MOLECULAR GENETICS
OF INHERITED EYE DISORDERS

Modern Genetics
A series of books covering new developments across the entire field of genetics
Edited by H. J. Evans, MRC Human Genetics Unit, Western General Hospital, Edinburgh, UK

Volume 1
Embryonal Stem Cells: Introducing Planned Changed into the Animal Germline
Martin L. Hooper

Volume 2
Molecular Genetics of Inherited Eye Disorders
Edited by Alan F. Wright and Barrie Jay

Aditional volumes in preparation

Volume 3
Molecular Genetics of Drug Resistance
edited by C. Roland Wolf and John D. Hayes

This book is part of a series. The publisher will accept continuation orders which may be cancelled at any time and which provide for automatic billing and shipping of each title in the series upon publication. Please write for details.

MOLECULAR GENETICS

OF INHERITED EYE DISORDERS

edited by

Alan F. Wright

MRC Human Genetics Unit, Western General Hospital
Crewe Road, Edinburgh EH4 2XU, UK

Barrie Jay

Emeritus Professor of Clinical Ophthalmology,
University of London, UK

informa
healthcare

New York London

First published in 1994 by Harwood Academic Publishers GmbH .

This edition published in 2011 by Informa Healthcare, Telephone House, 69-77 Paul Street, London EC2A 4LQ, UK.

Simultaneously published in the USA by Informa Healthcare, 52 Vanderbilt Avenue, 7th Floor, New York, NY 10017, USA.

Informa Healthcare is a trading division of Informa UK Ltd. Registered Office: 37–41 Mortimer Street, London W1T 3JH, UK. Registered in England and Wales number 1072954.

A CIP record for this book is available from the British Library.

Library of Congress Cataloging-in-Publication Data available on application

ISBN-13: 9783718654932

Orders may be sent to: Informa Healthcare, Sheepen Place, Colchester, Essex CO3 3LP, UK
Telephone: +44 (0)20 7017 5540
Email: CSDhealthcarebooks@informa.com
Website: http://informahealthcarebooks.com/

For corporate sales please contact: CorporateBooksIHC@informa.com
For foreign rights please contact: RightsIHC@informa.com
For reprint permissions please contact: PermissionsIHC@informa.com

CONTENTS

Preface to the Series vii

Preface ix

Contributors xi

1 Background to Molecular Genetic Principles and Techniques 1
M. B. Gorin and A. F. Wright

(1) INVERTEBRATES

2 Retinal Degeneration Mutants of *Drosophila* 29
W. L. Pak

(2) VERTEBRATES

A. CHOROID AND RETINA

3 Diagnosis in Inherited Retinal Disorders 53
A. C. Bird and B. Jay

4 Human Albinism and Mouse Models 89
R. A. King, I. J. Jackson and W. S. Oetting

5 Inherited Retinal Degenerations in the Mouse 123
D. B. Farber and M. Danciger

6 Autosomal Dominant Retinitis Pigmentosa 153
P. Humphries, P. Kenna and G. J. Farrar

7 The Molecular Genetic Approach to Macular Degeneration 173
E. M. Stone and V. C. Sheffield

8 Molecular Genetics of Retinoblastoma 197
J. K. Cowell

9 Color Vision Defects 217
J. Neitz and M. Neitz

10 X-linked Retinitis Pigmentosa 259
M. A. Aldred, M. Jay and A. F. Wright

11 Congenital Stationary Night Blindness: A Critical Review for Molecular Approaches 277
E. Héon and M. A. Musarella

12 Choroideremia 303
F. P. M. Cremers and H.-H. Ropers

13 Norrie Disease 321
A. A. B. Bergen, W. Berger, Z.-Y. Chen, P. J. Diergaarde, E. M. Bleeker-Wagemakers, E. M. Battinelli, M. Warburg, H.-H. Ropers and I. Craig

14 X-linked Juvenile Retinoschisis 339
A. de la Chapelle, T. Alitalo and H. Forsius

15 Clinical and Genetic Heterogeneity of Usher Syndrome 359
W. J. Kimberling, M. Weston and C. Moller

B. VITREOUS/LENS

16 Marfan Syndrome 383
P. Tsipouras and M. W. Kilpatrick

17 Lowe Oculocerebrorenal Syndrome 399
I. Okabe and R. L. Nussbaum

18 Crystallin Genes and Cataract 413
M. Cartier, L.-C. Tsui, S. P. Ball and N. H. Lubsen

C. ANTERIOR SEGMENT

19 Aniridia 445
I. Hanson, T. Jordan and V. van Heyningen

D. METABOLIC

20 Mitochondrial DNA Mutations and the Eye 469
M. D. Brown, M. T. Lott and D. C. Wallace

GLOSSARY 491

INDEX 515

PREFACE TO THE SERIES

Developments in cell and molecular biology have been responsible for enormous advances in our understanding of the structure and functions of genetic factors at the molecular, cell, tissue, organism and population levels. These advances are wide-ranging, impinging upon virtually all fields in the biological and biomedical sciences, and are published in a wide variety of journals. It seemed appropriate therefore to provide a means whereby experts in a field could be invited to review recent advances in their own areas of research, which would provide the interested reader with a comprehensive and critical appraisal of progress and prospects. Our general aim is to cover new developments across the entire field of genetics of plants and animals, including man, and at all levels from the molecule to the population. An important feature of the series will be its flexibility. Individual titles will range from short monographs on a narrow topic to larger multiauthor works covering several related developments in a single area.

The editors and contributors to this volume on "The Molecular Genetics of Inherited Eye Disorders" are to be congratulated, for this is a lucid, authoritative and up-to-date compilation of progress in an exciting and important field.

H. J. Evans

PREFACE

Abnormalities of structure relating to the eye, such as to the eyelids, lens, iris, etc. are rather apparent ophthalmological features as, of course, are abnormalities of function of this very important sense organ. It is not surprising therefore that the familial nature of a range of eye defects has been documented over the centuries and attracted the attention of the early human geneticists. It is worth noting that it is almost exactly 200 years since Dalton the chemist described his own, and his brother's, colour blindness (1794) and in 1838 Cuniers described a family, well known to most students, with an autosomal dominant form of congenital stationary night blindness which could be traced back to a French man born in the early seventeenth century. The importance of inheritance in a range of eye disorders is therefore well established and in the Western world around one in four of the blind in the community, and roughly one-half of individuals who are blinded as children, suffer their disability as a consequence of the inheritance of one or more mutated genes.

The paramount importance of the sense of vision, the range of variation in structure and function of human and other animal eyes, and the wealth of information on the Mendelian inheritance of eye abnormalities in humans, has therefore provided both a platform and a spur for the application of modern molecular approaches in the field of ophthalmological genetics. The present volume brings together a rich collection of papers by internationally distinguished researchers documenting the many remarkable and recent advances in our understanding at the molecular level of the development, structure and function of the eye in health and disease. We learn that in many cases we are now able to relate a particular structural or functional disorder to a particular mutation in a particular gene. We also now have some understanding of the mechanisms involved in Dalton's colour blindness and of the range of mutational changes responsible for example for retinal degeneration, cataract and other important and disabling ophthalmic abnormalities. Studies on the genetics of retinoblastoma led to the isolation and characterisation of the first human tumour suppressor gene and have essentially laid the foundations for our understanding of certain inherited cancer predispositions, whereas those on Leber's hereditary optic neuropathy have highlighted the importance of mitochondrial mutations in human disease. Some of the advances that have been made have relied on studies on other vertebrates, and indeed invertebrates, and relevant contributions on the use of animal models in investigating retinal dystrophies and abnormalities of eye structure and development in humans are discussed.

The Editors

HEREDITY

I am the family face;
Flesh perishes, I live on,
Projecting trait and trace
Through time to times anon,
And leaping from place to place
Over oblivion.

The years-heired feature that can
In curve and voice and eye
Despise the human span
Of durance – that is I;
The eternal thing in man,
That heeds no call to die.

Thomas Hardy (1917)

Charles Darwin (1859) from "The Origin of Species"

To suppose that the eye, with all its inimitable contrivances for adjusting the focus to different distances, for admitting different amounts of light, and for the correction of spherical and chromatic aberration, could have been formed by natural selection, seems, I freely confess, absurd in the highest degree. Yet reason tells me, that if numerous gradations from a perfect and complex eye to one very imperfect and simple, each grade being useful to its possessor, can be shown to exist; if further, the eye does vary ever so slightly, and the variations be inherited, which is certainly the case; and if any variation or modification in the organ be ever useful to an animal under changing conditions of life, then the difficulty of believing that a perfect and complex eye could be formed by natural selection, though insuperable by our imagination, can hardly be considered real.

CONTRIBUTORS

M. A. Aldred, MRC Human Genetics Unit, Western General Unit, Crewe Road, Edinburgh EH4 2XU, United Kingdom

Tiina Alitalo, Department of Medical Genetics, University of Helsinki, Haartmaninkatu 3, 00290 Helsinki, Finland

Sarah P. Ball, Department of Biological Sciences, Washington Singer Laboratories, University of Exeter, Exeter EX4 4QG, United Kingdom

E. M. Battinelli, Department of Biochemistry, Genetics Laboratory, University of Oxford, South Parks Road, Oxford OX1 3QU, United Kingdom

A. A. B. Bergen, The Netherlands Ophthalmic Research Institute, P O Box 12141, NL-1100 AC Amsterdam, The Netherlands

W. Berger, Department of Human Genetics, University Hospital Nijmegen, P O Box 9101, 6500 HB Nijmegen, The Netherlands

Alan C. Bird, Department of Clinical Ophthalmology, Moorfields Eye Hospital, City Road, London EC1V 2PD, United Kingdom

E. M. Bleeker-Wagemakers, The Netherlands Opthalmic Research Institute, P O Box 12141, NL-1100 AC Amsterdam, The Netherlands

Michael D. Brown, Department of Genetics and Molecular Medicine, Emory University School of Medicine, 403 Dental Building, 1462 Clifton Road, Atlanta, GA 30322, United States of America

Mireille Cartier, Department of Genetics, Research Institute, The Hospital for Sick Children, 555 University Avenue, Toronto, Ontario M5G 1X8, Canada

Albert de la Chapelle, Department of Medical Genetics, University of Helsinki, Haartmaninkatu 3, 00290 Helsinki, Finland

Z.-Y. Chen, Department of Biochemistry, Genetics Laboratory, University of Oxford, South Parks Road, Oxford OX1 3QU, United Kingdom

John K. Cowell, ICRF Oncology Group, Institute of Child Health, 30 Guilford Street, London WC1N 1EH, United Kingdom

I. Craig, Department of Biochemistry, Genetics Laboratory, University of Oxford, South Parks Road, Oxford OX1 3QU, United Kingdom

Frans P. M. Cremers, Department of Human Genetics, University Hospital Nijmegen, P O Box 9101, 6500 HB Nijmegen, The Netherlands

Michael Danciger, Loyola-Marymount University, Los Angeles, CA 90045, United States of America

P. J. Diergaarde, The Netherlands Ophthalmic Research Institute, P O Box 12141, NL-1100 AC Amsterdam, The Netherlands

Debora B. Farber, Jules Stein Eye Institute, UCLA School of Medicine, Los Angeles, CA 90024-7008, United States of America

G. Jane Farrar, Department of Genetics, Trinity College, Dublin 2, Ireland

Henrik Forsius, Department of Medical Genetics, University of Helsinki, Haartmaninkatu 3, 00290 Helsinki, Finland

Michael B. Gorin, Eye and Ear Institute of Pittsburgh, Departments of Ophthalmology and Human Genetics, University of Pittsburgh, PA 15213, United States of America

Isabel Hanson, MRC Human Genetics Unit, Western General Hospital, Crewe Road, Edinburgh EH4 2XU, United Kingdom

Elise Héon, University of Iowa Hospitals and Clinics, Department of Ophthalmology, 200 Hawkins Drive, Iowa city, Iowa 52242, United States of America

Veronica van Heyningen, MRC Human Genetics Unit, Western General Hospital, Crewe Road, Edinburgh EH4 2XU, United Kingdom

Peter Humphries, Department of Genetics, Trinity College, Dublin 2, Ireland

Ian J. Jackson, MRC Human Genetics Unit, Western General Hospital, Crewe Road, Edinburgh EH4 2XU, United Kingdom

Barrie Jay, *Retired from* Institute of Ophthalmology, University of London, London, United Kingdom

Marcelle Jay, Department of Clinical Ophthalmology, Moorfields Eye Hospital, City Road, London ECIV 2PD, United Kingdom

Tim Jordan, MRC Human Genetics Unit, Western General Hospital, Crewe Road, Edinburgh EH4 2XU, United Kingdom

Paul Kenna, Department of Genetics, Trinity College, Dublin 2, Ireland

Michael W. Kilpatrick, Department of Clinical Genetics, University of Birmingham, Birmingham B15 2TT, United Kingdom

W. J. Kimberling, Boys Town National Research Hospital, Omaha, Nebraska, NE 68131, United States of America

Richard A. King, University of Minnesota, Box 485 UMHC, 420 Delaware Street SE, Minneapolis, Minnesota 55455, United States of America

Marie T. Lott, Department of Genetics and Molecular Medicine, Emory University School of Medicine, 403 Dental Building, 1362 Clifton Road, Atlanta, GA 30322, United States of America

Nicolette H. Lubsen, Department of Molecular Biology, University of Nijmegen, Toernooiveld, 6525 ED Nijmegen, The Netherlands

C. Moller, University of Linkoping, Linkoping, Sweden

Maria A. Musarella, The Hospital for Sick Children, Research Institute, Departments of Ophthalmology and Genetics, 555 University Avenue, Toronto, Ontario M5G 1X8, Canada

Jay Neitz, Medical College of Wisconsin, Department of Cellular Biology and Anatomy, and Department of Ophthalmology, 8701 Watertown Plank Road, Milwaukee, WI 53226, United States of America

Maureen Neitz, Medical College of Wisconsin, Department of Cellular Biology and Anatomy, 8701 Watertown Plank Road, Milwaukee, WI 53226, United States of America

Robert L. Nussbaum, Chief, Laboratory for Genetic Disease Research, National Center for Human Genome Research, NIH, Building 49/4A72, 9000 Rockville Pike, Bethesda, Maryland 20892, United States of America

William S. Oetting, University of Minnesota, Box 485 UMHC, 420 Delaware Street, SE Minneapolis, Minnesota 55455 United States of America

I. Okabe, Howard Hughes Medical Institute Laboratory, University of Pennsylvania School of Medicine, 422 Curie Boulevard, Philadelphia, PA 19104, United States of America

William L. Pak, Department of Biological Sciences, Lilly Hall of Life Sciences, Purdue University, West Lafayette, Indiana 47907, United States of America

Hans-Hilger Ropers, Department of Human Genetics, University Hospital Nijmegen, P O Box 9101, 6500 HB Nijmegen, The Netherlands

Val C. Sheffield, Department of Pediatrics, University of Iowa College of Medicine, Iowa City, Iowa 52242, United States of America

Edwin M. Stone, Department of Ophthalmology, University of Iowa College of Medicine, Iowa City, Iowa 52242, United States of America

Petros Tsipouras, Department of Pediatrics, University of Connecticut Health Center, Farmington, CT 06030, United States of America

Lap-Chee Tsui, Department of Genetics, Research Institute, The Hospital for Sick Children, 555 University Avenue, Toronto, Ontario M5G 1X8, Canada

Douglas C. Wallace, Department of Genetics and Molecular Medicine, Emory University School of Medicine, 403 Dental Building, 1462 Clifton Road, Atlanta, GA 30322, United States of America

M. Warburg, Gentofte Hospital, Department of Ophthalmology, Division of Paediatric Ophthalmology and Handicaps, 40 Sognevej, DK-2820 Gentofte, Denmark

M. Weston, Boys Town National Research Hospital, Omaha, Nebraska, NE 68131, United States of America

Alan F. Wright, MRC Human Genetics Unit, Western General Hospital , Crewe Road, Edinburgh EH4 2XU, United Kingdom

1. BACKGROUND TO MOLECULAR GENETIC PRINCIPLES AND TECHNIQUES

MICHAEL B. GORIN[1] and ALAN F. WRIGHT[2]

[1] *The Eye and Ear Institute of Pittsburgh, Departments of Ophthalmology and Human Genetics, University of Pittsburgh, Pittsburgh, PA 15213, USA*
[2] *MRC Human Genetics Unit, Western General Hospital, Crewe Road, Edinburgh EH4 2XU, UK*

Throughout this book, there are numerous examples of the genetic and physical mapping of genes related to ophthalmic disorders and of the unravelling of their genetic functions. The techniques that are used cover a range of disciplines including molecular biology, genetics, biochemistry and cytology and, to the novice, the terms and methods can be overwhelming. It is the objective of this chapter to provide a simple "guide for the perplexed" to assist the reader. The descriptions and historical origins are kept to a minimum, so that material can be rapidly accessed. The definitions are by no means comprehensive and some excellent and complete molecular genetics guidebooks are available (Davies and Read, 1988; King and Stansfield, 1990; Thompson et al., 1991; Casey, 1992).

There are three sections to this introduction. The first covers the anatomy of the mammalian genome and of the gene. It is important to recognize the extent of the problem of identifying a single base pair mutation within a genome that spans over three billion base pairs and contains up to 100,000 genes. While the basic alphabet of the genetic code – A,C,G and T (U in RNA) – has been recognized since Watson and Crick's classic paper in 1953, it is still not clear what is the full significance of the 95–98% of DNA that does not directly code for genes. This material is full of highly repetitive elements that make the task of the gene hunter considerably more difficult and genes themselves are split up into short coding stretches (exons), each only a few hundreds of base pairs in length, that may be scattered over a region that covers tens or hundreds of thousands of base pairs.

The second section addresses the concepts of linkage analysis and related genetic, as opposed to physical, mapping techniques. Genetic mapping is concerned with that most fundamental of all geneticists' techniques, pedigree analysis, in which parental and recombinant type offspring are counted and from which genetic distances can be computed. This area has developed very rapidly during the past ten years to the extent that a trickle of genes mapped using restriction fragment length polymorphisms in the first half of the 1980s has now turned into a flood using new types of genetic marker such as the microsatellite repeat. The next five years will almost certainly see the automation of such genetic mapping procedures and the introduction of new methods, making it feasible to elucidate the genetics of more complex ophthalmic disorders such as glaucoma and age-related macular dystrophies.

There is a vast array of techniques that the molecular geneticist employs for the analysis of nucleic acids and gene expression, many of which are referred to in the succeeding chapters. Essentially all of these methods are built upon a fundamental set of tools. The third section provides a brief overview of these tools. The revolution in genetics – the "new genetics" – is underpinned by basic advances in fundamental molecular biological techniques such as the ability to clone DNA in microorganisms, to identify specific nucleic acids within complex mixtures by filter hybridisation, to exponentially amplify small segments of DNA by thermal cycling, to identify the sequence of nucleotide bases on a linear stretch of DNA and so on. Rather than reviewing these essential advances in a historical context, the tools that have emerged from them will be summarised. This will give some insight into the everyday techniques used by the gene mapper.

In addition, a glossary is included at the end of the book that may help to clarify both the basic language of molecular genetics and the more technical terms and abbreviations that make the subject less accessible than it should be.

I(a) ANATOMY OF THE GENOME

The mammalian genome consists of an estimated 3 billion base pairs of DNA organized into 23 pairs of chromosomes in man and 20 pairs in the mouse. The invertebrate genome is considerably smaller, so that for example the fruit fly *Drosophila melanogaster* has only four pairs of chromosomes and 300 million base pairs of DNA. This smaller genome size, very high resolution cytological and genetic maps and rather special techniques both for mutating and restoring the function of genes, has made Drosophila a powerful model for genetic analysis – not least in the analysis of mutations affecting the eye and its development (Chapter 2). By contrast, the mammalian genome has greater complexity, not only in size but also in organization. The basic apparatus for genetic replication and segregation are similar however. Each **chromosome** contains, firstly, a double stranded linear DNA molecule with two "antiparallel" α-helical strands –they are in opposite orientation with respect to the 5´ ("head") and 3´ ("tail") carbon atoms of the deoxyribose in the sugar–phosphate "backbone" which is oriented as a 5´-3´-5´-3´ array. Histone and non-histone chromosomal proteins bind and contribute to the folding of DNA such that it undergoes a linear contraction of at least three orders of magnitude within the chromosome. Chromosomes contain a **centromere** for attachment to the mitotic or meiotic spindle, ensuring proper segregation into daughter cells, and specialised chromosomal structures called **telomeres** that facilitate the replication of chromosomal ends. One of the chromosomal pairs is the sex chromosomes, which in the heterogametic sex (males in mammals, with an X and Y chromosome) share only a small region of homology (pseudo-autosomal region); the majority of the X and Y chromosomes contain different genes. Recessive mutations occurring on the X chromosome are more readily identified than on non-sex chromosomes (autosomes) since in the hemizygous (XY) state they lack a second, normal copy of the affected gene. The expression of the X chromosome is also different from the autosomes in the homogametic sex (XX), the female in mammals, because of the process of random functional inactivation of one of the two chromosomes in each cell in early embryogenesis (**X-inactivation**). This process takes place at a stage of embryogenesis where cell numbers are small (e.g. ≤5 day old mouse embryos) so that chance fluctuations in the distribution of cells with one or other X chromosome active can have large effects on the expression of deleterious genes in heterozygotes – particularly where the number of primordial cells giving rise to a tissue, such as the retina, are relatively small. This process is responsible for the variable and often subtle manifestations of disease in heterozygous carriers of X-linked disorders.

Autosomal genes are expressed in every cell although it is now known that the parental source (father or mother) of the chromosome can have a profound effect on the expression of certain genes. This process has been termed **imprinting** and, until recently, most examples came from studies of genes expressed in early mammalian development. Recent studies on the parental origin of chromosome 15 deletions associated with the Prader–Willi and Angelman syndromes in man has now established a significant role for this process not only in early development but also in the expression of disease.

Genetic diversity is achieved by two major mechanisms among diploid species. The first is by **mutation**, which occurs at a slow but finite rate throughout the genome. The evolutionary constraints on altering the function of genes are such that relatively few mutational differences survive in regions that code for proteins or control gene expression. The "silent" regions of the genome, on the other hand, maintain a high level of diversity. The advantage of this high polymorphism content became evident to gene mappers in the early 1980s, who proposed the use of such sequence variants as markers to follow the segregation of genes in kindreds. The variants were identified initially by their ability to alter restriction endonuclease cleavage sites (**restriction fragment length polymorphisms**, RFLPs), which could be detected by hybridisation of specific radiolabelled "probes" against genomic DNA after digestion with the appropriate endonuclease (Figure 1). Such variants were found to occur once every 300–500 base pairs in non-coding regions of the genome and to be present as frequent alternative forms (alleles) in the general population. They have been exploited to the full in mapping genetic traits, where only a limited range of blood group and serum protein polymorphisms existed previously for this purpose.

In contrast to the nuclear genome, the **mitochondrial genome** consists of a small, circular DNA molecule (16.5×10^3 base pairs long) that is maintained in the cytoplasm as hundreds or thousands of copies per cell. It encodes several subunits of oxidative phosphorylation enzymes that are vital to the function of respiring cells, the others being coded by the nuclear genome. There are important differences between these two genomes (Chapter 21), including a significantly higher rate at which mutations are incorporated into the mitochondrial genome (fixa-

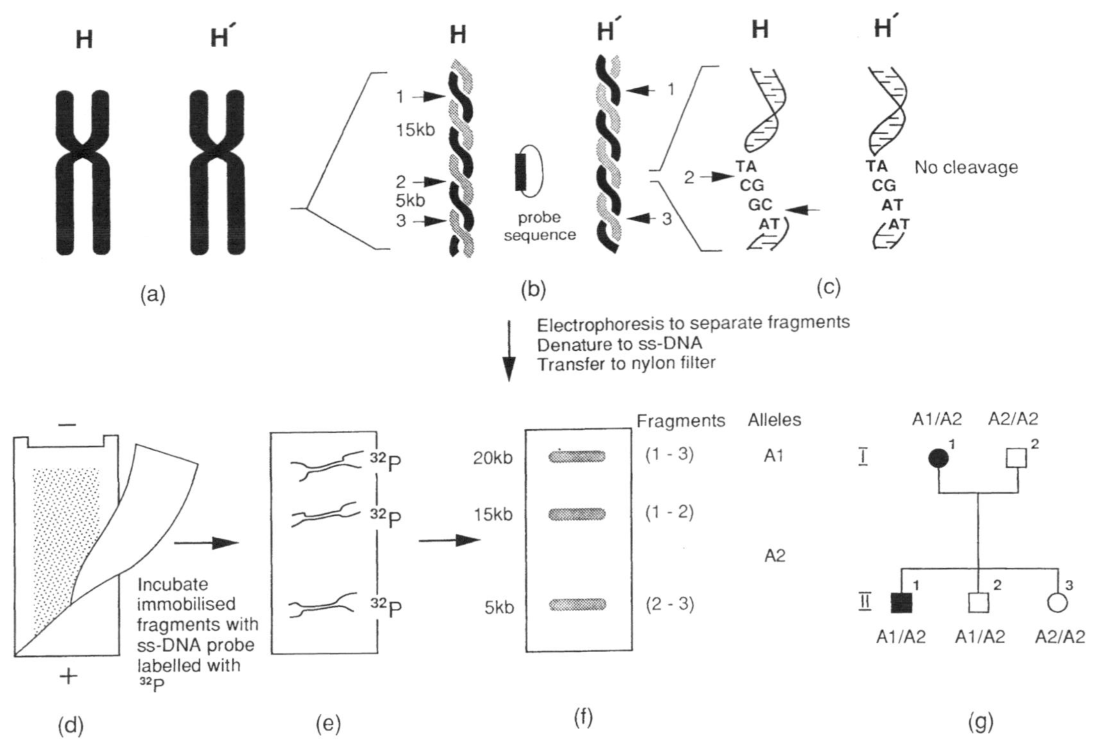

Figure 1 Restriction fragment length polymorphisms (RFLPs) detected using a Southem blot. The process of restriction endonuclease digestion cleaves genomic DNA into millions of fragments, three of which (20 kb, 15 kb, 5 kb) are detected using a specific hybridisation probe. One fragment (allele A1, 20 kb) corresponds to the allele on chromosome H´, the other two (allele A2, 15/5 kb) come from chromosome H . The RFLP results from the presence of a previous mutation within the restriction endonuclease cleavage site (G → A) on H' that results in failure to cleave at that point. The digested fragments are separated electrophoretically, denatured, transferred to a membrane (Southern transfer) and hybridised with a single stranded ^{32}P labelled probe sequence that specifically anneals to the fragments containing the RFLP. The segregation of the two alleles are shown within a small pedigree on the right. The RFLP alleles do not segregate with the disease phenotype (shaded symbols) in this family.

tion), some of which give rise to genetic eye disorders. Mutation also arises as a result of chromosomal aberrations such as translocations, inversions and deletions. These are important both in creating evolutionary diversity and, in the context of gene mapping, provide an important means of identifying genes that have been disrupted by translocation or deletion events (Chapters 8, 9, 10, 12, 13, 17, 19, 21).

The second mechanism for generating diversity occurs during meiosis when homologous parental chromosomes pair, exchange genetic material by **recombination** and segregate into the germ cells. The average number of exchanges per chromosome during meiosis in a mammalian cell is only 2–4, so that the daughter chromatids are composites made up of relatively long segments of either maternal or paternal chromosomes. The recombination process ensures that the probability of an individual inheriting two genes or markers from the chromosome originally contributed by a grandparent decreases rapidly to the level of chance, depending on how far apart they are. The paucity of recombinational "crossovers" makes for a relatively efficient mapping procedure in which the rate at which the two sites (genes, markers) recombine is a measure of the **genetic distance** between them (section II). Mistakes that occur either during chromosomal replication or during meiotic pairing can give rise firstly to duplication of genes and later to multiple copies of genes, providing scope for the divergence of genes with related functions, as discussed in Chapter 9 on color vision.

The **number of genes** in the mammalian genome has been estimated to be in the region of 50,000–100,000. The average cell expresses one-quarter to one half of this number and some gene transcripts are very abundant, others make only a few copies per cell. This highly skewed distribution of gene expression can be an important consideration in identifying candidate genes for genetic disorders, particularly with regard to the construction or screening of expressed sequence (cDNA) libraries. If an average mammalian gene product contains about 300 amino acids, the equivalent amount of DNA required to code and express such a gene is only about one kilobase pair (kb; 10^3 base pairs). When extrapolated to the size of the whole genome, it is readily apparent that only 2–4% of the genome actually codes for proteins. Nucleic acid reassociation studies show that as much as 60% of mammalian DNA contains unique or low copy number (<10) sequences, while the remainder is repetitive. Within the latter group, three classes of **repetitive DNA** can be distinguished – high, intermediate and low abundance classes. The first class of highly repetitive sequences includes those with reiteration frequencies above about 100,000 copies, such as the tandemly repetitive "satellite" sequences near the centromeres of some chromosomes. Another important example of this group is the Alu repetitive element, which is 300 base pairs in length and interspersed throughout the genome, occurring on average every 3–4 kb. These and other interspersed repeats are useful for distinguishing the DNA of different mammalian species but in general are a geneticists' nightmare. They make the task of identifying specific sequences such as genes considerably more difficult, so that radiolabelled probe sequences often require competition with high concentrations of repetitive DNA in order to prevent non-specific hybridisation signals.

Other repetitive elements have been exploited to great advantage. The microsatellite repeat class consists of two to four bases repeated in tandem up to 50-fold and again widely dispersed throughout the genome. Microsatellites are thought to occur, on average, every 30 kb and are increasingly exploited as genetic markers. Most categories of repetitive DNA accumulate mutations, some of which become genetically *fixed* in the population, but the sequence variation within microsatellites is particularly useful. Towards the end of the 1980s, it emerged that microsatellites show a high degree of variation in the *number* of repeat units within different members of most outbred populations. These are relatively easily typed after sequence amplification by means of the polymerase chain reaction (see below) and now constitute a major class of genetic marker for mapping studies. The most widely used microsatellite is the CA repeat polymorphism, in which the CA motif is reiterated from 10 to 50 times *en bloc*.

If most of the mammalian genome does not code for genes, what is this "junk" DNA doing? The repetitive elements do, in certain cases, have the ability to replicate and re-insert themselves in new regions of the genome, but for the most part this ability is lost or very rare. Some non-coding DNA may simply provide an evolutionarily useful strategy of spacing out the regions (exons) coding for small domains within proteins (see below), providing scope for chromosomal rearrangements to create new genetic combinations and functions. Some DNA is certainly concerned with the ability to replicate itself, providing recognition sequences for replication proteins and chromosomal scaffold attachment sites. Other sites contribute to the regulation of gene expression either directly, as with promoter and enhancer sequences (see below), or indirectly by means of sequences necessary for the packaging of DNA into chromosomes. However, most sequences known to be required for replication or expression are short and account for only a small proportion of the genome. The majority of the genome appears to retain mutations at a high rate, which itself implies a non-specific but perhaps evolutionarily important role. Specificity may not always be obvious however, since the secondary and higher-order structure of DNA is much less well understood than its primary structure and there may be considerable constraints on the sequence composition of what is currently regarded as "junk" DNA.

I (b) ANATOMY OF THE GENE

The eukaryotic gene can be considered in two parts. The first part contains the information required for a ribonucleic acid (RNA) gene product. Transfer and ribosomal RNA gene transcripts are required for the addition of charged amino acids to the nascent polypeptide chain and for ribosomal binding and translation of the messenger RNA (mRNA) respectively. Messenger RNA contains all the sequence information required for translation into protein although it first undergoes considerable modification within the nucleus. Unlike bacteria or mitochondria, the parts of eukaryotic genes that encode a protein are typically broken up into discontinuous coding segments contained within **exons**, separated by intervening sequences or **introns**, which are removed from the initial transcript. The RNA polymerase transcript commences about 30 base pairs (three DNA helix turns) downstream of a promoter element (e.g. TATA box), at the **cap site**, which marks the 5´ end of the transcript and is modified by the enzymatic addition of a guanosine residue in the opposite orientation (5´ to 5´) to the mRNA chain (5´ to 3´): this is then methylated to form the "cap" (Figure 2). The cap may protect the mRNA from nuclease attack or assist in ribosome binding. Transcription proceeds in a 5´ to 3´ direction through the **5´ untranslated region** of the gene, past the protein coding regions and introns into the **3´ untranslated region** of the gene and finishes at one or more stop sequences, which are usually in mammalian genes. The initial transcript, which is called **heterogeneous nuclear (hn) RNA**, is rapidly processed in the nucleus. The 3' terminal RNA contains a polyadenylation signal sequence (Figure 2), to which are added 50–200 adenosine residues (polyA^+ tail), which serve to stabilise the transcript and to effect its translocation to the cytoplasm.

Hn RNA is also processed by **splicing** – enzymatic cleavage at partially conserved residues flanking the exons. Some genes are spliced at different points in different cell types or even within the same cell type – differential splicing. This may result from cell-specific proteins stabilising certain splice complexes at the expense of others. Differential splicing can result in functionally significant differences in gene expression.

Exons may be tens to thousands of base pairs in length and are separated by introns which are typically much longer – anything up to hundreds of kilobases long. There may be from one to over one hundred exons in a eukaryotic gene. During translation of the mature, processed eukaryotic mRNA, the 40S ribosomal subunit, which contains the initiator $tRNA^{met}$, binds to the 5´ end and tracks along the transcript until it reaches the first **initiation codon**, AUG (methionine), when the 60S ribosomal subunit attaches and translation begins.

Mutations within the coding or associated regions required for processing the transcript (e.g. splice sites) are the commonest cause of genetic disorder. These may take the form of **missense mutations**, in which the triplet codon for one amino acid is changed to that of another (e.g. CAC → TAC changes a histidine to a tyrosine). The resultant change may be functionally benign or conservative if there is substitution of a similar amino acid e.g. both residues have no net charge and share similar biophysical properties. Alternatively, a change from an acidic to a basic or uncharged amino acid is non-conservative and is more likely to be deleterious. **Nonsense mutations** introduce a premature translation stop codon (UAA, UAG) into the transcript, producing a truncation of the resultant polypeptide. **Frameshifts** are small insertions or deletions that alter the codon reading frame and so both change the subsequent sequence and often lead to truncation of the protein due to introduction of a stop codon. **Splice site mutations** result in abnormal processing of the transcript leading to an abnormal protein product. Mutations can occur within any functionally important part of the gene and the

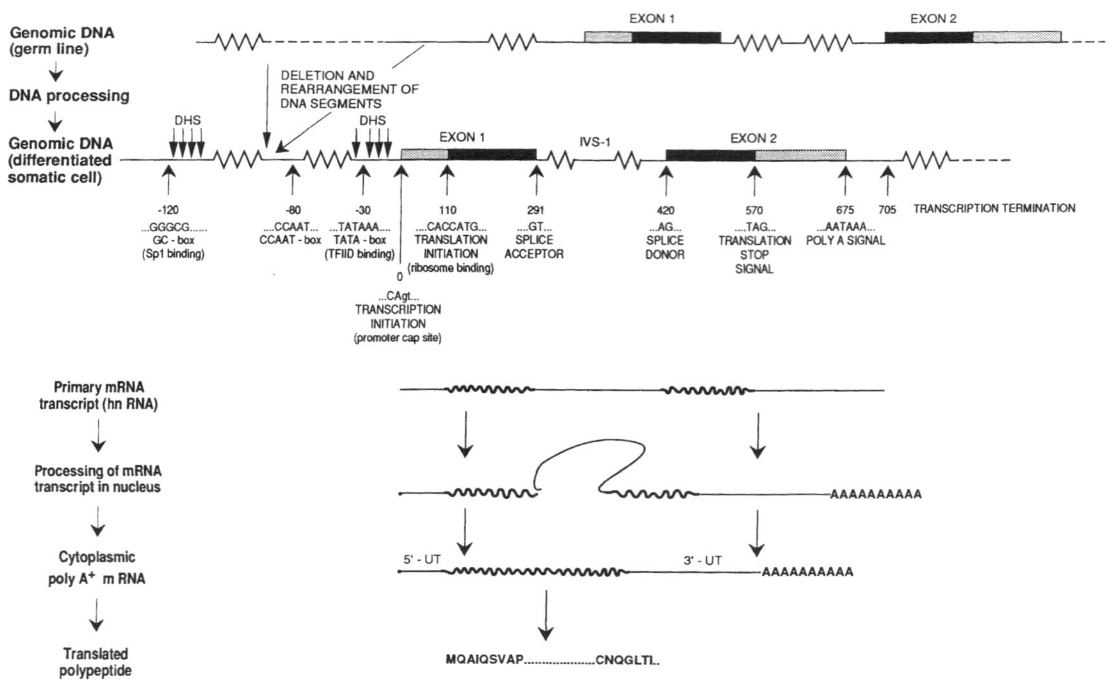

Figure 2 Major structural and regulatory features of a mammalian gene that determine the expression of its mRNA and protein products. Only certain genes are known to involve genomic deletion and rearrangement (DNA processing) during cellular differentiation as a means of regulating expression (e. g. immunoglobulin and T cell receptor genes). The upstream elements include deoxyribonuclease hypersensitive sites (DHS), which probably indicate the "openness" of a chromatin region undergoing active expression; and promoter regions (GC-box, CCAAT-box, TATA-box) that bind protein complexes, including RNA polymerase or transcription factors. The sequence surrounding the transcription initiation site (cap site) shows limited sequence conservation between genes but marks the start of the first exon. The coding regions of the first two exons are shown in bold with translation initiation (ATG) and stop (TAG) signals. The splice donor and acceptor sites (AG, GT) flanking each intervening sequence (IVS) or intron are strongly conserved across genes. Transcription termination sites are rarely known in detail but polyadenylation signal sequences are associated with polyA addition near the 3′ end of the transcript. The sequential processing (splicing) of the primary transcript (hnRNA) into mature mRNA and the amino acids (using the single letter code) of its translated peptide are shown schematically. The jagged lines indicate the position of repetitive sequences.

relative distribution of mutational types may vary with the gene structure and the nature of the phenotype. Genes containing repeated sequences may be unstable either because of a tendency to **expansion** of the region (as in Huntington's disease) or because they may be more prone to **deletion** as a result of mispairing between repeated sequences in meiosis (as in some LDL receptor mutations). Deletions are a very important means of identifying disease genes and may represent a high proportion of all mutations in the case of large genes such as factor VIII (10%) or dystrophin (65%), which cause haemophilia A and Duchenne muscular dystrophy respectively. In most genes, however, deletions only comprise a small percentage of mutations.

The second part of the eukaryotic gene is concerned with the control of gene expression and contains conserved sequences regulating the level of transcription and tissue or developmental expression. These sequences are not transcribed into RNA. They include promoter and enhancer sequences. The **promoter region** is located upstream or 5' to the transcription start site and includes one or more of several sites with relatively short sequence motifs, many of them binding sites for transcription activating or repressing proteins. Most genes contain a **TATA box,** a seven base pair AT-rich segment located 20–30 bp upstream from the transcription start site (Figure 2), which binds part of the TFIID transcription factor complex. This appears to orientate the start of transcription by RNA polymerase II. Some genes contain a variety of upstream elements located 80–200 bp upstream of the cap site, such as the **CCAAT box** and **GC box,** whose precise functions in regulating the level of transcription are less clear. These **upstream elements** may be present in multiple copies and are often gene family or promoter – specific. Genes that are constitutively expressed in all tissues – "housekeeping genes" – tend to lack both TATA and CCAAT boxes but have GC boxes, which bind the Sp1 transcription factor. Some transcription factors are thought to bind cooperatively to the more general promoter sequences, helping to produce an active transcription initiation complex. Others are more truly regulatory for expression of specific genes or gene families. The promoter regions of housekeeping genes also tend to have small 1–2 kb "islands" of CpG-rich sequences that are undermethylated, compared with CpG sequences elsewhere in the vertebrate genome that are methylated at the C5 position. These **CpG islands** are found around the 5´ end of a significant proportion of all genes and although their role is still unclear, they provide important landmarks for identifying the location of genes. An island can be identified by a clustering of sites for methylation-sensitive restriction enzymes that cleave unmethylated CpG-rich sequences.

In addition to the more general upstream regulatory elements discussed above, there are important **enhancer sequences** that are essential for full or tissue-specific expression and act further upstream or even within introns or downstream regions. Unlike promoter elements, enhancers can act in an orientation and position-independent manner. These are often rather short (10–20 bp) sequences but are often present in several copies, as with GC boxes. Enhancers generally activate a gene but they may also inhibit it, depending on the cellular conditions.

Because both promoter and enhancer sequences are short and, in some cases, present in multiple copies, they are less vulnerable to point mutation than exon or splice sequences, so that comparatively few regulatory mutants are found in (heritable) genetic diseases. In the case of somatic disorders such as cancer, changes in genetic regulatory regions are of great importance.

II GENETIC MAPPING

The concept of genetic linkage was first introduced in 1910 by TH Morgan who realised that the tendency of some allelic pairs to segregate together (coupling) and of others to segregate apart in the offspring (repulsion) were two aspects of the same phenomenon, in which genes reside on the same chromosome (Figure 3).

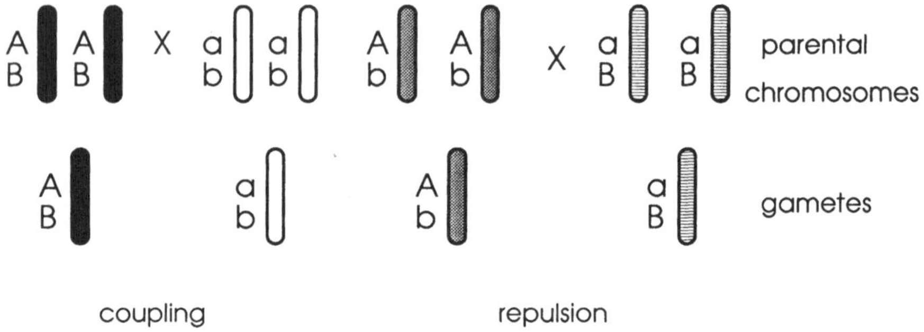

Figure 3 Coupling and repulsion. Two alleles at each of two loci (A,a and B,b) are shown, two of which are in coupling (the same chromosome), two of which are in repulsion (on opposite chromosomes) and segregate together or apart in the gametes.

Morgan hypothesised that the strength of the observed linkage between paired genes was dependent on the separation of the genes on the chromosome. Non-parental combinations of gametes in the offspring or **recombinants** were thought to result from exchange (**crossing-over**) between the homologous parental chromosomes during meiosis (Figure 4).

This hypothesis has since been fully confirmed. Genetic linkage is therefore present when two loci occur on the same chromosome and/or are inherited together more often than expected from independent assortment of the chromosomes. Genetic distances are measured either in terms of the fraction of recombinant to total offspring or gametes – the **recombination fraction** (usually designated by the symbol θ) – or in genetic map units such as centiMorgans. The early Drosophila geneticists soon observed that genetic distances measured in terms of recombination fractions were approximately additive, which led to the hypothesis

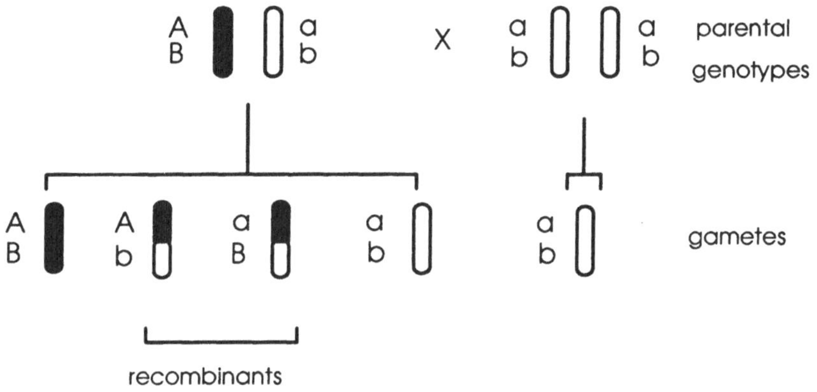

Figure 4 Recombination. Two loci on the same chromosome are shown, each with two alleles (A,a and B,b) which segregate in the gametes either as parental (AB, ab) or recombinant (Ab and aB) combinations. Loci that are far apart on the same chromosome will assort independently of each other in the gametes (and offspring) as a result of recombination between homologous chromosomes in meiosis.

of colinearity of genes along the chromosome and to the development of the first genetic maps of chromosomes. Since the occurrence of double cross-overs between two genetic loci will not be observed as recombinant offspring, the occurrence of all categories of multiple crossover between loci will lead to underestimation of the true genetic distance. This leads to non-linearity of the genetic map when markers are separated by longer distances. This effect is offset to some extent by the widespread occurrence of **interference,** the tendency of one crossover to suppress the occurrence of adjacent ones. Genetic maps are therefore usually expressed in terms of genetic map units, which are additive, even over long distances. The commonly used genetic map unit is the **centiMorgan**, defined as the length of chromosome that shows one crossover *per chromatid strand* in 100 meioses. The relationship between the recombination fraction, θ, and map units is expressed in terms of **mapping functions,** such as that of Kosambi:

$$\theta = \frac{1}{2}\frac{(e^{4x} - 1)}{(e^{4x} + 1)}$$

where x is in map units (Morgans). The relationship is approximately linear over short distances, so that in most mammals a recombination fraction of 0.1 (or 10% frequency of recombination) corresponds to a genetic distance of 10 centiMorgans, due to the lack of double crossovers over this distance. Above about 20 centiMorgans (cM) the two measures of recombination diverge increasingly until at a recombination fraction of 0.5, this corresponds to an infinite distance between markers. Two markers separated by a θ value of 0.5 may be on different chromosomes and therefore assort independently at meiosis or may be on the same chromosome but separated by a large genetic distance. In the latter case, a recombination fraction of 0.5 is the maximum value that can be expected, because each of a pair of chromatid strands has only a 50% chance of being involved in any one exchange, provided that each strand is equally likely to be involved in a crossover.

The total **genetic map lengths** in man and mouse are known quite precisely from genetic marker studies (4000 cM in man and 1600 cM in mouse), but in both cases there is a significant sex difference in recombination frequencies. In man, for example, the total autosomal length in males is about 2800 cM and in females 4800 cM. The reason for the increased recombination in females[1], which is generally true in mammals, is unknown, but genetic maps are therefore best expressed as separate male and female maps. Recombination rates vary in different parts of the genome, with some "hot spots" and other "cold spots", but over longer distances they tend to equalise. Rates tend to be increased towards the ends of chromosomes and decreased near the centromeres. There is a rough correspondence between the number of chiasmata – the presumed physical counterpart of the crossover – observed in male meioses and the number of crossovers measured by genetic mapping studies. Finally, there is an approximate relationship between genetic and physical distance such that 1 centiMorgan in man corresponds to about 1 Megabase (10^6 base pairs) of DNA (about 1.6 Mb in mouse)[2], although this may vary markedly in different parts of the genome depending on whether recombination is occurring above or below the average rate.

Genetic linkage is most commonly evaluated by means of the lod score statistic devised by Morton in 1951. In order to calculate a **lod score** for a given set of data, first the probability (likelihood) of observing r recombinants out of n informative meioses is calculated for each possible value of the recombination fraction (i.e. of the genetic distance between two markers), using binomial theorem:

$$L_{(\theta)} = (\theta)^r (1-\theta)^{n-r} \frac{n!}{r!\,(n-r)!} \cdot$$

For example, when 1 recombinant out of 6 offspring are observed, the likelihood for $\theta = 0.1$ is:

$$L_{(\theta = 0.1)} = (0.1)^1 (0.9)^5 \frac{6!}{1!5!} \cdot$$

If the number of recombinants cannot be inferred unambiguously (e.g. two generation families) since the **linkage phase** of the parents is unknown (i.e. whether the alleles are in coupling or repulsion), the average likelihood of the two alter-

[1]There are exceptions to this, particularly towards the ends of some chromosomes.
[2]The total amount of DNA per haploid genome is about 3×10^9 base pairs (bp) in mammals, these values are therefore obtained by dividing the genetic length for each species by 3×10^9bp.

native phases is computed. Next, the probability of the family data given the null hypothesis of random assortment ($\theta = 0.5$) is calculated in the same way. In order to assess the overall likelihood of linkage, the likelihood ratio (LR) is obtained:

$$LR_{(\theta i)} = \frac{L_{(\theta i)}}{L_{(\theta = 0.5)}} .$$

The $\log_{10}$ of the LR is conventionally used, which is the **lod score** or log of the odds ratio. This calculation is then repeated over several values of θ and the θ value corresponding to the **maximum likelihood value of the lod score** (Z_{max}) is generally taken to be an appropriate summarising statistic, since it minimises the variance of the estimate. The 95% confidence limits of θ_{max} are most conveniently approximated by reading the θ values corresponding to $Z_{max} \pm 1.00$, provided that the sample size is large enough. The lod score is a sequential test, in that lods can be added together from different families and if necessary over different studies until the required degree of statistical significance is obtained. In human genetics, a lod score of 3 or greater (LR $\geq$1000:1) is conventionally used, for Mendelian traits, to demonstrate a statistically significant linkage, while linkage is excluded at θ values for which the lod score is –2 or less. For X-linked loci, the upper critical limit is reduced from 3 to 2. The derivation of these limits takes into account the fact that the prior probability of linkage is about 1 in 20 for autosomal loci, and reduces the posterior probabilities both that linkage is actually present, given that it has been excluded (type 2 error), and that it is absent, given that it has been confirmed (type 1 error) to suitably low values. Under the null hypothesis ($\theta = 0.5$), the function 2 ln (LR) is distributed as χ^2 with one degree of freedom, so that a lod score of 3 corresponds to a χ^2 of 13.8.

Lod scores from small nuclear families can be checked manually but for most purposes they are calculated using computer programmes designed for the purpose. Linkage to an unmapped disease is usually established by screening a large number of marker loci. About 200 markers, evenly spaced along the human genetic map, would be sufficient to detect linkage to any locus within 10 cM on either side of a marker. Typing a large set of individuals with this number of markers is clearly a very large undertaking but the strategy has proved remarkably successful (Chapters 6, 7, 15, 16). Once linkage is established between a disease gene and a marker, other markers from the region are typed and a **multipoint linkage analysis** is carried out. Here the position of the disease locus is usually varied with respect to a fixed series of marker loci, whose order and separations are known. These computations are considerably more demanding than two-point lod scores and may take a lot of computer time. The output is again usually in the form of (multipoint) lod scores corresponding to different disease gene locations over the entire map. The relative likelihood of various gene orders can then be estimated.

In human linkage, there are many potential problems, most notably small sample size, resulting in wide confidence limits. Genetic heterogeneity is also a major problem in genetic eye disease with many loci producing virtually indistinguishable phenotypes. Lack of information on linkage phase, the need to infer the genotypes of missing inividuals, phenocopies, reduced penetrance and non-paternity all compound the difficulties. Replication of a human genetic linkage result is therefore particularly important.

In Drosophila, the order of linked loci was first inferred by pairwise comparisons and minimisation of the number of double or multiple recombinants to produce a linear map or linkage group. In the mouse, linkage maps are constructed along similar lines. The mouse data provide further justification for this simple approach since the number of double crossovers has been found to be even less frequent than expected, as a result of interference. The power of this approach increases as the data accumulate until the point is reached where all crossover sites can be inferred accurately. A means of calculating the relative probabilities of different gene orders and a best estimate of the likely location of a gene may still be required however.

Genetic markers

What markers are used for genetic mapping? Initially genes were mapped in man and mouse using either blood group variants or protein electrophoresis polymorphisms. There are comparatively few such markers and they are not highly **informative** for mapping studies – they show low heterozygosity. Since only the offspring of doubly or multiply heterozygous individuals provide information about the segregation of gametes in the offspring, these are said to be informative for linkage. The first RFLPs (Figure 1) were used in linkage studies in the early 1980s, which marked a revolution in gene mapping, since it became clear that there was now a limitless supply of polymorphic DNA markers. Increasingly, the labour intensive nature of RFLP typing and their comparatively low heterozygosity compared with newer markers, has led to their decline. The method that is currently in favour is to use microsatellite repeat polymorphisms. These markers can be selected to have a high heterozygosity (>0.5), and are relatively evenly spaced throughout the genomes of most mammalian species. They are easily and rapidly typed using the polymerase chain reaction (section III(3) and Figure 6) and have a low enough mutation rate to be suitable for linkage mapping. Recently, human genetic maps containing over 1000 microsatellites have been published. Standardised human genetic linkage maps are obtained by typing markers against a set of 40 large three-generation reference (CEPH) families.

The mouse genetic map

The laboratory mouse (*Mus musculus*) is one of the most intensively studied mammalian species, which has led to the identification of about 500 mutant phenotypes, including many that affect the eye, and close to 1000 genetically mapped loci. The cloning of the *small eye, rd* and *rds* loci in the mouse led directly to the identification of the corresponding human disease genes (discussed in Chapters

5, 6, 19). The powerful genetic and analytical resources that are available in the mouse together with the strong evolutionary conservation of large genomic regions between mouse and man has led to the increasing importance of mouse genetics. More recently, this has been further enhanced by the ability to create transgenic mice.

The primary resource for genetic mapping of an unknown locus in the mouse used to be **recombinant inbred (RI) strains** in which closely related laboratory strains were crossed and backcrossed to genetically fix the recombinant chromosomes on an otherwise inbred background (intraspecific crosses). In this way, the segregation of marker alleles that are distinguishable between the parental strains can be followed by typing the RI offspring. Unfortunately, many marker alleles are not distinguishable between the parental strains so that the RI offspring are uninformative - particularly within more interesting evolutionarily conserved regions such as genes. This problem has to a large extent been overcome by the introduction of inter-specific crosses between *M. musculus* and *M. spretus*, one of the most distantly related *Mus* species that still breeds with *M. musculus*. These species have diverged sufficiently so that the majority of markers differ between them. The offspring of the (*M. musculus* x *M. spretus*) F1 hybrids can therefore be typed directly and analysed for linkage. There are some minor drawbacks with the method, for example the F1 females but not the males are fertile so that only a female genetic map is produced. In this way, over 600 loci have been genetically mapped, producing an average map resolution of <3 cM.

In mice, the total genetic length is in the region of 1600 cM compared with 4000 cM in man despite the fact the the two species have almost identical amounts of DNA per cell (3×10^9 bp per haploid genome). The comparative map of man and mouse is increasingly important, since closely linked genes in one species tend to remain so in the other. The mean length of autosomal segments conserved between man and mouse is 10.4 ± 2.4 cM and approximately 90 conserved segments have been identified to date. Identification of linkage to a disease gene in man results in a search both for 'candidate' genes and for mouse mutants with a related phenotype either of which map to the corresponding conserved region of the mouse genome. This has proved useful in moving from mouse to man and vice versa. Several examples are discussed in detail in other chapters (Chapters 4, 5, 18, 19).

Transgenics and ES cell technology

The ability to create transgenic mice, incorporating foreign genes either with mutations or showing abnormal expression, has enormously enriched the field of molecular genetics. The function of genes can be investigated by producing mice with aberrant expression, overexpression or by mutating the normal gene. A mutant phenotype can be rescued by insertion of a "candidate" gene, providing the final proof that the correct gene has been identified. There are two main strategies for creating transgenic mice that are illustrated in Figure 5.

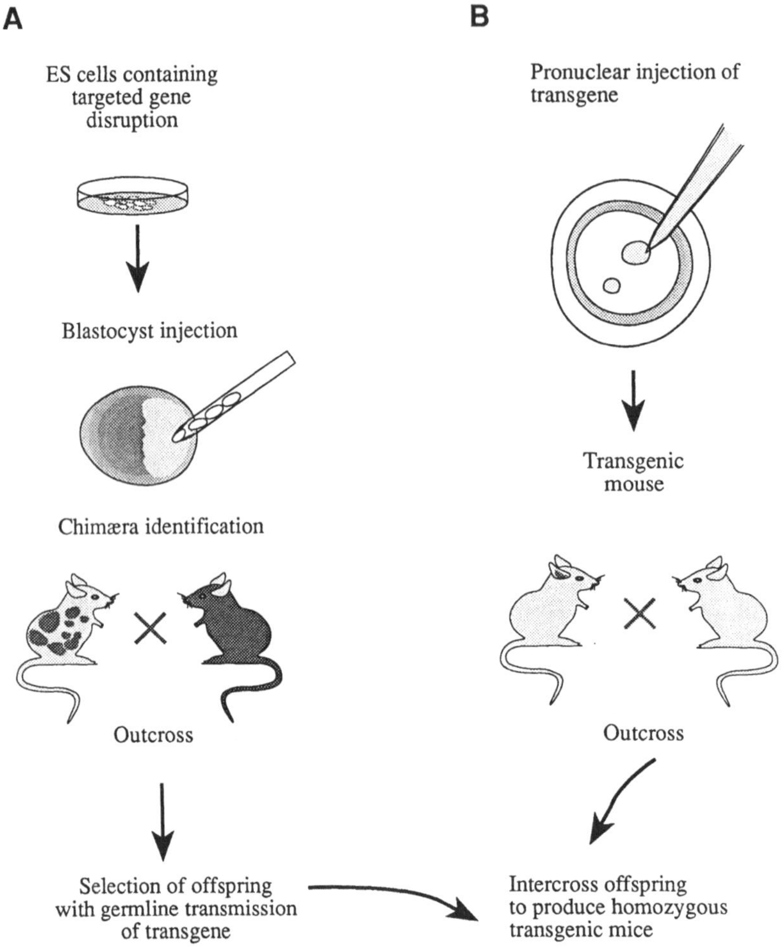

Figure 5 Transgenic experiments. Transgenic mice can be produced in two main ways. Mouse embryonic stam (ES) cells may be targeted by homologous recombination into a gene in order to introduce a specific mutation (e.g. "knockout"). (A) The targeted ES cells are microinjected into a blatocyst, reimplanted and chimaeric progeny identified by means of coat colour markers. These can be outcrossed to identify those capable of germline transmission of the transgene and intercrossed to produce homozygotes. Alternatively the male pronucleus within fertilised oocytes is microinjected with the transgene and after reimplantation into a pseudopregnant female, the progeny are outcrossed to identify those showing hemizygous expression of the transgene and then intercrossed to produce homozygous expression (B).

The first approach involves the use of totipotential mouse embryonic stem cells (ES cells), which can be modified by the introduction of gene constructs aimed at disrupting a gene either specifically by homologous recombination or non-specifically by insertional inactivation (Figure 5A). The gene construct can be introduced into ES cells by electroporation or microinjection. Disruption of a gene by homologous recombination in mitotically dividing cells is a low frequency event so that tedious selection and screening procedures are often required to detect the rare clones that have integrated the gene construct at the homologous site. Specific mutations can be introduced in this way or the gene disrupted by insertion of the construct. Once the transgene has been introduced into an ES cell, which is usually recognised by the acquisition of a drug-resistance phenotype, the cells are introduced into mouse blastocysts by microinjection and after reimplantation into a foster mother, chimaeric offspring can be identified by coat colour or other genetic markers. Further problems arise in finding chimaeras that express the transgene in the germline and so are able to transmit it to the offspring after outcrossing to an appropriate strain. Finally, intercrossing of germline chimaeras may produce identifiable homozygous offspring, unless the mutation is a homozygous lethal.

The second and often simpler strategy for producing transgenic animals involves microinjection of pronuclei within fertilised mouse oocytes with the gene construct, which are reimplanted into pseudo-pregnant females and lead to the production of mice that are hemizygous for the transgene (Figure 5B). The resultant progeny can again be outcrossed and intercrossed to obtain homozygotes. In this way, a wide variety of investigations can be carried out on gene expression and function including the "rescue" of mutant phenotypes. The homozygous *rd and rds* mutants, which are both associated with retinal degeneration, were partially rescued using pronuclear injection of the cloned β-subunit of cGMP phosphodiesterase and peripherin/RDS genes respectively (Chapters 5, 6). The technique can also be used to see if a dominant mutation in a cloned gene (e.g. rhodopsin, Chapter 6) confers an abnormal phenotype even in the presence of two normal copies of the gene. The final proof that a "disease gene" has been cloned, is to recreate the phenotype in the mouse with a mutant candidate gene or to rescue the normal phenotype in the mutant mouse by introduction of the candidate gene. In both cases, upstream regulatory regions must also be introduced in order to retain appropriate expression within the eye.

The use of the mouse for studying human genetic disorders does have limitations, but they are far fewer than was generally appreciated ten years ago. The *mdx* mouse has a remarkably mild phenotype compared with mutations in the human homologue of mouse dystrophin which causes the fatal disorder of Duchenne muscular dystrophy. Similarly, some of the transgenic mice developed as models of cystic fibrosis show differences from the human disorder. In general, there is still so much to be learnt from studying mouse mutants and why they may differ from the human disorder that their increasing use is easily justified. This is particularly true as gene therapy starts to come of age in the 1990s.

III THE TOOLS OF THE MOLECULAR GENETICIST

These methods are not unlike the simple machines, such as the lever or wheel, that underlie the workings of complex mechanical instruments. The basic methods or tools of molecular genetics can be considered under five headings:

(1) The **separation of nucleic acids** on the basis of size and structure.
(2) The **manipulation of nucleic acids** with specific enzymes or chemicals that enable molecules to be modified, cleaved, joined (ligated) or copied.
(3) The **production of specific nucleic acids** by enzymatic or chemical synthesis or by introduction into microorganisms (cloning).
(4) The **visualisation and quantification of nucleic acids** by non-covalent binding of stains or dyes, by covalent attachment of detector molecules and by ultraviolet absorbance.
(5) The **hybridisation of nucleic acids** to complementary single-stranded nucleotide probes in order to detect specific sequences within complex mixtures.

Under each of these headings there are numerous variations and modifications that allow a broad range of tasks to be carried out on a variety of nucleic acid sources. The resultant analyses can involve the discrimination of single nucleotide differences between two genes but are also capable of assessing enormous DNA structures such as chromosomes. All routine laboratory methods in molecular genetics use these five principles. A brief overview will be given of each tool and its principal applications, followed by a more detailed table of specific methods within each category.

1) Separation of nucleic acids on the basis of size and structure

Nucleic acids are routinely separated by gel electrophoresis. For relatively small, linear double-stranded (ds) DNA molecules (<50 kb), migration within an agarose gel in the presence of an electric field is a function of molecular size rather than conformation or sequence. The number of negatively charged groups on the phosphate backbone is proportional to the number of base pairs, so that the charge density is constant. The different nucleotide residues have little effect on the charge behaviour of DNA, although this is less so for electrophoresis in polyacrylamide. DNA sequence analysis and high resolution separation of molecules up to about 1000 nucleotides in length (e.g. sequencing or microsatellite analysis) is generally carried out using polyacrylamide gel electrophoresis under denaturing conditions. In the case of single-stranded (ss) DNA, the different nucleotides can have a major effect on the conformation of the molecule, which can alter its behaviour in a gel unless denaturants are used to eliminate the secondary structure. This is exploited in the technique of single strand conformational polymorphism analysis (SSCP) in order to detect single base changes for mutation screening.

Large molecules of DNA (>50 kb) would normally be unable to migrate within an agarose gel, but the use of pulsed-field and field-inversion gel electrophoresis (PFGE, FIGE) exploit a changing electric field to alter their secondary structure so that they can migrate at a rate that is inversely proportional to size. In contrast to yeast chromosomes, human chromosomes are much too large to be separated by PFGE and have first to be cleaved with "rare cutter" restriction endonucleases into fragments below about 2×10^6 base pairs (2 Mb) in size. Yeast or bacterial artificial chromosomes (YACs, BACs) and P1 vectors have been used to clone human, mouse or Drosophila DNA fragments in the size range that can be separated and purified by PFGE.

Flow cytometry or "sorting" is based on a different principle than electrophoresis since it distinguishes chromosomes (or cells) on the basis of their ability to bind fluorescent dyes (e.g. bisbenzimide, DAPI) that are detected after laser excitation. A dilute solution of chromosomes is passed through a laser and the emitted fluorescent signal is detected and a magnetic field is activated to divert the electrically charged water droplet containing the chromosome of interest into a tube. The separation of chromosomes is therefore based on a combination of size and the structural properties that allow the binding of dyes.

Table 1 illustrates the different methods of nucleic acid separation, the size range and sources of the molecules separated and special applications of each technique.

2) Manipulation of nucleic acids with specific enzymes or chemicals

DNA can be manipulated by a wide variety of specific enzymes, by chemical reagents, radiation and physical shearing. These methods allow nucleic acids to be specifically cleaved, modified by methylase, phosphatase or kinase enzymes, joined and synthesised. The following table summarises the major methods by means of which nucleic acids are altered. Only a small number of the large repertoire of enzymes and chemicals is listed but table 2 gives some indication of the scope of these tools.

3. Production of pure nucleic acids – synthesis, cloning

Specific pieces of DNA can be produced in large quantities to allow for direct study or use as molecular tools. This amplification process can be carried out by **chemical synthesis** (as in the manufacture of oligonucleotides), by enzymatic amplification using the **polymerase chain reaction (PCR)**, or by **cloning into vectors** such as plasmids, cosmids, bacteriophage, P1 or into yeast artificial chromosomes (YACs) (Table 3).

Chemical synthesis of DNA is limited to molecules of less than about 100 nucleotides in length and is widely used for synthesising specific oligonucleotide sequences (primers) for priming PCR and sequencing reactions. Small synthetic oligonucleotides that act as adaptors or linkers for modifying DNA ends are also useful.

The production of DNA by enzymatic amplification using PCR is central to a wide range of molecular genetic experiments. It is illustrated and explained in Figure 6. PCR synthesis of DNA uses the other fundamental tools, for example annealing of primers to the DNA template, replication of DNA using heat-stable polymerase, visual detection (as part of screening and analysis) and purification of the product by electrophoresis but it is of such fundamental importance that it almost amounts to a basic tool in itself.

Table 1 Fractionation of DNA by Size

DNA Type	*Method*	*Size Range*	*Special Uses*
Chromosomes, human	Flow sorting Metaphase spreads	30–200 Mb	Chromosome-specific libraries FISH, microdissected libraries
Chromosomes, yeast	PFGE	0.1–10 Mb	Purification of YAC DNA
Genomic DNA	PFGE FIGE Agarose	0.1–10 Mb 25 kb–2 Mb 100 b–50 kb	Large-scale genomic mapping DNA purification, mapping Mapping, RFLPs
RNA	Agarose, denaturing	1–20 kb	Sizing and analysis of RNA
dsDNA fragments partial dsDNA fragments	Polyacrylamide Polyacrylamide or Hydrolink	15–1500 b	Sizing and analysis of dsDNA SSCP analysis Heteroduplex analysis DGGE
ssDNA fragments	Polyacrylamide, denaturing	1–500 b, single base resolution	Sequencing, S1 nuclease, microsatellite analysis Gel retardation assays
RNA fragments (small)	Polyacrylamide, denaturing	50–1500 b	RNase protection assays – to define transcribed regions

Table 2 Methods for Modifying Nucleic Acids

Process	*Type*	*Reagents*	*Comments*
Cleavage of DNA	<u>Nonspecific</u>		
	Enzymatic	DNaseI	Use for radio-labelling DNA by nick translation
	Chemical	1,10 phenanthroline-Cu, EDTA-Fe^{+2}	
	Radiation	UV, Xray	Fragment size determined by dose
	Physical	Shearing, ultrasound	
	<u>Specific for structure</u>		
	a) Single-stranded		
	Non-specific	S1 nuclease, micrococcal nuclease	DNA/RNA hybrid analysis for transcript and intron mapping
	Ends only	Mung bean nuclease	
	b) Double-stranded		
	Ends only	λ exonuclease, Bal31 nuclease	
	<u>Specific for sequence</u>		
	Enzymatic	Restriction endonucleases	4-8 b recognition sequences
	Chemical	OsO_4, hydroxylaminine + piperidine	Mismatch cleavage used in mutation detection

Table 2 Continued.

Process	*Type*	*Reagents*	*Comments*
Cleavage of RNA	Non-specific	RNase A	Transcript mapping
	Specific for structure		
	Single-stranded	S1 nuclease	RNase protection
	Double-stranded	RNaseH	
	RNA/DNA hybrid	RNaseH	cDNA synthesis
	Specific for sequence	RNase M1, T1, T2	RNA sequencing
Modification	Methylation	Methylase	Protects against R.E.
	Removal of 5' phosphate	Phosphatase	Prevents ligation
	Addition of 5' phosphate	Polynucleotide kinase	Used to label with ^{32}P
Ligation (joining)	Blunt ends	T4 ligase	Covalent joining of
	Overlapping ends	T4 ligase	DNA molecules
	Single-stranded	RNA ligase	
Synthesis	Chemical	Oligonucleotide synthesis	Up to 100 bases
	Enzymatic		
	Not sequence-specific		
	3' Tailing	Terminal deoxynucleotide transferase (Tdt)	Labelling, cloning
	Sequence-specific		
	DNA template		
	DNA product	DNA polymerase I, Klenow enzyme, T4 polymerase, Taq polymerase	Labelling, cloning, sequencing, thermostable enzyme used for PCR
	RNA product	RNA polymerase	RNase protection, *in vitro* transcription, cosmid walking
	RNA template		
	DNA product	Reverse transcriptase	cDNA cloning, RT/PCR

Cloning of DNA results in the ability to produce large amounts of a specific DNA as well as of its gene products – RNA and protein. The essence of cloning is that DNA can be introduced into foreign cells and stably propagated. The DNA of interest is joined to vector sequences that allow the cell to replicate the foreign genetic material and that contain selectable markers (e.g. antibiotic resistance genes) which can be used to select recombinant clones – those that contain a foreign insert. The production of pure DNA of known origin or sequence is of such fundamental importance that it must be considered a primary tool.

4. Nucleic acid visualisation and quantification

DNA can be visualised by a variety of methods, both specific and non-specific, that include colorimetric, radioactive, fluorescent, chemiluminescent and cyto-

Table 3 Production of Nucleic Acids: Synthesis and Cloning

	Size Range	*Method*	*Vector*	*Host*	*Product*	*Purification*
Deoxynucleotide sequence	2–100 nt	Chemical	NA	NA	Oligonucleotide	Gel, HPLC, Column
RNA						
Known short sequence	2–100 nt	Chemical	NA	NA	Oligoribonucleotide	Gel, HPLC, Column
Cellular RNA	100nt–5 kb	RT/PCR	NA	NA	ds-DNA (cDNA)	Gel
	100nt–20 kb	Reverse transcription	NA	NA	cDNA	Column, Gel
cDNA (made from RNA)	10 nt–20 kb	Cloning	Plasmid, phage	E.coli	Vector with insert	Column, Banding, Gel
Cloned DNA	10 nt–20 kb	In vitro transcription	NA	NA	RNA	DNaseI, Gel
	10 nt–40 kb	Subcloning	Plasmid, phage, cosmid	E. coli	Vector with insert	Column, Banding, Gel
Genomic DNA	100 nt–5 kb	PCR	NA	NA	ds-DNA	Gel, Filter, Column
	100 nt–20 kb	Cloning	Plasmid, phage	E.coli	Vector with insert	Column, Banding, Gel
	15 kb–40 kb	Cloning	Cosmids	E.coli	Vector with insert	Column, Banding, Gel
	15 kb–150 kb	Cloning	P1/BAC	E. coli	Vector with insert	FIGE, PFGE
	100 kb–1 Mb	Cloning	YAC	Yeast	Vector with insert	PFGE

Table 3 (Continued).

	Size Range	*Method*	*Vector*	*Host*	*Product*	*Purification*
Chromosomal fragments	100nt–3 kb	Microdissected/ PCR	Plasmid/ phage	E.coli	Vector with insert	Column, Banding, Gel
	200 nt–200 kb	Flow-sorted/ Cloning	Plasmid/ phage/ cosmid/YAC	E.coli Yeast	Vector with insert	Column, Banding, Gel
	25 kb–100 Mb	Radiation Hybrids	NA	Rodent cells	Complex mixture of limited human fragments and normal rodent chromosomes	Not usually purified for use, although occasionally flow-sorted
Chromosomes	50–200Mb	Somatic Cell Hybrids-cell fusion	NA	Rodent cells	Complex mixture of one or more donor chromosomes and rodent genome	Not usually purified for use, though occasionally flow-sorted

logical stains or dyes. Selection of the method depends on the requirements for sensitivity, specificity and convenience. Ethidium bromide staining of DNA within gels is routinely used to monitor the concentration of specific bands by examining the fluorescence emitted after ultraviolet (UV) irradiation. Quantification of nucleic acids can be carried out more accurately by measuring the absorbance of UV light at 260 nm or by fluorimetry after binding of fluorescent dye. Table 4 is not comprehensive but highlights the choices available for different types of experiment.

5. Specific sequences of DNA can be identified by hybridisation

Complementary strands of nucleic acid will anneal or hybridise to each other and form stable structures even under unfavourable conditions such as elevated temperature and low salt (high stringency). Hybridisation reactions in which one strand (the "probe") is radiolabelled and present at high concentration can be used to recognise specific sequences of immobilised DNA within complex mixtures with very high sensitivity. As little as a few picograms of nucleic acid can be detected using these methods (Table 5, Figure 1). For example 10μg of mammalian genomic DNA contains only 3pg of a 1 kb single copy sequence (1 in 3×10^6) which can be readily detected by Southern hybridisation using a high specific activity probe.

When double-stranded molecules are formed, they align so that the "head" (5′ end) of one strand is matched with the "tail" (3′ end) of the other strand. The nucleotide bases are paired so that adenine and thymine form a stable structure, held together by two hydrogen bonds, while guanine and cytosine are more tightly bound by three hydrogen bonds. Since they do not form covalent bonds, the two strands can be relatively easily denatured and separated by heat, low salt and chemicals such as alkali, urea and formamide. By varying the hybridisation or washing conditions, hybrids can be selected that are either very well matched or contain a significant proportion of mismatched bases. Similar considerations apply to the use of primer annealing reactions during amplification of DNA by PCR (Figure 6).

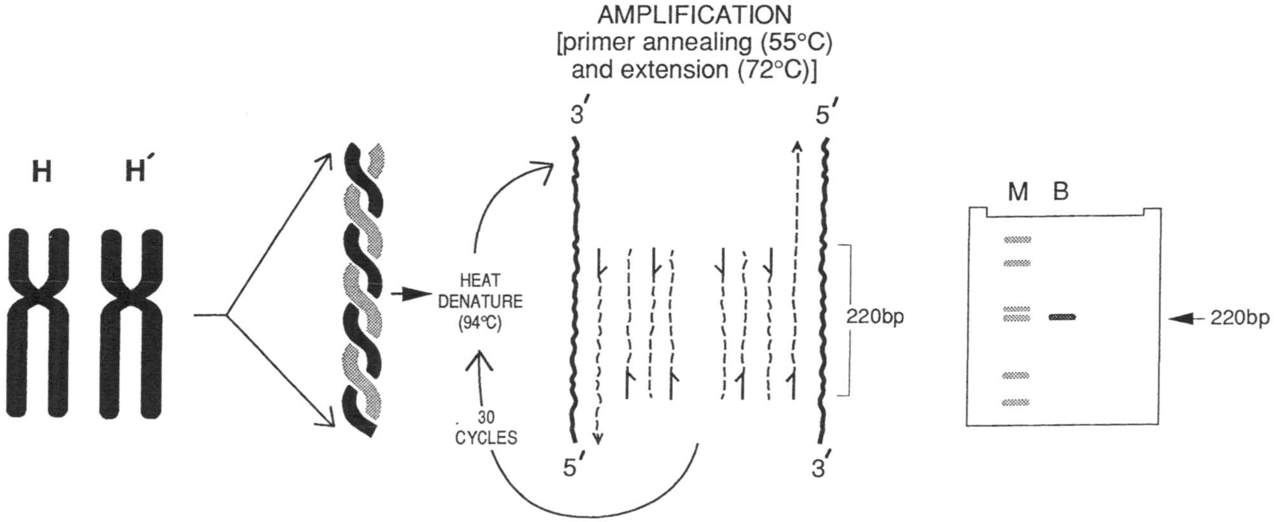

Figure 6 Use of primer annealing reactions during amplification of DNA by PCR.

Table 4 Detection of Nucleic Acids

Type of Indicator	*Specificity/ Incorporation*	*Reagents*	*Detection*	*Uses*
NONSPECIFIC DETECTION				
Electron structures microscopy (shadowing) EM	Can distinguish ss and ds DNA	Carbon	Electron Microscopy	For DNA rarely used
Stains				
Nonfluorescent	Chromosome banding	Giemsa	Light photomicroscopy	Cytogenetics
	Nonspecific	Silver Gold	Direct visualisation	DNA/RNA on gels (very sensitive)
Fluorescent	Chromosome banding	Quinacrine DAPI	Fluorescence microscopy	Cytogenetics
	Nonspecific (ds > ss) (ds + ss)	Ethidium Acridine Orange	Fluorescence photography	DNA/RNA on gels
DETECTING SPECIFIC DNA FRAGMENTS				
Direct Detection				
Radioisotopes				
End-labelled	Dephosphorylated 5' end 3' tailing reaction	γ^{32}P-ATP α^{32}PdNTP, ^{35}S-dNTP	Autoradiography	Sequencing, Hybridisation
Incorporation by synthesis	PCR, nick-translation, random primer synthesis, in vitro transcription	α^{35}P-33P-, ^{35}S-dNTP ^{32}P-,^{33}P-, ^{35}S-NTP	Autoradiography	Sequencing, Hybridisation
Fluorescent adducts				
Chemically incorp.	Oligonucleotide synthesis	Fluorescein, rhodamine derivatives	Fluorescence	Sequencing Hybridisation
Internal incorporation	PCR, nick translations, random primer synthesis	Fluor-dUTP	Fluorescence	Hybridisation FISH
Incorporation with labelled primer	PCR	Derivatized oligonuc-leotide	Fluorescence	Sequencing

Table 4 Continued.

Type of Indicator	*Specificity/ Incorporation*	*Reagents*	*Detection*	*Uses*
Secondary Detection				
Enzyme adducts to DNA				
Chemically linked	glutaraldehyde crosslink amino-modified oligo	HRP, alkaline phosphatase	Stains, Chemilum	Hybridisation In situ hybrid
Avidin-conjugates				
Biotinylated probe	Internal incorporation	avidin-FITC	Fluorescence microscopy	FISH Hybridization
	End-labelling Photobiotinylation	avidin-HRP	Stains, Chemilum	
Antibody-enzyme conjugates				
Incorp. by synthesis	PCR, nick translation, random primer synthesis in vitro transcription	Alk. ptase-, HRP- β-galacto-sidase-conjugates	Chemiluminescence, colorimetric stains	Hybridization
Digoxigenin Fluorescein	digoxigenin-dUTP fluor-dUTP	Substrates with. chemilum. or colored products		

Table 5 Hybridisation of Nucleic Acids

Name of method	*Probe*	*Target Component*	*Immobilized*	*Separation*	*Support*	*Detection*
Southern	Oligo, PCR product, Cloned DNA	DNA	Target	Electro-phoresis	Membrane	Probe
Northern	Oligo, PCR product, Cloned DNA	RNA	Target	Electro-phoresis	Membrane	Probe
DNA or RNA dot blots	Oligo, PCR product, Cloned DNA	DNA or RNA	Target		Membrane	Probe

Table 5 Continued.

Name of method	*Probe*	*Target Component*	*Immobilized*	*Separation*	*Support*	*Detection*
In situ hybridisation of RNA transcripts	Oligo-nucleotides, PCR products, Cloned DNA, RNA	RNA		Target	Glass slides	Probe
PolyA$^+$ mRNA purification	oligo dT	RNA	Probe	Washing	Membrane, column, magnetic beads	cDNA
Reverse transcription (RT)	Primer or oligo dT	RNA	None	Primer-directed synthesis	NA	Nonspecific or direct detection
Polymerase Chain Reaction (PCR)	Primers	DNA	Product	Amplify specific fragments	Electro-phoresis	Nonspecific or direct detection
RNAse protection	RNA	RNA	Product	Resistance to RNAse	NA	Nonspecific or direct detection
Allele-specific oligonucleotide (ASO)	Oligo-nucleotide	DNA	Probe	Retention of target to bound probe	Membrane	Bound target
In situ hybridisation	Cloned DNA	Chromo-somes	Target	Retention of probe	Glass slides	Bound probe
FISH, Chromosome painting	YACs, PCR products					

ACKNOWLEDGEMENTS

We would like to thank Andrew Carothers and Ian Jackson for critically reading parts of the manuscript. M.B.G is supported in part by unrestricted funds from Research to Prevent Blindness, Inc.

REFERENCES

Casey, D. (1992) Primer on Molecular Genetics In: *Human Genome 1991–92 Program Report.* U.S. Department of Energy, Washington, D.C.

Davies, K. E. and Read, A. P. (1988) *Molecular Basis of Inherited Disease.* IRL Press, Oxford, 77 pp.

King, R. C. and Stansfield, W. D. (1992) *A Dictionary of Genetics,* (Fourth ed.) Oxford University Press, Inc, New York, 406 pp.

Thompson, M. W., McInnes, R. R., Willard, H. F. (1991) *Thompson & Thompson Genetics in Medicine,* (Fifth ed.) W. B. Saunders Co., Philadelphia, 500 pp.

2. RETINAL DEGENERATION MUTANTS OF *Drosophila*

WILLIAM L. PAK

Department of Biological Sciences, Purdue University, West Lafayette, Indiana 47907, USA

KEY WORDS: *Drosophila*, retinal degeneration mutants

ABSTRACT

Six genes which, when defective, cause photoreceptor degeneration have now been cloned and analyzed in *Drosophila*. Most of these genes were identified from mutants generated by chemical mutagenesis and isolated on the basis of their defects in the electroretinogram or visual behavior. The first gene to be cloned was that encoding opsin in R1-6 photoreceptors, which roughly correspond to rod photoreceptors of vertebrates. The first mutation to be isolated in this gene was a nonsense mutation that causes severe photoreceptor degeneration. Recently, a human counterpart of this mutation has been identified. Most of the six genes encode proteins that are directly or indirectly implicated in phototransduction in this organism. The protein product of one of them, *rdgC*, however, has been implicated in a rhodopsin-activated pathway distinct from phospholipase C-mediated phototransduction. Where multiple alleles of a given gene have been studied, strong allele dependence of retinal degeneration has been found. The severity and the time course of retinal degeneration can vary greatly among different mutants in a given gene. Retinal degeneration in mutants of all genes studied except those in the rhodopsin gene show strong light dependence. In these mutants, structural deterioration can be prevented from occurring or considerably slowed down by raising the flies in the dark. In most of these mutants, however, functional defects are already present before the onset of degeneration. In mutants of all six genes, R1-6 photoreceptors are more susceptible to degeneration than either just R7 or both R7 and R8 photoreceptors. Eventually, however, R7 and R8 photoreceptors, which roughly correspond to cones, also degenerate. This is true whether the defective gene is expected to be expressed just in R1-6 photoreceptors, as in R1-6 opsin gene mutants, or in both R1-6 and R7/8 classes of photoreceptors, as in mutants of the other genes.

I INTRODUCTION

Many *Drosophila* mutations are known to affect the gross morphology of the compound eye through their effects on eye development (e.g. Renfranz and Benzer,

1989). This review, however, will focus on mutants characterized by age-dependent degeneration of the retina. Moreover, the discussion will be largely limited to those mutants for which the corresponding genes have been identified and cloned. By and large, in these mutants, the gross morphology of the eye is normal, and nearly or completely normal adult retinal structures form initially but degenerate with age.

Previous to the late 1960s the only eye mutants available in *Drosophila* were those that affect the eye color or the external morphology of the eye. Beginning in the late 1960s, several laboratories began chemically mutagenizing *Drosophila* and isolating mutants that are defective in visual behavior or light-evoked electrical responses of the eye (electroretinogram, ERG) (Hotta and Benzer, 1969; Pak et al., 1969; Heisenberg, 1971; Koenig and Merriam, 1977). Each of these laboratories had different experimental objectives and employed different mutant selection schemes. For example, the Benzer laboratory was interested in dissecting fly behavior and isolated mutants defective in phototaxis (Benzer, 1973); the Pak laboratory was looking for mutants defective in the ERG for use in dissection of the phototransduction process (Pak et al., 1969); and M. Heisenberg, then working in K. Gotz' laboratory in Tübingen, Germany, isolated mutants defective in the optomotor behavior for use in the analysis of the integration of visual information (Heisenberg, 1972). The initial mutagenesis schemes by these three laboratories targeted the X chromosome. Mutagenesis of the autosomes for the isolation of ERG-defective mutants was carried out by the Merriam laboratory (Koenig and Merriam, 1977) and the Pak laboratory (reviewed by Pak, 1979). All these approaches, however, yielded many mutants of the same genes, and some of the mutants isolated were found to display retinal degeneration. In addition, Cosens and Manning (1969) described a spontaneously arising ERG-defective mutant, which turned out to be a retinal degeneration mutant (Cosens and Perry, 1972). More recently, Steele and O'Tousa (1990) have isolated retinal degeneration mutants by looking for structural deterioration with age using an optical technique (see next section).

Described in Section III are retinal degeneration mutants of six genes, all of which have now been cloned. Preliminary analysis of other ERG-defective mutants suggests that there are many more genes which, when defective, cause retinal degeneration. Although detailed information is available only on a small fraction of mutants that display retinal degeneration, the information that is becoming available is beginning to provide insights into some common properties shared by many of these mutants.

II STRUCTURE OF THE *Drosophila* COMPOUND EYE AND METHODS OF MONITORING RETINAL DEGENERATION

The *Drosophila* compound eye consists of approximately 800 repeating units called the ommatidia. Each ommatidium contains 20 cells, eight of which are photoreceptors (Figure 1). Each photoreceptor has a membrane structure, called the rhabdomere, consisting of tightly packed microvilli that project laterally toward the center of the ommatidium. These structures are specialized for light reception and phototransduction, contain visual pigment molecules as integral membrane proteins of the microvillar membrane, and thus correspond to the outer segments of vertebrate photoreceptors. Six of the photoreceptors, R1-6, have rhabdomeres that extend the full depth of the retina and that are arranged in a trapezoidal pattern in cross-section (Figure 1). R1-6 photoreceptors all contain the same visual pigment, R1-6 rhodopsin, and roughly correspond to rods of vertebrates. The remaining two photoreceptors, R7 and 8, have rhabdomeres near the center of the ommatidium, with R7 rhabdomere occupying a position directly above the R8 rhabdomeres. R7/8 photoreceptors have different visual pigments and roughly correspond to vertebrate cones. Pigment cells, which contain screening pigments, pterins and ommachromes, surround each ommatidium (Figure 1), optically isolating individual ommatidia. For many applications, it is convenient to eliminate these pigments genetically. In fact, many of the studies summarized in Section III were done in the absence of these pigments. Although these pigments do not affect the transduction process, they do affect the amount of light reaching the rhabdomeres and can affect the severity and time course of degeneration in light-dependent degeneration mutants.

Retinal degeneration in *Drosophila* can be monitored in a number of different ways. Although light and electron microscopy of the retina carried out as a function of age is the most direct and reliable method of studying degeneration, the method is time-consuming and labor-intensive. It is, therefore, desirable to have other, simpler means of monitoring degeneration. Monitoring the ERG or the deep pseudopupil as a function of age often yields reliable indications of degeneration, although details of structural deterioration are lacking. In most retinal degeneration mutants, the ERG is already abnormal at eclosion before any structural deterioration is apparent. As the retina degenerates, the ERG deteriorates further with age. Thus, deterioration of the ERG with age is a reliable initial indication of retinal degeneration. The deep pseudopupil is a microscopically observed superposition of virtual images of about 20 neighboring ommatidia (Franceschini and Kirschfeld, 1971). As the retina degenerates, the deep pseudopupil becomes indistinct and disappears. Because superposition of virtual images is dependent on the correct structural relationship of neighboring ommatidia, the technique is sensitive to any structural misalignment of ommatidia. Consequently, this technique tends to detect defects before they are seen histologically, and the time course of degeneration determined by this method tends to be faster than that obtained histologically.

In monitoring retinal degeneration, the life span of *Drosophila* is measured in days or weeks post-eclosion. While it is difficult to define a *Drosophila* time scale corresponding to that of human, convenient yard sticks to compare are the generation time and the age at which *Drosophila* start dying off. At 25° C, the generation time of *Drosophila* is approximately 10 d, and they start dying off at about 5 wk post-eclosion under laboratory conditions.

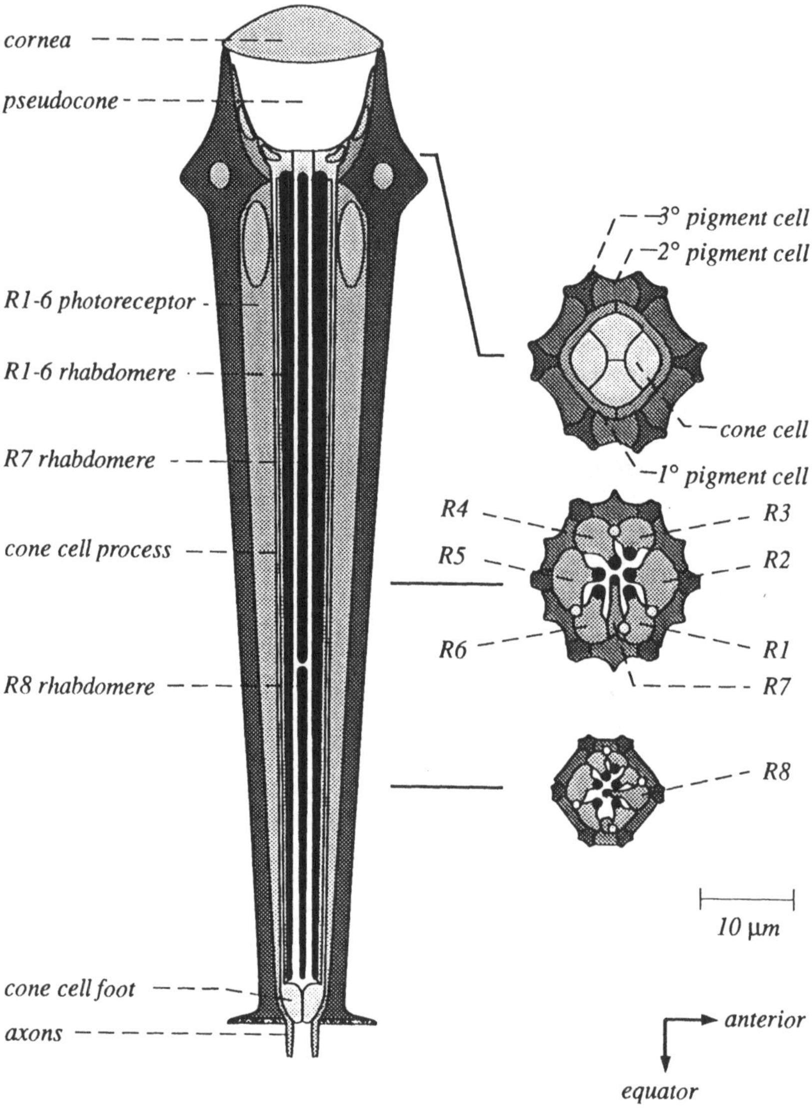

Figure 1 Structure of *Drosophila* ommatidium. A repeating unit in the compound eye, the ommatidium, contains eight photoreceptors, R1 to R8. Each photoreceptor has a rhabdomere, consisting of tightly packed microvilli, projecting laterally into the center of the ommatidium (shown in black). Each ommatidium is surrounded by primary (1°), secondary (2°), and tertiary (3°) pigment cells. R1-6 photoreceptors roughly correspond to rods, R7/8 photoreceptors to cones, and the rhabdomeres to the outer segments of vertebrate photoreceptors. Adapted from Wolff and Ready (1993). Reproduced with permission from the Cold Spring Harbor Press.

III DESCRIPTION OF MUTANTS

1 Mutants of the Rhodopsin Gene, *ninaE*

The R1-6 opsin gene of *Drosophila, ninaE,* which corresponds to the rod opsin gene of vertebrates, was identified from a class of mutants in which the amount of R1-6 rhodopsin is greatly reduced. These mutants, and the corresponding genes, were named *nina* (*neither inactivation nor afterpotential*) because of the characteristic electrophysiological phenotype displayed by the mutants (Pak, 1979; Stephenson et al., 1983). Since these mutations apparently disrupt either rhodopsin synthesis or its maintenance, it was reasoned that one of the five *nina* genes then identified, *ninaA, B, C, D,* or *E,* might code for the R1-6 opsin protein itself. Two criteria were used to identify the R1-6 opsin structural gene among the five *nina* genes: cell line specificity and the gene dosage effect. If different classes of opsin are encoded by different genes, as seemed likely, one would expect that mutations in the R1-6 opsin gene would affect the visual pigment content only in R1-6 photoreceptors and not in R7/8 photoreceptors. The gene dosage effect refers to the approximately linear relationship found between the number of copies of a gene (gene dosage) and the amount or activity of the protein encoded by the gene (e.g. review by O'Brien and MacIntyre, 1978). Of the five *nina* genes tested, only the *ninaE* gene satisfied both these criteria, identifying it as the prime candidate for the R1-6 opsin gene in *Drosophila* (Scavarda et al., 1983). The *ninaE* gene was mapped using deletions to the 92B region on the right arm of the third chromosome (O'Tousa et al., 1985) and was cloned by interspecies hybridization using bovine rhodopsin cDNA or RNA as probe (O'Tousa et al., 1985; Zuker et al., 1985).

Comparison of the deduced *Drosophila* protein sequence with the vertebrate rod opsin sequences then available, bovine (Ovchinnikov et al., 1982; Nathans and Hogness, 1983) and human (Nathans and Hogness, 1984), showed that it had all the structural features of opsin. For example, the protein had seven transmembrane segments (Figure 2), the chromophore-binding lysine residue near the middle of the seventh transmembrane segment, several C-terminal serine and threonine residues that could serve as phosphorylation sites, and several helix-breaking proline residues scattered throughout the transmembrane segments. When aligned for maximum homology, the *Drosophila* protein showed about 36% overall sequence identity with the mammalian proteins. Most of the sequence identity could be accounted for by several clusters of very strong sequence identity. The work, thus, established the basic structural similarity of *Drosophila* R1-6 opsin to mammalian opsins and identified several regions of strong sequence homology which could be functionally important.

The first indication that a mutation in the rhodopsin gene might cause structural deterioration of photoreceptors was obtained in *Drosophila*. In the early seventies, J. Koenig working in J. Merriam's lab at UCLA, isolated a mutant, later named *ora*^{JK84} (*outer rhabdomeres absent*), that seemed to lack R1-6 rhabdomeres (Harris

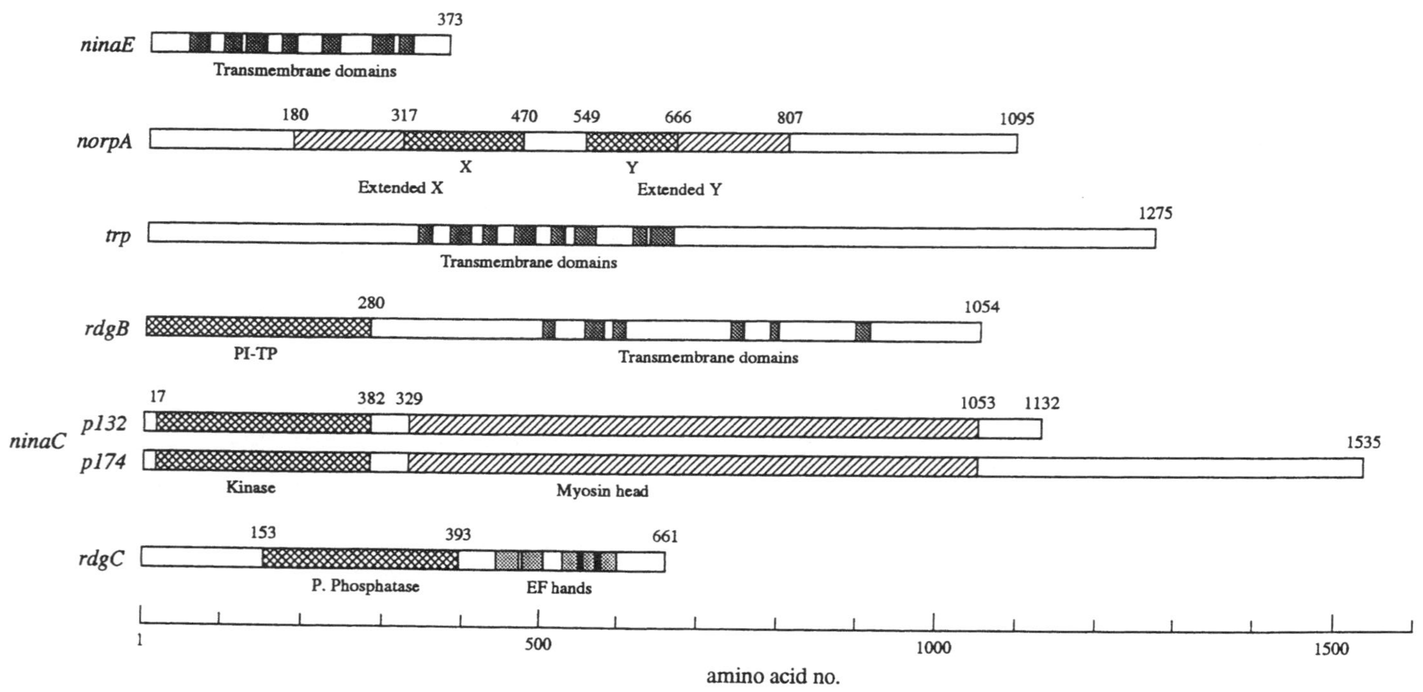

Figure 2 Structure of the proteins encoded by retinal degeneration mutant genes. Proteins encoded by the *ninaE*, *trp*, and *rdgB* genes contain potential transmembrane domains. Other regions of interest are also indicated. X and Y: Boxes X and Y, regions of highest sequence identity among known phospholipase Cs. Extended X and Y include additional regions that also share high sequence identity among phospholipase Cs (R. Shortridge, unpublished). PI-TP: domain homologous to phosphatidylinositol transfer protein. Kinase: protein kinaselike domain. Myosin head: domains with homology to the head region of myosin heavy chain. P. phosphatase: protein phosphatase-like domain. EF hands: domain containing EF hand motifs.

et al., 1976; Koenig and Merriam, 1977). Subsequent studies showed that small and deformed but recognizable R1-6 rhabdomeres are present in *ora*^{JK84} mutants at eclosion but quickly disappear with age (Schinz et al., 1982; Stark and Sapp, 1987; O'Tousa et al., 1989). It was soon realized from complementation tests that this mutant carries a lesion in the R1-6 opsin gene, ninaE (Scavarda et al., 1983). However, complementation tests also revealed that the mutant seemed to be carrying another mutation. O'Tousa et al. (1989) showed that (1) *ora*^{JK84} is, in fact, a double mutant carrying a mutation in the *ninaE* gene and another one in a closely linked gene and (2) the lesion in the *ninaE* gene is solely responsible for the structural defect in R1-6 photoreceptors.

Since then, photoreceptor degeneration has been found in all *ninaE* mutants examined to date. Leonard and Pak (1984) examined, at the light microscope level, the photoreceptor morphology of nine different *ninaE* mutants up to the age of nine weeks post-eclosion. Subsequently, Leonard et al. (1992) carried out detailed electron microscopic studies on five of these mutants for which the molecular defects in the *ninaE* gene had been elucidated (Washburn and O'Tousa, 1989; summarized in Figure 3 insert). The defects vary from a small deletion in the 5' region resulting in no transcripts (*ninaE*^{oI17}) to missense mutations resulting in single amino acid substitutions (*ninaE*^{P318} and *ninaE*^{P332}). The nonsense mutation carried by *ninaE*^{P352} is identical to that carried by *ora*^{JK84} in the *ninaE* gene except that the latter is a double mutant carrying another mutation. Degeneration of photoreceptors was found in all these mutants. It is light-independent in that illumination cycles have little or no effect on the course of degeneration. The time course of degeneration varies strongly among mutants, with mutants having the most severe defects showing the fastest degeneration (Figure 3). In strong *ninaE* mutants, R1-6 rhabdomeres almost completely disappear by 3 d post-eclosion, while in weaker mutants, many R1-6 rhabdomeres are still present at 9 wk post-eclosion at room temperature. The degeneration first affects R1-6 rhabdomeres, but eventually R1-6 cell bodies also become affected. By the time the degeneration of R1-6 cells is well under way, even the R7/8 rhabdomeres, in which different classes of opsin are expressed, also start degenerating. An observation of some import is the occurrence of membrane proliferations at the base of R1-6 rhabdomeres in many of these mutants at about the time these rhabdomeres are degenerating. Each of these structures consists of a closely apposed pair of membranes, one end of which is continuous with the plasma membrane at the base of R1-6 rhabdomeres and the other end of which extends into the cytoplasm and is closed. As a working hypothesis, Leonard et al. (1992) proposed that this accumulation of membranes might be caused by the failure of newly inserted membranes to be assembled into R1-6 rhabdomeres. A similar suggestion was made by Stark and Carlson (1983) to explain the morphological defect of *ora*^{JK84}.

An important recent development in retinitis pigmentosa (RP) research is the finding that a surprisingly large fraction of RP is caused by various mutations in the rhodopsin gene (Dryja et al., 1990; 1991; Sung et al., 1991). In fact, Rosenfeld et al. (1992) have recently reported on an autosomal recessive patient with a defect almost identical to that carried by *ora*^{JK84} or *ninaE*^{P352} suggesting an incred-

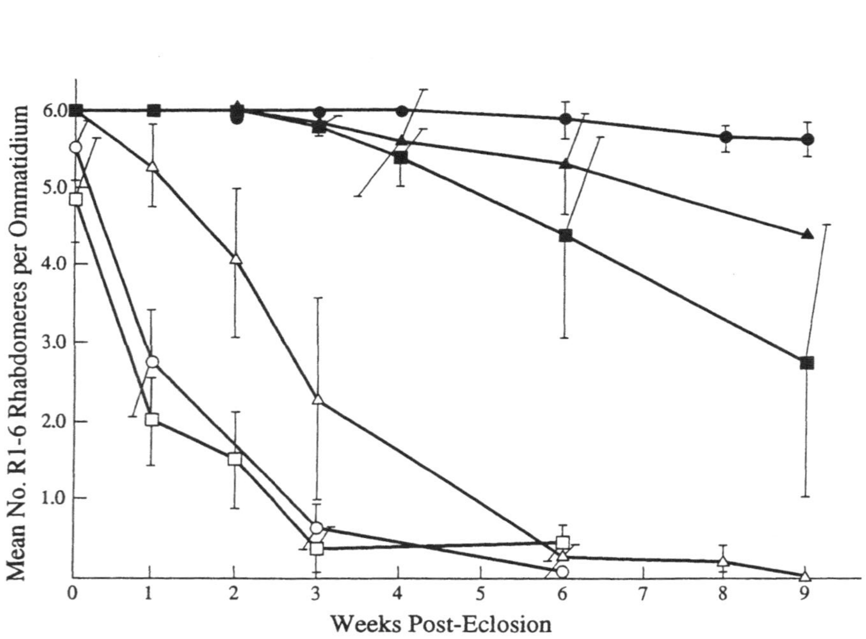

	Molecular Defects	Protein or DNA Region Affected
P318	Pro (120) Leu	Interface between 1st extracellular loop and 3rd transmembrane segment
P322	Gly (128) Arg	3rd transmembrane domain
P334	Thr (283) Met Trp (289) Arg Cys (297) Ser	6th transmembrane domain
P352	Gln (251) nonsense	Truncation in the 3rd cytoplasmic domain
ol17	Deletion	5' flanking region and 5' coding region

Figure 3 Allele dependence of the time course of photoreceptor degeneration in *Drosophila* rhodopsin gene (*ninaE*) mutants and the molecular defects carried by the mutants. Mean number of R1-6 rhabdomeres per ommatidium, determined in light microscopy, is plotted as a function of age. Adapted from Figure 1 and Table 1 of Leonard et al. (1992). Reproduced with permission from the Journal of Neurobiology.

ible degree of conservation of degeneration mechanisms across species in at least some forms of photoreceptor degeneration. However, most of the human RP caused by rhodopsin gene mutations reported to date is autosomal dominant, while no dominant rhodopsin gene mutation causing retinal degeneration has yet been found in *Drosophila.* It is possible that mammalian photoreceptors are much more sensitive to the effects of single copies of rhodopsin gene mutations than *Drosophila* photoreceptors. It is also likely that there are many more recessive mutations, which cause photoreceptor degeneration only when homozygous, in humans, but most of them remain unrecognized because homozygous expression of recessive mutations is not a frequent event in a natural population.

2 Mutants of the Phospholipase C Gene, *norpA*

Isolation of *norpA* (*no receptor potential A*) mutants was first reported independently by Hotta and Benzer (1970) and Pak et al. (1970). Because strong mutations in this gene abolish the receptor potential, it had long been felt that the protein product of this gene plays a critical role in phototransduction in *Drosophila.* The first pieces of evidence that *norpA* mutants might be defective in inositol lipid metabolism were obtained by Inoue et al. (1985) and Yoshioka et al. (1984), who showed that both phospholipase C (PLC) and diacylglycerol kinase (DGK) activities are greatly reduced in these mutants. Selinger and Minke (1988) helped focus on PLC as a potential target of *norpA* mutation by showing that, in a temperature-sensitive *norpA* mutant, the PLC, but not DGK, activity is reduced only when the mutant is exposed to restricted temperatures. In the mean time, a body of evidence started emerging to suggest that the invertebrate phototransduction cascade may involve PLC activation (Figure 4). For example, injection of inositol 1,4,5-trisphosphate (IP_3) into the ventral photoreceptors of the horseshoe crab *Limulus* was shown to produce electrical activities resembling those generated by light (Brown et al., 1984; Fein et al., 1984), and light-dependent activation of PLC was demonstrated in a number of invertebrate eye preparations (Brown et al., 1984, 1987; Szuts et al., 1986; Devary et al., 1987; Baer and Saibil, 1988; Wood et al., 1989).

Molecular isolation and analysis of the *norpA* gene provided the most compelling evidence for the involvement of PLC in invertebrate phototransduction. The *norpA* gene was mapped to the 4B6-4C1 region of the X chromosome (Banga et al., 1986) and was cloned by a combination of chromosomal walking and transposon mutagenesis (Bloomquist et al., 1988). Sequence analysis of cognate cDNAs showed that the gene encodes a polypeptide of 1095 amino acids (Figure 2) with strong sequence similarity with mammalian PLC sequences that became known from cDNA cloning at about the same time (Katan et al., 1988; Stahl et al., 1988; Suh et al., 1988a, b). These PLCs fall into three structurally distinct classes, PLCα, β, and γ, and are characterized by the presence of two, 120–200 amino acid domains of strong sequence similarity, Boxes X and Y (Rhee et al., 1989; Figure 2) Comparison of the deduced *norpA* protein sequence with mammalian sequences

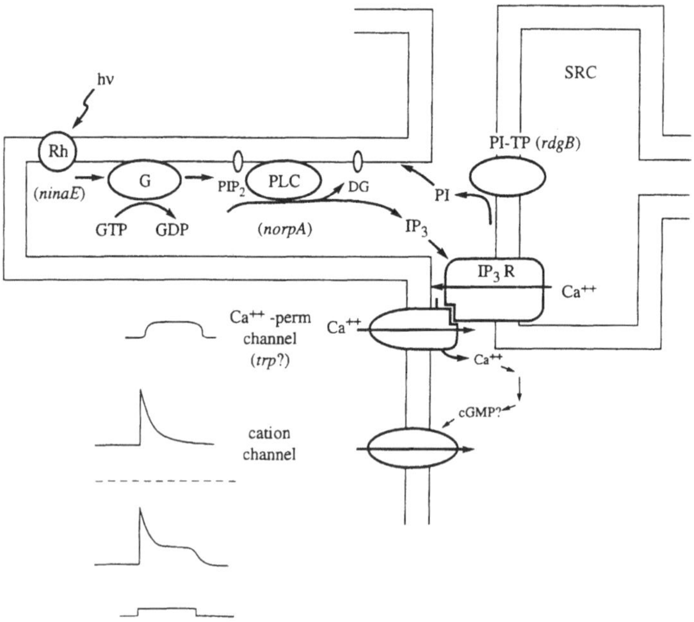

Figure 4 Hypothetical phototransduction scheme in *Drosophila* photoreceptors. Light-activated rhodopsin (Rh), located in the rhabdomeric membrane, activates G protein (G) by catalyzing the exchange of GTP for GDP. The G protein, in turn, activates phospholipase C (PLC) to catalyze the hydrolysis of phosphatidylinositol 4,5-bisphosphate (PIP_2) into diacylglycerol (DG) and inositol 1,4,5-trisphosphate (IP_3). IP_3 binds to IP_3 receptor (IP_3R) on the subrhabdomeric cisternae (SRC) membrane to release Ca^{2+} from the intracellular compartments. The released Ca^{2+}, through a number of unknown steps, opens a class of cation channels with relatively low Ca^{2+} permeability. This class of channels is responsible for the transient phase of the receptor potential. In the meantime, the activated IP_3R also opens a different class of ion channels with high Ca^{2+} permeability through direct interaction. This class of channels is responsible for the steady-state phase of the receptor potential, and the Ca^{2+} influx through these channels helps replenish Ca^{2+} in the intracellular compartments. Phosphatidylinositol transfer protein (PI-TP) encoded by the *rdgB* gene located in the SRC membrane transports PI into the rhabdomeric membrane to replenish it for use in phototransduction. Except for the first few steps, much of the scheme is still largely hypothetical. Adapted from Minke and Selinger (1991). Reproduced with permission from the Pergamon Press.

showed that the former is structurally similar to mammalian PLCβs and has 49 and 53% sequence identity with mammalian PLCβs within Boxes X and Y, respectively; 35 and 44% identity within Extended Boxes X and Y, respectively; larger regions of sequence homology were revealed in a subsequent analysis (R. Shortridge and Pak, unpublished), all strongly indicating that the *norpA* gene product is a PLCβ.

Toyoshima et al. (1990) provided proof that identification of the *norpA* protein as a PLC is correct. They purified the major PLC component present in wild-type but absent in *norpA* heads and showed that sequences of peptide fragments derived from the protein matched the deduced *norpA* protein sequence (Bloomquist et al., 1988). Ultrastructural localization of the *norpA* protein by immuno-gold labeling showed that the protein is localized in the lumen of the microvilli of the rhabdomeres in close association with the microvillar membrane (Schneuwly et al., 1991). It has, thus, been established unequivocally that the *norpA* gene encodes a phospholipase C. Since genetic disruption of the *norpA* gene almost completely abolishes the receptor potential, PLC is an essential component of the phototransduction machinery in *Drosophila* (Figure 4). This conclusion is strongly supported by the localization of the *norpA* protein in the rhabdomere, the structure specialized for photoreception and phototransduction.

Meyertholen et al. (1987) studied the rhodopsin content, deep pseudopupil, and rhabdomere structure in three different *norpA* allelic mutants up to the age of 3 wk and first reported that *norpA* rhabdomeres degenerate with age. Stark et al. (1989) and Zinkl et al., (1990) also studied age-dependent deterioration of rhabdomeres in *norpA* mutants by electron microscopy and spectrophotometry of rhodopsin content up to the age of 10 d. At eclosion, rhabdomeres are largely normal even in the strongest mutants studied, even though a severe electrophysiological defect is present. If grown on a 12 hr light/12 hr dark illumination cycle, the rhabdomeres become smaller and distorted (Zinkl et al., 1990) and finally disappear (Meyertholen et al., 1987), with the rhodopsin content decreasing roughly in parallel. The time course of degeneration is allele-dependent, temperature-dependent, and light-dependent (Meyertholen et al., 1987). The mutants carrying the strongest alleles display the fastest degeneration, with R1-6 rhabdomeres disappearing within about 10 d at 24° C (Meyertholen et al., 1987). The time course is slower for mutants carrying weaker alleles and at lower temperatures. The rhabdomeres of R1-6 photoreceptors are much more susceptible to degeneration than those of R7/8 photoreceptors (Meyertholen et al., 1987), even though the *norpA* product is expressed in all adult photoreceptors (Schneuwly et al., 1991). Because rhabdomere degeneration is strongly light-dependent, no degeneration is apparent even in severely affected mutants if grown in complete darkness (Meyertholen et al., 1987; Zinkl et al., 1990). Most of the above studies were carried out on mutants from which the screening pigments in the eye were genetically eliminated. If the screening pigments are left intact and thus the amount of light reaching the rhodopsin molecules is reduced, rhabdomere degeneration is not apparent up to 10 d post-eclosion (Stark et al., 1991).

The phototransduction cascade in mammalian photoreceptors does not utilize PLC. Thus, a question of some importance is whether photoreceptor degeneration in *norpA* mutants has any relevance to mammalian photoreceptors. The question may be answered along two lines. The first of these is to note that, in vertebrate phototransduction, cGMP phosphodiesterase (PDE) occupies a position analogous to that occupied by PLC in invertebrate phototransduction in that it is the target of transducin interaction. Thus, by analogy, one might expect to find that a defect in the genes encoding cGMP PDE subunits might cause photoreceptor degeneration in mammals. In fact, the mouse retinal degeneration mutant *rd* has been shown to carry a mutation in the gene encoding the PDE β subunit (Bowles et al., 1990; Pittler and Baehr, 1991).

Another way to answer the question is to ask if there are any mammalian PLCs homologous to the *norpA* protein expressed predominantly or exclusively in the retina. Although PLC has no role in mammalian phototransduction, it could have a role in other signal transduction carried out by photoreceptors. PLCs involved in retina-specific functions could be highly homologous to the *norpA* protein in both structure and function, and thus a defect in the genes encoding these proteins could have consequences similar to those caused by *norpA* mutations. Search for mammalian homologs of the *norpA* protein has led to the isolation of bovine cDNAs encoding PLCβ isoforms, expressed almost exclusively in the retina and having sequence identity to the *norpA* protein at least 10% higher than any other PLCβs identified to date (Ferreira et al., 1991).

3 Mutants of the *trp* Gene

The first mutant to be isolated in the *trp* (*transient receptor potential*) gene was a spontaneously arising one that behaved as though blind in bright ambient light (Cosens and Manning, 1969; Cosens, 1971). Other mutant alleles of this gene were isolated by chemical mutagenesis (review by Pak, 1979). Mutations in this gene cause the receptor potential to decay to the baseline during a bright and/or prolonged light stimulus, thereby causing blindness, while in wild type a steady-state response is present throughout the duration of stimulus (Cosens and Manning, 1969; Cosens, 1971; Minke et al., 1975). The receptor potential is generated by a summation of a large number of small, discrete potentials known as bumps, each of which arises in response to a single photon (Rushton, 1961; Fuortes and Yeandle, 1964; Dodge et al., 1968). From noise analysis, Minke et al. (1975) and Minke (1982) concluded that the decay of receptor potential during sustained stimulus in *trp* mutants is caused by a severe reduction in the rate of bumps, i.e. the absorbed photons become ineffective in generating bumps.

Wong et al. (1989) mapped the gene to the 99C5-6 region of the third chromosome, and Montell and Rubin (1989) and Wong et al. (1989) cloned the gene by a short chromosomal walk using a head-specific genomic clone that maps to the 99C region, isolated by Levy et al. (1982). A genomic DNA fragment obtained in the walk was used to rescue a *trp* mutant by P-element-mediated germ line transformation (Montell et al., 1985), providing definitive proof that the DNA fragment contained the entire *trp* gene. This is probably the first example of gene therapy applied to a retinal degeneration mutant in any organism. The gene encodes a novel protein of 1275 amino acid residues with at least four, and possibly as many as eight, potential transmembrane domains (Figure 2). However, because the protein has no homology to any known protein, gene cloning provided little clue to the function of the protein.

The following two sets of observations led Minke and Selinger (1991) to suggest that the *trp* protein is a Ca^{+2} transporter localized in the plasma membrane near the base of the rhabdomere: (1) The phenotypes of *trp* and an analogous mutant *nss* (no steady state) of the sheep blowfly *Lucilia* are closely mimicked by the application of La^{+3}, a known inhibitor of Ca^{+2}-binding proteins including Ca^{+2} transporters, to the extracellular space of wild-type photoreceptors (Hochstrate, 1989), and (2) several lines of indirect evidence exist to suggest that the *trp* and *nss* mutant photoreceptors do not display an increase in intracellular [Ca^{+2}] that normally occurs during illumination (summarized by Minke and Selinger, 1991). According to this hypothesis, IP_3 generated by light-activated PLC binds to the IP_3 receptors on the intracellular Ca^{+2} storage compartment to release Ca^{+2}, and, at the same time, the IP_3-bound IP_3 receptor directly interacts with the *trp*-encoded Ca^{+2} transporter in the plasma membrane to cause transport of Ca^{+2} into the cytoplasm and thus to replenish Ca^{+2} in the storage compartment (Figure 4). The Ca^{+2} pool in the storage compartment is sufficient to generate only the transient phase of the receptor potential during a prolonged stimulus, and the replenishment of the pool through the action of the Ca^{+2} transporter is needed to maintain the steady-state response. More recently, whole-cell recordings from dissociated photoreceptors of both wild type and *trp* mutants led Hardie and Minke (1992) to propose that the receptor potential in *Drosophila* is mediated by two classes of light-activated channels: one consisting of those with high Ca^{+2} permeability mediating the steady-state phase of the receptor potential and the other consisting of those with much lower Ca^{+2} permeability mediating the transient phase (see Figure 4). They further (1) suggested that the *trp* gene encodes either the subset of channels with high Ca^{+2} permeability or a protein product essential for their function and (2) identified this subset of high Ca^{+2} permeability channels with the Ca^{+2} transporter proposed earlier by Minke and Selinger (1991), unifying the two hypotheses. Since the hypothesis requires that the Ca^{+2} channel be located in the plasma membrane near the base of the rhabdomeral microvilli (Figure 4), immunolocalization of the *trp* protein to the base of the rhabdomere by J. A. Pollock (cited by Hardie and Minke, 1992) was taken as strong support for their hypothesis.

Cosens and Perry (1972) carried out EM studies of *trp* photoreceptors and reported that many of the R1-6 rhabdomeres show degenerative changes with age. These rhabdomeres are structurally normal at eclosion, but by 2 to 4 wk post-eclosion, many of them are reduced in cross-sectional area and are disrupted by vesiculation. R8 rhabdomeres also undergo similar degenerative changes, but R7 rhabdomeres appear to be largely free of these changes at the ages tested. Although no other published studies of rhabdomere degeneration in *trp* are available, our own unpublished observations (K. Isono and D. Leonard) are consistent with Cosens and Perry's early report. As in the case of *norpA*, the degeneration is light-dependent. It does not manifest itself if the flies are maintained in complete darkness up to the age of 3 1/2 wk post-eclosion. The presence of screening pigments seems to considerably slow down the degeneration time course, and R1-6 photoreceptors are more susceptible to degeneration than R7 cells.

4 Mutants of the *rdgB* Gene

Hotta and Benzer (1970) first reported that mutations in two genes, later named *rdgA* and *B* (*retinal degeneration A* and *B*), cause age-dependent degeneration of rhabdomeres. Independently, others have also isolated mutations in these genes (Heisenberg, 1971; review by Pak, 1975). Harris and Stark (1977) reported that degeneration caused by *rdgA* mutations proceeds independently of light while that caused by *rdgB* mutations is light dependent. Since the *rdgA* gene has not yet been cloned, no further mention will be made of this gene in this review.

The first detailed studies of *rdgB* mutants were carried out by Harris and Stark (1977). They showed that rhabdomere degeneration is not only light-dependent but also temperature- and allele-dependent. Thus, if exposed to a 12 hr light/12 hr dark cycle at 25°C, the R1-6 rhabdomeres, which are histologically normal at eclosion, disappear within about a week. The time course is faster if the flies are raised at higher temperatures or on continuous illumination. R7/8 rhabdomeres are much less sensitive to degeneration than R1-6 rhabdomeres, the extent to which R7/8 rhabdomeres degenerate being allele-dependent. Some alleles cause virtually no degeneration of R7/8 rhabdomeres, while others cause almost complete degeneration of R7/8 rhabdomeres within a week. The allele dependence of R7/8 degeneration, however, was disputed later by Stark et al. (1983).

Degeneration in *rdgB* mutants can be largely prevented if the mutants are raised in the dark or deprived of vitamin A (Harris and Stark, 1977). Since vitamin A deprivation drastically reduces the rhodopsin content, it would be equivalent to dark rearing. Similarly, *norpA* mutations, if combined with an *rdgB* mutation in double mutants, inhibit *rdgB*-caused degeneration (Harris and Stark, 1977; Stark et al., 1983), presumably because the *norpA* mutation prevents the generation of the light-evoked potential. (*rdgB*-caused degeneration can be distinguished from *norpA*-caused degeneration because of their different time courses.) Harris and Stark (1977), however, isolated a *norpA* allele which does not abolish the receptor potential and yet prevents retinal degeneration caused by a specific *rdgB* allele. The allele-specific rescue led them to propose that the protein product of *norpA* interacts with that of *rdgB*.

Stark and Carlson (1982) showed by electron microscopy that *rdgB*-induced degeneration is not confined to the rhabdomeres. In fact, R1-6 photoreceptor axons are among the first structures to be affected. In addition, by 7 d on a 12 hr light/12 hr dark cycle, the photoreceptor cytoplasm becomes electron dense and begins to show various structures associated with degeneration, such as vacuoles, electron dense reticulum, and degenerate mitochondria. Moreover, photoreceptor axon terminals lack synaptic vesicles and presynaptic structures.

Minke, Selinger and colleagues (reviewed by Minke and Selinger, 1991) showed that a number of chemicals applied to the retina of *rdgB* mutants cause photoreceptor degeneration in the dark, while the same treatment has no effect on wild type flies. In particular, phorbol ester was found to cause photoreceptor degeneration indistinguishable from that caused by light (Minke et al., 1990). Since phorbol ester is a powerful activator of protein kinase C (Nishizuka, 1988), the above result suggested that photoreceptor degeneration in *rdgB* might be caused by excessive phosphorylation of photoreceptor proteins. In fact, these investigators showed that, in pulse labeling experiments with radiolabelled phosphate, much greater incorporation of label is found in the eye proteins of *rdgB* mutants than wild type. More direct evidence for possible involvement of protein kinase C (PKC) was provided by Smith et al. (1991), who showed that *rdgB*-caused photoreceptor degeneration is suppressed by a mutation in the structural gene for eye-specific PKC.

One of the early electrophysiological manifestations of photoreceptor degeneration in *rdgB* is the appearance of spikes superimposed on the ERG (Rubinstein et al., 1989). The spikes have regenerative properties and arise from the normally non-spiking axons and axon terminals of photoreceptors. The voltage-dependent Ca^{2+} channel blockers, Cd^{2+}, verapimil chloride, and diltiazem, eliminate the spikes without affecting the light-induced response. In fact, application of diltiazem to the eyes of *rdgB* mutants over a 7-day period largely prevents light-induced photoreceptor degeneration. Combining these results, Minke and Selinger (1991) and Sahly et al. (1992) proposed that photoreceptor degeneration in *rdgB* mutants is caused by a toxic increase in intracellular Ca^{2+} concentration through voltage-gated Ca^{2+} channels, possibly as a result of excessive phosphorylation.

The *rdgB* gene had been cytogenetically mapped to the 12A-E region of the X chromosome (Harris and Stark, 1977). Vihtelic et al. (1991a) refined the mapping to 12C1 using X-ray induced *rdgB* alleles containing detectable cytological lesions and isolated genomic DNA corresponding to the region by a chromosomal walk. They found a single transcription unit affected by each of the new *rdgB* alleles within the cloned DNA fragment and thus identified it as the *rdgB* gene. Immunolocalization of the *rdgB* protein showed that, in photoreceptors, it is found in the cytosolic base of the rhabdomeres of R1-6 photoreceptors, near the subrhabdomeric cisternae (SRC) (Figure 4) and the axons of all photoreceptors (Vihtelic et al., 1991b). The SRC are a network of smooth endoplasmic reticulum located just beneath the rhabdomeres of rhabdomere-bearing photoreceptors (e.g. Walz, 1982; Matsumoto-Suzuki et al., 1989). The protein is also found in the ocelli, optic lobes, antennae, antennal lobes, and mushroom bodies of the central brain.

Sequence analysis of the transcription unit revealed an open reading frame capable of encoding a protein of 1054 amino acids (Vihtelic et al., 1991a), the amino terminal 280 residues of which are more than 40% identical to the entire length of the rat brain phosphatidylinositol transfer protein (PI-TP; Dickeson et al., 1989) (Figure 2). The PI-TP-like domain is followed by six potential transmembrane domains. Thus, it has been proposed that the *rdgB* protein is a novel, membrane-bound member of the PI-TP family localized in or near the SRC, involved in shuttling PI between the SRC and the rhabdomeres for use in phototransduction (Vihtelic et al., 1991b) (Figure 4). The relationship between this finding and the previous proposals of excessive phosphorylation and Ca^{2+} channel defect has not yet been clarified.

5 Mutants of the *ninaC* Gene

Along with the *ninaE* gene (Section III 1), the *ninaC* gene was one of the original five *nina* genes identified from mutations that reduce the rhodopsin content (Pak, 1979; Stephenson et al., 1983). From their studies on several alleles of *ninaC*, Matsumoto et al. (1987) suggested (1) that the *ninaC* gene likely encodes two retina-specific proteins of 130 and 170 kD identified in SDS polyacrylamide gel electrophoresis and (2) that these proteins are probably cytoskeletal components of rhabdomeral microvilli. The first suggestion was based on strong *ninaC* gene dosage dependence displayed by the two proteins, and the second suggestion was based on electron micrographs showing reduced or absent central cytoskeletal cores of rhabdomeral microvilli in *ninaC* mutants. Noting that the rhabdomere size is reduced in *ninaC* mutants, they further suggested that the reduction in rhodopsin content is due to the reduced size of rhabdomeres resulting from defective microvillar cytoskeleton.

Montell and Rubin (1988) isolated the *ninaC* gene using a "head-specific" genomic clone (Levy et al., 1982) mapping to the 28A region of the second chromosome to which the *ninaC* gene had also been mapped (Matsumoto et al., 1987). To prove that a DNA fragment derived from the Levy clone contains the *ninaC* gene, Montell and Rubin (1988) rescued a *ninaC* mutant by introducing the DNA fragment into the mutant by P-element-mediated germ line transformation. Molecular analysis of the gene by Montell and Rubin (1988) showed that it is expressed as two highly overlapping, alternatively spliced mRNAs which encode two very closely related, photoreceptor-specific proteins of 1135 (p132) and 1501 (p174) amino acids (Figure 3). Each isozyme consists of two adjacent domains homologous to protein kinases and the head region of the myosin heavy chain, respectively (Figure 2). Thus, the *ninaC* protein products appear to be novel photoreceptor-specific cytoskeletal proteins containing a protein kinase domain.

Porter et al. (1992) showed that a strong *ninaC* mutation causes light-dependent and age-dependent degeneration of rhabdomeres. When maintained on a 12 hr light/12 hr dark cycle, the rhabdomeres, which are essentially normal at eclosion, almost disappear by 3 wk post-eclosion. R7 rhabdomeres are somewhat less affected than either R1-6 or R8 rhabdomeres. If grown in the dark, on the other hand, degenerative changes, if present at all, are only slight at 3 wk post-eclosion.

To study the effects of eliminating one or the other of the two isozymes encoded by *ninaC*, Porter et al. (1992) generated two *ninaC* mutants expressing, respectively, only p174 and only p132 by appropriate oligonucleotide-directed mutagenesis of the *ninaC* gene and introduction of the altered genes into a *ninaC* null mutant. They found that elimination of p174 alone is sufficient to cause both morphological and physiological phenotypes indistinguishable from those of *ninaC* null mutants, while elimination of p132 has little or no effect. EM immunolocalization of the two isozymes showed that the p174 and p132 proteins are located primarily in the rhabdomeres and cytoplasm, respectively. From these observations, these authors have suggested that the p174 protein associates with actin in the central cores of rhabdomeral microvilli through its myosin-I-like tail and is involved in phosphorylating rhabdomeric proteins important in phototransduction.

6 Mutants of the *rdgC* Gene

Mutations in the *rdgC* gene were isolated from chemical mutagenesis by looking for their degenerative effects on the deep pseudopupil with age (Steele and O'Tousa, 1990). Rhabdomeres of *rdgC* mutants are normal when young but undergo age- and light-dependent degeneration. As monitored by examining the deep pseudopupil, degeneration is essentially complete by 8 d post-eclosion when reared on a 12 hr light/12 hr dark cycle. R7 and R8 rhabdomeres are much more resistant to degeneration than R1-6 rhabdomeres and just begin showing signs of degeneration when the latter have almost completely degenerated. Degeneration is not observed if the flies are reared in complete darkness. Unlike most other known retinal degeneration mutants, prior to the onset of degeneration, no obvious defect in the receptor potential is evident in *rdgC* mutants. Degeneration requires photoactivation of rhodopsin because it does not occur if the rhodopsin level is reduced by either vitamin A deprivation or the use of a rhodopsin (*ninaE*) gene mutation. However, unlike *rdgB*-induced degeneration, *rdgC*-induced degeneration does not require the *norpA*-encoded PLC, suggesting that it is mediated by a rhodopsin activated pathway different from the PLC-mediated phototransduction pathway.

The gene was mapped to 77B1 of the third chromosome and was isolated by chromosomal walking (Steele et al., 1992). A DNA fragment thought to contain the gene, isolated from the walk, was used to rescue *rdgC* mutant flies by germ line transformation, providing proof that the gene had been cloned in the fragment. The gene encodes a protein of 661 amino acids which contains a domain homologous to the catalytic domains of types 1, 2A, and 2B serine/threonine protein phosphatases (Figure 2). The putative catalytic domain is followed by a domain containing several EF hand motifs, or potential Ca^{2+}-binding sites (Figure 2). Steele et al. (1992), thus, suggested that the *rdgC* protein might be a novel type of serine/threonine phosphatase regulated directly by Ca^{2+} levels. The protein is expressed in the retina, ocelli, and mushroom bodies of the brain. The function of the protein has not yet been elucidated.

IV DISCUSSION

Although only a few genes which, when defective, cause photoreceptor degeneration have been isolated and analyzed in *Drosophila* to date, some tentative conclusions may be advanced regarding retinal degeneration in this organism. First of all, as we have seen in the previous section, there is a strong correlation between retinal degeneration and defective phototransduction. Almost all mutations that affect phototransduction may also cause retinal degeneration (see also Minke and Selinger, 1991). However, mutations that cause retinal degeneration do not necessarily all affect phototransduction. For example, retinal degeneration caused by *rdgC* mutations proceeds in the absence of *norpA*-encoded PLC (Steele and O'Tousa, 1990), suggesting that this mode of degeneration is not consequent on a defect in PLC-mediated phototransduction. Rather, it appears to be a consequence of a defect in another light-activated, rhodopsin-mediated pathway (Steele and O'Tousa, 1990). The implication is that defects not only in phototransduction but also in any other light-activated signaling pathway of photoreceptors may cause photoreceptor degeneration.

Allele dependence of degeneration has been studied in detail only for *ninaE* mutants to date (Section III I; Leonard et al., 1992). As shown in Figure 3, the time course of degeneration is strongly allele-dependent. The results, though striking, are not unexpected, since different mutant alleles can have very different effects on the function of the protein encoded by the gene. Thus, although allele dependence has been reported only for *ninaE* (Leonard et al., 1992), *rdgB* (Harris and Stark, 1977), and *norpA* (Meyertholen et al., 1987) mutants to date, it is expected to be a property common to all degeneration mutants.

Retinal degeneration in most mutants studied to date is strongly light-dependent (Section III). The only exception appears to be *ninaE* mutants (Section III I; Leonard et al., 1992). Mutants of another gene, *rdgA*, which are not discussed in this review, have also been reported to undergo light-independent retinal degeneration (Harris and Stark, 1977). Indeed, degeneration proceeds in the dark in this mutant. However, our own unpublished observations indicate that degeneration has a substantial light-dependent component in *rdgA* in that it is considerably accelerated by illumination (D. Leonard and V. Bowman, unpublished data). Thus, in most mutants, retinal degeneration is either completely blocked or substantially retarded by raising the mutants in the dark. However, dark rearing does not result in functional rescue in most light-dependent degeneration mutants because functional defects are already present before any evidence of degeneration is apparent. As the flies age and degeneration proceeds, electrophysiological responses deteriorate further. Thus, in these mutants, structural degeneration does not initiate a functional defect, but rather it further exacerbates a defect that is already present. There are, however, exceptions. In *rdgC* mutants, for example, the ERG is initially normal (Steele and O'Tousa, 1990) but deteriorates with age apparently as a consequence of retinal degeneration. Therefore, dark rearing would result in both functional and structural rescue of *rdgC* mutants.

In all retinal degeneration mutants studied to date, R1-6 photoreceptors tend to degenerate before R7 or R8 photoreceptors, although eventually the latter degenerate also. In the case of *ninaE* mutants, the greater susceptibility of R1-6 photoreceptors to degeneration is understandable, since the *ninaE* gene encodes R1-6 opsin. In fact, the question in this case is why R7 and R8 photoreceptors degenerate at all. However, R7 and R8 photoreceptors do degenerate once degeneration of R1-6 photoreceptors is well under way (Leonard et al., 1992). A part of the explanation might be that R7/8 photoreceptors are trophically dependent on the presence of R1-6 photoreceptors. Thus, when R-6 photoreceptors degenerate, R7/8 photoreceptors can no longer maintain themselves and degenerate also.

Even in mutants in which the gene products are expressed in both R1-6 and R7/8 cells, R1-6 photoreceptors tend to degenerate first. Since degeneration in these mutants is light-dependent, greater susceptibility of R1-6 photoreceptors to degeneration may be due, in part, to greater sensitivity of R1-6 photoreceptors to light because of their larger rhabdomeres. However, it has been reported that in *trp* and *ninaC*, R8 photoreceptors degenerate as fast as R1-6 photoreceptors, while R7 photoreceptors are more resistant to degeneration (Cosens and Perry, 1972; Porter et al., 1992). Porter et al. (1992) have suggested that R7 photoreceptors are resistant to degeneration because they are UV receptors and the ambient light contains relatively little UV. In other light-dependent retinal degeneration mutants, on the other hand, differential susceptibility to degeneration between R7 and R8 cells has not been reported. Thus, neither of the above explanations is adequate to explain both sets of observations: that only R7 photoreceptors are resistant to degeneration in some classes of mutants and that both R7 and R8 photoreceptors are resistant to degeneration in other classes of mutants. Thus, while these explanations may play a part in the observed phenomena, the whole explanation is still to be obtained in future experiments.

Finally, mutations in a rather large number of genes can cause retinal degeneration in *Drosophila*. Estimating from preliminary analysis of a collection of ERG-defective mutants, more than a dozen other genes, in addition to those described in this review, are likely to cause retinal degeneration when defective. Thus, the genes described in Section III very likely represent only a fraction of "retinal degeneration genes" in this organism. Moreover, analyses of degeneration phenotypes are still largely incomplete even for those described in Section III. Isolation and analysis of these other genes and more complete knowledge of the known degeneration mutants are likely to be important in the understanding of retinal degeneration not only in this organism but also in mammals as well.

V ACKNOWLEDGMENTS

I thank Martin Burg, Paulo Ferreira, Lydia L. Randall, and Bharath Srinivasan for their comments on the manuscript, Barbara Petty and Lydia L. Randall for editorial and stenographic assistance, Donald Ready for help with Figure 1. The preparation of the manuscript was supported in part by a grant from the National Eye Institute (EY00033). I gratefully acknowledge the work of many colleagues whose contributions are reviewed in the manuscript and apologize for any inadvertent omissions.

VI REFERENCES

Baer, K. M. and Salbil, H. R. (1988). Light- and GTP-activated hydrolysis of phosphatidylinositol bisphosphate in squid photoreceptor membranes. *Journal of Biological Chemistry* **263**, 17–20.

Banga, S. S., Bloomquist, B. T., Brodberg, R. K., Pye, Q. N., Larrivee, D. C., Mason, J. M., et al. (1986). Cytogenetic characterization of the 4BC region on the X chromosome of *Drosophila melanogaster.* Localization of the mei-9, *norpA* and omb genes. Chromosoma **93**, 341–346.

Benzer, S. (1973). Genetic dissection of behavior. *Scientific American* **229**, 24–37.

Bloomquist, B. T., Shortridge, R. D., Schneuwly, S., Perdew, M., Montell, C., Steller, H., et al. (1988). Isolation of a putative phospholipase C gene of *Drosophila, norpA,* and its role in phototransduction. Cell **54**, 723–733.

Bowes, C., Li, T., Danciger, M., Baxter, L. C., Applebury, M. L. and Farber, D. B. (1990). Retinal degeneration in the *rd* mouse is caused by a defect in the subunit of rod cGMP-phosphodiesterase. *Nature* **347**, 677–680.

Brown, J. E., Rubin, L. J., Ghalayini, A. J., Tarver, A. P., Irvine, R. F., Berridge, M. J., et al. (1984). *myo*-inositol polyphosphate may be a messenger for visual excitation in Limulus photoreceptors. *Nature* **311**, 160–163.

Brown, J. E., Watkins, D. C. and Malbon, C. C. (1987). Light-induced changes in the content of inositol phosphates in squid *Loligo pealei* retina. *Biochemical Journal* **247**, 293–297.

Cosens, D. (1971). Blindness in a *Drosophila* mutant. *Journal of Insect Physiology* **17**, 285–302.

Cosens, D. and Manning, A. (1969). Abnormal electroretinogram from a *Drosophila* mutant. *Nature* **224**, 285–28?.

Cosens, D. J. and Perry, M. M. (1972). The fine structure of the eye of a visual mutant, A-type, of *Drosophila* melanogaster. *Journal of Insect Physiology* **18**, 1773–1786.

Devary, 0., Heichal, O., Blumenfeld, A., Cassel, D., Suss, E., Barash, S., et al. (1987). Coupling of photoexcited rhodopsin to inositol phospholipid hydrolysis in fly photoreceptors. *Proceedings of the National Academy of Sciences USA* **84**, 6939–6943.

Dickeson, S. K., Lim, C. N., Schuyler, G. T., Dalton, T. P., Helmkamp, J. G. M. and Yarbrough, L. R. (1989). Isolation and sequence of cDNA clones encoding rat phosphatidylinositol transfer protein. *Journal of Biological Chemistry* **264**, 16557–16564.

Dodge, F. A., Jr., Knight, B. W. and Toyoda, J. (1968). Voltage noise in *Limulus* visual cells. *Science* **160**, 88–96.

Dryja, T. P., McGee, T. L., Reichal, E., Hahn, L. B., Cowley, G. S., Yandell, D. W., et al. (1990). A point mutation of the rhodopsin gene in one form of retinitis pigmentosa. *Nature* **343**, 364–366.

Dryja, T. P., Hahn, L. B., Cowley, G. S., McGee, T. L. and E. L. Berson. (1991). Mutation spectrum of the rhodopsin gene among patients with autosomal dominant retinitis pigmentosa. *Proceedings of the National Academy of Sciences USA* **88**, 9370–9374.

Fein, A., Payne, R., Corson, D. W., Berridge, M. J. and Irvine, R. F. (1984). Photoreceptor excitation and adaptation by inositol 1,4,5-trisphosphate. *Nature* **311**, 157–160.

Ferreira, P. A., Shortrldge, R. D. and Pak, W. L. (1991). Identification and characterization of *norpA*-like retina-specific bovine cDNA. Molecular Neurobiology of *Drosophila* (Abstract), Cold Spring Harbor Laboratory, Cold Spring Harbor, N.Y.

Franceschini, N. and Kirschfeld, D. (1971). Les phenomenes de pseudopupille dans l'oeil compose de *Drosophila. Kybernetik* **9**, 159–182.

Fuortes, M. G. F. and Yeandle, S. (1964). Probability of occurrence of discrete potential waves in the eye of *Limulus. Journal of General Physiology* **47**, 443–463.

Hardie, R. C. and Minke, B. (1992). The *trp* gene is essential for a light activated Ca^{2+} channel in *Drosophila* photoreceptors. *Neuron* **8**, 643–651.

Harris, W. A. and Stark, W. S. (1977). Hereditary retinal degeneration in *Drosophila melanogaster*: a mutant defect associated with the phototransduction process. Journal of General Physyslology **69**, 261–291.

Harris, W. A., Stark, W. S. and Walker, J. A. (1976). Genetic dissection of the photoreceptor system in the compound eye of *Drosophila melanogaster*. *Journal of Physiology* **256**, 415–439.

Heisenberg, M. (1971). Isolation of mutants lacking the optomotor response. *Drosophila Information Service* **46**, 68.

Heisenberg, M. (1972). Behavioral diagnostics: a way to analyze visual mutants of *Drosophila*, pp. 265–268. *Information Processing in the Visual Systems of Arthropods*. Ed. R. Wehner, Springer-Verlag, Berlin.

Hochstrate, P. (1989). Lanthanum mimics the trp photoreceptor mutant of *Drosophila* in the blowfly *Calliphora*. *Journal of Comparative Physiology* A **166**, 179–188.

Hotta, Y. and Benzer, S. (1969). Abnormal electroretinograms in visual mutants of *Drosophila*. *Nature* **222**,354–356.

Hotta, Y. and Benzer, S. (1970). Genetic dissection of the *Drosophila* nervous system by means of mosaics. *Proceedings of the National Academy of Sclences USA* **67**, 1156–1163.

Inoue, H., Yoshioka, T. and Hotta, Y. (1985). A genetic study of inositol trisphosphate involvement in phototransduction using *Drosophila* mutants. *Biochemical and Biophysical Research Communications* **132**, 513–519.

Katan, M., Kriz, R. W., Totty, N., Phllip, R., Meldrum, E., Aldape, R. A., et al. (1988). Determination of the primary structure of PLC-154 demonstrates diversity of phosphoinositide-specific phospholipase C activities. *Cell* **54**, 171–177.

Koenig, J. and Merriam, J. R. (1977). Autosomal ERG mutants. *Drosophila Information Service* **52**, 50–51.

Leonard, D. S. and Pak, W. L. (1984). Photoreceptor degeneration associated with mutations in presumptive opsin structural gene in *Drosophila*. Society of Neuroscience Abstracts 10, 1032.

Leonard, D. S., Bowman, V. D., Ready, D. F. and Pak, W. L. (1992). Degeneration of photoreceptors in rhodopsin mutants of *Drosophila*. *Journal of Neurobiology* **23**, 605–626.

Levy, L. S., Ganguly, R., Ganguly, N. and Manning, J. E. (1982). The selection, expression, and organization of a set of head-specific genes in *Drosophila*. *Developmental Biology* **94**, 451–464.

Matsumoto, H., Isono, K., Pye, Q. and Pak, W. L. (1987). Gene encoding cytoskeletal proteins in *Drosophila* rhabdomeres. *Proceedings of the National Academy of Sciences USA* **84**, 985–989.

Matsumoto-Suzuki, E., Hirosawa, K. and Hotta, Y. (1989). Structure of the subrhabdomeric cisternae in the photoreceptor cells of *Drosophila* melanogaster. *Journal of Neurocytology* **18**, 87–93.

Meyertholen, E. P., Stein, P. J., Williams, M. A. and Ostroy, S. E. (1987). Studies of the *Drosophila norpA* phototransduction mutant, II: photoreceptor degeneration and rhodopsin maintenance. *Journal of Comparative Physiology A* **161**, 793–798.

Minke, B.(1982). Light-induced reduction in excitation efficiency in the trp mutant of *Drosophila*. *Journal of General Physiology* **79**, 361–385.

Minke, B. and Selinger, Z. (1991). Inositol lipid pathway in fly photoreceptors: excitation, calcium mobilization and retinal degeneration, pp. 100–121. *Retinal Research, Vol. II*. Ed. N. N. Osborne and G. J. Chader, Pergamon Press, Oxford.

Minke, B., Wu, C. -F. and Pak, W. L. (1975). Induction of photoreceptor voltage noise in the dark in *Drosophila* mutant. *Nature* **258**, 84–87.

Minke, B., Rubinstein, C. T., Sahly, I., Bar-Nachum, S., Timberg, R. and Z. Selinger, Z. (1990). Phorbol ester induces photoreceptor-specific degeneration in a *Drosophila* mutant. *Proceedings of the National Academy of Sclences USA* **87**, 113–117.

Montell, C. and Rubin, G. M. (1988). The *Drosophila ninaC* locus encodes two photoreceptor cell specific proteins with domains homologous to protein kinases and the myosin heavy domain head. *Cell* **52**, 757–772.

Montell, C. and Rubin, G. M. (1989). Molecular characterization of the *Drosophila trp* locus: a putative integral membrane protein required for phototransduction. *Neuron* **2**, 1313–1323.

Montell, C., Jones, K., Hafen, E. and Rubin, G. (1985). Rescue of the *Drosophila* phototransduction mutation *trp* by germline transformation. *Science* **230**, 1040–1043.

Nathans, J. and Hogness, D. S. (1983). Isolation, sequence analysis and intron-exon arrangement of the gene encoding bovine rhodopsin. *Cell* **34**, 807–814.

Nathans, J. and Hogness, D. S. (1984). Isolation and nucleotide sequence of the gene encoding human rhodopsin. *Proceedings of the National Academy of Sciences USA* **81**, 4851–4855.

Nishizuka, Y. (1988). The molecular heterogeneity of protein kinase C and its implications for cellular regulation. *Nature* **334**, 661–665.

O'Brien, S. J. and MacIntyre, R. J. (1978). Genetics and biochemistry of enzymes and specific proteins of *Drosophila*, pp. 396–551. *The Genetics and Biology of Drosophila*, Vol. 2a., M. Ashburner and T. R. F. Wright, Academic Press, New York.

O'Tousa, J., Hirsh, J., Applebury, M. L. and Pak, W. L. (1985). Isolation of ninaE, the major *Drosophila* opsin gene. *Supplement of Investigative Ophthalmology and Visual Science* **26**, 290.

O'Toursa, J. E., Leonard, D. S. and Pak, W. L. (1989). Morphological defects in oraJK84 photoreceptors caused by mutation in R1-6 opsin gene of *Drosophila. Journal of Neurogenetics* **6**, 41–52.

Ovchinnikov, Y. A., Abdulaev, N. G., Feigina, M. Y., Artamonov, I. D., Zolotarev, A. S., Miroshnikov, A. I., et al. (1982). The complete amino acid sequence of visual rhodopsin. *Bioorganicheskaya Khimia* **8**, 1011–1014.

Pak, W. L. (1975). Mutations affecting the vision of *Drosophila melanogaster*, pp. 703–733. *Handbook of Genetics, Vol. 3.* Ed. R. C. King, Plenum Publishing Corp., New York.

Pak, W. L. (1979). Study of photoreceptor function using *Drosophila* mutants, pp. 67–99. Neurogenetics: *Genetic Approaches to the Nervous System.* Ed. X. O. Breakefield, Elsevier North Holland, New York.

Pak, W. L., Grossfield, J. and White, N. V. (1969). Nonphototactic mutants in a study of vision of *Drosophila. Nature* **222**, 351–354.

Pak, W. L., Grossfield, J. and Arnold, K. (1970). Mutants of the visual pathway of *Drosophila* melanogaster. *Nature* **227**, 518–520.

Pittler, S. J. and Baehr, W. (1991). Identification of a nonsense mutation in the rod photoreceptor cGMP phosphodiesteras β–subunit gene of the *rd* mouse. *Proceedings of the National Academy Sciences USA* **88**, 8322–8326.

Porter, J. A., Hicks, J. L., Williams, D. S. and Montell, C. (1992). Differential localizations of and requirements for the two *Drosophila ninaC* kinase/myosins in photoreceptor cells. *Journal of Cell Biology* **116**, 683–693.

Renfranz, P. J., and Benzer, S. (1989). Monoclonal antibody probes discriminate early and late mutant defects in development of the *Drosophila* retina. *Developmental Biology* **136**, 411–429.

Rhee, S. G., Suh, P.-G., Ryu, S.-H., Lee, S. Y. (1989). Studies of inositol phospholipid-specific phospholipase C. *Science* **244**, 546–550.

Rosenfeld, P. J., Cowley, G. S., McGee, T. L., Sandberg, M. A., Berson, E. L., Dryja, T. P. (1992). A *null* in the rhodopsin gene causes rod photoreceptor dysfunction and autosomal recessive retinitis pigmentosa. *Nature Genetics* **1**, 209–213.

Rubinstein, C.-T., Bar-Nachum, S., Selinger, Z. and Minke, B. (1989). Light-induced retinal degeneration in *rdgB* (retinal degeneration B) mutant of *Drosophila*: Electrophysiological and morphological manifestations of degeneration. *Visual Neuroscience* **2**, 529–539.

Rushton, W. A. H. (1961). The intensity factor in vision, pp. 706–722. *Light and Life.* Ed. W. D. McElroy and H. B. Glass, The Hopkins University Press, Baltimore.

Sahly, I., Bar-Nachum, S., Suss-Toby, E., Rom, A., Peretz, A., Kleiman, J., et al. (1992). Calcium channel blockers inhibit retinal degeneration in the retinal-degeneration-B mutant of *Drosophila. Proceedings of the National Academy of Sciences USA* **89**, 435–439.

Scavarda, N. J., O'Tousa, J. and Pak, W. L. (1983). *Drosophila* locus with gene dosage effects on rhodopsin. *Proceedings of the National Academy of Sciences USA* **80**, 4441–4445.

Schinz, R. H., Lo, M.-V. C., Larrivee, D. C. and Pak, W. L. (1982). Freeze-fracture study of the *Drosophila* photoreceptor membrane: mutations affecting membrane particle density. *Journal of Cell Biology* **93**, 961–969.

Schneuwly, S., Burg, M. G., Lending, C., Perdew, M. H. and Pak, W. L. (1991). Properties of photoreceptor-specific phospholipase C encoded by the *norpA* gene of *Drosophila* melanogaster. *Journal of Biological Chemistry* **266**, 24314–24319.

Selinger, Z. and Minke, B. (1988). Inositol lipid cascade of vision studied in mutant flies, pp. 333–341. *The Molecular Biology of Signal Transduction, Cold Spring Harbor Symposium on Quantitative Biology, Vol. 53.* Ed. J. D. Watson, M. Wigler, and J. Feramisco, Cold Spring Harbor Laboratory, Cold Spring Harbor, New York.

Smith, D. P., Ranganathan, R., Hardy, R. W., Marx, J., Tsuchida, T. and Zuker, C. (1991). Photoreceptor deactivation and retinal degeneration mediated by a photoreceptor-specific protein kinase C. *Science* **254**, 1478–1484.

Stahl, M. L., Ferenz, C. R., Kelleher, K. L., Kriz, R. W. and Knopf, J. L. (1988). Sequence similarity of phospholipase C with the non-catalytic region of src. *Nature* **332**, 269–272.

Stark, W. S. and Carlson, S. D. (1982). Ultrastructural pathology of the compound eye and optic neuropiles of the retinal degeneration mutants (*w rdgB*^{KS222}) *Drosophila melanogaster. Cell Tissue Research* **225**, 11–22.

Stark, W. S. and Carlson, S. D. (1983). Ultrastructure of the compound eye and first optic neuropile of the photoreceptor mutant oraJK84 of *Drosophila. Cell Tissue Research* **233**, 305–317.

Stark, W. S. and Sapp, R. (1987). Ultrastructure of the retina of *Drosophila* melanogaster: the mutant ora (outer rhabdomeres absent) and its inhibition of degeneration in *rdgB* (retinal degeneration-B). Journal of Neurogenetics **4**, 227–240

Stark, W. S., Chen, D. -H., Johnson, M. A. and Frayers, K. L. (1983). The *rdgB* gene in *Drosophila*: Retinal degeneration in different mutant alleles and inhibition of degeneration by *norpA*. *Journal of Insect Physiology* **29**, 123–131.

Stark, W. S., Sapp, R. and Carlson, S. D. (1989). Photoreceptor maintenance and degeneration in the *norpA* (no receptor potential-A) mutant of *Drosophila* melanogaster. *Journal of Neurogenetics* **5**, 49–59.

Stark, W. S., Christianson, J. S., Maier, L. and Chen, D. -M. (1991). Inherited and environmentally induced retinal degeneration in *Drosophila*, pp. 61–75. *Retinal Degenerations.* Ed. R. E. Anderson, J. G. Hollyfield, and H. M. Lavail, CRC Press, Boca Raton.

Steele, F. and O'Tousa, J. E. (1990). Rhodopsin activation causes retinal degeneration in *Drosophila rdgC* mutant. *Neuron* **4**, 883–890.

Steele, F., Washburn, T., Rieger, R. and O'Tousa, J. E. (1992). *Drosophila* retinal degeneration C (*rdgC*) encodes a novel serine/threonine protein phosphatase. *Cell* **69**, 669–676.

Stephenson, R. S., O'Tousa, J., Scavarda, N. J., Randall, L. L. and Pak, W. L. (1983). *Drosophila* mutants with reduced rhodopsin content, pp. 447–501. *The Biology of Photoreception.* Ed. D. J. Cosens and D. Vince-Price, Cambridge University Press, Cambridge, England.

Suh, P. G., Ryu, S. H., Moon, K. H., Suh, H. W. and Rhee, S. G. (1988a). Cloning and sequence of multiple forms of phospholipase C. *Cell* **54**, 161–169.

Suh, P. G., Ryu, S. H., Moon, K. H., Suh, H. W. and Rhee, S. G. (1988b). Inositol phospholipid-specific phospholipase C: complete cDNA and protein sequences and sequence homology to tyrosine kinase-related oncogene products. *Proceedings of the National Academy of Sciences USA* **85**, 5419–5423.

Sung, C.-H., Davenport, C. M., Hennessey, J. C., Maumenee, I. H., Jacobson, S. G., Heckenlively, J. R., et al. (1991). Rhodopsin mutations in autosomal dominant retinitis pigmentosa. *Proceedings of the National Academy of Sciences USA* **88**, 6481–6485.

Szuts, E. Z., Wood, S. F., Reid, M. A. and Fein, A. (1986). Light stimulates the rapid formation of inositol trisphosphate in squid retinae. *Biochemical Journal* **240**, 929–932.

Toyoshima, S., Matsumoto, N., Wang, P., Inoue, H., Yoshioka, T., Hotta, Y., et al. (1990). Purification and partial amino acid sequences of phosphoinositide-specific phospholipase C of *Drosophila* eye. *Journal of Biological Chemistry* **265**, 14842–14848.

Vihtelic, T. S., Hyde, D. R. and O'Tousa, J. E. (1991a). Isolation and characterization of the *Drosophila retinal degeneration B (rdgB)* gene. *Genetics* **127**, 761–768.

Vihtelic, T., O'Tousa, J. and Hyde, D. (1991b). Localization of the *Drosophila rdgB* protein: a phospholipid transfer protein. *Molecular Neurobiology of Drosophila* (Abstract), Cold Spring Harbor Laboratory, Cold Spring Harbor, New York, p184.

Walz, B. (1982). Calcium-sequestering smooth endoplasmic reticulum in retinula cells of the blowfly. *Journal of Ultrastructure Research* **81**, 240–248.

Washburn, T. and O'Tousa, J. E. (1989). Molecular defects in *Drosophila* rhodopsin mutants. *Journal of Biological Chemistry* **264**, 1564–1566.

Wolff, T. and Ready, D. F. (1993). Pattern formation in the *Drosophila* retina. *Development of Drosophila.* Ed. M. Bate and A. Martinez-Arias, Cold Spring Harbor Press, Cold Spring Harbor, N. Y.

Wong, F., Schaefer, E. L., Roop, B. C., LaMendola, J. N., Johnson-Seaton, D., and Shao, D. (1989). Proper function of the *Drosophila* trp gene product during pupal development is important for normal visual transduction in the adult. *Neuron* **3**, 81–94.

Wood, S. E., Szuts, E. Z. and Fein, A. (1989). Inositol trisphosphate production in squid photoreceptors. Activation by light, aluminum fluoride, and guanine nucleotides. *Journal of Biological Chemistry* **264**, 12970–12976.

Yoshioka, T., Inoue, H. and Hotta, Y. (1985). Absence of diglyceride kinase activity in the photoreceptor cells of *Drosophila* mutants. *Biochemical and Biophysical Research Communications* **119**, 389–395.

Zinkl, G. M., Maier, L., Studer, K., Sapp, R., Chen, D.-M., and Stark, W. S. (1990). Microphotometric, ultrastructural, and electrophysiological analyses of light-dependent processes on visual receptors in white-eyed wild-type and *norpA* (no receptor potential) mutant *Drosophila. Visual Neuroscience* **5**, 429–439.

Zuker, C. S., Cowman, A. F. and Rubin, G. M. (1985). Isolation and structure of a rhodopsin gene from D. melanogaster. *Cell* **40**, 851–858.

3. DIAGNOSIS IN INHERITED RETINAL DISORDERS

ALAN C. BIRD and BARRIE JAY

Department of Clinical Ophthalmology, Moorfields Eye Hospital, City Road, London EC1V 2PD, UK

INTRODUCTION

Retinal receptor dystrophies comprise a variety of disparate genetically determined conditions that differ from one to another in their mode of inheritance, their pattern of visual loss and their ophthalmoscopic appearances. Over the past few years an increasing number of distinct disorders have been recognised within this heterogeneous group, and in some a clue to their pathogenesis has started to emerge. It is possible to subdivide patients with receptor dystrophies into several groups depending on their clinical features. Some patients have symptoms early in their disease, indicating primary loss of rod function, and examination reveals defective vision in the mid zone of the visual field and morphological changes in the postequatorial fundus; the diseases in such cases are known collectively as retinitis pigmentosa. Other disorders, with loss of cone function and morphological changes in the central fundus, are known as macular degenerations or cone dystrophies.

This subdivision into "peripheral degenerations" in which the rods may be the primary target of disease and "central degenerations" in which the cones may be the cells initially affected is superficially attractive. It is tempting to assume that the pathogenesis of conditions within the retinitis pigmentosa group is related to a disorder of a metabolic function or structural protein peculiar to rods and conversely that macular dystrophies are due to similar defects of cones. However, many patients have combined disorders of both rod and cone systems, even in the early stages of their disease, and all patients with retinitis pigmentosa have some degree of cone loss. The receptor dystrophies, therefore, comprise a spectrum of disorders ranging from pure rod dystrophies, through rod/cone and cone/rod dystrophies, to pure cone dystrophies. It is probable that different forms of disease usually result from the effect of different *genes* but in some instances the differences may be due to observation of the same disorder at different stages of evolution, of different phenotypic expression of the same genetic disorder or even of different mutations within the same gene. Despite these reservations, receptor dystrophics will be considered in two broad categories: peripheral receptor dystrophies typified by retinitis pigmentosa; and central dystrophies. In addition, retinal dystrophies may also occur as part of multisystem disease in a variety of well-described syndromes, two of which, Refsum disease and abetalipoproteinaemia, will also be discussed here.

PERIPHERAL RECEPTOR DYSTROPHIES

Retinitis pigmentosa

Retinitis pigmentosa (RP) is a solitary manifestation of several genetically determined disorders characterised by loss of dark adaptation and progressive reduction of peripheral visual fields early in the disease, leading eventually to impairment of central vision. It is becoming increasingly obvious that there are many different genetic defects within this family of disorders.

Genetic studies

Retinitis pigmentosa may be inherited as an autosomal dominant, an autosomal recessive or an X-linked trait. When transmission has occurred through several generations the recognition of X-linked or autosomal dominant inheritance is usually simple. As with other disorders, parental consanguinity may indicate the possibility of autosomal recessive inheritance. When faced with a sporadic case, however, the problem of identifying the genetic form of the disease may be more difficult. In the large series of cases at Moorfields Eye Hospital, the frequency of simplex (sporadic) cases in families with retinitis pigmentosa is at present 52 per cent (Jay M, personal communication), a figure a little larger than that published 10 years ago (42% – Jay, 1982). Other authors have found a similar high frequency of simplex cases (63% – Pearlman, 1979; 48% – Hu, 1982; 50% – Boughman & Fishman, 1983; 42% – Heckenlively, 1987). Until fairly recently it was commonly assumed that patients with simplex RP had autosomal recessive disease but there is increasing clinical evidence that many of these patients do not express the disease in the same manner as patients with proven autosomal recessive RP (Heckenlively, 1988). There is also genetic evidence that some cases of simplex RP carry a mutation of the rhodopsin gene (Dryja et al, 1991). Simplex cases may be X-linked if heterozygous females are asymptomatic or only very mildly affected and affected male relatives are not known to the proband, or the disease may be autosomal dominant as a result of a new mutation.

The frequency of the different genetic forms of retinitis pigmentosa varies in different series and in different communities (Table 1).

Further genetic information may be obtained by examining asymptomatic relatives. In particular, it is useful to examine the mothers of severely affected males because they frequently show mild retinal involvement, indicating the heterozygous state of X-linked disease (Bird, 1975).

The severity of the disease may be used as a guide to the likely inheritance (Jay & Bird, 1973). In general X-linked and autosomal recessive retinitis pigmentosa cause severe disease of early onset, while dominant disease is milder. Severe disease in a female usually indicates autosomal recessive and mild disease in a male autosomal dominant disease. Mild disease in a female is compatible with either the

Table 1

Author	No of Cases	AD %	AR %	XL %	Sp %
Ammann et al., 1961	153	9	90	1	
François & Verriest, 1962	153	19	35	4.1	41.3
Imaizumi, 1971	1091	2.1	40.1		43.2
Fishman, 1978	173	26	19	16	38
Pearlman, 1979	250	11	21	5	63
Boughman et al., 1980	670	10	69	6	15
Hu, 1982	151	11	33.1	7.7	48.3
Jay, 1982	426	24.4	15.5	18	42.1
Boughman & Fishman, 1983	300	21.7	16	9	50
Heckenlively 1987	609	16.9	30.8	10.1	42.2

(AD = autosomal dominant; AR = autosomal recessive; XL = X-linked; Sp = sporadic

heterozygous state of X-linked disease or autosomal dominant disease, whereas severe disease in a male suggests X-linked or autosomal recessive disease.

Functional studies

By careful studies of visual function, considerable success has been achieved in establishing a subdivision of RP. In some disorders, the target cells have been identified, and the characteristics of functional loss indicate the nature of the disease mechanism.

Two broad categories of autosomal dominant RP have been identified which are designated as type I or "diffuse" and type II or "regional" forms (Massof & Finkelstein, 1981; Arden et al., 1983; Lyness et al., 1985). The functional characteristics are consistent within families indicating genetic heterogeneity. In families with diffuse RP, affected members show widespread loss of rod function with relatively well preserved cone function at some stage in the evolution of their disease. Those with regional RP show variation in the state of rod and cone function with severe losses of both systems in some regions and near normal function in others. In regional RP, loss of rhodopsin accounts for the reduction of sensitivity, but in diffuse RP there is more rhodopsin than would be predicted if reduced light absorption by visual pigment was responsible for sensitivity loss (Kemp et al., 1988). These findings suggest that in regional RP loss of rod function is due to photoreceptor cell death, short outer segments or a combination of the two, while in diffuse RP this cannot be the case.

Additional variants have been identified in autosomal dominant disease. Sector RP (Bietti, 1937; Haase & Hellner, 1965) is characterised by retinal atrophy restricted to part of the fundus, and gross field loss is confined to the area of visual field corresponding to the involved retina. When this occurs, retinitis pigmentosa usually affects the lower half of the fundus with loss of visual function in the upper field (Fledelius & Simonsen, 1970), but rarely the disease affects only the superior (Ragnetti, 1962), nasal (Vukovich, 1959) or temporal fundus (Alezzandrini, 1965). In most reported cases with limited sector involvement the inheritance was autosomal dominant and the regional distribution of disease was common to affected members of the family (Hommer, 1959; Kuper, 1960; Lisch, 1960; Haase & Hellner, 1965; Fledelius & Simonsen, 1970). However, similar cases also appear in families with relatives who have involvement of the whole fundus (Krill et al., 1970) and some are heterozygous for X-linked retinitis pigmentosa (Krill et al., 1970; Bird, 1975). By fluorescein angiography, minor changes are often found to be more widespread than might be appreciated by ophthalmoscopy with white light, and the sector of retinitis pigmentosa may be an area of maximum disease rather than of exclusive involvement (Krill et al., 1970; Bird, 1975). In typical sector RP, rod and cone ERGs show mild reduction in amplitude with normal cone implicit times (Berson & Howard, 1971; Massof & Finkelstein, 1979; Fulton & Hansen, 1988). This pattern of disease is common to all affected members within a family irrespective of age, suggesting that the disease, in contrast to other forms of autosomal dominant RP, is non-progressive or progresses very slowly.

In some families there is slow recovery from bleach (Alexander & Fishman, 1984). Cone adaptation and the early part of the recovery of rod sensitivity follow the normal time course, but the later phase of rod adaptation is markedly prolonged.

Unilateral retinitis pigmentosa was reported as early as 1865 by Pedralgia and the ophthalmoscopic observations were corroborated by histological examination 25 years later. In 1952, François and Verriest reviewed the 56 unilateral cases reported up to that time and concluded that only 10 cases had the typical ophthalmoscopic appearance of genetically determined retinitis pigmentosa and were strictly unilateral. Subsequent reports (Kolb & Galloway, 1964; Carr & Siegel, 1973) have shown that, in some cases, minor changes can be demonstrated in the unaffected fellow eye by sophisticated examination. Furthermore, Carr and Siegel (1973) thought that the lesions in cases described by them and by some previous authors were due to vascular disease and were not heritable. To illustrate the difficulty in assessing the significance of such cases, Pearlman and colleagues (1976) described a patient with mild but widespread unilateral outer retinal disease, but they presented no evidence as to the inheritance of the disease. It should be emphasised that in no case of unilateral retinitis pigmentosa has a family history of eye disease been identified although, in one patient, the parents were consanguineous (Cordier et al., 1966). Most workers take the view that there is no good evidence that unilateral retinitis pigmentosa represents a distinct heritable disease.

The expressivity in some families is highly variable such that about 70% of members with the abnormal genotype have moderate to severe disease while the remainder are asymptomatic, but with mild fundus abnormalities and electrophysiological changes indicating the presence of the abnormal gene (Ernst & Moore, 1988). In these families, there appears to be bimodal expression of disease although the factors that modulate gene expression are unknown.

Specific clinical subtypes of RP have also been recognised with autosomal recessive disease. RP has been described with preservation of the retinal pigment epithelium adjacent to the retinal arterioles, and unlike many patients with RP they are consistently hypermetropic (Heckenlively, 1982; Porta et al., 1992).

A further variant comprises night blindness and variably reduced visual acuity associated with restricted pigmentation in the ocular fundus, cystoid macular changes resembling macular schisis, and distinctive electroretinographic abnormalities (Marmor et al., 1990). The rod ERG is unrecordable, maximal stimulus evokes a large response with very prolonged implicit time of the b-wave, photopic and scotopic responses are of similar magnitude, and the response is much larger when produced by short than by long wavelengths. The last attribute is known as the enhanced S cone response.

As early as 1914, Diem reported abnormal fundus reflexes in females heterozygous for X-linked retinitis pigmentosa, but during subsequent years similar phenomena were reported in other genetic forms of retinitis pigmentosa and were referred to as "tapetal reflexes" (Mann, 1937). An abnormal tapetal reflex has been reported to be the most common expression of the heterozygous state in X-linked disease (Falls & Cotterman, 1948; Wiener & Falls, 1955; Francois, 1962; Ricci et al., 1963; Hussels, 1967; Warburg & Simonsen 1968; Krill 1969). Ricci and his colleagues (1963) found that this reflex could not be seen after exposure of the fundus to light but reappeared after a period of darkness; they called this the "phénoméne de Mizuo inverse". However, Schappert-Kimmijser (1963) could identify this abnormal reflex in females of only one family out of eight and concluded that although this sign is useful when present, its absence does not indicate a normal genotype.

Attempts have been made to subdivide X-linked retinitis pigmentosa on the basis of differential affection of heterozygotes. Only a few of the early reports of families with X-linked disease described retinal degeneration in females (McQuarrie, 1935; Janssen, 1938) but profound visual loss in heterozygotes was reported by McKenzie (1951) in New Zealand. In 1960 Kobayashi described a family with even more severe involvement of the female members, five heterozygous females having retinitis pigmentosa, 11 having some stigmata of the disease and at least four being normal. No accurate comparative analysis was made of the severity of the disease between affected men and women, although it was stated that the females were more mildly affected than the males. From this information, it was concluded that there were three separate X-linked conditions: X-linked recessive, X-linked intermediate and X-linked dominant retinitis pigmentosa. However, subsequent experience indicates that there is considerable intrafamilial variation in the severity of the disease in heterozygotes, and that there is no justification for separating X-linked retinitis pigmentosa into different categories on the basis of the severity of the disease in women (Berson et al., 1969; Bird, 1975).

It does appear that the presence of a tapetal reflex in women heterozygous for the abnormal gene characterises some families but not others (Schappert-Kimmijser, 1963), such that this sign appears to differentiate X-linked RP into two categories on clinical grounds, an observation in keeping with recent genetic studies that indicate the existence of two loci for X-linked RP.

Molecular genetic studies

The first RP gene to be localised (RP2) was one for X-linked RP using a probe L1.28 (Bhattacharya et al., 1984) which maps to the proximal X chromosome short arm in band Xp11. Subsequent studies have supported the original observation (Friedrich et al., 1985; Mukai et al., 1985; Wright et al., 1987). The first indication of a second locus resulted from the identification of a deletion at Xp21 in a child with retinitis pigmentosa, Duchenne muscular dystrophy and X-linked chronic granulomatous disease (Francke et al., 1985); this locus is distal (telomeric) to that of RP2 and is designated RP3. Localisation of an X-linked retinitis pigmentosa gene to this distal site was supported by other linkage studies (Nussbaum et al., 1985a; Denton et al., 1988). There is no wide agreement on the clinical differences between diseases transmitted at the two loci, although all heterozygotes in whom a tapetal reflex is seen appear to belong to families in which mutation is at the RP3 locus. The possibility of more than one abnormal allele at each locus has not been excluded; that not all families with disease transmitted at the RP3 site have a tapetal reflex would support this view.

In 1989, a locus for autosomal dominant retinitis pigmentosa (adRP) was identified on the long arm of chromosome 3 (McWilliam et al., 1989). Three candidate genes exist close to the proposed site, and within a short time a proline/histidine mutation was detected at codon 23 of the rhodopsin molecule (Dryja et al., 1990a). Since then, more than 30 different mutations have been detected in the rhodopsin gene and it appears that this gene accounts for about 25% of autosomal dominant RP (Dryja et al., 1990b; Sung et al., 1991; Keen et al., 1991; Inglehearn et al., 1992). Most of the substitutions appear to be peculiar to a single family, although some, notably the mutation at codon 347, have been found in many parts of the world. The proline/histidine mutation at codon 23 was identified in a proportion of subjects in the United States but not in other countries. There is now some evidence that the RP patients with this mutation may form part of one large pedigree.

In several families with autosomal dominant RP, there is no linkage with markers in the long arm of chromosome 3 (Lester et al., 1990), confirming heterogeneity of dominant disease, and it has now been shown that genes other than that for rhodopsin may be responsible for autosomal dominant disease (Blanton et al., 1990, 1991; Farrar et al., 1991a,b; Kajiwara et al., 1991). One of these is the gene for rds peripherin that is of particular interest since it is a mutation in this gene that is responsible for the retinal degeneration in the *rds* mouse (Travis et al., 1989; Connell et al., 1991). Peripherin is localised to the outer segment disc membranes of both rods and cones, and is thought to be essential for the assembly, orientation and physical stability of the outer segment discs of the retinal photoreceptors (Connell & Molday, 1990; Travis et al., 1991a; Arokawa et al., 1992).

Of great interest is the identification that macular dystrophy may also be due to mutations in the *RDS* gene. Two families with a form of macular dystrophy have been shown to have a tryptophan substitution for arginine at codon 172, and another family with almost identical disease has a glutamine substitution for arginine at the same codon. A mutation resulting in a stop codon at position 258 has been identified in another family with a retinal degeneration similar in appearance to adult vitelliform macular dystrophy. These findings demonstrate that some forms of autosomal dominant retinitis pigmentosa and dominantly inherited macular dystrophies are caused by mutations in the same (RDS peripherin) gene. The variability of phenotype caused by mutations at different codons suggests that the functional significance of certain amino-acids to cones and rods may be different (Wells et al., 1993).

Functional consequences of mutations in adRP genes

In the families with autosomal dominant RP showing mutations of the rhodopsin gene on chromosome 3, different patterns of retinal dysfunction have been demonstrated with different mutations (Fishman et al., 1991; Heckenlively et al., 1991; Jacobson et al., 1991; Kemp et al., 1992; Moore et al., 1992).

Functional loss compatible with diffuse (type I) autosomal dominant RP was found in patients with three types of rhodopsin mutation in whom rod and cone function was tested; proline to leucine at codon 347, lysine to glutamic acid at codon 296 (which is the binding site for retinal), and an isoleucine deletion at codon 255 or 256. With all three mutations, poor night vision was consistently identified in early life and rod ERGs were severely reduced or unmeasurable. In those with measurable visual function, psychophysical testing showed rod function to be severely affected throughout the retina (with threshold elevations of more than 3 log units), even in younger individuals, while loss of cone function varied widely between families, and was less widespread and less severe than loss of rod function. In the most severely affected family (lys296glu), there was little visual function after the age of 30 years in most members (Keen et al., 1991). With the isoleucine deletion at codon 256, cone function was limited to the central 10° by the age of 25 years but little further loss occurred thereafter. Cone function was retained over most of the visual field until middle life in the codon 347 mutation despite widespread and severe early loss of rod function.

The functional abnormality in families with mutations threonine to arginine at codon 17, proline to histidine at codon 23, threonine to arginine at codon 58, glycine to arginine at codon 106 and guanine to adenine at codon 182 is qualitatively similar with altitudinal or sectorial distribution of disease and this attribute appears to be constant within the families (Fishman et al., 1991, 1992; Heckenlively et al., 1991; Richards et al., 1991; Stone et al., 1991; Kemp et al., 1992; Moore et al., 1992). This characteristic of disease is in marked contrast with that seen in patients with other rhodopsin mutations examined to date. Rod function is severely depressed in the superior field but is nearly normal in the inferior field. Loss of cone sensitivity closely follows that of the rods. In this respect the pattern of disease resembles regional (type II) RP.

An additional striking finding in these families with sectorial type RP is a characteristic abnormality in one component of the kinetics of dark adaptation following exposure to a bright light. Prior to bleaching, measurements made in the relatively intact portion of the visual field show mild threshold elevations of approximately one log unit. Following light adaptation there is a marked delay in the recovery of sensitivity. The initial portion of recovery mediated by cones and rods is normal. However by one hour, when in the normal subjects recovery of sensitivity is complete, these patients show residual threshold elevations of 1–2 log units from the prebleach values. Even after nearly two hours, thresholds are still elevated by more than 1.0 log unit. In two subjects, it was found that the time course of this slow recovery of sensitivity was of the order at least 80–120 hours (Moore et al., 1992).

Using a model based upon primate data of rod outer segment length and turnover, it has been calculated that the delayed phase of the recovery of rod sensitivity following strong light adaptation could be due in part to the formation of new disc membranes with a normal concentration of rhodopsin rather than *in situ* regeneration of photopigment (Moore et al., 1992). The model requires that the outer segments are short as a result of RP, and that a major portion of the outer segment is shed following strong light adaptation.

These observations show that there are both quantitative and qualitative differences in the RP consequent upon different mutations in the rhodopsin gene. Also important is that it has been shown for the first time that some forms of RP are due to defects of metabolic systems that are limited to rod photoreceptors, and yet it is evident that loss of photopic function is consistently found in these disorders. No explanation exists to account for cone cell death. It is possible that cone loss occurs as a result of either a release of toxic products by dying rods or the cones being metabolically dependent upon the presence of rods. It follows that loss of cone function, although important to the patient, is a secondary effect. It is also clear that these studies are incomplete and that further work will produce more information.

The current knowledge of the putative function of rds peripherin provides an explanation for different phenotypes being associated with different mutations in the *RDS* gene. The photoreceptor outer segment protein, rds peripherin, has 346 amino acids with four hydrophobic transmembrane domains and two putative N-linked glycosylation sites, one of which is conserved across four species and is thought to be important to the protein's function in stabilising the parallel array of photoreceptor outer segment membranes (Travis et al., 1991a; b). Immunohistochemical studies have shown that the protein is limited to the membranes of outer segments of both rods and cones although there is disagreement over its precise localisation. One study using a polyvalent antibody to a short peptide sequence near the carboxyl terminus implied that the protein was distributed over the entire length of the outer segment disc membrane (Travis et al., 1991a). It was suggested that covalent bonding between rds peripherin molecules was responsible for the maintenance of the parallel arrangement of the outer segment membranes. By contrast, a separate study using a monoclonal antibody to purified photoreceptor discs showed that labelling was confined to the rims of the outer segment, implying that the primary function of the protein was to stabilise the thermodynamically unfavourable bend at the disc rim (Kajiwara et al., 1991). To explain these different findings it was hypothesised that the epitopes available for antibody binding may differ at the disc rim when compared with other parts of the disc membrane. Regardless of the precise mechanism, the general belief is that rds peripherin is important to the structural stability of the outer segment disc membrane. Studies of the *rds* mouse support this belief. In the homozygous *rds/rds* genotype, outer segment discs are not formed, the disc membrane being discharged as small vesicles into the subretinal space (Sanyal & Jansen, 1981; Cohen, 1983; Jansen & Sanyal, 1984; Usukura & Bok, 1987). In the heterozygous rds/+ mouse, outer segment disc membranes are long and distorted (Hawkins et al., 1985; Sanyal & Hawkins, 1988).

Recently it has been shown that rds peripherin may be non-covalently associated with ROM1, a protein structurally related to peripherin. ROM1 has been localised to the disc rims of rod outer segments but has not been identified in cones (Bascom et al., 1992). It has been proposed that the formation and stability of the bend in the disc membrane in rods are dependent upon the association between rds peripherin and ROM1. However, the absence of ROM1 in cones implies that differences exist between the precise mechanisms by which rds peripherin stabilises outer segment membranes in the two classes of photoreceptor. In rods, the association between rds peripherin and ROM1 may be important, whereas in cones, rds peripherin may bind to a different membrane protein or act alone. If the binding sites are different in rods and cones, conservation of one amino acid in rds peripherin may be important to rods only, and a mutation at this site could cause a dystrophy falling within the category of retinitis pigmentosa, in which rods were the target cell of disease with relative preservation of cones. Conversely, a different mutation in the *RDS* gene may disrupt the metabolism or structure of either cones alone causing macular dystrophy or of both rods and cones. From our data, it appears that the presence of arginine at position 172 is important to the structure and function of cones but not rods (Wells et al., 1993).

Choroideremia

Choroideremia, a diffuse photoreceptor dystrophy that may be difficult to distinguish from RP, particularly in young males, was recorded as a distinct condition by Mauthner (1871) and the inheritance was recognised as X-linked in 1942 (Goedbloed, 1942; Waardenburg, 1942). Apart from its inheritance, the distinctive feature of this disorder is diffuse progressive atrophy of the retinal pigment epithelium and choriocapillaris.

The symptoms in affected men are identical to retinitis pigmentosa, with loss of dark adaptation and peripheral fields early, which progresses to leave a small central field. Visual acuity is maintained at a good level until late in the disorder, often well into the 50s. Electrophysiological responses are severely depressed early in the disease, as they are in RP.

In males, the fundus shows a generalised loss of the retinal pigment epithelium, sometimes resulting in a blond fundus as early as two years of age, followed by the appearance of fine granular subretinal pigment deposits in the midequatorial region. By the third decade of life, there are multifocal areas of loss of the retinal pigment epithelium with concurrent underlying choriocapillaris atrophy in a typically scalloped pattern, becoming more widespread and slowly progressing to a generalised confluent loss of retinal pigment epithelium and choriocapillaris. The disease progresses until a small area of choriocapillaris and pigment epithelium survives at the fovea (McCulloch & McCulloch, 1948), and even this is eventually lost. Fluorescein angiographic studies demonstrate well the loss of pigment in the pigment epithelium and the associated loss of choriocapillaris.

Histopathological examination confirmed the extensive chorioretinal atrophy (Rafuse & McCulloch, 1968; McCulloch, 1969; Cameron et al., 1987), as well as epiretinal membrane formation and vascular endothelial cell abnormalities (Cameron et al., 1987).

The heterozygous females almost always have abnormal fundi with changes in the pigment epithelium, but functional loss is unusual in the young and is mild later in life. The most common appearance is of midperipheral pigmentation localised deep in the retina and associated with spotty areas of pigment epithelial atrophy. There may be pigment clumping that may be linear in distribution. Occasionally, focal areas of choroidal atrophy occur, around the optic disc or in the midperiphery. These fundus changes are usually stationary, but progression has been noted (Krill, 1977). Microscopic examination demonstrated abnormalities of the pigment epithelium only (Krill, 1977), while other reports are of widespread malformation of the outer receptor segments and patches of retinal atrophy (Ghosh et al., 1988), or of short or absent photoreceptor outer segments in much of the equatorial region, irregular thickness and pigmentation of the retinal pigment epithelium, and areas of profound atrophy in the equatorial region (Flannery et al., 1990).

Choroideremia is a distinct entity within the group of outer retinal dystrophies, in which it is the retinal pigment epithelium that is primarily affected by the disorder, rather than the choroid as the name implies. The choroideremia locus has been mapped to band q21 on the X chromosome (Nussbaum et al., 1985b) and DNA clones that span Xq21 deletions in patients with the disease have been isolated (Cremers et al., 1990). In some of the patients with microdeletions, mental retardation and/or deafness have been reported (Hodgson et al., 1987; Rosenberg et al., 1987; Cremers et al., 1989; Merry et al., 1989), but despite the size of the deletion, the choroideremia remains remarkably constant in clinical expression.

It is likely that the prominent choroidal atrophy is a secondary response to pigment epithelial cell loss. Choroidal atrophy is not unique to choroideremia, since it has been identified in about half the eyes with retinitis pigmentosa subject to histopathological study, but profound atrophy early is unlike retinitis pigmentosa and signifies a pathogenic process different from those in other outer retinal diseases.

Refsum disease

This rare autosomal recessive disorder, characterised by retinitis pigmentosa, chronic polyneuritis and cerebellar signs, was first described in 1945 (Refsum, 1945) and in 1963 it was shown that affected individuals had elevated levels of serum phytanic acid (Klenk & Kahlke, 1963). This fatty acid is exogenous, being derived from the breakdown of chlorophyll. Cultured fibroblasts from skin have defective alpha-hydroxylation of phytanate and in heterozygotes the level of activity is about half the normal level although they have no symptoms (Steinberg et al., 1967; Herndon et al., 1969; Hutton & Steinberg, 1973).

Loss of night vision is usually the first symptom occurring before the age of 30 years. The progression and fundus changes are similar to those of more classical forms of retinitis pigmentosa, although loss of central vision and macular changes occur earlier than is usual. Miosis (small poorly reacting pupils) is common and contributes to the reduced vision (Gibberd et al., 1985).

Hypertrophic interstitial peripheral neuropathy is a constant feature of the disorder although it is usually not clinically apparent until after the onset of ocular disease, while anosmia is usually apparent from an early age. Progressive neurosensory hearing loss, abnormal vestibular responses, ichthyosis and cardiac conduction defects are commonly found.

Histopathological changes of the eye show accumulation of an abnormal deposition, presumed to be phytanic acid in the pigment epithelium of iris and retina (Levy, 1970; Toussaint & Danis, 1971; Dick et al., 1990).

Treatment consists of reduction of intake of all foods containing chlorophyll, which includes all fat products of herbivores. The principal foods allowed are fish, lean meat, poultry, margarine and grain products. Plasmapheresis performed once or twice a month effectively removes phytanic acid from the body and reduces the dietary restriction (Gibberd et al., 1979; Moser et al., 1980). Reduction in plasma phytanic acid produces a considerable clinical improvement in the neuropathy but not in the retinitis pigmentosa (Gibberd et al., 1985).

Abetalipoproteinaemia

This rare autosomal recessive disorder, which was first described in 1950 (Bassen & Kornzweig, 1950), presents initially with malabsorption syndrome and fat intolerance, and has a pigmentary retinopathy, progressive ataxic neuropathy and acanthocytosis as other common features. The retinopathy is indistinguishable from that in a separate genetic disorder, hypobetalipoproteinaemia. It has become clear that the major abnormality is defective fat absorption causing deficiency of vitamins A and E, undetectable beta-lipoproteins (Salt et al., 1960) and low cholesterol in serum, reduced plasma phospholipids, total lipids, triglycerides and alpha-1-lipoproteins. The intestinal mucosa shows characteristic changes, including packing of the villa with lipid droplets. Some erythrocytes have a curious crenated appearance, termed acanthocytosis (Isselbacher et al., 1964).

Fundus abnormalities may become apparent in the first year of life with white dots at the level of the pigment epithelium that is reminiscent of changes seen in dietary vitamin A deficiency. With time, the characteristic fundus appearances and histopathological changes of retinitis pigmentosa appear (Jampel & Falls, 1958; Von Sallman et al., 1969; Cogan et al., 1984).

Progressive neurological disease develops which is usually manifest initially as abnormal gait. Abnormalities include cerebellar ataxia, muscle wasting and sensory loss.

Treatment by restriction of dietary fats and supplementation with large doses of vitamins A and E appears to be effective in preventing the advance of the disease (Wolff et al., 1964; Muller et al., 1977; Bishara et al., 1982) and reversal of the fundus changes has been reported (Gouras et al., 1971). Prophylactic vitamins in the first 12 to 18 years of life appear to prevent or modify the occurrence of retinopathy (Runge et al., 1986).

CENTRAL RECEPTOR DYSTROPHIES

A number of genetically determined disorders cause progressive loss of visual functions associated with cones: losses of visual acuity, colour vision and central visual field, and diffuse, poor vision in bright light. These patients also remark upon difficulty with night vision, but on closer questioning this often relates to a short transient period on first entering a dark environment.

Specific disorders, such as Best's disease, can be identified but the remainder include a group of entities that cannot be clearly distinguished one from another. These include the recently described mutations in the *RDS* gene causing macular dystrophy (Wells et al., 1993).

Vitelliform macular dystrophy (Best's disease)

The typical lesion of Best's disease is a round yellow deposit at the macula that may be identified within a short time of birth (Barkman, 1961) or which may develop later in a previously normal fundus (Deutman, 1971; Barricks, 1977). On rare occasions, the macular lesion is not seen until later life (Deutman, 1971); the lesion may be extramacular, the disease may be asymmetrical (Maloney et al.,

1977) or the fundus may be entirely normal (Deutman, 1971; Krill, 1977; Mohler & Fine, 1981; Godel, et al., 1986). Godel and colleagues (1986) state that about half of those with the abnormal genotype have a normal or near normal fundus appearance and normal vision. The abnormal material appears to accumulate at the level of the pigment epithelium, histopathological evidence indicating that it is probably between Bruch's membrane and the pigment epithelium (O'Gorman et al., 1988). Although flicker fusion studies indicate cone dysfunction much earlier (Massof et al., 1977), central cone function remains relatively good in most patients until the second decade of life, at which time there is progressive loss of central vision, associated with disintegration of the central yellow area. The evolution of the lesion during this period is determined by whether or not new blood vessels invade the abnormal deposit from the choroid. If new vessels invade the lesion, there may be subretinal haemorrhages and fibrosis (disciform response), and the visual outcome is poor (Miller et al., 1976). In other patients, the yellow material may become partially liquefied, forming fluid levels within the subretinal space and eventually the material resorbs, leaving confluent atrophy of the central pigment epithelium and choriocapillaris. The appearances of patients with normal maculae, minimal pigment epithelial changes, atrophic maculae or fibrous macular scars tended not to change; those with vitelliform or pseudohypopyon lesions progressed to later atrophic macular stages (Mohler & Fine, 1981).

The autosomal dominant mode of inheritance of this disorder was first suggested by Best (1905) and was substantiated later by further observations on the same family (Vossius, 1921; Weisel, 1922; Jung, 1936).

The rise in ocular potential induced by light,as seen in the electro-oculogram, is always reduced in patients with the abnormal genotype for Best's disease, whatever their clinical status, but ERG potentials are normal (Deutman, 1969, 1971). The universal reduction of the light-induced rise in ocular potential even in those patients with normal fundi, indicates a widespread dysfunction of the receptor pigment epithelial complex that has possibly been present from birth.

In patients with advanced disease, histopathology showed extensive atrophy of photoreceptors in the macular area (McFarland, 1955; Anderson, 1977). In slightly less advanced cases, there was also widespread abnormality of the retinal pigment epithelial cells, in which an excessive amount of lipofuscin had accumulated (Frangieh et al., 1982; Weingeist et al., 1982; O'Gorman et al., 1988). In one case, an accumulation of heterogeneous material between Bruch's membrane and the pigment epithelium at the fovea was believed to represent the location of a previtelliform lesion. This material appeared to be derived from degenerating pigment epithelial cells (O'Gorman et al., 1988).

Adult vitelliform macular dystrophy

Adult vitelliform macular dystrophy was first described by Gass (1974) as a peculiar foveomacular dystrophy. It is characterised by a focal, round or oval shaped, subretinal yellowish foveal lesion, often with one or more pigment spots on the anterior surface at the level of the pigment epithelium. The lesions may vary in size but are typically one third to one half of a disc diameter, and are usually bilateral and symmetrical. Patients usually present in the fourth or fifth decade of life and tend to have minimal visual symptoms (Brecher & Bird, 1990).

The disorder differs from vitelliform macular dystrophy (Best's disease) in that the foveal lesions are smaller, it presents at a later age, it does not demonstrate evolutionary changes of the foveal lesion, and the light induced rise in ocular potential is rarely absent.

Although series have been reported where there was no evidence of familial involvement (Kingham & Lochen, 1977), this condition is now generally regarded as being transmitted as an autosomal dominant trait (Gass, 1974; Vine & Schatz, 1980; Brecher & Bird, 1990). It is interesting that a mutation in the peripherin RDS gene, in a stop codon at position 258, has been found in one patient with a retinal degeneration similar in appearance to adult vitelliform macular dystrophy (Wells et al., 1993).

Butterfly-shaped dystrophy

In 1970, five members of a family were described from Holland with a unique fundus appearance consisting of patterned hyperpigmentation at the macula (Deutman et al., 1970). The pigment appeared to be in the pigment epithelium, and fluorescein angiography showed granular hyperpigmentation of the remaining pigment epithelium at the posterior pole. The disorder caused few symptoms since the worst visual acuity was 0.8. The most striking feature was the universal reduction of the light-induced rise in ocular potential to 130% or lower in the presence of good visual function.

Deutman and colleagues (1970) concluded that there was diffuse dysfunction of the retinal pigment epithelium with little associated visual deficit since there was no indication of rod dysfunction as gauged by the electroretinogram and no abnormality of colour vision. In this respect butterfly-shaped dystrophy is similar to Best's disease. The original pedigree is suggestive of autosomal dominant inheritance, although the disease was identified in only two generations.

Bietti's crystalline dystrophy

A unique fundus appearance was reported by Bietti in three patients, two of whom were brothers, who had crystals in the retina and in the peripheral cornea (Bietti, 1937). There is progressive atrophy of the retina with prominent atrophy of the pigment epithelium and choroid. As the atrophy supervenes, the intraretinal crystals disappear. The pattern of retinal functional loss is similar to that of retinitis pigmentosa with the initial symptoms usually appearing in the third or fourth decade of life. Autosomal recessive inheritance was established in this condition by Hu (1983). Patients with a similar fundus appearance but without corneal changes have been described and considered to represent the same disorder (François and De Laey 1970; Welsh 1977; Hayasaka & Okoyuma 1984). No systemic biochemical disorder has been identified, although the corneal crystals resemble cholesterol or cholesterol ester, suggesting that Bietti's crystalline dystrophy may be due to a systemic abnormality of lipid metabolism (Wilson et al., 1989).

Other central dystrophies

The remaining cone dystrophies have been subdivided with respect to inheritance and the appearance of the fundus, but no satisfactory categorisation of these conditions has yet been devised, and it is likely that each subdivision contains more than one condition. Two broad subdivisions, fundus flavimaculatus and bull's eye maculopathy, can be identified on the basis of fundus morphology, although, the distinction may not be absolute.

Fundus flavimaculatus

This broad category comprises diseases in which there is deposition of white material at the level of the pigment epithelium; Stargardt's disease (Stargardt, 1909, 1913) and fundus flavimaculatus (Franceschetti, 1963; Franceschetti & Francois, 1965) have been used to denote these disorders, but there is no evidence that these two terms describe separate conditions. Most cases show autosomal recessive inheritance although autosomal dominant forms are occasionally seen (Cibis et al., 1980). These conditions usually cause rapid loss of central vision during a 6 month period in the first 15 years of life, although in some cases good visual acuity is maintained until the age of 40 years.

At the time of visual loss, confluent atrophy of the outer retina, pigment epithelium and choriocapillaris occurs at the fovea, and this area grows slowly during the rest of the patient's life. The white "fish tail" flecks occupy the remaining part of the posterior pole with characteristic sparing of the peripapillary region. These lesions can be identified at the time of the initial visual loss and resolve as additional ones appear elsewhere (Hadden & Gass, 1976).

On fluorescein angiography, the choroid is not seen in the majority of patients although it appears normal in some (Bonin, 1971; Fish et al., 1981; Fishman et al., 1987). It seems likely that the non-appearance of the choroid signifies an even deposition of abnormal material at the level of the pigment epithelium that absorbs blue-green light. This has been confirmed by histopathological studies in which it has been demonstrated that the retinal pigment epithelial cells were packed with lipofuscin and melanolipofuscin (Eagle et al., 1980; Lopez et al., 1990); these changes start to develop in childhood (Steinmetz et al., 1991). At the site of previous white lesions, atrophy of the pigment epithelium occurs and fluorescein angiography shows multifocal hyperfluorescence, corresponding to the areas of pigment epithelial atrophy.

On electroretinography, all patients have some significant abnormality of cone function while many have abnormal rod function (Moloney et al., 1983). The EOG is initially normal but frequently becomes subnormal late in the disease (Noble & Carr, 1979; Moloney et al., 1983).

Bull's eye dystrophies (including cone and cone–rod dystrophies)

The typical fundus changes in this group of cone and cone–rod dystrophies consist of one or more concentric rings of pigment epithelial change around the fovea, which gives rise to a characteristic appearance on fluorescein angiography. Visual loss may occur at any time during the second to the fifth decades of life and once started progresses slowly (Grey et al., 1977). There may or may not be white deposits at the level of the pigment epithelium. The phenomenon of dark choroid is also seen in some families with bull's eye dystrophy (Fish et al., 1981).

The heterogeneity of this group is indicated by the inheritance in different families, which may be autosomal dominant, autosomal recessive or X-linked. Within this group are the progressive cone dystrophies that usually present in adolescence or early adult life with reduced vision, photophobia, and defects of colour vision and which usually precede the macular changes. These macular changes result from atrophy of the retinal pigment epithelium, giving rise to a "bull's eye" appearance.

The cone dystrophies are at present classified by their mode of inheritance, which may be autosomal dominant (Goodman et al., 1963; Berson et al., 1968; Krill & Deutman, 1972; Pearlman et al., 1974; van Schooneveld et al., 1991; Went et al., 1992), autosomal recessive (Gouras et al., 1983; Yagasaki & Jacobson, 1989) or X– linked (Heckenlively & Weleber, 1986; Jacobson et al., 1989; Reichel et al., 1989; van Everdingen et al., 1992). Even within these different genetic forms of cone dystrophy there is heterogeneity. In the group of dominant cone dystrophies there is one recently described form with early and near complete absence of blue cone function, but with a blue-sensitive visual pigment gene indistinguishable from normal (Went et al., 1992). In a family with X-linked cone dystrophy accompanied by loss of red cone function, a 6.5 kilobase deletion within the red cone pigment gene was found (Reichel et al., 1989).

Heterogeneity is further indicated by the presence of peripheral retinal degeneration in some cases, and these cone-rod dystrophies, with rod involvement, must be distinguished from cone dystrophies, where rod function is normal, by means of electrophysiological tests of rod function. It must also be remembered that many patients with each of the genetic forms of retinitis pigmentosa (rod-cone dystrophies) also have a bull's-eye appearance at their maculae (Pruett, 1983). Drug induced phenocopies occur (Kearns & Hollenhorst, 1966; Krill et al., 1971) and comparable dystrophies have also been described in the Pierre-Marie type of hereditary ataxia (Bjork et al., 1956), in fucosidosis (Snodgrass, 1976) and in amelanogenesis imperfecta (Jalili & Smith, 1988).

There is considerable doubt as to whether the fundus changes of fundus flavimaculatus and bull's eye dystrophy indicate that the disorders are clearly separated into two broad groups. Studies of fundus appearance show that in some families the changes are constant, whilst in others bull's eye dystrophy may be seen in early disease and flavimaculatus lesions seen later.

Sorsby's fundus dystrophy (pseudoinflammatory macular dystrophy)

Macular dystrophies have been described in which there are features other than atrophy of the outer retinal layers. Sorsby and his colleagues (1949) first reported an autosomal dominant disorder in which there is bilateral central visual loss in the fifth decade of life from subretinal neovascularisation, and progressive atrophy of the peripheral retina and choroid leading to loss of ambulatory vision by the seventh decade in most cases.

Three of the original families reported by Sorsby (Sorsby et al., 1949) have been extensively studied more recently and have confirmed the three key features of the dystrophy described by Sorsby: autosomal dominant inheritance, loss of central vision occurring in the fifth decade of life, and loss of ambulatory vision occurring during the eighth decade. In one of these families (E family), prior to any reduction of central vision, the authors found fine drusen in some patients, and in others a confluent deposit of yellow subretinal material at the macula (Hoskin et al., 1981). In another family (K family), peripheral retinal dysfunction occurred, with a deposit of a yellow subretinal material throughout the fundus and a tritan colour defect, all prior to the loss of central vision. In some of these patients there was loss of central vision as a result of atrophic disease, rather than from ingrowth of subretinal new vessels (Capon et al., 1988). In a third family (C family), as in the other two, yellow material was present at the level of Bruch's membrane early in the course of the disease, although the earliest phenotypic marker was delayed filling of the choriocapillaris. The loss of central vision was commonly due to atrophy of the outer retina and choroid; subretinal neovascularisation was a rare occurrence (Polkinghorne et al., 1989). In another family (not one of Sorsby's), three variations in the fundus appearance were seen: firstly and most commonly, white to yellow fundus spots accompanied a disciform macular degeneration; in the second, the fundus spots were absent; in the third, the yellow deposits were associated with atrophic macular degeneration (Hamilton et al., 1989).

Electrophysiological responses were originally thought to be usually within normal limits, suggesting that the disorder was localised to the macular region (Carr et al., 1975), although a reduced light rise on electro-oculography has more recently been reported (Capon et al., 1988; Dreyer & Hidayat, 1988).

A light and electron microscopic study of the eyes of one patient showed a 30 micron thick deposit within Bruch's membrane that stained positive for lipids. In addition, there was gross loss of the outer retina, a discontinuous retinal pigment epithelium and atrophy of the choriocapillaris (Capon et al., 1989).

It has to be appreciated that atrophy of the choriocapillaris or of the whole thickness of the choroid appears to be a common phenomenon at the terminal phase in all outer retinal dystrophies. It has been described histologically in half the cases of retinitis pigmentosa studied and is a common late feature of most, if not all, central dystrophies, such as Stargardt's disease and bull's eye dystrophy. These observations suggest that the choroidal atrophy is a response to absence of the retinal pigment epithelium and retinal photoreceptors. Initially, the choriocapillaris was thought to be primarily affected in central areolar choroidal atrophy (Sorsby, 1939; Sorsby & Crick, 1953; Noble, 1977) because atrophy of the retinal pigment epithelium and choroid at the macula was apparently the initial or at least the most prominent ophthalmoscopic change when the patient was first seen. However, Ashton (1953) demonstrated that the major choroidal blood vessels were normal in a case of central areolar choroidal sclerosis by histopathological examination. It is also evident that Noble (1977) could not have been able to comment on the state of the choriocapillaris had the pigment epithelial pigment content been normal. Clearly, there is doubt concerning the primary site of disease in so-called choroidal dystrophies. The two possibilities are primary atrophy of the choroid or primary atrophy of the pigment epithelium, each causing secondary atrophy of the neighbouring tissues.

It is possible that early choriocapillaris closure distinguishes disorders with early pigment epithelial cell death from those in which the disease affects primarily the retinal receptors.

CONGENITAL RECEPTOR DEFECTS

A series of conditions has been recognised in which there is a genetically determined nonprogressive visual defect that may be related to either the cone or the rod systems.

Cone defects

Colour vision defects

Defects in colour vision have been classified on the basis of the concept that colour vision is determined by three classes of cones: red-sensitive, green-sensitive and blue-sensitive. The basis for this classification has been confirmed by the sequencing of the three cone pigment genes (Nathans et al., 1986a,b; Vollrath et al., 1988). If colour matching tests demonstrate that one of the three systems is defective but present (trichromats), the terms protanomaly (red), deuteranomaly (green) and tritanomaly (blue) are used and if one is functionally absent (dichromats), the suffix -anopia replaces the suffix -anomaly.

Deuteranomaly, deuteranopia, protanomaly and protanopia are X-linked conditions: deuteranomaly is found in about 5% of the population of Western Europe and North America; deuteranopia, protanomaly and protanopia are each found in about 1% (Waardenburg et al., 1963). Tritanomaly and tritanopia, which are inherited as autosomal dominant disorders, are much less common affecting between 0.002 and 0.007% of the population (Verriest, 1974). Individuals with these abnormalities of colour vision have normal visual acuities, and anomalous trichromats are often unaware of this condition. Apart from the colour defect, the eye is otherwise normal. The molecular genetic basis of red/green colour vision defects is discussed in detail in Chapter 9.

Monochromatism

Patients suffering from monochromatism have absent or markedly impaired colour vision. In complete rod monochromatism there is little evidence of cone function, so that visual acuity is poor, the patient has nystagmus and extreme photophobia is characteristic. The fundi show no gross abnormality, although, the foveola may appear abnormal and changes in the central pigment epithelium are sometimes identified (Waardenburg et al., 1963). The dark adaptation curve is typically monophasic, fusion frequency is very low and the photopic (cone) ERG is absent but the scotopic (rod) ERG is normal. Rod monochromatism is an autosomal recessive trait. An incomplete form of rod monochromatism has been described where the symptoms are less severe, visual acuity is better and photophobia and nystagmus may be absent (Sloan & Newhall, 1942; Siegel et al., 1966). In another form of monochromatism there is absence of central cones (Krill, 1977).

Blue cone monochromatism presents in a similar fashion to complete rod monochromatism, except that there appears to be an intact blue cone system and the inheritance is X-linked (Alpern et al., 1965). The gene for this form of monochromatism maps to Xq28, and this rare disorder results from alterations of the red and green pigment genes, or from their deletion (Lewis et al., 1987; Nathans et al., 1989).

Four eyes from individuals with monochromatism have been examined histopathologically (Larsen, 1921; Harrison et al., 1960; Falls et al., 1965; Glickstein & Heath, 1975). These showed a change at the fovea with a reduced number of cones and an abnormality of those which remained.

Rod defects

Congenital stationary night blindness

Congenital stationary night blindness (CSNB) is characterised by night blindness, normal fundus appearance, the absence of rod dark adaptation (monophasic dark adaptation curve) and lack of progression. It may be inherited as an autosomal dominant, an autosomal recessive or an X-linked trait.

The most widely reported form of CSNB is an autosomal dominant disorder (Dejean & Gassenc, 1949). These patients appear to have normal cone function but little rod function. Visual acuity, colour vision and photopic visual fields are normal or are at most mildly abnormal and dark adaptation shows only a cone segment that may be abnormally prolonged (Krill & Martin, 1971); there is no shift from cone to rod characteristics in the dark adaptation curve, and the ERG shows no prolongation of the b wave implicit time between the photopic and scotopic records. However, fundus reflectometry indicates a normal concentration of bleachable rhodopsin in one case (Carr et al., 1966) and histological studies have shown no structural abnormalities either in the retina as a whole or in the rods in particular (Vaghefi et al., 1978).

The majority of patients with autosomal recessive and X-linked CSNB, and occasional patients with autosomal dominant CSNB, have an ERG with a near normal a wave and a substantially reduced b wave on testing under scotopic conditions (negative ERG) (Noble et al., 1990). With increasing intensity of the test stimulus, the amplitude of the a wave increases but that of the b wave is unchanged (Schubert & Bornschein, 1952). These patients could be further divided into two groups: one group lacked rod function (complete type), the other retained some rod function (incomplete type) (Miyake et al., 1986). Patients with the complete type are myopic; those with the incomplete type may be hyperopic or myopic (Khouri et al., 1988). Although complete and incomplete CSNB did not coexist in any of the families reported by Miyake et al. (1986), others have found patients with both complete and incomplete CSNB within the same X-linked pedigrees (Khouri et al., 1988; Pearce et al., 1990). This led Pearce et al. (1990) to propose that X-linked CSNB is a single clinical entity manifesting a wide variation in clinical expression. It has also been suggested that incomplete CSNB and Aland Island eye disease may be the same condition (Weleber et al., 1989; Alitalo et al., 1991).

Myopia is almost always associated with the X-linked form of CSNB in which visual acuity is reduced and nystagmus may be present. Carriers of X-linked CSNB are not night blind but may show abnormal oscillatory potentials on ERG (Miyake & Kawase, 1984; Young et al., 1989). Myopia also occurs in those cases of autosomal recessive CSNB that have abnormal vision and, again, these patients may have nystagmus (Krill, 1977). These forms of CSNB may present in infancy with blindness (Weleber & Tongue, 1987).

The complete X-linked CSNB gene has been assigned to Xp11.2-p11.3 (Musarella et al., 1989; Gal et al., 1989; Bech-Hansen et al., 1990).

Light and electron microscopic studies of one eye with CSNB and a negative ERG showed a normal arrangement of rod outer segment discs, normal synaptic ends of the photoreceptors and complete loss of ganglion cells. It was suggested that the cause of CSNB in this case may be related to mechanisms inhibitory to cells of the bipolar layer (Watanabe et al., 1986). The absence of rod–cone interaction, together with an absent scotopic b wave, also implies that the defect is in the midretinal layers (Siegel et al., 1987).

Prolonged Dark Adaptation

Three distinct conditions have been described in which the final threshold of dark adaptation is normal but in which the rod phase is abnormally prolonged; the defect in each appears to be static.

Oguchi's disease

This autosomal recessively inherited disorder with prolonged dark adaptation was first reported in Japan in 1907 by Oguchi and most subsequent reported cases have come from the same country, although some non-Japanese patients have been described (Klein, 1939; Winn et al., 1939). In most patients, defective night vision is the only complaint, visual acuity being normal. The characteristic feature of this condition is the abnormal coloration of the light-adapted fundus, the abnormal white or cream-coloured appearance being derived from the inner limiting membrane and presumably from the foot plates of Muller cells. The abnormal colour resolves over a period of 0.5 – 8 hours in darkness and has been termed the Mizuo–Nakamura phenomenon after the authors who provided the first description (Mizuo, 1913; Nakamura. 1920). A similar phenomenon has been described in X-linked cone dystrophy (Heckenlively & Weleber, 1986) and in X-linked retinoschisis (de Jong et al., 1991). Dark adaptation is characteristically slow and a final rod threshold may be attained only after several hours, and even then may be slightly elevated (François & Verriest, 1954).

Histopathological examination has been undertaken on three eyes. An excess of cones compared with rods together with an abnormal layer of material between the receptors and pigment epithelium was found by Oguchi (1925). Parallel histological studies on the other half of the same eye by another investigator (Yamanaka, 1924) failed to confirm this additional layer. He considered the abundance of round fuscin granules confined to the apical portion of the pigment cells as the characteristic feature. Histopathological and electron microscopic study of another eye more recently (Kuwabara et al., 1963) confirmed the existence of an abnormal layer between the outer segments of the photoreceptors and the pigment epithelial cells. However, the constituents were normal components of the retina consisting of fuscin granules and protrusions of the pigment epithelium with complex interdigitations of the outer segments. There was no abnormal cone distribution. A third histological study (Yamanaka, 1969) is open to question as the patient had reduced vision with retinal pigmentary changes and both parents had retinitis pigmentosa. On electrophysiology, both the a-wave of the ERG and the light rise in the standing potential of the eye are normal, but the scotopic b-wave of the ERG is severely depressed even in the fully dark adapted eye (Carr & Ripps, 1967).

The pathogenesis of Oguchi's disease is not understood and although it is tempting to speculate that it is related to a slow regeneration of rhodopsin, this is not substantiated by a single study of rhodopsin kinetics in this disorder, in which rhodopsin regeneration was normal (Carr & Ripps, 1967). These observations imply that the primary abnormality is unrelated to light catch and rhodopsin bleaching but is related to other systems of transduction. From the electrophysiology results, the region of bipolar cells appears to be the earliest stage in the visual pathway exhibiting signs of defective function. However, the presence of an abnormal layer either on the vitreal side of the pigment epithelial cells or within the apical cytoplasm incites some abnormality of outer-segment shedding or of the pigment epithelium, whether due to slow phagocytosis or inadequacy of the lysosomal system. From their observation in X-linked retinoschisis, de Jong and his colleagues (1991) suggested that the Mizuo–Nakamura phenomenon is caused by an excess of extracellular potassium in the retina as a result of a decreased potassium scavenging capacity of retinal Muller cells.

Fundus albipunctatus

Fundus albipunctatus is a static autosomal recessive condition in which the only symptoms are related to defective dark adaptation (Lauber, 1924). The condition should not be confused with albipunctate dystrophy (retinitis punctata albescens), which is progressive and which represents a variant of retinitis pigmentosa.

The fundus shows widespread distribution of uniform-sized almost white dots at the level of the pigment epithelium, which are most dense in the post-equatorial fundus; the macula may or may not be involved. Changes in the distribution of the white dots have been described (Marmor, 1977), as have their change from flecks in childhood to relatively permanent punctate dots that increase in number over the years (Marmor, 1990). Diffuse pigmentary changes in the pigment epithelium are unusual. Fluorescein angiography shows punctate hyperfluorescence that does not, however, correspond with the punctate white dots (Gass, 1970).

Although fundus albipunctatus usually involves both eyes, unilateral disease has been reported (Henkes, 1963). Typically, the visual acuity and visual fields are normal, but minor loss of field has been described (Marmor, 1977). In a majority of cases, the dark adaptation of both cones and rods is markedly prolonged (Krill & Martin, 1971; Carr & Siegel, 1973; Marmor, 1977; Margolis et al., 1987) and there is delay in the acquisition of scotopic ERG thresholds (Smith et al., 1959; Krill & Martin, 1971; Marmor, 1977; Margolis et al., 1987). Variation from this pattern has been described in which the dark adaptation and ERG are normal (François et al., 1956; Franceschetti et al., 1963) or dark adaptation shows a cone segment only (Mandelbaum, 1941). It is not clear whether this variation implies various degrees of severity of a single disease or that several disorders have been described which share this fundus abnormality.

Studies of rhodopsin kinetics showed slow rhodopsin regeneration that parallels dark adaptation (Carr et al., 1974). This implies that, by contrast with Oguchi's disease, the sensory defect is due to abnormal photopigment kinetics in fundus albipunctatus.

Fleck retina of Kandori

This rare condition in which there is prolonged dark adaptation giving rise to difficulty with night vision but no other symptoms, has been described only in Japan (Kandori et al., 1972). Dark adaptation shows a prolonged rod phase reaching normal thresholds within 40 minutes. The fundus presents large irregular white lesions at the level of the pigment epithelium that are most concentrated in the equatorial region. The photopic ERG is normal and a prolonged interval of dark adaptation is needed to reach scotopic potentials.

CONCLUSION

The recent advances in molecular biology whereby specific genetic defects have been identified, and in cell biology demonstrating the influence of genetic defects on cell function, render the findings of clinical investigation much more significant. In the past, the functional deficits in retinitis pigmentosa were of value in distinguishing one disorder from another or at least identifying categories of disease. It is now possible to relate these characteristics to putative disease mechanisms and to generate hypotheses that are amenable to testing in the laboratory. It is also the case that observations are possible in man, particularly with respect to detailed recording of functional loss in those with retinal dystrophies, which cannot be made in animals. Thus the results of clinical and laboratory studies are relevant one to the other.

The initial subdivision of retinitis pigmentosa was achieved on the basis of its inheritance; functional studies added to our understanding of this complex group of disorders. The recent identification of a number of different mutations in the rhodopsin and peripherin RDS genes are at present being correlated with clinical and functional abnormalities, and we can expect, in the near future, to find other genes responsible for further members of this group of disorders.

The mechanism by which alteration of the amino acid sequence in rhodopsin or rds peripherin molecules may influence photoreceptor function is unclear, but much may depend upon the site of expression of the abnormal protein. Further studies should clarify this aspect of the problem.

Although there has been a very rapid advance in the understanding of disease mechanisms in retinitis pigmentosa, there is every prospect that the expansion of knowledge will accelerate in the next few years, particularly if there is continued and, if possible, increasing collaboration between clinicians and basic scientists. The considerable advances so far achieved have not yet brought any great advance in therapy, although some effect on clinical management is already evident. Genetic counselling in a family is undoubtedly simplified if the genetic abnormality is known, and advice on visual prognosis may be made on a firmer basis if single nosological entities are identified by genetic studies. In addition, it is to be hoped that with the knowledge that will result from extension of current studies, it may eventually be possible to offer effective treatment to the patient with retinitis pigmentosa.

REFERENCES

Alexander, K. R. and Fishman, G. A. (1984) Prolonged rod adaptation in retinitis pigmentosa. *British Journal of Ophthalmology*, **68**, 561–569.

Alezzandrini, A. (1965) Retinitis pigmentosa in symmetric quadrants. *American Journal of Ophthalmology*, **60**, 1160.

Alitalo, T., Kruse, T. A., Forsius, H., Eriksson, A. W. and de la Chapelle, A. (1991) Localization of the Aland Island eye disease locus to the pericentromeric region of the X chromosome by linkage analysis. *American Journal of Human Genetics*, **48**, 31–38.

Alpern, M., Lee, G. B. and Spivey, B. E. (1965) Pi cone monochromatism. *Archives of Ophthalmology*, **74**, 334–337.

Anderson S. (1977) Quoted in Krill, A. E. *Hereditary and Choroidal Diseases*, p. 697. New York: Harper & Row.

Arden, G. B., Carter, R. M., Hogg, C. R., Powell, D. J., Ernst, W. J. K., Clover, G. M. et al. (1983) Rod and cone activity in patients with dominantly inherited retinitis pigmentosa: comparison between psychophysical and electroretinographic measurements. *British Journal of Ophthalmology*, **67**, 405–418.

Arokawa, K. Molday, M. M., Molday, R. S. and Williams, D. S. (1992) Localization of peripherin/rds in the disk membranes of cone and rod photoreceptors; relationship to disk membrane morphogenesis and retinal degeneration. *Journal of Cell Biology*, **116**, 659–667.

Ashton, N. (1953) Central areolar choroidal sclerosis. A histopathological study. *British Journal of Ophthalmology*, **37**, 140–147.

Barkman, Y. (1961) A clinical study of a central tapeoretinal degeneration. *Acta Ophthalmologica*, **39**, 663–671.

Barricks, M. E. (1977) Vitelliform lesions developing in normal fundi. *American Journal of Ophthalmology*, **83**, 324–327.

Bascom, R. A., Manara, S., Collins, L., Molday, R. S., Kalnins, V. I. and McInnes, R. R. (1992) Cloning of the cDNA for a novel photoreceptor membrane (rom-1) identifies a disk rim protein family implicated in human retinopathies. *Neuron*, **8**, 1171–1184.

Bassen, F. A. and Kornzweig, A. L. (1950) Malformation of the erythrocytes in a case of atypical retinitis pigmentosa. *Blood*, **5**, 381–387.

Bech-Hansen, N. T., Field, L. L., Schramm, A. M., Reedyk, M., Craig, I. W., Fraser, N. J. and Pearce, W. G. (1990) A locus for X-linked congenital stationary night blindness is located on the proximal protion of the short arm of the X chromosome. *Human Genetics*, **84**, 406–408.

Berson E. L. and Howard, J. (1971) Temporal aspects of the electroretinogram in sector retinitis pigmentosa. *Archives of Ophthalmology*, **48**, 653–665.

Berson, E. L., Gouras, P. G. and Gunkel, R. D. (1968) Progressive cone degeneration, dominantly inherited. *Archives of Ophthalmology*, **80**, 77–83.

Berson, E. L., Gouras, P., Gunkel, R. D. and Myrianthopoulos, N. C. (1969) Rod and cone responses in sex-linked retinitis pigmentosa. *Archives of Ophthalmology*, **81**, 215–225.

Best, F. (1905) Uber eine hereditäre Maculaaffektion: Beitrag zur Vererbungslehre. *Zeitschrift fur Augenheilkunde*, **13**, 199–212.

Bhattacharya, S. S., Wright, A. F., Clayton, J. F., Price, W. H., Phillips, C. T., McKeown, C. M. E. et al. (1984) Close genetic linkage between X-linked retinitis pigmentosa and a restriction fragments length polymorphism identified by recombinant DNA probe L1.28. ***Nature***, **309**, 253–255.

Bietti, G. B. (1937) Su alcune forme atipiche o rare di degenerazione retinica (degenerazioni tappeto-retiniche e quadri morbosi similari). *Bollettino di Oculistica*, **16**, 1159–1241.

Bird, A. C. (1975) X-linked retinitis pigmentosa. *British Journal of Ophthalmology*, **59**, 177–199.

Bishara, S., Merin, S., Cooper, M., Azizi, E., Delpre, G. and Deckelbaum, R. J. (1982) Combined vitamin A and E therapy prevents retinal electrophysiological deterioration in abetalipoproteinaemia. *British Journal of Ophthalmology*, **66**, 767–770.

Bjork, A., Lindbalm, V. and Wadanstein, L. (1956) Retinal degeneration in hereditary ataxia. *Journal of Neurology, Neurosurgery and Psychiatry*, **19**, 186–193.

Blanton, S. H., Cottingham, A. W., Giesenschlag, N., Heckenlively, J. R., Humphries, P. and Daiger, S. P. (1990) Further evidence of exclusion of linkage between type II autosomal dominant retinitis pigmentosa (ADRP) and D3S47 on 3q. *Genomics*, **8**, 179–181.

Blanton, S. H., Heckenlively, J. R., Cottingham, A. W., Friedman, J., Sadler, L. A., Wagner, M. et al. (1991) Linkage mapping of autosomal dominant retinitis pigmentosa (RP1) to the pericentric region of human chromosome 8. *Genomics*, **11**, 857–869.

Bonin, P. (1971) Le signe du silence choroidien dans les dégénérescences tapéto- rétiniennes centrales examinïes sous fluoresceïne. *Bulletin des Sociétiés d'Ophtalmologie de France*, **71**, 348–351.

Boughman, J. A. and Fishman, G. A. (1983) A genetic analysis of retinitis pigmentosa. *British Journal of Ophthalmology*, **66**, 405–416.

Brecher, R. and Bird, A. C. (1990) Adult vitelliform macular dystrophy. *Eye*, **4**, 210–215.

Bresnick, G. H., Smith, V. C. and Pokorny, J. (1989) Autosomal dominantly inherited macular dystrophy with preferential short wavelength sensitive cone involvement. *American Journal of Ophthalmology*, **108**, 265–276.

Cameron, J. D., Fine, B. S. and Shapiro, I. (1987) Histopathological observations in choroideremia with emphasis on vascular changes of the uveal tract. *Ophthalmology*, **94**, 187–196.

Capon, M. R. C., Polkinghorne, P. J., Fitzke, F. W. and Bird, A. C. (1988) Sorsby's pseudoinflammatory macula dystrophy – Sorsby's fundus dystrophies. *Eye*, **2**, 114–122.

Capon, M. R. C., Marshall, J., Krafft, J. I., Alexander, R. A., Hiscott, P. S. and Bird, A. C. (1989) Sorsby's fundus dystrophy: a light and electron microscopic study. *Ophthalmology*, **96**, 1769–1777.

Carr, R. E., Ripps, H., Siegel, I. M. and Weale, R. A. (1966) Rhodopsin and the electrical activity of the retina in congenital night blindness. *Investigative Ophthalmology*, **5**, 497–507.

Carr, R. E. and Ripps, H. (1967) Rhodopsin kinetics and rod adaptation in Oguchi's disease. *Investigative Ophthalmology*, **6**, 426–436.

Carr, R. E. and Siegel, I. M. (1973) Unilateral retinitis pigmentosa. *Archives of Ophthalmology*, **90**, 21–26.

Carr, R. E., Ripps, H. and Siegel, I. M. (1974) Visual pigment kinetics and adaptation in fundus albipunctatus. *Documenta Ophthalmologica Proceedings Series X^{th} ISCERG Symposium*, **4**, 193–204.

Carr, R. E., Mittl, R. N. and Noble, K. G. (1975) Choroidal abiotrophies. *Transactions of the American Academy of Ophthalmology and Oto-laryngology*, **79**, 796–816.

Cibis, G. N., Morey, M. and Harris, D. J. (1980) Dominantly inherited macular dystrophy with flecks (Stargardt). *Archives of Ophthalmology*, **98**, 1785–1789.

Cogan, D. G., Rodrigues, M., Chu, F. C. and Schaefer, E. J. (1984) Ocular abnormalities in abetalipoproteinemia. A clinicopathologic correlation. *Ophthalmology*, **91**, 991–998.

Cohen, A. I. (1983) Some cytological and initial biochemical observations on photoreceptors in retinas of rds mice. *Investigative Ophthalmology and Visual Science*, **24**, 832–843.

Connell, G. and Molday, R. S. (1990) Molecular cloning, primary structure and orientation of the vertebrate photoreceptor cell protein peripherin in the rod disc membrane. *Biochemistry*, **29**, 4691–4698.

Connell, G., Bascom, R., Molday, L., Reid, D., McInnes, R. R. and Molday, R. S. (1991) Photoreceptor peripherin is the normal product of the gene responsible for retinal degeneration in the rds mouse. *Proceedings of the National Academy of Sciences of the USA*, **88**, 723–726.

Cordier, J., Reny, A. and Seigneur, J.-B. (1966) Rétinite pigmentaire unilatérale. *Bulletin des Sociétés d'Ophtalmologie de France*, **66**, 224–227.

Cremers, F. P., van de Pol, D. J., Diergaarde, P. J., Wieringa, B., Nussbaum, R. L., Schwartz, M. and Ropers, H-H. (1989) Physical fine mapping of the choroideremia locus using Xq21 deletions associated with complex syndromes. *Genomics*, **4**, 41–46.

Cremers, F. P. M., van de Pol, D. J. R., van Kerkhoff, L. P. M., Wieringa, B. and Ropers, H.-H. (1990) Cloning of a gene that is rearranged in patients with choroideraemia. *Nature*, **347**, 674–677.

Dejean, C. and Gassenc, R. (1949) Note sur la généalogie de la famille Nougaret, Véndemian. *Bulletin des Sociétés d'Ophtalmologie de France*, **1**, 96–99.

Denton, M. J., Chen, J.-D., Serravalle, S., Colley, P., Halliday, F. B. and Donald, J. (1988) Analysis of linkage relationships of X-linked retinitis pigmentosa with the following Xp loci: L1.28, OTC, 754, XJ1.1, pERT 87 and C7. *Human Genetics*, **78**, 60–64.

Deutman, A. F. (1969) Electro-oculography in families with vitelliform dystrophy of the fovea. *Archives of Ophthalmology*, **81**, 305–316.

Deutman, A. F., van Blommestein, J. D. A., Henkes, H. E., Waardenburg, P. J. and Solleveld– van Dreist, E. (1970) Butterfly shaped pigment dystrophy of the fovea. *Archives of Ophthalmology*, **83**, 558–569.

Deutman, A. F. (1971) *The Hereditary Dystrophies of the Posterior Pole of the Eye.* Assen: Van Gorcum.

Dick, A. D., Jagger, J. and McCartney, A. C. (1990) Refsum's disease: electron microscopy of an iris biopsy. *British Journal of Ophthalmology*, **74**, 370–372.

Diem, M. (1914) Retinitis punctata albescens et pigmentosa. *Klinische Monatsblatter fur Augenheilkunde*, **53**, 371–379.

Dreyer, R. F. and Hidayat, A. A. (1988) Pseudoinflammatory macular dystrophy. *American Journal of Ophthalmology*, **106**, 154–161.

Dryja, T. P., McGee, T. L., Reichel, E., Hahn, I. B., Cowley, G. S., Yandel, D. W. et al. (1990a) A point mutation of the rhodopsin gene in one form of retinitis pigmentosa. *Nature*, **343**, 364–366.

Dryja, T. P., McGee, T. L., Hahn, L. B., Cowley, C. G., Olsson, J. E., Reichel, E. et al. (1990b) Mutations within the rhodopsin gene in patients with autosomal dominant retinitis pigmentosa. *New England Journal of Medicine*, **323**, 1302–1307.

Dryja, T. P., Hahn, L. B., Cowley, G. S., McGee, T. L. and Berson, E. L. (1991) Mutation spectrum of the rhodopsin gene among patients with autosomal dominant retinitis pigmentosa. *Proceedings of the National Acadamy of Sciences of the USA*, **88**, 9370–9374.

Eagle, R. C., Lucier, A. C., Bernardino, J. R. and Yanof, M. (1980) Retinal pigment epithelial abnormalities in fundus flavimaculatus; a light and electron microscopic study. *Ophthalmology*, **87**, 1189—1200.

Ernst, W. and Moore, A. T. (1988) Heterogeneity, anomalous adaptation and incomplete penetrance in autosomal dominant retinitis pigmentosa. In *Research in Retinitis Pigmentosa*, edited by E. Zrenner, H. Krastel and H.-H. Goebel, pp. 115–120. Oxford: Pergamon Press.

van Everdingen, J. A. M., Went, L. N., Keunen, J. E. E. and Oosterhuis, J. A. (1992) X linked progressive cone dystrophy with specific attention to carrier detection. *Journal of Medical Genetics*, **29**, 291–294.

Falls, H. F. and Cotterman, C. W. (1948) Choroido-retinal degeneration. A sex-linked form in which heterozygous women exhibit a tapetal-like reflex. *Archives of Ophthalmology*, **40**, 685–703.

Falls, H. F., Wolter, J. R. and Alpern, M. (1965) Typical total monochromacy. *Archives of Ophthalmology*, **74**, 610–616.

Farrar, G. J., Jordan, S. A., Kenna, P. et al. (1991a) Autosomal dominant retinitis pigmentosa; localization of a disease gene (RP6) to the short arm of chromosome 6. *Genomics*, **11**, 870–874.

Farrar, G. J., Kenna, P., Jordan, S. A., Kumar-Singh, R., Humphries, M. M., Sharp, E. M. et al. (1991b) A three-base-pair deletion in the peripherin-RDS gene in one form of retinitis pigmentosa. *Nature*, **354**, 478–480.

Fish, G., Grey, R. H. B., Sehmi, K. S. and Bird, A. C. (1981) The dark choroid in posterior retinal dystrophies. *British Journal of Ophthalmology*, **65**, 359–363.

Fishman, G. A., Farber, M., Patel, B. S. and Derlacki, D. J. (1987) Visual acuity loss in patients with Stargardt's macular dystrophy. *Ophthalmology*, **94**, 809–814.

Fishman, G. A., Stone, E. M., Gilbert, L. D., Kenna, P. and Sheffield, V. C. (1991) Ocular findings associated with a rhodopsin gene codon 58 transversion mutation in autosomal dominant retinitis pigmentosa. Archives of *Ophthalmology*, **109**, 1387–1393.

Fishman, G. A., Stone, E. M., Sheffield, V. C., Gilbert, L. D. and Kimura, A. E. (1992) Ocular findings associated with rhodopsin gene codon 17 and codon 182 transition mutations in dominant retinitis pigmentosa. *Archives of Ophthalmology*, **110**, 54–62.

Flannery, J. G., Bird, A. C., Farber, D. B., Weleber, G. R. and Bok, D. (1990) A histopathologic study of a choroideremia carrier. *Investigative Ophthalmology and Visual Science*, **31**, 229–236.

Fledelius, H. and Simonsen, S. E. (1970) A family with bilateral symmetrical sectorial pigmentary retinal lesion. *Acta Ophthalmologica*, **48**, 14–22.

Franceschetti, A. (1963) Uber Tapeto-retinale degeneration in Kindesalter. In *Entwicklung ud Fortschritt in der Augenheilkhunde*, p. 107. Stuttgart: Enke Verlag.

Franceschetti, A., Dieterle, P., Amman, P. and Marty, F. (1963) Une nouvelle forme de fundus albipunctatus cum hemeralopia. *Ophthalmologica*, **145**, 403–410.

Franceschetti, A. and François, J. (1965) Fundus flavimaculatus. *Archives d'Ophtalmologie (Paris)*, **25**, 505–530.

Francke, U., Ochs, H. D., DeMartinville, B., Giacalone, J., Lindergren, V., Disteche, C. et al. (1985) Minor Xp21 chromosome deletion in a male associated with expression of Duchenne muscular dystrophy and McLeod syndrome. *American Journal of Human Genetics*, **37**, 250–267.

François, J. and Verriest, G. (1952) Rétinopathie pigmentaire unilatérale. *Ophthalmologica*, **124**, 65–88.

François, J. and Verriest, G. (1954) La Maladie d'Oguchi. *Bulletin de la Societe Belge* d'Ophtalmologie, **108**, 465–506.

François, J., Verriest, G. and De Rouck, A. (1956) Les fonctions visuelles dans les dégénérescences tapéto-rétiniennes. *Ophthalmologica*, **131** (Suppl. 43), 1–40.

François, J. (1962) Chorioretinal degeneration of retinitis pigmentosa of intermediate sex–linked heredity. *Documenta Ophthalmologica*, **16**, 111–127.

François, J. and De Laey, J. J. (1970) Bietti's crystalline fundus dystrophy. *Klinische Monatsblatter fur Augenheilkunde*, **170**, 353–362.

Frangieh, G. T., Green, R. and Fine, S. L. (1982) A histopathological study of Best's macular dystrophy. *Archives of Ophthalmology*, **100**, 1115–1121.

Friedrich, U., Warburg, M., Wieacker, P., Wienker, T. F., Gal, A. and Ropers, H.-H. (1985) X-linked retinitis pigmentosa: linkage with the centromere and a cloned DNA sequence from the proximal short arm of the X chromosome. *Human Genetics*, **71**, 93–99.

Fulton, A. B. and Hansen, R. M. (1988) The relationship of rhodopsin and scotopic retinal sensitivity in sector retinitis pigmentosa. *American Journal of Ophthalmology*, **105**, 132–140.

Gal, A., Schinzel, A., Orth, U., Fraser, N. A., Mollica, F., Craig, I. W., Kruse, T., Machler, M., Neugebauer, M. and Bleeker-Wagemakers, L. M. (1989) Gene of X-chromosomal congenital stationary night blindness is closely linked to DXS7 on Xp. *Human Genetics*, **81**, 315–318.

Gass, J. D. M. (1970) *Stereoscopic Atlas of Macular Diseases*, p. 124. St Louis: CV Mosby.

Gass, J. D. M. (1974) A clinicopathologic study of a peculiar foveomacular dystrophy. *Transactions of the American Ophthalmological Society*, **72**, 139–156.

Ghosh, M., McCulloch, C. and Parker, J. A. (1988) Pathological study in a female carrier of choroideremia. *Canadian Journal of Ophthalmology*, **23**, 181–186.

Gibberd, F. B., Page, N. G. R., Billimoria, J. D. and Retsas, S. (1979) Heredopathia atactica polyneuritiformis (Refsum's disease) treated by diet and plasma-exchange. *Lancet*, i, 575–578.

Gibberd, F. B., Billimoria, J. D., Goldman, J. M., Clemens, M. E., Evans, R., Whitelaw, M. N. et al. (1985) Heredopathia atactica polyneuritiformis: Refsum's disease. *Acta Neurologica Scandinavica*, **72**, 1–17.

Glickstein, M. and Heath, G.G. (1975) Receptors in the monochromat eye. *Vision Research*, **15**, 633–636.

Godel, V., Chaine, G, Regenbogen, L. and Coscas, G. (1986) Best's vitelliform macular dystrophy. *Acta Ophthalmologica Supplement (Copenhagen)*, **175**, 1–31.

Goedbloed, J. (1942) Mode of inheritance in choroideraemia. *Ophthalmologica*, **104**, 308–315.

Goodman, G., Ripps, H. and Siegel, I. M. (1963) Cone dysfunction syndromes. *Archives of Ophthalmology*, **70**, 214–231.

Gouras, P., Carr, R. E. and Gunkel, R. D. (1971) Retinitis pigmentosa in a–beta– lipoproteinemia: effects of Vitamin A. *Investigative Ophthalmology*, **10**, 784–793.

Gouras, P., Eggars, H. M. and MacKay, C.J. (1983) Cone dystrophy, nyctalopia and supernormal rod responses. A new retinal degeneration. *Archives of Ophthalmology*, **101**, 718–724.

Grey, R. H. B., Blach, R. K. and Barnard, W. M. (1977) Bull's eye maculopathy with early cone degenertion. *British Journal of Ophthalmology*, **61**, 702–718.

Haase, W. and Hellner, K. A. (1965) Uber familiare bilaterale sektorenformige Retinopathia pigmentosa. *Klinische Monatsblatter fur Augenheilkunde*, **147**, 365–375.

Hadden, O. B. and Gass, J. D. M. (1976) Fundus flavimaculatus and Stargardt's disease. *American Journal of Ophthalmology*, **82**, 527–539.

Hamilton, W. K., Ewing, C. C., Ives, E. J. and Carruthers, J. D. (1989) Sorsby's fundus dystrophy. *Ophthalmology*, **96**, 1755–1762.

Harrison, R., Hoeffnagel, D. and Hayward, J. N. (1960) Congenital total color blindness: A clinicopathological report. *Archives of Ophthalmology*, **64**, 685–692.

Hawkins, R. K., Jansen, H. G. and Sanyal, S. (1985) Development and degeneration of retina in rds mutant mice: photoreceptor abnormalities in the heterozygotes. *Experimental Eye Research*, **41**, 701–720.

Hayasaka, S. and Okoyuma, S. (1984) Crystalline retinopathy. *Retina*, **4**, 177–181.

Heckenlively, J. R. (1982) Preserved para-arteriole retinal pigment epithelium (PPRPE) in retinitis pigmentosa. *British Journal of Ophthalmology*, **66**, 26–31.

Heckenlively, J. R. and Weleber, R. G. (1986) X-linked recessive cone dystrophy with tapetal like sheen. A newly recognised entity with Mizuo-Nakamura phenomenon. *Archives of Ophthalmology*, **104**, 1322–1328.

Heckenlively, J. R. (1987) Quoted in Heckenlively, J. R. (1988) *Retinitis Pigmentosa* p. 21. Philadelphia: J. B. Lippincott Company.

Heckenlively, J. R. (1988) *Retinitis Pigmentosa* p. 21. Philadelphia: J. B. Lippincott Company.

Heckenlively, J. R., Rodriguez, J. A. and Daiger, S. P. (1991) Autosomal dominant sectoral retinitis pigmentosa. Two families with transversion mutation in codon 23 of rhodopsin. *Archives of Ophthalmology*, **109**, 84–91.

Henkes, H. E. (1963) Unilateral fundus albipunctatus. *Ophthalmologica*, **145**, 470–480.

Herndon, J. H. Jr., Steinberg, D. and Uhlendorf, B. W. (1969) Refsum's disease: defective oxidation of phytanic acid in tissue cultures derived from homozygotes and heterozygotes. *New England Journal of Medicine*, **281**, 1034–1038.

Hodgson, S. V., Robertson, M. E., Fear, C. N., Goodship, J., Malcolm, S., Jay, B. et al. (1987) Prenatal diagnosis of X-linked choroideremia with mental retardation, associated with a cytologically detectable chromosome deletion. *Human Genetics*, **75**, 286–290.
Hommer, K. (1959) Das Elektroretinogramm bei sektorenförmiger Retinitis pigmentosa (Retinopathia pigmentosa). *Albrecht von Graefes Archiv fur Klinische und Experimentelle Ophthalmologie*, **161**, 16–26.
Hoskin, A., Bird, A. C. and Sehmi, K. (1981) Sorsby's pseudoinflammatory macular dystrophy. *British Journal of Ophthalmology*, **65**, 859–865.
Hu, D. (1982) Genetic aspects of retinitis pigmentosa in China. *American Journal of Medical Genetics*, **12**, 51–56.
Hu, D-N. (1983) Ophthalmic genetics in China. *Ophthalmic Paediatrics and Genetics*, **2**, 39–45.
Hutton, D. S. and Steinberg, D. (1973) Localization of the enzymatic defect in phytanic acid storage disease (Refsum's syndrome). *Neurology*, **23**, 1333.
Hussels, I. (1967) Une famille atteinte de rétinopathie pigmentaire liée au sexe, de maladie de Parkinson et d'autres troubles neuro-psychiatriques. *Journal de Génétique Humaine*, **16**, 106–155.
Inglehearn, C. F., Keen, T. J., Bashir, R., Jay, M., Fitzke, F., Bird, A. C. et al. (1992) A completed screen for mutations of the rhodopsin gene in a panel of patients with autosomal dominant retinitis pigmentosa. *Human Molecular Genetics*, **1**, 41–45.
Isselbacher, K. J., Scheig, R., Plotkin, G. R. and Caulfield, J. B. (1964) Congenital beta– lipoprotein deficiency: an hereditary disorder involving a defect in the absorption and transport of lipids. *Medicine*, **43**, 347–361.
Jacobson, D. M., Thompson, H. S. and Bartley, J. A. (1989) X-linked progressive cone dystrophy. Clinical characteristics of affected males and female carriers. *Ophthalmology*, **96**, 885–895.
Jacobson, S.G., Kemp, C. M., Sung, C. H. and Nathans, J. (1991) Retinal function and rhodopsin levels in autosomal dominant retinitis pigmentosa with rhodopsin mutations. *American Journal of Ophthalmology*, **112**, 256–271.
Jalili, I. K. and Smith, N. J. D. (1988) A progressive cone rod dystrophy and amelanogenesis imperfecta: a new syndrome. *Journal of Medical Genetics*, **25**, 738–740.
Jampel, R. S. and Falls, H. F. (1958) Atypical retinitis pigmentosa, acanthocytosis and rative neuromuscular disease. *Archives of Ophthalmology*, **59**, 818–820degene.
Jansen, H. G. and Sanyal, S. (1984) Development and degeneration of retina in rds mutant mice: electron microscopy. *Journal of Comparative Neurology*, **224**, 71–84.
Janssen, O. (1938) *Zur Erbbiologie der Retinitis pigmentosa.* Inaugural dissertation. Munster in Westfalen.
Jay, B. and Bird, A. C. (1973) X-linked retinitis pigmentosa. *Transactions of the American Academy of Ophthalmology and Otolaryngology*, **77**, 641–651.
Jay, M. (1982) On the hereditary of retinitis pigmentosa. *British Journal of Ophthalmology*, **7**, 405–416.
de Jong, P. T., Zrenner, E., van Meel, G. J., Keunen, J. E. and van Norren, D. (1991). Mizuo phenomenon in X-linked retinoschisis. Pathogenesis of the Mizuo phenomenon. *Archives of Ophthalmology*, **109**, 1104–1108.
Jung, E. E. (1936) Quoted in Duke-Elder, S. *System of Ophthalmology, Volume X*, p. 632. London: Henry Kimpton.
Kajiwara, K., Hahn, L. B., Mukai, S., Travis, G. H., Berson, E. L. and Dryja, T. P. (1991) Mutations in the human retinal degeneration slow gene in autosomal dominant retinitis pigmentosa. *Nature*, **354**, 480–483.
Kandori, F., Tamai, A., Kurimoto, S. and Fukunaga, K. (1972) Fleck retina. *American Journal of Ophthalmology*, **73**, 673–685.
Kearns, T. P. and Hollenhorst, R. W. (1966) Chloroquine retinopathy. Evaluation by fluorescein angiography. *Archives of Ophthalmology*, **76**, 378–384.

Keen, T. J., Inglehearn, C. F., Lester, D. H., Bashir, R., Jay, M., Bird, A. C. et al. (1991) Autosomal dominant retinitis pigmentosa: four new mutations in rhodopsin, one of them in the retinal attachment site. *Genomics*, **11**, 199–205.

Kemp, C. M., Jacobson, S. G. and Faulkner, D. J. (1988) Two types of visual disfunction in autosomal dominant retinitis pigmentosa. *Investigative Ophthalmology and Visual Science*, **29**, 1235–41.

Kemp, C. M., Jacobson, S. G., Roman, A. J., Sung, C. H. and Nathans, J. (1992) Abnormal rod adaptation in autosomal dominant retinitis pigmentosa with Pro-23-His rhodopsin mutation. *American Journal of Ophthalmology*, **113**, 165–174.

Khouri, G., Mets, M. B., Smith, V. C., Wendell, M. and Pass, A. S. (1988) X-linked congenital stationary night blindness. Review and report of a family with hyperopia. *Archives of Ophthalmology*, **106**, 1417–1422.

Kingham, J. D. and Lochen, G. P. (1977) Vitelliform macular degeneration. *American Journal of Ophthalmology*, **84**, 526–531.

Klein, B. A. (1939) A case of so-called Ouguchi's disease in the USA. *American Journal of Ophthalmology*, **22**, 953–955.

Klenk, E. and Kahlke, W. (1963) Ueber das Vorkommen der 3.7.11.15-tetramethyl– hexadecansaeure (Phytansaeure) in den Cholesterinestern und anderen Lipoidfraktionen der Organe bei einem Krankheitsfall unbekannter Genese (Verdacht auf heredopathia atactica polyneuritiformis Refsum-syndrom). *Hoppe-Seylers Zeitschrift fur Physiologische Chemie*, **333**, 133–142.

Kobayashi, V. A. (1960) Genetic study on retinitis pigmentosa. *Japanese Journal of Ophthalmology*, **7**, 82–88.

Kolb, H. and Galloway, N. R. (1964) Three cases of unilateral pigmentary degeneration. *British Journal of Ophthalmology*, **48**, 471–479.

Krill, A. E. (1969) X-chromosomal linked diseases affecting the eye: status of the heterozygote female. *Transactions of the American Ophthalmological Society*, **67**, 535–608.

Krill, A. E., Archer, D. B. and Martin, D. (1970) Sector retinitis pigmentosa. *American Journal of Ophthalmology*, **69**, 977–987.

Krill, A. E. and Martin, D. (1971) Photopic abnormalities in congenital stationary night blindness. *Investigative Ophthalmology*, **10**, 625–636.

Krill, A. E., Potts, A. M. and Johanson, C. E. (1971) Chloroquine retinopathy. Investigation of discrepancy between dark adaptation and electroretinographic findings in advanced stages. *American Journal of Ophthalmology*, **71**, 530–543.

Krill, A. E. and Deutman, A. F. (1972) Dominant macular degenerations. The cone dystrophies. *American Journal of Ophthalmology*, **73**, 352–369.

Krill, A. E. (1977) *Hereditary Retinal and Choroidal Diseases*, p. 1036. New York: Harper & Row.

Kuper, J. (1960) Familiäre sektorenförmige Retinitis pigmentosa. *Klinische Monatsblatter fur Augenheilkunde*, **136**, 97–102.

Kuwabara, Y., Ishikara, K. and Akiyas, S. (1963) Histologic and electron microscopic studies of the retina in Oguchi's disease. *Acta Societatis Ophthalmologicae Japonicae*, **67**, 1323–1351.

Larsen, H. (1921) Demonstration mikroskopischer Präparate von einem monochromatischen Auge. *Klinische Monatsblatter fur Augenheilkunde*, **67**, 301–302.

Lauber, H. (1924) The origin of hyalin formations within the eye. *Berliner Deutsche Ophthalmologische Gesundheitwesen*, **44**, 216–220.

Lester, D. H., Inglehearn, C. F., Bashir, R., Ackford, H., Esakowitz, L., Jay, M. et al. (1990) Linkage to D3S47 (C17) in one large dominant retinitis family and exclusion in another: confirmation of genetic heterogeneity. *American Journal of Human Genetics*, **47**, 536–541.

Levy, I. S. (1970) Refsum's syndrome. *Transactions of the Ophthalmological Societies of the United Kingdom*, **90**, 181–185.

Lewis, R. A., Holcomb, J. D., Bromley, W. C., Wilson, M. C., Roderisk, T. H. and Hejtmancik, J. F. (1987) Mapping X-linked ophthalmic diseases. III. Provisional assignment of the locus for blue cone monochromacy to Xq28. *Archives of Ophthalmology*, **105**, 1055–1059.

Lisch, K. (1960) Isolierte Entwicklungsstorungen. *Medizinische Wochenschrift*, **14**, 720–725.

Lopez, P. F., Maumenee, I. H., de la Cruz, Z. and Green, W. R. (1990) Autosomal-dominant fundus flavimaculatus. Clinicopathologic correlation. *Ophthalmology*, **97**, 798–809.

Lyness, A. L., Ernst, W., Quinlan M. P., Clover, G. M., Arden, G. B., Carter, R. et al. (1985) A clinical, psychophysical, and electroretinographic survey of patients with autosomal dominant retinitis pigmentsa. *British Journal of Ophthalmology*, **69**, 326–339.

Mandelbaum, J. (1941) Dark adaptation: some physiologic and clinical observations. *Archives of Ophthalmology*, **26**, 203–239.

Mann, I. (1937) *Developmental Abnormalities of the Eye*, Figure 105. Cambridge: Cambridge University Press.

Margolis, S., Siegel, I. M. and Ripps, H. (1987) Variable expressivity in fundus albipunctatus. *Ophthalmology*, **94**, 1416–1422.

Marmor, M. F. (1977) Defining fundus albipunctatus. *Documenta Ophthalmologica*, **13**, 227–234.

Marmor, M. F. (1990) Long-term follow-up of the physiologic abnormalities and fundus changes in fundus albipunctatus. *Ophthalmology*, **97**, 380–384.

Marmor, M. F., Jacobson, S. G., Foerster, M. H., Kellner, U. and Weleber, R. G. (1990) Diagnostic findings of a new syndrome with night blindness, maculopathy, and enhanced s-cone sensitivity. *American Journal of Ophthalmology*, **110**, 124–134. Massof, R. W., F.

Fleishman, J. A., Fine, S. L. and Yoder, F. (1977) Flicker fusion thresholds in Best's macular dystrophy. *Archives of Ophthalmology*, **95**, 991–994.

Massof, R. W. and Finkelstein, D. (1981) Two forms of autosomal dominant primary retinitis pigmentosa. *Documenta Ophthalmologica*, **51**, 289–346.

Mauthner, L. (1871) Quoted in Duke-Elder, S. (1964) System of Ophthalmology. *Vol. III. Part 2.* p. 619. London: Henry Kimpton.

McCulloch, C. and McCulloch, R. J. P. (1948) A hereditary and clinical study of choroideremia. *Transactions of the American Academy of Ophthalmology and Oto-laryngology*, **52**, 160–190.

McCulloch, C. (1969) Choroideraemia: a clinical and pathological review. *Transactions of the American Ophthalmological Society*, **67**, 142–195.

McFarland, C. B. (1955) Heredodegeneration of macula lutea; study of clinical and pathological aspects. *Archives of Ophthalmology*, **53**, 224–228.

McKenzie, D. S. (1951) The inheritance of retinitis pigmentosa in one family. *Transactions of the Ophthalmological Society of New Zealand*, 79–82.

McQuarrie, M. D. (1935) Two pedigrees of hereditary blindness in man. *Journal of Genetics*, **30**, 147–153.

McWilliam, P., Farrar, G. J., Kenna, P., Bradley, D. G., Humphries, M. M. et al. (1989) Autosomal dominant retinitis pigmentosa (ADRP): localization of an ADRP gene to the long arm of chromosome 3. *Genomics*, **5**, 619–622.

Merry, D. E., Lesko, J. G., Sosnoski, D. M., Lewis, R. A., Lubinsky, M., Trask, B., et al. (1989) Choroideremia and deafness with stapes fixation: a contiguous gene deletion syndrome in Xq21. *American Journal of Human Genetics*, **45**, 530–540.

Miller, S. A., Bresnik, G. H. and Chandra, S. R. (1976) Choroidal neovascular membrane in Best's vitelliform macular dystrophy. *American Journal of Ophthalmology*, **82**, 252–255.

Miyake, Y. and Kawase, Y. (1984) Reduced amplitude of oscilatory potentials in female carriers of X-linked recessive congenital stationary night blindness. *American Journal of Ophthalmology*, **98**, 208–15.

Miyake, Y., Yagasaki, K., Horiguchi, M., Kawase, Y. and Kanda, T. (1986) Congenital stationary night blindness with negative electroretinogram. A new classification. *Archives of Ophthalmology*, **104**, 1013–1020.

Mizuo, A. (1913) On new discovery in dark adaptation in Oguchi's disease. *Acta Societatis Ophthalmologicae Japonicae,* **17**, 1148–1150.

Mohler, C. W. and Fine, S. L. (1981) Long-term evaluation of patients with Best's vitelliform dystrophy. *Ophthalmology,* **88**, 688–692.

Moloney, J. B., Mooney, D. J. and O'Connor, M. A. (1983) Retinal function in Stargardt's disease and fundus flavimaculatus. *American Journal of Ophthalmology,* **96**, 57–65.

Moore, A. T., Fitzke, F. W., Kemp, C. M., Arden, G. B., Keen, T. J., Inglehearn, C. F. et al. (1992) Abnormal dark adaptation kinetics in autosomal dominant sector retinitis pigmentosa due to rhodopsin mutation. *British Journal of Ophthalmology,* **76**, 465–469.

Moser, H. W., Braine, H., Pyeritz, R. E., Ullman, D. D., Murray, C. and Asbury, A. K. (1980) Therapeutic trial of plasmapheresis in Refsum disease and in Fabry disease. *Birth Defects Original Article Series,* **16**, 491–497.

Mukai, S., Dryja, T. P., Bruns, G. A. P., Alridge, J. F. and Berson, E. L. (1985) Linkage between the X-linked retinitis pigmentosa locus and the L1. 28 locus. *American Journal of Ophthalmology,* **100**, 225–229.

Muller, D. P. R., Lloyd, J. K. and Bird, A. C. (1977) Long-term management of a-beta– lipoproteinaemia: possible role for vitamin E. *Archives of Disease in Childhood,* **52**, 209–214.

Musarella, M. A., Weleber, R. G., Murphey, W. H., Young, R. S., Anson-Cartwright, L., Mets, M., Kraft, S. P., Polemeno, R., Litt, M. and Worton, R. G. (1989) Assignment of the gene for complete X-linked congenital stationary night blindness (CSNB1) to Xp11. 3. *Genomics,* **5**, 727–737.

Nakamura, B. (1920) Uber ein neues Phänomen der Farberveränderung des menschlichen Augenhintergrundes im Zusammenhang mit der fortshreitenden Dunkeladaptation. *Klinische Monatsblatter fur Augenheilkunde,* **65**, 83–85.

Nathans, J., Thomas, D. and Hogness, D. S. (1986a) Molecular genetics of human color vision: the genes encoding blue, green, and red pigments. *Science,* **232**, 193–202.

Nathans, J., Piantandida, T. P., Eddy, R. L., Shows, T. B. and Hogness, D. S. (1986b) Molecular genetics of inherited variation in human color vision. *Science,* **232**, 203–210.

Nathans, J., Davenport, C. M., Maumenee, I. H., Lewis, R. A., Hejtmancik, J. F., Litt, M., Lovrien, E., Weleber, R., Bachynski, B., Zwas, F., Klingaman, R. and Fishman, G. (1989) Molecular genetics of blue cone monochromacy. *Science,* **245**, 831–838.

Noble, K. G. (1977) Central areolar choroidal dystrophy. *American Journal of Ophthalmology,* 84, 310–318.

Noble, K. G. and Carr, R. E. (1979) Stargardt's disease and fundus flavimaculatus. *Archives of Ophthalmology,* **97**, 1281–1285.

Noble, K. G., Carr, R. E. and Siegel, I. M. (1990) Autosomal dominant congenital stationary night blindness and normal fundus with an electronegative electroretinogram. *American Journal of Ophthalmology,* **109**, 44–48.

Nussbaum, R. L., Lewis, R. A., Lesko, J. G. and Ferrell, R. (1985a) Mapping ophthalmological disease. II Linkage of relationship of X-linked retinitis pigmentosa to X chromosome short arm markers. *Human Genetics,* **70**, 45–50.

Nussbaum, R. L., Lewis, R. A., Lesko, J. G. and Ferrell, R. (1985b) Choroideremia is linked to the restriction fragment length polymorphism DXYS1 at Xq13-21. *American Journal of Human Genetics,* **37**, 473–481.

O'Gorman, S., Flaherty, W. A., Fishman, G. A. and Berson, E. L. (1988) Histopathologic findings in Best's vitelliform macular dystrophy. *Archives of Ophthalmology,* **106**, 1261–1268.

Oguchi, C. (1907) Uber einen Fall von eigenartiger Hemeralopie. *Nippon Ganka Gakkai Zasshi,* **11**, 123.

Oguchi, C. (1925) Zur Anatomie der sogenannten. Oguchi'schen Krankheit. *Albrecht von Graefes Archiv fur Klinische und Experimentelle Ophthalmologie,* **115**, 234–245.

Pearce, W. G., Reedyk, M. and Coupland, S. G. (1990) Variable expressivity in X-linked congenital stationary night blindness. *Canadian Journal of Ophthalmology*, **25**, 3–10.

Pearlman, J. T. Owen, G. W., Brounley, D. W. and Sheppard, J. J. (1974) Cone dystrophy with dominant inheritance. *American Journal of Ophthalmology*, **77**, 293–303.

Pearlman, J. T., Saxton, J., Hoffman, G. and Carson, S. (1976) Unilateral retinitis pigmentosa sine pigmento. *British Journal of Ophthalmology*, **60**, 354–360.

Pearlman, J. T. (1979) Mathematical models of retinitis pigmentosa: a study of the rate of progress in the different genetic forms. *Transactions of the American Ophthalmological Society*, **77**, 643–656.

Pedralgia, C. (1865) Klinische Berobachtungen. Retinitis pigmentosa. *Klinische Monatsblatter fur Augenheilkunde*, **3**, 114–117.

Polkinghorne, P. J., Capon, M. R. C., Berninger, T., Lyness, A. L., Sehmi, K. and Bird, A. C. (1989) Sorsby's fundus dystrophy: a clinical study. *Ophthalmology*, **96**, 1763–1768.

Porta, A., Pierrottet, C., Aschero, M. and Orzalesi, N. (1992) Preserved para-arteriole retinal pigment epithelium retinitis pigmentosa. *American Journal of Ophthalmology*, **113**, 161–164.

Pruett, R. C. (1983) Retinitis pigmentosa: clinical observations and correlations. *Transactions of the American Ophthalmological Society*, **81**, 693–735.

Rafuse, E. V. and McCulloch, C. (1968) Choroideraemia, a pathological report. *Canadian Journal of Ophthalmology*, **3**, 347–352.

Ragnetti, E. (1962) An atypical form of retinitis pigmentosa. *Bollettino di Oculistica*, **41**, 617–625.

Refsum, S. (1945) Heredoataxia hemeralopia polyneuritiformis, a familial syndrome not previously described: preliminary report. *Nordisk Medicin*, **28**, 2682.

Reichel, E., Bruce, A. M., Sandberg, M. A. and Berson, E. L. (1989) An electroretinographic and molecular genetic study of X-linked cone degeneration. *American Journal of Ophthalmology*, **108**, 540–547.

Ricci, A., Ammann, F. and Franceschetti, A. (1963) Reflet tapétoïde reversible (phénomene de Mizuo inverse) chez des conductrices de rétinopathie pigmentaire récessive liée au sexe. *Bulletins et Memoires de la Société Française d'Ophtalmologie*, **76**, 31–35.

Richards, J. E., Kuo, C. Y., Boehnke, M. and Sieving, P. A. (1991) Rhodopsin Thr58Arg mutation in a family with autosomal dominant retinitis pigmentosa. *Ophthalmology*, **98**, 1797–1805.

Rosenberg, T., Niebuhr, E., Yang, H. M., Parving, A. and Schwartz, M. (1987) Choroideremia, congenital deafness and mental retardation in a family with an X chromosomal deletion. *Ophthalmic Paediatrics and Genetics*, **8**, 139–143.

Runge, P., Muller, D. P., McAllister, J., Calver, D., Lloyd, J. K. and Taylor, D. (1986) Oral vitamin E supplements can prevent the retinopathy of abetalipoproteinaemia. *British Journal of Ophthalmology*, **70**, 166–173.

Salt, H. B., Wolff, O. H. and Lloyd, J. K. (1960) On having no beta-lipoprotein: a syndrome comprising a-beta-lipoproteinaemia, acanthocytosis and steatorrhoea. *Lancet*, ii, 325–329.

Sanyal, S. and Jansen, H. (1981) Absence of receptor outer segments in the retina of rds mutant mice. *Neurosciences Letter*, **21**, 23–26.

Sanyal, S. and Hawkins, R. K. (1988) Development and degeneration of retina in rds mutant mice. Altered disc shedding pattern in the albino heterozygotes and its relation to light exposure. *Vision Research*, **28**, 1171–1178 .

Schappert-Kimmijser, J. (1963) Les dégénérescences tapéto-rétiniennes du type X chromosomal aux Pays-Bas. *Bulletins et Memoires de la Société Française d'Ophtalmologie*, **76**, 122–129.

van Schooneveld, M. J., Went, L. N. and Oosterhuis, J. A. (1991) Dominant cone dystrophy starting with blue cone involvement. *British Journal of Ophthalmology*, **75**, 332–336.

Schubert. G. and Bornschein, H. (1952) Beitrag zur Analyse des menschlichen Elektroretinogramms. *Ophthalmologica*, **123**, 396–412.

Siegel, I. M., Graham, C. H., Ripps, H. and Hsia, Y. (1966) Analysis of photopic and scotopic function in an incomplete achromat. *Journal of the Optical Society of America*, **56**, 699–704.

Siegel, I. M., Greenstein, V. C., Seiple, W. H. and Carr, R. E. (1987) Cone function in congenital nyctalopia. *Documenta Ophthalmologica*, **65**, 307–318.

Sloan, L. L. and Newhall, S. M. (1942) Comparison of cases of atypical and typical achromatopsia. *American Journal of Ophthalmology*, **25**, 945.

Smith, B. F., Ripps, H. A. and Goodman, G. (1959) Retinitis punctata albescens. A functional and diagnostic evaluation. *Archives of Ophthalmology*, **61**, 93–101.

Snodgrass, N. B. (1976) Ocular findings in fucosidosis. *British Journal of Ophthalmology*, **60**, 508–511.

Sorsby, A. (1939) Choroidal angiosclerosis with special reference to its hereditary character. *British Journal of Ophthalmology*, **23**, 433–444.

Sorsby, A., Mason, M. E. J. and Gardener, N. (1949) A fundus dystrophy with unusual features. *British Journal of Ophthalmology*, **33**, 67–97.

Sorsby, A. and Crick, R. P. (1953) Central areolar choroidal sclerosis. *British Journal of Ophthalmology*, **37**, 129–139.

Stargardt, K. (1909) Uber familiäre, progressive Degeneration in der Makulagegend des Auges. *Albrecht von Graefes Archiv fur Klinische und Experimentelle Ophthalmologie*, **71**, 534–550.

Stargardt, K. (1913) Uber familiäre, progressive Degeneration in der Maculagegend des Auges. *Zeitschrift Augenheilkunde*, **30**, 95–116.

Steinberg, D., Herndon, J. H. Jr., Uhlendorf, B. W., Mize, C. E., Avigan, J. and Milne, G. W. (1967) Refsum's disease: nature of the enzyme defect. *Science*, **156**, 1740–1742.

Steinmetz, R. L., Garner, A., Maguire, J. I. and Bird, A. C. (1991) Histopathology of incipient fundus flavimaculatus. *Ophthalmology*, **98**, 953–956.

Stone, E. M., Kimura, A. E., Nichols, B. E., Khadivi, P., Fishman, G. A. and Sheffield, V. C. (1991) Regional distribution of retinal degeneration in patients with the proline to histidine mutation in codon 23 of the rhodopsin gene. *Ophthalmology*, **98**, 1806–1813.

Sung, C. H., Davenport, C. M., Hennessey, J. C., Maumenee, I. H., Jacobson, S. G., Heckenlively, J. R. et al. (1991) Rhodopsin mutations in autosomal dominant retinitis pigmentosa. *Proceedings of the National Academy of Sciences of the USA*, **88**, 6481–6485.

Toussaint, D. and Danis, P. (1971) An ocular pathological study of Refsum's syndrome. *American Journal of Ophthalmology*, **72**, 342–347.

Travis, G. H., Brennan, M. B., Danielson, P. E., Kozak, C. A. and Sutcliffe, J. G. (1989) Identification of a photoreceptor-specific mRNA encoded by the gene responsible for retinal degeneration slow (rds). *Nature*, **338**, 70–73.

Travis, G., Sutcliffe, J. G. and Bok, D. (1991a) The retinal degeneration slow (rds) gene product is a photoreceptor disc membrane associated glycoprotein. *Neuron*, **6**, 61–70.

Travis, G. H., Christerson, L., Danielson, P. E., Klisak, I., Sparkes, R. S., Hahn, L. B. et al. (1991b) The human retinal degeneration slow (RDS) gene: chromosome assignment and structure of the mRNA. *Genomics*, **10**, 733–739.

Usukura, J. and Bok, D. (1987) Changes in the localization and content of opsin during retinal development in the rds mutant mouse: immunocytochemistry and immunoassay. *Experimental Eye Research*, **45**, 501–515

Vaghefi, H. A., Green, R., Kelly, J. S., Sloane, L. L. and Patz, A. (1978) Correlation of clinicopathological findings in a patient: congenital night blindness, branch retinal vein occlusion, cilioretinal artery, drusen of the nerve head and intraretinal pigmented lesion. *Archives of Ophthalmology*, **96,** 2079–2104.

Verriest, M. G. (1974) Recent progress in the study of acquired deficiencies of colour vision. *Bulletin des Sociétés d'Ophtalmologie de France*, **74**, 595–620.

Vine, A. K. and Schatz, H. (1980) Adult-onset foveomacular pigment epithelial dystrophy. *American Journal of Ophthalmology*, **89**, 680–691.

Vollrath, D., Nathans, J. and Davis, R. W. (1988) Tandem array of human visual pigment genes at Xq28. *Science*, **240**, 1669–1672.

Von Sallman, L., Gelderman, A. H. and Laster, L. (1969) Ocular histopathological changes in a case of a–beta-lipoproteinaemia. *Documenta Ophthalmologica*, **26**, 451–460.

Vossius, A. (1921) Quoted in Duke-Elder, S. *System of Ophthalmology, Volume X, p. 632. London: Henry Kimpton.*

Vukovich, V. (1959) Das ERG bei Retinitis pigmentosa (retinopathia pigmentosa) mit bitemporalen Gesichstfeld ausfall. *Albrecht von Grafes Archiv fur Klinische und Experimentelle Ophthalmologie*, **161**, 27–32.

Waardenburg, P. J. (1942) Choroideremia als Erbmerkmal. *Acta Ophthalmologica (Copenhagen)*, **20**, 235–274.

Waardenburg, P. J., Franceschetti, A. and Klein, D. (1963) *Genetics and Ophthalmology, Vol 2*, p. 1736. Springfield Ill: Charles C Thomas.

Warburg, M. and Simonsen, S. E. (1968) Sex-linked recessive retinitis pigmentosa. *Acta Ophthalmologica (Copenhagen)*, **46**, 494–499.

Watanabe, I., Taniguchi, Y., Morioka, K. and Kato, M. (1986) Congenital stationary night blindness with myopia: a clinico-pathologic study. *Documenta Ophthalmologica*, **63**, 55–62.

Weiner, R. L. and Falls, H. F. (1955) Intermediate sex-linked retinitis pigmentosa. *Archives of Ophthalmology*, **53**, 539–553.

Weingeist, T. A., Kobrin, J. L. and Watzke, R. C. (1982) Histopathology of Best's macular dystrophy. *Archives of Ophthalmology*, **100**, 1108–1114.

Weisel, G. (1922) Quoted in Duke-Elder, S. *System of Ophthalmology, Volume X*, p. 632. London: Henry Kimpton.

Welch, R. B. (1977) Bietti's tapetoretinal degeneration with marginal corneal dystrophy. Crystaline retinopathy. *Transactions of the American Ophthalmological Society*, **75**, 164–179.

Weleber, R. G. and Tongue, A. C. (1987) Congenital stationary night blindness presenting as Leber's congenital amaurosis. *Archives of Ophthalmology*, **105**, 360–365.

Weleber, R. G., Pillers, D. A., Powell, B. R., Hanna, C. E., Magenis, R. E. and Buist, N. R. (1989) Aland Island eye disease (Forsius-Eriksson syndrome) associated with contiguous deletion syndrome at Xp21. Similarity to incomplete congenital stationary night blindness. *Archives of Ophthalmology*, **107**, 1170–1179.

Wells, J., Wroblewski, J., Keen, J., Inglehearn, C., Jubb, C., Eckstein, A. et al. (1992) Mutations in the human retinal degeneration slow (rds) gene can cause either retinitis pigmentosa or macular dystrophy. *Nature*, submitted

Went, L. N., van Schooneveld, M. J. and Oosterhuis, J. A. (1992) Late onset dominant cone dystrophy with early blue cone involvement. *Journal of Medical Genetics*, **29**, 295–298.

Wilson, D. J., Weleber, R. G., Klein, M. L., Welch, R. B. and Green, W. R. (1989) Bietti's crystalline dystrophy: a clinicopathologic correlative study. *Archives of Ophthalmology*, **107**, 213–221.

Winn, S., Tasman, W., Spaeth, G., McDonald, P. R. and Justice, J. (1969) Ouguchi's disease in Negroes. *Archives of Ophthalmology*, **81**, 501–507.

Wolff, O. H., Lloyd, J. K. and Tonks, E. L. (1964) A-beta-lipoproteinaemia: with special reference to the visual defect. *Experimental Eye Research*, **3**, 439–442.

Wright, A. F., Bhattacharya, S. S., Clayton, J. F., Dempster, M., Tippett, P., McKeown, C. M. E. et al. (1987) Linkage relationships between X-linked retinitis pigmentosa and six short arm markers: exclusion of the disease locus from Xp21. *American Journal of Human Genetics*, **41**, 635–644.

Yagasaki, Y. and Jacobson, S. G. (1989) Cone-rod dystrophy. Phenotypic diversity by retinal function testing. *Archives of Ophthalmology*, **107**, 701–708.

Yamanaka, J. (1924) Existiert die Pigmentverschiebung im Retinalepithel im menschlichen Auge? Der erste Sektionsfall von sogenannter Oguchischer Krankheit. *Klinische Monatsblatter fur Augenheilkunde*, **73**, 742–752.

Yamanaka, M. (1969) Histologic study of Oguchi's disease: its relationship to pigmentary degeneration of the retina. *American Journal of Ophthalmology*, **68**, 19–26.

Young, R. S., Chaparro, A., Price, J. and Walters, J. (1989) Oscillatory potentials of X-linked carriers of congenital stationary night blindness. *Investigative Ophthalmology and Visual Science*, **30**, 806–812.

4. HUMAN ALBINISM AND MOUSE MODELS

RICHARD A. KING, IAN J. JACKSON* and WILLIAM S. OETTING

University of Minnesota, BOX 485 UMHC 420 Delaware Street SE, Minneapolis, Minnesota 55455, USA

**MRC Human Genetics Unit, Western General Hospital, Crewe Road, Edinburgh, EH4 2XU, UK*

ABBREVIATIONS USED

AS	Angelman Syndrome
c	mouse *albino* locus
CHS	Chediak Higashi Syndrome
HPS	Hermansky Pudlak Syndrome
OA	Ocular Albinism
OA1	X-Linked OA
OCA	Oculocutaneous Albinism
OCA1	Tyrosinase-related OCA
OCA2	Tyrosinase-positive OCA
p	mouse *pink-eyed dilution* locus
P	Human P gene locus which is homologous to mouse *pink-eyed dilution* locus
PWS	Prader-Willi Syndrome
TRP1	Tyrosinase related protein 1; gp75; *brown* locus
TRP2	Tyrosinase related protein 2; dopachrome tautomerase; *slaty* locus
TYR	Human tyrosinase gene locus

INTRODUCTION

Inherited abnormalities of melanogenesis that present with a congenital reduction in melanin synthesis in all melanocytes of the skin, hair and eyes are termed oculocutaneous albinism (OCA), and those in which the hypopigmentation primarily involves the melanocytes of the eye are termed ocular albinism (OA). The term "albino" is derived from the Latin *albus* for white, and is attributed to Bathazar Tellez with his 1660 description of African individuals with white skin (Pearson et al.,1911). Sir Archibald Garrod included OCA in his original description of inborn errors of metabolism (Garrod, 1908). OCA is universal in distribution and has been described in all human ethnic groups.

A similar genetic variant, involving white skin and hair and pink eyes is seen in many other mammalian species, including cats, dogs, cattle, mice, rats and rabbits (Searle, 1990). Pigmentation defects have been particularly well studied in the mouse, where there are more than 80 different genetic loci at which mutations affect hair or eye pigmentation. Some of these mutations, in addition to providing useful tools by which the development of melanocytes can be dissected, are also models for the understanding of homologous human diseases. They are particularly informative for examining the effect of different mutations of the same gene. The term "albino" in mouse is used for animals mutant at one particular locus, the *c*-locus on chromosome 7 and the model for human OCA1. Other forms of human albinism correspond to other mouse mutations.

MELANIN SYNTHESIS

Biochemical Process

Melanin biosynthesis has been studied for over seventy years (Raper, 1928; Mason, 1948; Lerner and Fitzpatrick, 1950), and for more than 50 years was thought to be regulated solely by the enzyme tyrosinase with all subsequent reactions occurring non-enzymatically (Coleman, 1962). In 1978, Logan and Weatherhead reported that hair follicle tyrosinase activity of the Chinese hamster was high with the winter and summer molt, while pigmented hairs were produced only with the summer molt (Logan and Weatherhead, 1978). This showed that tyrosinase activity in itself was not sufficient for pigmentation and other mechanisms for control of the process must exist. This observation led to a search for other enzymes and factors that regulate melanin synthesis, and several have now been identified, including tyrosinase related protein 1 or gp75 (TRP1) (Shibahara et al.,1986; Jackson, 1988; Jackson et al.,1990), tyrosinase related protein 2 or dopachrome tautomerase (TRP2) (Jackson, 1988; Jackson et al.,1992; Tsukamoto et al., 1992) and Pmel 17-1 (Kwon et al., 1987). The specific role of these newly described factors in normal melanogenesis is unknown but a great deal of work is being devoted to this problem, and human and mouse mutants are providing important insights.

The biosynthesis of melanin starts with the oxidation of tyrosine and either red-yellow pheomelanin or black-brown eumelanin can form (Prota, 1988a; Prota and Thomson, 1976; Prota, 1986; Prota, 1988b; Prota, 1980). The common steps to both types of melanin are the oxidation of tyrosine to dopa and the dehydrogenation of dopa to dopaquinone, the two steps being catalyzed by tyrosinase (Lerner and Fitzpatrick, 1950; Pomerantz and Warner, 1967). If sulfhydryl compounds are present, they react with dopaquinone and commit the process to phaeomelanization through a series of cysteinyldopa and benzothiazine intermediates that polymerize to acid-soluble pheomelanin (Jergil et al., 1983; Agrup et al., 1979, 1982; Hansson et al., 1980; Rorsman et al., 1973; Jara et al., 1988). The production of pheomelanin appears to be controlled by the availability of sulfhydryl compounds in the melanosome (Jara et al.,1988). In the absence of sulfhydryl compounds, dopa is converted to dopachrome (Cabanes et al., 1987) which is then converted to 5,6-dihydroxy indole-2-carboxylic acid by the catalytic action of dopachrome tautomerase (Leonard et al., 1988; Hearing et al., 1982; Körner and Pawelek, 1980; Pawelek, 1985, 1990, 1991; Pawelek et al., 1980; Chakraborty et al. 1992) or to 5,6-dihydroxyindole by spontaneous reaction (Leonard et al., 1988). Tyrosinase may also play a role in controlling the oxidation of dihydroxyindole-2-carboxylic acid to the corresponding indolequinone (Körner and Pawelek, 1982). The steps beyond the dihydroxyindole compounds are not well elucidated, but generally involve the formation of indolequinones which then polymerize to acid-insoluble eumelanin (Prota, 1988b; Prota, 1980; Ito and Wakamatsu, 1989). In a sense, the formation of melanin can be viewed, in part, as a detoxification mechanism of the cell by converting relatively toxic intermediate compounds such as quinones into inert melanin that cannot diffuse out of the cell (Prota, 1980).

Tyrosinase is a multifunctional copper containing glycoprotein with a *de novo* molecular weight of about 65kDa and a glycosylated weight of 75kDa (Nishioka, 1978; Miyazaki and Ohtaki, 1976; Hearing et al., 1978). The tyrosinase polypeptide contains 528 amino acids including an 18 amino acid leader sequence that is missing in the mature enzyme (Wittbjer et al., 1990). Tyrosinase contains two copper atoms as a binuclear complex. The copper complex forms a peroxide with dioxygen which then interacts with tyrosine as the first step in melanin formation (Jackman et al., 1992; Ross and Solomon, 1991). The amino acid sequence of the copper binding sites in human tyrosinase are similar to the amino acid sequence of hemocyanin (Volbeda and Hol, 1989a; Ross and Solomon, 1991), a compound for which X-ray defraction data are available, and this has permitted detailed tertiary structural analysis of the copper binding site of human tyrosinase (Oetting and King, 1992b). Tyrosinase has a hydrophobic transmembrane region that anchors the protein to the membrane and a cytoplasmic tail, both at the carboxy end of the protein.

Dopachrome tautomerase catalyzes the conversion of dopachrome to 5,6-dihydroxy indole-2-carboxylic acid (Leonard et al., 1988; Hearing et al.,1982; Körner and Pawelek, 1980; Pawelek, 1990, 1991). The murine or human enzyme has not been isolated and purified, but the gene has been cloned as cDNA. The predicted amino acid sequence of dopachrome tautomerase has a moderate degree of similarity to that of tyrosinase (40%) and includes areas that may be a transmembrane region and two copper binding sites as well as having two conserved cysteine-rich domains. The function of these cysteine-rich domains is unknown, but one of them is homologous to the EGF motif family of amino acid sequences which are found in many membrane-bound or secreted proteins from nematode and *Drosophila* developmental genes, mammalian growth factors and extracellular matrix proteins. (Jackson et al., 1992; Davis, 1990) The motif may function to mediate protein–protein interaction between the melanogenic enzymes and other proteins in the melanosome. Although dopachrome tautomerase has a specific action in the conversion of dopachrome and is encoded by a recognized coat color locus in the mouse (*slaty* locus), the precise role of this enzyme in normal melanogenesis remains largely unknown.

Molecular Biology

Tyrosinase

The human tyrosinase gene (TYR) is located on chromosome 11q14- q21(Barton et al.,1988). There is also a tyrosinase pseudogene (TYRL) that is a truncated gene on chromosome 11p11.2- cen (Barton et al., 1988; Takeda et al., 1989, 1991). The TYR gene is approximately 50 kb in length and the coding region is divided into 5 exons, with almost half the coding region in exon 1 (Giebel et al. 1991b). The mouse gene, on chromosome 7, has an identical exon structure to the human, but at 70kb is a little longer (Ruppert et al., 1988). Expression studies using the mouse tyrosinase gene have shown that only the first 270 bp of the promoter region are necessary for both melanocyte specific expression and the correct temporal expression in transgenic mice (Kluppel et al.,1991), although expression studies using the entire gene including 50 kb of the promoter region is necessary to completely mimic wild type expression of the gene (Schedl et al., 1992, 1993).

TRP1

A second pigment cDNA was identified that encodes a protein with about 40% amino acid identity with tyrosinase and is currently called tyrosinase related protein-1 or TRP1, initially named TRP (Shibahara et al., 1986, Cohen et al., 1990). The encoded protein has the characteristic copper-binding sites, membrane spanning domain and cysteine-rich features seen in tyrosinase and dopachrome tautomerase. The mouse gene maps to the *brown* (*b*) locus on chromosome 4 (Jackson, 1988). Mutations at this locus that affect the protein sequence or expression levels alter the type of eumelanin made from black to brown (Jackson et al., 1990). Interestingly, despite the clear common evolutionary history, TRP1 has an exon structure quite different from tyrosinase (Jackson et al., 1991), which is a characteristic of gene families separated by a very long history. The implication is that the family of related proteins should be found in a wide range of species.

The human TRP1 shares about 93% amino acid identity with mouse TRP1, being a polypeptide of 527 amino acids with a molecular weight of 60,000. The human gene has been mapped to human chromosome 9p23 using *in situ* hybridization (Murty et al., 1992). The TRP1 locus has been shown to be the gene for the human melanoma autoantigenic glycoprotein 75 (gp75) (Halaban and Moellmann, 1990; Vijayasaradhi and Houghton, 1991). The glycoprotein gp75 is the most abundant glycoprotein in melanoma cell lines, but absent in nonpigmented melanoma cell lines (Tai et al., 1983; Thomson et al., 1985).

Biochemical analysis of mouse TRP1, purified by immune affinity chromatography, has shown that the protein exhibits low levels of tyrosine hydroxylase activity (Jiménez et al., 1991) but column purified human gp75 from melanoma has been shown not to contain any tyrosine hydroxylase activity (Vijayasaradhi and Houghton, 1991). This discrepancy may be due to the high similarity of tyrosinase to gp75 and the immunoprecipitated brown protein may have been contaminated with tyrosinase. Others have ascribed catalase activity to this protein and have given it the locus name CAS2 for the human gene but this activity may not be the primary catalytic function of this protein (Halaban and Moellmann, 1990; Chintamaneni et al., 1991). It has been shown that both tyrosinase and gp75 contain catalase-like activity indicating that this activity may be artifactual.

TRP2

A third member of the tyrosinase-related family, TRP2, was initially cloned as a candidate for mouse tyrosinase, and has about 40% amino acid identity to both tyrosinase and TRP1, including the various characteristic motifs (Jackson et al., 1992). Immunopurification has subsequently shown that TRP2 is dopachrome tautomerase (Tsukamoto et al., 1992). The human cDNA has recently been cloned (B. Bouchard, personal communication) and is 80% identical to its mouse homologue. The mouse gene maps to and is mutated at the *slaty* locus on chromosome 14, (Jackson et al., 1992) whilst the human gene is at 13q32 (B. Bouchard, personal communication).

DEFINITION OF HUMAN ALBINISM

Our current understanding of the pathophysiology of hypopigmentation allows a precise definition of albinism. First, there are no characteristic changes associated with the skin hypopigmentation other than a generalized reduction of melanin in OCA. The hypopigmentation of the skin is associated with sensitivity to ultraviolet radiation and a predisposition to skin cancer, but the degree of the sensitivity in albinism is the same as in normally pigmented individuals with light or fair pigmentation. The specific and defining changes in albinism occur in the optic system, and are part of the diagnostic criteria (Creel et al., 1974, 1990). Melanin is involved in the normal development of the eye and the optic system through unknown mechanisms, and reduced melanin in the eye during development is associated with lifelong nystagmus, reduced retinal pigment, foveal hypoplasia with reduced visual acuity, and abnormal routing of the neural fibers from the retina to the brain resulting in strabismus and the loss of binocular vision. The optic system changes are found in all types of albinism, suggesting that it is the reduction in melanin in the involved tissue rather than a pleiotropic effect of the responsible gene.

Nystagmus

Almost all individuals with albinism have nystagmus (Collewijn et al., 1985; van Dorp, 1987; Creel et al., 1990). Several recently reported families contain individuals who appeared to have albinism without nystagmus, but this is quite rare (Summers et al.,1991; Cheong et al.,1992; Castronuovo et al.,1991). Nystagmus is present at birth or develops within the first few months of life, and the amplitude is usually greater early in life and becomes less marked with time, to the point where it can be minimal in an adult. There can be an associated head tilt or turn that appears to dampen the amplitude of the nystagmus.

Photophobia

Most individuals with albinism have some degree of photophobia, the degree related to the amount of pigment present in the iris and retina. Photophobia is usually most marked at a young age, and babies with albinism, particularly the tyrosinase-negative type of OCA1 will not want to go outside and will cry when they are in the sun. Some individuals with albinism may have no photophobia.

Visual Acuity

Affected individuals usually have diminished visual acuity, typically in the range of 20/100 to 20/200, with a range of 20/40–50 to 20/400+ (Edmunds, 1949; Fonda, 1962; Summers et al., 1988). Several recent reports have identified individuals with albinism who have normal or near-normal visual acuity, but this is thought to be rare (Summers et al., 1991). Visual development in an affected infant is slower than normal, with attainment of visual acuity delayed in the first years of life. Refractive errors are common, and many affected individuals have improvement of their acuity with proper correction (Taylor, 1978; Fonda, 1962). Color vision is essentially normal.

Iris Pigment

Iris pigment is normally found in the stroma and in the posterior iris epithelium, and both compartments have reduced amounts in albinism. Stromal pigment changes the iris from a blue color to one that is partially or completely brown, while pigment in the posterior iris epithelium reduces the translucency of the iris. Stromal pigment in albinism can be absent, minimal (light brown color to central portions of iris), or moderate (light brown iris), and often increases with age (Summers et al., 1988). When the posterior epithelial pigment is absent, the iris is fully translucent on globe transillumination. Increasing amounts of stromal and epithelial pigment reduce translucency to a "cartwheel" or "wagon-wheel", segmental or spotty pattern on globe transillumination (Summers et al., 1988).

Retinal Pigment and Foveal Development

The amount of pigment in the pigment epithelium of the retina is reduced or absent, allowing the choroidal vessels to be seen beneath the retina. The normal foveal depression and light reflex in the center of the fovea are absent (Usher, 1920; Summers et al., 1988; Abadi and Dickinson, 1983; Adler, 1910), and this developmental abnormality is largely responsible for the reduced acuity. The macular area can be translucent or opaque, the latter produced by luteal rather that melanin pigment (Summers et al., 1988).

Optic Tract Development

All individuals with albinism have excessive decussation of the optic fibers at the chiasm (Creel et al., 1974, 1990; Guillery et al., 1975). The ipsilateral projections from the retina to the lateral geniculate are reduced and the structure of the lateral geniculate is abnormal (Guillery et al., 1975). The demonstration of optic tract misrouting by visual evoked potential studies is the critical diagnostic procedure for questionable cases (Creel et al., 1974, 1978, 1981, 1990; Apkarian et al., 1990). Misrouting of the optic fibers can be demonstrated with visual evoked potentials recorded with monocular stimulation (Leventhal and Creel, 1985; Creel et al., 1974, 1978, 1981; Apkarian et al., 1990). The altered development of the optic tracts contributes to the strabismus and diminished stereoacuity that is usually present.

OCULOCUTANEOUS ALBINISM: PRIMARY DEFECT SPECIFIC FOR MELANIN SYNTHETIC PATHWAY

Overview

Oculocutaneous albinism can be divided into those types in which the primary defect is in the melanin synthetic system in the melanocyte and those presenting with albinism and associated defects in which the primary defect is not specific for the melanin synthetic system (Table 1). All types are autosomal recessive in inheritance, and obligate heterozygotes do not regularly manifest cutaneous or ocular hypopigmentation. The melanocyte structure and number in the skin and hair follicle are normal in all types of OCA, although the melanosomal structure may be altered, particularly in those types in which the primary defect is not specific for the melanin synthetic pathway (i.e. CHS). Primary OCA can be divided into OCA1 or tyrosinase-related OCA, OCA2 or tyrosinase-positive OCA (using the most recent terminology from McKusick (McKusick, 1992)) and several types that are unclassified because of incomplete clinical and biochemical data.

Table 1 Classification Of Albinism

OCULOCUTANEOUS ALBINISM (OCA)

Primary Defect Specific for Melanin Synthetic Pathway
- OCA1 Tyrosinase related OCA
  - *Mutations associated with no residual enzyme activity*
    - Tyrosinase negative OCA (OCA1A)
  - *Mutations associated with residual enzyme activity*
    - Yellow OCA (OCA1B)
    - Minimal Pigment OCA (OCA1MP)
  - *Mutation associated with unusual enzyme activity*
    - Temperature sensitive OCA (OCA1TS)
- OCA2 Tyrosinase positive OCA
  - Prader-Willi Syndrome/Angelman Syndrome
- Unclassified types
  - Brown OCA
  - Rufous/red OCA

Primary Defect not Specific for Melanin Synthetic Pathway

Hermansky-Pudlak Syndrome
Chediak-Higashi Syndrome

OCULAR ALBINISM
- OA1 X-Linked OA
- OA2 Autosomal Recessive OA

Tyrosinase Related OCA (OCA1)

Clinical phenotype

Tyrosinase related OCA is produced by mutations of the structural gene for the tyrosinase enzyme. Affected individuals have reduced or absent hairbulb tyrosinase activity, and heterozygous individuals generally have very low activity (King and Olds, 1985; King et al., 1978; King and Witkop, 1977). The phenotype of tyrosinase-related OCA varies from having no melanin pigment in the skin and eyes to having an unusual pattern of cutaneous pigment. An important characteristic of OCA1 is the marked hypopigmentation at birth, with most affected individuals having white hair, milky white skin (in comparison to the creamy white skin often present in individuals with tyrosinase-positive OCA), and blue eyes at birth. The irides can be very light blue and translucent such that the whole iris has a pink or red color in ambient light. With time, the irides often become a darker blue and remain translucent or become lightly pigmented with reduced translucency. The skin remains white but may appear to have more color with time. Sun exposure produces erythema and a burn if the skin is unprotected, and a generalized tan is absent or minimal in most, although an occasional individual with OCA1 can tan well, particularly if the hair and eye are pigmented. Few pigmented lesions (nevi, freckles, lentigines) develop in the skin. The clinical phenotype is dependent on the type of mutation present, and the variation in phenotype is more extreme than previously recognized. It is expected that future studies will broaden this group to include individuals who have nearly normal pigment associated with mutations of the tyrosinase gene that only minimally reduce activity of the enzyme.

Individuals with tyrosinase negative OCA are born with white hair and skin and blue eyes, and there is no melanin synthesis in these tissues throughout life (Witkop et al., 1989; King and Olds, 1985; King and Witkop, 1976). The phenotype is the same in all ethnic groups. With time, the hair may develop a darker white or a slight yellow tint that varies with different shampoos. The irides are translucent and appear pink early in life and often turn a gray-blue color with time. No pigmented lesions develop in the skin, although amelanotic nevi can be present. In the U.S., the frequency of tyrosinase negative OCA is estimated to be 1 in 39,000 in the Caucasian and 1 in 28,000 in the African-American population (Witkop et al., 1989).

Yellow OCA was originally thought to represent a discrete type of OCA with a characteristic phenotype but molecular studies now show that this is part of OCA1. The term yellow OCA refers to those types of tyrosinase related OCA produced by mutations associated with residual tyrosinase function (leaky mutations). The amount of residual enzyme function is sufficient to produce small to moderate amounts of iris, hair and perhaps skin pigment.

Yellow OCA was first recognized in the Amish (Nance et al., 1970) and subsequently identified in other groups. Family studies suggested the responsible gene was allelic to that for tyrosinase-negative OCA (Hu et al., 1980) and molecular studies eventually demonstrated several different mutations of the tyrosinase gene in this type of OCA. The phenotype of yellow OCA varies, but all affected individuals have white hair and skin and blue eyes at birth. Hair and iris pigment develop in the first few years of life. The hair color changes to light yellow, light blond or golden blond first, and eventually can turn dark blond in the adult (Giebel et al., 1991d; Nance et al., 1970). The irides can develop light tan pigment, particularly in the inner third of the iris, and the skin may tan with sun exposure. Ethnic and family pigment patterns can influence the phenotype, and hair color can turn light red or brown. The frequency of this type of mutation is unknown, but does not appear to be rare (Giebel et al., 1991d; Witkop et al., 1989).

Another OCA1 type associated with some melanin formation is minimal pigment OCA (King et al., 1986). The original description of minimal pigment OCA included a phenotype similar to tyrosinase negative OCA except for the development of minimal amounts of iris pigment in the first decade of life, and biochemical evaluation suggested that this was a tyrosinase related OCA. Molecular analysis has identified mutations of the tyrosinase gene in individuals with this type (Oetting and King, 1993). It is hypothesized that the responsible mutant allele is associated with less residual enzyme activity than a yellow allele, and the pigment is seen primarily in the iris because this is a stable rather than a deciduous pigment tissue (i.e. the pigment can accumulate without being shed with turnover of the tissue). It may be more appropriate to include minimal pigment OCA within the broad range of yellow OCA

Affected individuals have white skin and hair and blue eyes at birth. Iris pigment is present at birth or develops in the first decade. The hair remains white or develops a very slight yellow tint. The skin remains milky white and tanning has not been observed. To date, all individuals recognized with this type of OCA have been Caucasian, and the phenotype in other ethnic groups is unknown. The frequency is not known. Platinum OCA (Witkop et al., 1989) most likely represents this same phenotype.

A very interesting and unusual tyrosinase related OCA has recently been described (King et al., 1991b). The proband of the original family appeared to have tyrosinase negative OCA through the first years of her life, with white hair and skin, and blue eyes. With puberty her body hair color changed. The axillary hair remained white, the scalp hair also remained white but with a slight yellow tint, the arm hair turned light reddish brown, and the leg hair turned dark brown. The eyes remained blue and the skin remained white. Analysis of tyrosinase from scalp and leg hairbulbs showed that the enzyme was temperature sensitive, losing activity above 35°C. As a result, pigment could form in the cooler but not the warmer parts of the body. Molecular analysis showed that the proband was a compound heterozygote at the tyrosinase locus, having two different missense mutations that included the temperature sensitive allele and a tyrosinase-negative allele (Giebel et al.,1991c). There were three similarly affected siblings in the family and no other family of this type has been published to date. This type of tyrosinase related OCA is analogous to the Siamese cat and the Himalayan mouse.

Molecular phenotype

Analysis has shown that mutations in the tyrosinase gene are responsible for OCA1. DNA sequence analyses of the genomic DNA from individuals with OCA1 has revealed that many different mutations are responsible for this single gene disorder (Oetting and King, 1993). At present 25 missense, 4 nonsense and 9 frameshift mutations have been reported for a total of 38 different mutations. This analysis has shown that most individuals with OCA1 are compound heterozygotes with different mutant maternal and paternal alleles. These individuals have all combinations of missense, nonsense and frameshift mutations. Approximately 15% of all individuals in which the entire coding region and flanking intron sequences have been sequenced have shown only one mutation and in these cases the correct nucleotide is displayed on the homologous allele to that of the mutation showing that they are compound heterozygotes. The second mutation is thought to be within a regulatory region that is outside of the region sequenced or in the intron sequences producing a cryptic splice site or affecting an intronic transcriptional regulator. Most of these mutations have been found in individuals with the classic tyrosinase negative type of OCA1.

The majority of tyrosinase mutations have been found in Caucasian individuals but analysis of individuals with different racial backgrounds are revealing mutations that seem to be unique to that group of people. Two different mutations have been found in Japanese individuals, a missense mutation at codon 77 (R77Q) resulting in an arginine to glutamine substitution and a single base insertion of a cytosine at codon 310 (+ΔC310) resulting in a premature termination at codon 316 (Tomita et al., 1989; Kikuchi et al., 1990; Takeda et al., 1990). There have also been reported a missense mutation in an American black at codon 89 (cysteine to arginine, C89R), a nonsense mutation at codon 178 in an Afghan individual (tryptophan to termination, T178X) and a dinucleotide deletion in a Pakastani individual (-ΔGA115) (Giebel et al., 1991a; Spritz et al., 1991; Oetting et al., 1993a).

Mapping the location of the missense mutations shows that they cluster into three distinct regions of the tyrosinase coding region (Figure 1). Two of these clusters are within the putative copper A or copper B binding regions while the third is located at the 5' end of the coding sequence. It is thought that these clusters define functional areas of the tyrosinase enzyme (King et al., 1991a; Oetting and King, 1992a). There are several possibilities as to how these mutations could cause their effect on tyrosinase activity. For the putative copper A and copper B binding region the most obvious explanation is that the mutations alter the protein structure in such a way that the copper atoms are not correctly coordinated with the histidine ligands that complex with the copper atoms (Oetting and King, 1992b). In the copper A binding region, two of these mutations are within conserved amino acid motifs. The mutation F176I alters a phenylalanine in the motif Phe-X-X-X-His that is conserved in the copper binding centers found in tyrosinase and hemocyanin (Volbeda and Hol, 1989a,b). A second mutation, A206T, is within the motif Pro-X-Phe-X-X-X-His that is conserved in all tyrosinase polypeptides including fungal tyrosinase molecules (Huber et al., 1985). Four of the missense mutations, at codons 371, 373, 382 and 383, are in a loop structure of the copper B binding region between the α-helices responsible for the correct placement of the three histidine copper ligands. These missense mutations in the copper A and copper B binding regions are next to histidine residues which are responsible for the correct orientation of the two copper atoms in the catalytic site of the enzyme. Computer modeling of the copper B binding region shows perturbations in the secondary structure when these mutations are substituted for the normal sequence. The juxtaposition of the two copper atoms is critical because of the need to form a peroxide with dioxygen which is necessary for catalytic function and any alteration in the copper to copper distance would interfere with the peroxide bridge and render the enzyme inactive. These mutations may eliminate enzymatic activity by altering the location of the histidine ligands and altering copper-protein binding (Oetting and King, 1992b). A second possible mechanism is that these domains also contain a substrate binding region or another catalytic site affected by these mutations. The binding of tyrosine or DOPA should occur next to the copper atoms and if these amino acids are involved in this function, then even a slight change in the structure of this region or the chemical nature of an amino acid side chain might affect enzymatic activity. The third cluster, at the amino ter minal of the enzyme (codons 42-89), could represent a structural domain that involves a catalytic site of the enzyme. One mutation in this cluster, R77Q, results in the removal of a positively charged amino acid (arginine) and is substituted by a uncharged amino acid (glutamine) (Shibahara et al., 1990; Shibahara, 1992). The role of arginine in enzyme active sites as an anionic binding site has been reported in over one hundred enzymes and chemical modification of this residue results in inactivation of the enzyme. The arginine at codon 77 may play a role in the catalytic site of tyrosinase and removal of this amino acid could disrupt the interaction of the tyrosine substrate or dopa co-factor with the active site of this enzyme (Riordan, 1979). The cluster at the copper B binding site extends 3 of this region into exon 4. This area of the cluster includes a yellow OCA1 mutation (P406L) and the temperature sensitive OCA1 mutation. This region is removed from the copper B binding site and may be a separate functional domain, possibly involved in the substrate binding region of the catalytic site.

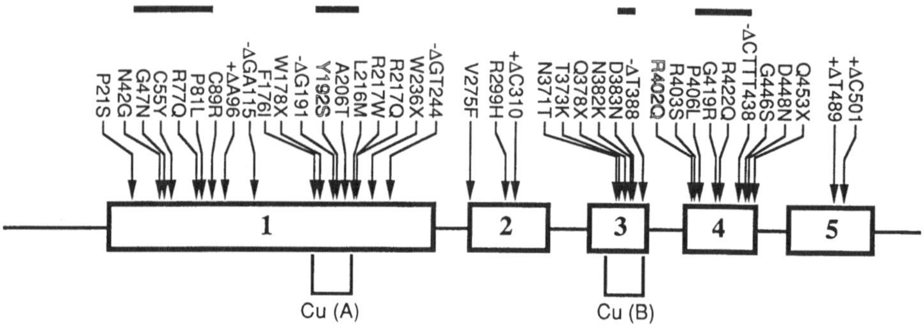

Figure 1 Location of the Mutations in the Tyrosinase Gene Associated with OCA1
The boxes represent the 5 exons of the tyrosinase coding region and the Cu(A) and Cu(B) labels represent the copper binding regions. Arrows show the location of the missense, nonsense and frameshift mutations associated with OCA1. The outlined text represents nonpathogenic polymorphisms. The black bars show the four regions where the missense mutation tend to cluster.

There are mutations of the tyrosinase gene that reduce but do not completely eliminate the level of tyrosinase activity, and mutations that confer temperature sensitivity to the enzymatic activity. These mutations were found in individuals with yellow OCA1, who have partial pigmentation in their hair and eyes (Giebel et al., 1991d). Analysis of the Amish families in which this was originally described has shown that the affected individuals are homozygous for a codon 406 missense mutation (P406L) resulting in a proline to leucine substitution (Giebel et al., 1991d; Nance et al., 1970). Expression studies in HeLa cells show that this mutation reduces enzymatic activity to 7% of the wild type levels. A codon 275 missense mutation (valine to phenylalanine; V275F) has been found in individuals with this phenotype who are compound heterozygotes, with the mutation on the homologous allele (codon 81 proline to leucine; P81L) found in tyrosinase negative OCA1 (supposedly an allele associated with no tyrosinase function). Since the yellow OCA1 phenotype varies in families containing individuals homozygous for the P406L mutation as well as in individuals who are compound heterozygotes, it appears that other familial pigment genes can influence this pheno-type (Giebel et al., 1991d). In the case of reduced tyrosinase activity, red/yellow pheomelanin biosynthesis is favored over black/brown eumelanin biosynthesis.

In one individual, a temperature-sensitive form of tyrosinase was observed (see above) (King et al., 1991b). This individual was found to have a mutation at codon 422 (R422Q) resulting in an arginine to glutamine substitution with a tyrosinase negative OCA1 mutation on the homologous allele (Giebel et al., 1991c). Expression of the 422 codon mutation produced a tyrosinase that was inactivated above 35°C. It is interesting that the mouse Himalayan mutation, also a temperature sensitive tyrosinase mutation, is located in the same region of the peptide, codon 420 (histidine to arginine) (Kwon et al., 1989). There has also been reported a temperature sensitive mutation at codon 402 (R402Q) (Tripathi et al., 1991). It is now obvious that there is a continuous spectrum of reduced tyrosinase activities resulting from mutations of the tyrosinase gene (King et al., 1988; King and Oetting, 1992). These altered tyrosinase molecules produce phenotypic variation from a complete absence of pigment to a pigment phenotype indistinguishable from normal. In all these cases of OCA1 there is ocular involvement. This spectrum of pigment has also been observed in the *c*-locus mutations of the mouse (Silvers, 1979).

Several non pathogenic polymorphisms have also been detected. Two of these result in amino acid substitutions. A polymorphism at codon 192 contains either a serine (TAT) or tyrosine (TCT) and affects a *Mbo* I restriction endonuclease site (Giebel and Spritz, 1990). It is interesting that this mutation at codon 192 is within the copper A binding site but does not appear to affect the phenotype of the individual. The distribution of this polymorphism is equal in Caucasian populations, but only the TCT codon is found in the Oriental population. The second polymorphism is at codon 402 and is either an arginine or glutamine. This polymorphism does affect tyrosinase activity by reducing enzymatic activity by 75% when expressed in cultured cells but the clinical phenotype is not altered (Tripathi et al., 1991). A third polymorphism is at the CCAATT box of the promoter and affects a *Taq* I restriction endonuclease site (Oetting et al., 1991). Another polymorphism located in the promoter region is a GA-cluster 300 bp upstream of the start site; four alleles have been detected using PCR amplification (Morris et al., 1991). Two other polymorphisms are found by means of Restriction Fragment Length Polymorphisms (RFLP) analysis with either *Taq* I or *Bgl* II and using the tyrosinase cDNA as a probe (Spritz et al., 1988; Spritz and Strunk, 1990). Haplotype analysis can be performed with analysis of these polymorphic sites. The missense mutation at codon 47 (G47D) has been observed in three different populations, United States, Puerto Rico and the Canary Islands (Oetting et al., 1993b). All individuals had the G47D mutation associated with haplotype 1 showing a common founder for this mutation. On the other hand, the mutation at codon 81 (P81L) was found in individuals from the United States and the Canary Islands, but the two regions had two different haplotypes, haplotype 1 being associated with the P81L mutation in individuals from the Canary Islands and haplotype 4 associated with the United States, showing the possibility of two different founders for the same mutation.

Mouse Tyrosinase Mutations

The recessive mouse *albino* mutation was one of the first genes to be studied after the rediscovery of Mendel's Laws (Bateson, 1903). The phenotype of the mutant animal is a clear homologue of human OCA1; there is no hair, skin or eye pigment synthesised, and no tyrosinase activity is detectable. The iris is not normally externally visible in the mouse, so the dark pigment seen in the eye is the pigmented retinal epithelium. Lack or reduction of eye pigment results in the eye appearing pink, red or ruby-coloured, depending on severity. Mutants with absent or much reduced eye pigment also have, like their human counterparts, defective vision due to a reduced number of ipsilateral optic nerve connections to the brain.(Guillery, 1974; LaVail et al., 1978)

The tyrosinase gene from the classical *albino* mutation has been sequenced, and in fact encodes a protein with two amino acid differences from wild-type (Shibahara et al., 1990). These are C103S and R405H (numbering is the same as the human enzyme). Several lines of evidence indicates that only one of these changes, C103S is causative of the *albino* mutation. Firstly, a tyrosinase gene with only this change is inactive in expression in cultured cells (Shibahara, 1990) or in transgenic mice (Yokoyama et al., 1990). Second, a pigmented clone, apparently a revertant to wild-type, was identified in a cultured *albino* melanocyte line. Sequencing of the tyrosinase gene showed that the revertant cells had changed the S103 back to the wild-type cysteine, but only on one chromosome (Jackson and Bennett, 1990). All commonly used *albino* mouse lines have this same mutation, indicating that they all derive from a common ancestor (Jackson and Bennett, 1990; Yokoyama et al., 1990). Cysteine-103 is an essential part of the EGF-motif and is conserved in human tyrosinase, but has not been seen to be mutant in any OCAI patients so far. Other mutations have been seen in the EGF motif; cysteine-89 is part of the motif, which is mutated to arginine in human tyrosinase causing OCAI (Spritz et al., 1991) and the C111Y change which causes the *brown* mutation in mouse TRP-1 is also in the motif (Zdarsky et al., 1990).

There are numerous alleles of the classic *albino* mutation. Some have the same lack of function phenotype, and are due to other (as yet uncharacterised) point mutations or to deletions of the gene (Yokoyama et al., 1990; Rinchik et al., 1993a). More interesting from a comparative genetic point of view are the alleles giving phenotypes intermediate between albino and full pigmentation. A number, such as *chinchilla, ruby-eyed dilute, extreme dilution,* and *platinum* form a series, with decreasing amounts of pigment, resulting in progressively paler coats going along the series, and eyes ranging from almost black to pink (Silvers, 1979). The phenotypic series corresponds to decreasing levels of detectable tyrosinase activity, and these mutants are likely to be mouse models of yellow OCA. Another, *c-44H* has almost no coat pigmentation, but eyes which darken after birth to become nearly black at a few months of age. Possibly this allele is a model of minimal pigment OCA or platinum OCA.

The molecular nature of the changes in the tyrosinase gene of these partial or "leaky" mutations is likely to be very informative, but with the exception of *chinchilla* these have not been studied in detail. The *chinchilla* tyrosinase gene contains a point mutation A482T (Beerman et al., 1990), located in the transmem-brane region of the encoded enzyme, which may be the cause of decreased activity, although an effect of the mutation on mRNA levels has not been rigourously ruled out. The temperature-sensitive tyrosinase of the *himalayan* mutation was referred to above. The animals are largely very lightly pigmented, but have dark ears, nose, tail and scotum and red eyes. The temperature senstivity of the enzyme results from an H420R change (Kwon et al., 1989). One other mutation has been subject to molecular study, *chinchilla-mottled*, animals which have patches of chinchilla fur and patches of lighter fur. The *chinchilla-mottled* tyrosinase structural gene seems normal, but there is a DNA rearrangement, insertion or deletion located several kilobases upstream. (Porter et al., 1991) The rearrangement affects transcription of the gene, but to a variable extent in different clones. Transgenic animals made with a tyrosinase minigene, and with 5kb of upstream DNA do not always have normal levels of gene expression (Beerman et al., 1990). Transgenes carrying the full tyrosinase gene, plus 150kb upstream result in position-independent expression (Schedl et al., 1993), indicating a requirement for sequence elements between -5kb and -150kb for full-level expression.

Tyrosinase Positive OCA (OCA2)

Overview

A number of observations have been made that broadly define this type of OCA. First, fresh scalp anagen hairbulbs from an affected individual form visible pigment when incubated in tyrosine or dopa (Witkop et al., 1970) and the melanin in these hairbulbs is synthesized in the melanosomes when the incubated hairbulbs are observed by electron microscopy (Witkop et al., 1970). Historically, this observation was the first to separate human OCA into two types, as hairbulbs from an individual with tyrosinase negative OCA do not form pigment when incubated in tyrosine or dopa under identical conditions (Witkop et al., 1970). Second, the activity of tyrosinase in hairbulbs from affected individuals is nomal or nearly normal when measured by a quantitative assay or evaluated by electrophoresis (King and Olds, 1985; King et al., 1978; King and Witkop, 1976). Third,

Table 2 Mouse Mutations as Models for Human Albinism

OCA1		*albino* (c) locus mutations
	OCA1A	*albino*
	OCA1B, OCA1MP	*chinchilla*
		ruby-eyed dilute
		extreme dilution
		platinum
	OCA1TS	*himalayan*
OCA2		*pink-eye dilution locus*
		multiple mutations
Hermansky-Pudlak Syndrome		Mutations at 12 loci:
Chediak-Higashi Syndrome		*beige, cocoa, pale ear, light ear, mocha, muted, pallid, pearl, reduced pigmentation, ruby-eye, ruby-eye-2, sandy*

the ultrastructure of the melanocyte and the melanosome appear to be normal (Witkop et al., 1973). Fourth, the gene (or genes) responsible for this type of OCA is not genetically linked to the tyrosinase gene (Heim et al., 1988).

The general phenotypic features of this type of OCA include the presence of hair and iris pigment at birth or early in life, and the development of localized (nevi, freckles, and lentigines) skin pigment, often in sun exposed regions of the skin (Witkop et al., 1989). The ethnic and familiar pigment background of the affected individual has a more profound effect on tyrosinase positive OCA than on tyrosinase related OCA (Witkop et al., 1989). Most affected individuals accumulate pigment in their hair and eyes during their lifetime, and this is more marked in individuals from darker ethnic groups.

Broadly speaking, any or all of the many mouse coat-colour mutations could be candidates for mouse homologues of the human disorders. However, only a fraction of the mutations affect eye pigmentation to a significant degree. Leading amongst these is the *pink-eyed dilution (p)* mutation, on chromosome 7, which has numerous alleles which produce diluted hair pigmentation and pink eyes (Silvers, 1979).

Clinical phenotype

In Caucasian individuals, the amount of pigment present at birth or developing with time varies from minimal in northern European (particularly Scandinavian) to moderate in southern European or Mediterranean individuals. The hair can be very lightly pigmented at birth, having a light yellow or blond color, or more pigmented with a definite blond, golden blond or even red color. The skin is creamy white and pigmented spots may be present. The iris color is blue-gray or lightly pigmented, and the degree of translucency correlates with the amount of iris pigment present. With time, pigmented nevi and lentigines may develop and pigmented freckles are seen in exposed areas with repeated sun exposure. The skin does not tan. The hair slowly turns darker through the first two or more decades of life.

In most African-American individuals and those from west Africa, the phenotype of tyrosinase-positive OCA is distinct (King et al. 1980). The hair is yellow at birth and remains yellow with time, although the color may turn darker. The skin is creamy white at birth and changes little with time. Pigmented nevi, lentigines and freckles develop, but there is no tan. The irides are blue/gray or lightly pigmented. This phenotype is perhaps the most common in the world because of the high frequency of this type in west Africa (King et al., 1980; Okoro, 1975). In the African-American population, there is some phenotypic variation within this type, particularly in the hair color which may be brown, auburn, or red rather than yellow at birth or with time.

A newly recognized association with OCA2 is the hypopigmentation found with Prader-Willi syndrome (PWS) and Angelman Syndrome (AS) (King et al.,1993; Wiesner et al., 1987; Butler, 1989; Butler et al., 1986). PWS is a developmental syndrome that includes neonatal hypotonia, hyperphagia and obesity, hypogonadism, small hands and feet, and mental retardation associated with characteristic behavior (Cassidy, 1984; Bray et al., 1983). Approximately 70% of individuals with PWS have a deletion on the long arm of the paternal chromosome 15 (Butler et al.,1986; Ledbetter et al., 1982; Mutirangura et al., 1993), and most of those without a deletion have uniparental disomy for the maternal chromosome 15 (Magenis et al., 1990; Mascari et al., 1992).

Approximately 50% of individuals with PWS are hypopigmented but are without all of the typical features of albinism (Wiesner et al., 1987; Butler, 1989; Creel et al., 1986; Butler et al., 1986). Interestingly, a number of individuals with PWS and the typical features of OCA have been identified (Rinchik et al., 1993b). The hypopigmentation in PWS was originally described as oculocutaneous albinoidism to signify cutaneous without ocular involvement (Hittner et al., 1982), but subsequent studies have demonstrated hypopigmentation of skin, hair, and eyes, indicating that this is oculocutaneous hypopigmentation. For those without obvious OCA, hair and skin are lighter than unaffected family members, and childhood nystagmus and strabismus are common. The irides are pigmented but show some translucency on globe transillumination, and the amount of retinal pigment is reduced. Although foveal hypoplasia is not usually present, the fovea does not appear entirely normal (Creel et al., 1986). Visual evoked potential studies have revealed optic tract misrouting in a albinism pattern in some individuals with PWS and hypopigmentation (Creel et al., 1986); the presence of abnormalities of the optic tract implies pathologic hypopigmentation during development, indicating that the hypopigmentation in PWS is best described as a type of albinism. The individuals with PWS and OCA have not been fully described, but the features of the albinism are typical for OCA2.

Some individuals with AS are also hypopigmented, but the percentage showing this phenotypic characteristic is unknown (King et al., 1992; Fryburg et al., 1991).

AS is a complex developmental disorder that includes developmental delay and severe mental retardation, microcephaly, neonatal hypotonia, ataxic movements, and inappropriate laughter (Fryburg et al., 1991; Williams and Frias, 1982; Angelman, 1965). Those with hypopigmentation have light skin and hair and may have a history of nystagmus or strabismus (King et al., 1992). Iris translucency and reduced retinal pigment may be present. No analysis of the optic tract organization is available.

Molecular Phenotype

Recently, the human homologue to the mouse *pink-eyed-dilution* (*p*) gene has linked to a type of tyrosinase positive OCA, OCA2 (Gardner et al., 1992; Ramsay et al., 1992; Rinchik et al., 1993b). This linkage was foreshadowed by the identification of DN10, a gene associated with the PWS chromosomal region (Nakatsu et al.,1992), as the product of the mouse *p* gene. This made *P*, the human homologue of the murine p gene, a strong candidate for OCA2. The mouse *p* locus is located on chromosome 7 in a region that has conserved synteny with human chromosome 15q11.2-q12, the location of the critical region for PWS (Brilliant, 1992). The 3.4 kb mRNA transcript is expressed only in melanocytes (Rinchik et al., 1993b). The P poly-peptide contains 12 transmembrane domains with several *N*-glycosylation sites and potential protein kinase C phosphorylation sites (Gardner et al., 1992; Rinchik et al., 1993b). It has been hypothesized that the *P* polypeptide is a tyrosine transport protein bound to the membrane of the melanosome. Some support is lent to the idea by the observation that incubating the melanocytes from *p* mutant mice, or from OCA2 patients, in excess tyrosine increases their pigmentation (Sidman and Perlstein, 1965; Witkop et al., 1989)

Linkage analysis using the *p* locus cDNA has shown that this gene, which corresponds to the D15S12 marker, is within the chromosomal region at 15q11.2-q13 that is deleted in patients with PWS and AS (Gardner et al., 1992; Rinchik et al., 1993b). Linkage was also established between tyrosinase positive OCA and markers D15S10 and D15S13 which also map to 15q11.2-q12 (Ramsay et al., 1992). Evidence that the P locus is synonymous with the OCA2 gene has been strengthened by the finding of an individual with tyrosinase positive OCA with two different deletions of the P locus; a deletion of the entire P locus on the paternally derived chromosome 15 and a partial deletion of the P locus on the maternally derived allele (Rinchik et al., 1993b).

The mechanism for the hypopigmentation in PWS in those individuals without obvious OCA remains an enigma. The murine *p* locus is not imprinted and it is likely that the human locus is not (Gardner et al.,1992). The deletion on the paternal chromosome 15 in PWS and the maternal chromosome 15 in AS suggests that the hypopigmentation arises from a mechanism other than gene dose effect, since heterozygotes for tyrosinase positive OCA are normally pigmented. Further work will be necessary after the function of the *P* locus gene product is determined.

The multiple alleles of the mouse *p* mutation, as at *albino*, produce a range of phenotypes from only slight dilution of pigment of hair and eyes, to the full dilution seen in *p* animals. Some alleles, such as *p-6 Harwell* and *p-cleft palate* are complete deletions of the locus, others, for example *p-Jackson* have partial deletions whilst *p-25 Harwell* and *black-eyed sterile* have DNA rearrangements upstream of the gene. Most of these mutations, as well as the original *p*, produce no mRNA, but the *p-Jackson* mutant gene makes a smaller mRNA (presumably because it is an internal deletion) and *black-eyed sterile*, as well as others, have normal mRNA (which presumably encodes for a mutant protein) (Gardner et al., 1992). One further mutation, *p-unstable*, is particularly interesting as it is a large (70kb) internal duplication of part of the gene. The result is that the mRNA produced is larger by 1.5kb, resulting from addition of the duplicated exons into the mature mRNA, which does not make a functional protein (Gardner et al., 1992; Rinchik et al., 1993b). The DNA duplication causes the mutation to be unstable; it is able to revert at high frequency to wild-type by recombination, causing elimination of the duplicated segment (Gondo et al., 1993). The phenotype of *p-unstable* is normally characterised by small dark patches on a diluted pigment coat. Occasional animals have larger dark patches (indicating a reversion event earlier in development), and these mice often transmit the reverted gene to their offspring.

The expectation is that there will be great allelic heterogeneity of this gene in humans resulting in a spectrum of effects from severe to mild forms of albinism, as seen in the mouse and in the OCA1 locus in humans.

Unclassified Types

Brown OCA has been reported only in individuals in Africa or in African-American individuals, and the phenotype in Caucasian individuals is unknown (King et al., 1980, 1985). At birth, the hair and skin color are light brown, and the irides are gray to tan. With time there is little change in skin color, but the hair may turn darker and the irides may accumulate more tan pigment. The skin is resistant to the acute effects of sun exposure and does not burn. The iris has punctate and radial translucency, and moderate retinal pigment is present. The skin may darken with sun exposure. Visual acuity ranges from 20/60 to 20/150. The ultrastructure of the melanocyte and melanosome are normal, and hairbulb tyrosinase activity is normal (King et al., 1980, 1985). The frequency of Brown OCA is unknown; in one study, approximately 29% of indivi-duals with albinism in Nigeria had this type (King et al., 1980).

Individuals with rufous (Red) OCA having red hair and reddish-brown pigmented skin have been reported in Africa (Pearson et al., 1911; Stannus, 1913; Kromberg et al., 1990) and in New Guinea (Hornabrook et al., 1980; Walsh, 1971), but clinical descriptions are incomplete, no biochemical or ultrastructure data are available, and similar phenotypes in the US and European populations have not been recognized. Individuals with red hair who have either tyrosinase related OCA or tyrosinase positive OCA are recognized, but the reddish-brown skin pigment is usually not present. In New Guinea, the described phenotype includes reddish-brown skin, deep mahogany hair, reddish-brown to brown irides, and moderate amounts of retinal pigment.

OCULOCUTANEOUS ALBINISM: PRIMARY DEFECT NOT SPECIFIC FOR MEL SYNTHETIC PATHWAY

Several rare types of OCA are associated with clinical features that involve tissues outside of the pigment system, and the reduction in melanin synthesis in these conditions appears to be secondary to a defect that is not specific for the melanocyte or the melanin synthetic pathway. The phenotype of these conditions varies from obvious OCA to moderate hypopigmentation, and the associated medical problems can be severe and represent the major medical problem for the affected individual.

Hermansky-Pudlak Syndrome

Hermansky-Pudlak Syndrome (HPS) is an autosomal recessive condition that presents with OCA, a mild bleeding diathesis, and a ceroid storage disease affecting primarily the lungs and the gut. Hermansky and Pudlak first described this condition in two Czechoslovakian individuals in 1959 (Hermansky and Pudlak, 1959) and it has subsequently been recognized throughout the world, with the majority of affected individuals in the Puerto Rican population (Witkop et al., 1989). HPS is autosomal recessive in inheritance.

The albinism has a variable phenotype, ranging from marked hypopigmentation that is similar to tyrosinase-negative OCA to only moderate hypopigmentation with pigment present in the skin, hair and eyes. In the Puerto Rican population, the hair color varies from white to yellow to brown. The skin is creamy white and freckles are often present in the sun exposed regions (face, neck, arms and hands), often coalescing into large areas that look like normal dark skin pigment. Pigmented nevi are common. The iris color varies from blue to brown, and all of the ocular features of albinism are present. Visual acuity ranges from 20/60 to 20/400 (Simon et al., 1982; Summers et al., 1988). The presence of OCA may not be obvious in a Puerto Rican individual who has brown hair, pigment in the exposed areas of the skin, and brown eyes without comparison of the pigment pattern to unaffected family members (who are generally darker in pigment) or without recognizing the ocular features of albinism.

The most severe clinical manifestations of HPS are related to the pulmonary and intestinal changes. Interstitial pulmonary fibrosis has been described in many individuals with HPS, although the actual prevalence is unknown (Harmon et al., 1994; White et al., 1984). The fibrosis results in moderate to severe restrictive lung disease (Harmon et al., 1994). The development of granulomatous colitis, presenting with abdominal pain and bloody diarrhea in a child or an adult, has also been described in many individuals with HPS (Mahadeo et al., 1991; Schinella et al., 1980). The etiology is unknown, and immunologic studies do not show an abnormality (Shanahan et al., 1988). The presence of ceroid material in the epithelial cells of the gut suggests that this material may be involved in the development of the colitis, but this has not been proven.

The bleeding diathesis in HPS is related to a deficiency of storage granules in the platelets (i.e. storage pool-deficient platelets). The storage granules or dense bodies are reduced in number or are absent, and this is associated with a deficiency of serotonin, adenine nucleotides, and calcium in the platelet (Witkop et al., 1989). As a result, platelets in HPS do not show the irreversible secondary aggregation when stimulated. This deficiency produces mild hemorrhagic episodes in many affected individuals, including easy brusibility, epistaxis, hemoptysis, gingival bleeding with brushing or dental extraction, and postpartum bleeding. Occasional severe bleeding is observed. The basic defect in HPS is unknown. Hairbulb tyrosinase activity is present but the level is low (King and Olds, 1985). The structures of the melanocyte and melanosome structure are normal (Witkop et al., 1989).

Chediak-Higashi Syndrome (CHS)

The Chediak-Higashi Syndrome is a rare autosomal recessive syndrome that consists of increased susceptibility to bacterial infections, hypopigmentation, and the presence of giant peroxidase-positive lysosomal granules in peripheral blood granulocytes (Blume and Wolff, 1972). The skin, hair, and eye pigment is diluted in CHS but the affected individuals do not always have obvious albinism and the hypopigmentation may only be obvious in relationship to other family members. Hair color is light brown to blond, and the hair has a metallic silver-gray sheen. The skin is creamy white to slate gray. Iris pigment is present and nystagmus and photophobia may be present or absent (Benezra et al.,1980). Visual evoked potential studies show misrouting of the optic fibers in a pattern that is similar to individuals OCA. The primary defect in CHS is unknown. The susceptibility to bacterial infections appears to be the result of the abnormal granules in the neutrophils and other cells (White and Clawson, 1980; Tanaka, 1980). The hypopigmentation also arises from the formation of abnormal granules. Giant melanosomes form in the melanocyte and are unable to be transferred to the surrounding keratinocytes, leading to abnormal melanosome distribution and hypopigmentation.

Mouse Models for Platelet Storage Pool Deficiencies

The combination of pigmentation defects with a platelet storage pool deficiency has been found in a number of mouse mutants. The mutants (*beige, cocoa, pale ear, light ear, mocha, muted, pallid, pearl, reduced pigmentation, ruby eye, ruby eye-2 and sandy*) all have a combination of defects of lysosomal enzyme secretion in kidney and other tissues, a deficiency of platelet dense granules, and a pigmentation defect caused by melanosome abnormalities. There are differences between the 12 mutants in the range of pigmentation defects and the effects on the eye, but in general the eyes are lighter at birth than in adults. Three of the mutants, *pale ear, pearl,* and *ruby-eye* have been shown to have reduced ipsilateral projections from the retina to the lateral geniculate nucleus, as in ocular pigment-deficient humans. Any or all of these may be models for HPS or for CHS (Silvers, 1979; Green 1989).

This series of different mutant loci which all affect the same three processes suggests a common pathway linking the processes, in which the products of the 12 loci participate or interact. Melanosomes and lysosomes have a number of proteins in common and may share a common origin (Orlow et al., 1993; Zhou et al., 1993).

There has been promising progress towards elucidating the molecular basis of *pallid.* The gene encoding the mouse erythrocyte protein band 4.2 (Epb4.2) maps at or near the *pallid* locus, and expression of the Epb4.2 mRNA is abnormal in kidney (which is affected by the mutation) but not in liver or brain (which are not affected) (White et al., 1992). The protein 4.2 must be considered a strong candidate for the *pallid* locus, but the data is not, as yet, definitive. One unresolved issue is that human patients who lack protein 4.2 suffer from hereditary spherocy-tosis, but have none of the features of the pallid mice, whilst pallid mice do not have spherocytosis. There may be tissue-specific, or even organelle-specific, forms of the protein, and different diseases might manifest from mutations in the same gene. The identification of the product of one of the platelet storage pool deficiency mutations in mouse will greatly aid the identification of others, by locating the cellular processes affected and finding other interacting proteins within the process, which might be encoded at the other mutant loci.

OCULAR ALBINISM

X-Linked Recessive OA (OA1)

OA1 has an X-linked recessive pattern of inheritance. Affected males have normal skin and hair pigment. The irides are blue to brown, and the ocular changes are similar to OCA (Freed et al., 1989). In African-American males, iris color is often brown and there is little iris translucency (Charles et al., 1992). The skin of Caucasian individuals with OA1 appears normally pigmented, while many African-American individuals have scattered hypopigmented macules. Heterozygous females can be clinically detected because of ocular pigment changes that result from X-inactivation. A varigated pattern of retinal pigment and punctate areas of iris translucency are seen in approximately 80% of obligate heterozygotes females (Kinoshita et al., 1989; Freed et al., 1989; Smith et al., 1988; Charles et al., 1992). A small number of heterozygous females have ocular changes of albinism including nystagmus and reduced acuity that is thought to be the result of nonrandom patterns of x-inactivation (Jaeger and Jay 1981).

Melanocytes are normal in OA1 but there are changes in the melanosomes. Melanocytes in the skin and hair follicles as well as those of the iris and retina contain large melanosomes called giant melanosomes, macromelanosomes, or melanin macroglobules, along with normal melanosomes. The giant melanosomes are also found in the tissues of obligate heterozygous females. The systemic nature of the melanosome defect in OA1 suggests that this is really a type of OCA in which the major manifestations are in the eye. There may not be any type of albinism with changes limited to the eye, and all types probably represent variable expression of oculocutaneous albinism.

The gene for OA1 has been localized to the short arm of the X chromosome. Family studies placed the gene on the X chromosome. Initial linkage studies showed OA1 linked to the Xg blood group, now localized to Xp22.32 (Fialkow et al., 1967; Pearce et al., 1968). The gene has now been mapped more precisely to the Xp22.3-Xp22.2 region, (Schnur et al., 1991) and is thought to be between the markers DXS237 and DXS143 (Bergen et al., 1990, 1991). The gene has not been isolated and the function of its product is unknown.

Conservation of linkage between mouse and human X chromosomes readily allows identification of the region where the mouse homologue of OA1 should be. Curiously, there are no mutations which affect eye or hair melanocytes in this part of the chromosome in the mouse. It is possible that a mutation which has only a subtle effect on eye pigmentation would not have been identified in mouse stocks without a screen focused particularly on eye defects.

Autosomal Recessive Ocular Albinism (OA2)

This type of OA affects males and females equally (Kulkarni and Nathanson, 1989). The ocular features of the albinism are similar to those of OCA and OA1, except that most affected individuals have some retinal pigment in the posterior pole. Most published cases of OA2 have been in Caucasians, and they have frequently had light pigment at birth with a gradual increase in skin and hair pigment with time. Skin pigment appears to be normal in OA2 but affected African-American individuals have not been described.

The gene responsible of OA2 has not been identified or mapped. A recent report of a dysmorphic child having the features of OA and an absence of giant melanosomes in association with a deletion of chromosome 6(q13-q15) suggests that this may be the locus for OA2 (Rose et al., 1992). Another real possibility is that OA2 does not exist as a separate entity, and that the previously described cases actually have tyrosinase gene or P gene mutations associated with residual enzyme activity and relatively normal cutaneous pigmentation. Individuals having tyrosinase gene mutations associated with nearly normal cutaneous pigmentation have been described.

There are very few mouse mutations where eye pigmentation is affected without hair and skin involvement. One mutant allele of *ruby eye-2, maroon,* has been described in which some mutant animals can have lightened eyes but a normal coat color; although affected littermates might have much lightened coats. In support of the idea that OA2 humans might have tyrosinase mutations, the different mutant alleles of mouse *albino,* the tyrosinase gene, can have differential effects on hair versus eye pigmentation. The *extreme dilution* mutation has a very pale coat, but dark eyes, whereas the *ruby-eyed dilute* allele gives a darker coat with ruby eyes. On the whole, however, the reverse is true; mutations which have lightened hair but dark eyes occur much more frequently.

CLINICAL APPLICATION

Diagnosis of Albinism

Most individuals with OCA have an obvious phenotype, and extensive laboratory testing is unnecessary. The ocular examination is most important, and a diagnosis of OCA should not be made without the ocular features of albinism, particularly when cutaneous and iris pigment is present. It can be difficult to separate OCA1 and OCA2 with the clinical examination. The most reliable characteristic for this is the presence (OCA2) or absence (OCA1) of hair pigment at birth, but delayed hair growth in many children may make this difficult in the first six months of life.

The critical diagnostic issue is the presence or absence of albinism rather the specific type of albinism. Knowledge of the specific type can help in family counseling, but does not accurately predict the expected visual outcome. In general, individuals with OCA who form melanin in their hair and eyes will have a better visual outcome than those who form little or no pigment, but exceptions to this are common.

For OCA, the precise diagnosis of OCA1 or OCA2 requires molecular analysis of the tyrosinase or the P gene. The quantitative hairbulb tyrosinase assay has been used but is imprecise. Individuals with OCA1 who have alleles coding for an enzyme with moderate residual activity may demonstrate hairbulb tyrosinase activity on assay, and the incorrect interpretation may be OCA2 rather than OCA1. Heterozygote detection can only be performed with molecular analysis. At the present time, molecular analysis of the OA1 gene is not possible, and a skin or hair follicle biopsy to evaluate for giant melanosomes remains the most precise method of diagnosis. The OA1 gene is well localized on the X chromosome, however, and family counseling with linkage analysis may be possible. It is expected that the OA1 gene will be isolated in the very near future, and direct molecular analysis will then be possible.

Treatment

The two components to the treatment of OCA are skin protection and vision support. Most individuals with OCA do not tan and are susceptible to the ultraviolet rays of the sun. The use of hats with a brim, long-sleeve shirts and long pants, along with an avoidance of the sun at midday (10:00a.m. to 2:00p.m.) will often be enough, particularly in more northern climes. Sunscreen is necessary when additional protection is necessary or the above measures are impractical.

Ongoing ophthalmologic care is critical to support the normal development of an individual with albinism. This should include yearly eye examinations, full correction for myopia or hyperopia, and consultation with the appropriate local low vision program. Except for special circumstances, children with albinism should attend regular school classes and should be encouraged to participate in all aspects of school.

FUTURE

There are three exciting areas of future research in pigment biology and the eye. First, the genes responsible for all types of human albinism will be identified, allowing correct diagnosis and family counseling. Phenotypic variation of each type and correlation between genotype and phenotype, particularly the visual development, will be defined. Second, the isolation of a significant number of the genes involved in cutaneous and ocular melanogenesis, from both human and mouse, will allow the development and regulation of pigment in these tissues to be studied. For example, an analysis of the role of melanin in normal eye development will provide insight into the changes that occur in albinism, and may suggest methods for altering these changes. The mouse, as an experimental system, should play a major role in these analyses. Third, it may be possible to develop methods of gene therapy for the cutaneous and the ocular changes in albinism. This is already being discussed in several centers and the mouse models might be useful for testing gene therapy protocols.

REFERENCES

Abadi, R. V. and Dickinson, C. M. Monochromatic fundus photography of human *albinos. Arch.Ophthalmol.* **101**:1706–1711, 1983.

Adler, J. E. Histological examination of a case of albinism. *Biometrika* **7**:237–247, 1910.

Agrup, G., Hansson, C., Rorsman, H., Rosengren, A. M. and Rosengren, E. Intracellular distribution of DOPA and 5-S-cysteinylDOPA in pigment cells with minimal pigment formation. *Acta. Derm. Venereol. Suppl.* **59**:355–356, 1979.

Agrup, G., Hansson, C., Rorsman, H. and Rosengren, E. The effect of cysteine on oxidation of tyrosine, DOPA, and cysteinylDOPAs. *Arch. Dermatol. Res.* **272**:103–115, 1982.

Angelman, H. "Puppet" children: A report of three cases. *Develop. Med. Child. Neurol.* **7**:681–688, 1965.

Apkarian, P., Eckhardt, P. G. and van Schooneveld, M. J. Detection of optic pathway misrouting in the human albino neonate. *Neuropediatrics* **22**:211–215, 1990.

Barton, D. E., Kwon, B. S. and Francke, U. Human tyrosinase gene, mapped to chromosome 11 (q14->q21), defines second region of homology with mouse chromosome 7. *Genomics* **3**:17–24, 1988.

Bateson, W. The present state of knowledge of colour-heredity in mice and rats. *Proceedings of the Zoological Society of London* **2**:71–99, 1903.

Beerman, F., Ruppert, S., Hummler, E. et al. Rescue of the albino phenotype by introduction of a functional tyrosinase gene into mice. *EMBO J.* **9**:2819–2826, 1990.

Benezra, D., Mengitsu, F., Cividalli, G., Weizman, Z., Merin, S. and Auerbach, E. Chediak-Higashi syndrome: ocular features. *J. Ped. Ophthal. Strabismus* **17**:68–75, 1980.

Bergen, A. A. B., Samanns, C., van Dorp, D. B., Ferguson-Smith, M. A., Gal, A. and Bleeker-Wagemakers, E. M. Localization of the X-linked ocular albinism gene (OA1) between DXS278/DXS237 and DXS143/DXS16 by linkage analysis. *Ophthalmic Paediatr. Genet.* **11**:165–170, 1990.

Bergen, A. A. B., Samanns, C., Schuurman, E. J. M., et al. Multipoint linkage analysis in X-linked ocular albinism of the Nettleship-Falls type. *Hum. Genet.* **88**:162–166, 1991.

Blume, R. S. and Wolff, S. M. The Chediak-Higashi syndrome: studies in four patients and a review of the literature. *Medicine* **51**:247–280, 1972.

Bray, G. A., Dahms, W. T., Swerdloff, R. S., Fiser, R. H., Atkinson, R. L. and Carrel, R. E. The Prader-Willi syndrome: a study of 40 patients and a review of the literature. *Medicine* **62**:59–80, 1983.

Brilliant, M. H. The mouse pink-eyed dilution locus: a model for aspects of Prader-Willi syndrome, Angelman syndrome and a form of hypomelanosis of Ito. Mammal.Genome **3**:187–191, 1992.

Butler, M. G., Meaney, F. J. and Palmer, C. G. Clinical and cytogenetic survey of 39 individuals with Prader-Labhart-Willi syndrome. *Am. J. Med .Genet.* **23**:793–809, 1986.

Butler, M. G. Hypopigmentation: a common feature of Prader-Labhart-Willi syndrome. *Amer. J. Hum. Gen.* **45**:140–146, 1989.

Cabanes, J., García-Cánovas, F., Lozano, J. A. and García-Carmona, F. A kinetic study of the melanization pathway between l-tyrosine and DOPAchrome. 923:187–195, 1987.

Cassidy, S. B. Prader-Willi syndrome. *Curr. Prob. Pediat.* **14**:5–55, 1984.

Castronuovo, S., Simon, J. W., Kandel, G. L., et al. Variable expression of albinism within a single kindred. *Amer. J. Ophthalmol.* **111**:419–426, 1991.

Chakraborty, A. K., Orlow, S. J. and Pawelek, J. M. Evidence that DOPAchrome tautomerase is a ferrous iron-binding glycoprotein. *FEBS Lett.* **302**:126–128, 1992.

Charles, S. J., Moore, A. T., Grant, J. W. and Yates, J. R. W. Genetic counselling in X-linked ocular albinism. Clinical features of the carrier state. *Eye* **6**:75–79, 1992.

Cheong, P. Y. Y., Bateman, J. B. and King, R. A. Oculocutaneous albinism: variable expressivity of nystagmus in a sibship. *J. Ped. Ophthal. Strabismus* 1992 (in press).

Chintamaneni, C. D., Ramsay, M., Colman, M. A., Fox, M. F., Pickard, R. T. and Kwon, B. S. Mapping the human CAS2 gene, the homologue of the mouse brown (b) locus, to human chromosome 9p22-pter. *Biochem. Biophys. Res. Commun.* **178**:227–235, 1991.

Cohen, T., Muller, R. M., Tomita, Y. and Shibahara, S. Nucleotide sequence of the cDNA encoding human tyrosinase-related protein. *Nucleic Acids Res.* **18**:2807–2808, 1990.

Coleman, D. L. Effect of genic substitution on the incorporation of tyrosine into the melanin of mouse skin. *Arch. Biochem. Biophys.* **69**:562–568, 1962.

Collewijn, H., Apkarian, P. and Spekreijse, H. The oculomotor behaviour of human *albinos. Brain* **108**:1–28, 1985.

Creel, D., O'Donnell, F. E. and Witkop, C. J. Visual system anomalies in human ocular albinos. *Science* **201**:931–933, 1978.

Creel, D. J., Witkop, C. J. and King, R. A. Asymmetric visually evoked potentials in human albinos: evidence for visual system anomalies. *Invest. Ophthal.* **13**:430–440, 1974.

Creel, D. J., Spekreijse, H. and Reits, D. Evoked potentials in albinos: efficacy of pattern stimuli in detecting misrouted optic fibers. *Electroenceph. Clin. Neurophys.* **52**:595–603, 1981.

Creel, D. J., Bendel, C. M., Wiesner, G. L., Wirtschafter, J. D., Arthur, D. C. and King, R. A. Abnormalities of the central visual pathways in Prader-Willi Syndrome associate with hypopigmentation. *N. Engl. J. Med.* **314**:1606–1609, 1986.

Creel, D. J., Summers, C. G. and King, R. A. Visual anomalies associated with albinism. *Ophthalmic Paediatr. Genet.* **11**:193–200, 1990.

Davis, C. G. The many faces of epidermal growth factor repeats. *New Biologist* **2**:410–419, 1990.

Edmunds, R. T. Vision of albinos. *Arch. Ophthalmol.* **42**:755–767, 1949.

Fialkow, P. J., Giblett, E. R. and Motulsky, A. G. Measurable linkage between ocular albinism and Xg. *Am. J. Hum. Genet.* **19**:63–69, 1967.

Fonda, G. Characteristics and low-vision corrections in albinism. *Arch. Ophthalmol.* **68**:754–761, 1962.

Freed, W. J., Adinolfi, A. M., Laskin, J. D. and Geller, H. M. Transplantation of B16/C3 melanoma cells into the brains of rats and mice. *Brain. Res.* **485**:349–362, 1989.

Fryburg, J. S., Breg, W. R. and Lindgren, V. Diagnosis of Angelman syndrome in infants. *Amer. J. Hum. Gen.* **38**:58–64, 1991.

Gardner, J. M., Nakatsu, Y., Gondo, Y., et al. The mouse pink-eyed dilution gene: association with human Prader-Willi and Angelman syndromes. *Science.* **257**:1121–1124, 1992.

Garrod, A. E. Inborn Errors of Metabolism, Lecture II. *Lancet.* **2**:70–73, 1908.

Giebel, L. B. and Spritz, R. A. RFLP for MboI in the human tyrosinase (TYR) gene detected by PCR. *Nucleic Acids Res.* **18**:3103, 1990.

Giebel, L. B., Strunk, K. M., King, R. A., Hanifin, J. M. and Spritz, R. A. A frequent tyrosinase gene mutation in classic, tyrosinase-negative (Type IA) oculocutaneous albinism. *Proc. Natl. Acad. Sci. USA* **87**:3255–3258, 1990.

Giebel, L. B., Musarella, M. A. and Spritz, R. A. A nonsense mutation in the tyrosinase gene of Afghan patients with tyrosinase negative (type IA) oculocutaneous albinism. *J. Med. Genet.* **28**:464–467, 1991a.

Giebel, L. B., Strunk, K. M. and Spritz, R. A. Organization and nucleotide sequences of the human tyrosinase gene and a truncated tyrosinase-related segment. *Genomics* **9**:435–445, 1991b.

Giebel, L. B., Tripathi, R. K., King, R. A. and Spritz, R. A. A tyrosinase gene missense mutation in temperature-sensitive type I oculocutaneous albinism. *J. Clin. Invest.* **87**:1119–1122, 1991c.

Giebel, L. B., Tripathi, R. K., Strunk, K. M., et al. Tyrosinase gene mutations associated with type IB ("yellow") oculocutaneous albinism. *Amer. J. Hum. Gen.* **48**:1159–1167, 1991d.

Gondo, Y., Gardner, J., Nakatsu, Y. et al. High-frequency genetic reversion mediated by a DNA duplication: The mouse pink-eyed unstable mutation. *Proc. Natl. Acad. Sci. USA* **90**:297–301. 1993.

Green, M. C. in *Genetic Variants and Strains of the Laboratory Mouse* (Oxford University Press, Oxford). Pp. 12–403, 1989.

Guillery, R. W. Visual pathways in albinos. *Scientific American* **230**:44–54, 1974.

Guillery, R. W., Okoro, A. N. and Witkop, C. J. Abnormal visual pathways in the brain of a human albino. *Brain Research* **96**:373–377, 1975.

Halaban, R. and Moellmann, G. E. Murine and human b locus pigmentation genes encode a glycoprotein (gp75) with catalase activity. *Proc. Natl. Acad. Sci.USA* **87**:4809–4813, 1990.

Hansson, C., Rorsman, H. and Rosengren, E. 5-S-cysteinylDOPA as a substrate for tyrosinase. *Acta Derm. Venereol. Suppl.* **60**:399–402, 1980.

Harmon, K. R., Witkop, C. J., White, J. G., et al. Pathogenesis of pulmonary fibrosis – PDGF precedes structural alterations in the Hermansky-Pudlak syndrome. *J. Lab. Clin. Med.* **123**:617–627, 1994.

Hearing, V. J., Ekel, T. M., Montague, P. M., Hearing, E. D. and Nicholson, J. M. Mammalian tyrosinase: isolation by a simple new procedure and characterization of its steric requirements for cofactor activity. *Arch. Biochem. Biophys.* **185**:407–418, 1978.

Hearing, V. J., Körner, A. M. and Pawelek, J. M. Newregulators of melanogenesis are associated with purified tyrosinase isozymes. *J. Invest. Dermatol.* **79**:16–18, 1982.

Heim, R. A., Dunn, D. S., Candy, S. E., Zwane, E., Kromberg, J. G. and Jenkins, T. The tyrosinase-positive oculocutaneous albinism locus is not linked to the β-globin locus in man. *Hum.Genet.* **79**:89, 1988.

Hermansky, F. and Pudlak, P. Albinism associated with hemorrhagic diathesis and unusual pigmented reticular cells in the bone marrow: Report of two cases with histochemical studies. *Blood* **14**:162–169, 1959.

Hittner, H. M., King, R. A., Riccardi, V. M., et al. Oculocutaneous albinoidism as a manifestation of reduced neural crest derivatives in the Prader-Willi syndrome. *Amer. J. Ophthalmol.* **94**:328–337, 1982.

Hornabrook, R. W., McDonald, W. I. and Carroll, R. L. Congenital nystagmus among the redskins of the highlands of Papua New Guinea. *Br. J. Ophthal.* **64**:375–380, 1980.

Hu, F., Hanifin, J. M., Prescott, G. H. and Tongue, A. C. Yellow mutant albinism: cytochemical, ultrastructural, and genetic characterization suggesting multiple allelism. *Amer. J. Hum. Gen.* **32**:387–395, 1980.

Huber, M., Hintermann, G. and Lerch, K. Primary structure of tyrosinase from Streptomyces glaucescens. *Biochemistry* **24**:6038–6044, 1985.

Ito, S. and Wakamatsu, K. Melanin chemistry and melanin precursors in melanoma. *J. Invest. Dermatol.* **92**:261S–265S, 1989.

Jackman, M. P., Huber, M., Hajnal, A. and Lerch, K. Stabilization of the oxy form of tyrosinase by a single conservative amino acid substitution. *Biochem. J.* **282**:915–918, 1992.

Jackson, I. J. A cDNA encoding tyrosinase-related protein maps to the brown locus in mice. *Proc. Natl. Acad. Sci. USA* **85**:4392–4396, 1988.

Jackson, I. J., Chambers, D. M., Rinchik, E. M. and Bennett, D. C. Characterization of TRP-1 mRNA levels in dominant and recessive mutations at the mouse brown (b) locus. *Genetics* **126**:451–460, 1990.

Jackson, I. J. and Bennett, D. C. Molecular characterisation of the mouse *albino* mutation of tyrosinase and an in vitro revertant. *Proc. Natl. Acad. Sci. USA* **87**:7010–7014, 1990.

Jackson, I. J., Chamber, D. M., Budd, P. S. and Johnson, R. The tyrosinase-related protein-1 gene has a structure and promoter sequence very different from tyrosinase. *Nucleic Acids Res.* **14**:3799–3804, 1991.

Jackson, I. J., Chambers, D. M., Tsukamoto, K. et al. A second tyrosinase-related protein, TRP-2, maps to and is mutated at the mouse *slaty* locus. *EMBO J.* **11**:527–535, 1992.

Jaeger, C. and Jay B. X-linked ocular albinism. A family containing a manifesting heterozygote and an affected male married to a female with autosomal recessive ocular albinism. *Hum. Genet.* **56**:299–304, 1981.

Jara, J. R., Aroca, P., Solano, F., Martínez-Liarte, J. H. and Lozano, J. A. The role of sulfhydryl compounds in mammalian melanogenesis: the effect of cysteine and glutathione upon tyrosinase and the intermediates of the pathway. **967**:296–303, 1988.

Jergil, B., Lindbladh, C., Rorsman, H. and Rosengren, E. DOPA oxidation and tyrosine oxygenation by human melanoma tyrosinase. *Acta. Derm.Venereol.Suppl.* **63**:468–475, 1983.

Jiménez, M., Tsukamoto, K. and Hearing, V. J. Tyrosinases from two different loci are expressed by normal and by transformed melanocytes. *J. Biol. Chem.* **266**:1147–1156, 1991.

Kikuchi, H., Hara, S., Ishiguro, S., Tamai, M. and Watanabe, M. Detection of point mutation in the tyrosinase gene of a Japanese albino patient by a direct sequencing of amplified DNA. *Hum.Genet.* **85**:123–124, 1990.

King, R. A. and Witkop, C. J. Hairbulb tyrosinase activity in oculocutaneous albinism. *Nature* **263**:69–71, 1976.

King, R. A. and Witkop, C. J. Detection of heterozygotes for tyrosinase negative oculocutaneous albinism by hairbulb tyrosinase assay. *Amer. J. Hum. Gen.* **29**:164–168, 1977.

King, R. A., Olds, D. P. and Witkop, C. J. Characterization of human hairbulb tyrosinase: properties of normal and albino enzyme. *J. Invest. Dermatol.* **71**:136–139, 1978.

King, R. A., Creel, D. J., Cervenka, J., Okoro, A. N. and Witkop, C. J. Albinism in Nigeria with delineation of new recessive oculocutaneous type. *Clin.Genet.* **17**:259–270, 1980.

King, R. A., Lewis, R. A., Townsend, D., Zelickson, A., Olds, D. P. and Brumbaugh, J. A. Brown oculocutaneous albinism. Clinical, ophthalmological, and biochemical characterization. *Ophthalmology* **92**:1496–1505, 1985.

King, R. A. and Olds, D. P. Hairbulb tyrosinase activity in oculocutaneous albinism: suggestions for pathway control and block location. *Amer. J. Med. Gen.* **20**:49–55, 1985.

King, R. A., Wirtschafter, J. D., Olds, D. P. and Brumbaugh, J. A. Minimal pigment: a new type of oculocutaneous albinism. *Clin.Genet.* **29**:42–50, 1986.

King, R. A., Olds, D. P. and Townsend, D. Mechanisms of hypopigmentation in human oculocutaneous albinism. In: *Advances in Pigment Cell Biology*, edited by Bagnara, J. T. New York: Alan R. Liss, 1988, pp. 183–191.

King, R. A., Mentink, M. M. and Oetting, W. S. Non-random distribution of missense mutations within the human tyrosinase gene in Type 1 (tyrosinase-related) oculocutaneous albinism. *Mol. Biol. Med.* **8**:19–29, 1991a.

King, R. A., Townsend, D., Oetting, W. S., et al. Temperature-sensitive tyrosinase associated with peripheral pigmentation in oculocutaneous albinism. *J. Clin. Invest.* **87**:1046–1053, 1991b.

King, R. A. and Oetting, W. S. Molecular basis of type IA(tyrosinase negative) oculocutaneous albinism. *Pigment Cell Res. Supplement* **2**:249–253, 1992.

King, R. A., Wiesner, G. L., Townsend, D. and White, J. G. Hypopigmentation in Angelman syndrome. *Amer. J. Hum.Gen.* **46**:40–44, 1993.

Kinoshita, Y., Sato, S. and Takeuchi, T. Cellular sialic acid level and phenotypic expression in B16 melanoma cells: comparison of spontaneous variations and bromodeoxyuridine- and theophylline-induced changes. *Cell Struct. Funct.* **14**:35–43, 1989.

Kluppel, M., Beermann, F., Ruppert, S., Schmid, E., Hummler, E. and Schütz, G. The mouse tyrosinase promoter is sufficient for expression in melanocytes and in the pigmented epithelium of the retina. *Proc. Natl. Acad. Sci. USA* **88**:3777–3781, 1991.

Kromberg, J. G., Castle, D., Zwane, E. M., et al. Red or rufous albinism in southern Africa. *Ophthalmic.Paediatr. Genet.* **11**:229–235, 1990.

Kulkarni, G. A. and Nathanson, L. Specificity of growth inhibition of melanoma by 4-hydroxyanisole. *Pigment Cell Res.* **2**:40–43, 1989.

Kwon, B. S., Halaban, R., Kim, G. S., Usack, L., Pomerantz, S. H. and Haq, A. K. A melanocyte-specific complementary DNA clone whose expression is inducible by melanotropin and isobutylmethyl xanthine. *Mol. Biol. Med.* **4**:339–355, 1987.

Kwon, B. S., Halaban, R. and Chintamaneni, C. D. Molecular basis of the mouse Himalayan mutation. *Biochem. Biophys. Res. Commun.* **161**:252–260, 1989.

Körner, A. M. and Pawelek, J. M. DOPAchrome conversion: a possible control point in melanin biosynthesis. *J. Invest. Dermatol.* **75**:192–195, 1980.

Körner, A. M. and Pawelek, J. M. Mammalian tyrosinase catalyzes three reactions in the biosynthesis of melanin. *Science* **217**:1163–1165, 1982.

LaVail, J. H., Nixon, R. A. and Sidman, R. L. Genetic control of retinal ganglion cell projections. J. Comp. *Neurol.* **182**:399–422, 1978.

Ledbetter, D. H., Mascarello, J. T., Riccardi, V. M., Harper, V. D., Airhart, S. D. and Strobel, R. J. Chromosome 15 abnormalities and the Prader-Willi syndrome: a follow-up report of 40 cases. *Am. J. Hum. Genet.* **34**:278–285, 1982.

Leonard, L. J., Townsend, D. and King, R. A. Function of DOPAchrome oxidoreductase and metal ions in DOPAchrome conversion in the eumelanin pathway. *Biochemistry* **27**:6156–6159, 1988.

Lerner, A. B. and Fitzpatrick, T. B. Biochemistry of melanin formation. *Physiol. Rev.* **30**:91–126, 1950.

Leventhal, A. G. and Creel, D. J. Retinal projections and functional architecture of cortical areas 17 and 18 in the tyrosinase-negative albino cat. *J. Neurosci.* **5**:795–807, 1985.

Logan, A. and Weatherhead, B. Pelage color cycles and hair follicle tyrosinase activity in the Siberian hamster. *J. Invest. Dermatol.* **71**:295–298, 1978.

Magenis, R. E., Toth-Fejel, S., Allen, L. J., et al. Comparison of the 15q deletions in Prader-Willi and Angelman syndromes: Specific regions, extent of deletions, parental origin, and clinical consequences. *Am. J. Med. Genet.* **35**:333–349, 1990.

Mahadeo, R., Markowitz, J., Fisher, S. and Daum, F. Hermansky-Pudlak syndrome with granulomatous colitis in children. *J. Pediatr.* **118**:904–906, 1991.

Mascari, M. J., Gottlieb, W., Rogan, P. K., et al. The frequence of uniparental disomy in Prader-Willi syndrome: Implications for molecular diagnosis. *N. Engl. J. Med.* **326**:1599–1607, 1992.

Mason, H. S. The chemistry of melanin. *J. Biol. Chem.* **172**:83-99, 1948.

McKusick, V. A. *Mendelian Inheritance in Man. Catalogue of Autosomal Dominant, Autosomal Recessive, and X-Linked Phenotypes.* , Baltimore, MD:The Johns Hopkins University Press, 1992. Ed. 10 pp. -327–68.

Miyazaki, K. and Ohtaki, N. Tyrosinase as glycoprotein. Pigment Cell **3**:113–120, 1976.

Morris, S. W., Muir, W. and StClair, D. Dinucleotide repeat polymorphism at the human tyrosinase gene. *Nucleic. Acids. Res.* **19**:6968, 1991.

Murty, V. V. V. S., Bouchard, B., Mathew, S., Vijayasaradhi, S. and Houghton, A. N. Assignment of the Human TYRP (brown) locus to chromosome region 9p23 by nonradioactive in situ hybridization. *Genomics.* **13**:227–229, 1992.

Mutirangura, A., Greenberg, F., Butler, M. G., et al. Multiplex PCR of three dinucleotide repeats in the Prader-Willi/Angelman critical region (15q11-q13): molecular diagnosis and mechanism of uniparental disomy. *Hum. Mol. Gen.* **2**:143–151,1993.

Nakatsu, Y., Gondo, Y. and Brilliant, M. H. The *p* locus is closely linked to the mouse homolog of a gene from the Prader-Willi chromosomal region. *Mammalian Genome* **2**:69–71, 1992.

Nance, W. E., Jackson, C. E. and Witkop, C. J. Amish albinism: a distinctive autosomal recessive phenotype. *Amer. J. Hum. Gen.* **22**:579–586, 1970.

Nishioka, K. Particulate tyrosinase of human melanoma. Solubilization, purification following trypsin treatment, and characterization. *Eur. J. Biochem.* **85**:137–146,1978.

Oetting, W. S., Roed, C. M., Mentink, M. M. and King, R. A. PCR detection of a Taq1 polymorphism at the CCAATT box of the human tyrosinase (TYR) gene. *Nucleic. Acids. Res.* **19**:5800, 1991.

Oetting, W. S. and King, R. A. Molecular analysis of Type I-A (tyrosinase-negative) oculocutaneous albinism. *Hum. Genet.* **90**:258–262, 1992a.

Oetting, W. S. and King, R. A. Analysis of mutations in the copper B binding region associated with Type I (tyrosinase-related) oculocutaneous albinism. *Pigment. Cell. Res.* **5**:274–278, 1992b.

Oetting, W. S. and King, R. A. Molecular basis of type 1 (tyrosinase-related) oculocutaneous albinism: mutations and polymorphisms of the human tyrosinase gene. *Hum. Mutation.* **2**:1–6, 1993.

Oetting, W. S., Witkop, C. J., Brown, S. A., et al. A frequent mutation in the tyrosinase gene associated with type I-A (tyrosinase-negative) oculocutaneous albinism in Puerto Rico. *Amer. J. Hum. Gen.* **52**:17–23, 1993b.

Oetting, W. S., Fryer, J. P. and King, R. A. A dinucleotide deletion (-ΔGA115) in the tyrosinase gene responsible for type I-A (tyrosinase negative) oculocutaneous albinism in a Pakistani individual. *Hum. Mol. Gen.* **2**:1047–1048, 1993a.

Okoro, A. N. Albinism in Nigeria: a clinical and social study. *Br J Dermat* **92**:485–492, 1975.

Orlow, S. J., Boissy, R. E., Moran, D. J. and Pifko-Hirst, S. Subcellular distribution of tyrosinase and tyrosinase-related protein-1: Implications for melanosomal biogenesis. *J. Invest. Dermatol.* **100**:55–64, 1993.

Pawelek, J. M., Körner, A. M., Bergstrom, A. and Bolognia, J. New regulators of melanin biosynthesis and the autodestruction of melanoma cells. *Nature.* **286**:617–619, 1980.

Pawelek, J. M. A new enzyme in the pigment pathway. *J. Invest. Dermatol.* **84**:234, 1985.

Pawelek, J. M. DOPAchromeconversion factor functions as an isomerase. *Biochem. Biophys. Res.Commun.* **166**:1328–1333, 1990.

Pawelek, J. M. After DOPAchrome? Pigment Cell Res. 4:53-62, 1991.

Pearce, W. G., Sanger, R. and Race, R. R. Ocular albinism and *Xg. Lancet* **1**:1282–1283, 1968.

Pearson, K., Nettleship, E. and Usher, C. H. *A monograph on albinism in man: Drapers' company research memoirs, biometric series VI,* London:Department of Applied Mathematics, Dulau and Co., Limited, 1911, pp. 1–25.

Pomerantz, S. H. and Warner, M. C. 3,4-Dihydroxyphenylalanine as the tyrosinase cofactor. *J. Biol.Chem.* **242**:5308–5314, 1967.

Porter, S., Larue, L. and Mintz, B. Mosaicism of tyrosinase-locus transcription and chromatin structure in dark vs light clones of homozygous *chinchilla mottled* mice. *Dev. Genet.* **12**:393–402. 1991.

Prota, G. and Thomson, R. H. Melanin pigmentation in mammals. *Endeavor* **35**:32–38, 1976.

Prota, G. Recent advances in the chemistry of melanogenesis in mammals. *J. Invest. Dermatol.* **75**:122–127, 1980.

Prota, G. Pigment cell metabolism: chemical and enzymatic control. In: *Cutaneous Melanoma,* edited by Veronesi, U. and Cascinelli, N. NewYork: Academic Press, 1986, pp. 233–241.

Prota, G. Progress in the chemistry of melanins and related metabolites. *Med. Res. Rev.* **8**:525–556, 1988a.

Prota, G. Some new aspects of eumelanin chemistry. *Prog. Clin. Biol. Res.* **256**:101–124, 1988b.

Ramsay, M., Colman, M. A., Stevens, G., et al. The tyrosinase-positive oculocutaneous albinism locus maps to chromosome 15q11.2-q12. *Amer. J. Hum. Gen.* **51**:879–884, 1992.

Raper, H. S. The aerobic oxidases. *Physiol Rev.* **8**:245–248, 1928.

Rinchik, E. M., Bultman, S. J., Horsthemke, B., et al. A gene for the mouse pink-eyed dilution locus and for human type II oculocutaneous albinism. *Nature* **361**:72–76, 1993b.

Rinchik, E. M., Stoye, J. P., Frankel, W. N., Coffin, J., Kwom, B. S. and Russell, L. B. Molecular analysis of viable spontaneous and radiation-induced albino(c)-locus mutation in the mouse. *Mutation Res.* **286**:199–207, 1993a.

Riordan, J. F. Arginyl residues and anion binding sites in proteins. *Mol Cell Biochem* **26**:71–92, 1979.

Rorsman, H., Rosengren, A. M. and Rosengren, E. Determination of 5-S-cysteinlydopa in melanoma by a fluorimetric method. *Yale J. Biol. Med.* **46**:516–522, 1973.

Rose, N. C., Menacker, S. J., Schnur, R. E., et al. Ocular albinism in a male with del (6) (q13 - q15): candidate region for autosomal recessive ocular albinism? *Am. J. Med. Genet.* **42**:700–705, 1992.

Ross, P. K. and Solomon, E. I. An electronic structural comparison of copper-peroxide complexes of relevance to hemocyanin and tyrosinase active sites. *J. Am. Chem. Soc.* **113**:3246–3259, 1991.

Ruppert, S., Muller, G., Kwon, B and Schutz, G. Multiple transcripts of the mouse tyrosinase gene are generated by alternative splicing. *EMBO J.* **7**:2715–2722, 1988.

Schedl, A., Beermann, F., Thies, E., Montoliu, L., Kelsey, G. and Schütz, G. Transgenic mice generated by pronuclear injection of a yeast artificial chromosome. *Nucleic Acids Res.* **20**:3073–3077, 1992.

Schedl, A., Montollu, L., Kelsey, G. and Schütz, G. A yeast artificial chromosome covering the tyrosinase gene confers copynumber-dependent expression in transgenic mice. *Nature* **362**: 258–261, 1993.

Schinella, R. A., Greco, M. A., Cobert, B. L., Denmark, L. W. and Cox, R. P. Hermansky-Pudlak syndrome with granulomatous colitis. *Ann. Int. Med.* **92**:20–23, 1980.

Schnur, R. E., Nussbaum, R. L., Anson-Cartwright, L., McDowell, C., Worton, R. G. and Musarella, M. A. Linkage analysis in X-linked ocular albinism. *Genomics* **9**:605–613, 1991.

Searle , A. G. Comparative genetics of albinism. *Opthalmic Paed.* **11**:159–164, 1990.

Shanahan, F., Randolph, L., King, R., et al. The Hermansky-Pudlak syndrome: An immunological assessment of 15 cases. *Am. J. Med.* **85**:823–828, 1988.

Shibahara, S., Tomita, Y., Sakakura, T., Nager, C., Chaudhuri, B. and Muller, R. Cloning and expression of cDNA encoding mouse tyrosinase. *Nucleic Acids Res.* **14**:2413–2427, 1986.

Shibahara, S., Okinaga, S., Tomita, Y., et al. A point mutation in the tyrosinase gene of BALB/c albino mouse causing the cysteine to serine substitution at position 85. *Eur. J. Biochem.* **189**:455–461, 1990.

Shibahara, S. Mutations of the tyrosinase gene in oculocutaneous albinism. *Pigment Cell Research* **5**:279–283, 1992.

Sidman, R. L. and Perlstein, R. Pink-eyed dilution (p) gene in rodents: increased pigmentation in tissue culture. *Dev. Biol.* **12**:93–116, 1965.

Silvers, W. K. *The Coat Colors of Mice*, Basel: Springer-Verlag, 1979, pp. 1–380.

Simon, J. W., Adams, R. J., Calhoun, J. H., Shapiro, S. S. and Ingerman, C. M. Ophthalmic manifestations of the Hermansky-Pudlak syndrome (oculocutaneous albinism and hemorrhagic diathesis). *Am. J. Ophthal.* **93**:71–77, 1982.

Smith, C. J., O'Hare, K. B. and Allen, J. C. Selective cytotoxicity of hydroquinone for melanocyte-derived cells is mediated by tyrosinase activity but independent of melanin content. *Pigment Cell Res.* **1**:386–389, 1988.

Spritz, R. A., Strunk, K. M., Oetting, W. S. and King, R. A. RFLP for TaqI at the human tyrosinase locus. *Nucleic Acids Res.* **16**:9890, 1988.

Spritz, R. A. and Strunk, K. M. RFLP for BglII at the human tyrosinase(TYR) locus. *Nucleic Acids Res.* **18**:3672, 1990.

Spritz, R. A., Strunk, K. M., Hsieh, C. L., Sekhon, G. S. and Francke, U. Homozygous tyrosinase gene mutation in an American black with tyrosinase-negative (type IA) oculocutaneous albinism. *Amer. J. Hum. Gen.* **48**:318–324, 1991.

Stannus, H. S. Anomalies of pigmentation among natives of Nyasaland: a contribution to the study of albinism. *Biometrika* **9**:333–365, 1913.

Summers, C. G., Knobloch, W. H., Witkop, C. J. and King, R. A. Hermansky-Pudlak syndrome: ophthalmic findings. *Ophthalmology* **95**:545–554, 1988.

Summers, C. G., Creel, D. J., Townsend, D. and King, R. A. Variable expression of vision in sibs with albinism. *Am. J. Med. Genet.* **40**:327–331, 1991.

Tai, T., Eisinger, M., Ogata, S. and Lloyd, K. O. Glycoproteins as differentiation markers in human malignant melanoma and melanocytes. *Cancer Res.* **43**:2773–2779, 1983.

Takeda, A., Matsunaga, J., Tomita, Y., Tagami, H. and Shibahara, S. Molecular analysis of the DNA segments cross-hybridizable to the tyrosinase gene in patients affected with oculocutaneous albinism. *Tohoku J. Exp. Med.* **159**:333–340, 1989.

Takeda, A., Tomita, Y., Matsunaga, J., Tagami, H. and Shibahara, S. Molecular basis of tyrosinase-negative oculocutaneous albinism. A single base mutation in the tyrosinase gene causing arginine to glutamine substitution at position 59. *J. Biol. Chem.* **265**:17792–17797, 1990.

Takeda, A., Matsunaga, J., Tomita, Y., Tagami, H. and Shibahara, S. Nucleotide sequence of the putative human tyrosinase pseudogene. *Tohoku J. Exp. Med.* **163**:295–297, 1991.

Tanaka, A. Chediak-Higashi syndrome: abnormal lysosomal enzyme levels in granulocytes of patients and family members. *Pediatr. Res.* **14**:901–904, 1980.

Taylor, W. O. G. Visual disabilities of oculocutaneous albinism and their alleviation. *Trans. Ophthal. Soc. UK* **98**:423–445, 1978.

Thomson, T. M., Mattes, M. J., Roux, L., Old, L. J. and Lloyd, K. O. Pigmentation-associated glycoprotein of human melanoma and melanocytes: definition with a mouse monoclonal antibody. *J. Invest. Dermatol.* **85**:169–174, 1985.

Tomita, Y., Takeda, A., Okinaga, S., Tagami, H. and Shibahara, S. Human oculocutaneous albinism caused by a single base insertion in the tyrosinase gene. *Biochem. Biophys. Res. Commun.* **164**:990–996, 1989.

Tripathi, R. K., Giebel, L. B., Strunk, K. M. and Spritz, R. A. A polymorphism of the human tyrosinase gene is associated with temperature-sensitive enzymatic activity. *Gene Express.* **1**:103–110, 1991.

Tsukamoto, K., Jackson, I. J., Urabe, K., Montague, P. M. and Hearing, V. J. A second tyrosinase-related protein, TRP2, is a melanogenic enzyme termed DOPAchrome tautomerase. *EMBO J.* **11**:519–526, 1992.

Usher, C. H. Histological examination of an adult human albino's eyeball, with a note on mesoblastic pigmentation in foetal eyes. *Biometrika* **13**:46–56, 1920.

van Dorp, D. B. Albinism, or the NOACH Syndrome (The book of Enoch c.v. 1-20). *Clin. Genet.* **31**:228–242, 1987.

Vijayasaradhi, S. and Houghton, A. N. Purification of an autoantigenic 75-kDa human melanosomal glycoprotein. *Int. J. Cancer* **47**:298–303, 1991.

Volbeda, A. and Hol, W. G. Crystal structure of hexameric haemocyanin from Panulirus interruptus refined at 3.2 Å resolution. *J. Mol. Biol.* **209**:249–279, 1989a.

Volbeda, A. and Hol, W. G.J. Pseudo 2-fold symmetry in the copper-binding domain of arthropodan haemocyanins: possible implications for the evolution of oxygen transport proteins. *J. Mol. Biol.* **206**:531–546, 1989b.

Walsh, R. J. A distinctive pigment of the skin in New Guinea indigenes. *Ann. Hum. Genet.* **34**:379–385, 1971.

White, D. A,, Smith, G. J. W., Cooper, J. A. D., Glickstein, M. and Rankin, J. A. Hermansky-Pudlak syndrome and interstitial lung disease: report of a case with lavage findings. *Am. Rev. Resp. Dis.* **130**:138–141, 1984.

White, J. G. and Clawson, C. C. The Chediak-Higashi syndrome: the nature of the giant neutrophil granules and their interactions with cytoplasm and foreign particulates. *Am. J. Pathol.* **98**:151–196, 1980.

White, R. A., Peters, L. L., Adkinson, L. R., Korsgren, C., Cohen, C. M. and Lux, S. E. The murine pallid mutation is a platelet storage pool disease associated with the protein 4.2 (pallidin) gene. *Nature Genetics* **2**:80–83, 1992.

Wiesner, G. L., Bendel, C. M., Olds, D. P., et al. Hypopigmentation in the Prader-Willi syndrome. *Amer. J. Hum. Gen.* **40**:431–442, 1987.

Williams, C. A. and Frias, J. L. The Angelman ("happy puppet") syndrome. *Am. J. Med. Genet.* **11**:453–460, 1982.

Witkop, C. J., Nance, W. E., Rawls, R. F. and White, J. G. Autosomal recessive oculocutaneous albinism in man: evidence for genetic heterogeneity. *Am. J. Hum. Genet.* **22**:55–74, 1970.

Witkop, C. J., Hill, C. W., Desnick, S., et al. Ophthalmologic, biochemical, platelet and ultrastructural defects in the various types of oculocutaneous albinism. *J. Invest. Dermatol.* **60**:443–456, 1973.

Witkop, C. J., Quevedo, W. C., Fitzpatrick, T. B. and King, R. A. Albinism. In: *The Metabolic Basis of Inherited Diseases,* edited by Scriver, C. R., Beaudet, A. L., Sly, W. S. and Valle, D. New York: McGraw-Hill, 1989, pp. 2905–2947.

Wittbjer, A., Odh, G., Rosengren, A. M., Rosengren, E. and Rorsman, H. Isolation of soluble tyrosinase from human melanoma cells. *Acta. Derm. Venereol. Suppl.* **70**:291–294, 1990.

Yokoyama, T., Silversides, D. W., Waymire, K. G., Kwon, B. S., Takeuchi, T. and Overbeek, P. A. Conserved cysteine to serine mutation in tyrosinase is responsible for the classical albino mutation in laboratory mice. *Nucleic Acids Res.* **18**:7293–7298, 1990.

Zdarsky, E., Favor, J. and Jackson, I. J. The molecular basis of brown, an old mouse mutation, and of a revertant to wild-type. *Genetics* **126**:443–449, 1990.

Zhou, B.-K., Boissy, R. E., Pifko-Hirst, S., Moran, D. J. and Orlow, S. J. Lysosome-associated membrane protein-1 (LAMP-1) is the melanocyte vesicular membrane glycoprotein band II. *J. Invest. Dermatol.* **100**:110–114, 1993.

5. INHERITED RETINAL DEGENERATIONS IN THE MOUSE

DEBORA B. FARBER [1] and MICHAEL DANCIGER [1,2]

Jules Stein Eye Institute and Department of Ophthalmology, UCLA School of Medicine, Los Angeles, California 90024[1]; and Loyola-Marymount University, Los Angeles, California 90045, USA[2]

Until very recently, none of the genes responsible for the several inherited murine retinal degenerations[1] had been identified. However, in 1989 and 1990 the defective genes responsible for two of these diseases were isolated and characterized. In each case, a candidate gene was cloned by subtractive and differential hybridization and mapped to a particular mouse chromosome by somatic cell hybrid analysis. Both candidate genes were localized to the murine chromosome known to contain the respective disease locus. One of the genes identified is the *retinal degeneration* or *rd,* which causes a rapidly developing, early retinal degeneration mapping to mouse chromosome 5 (Sidman and Green, 1965) and encodes the β subunit of the rod-specific enzyme cyclic GMP-phosphodiesterase (Bowes et al., 1989; 1990). The other gene isolated is responsible for the *retinal degeneration slow* or *rds* mutation, which maps to mouse chromosome 17 (Demant et al., 1979; Travis et al., 1989). The *rds* gene encodes a rod photoreceptor outer segment membrane protein known as rds/peripherin (Connell et al., 1991). Further, the abortive development of the rod outer segments in *rds* is understandable in terms of a defective disc membrane protein. Similarly, the degenerative process in *rd* mouse photoreceptors closely correlates with the finding of increased levels of cyclic GMP due to a defective cyclic GMP-phosphodiesterase.

In this chapter, the studies that led to the identification of the *rd* gene will be discussed in detail; the work involved in the identification of the *rds* gene will be summarized and the use of the mouse as a means of identifying and characterizing human disease genes will be discussed.

---

[1]The following mouse mutants show signs of retinal degeneration (their symbols and chromosomal locations, where known, are shown in parentheses: retinal degeneration (*rd*, 5; *rd*-3, 1;*Rd*-4,4), retinal degeneration slow (*rds*, 17), purkinje cell degeneration (*pcd*, 13), cribriform degeneration (*cri*, 4), nervous (*nr*, 8), vitiligo (*vit*, 6) (Green, 1989; Heckenlively et al., 1992; Chang et al., 1993).

THE *rd* MOUSE

Retinal degeneration in the mouse (*rd* mutation) was first described in wild animals caught in different areas of Europe and is present in several inbred laboratory strains (Sidman and Green, 1965). Over the years, these mice have been the object of many genetic, morphological, electrophysiological, biochemical, immunological, molecular biological and behavioral studies.

The *rd* disease is inherited in an autosomal recessive fashion and results in the rapid and selective degeneration of the retinal photoreceptor cells (Sidman and Green, 1965; LaVail and Sidman, 1974). The morphology of the normal mouse retina is shown in Figure 1. The *rd* visual cells appear to form normally during the prenatal and early postnatal period and to achieve some degree of differentiation, as indicated by the development of rudimentary rod outer segments and a synaptic ribbon complex (LaVail and Sidman, 1974; Blanks et al., 1974). The morphology of a rod photoreceptor is shown in Figure 2. By 10 days of age, the photoreceptors of both the *rd/rd* mouse and the *rd* /+ heterozygote (which has a morphologically normal retina) look alike at the light microscope level, but by day 20, virtually all the *rd/rd* rod visual cells have degenerated (Noell, 1965). The single row of nuclei left from the outer nuclear layer at this time in the *rd* retina corresponds to cone nuclei (Carter-Dawson et al., 1978). Most of these nuclei disappear in the following few months (Figure 3). At the electron microscope level, the first signs of ultrastructural pathology in the *rd* photoreceptors are observed on the 7th to 8th postnatal day as 1) fewer ribosomes and swelling of the mitochondria of the inner segments, followed by 2) slower development and disorganization of the outer segment membranes and 3) absence of triad configurations at the photoreceptor terminals (Sonohara and Shiose, 1968; Caley, Johnson and Liebelt, 1972; Sanyal and Bal, 1973; Blanks, Adinolfi and Lolley, 1974).

The retinal pigment epithelium as well as the cells of the inner retina remain morphologically and functionally normal during the course of the dissolution of the outer nuclear layer (LaVail and Sidman, 1974). However, subtle and slow changes occur in the ganglion cells and in the inner nuclear and inner plexiform layers after several months of life (Grafstein, Murray and Ingoglia, 1972).

The *rd* retina is responsive to high intensity illumination and between 12 and 18 postnatal days it displays an attenuated electroretinogram which becomes completely flat by 28 postnatal days (Nöell, 1965). The rate of visual cell degeneration is not accelerated by the level of environmental light in which the mice are reared.

Biochemical and Molecular Biological Studies

The first clues regarding the cause of the *rd* mutation came from studies of cyclic nucleotide metabolism. Cyclic GMP levels become elevated in the *rd* retina when compared to control retina by the 6th postnatal day, at least two days prior to the onset of photoreceptor cell degeneration, and they are several-fold higher than normal by day 14 (Farber and Lolley, 1974). Cyclic GMP levels decline thereafter, correlating with the loss of visual cells from the *rd* retina. Analysis of microdis-

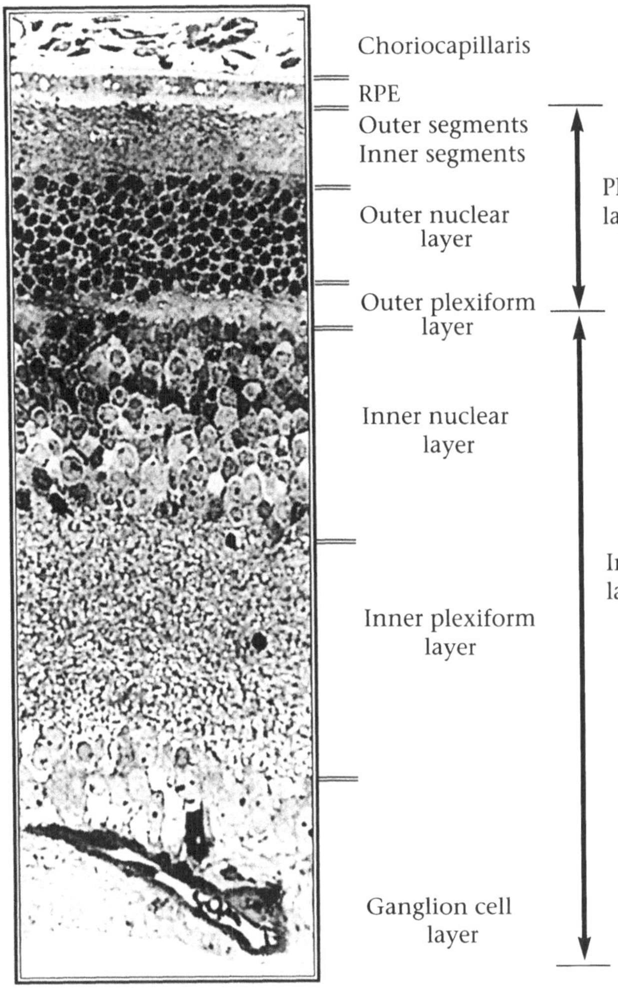

Figure 1 Light micrograph of the retina from a normal adult mouse, showing the major cell layers (courtesy of Dr. John Flannery). The retinal pigment epithelium (RPE) lies at the top and interdigitates with the outer segments of the photoreceptor cells. The nuclei of these cells are organized in several rows, forming the outer nuclear layer. Between this layer and the innermost retinal layer, formed by the ganglion cells and their axons, there is a cellular stratum, the inner nuclear layer, formed by horizontal, bipolar, interplexiform and amacrine neurons, and the glial like Muller cells. Synaptic contacts also occur in defined layers of the retina. The outer plexiform layer contains the synapses between the photoreceptors and horizontal and bipolar cells; and the inner plexiform layer is formed with the synaptic contacts between ganglion and bipolar cells, ganglion and amacrine cells, and amacrine cells interacting with each other. Interplexiform cells have processes extending into the inner plexiform and inner nuclear layers. The latter processes make synaptic contacts with horizontal and bipolar cells. Muller cells span the retina from the vitreal border through the outer nuclear layer.

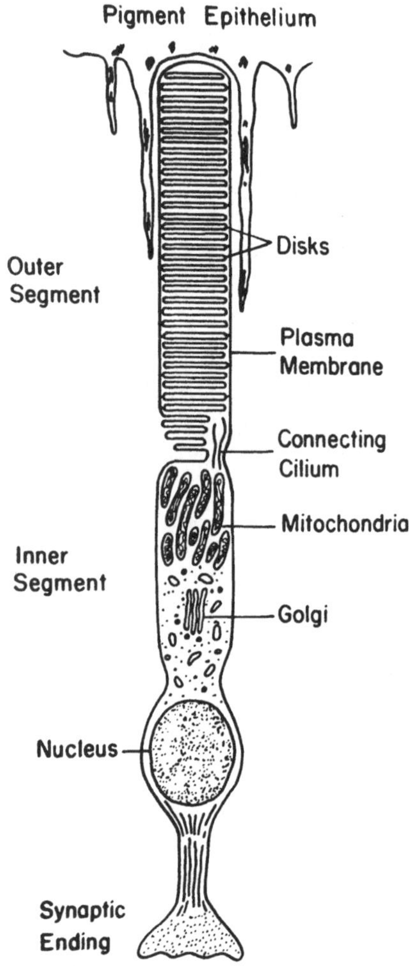

Figure 2 Diagram of a vertebrate rod photoreceptor cell. The outer segment contains stacks of membranous disks formed by lipid bilayers which provide a matrix for protein molecules. Most of these disks are not continuous with the plasma membrane of the photoreceptor although they are connected by filamentous structures. The visual pigment rhodopsin, the major protein of the rod outer segment, is an intrinsic glycoprotein which spans the disk membrane thickness crossing the membrane interface several times. A modified cilium connects the outer segment with the inner segment, where the cell synthesizes proteins, produces the energy-rich nucleoside triphosphates and assembles the disk membranes. Below the nucleus, the photoreceptor axon proceeds to the synaptic terminal, which forms synapses with the neurons of the inner retina and contains the specialized biochemical machinery needed for neural signaling.

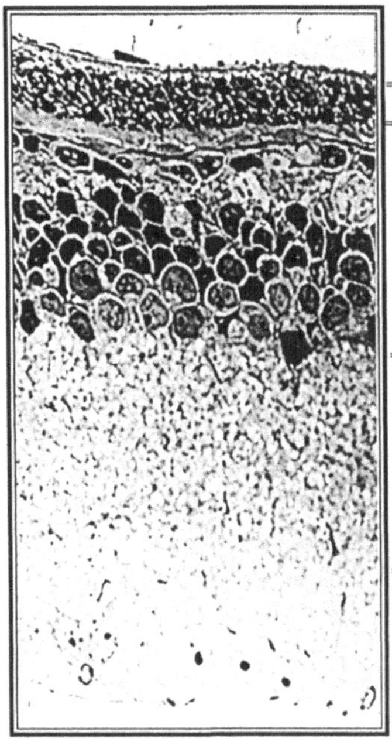

Figure 3 Light micrograph of the retina from a 30-day-old *rd* mouse (courtesy of Dr. John Flannery), showing its total lack of photoreceptor cells. As a consequence the inner nuclear layer is apposed to the retinal pigment epithelium (RPE).

sected samples consisting of the photoreceptor layer and the inner layers of the *rd* retina showed that cyclic GMP accumulates exclusively in the photoreceptor cells (Farber and Lolley, 1974). This elevation in cyclic GMP levels is the result of a deficiency in the activity of the rod photoreceptor cyclic GMP-phosphodiesterase, the enzyme that hydrolyzes cyclic GMP to 5'-GMP (Farber and Lolley, 1976). Cyclic GMP is synthesized normally by the *rd* visual cells via the guanylate cyclase reaction from the onset of differentiation (Farber and Lolley, 1976).

Cyclic GMP-phosphodiesterase activity could never be detected when it was assayed in homogenates of freshly dissected retinas obtained from *rd* mice of any age. But if kinetic studies were carried out using freeze-dried retinas of 12-day-old *rd* mice, the enzyme was found to have a Michaelis constant (Km) similar to that of photoreceptor cells of control retina, although its maximum velocity (Vmax) was well below normal (Farber and Lolley, 1977). The abnormality in cyclic nucleotide hydrolysis seems to be restricted to the retinal photoreceptors. Phosphodiesterases in other tissues of the *rd* mouse such as muscle, skin, brain and several blood components – plasma, serum, lymphocytes and erythrocytes – have activities comparable to those present in the corresponding tissues of control mice (Lolley and Farber, 1980).

Rod-specific cyclic GMP-phosphodiesterase is composed of four subunits. The two larger subunits (α: 88 kDa and β: 84 kDa) have catalytic activity, and the two identical, smaller γ subunits (11 kDa) have inhibitory activity (Baehr, Devlin and Applebury, 1979; Fung, Young, Yamane and Griswold-Prenner, 1990). This enzyme plays a key role in the phototransduction process of rod visual cells, a cascade of events which occurs in response to light. The cascade begins with the isomerization of the visual pigment rhodopsin (from 11-*cis* retinal to all-*trans* retinal), and ends with the degradation of cyclic GMP and the hyperpolarization of the rod photoreceptor membrane, along with the inhibition of neurotransmitter release at the synaptic terminal of the rod cell. The activation of cyclic GMP-phosphodiesterase by light is mediated by rhodopsin, GTP and the heterotrimeric protein transducin and its deactivation requires rhodopsin kinase, ATP and arrestin (also referred to as S-antigen or the 48-kDa protein).

Rod cyclic GMP-phosphodiesterase, which has been localized within the photoreceptor outer segments (Robb, 1974), can also be activated in the *rd* mutant by the phototransduction proteins. However, this activation is minimal and it occurs only during the postnatal period when the photoreceptor cells have morphologically intact outer segments. This is indicated by a small reduction in cyclic GMP content during light adaptation (Farber and Lolley, 1977).

Based on all of the above information, it was inferred either that 1) the *rd* retina had to have an abnormality in the cyclic GMP-phosphodiesterase molecule itself or 2) any of the proteins involved in its activation or deactivation could be dysfunctional (Farber, 1989). Other alternatives to consider were: 3) the *rd* enzyme could have an excess of γ subunit, in which case its activity could be greatly inhibited; 4) the biosynthesis of the *rd* enzyme could start later in *rd* than in normal mouse retina or it could be halted during the *rd* mouse early postnatal life; 5) there could be a greater degradation of cyclic GMP-phosphodiesterase in *rd* than in normal retina prior to the degeneration of the *rd* visual cells. Many different types of investigations were performed to clarify these possibilities.

To determine whether the γ subunit of cyclic GMP-phosphodiesterase had the correct amino acid sequence and was present in the right concentration in the *rd* retina, the cDNA encoding this subunit was first isolated and characterized from normal mouse retina (Tuteja and Farber, 1988). This cDNA was then used as a probe for the detection of the corresponding mRNAs from normal and *rd* retinas. The mRNAs in both types of retinas were found to have the same size (approximately 900 bp) and concentration (Tuteja, Tuteja and Farber, 1989). Furthermore, cloning and sequencing of the cDNA for the γ subunit of the *rd* enzyme confirmed that the deduced amino acid sequences of the *rd* and normal mouse retina γ subunits are identical (Tuteja, Tuteja and Farber, 1989). Thus, the γ subunit of cyclic GMP-phosphodiesterase is not the problem in the *rd* mouse disease.

Studies on the synthesis of cyclic GMP-phosphodiesterase indicated that it starts at the same time (postnatal day 6) both in normal and *rd* tissues and that the same amount of newly synthesized polypeptides – recognized by affinity purified antibodies to the combined α and β subunits of the enzyme – is made in both cases during all the developmental period in which the photoreceptor cells are viable (Farber et al., 1988). However, measurements of cyclic GMP-phosphodiesterase levels by radioimmunoassay showed that the concentration of the enzyme in the *rd* retina is already lower than normal at 6 postnatal days, when no signs of degeneration are discernible, and that it remains at about half the normal level until the photoreceptors begin to die (Farber et al., 1988). These results suggested that *rd* cyclic GMP-phosphodiesterase is abnormally labile and that it gets degraded very soon after it is made. In agreement with this idea is the finding that peptides from immature *rd* retinas that are recognized by antibodies against cyclic GMP-phosphodiesterase sediment on sucrose gradients more slowly than the corresponding proteins from normal mouse retina. These proteins migrate within the gradient with an apparent molecular mass of 105 kDa, whereas the proteins from the normal retina form a complex with an apparent molecular mass of 170 kDa. The 170 kDa complex is never detected in extracts from *rd* retina (Lee et al., 1988). Thus, the subunits of cyclic GMP-phosphodiesterase do not assemble into a functional complex in *rd* photoreceptors. This would make them susceptible to proteolysis once they are synthesized.

During the course of the studies focused on determining whether the proteins involved in the activation/deactivation of cyclic GMP-phosphodiesterase in the *rd* retina were functional, a defect in the process of dephosphorylation of *rd* rhodopsin was identified. Although the concentration of the visual pigment is comparable in normal and *rd* photoreceptors until 10–11 postnatal days (Caravaggio and Bonting, 1963; Shuster and Farber, 1986), *rd* rhodopsin cannot be phosphorylated *in vitro* at any age (Shuster and Farber, 1986). This lesion is not caused by an abnormality in rhodopsin kinase (the enzyme which catalyzes the transfer of phosphate from ATP to rhodopsin), since the activity of the rhodopsin kinases from normal and *rd* retinas is comparable during all the developmental period in which the *rd* photoreceptors are viable (Palczewski, Farber and Hargrave, 1991). Nor does the lesion result from a defective rhodopsin molecule, since at the mRNA level and in immunocytochemical reactions, the rhodopsins from normal and *rd* retinas are identical (Bowes and Farber, 1988; Bowes, van Veen and Farber, 1989). Lack of phosphorylation of *rd* rhodopsin stems from abnormal activity of protein phosphatase 2A, the enzyme that dephosphorylates rhodopsin (Palczewski, Farber and Hargrave, 1991). Protein phosphatase 2A activity is twice as high in *rd* than in normal retina between 5 and 10 postnatal days, although the *rd* enzyme seems to be structurally normal. Apparently, some regulatory component present in the *rd* retinal extract may be responsible for its increased activity. In previous studies carried out in our laboratory, rhodopsin phosphorylation in outer segments of normal rod photoreceptors was shown to be inhibited by high concentrations of cyclic GMP (Shuster and Farber, 1984). A possible explanation for this phenomenon is that raised cyclic GMP could enhance the activity of protein phosphatase 2A, which would remove the phosphate from rhodopsin as soon as the visual pigment is phosphorylated. Our findings with the *rd* retina are consistent with this hypothesis. The abnormality in the *rd* protein phosphatase 2A activity would then be secondary to the cyclic GMP-phosphodiesterase deficient activity which generates the high levels of cyclic GMP in the *rd* retina.

Further investigations on the components of the photoreceptor cell involved in the activation/deactivation of cyclic GMP-phosphodiesterase ruled out rhodopsin, the α, β and γ subunits of transducin, the α and γ subunits of cyclic GMP-phosphodiesterase, the 33-kDa protein (also referred to as phosducin) and arrestin as the cause of the *rd* disease. In the first place, the mRNAs encoding most of these proteins are qualitatively normal in the *rd* retina and are produced in the same amount and with the same onset of expression as in control retina (Bowes and Farber, 1987; Bowes, van Veen and Farber, 1988; Tuteja, Tuteja and Farber, 1989). In addition, each of the proteins translated from these mRNAs has the same time of appearance during development both in normal and *rd* retinas (Bowes, van Veen and Farber, 1988). In the second place, the genes encoding the cyclic GMP phosphodiesterase-related proteins are located on mouse chromosomes different from that to which the *rd* gene had been mapped, namely chromosome 5 (Sidman and Green, 1965). Screening of the candidate genes was carried out using somatic cell hybrid analysis and by linkage analysis of backcross progeny. The results are summarized in Table 1. These studies indicated that none of the proteins investigated is the primary lesion in the *rd* disorder.

Isolation of a Candidate cDNA for the *rd* Gene

Concomitant with the work described above, the search for the *rd* gene began by following a totally different strategic plan. This plan was based on the assumption that the mRNA encoded by the *rd* gene would be either absent or expressed in higher or lower levels in the *rd* photoreceptors compared to normal visual cells. The approach was to obtain the cDNAs present exclusively in the normal photoreceptors and then to compare these cDNAs with the same pool of cDNAs present in *rd* visual cells in retinas of 9- to 11-day-old mice. Nine to eleven day-old *rd*/+ and *rd/rd* littermates were used for the experiments, since the maximum amount of age-matched retinal tissue can be obtained at this age, before the *rd/rd* cells suffer advanced degeneration.

The adult mutant retina appeared to be the perfect starting material, since it is devoid of all the photoreceptors and is constituted solely by the inner retinal layers. Adult *rd/rd* retina cDNAs were subtracted from those of 9- to 11-day old normal, *rd*/+ mouse retina, and the photoreceptor-enriched, subtracted cDNAs were used to probe a library of the 9- to 11-day-old *rd*/+ retina. From approximately 100,000 clones tested, a pool of 588 photoreceptor-enriched cDNAs was obtained. These clones were screened by differential hybridization with single-stranded cDNAs from 9- to 11-day-old *rd*/+ and *rd/rd* retinas and from adult *rd/rd* retina. Comparison of the positive clones hybridizing to the young and the adult retinal cDNAs confirmed the photoreceptor specificity of 400 clones; the rest could be ruled out, since any clone able to hybridize to adult *rd/rd* mouse cDNAs had to belong to the inner retinal layers. Of these 400 clones, 3 clones hybridized differently with the 9- to 11-day-old *rd*/+ and *rd/rd* cDNAs. However, only one of these candi-

Table 1 Chromosomal loci for the genes of proteins involved in the visual cycle of rod photoreceptors.

Protein	*Gene symbol*	*Mouse Chromosome*	*Reference*
Rhodopsin	*Rho*	6	Elliott et al., 1990
Transducin α-subunit (also called Guanine nucleotide binding protein α-subunit, transducin-1)	*Gnat-1*	9	Danciger et al., 1989a
Tranducin β-subunit (also called Guanine nucleotide binding protein β-subunit-1 [Gβ1])	*Gnb-1*	4	Danciger et al., 1990a
Guanine nucleotide binding protein β-subunit-2 (Gβ2)	*Gnb-2*	5	Danciger et al., 1990a
Guanine nucleotide binding protein β-subunit-1 related sequence (Gβ1rs)	*Gnb-1rs1*	8	Danciger et al., 1990a
Guanine nucleotide binding protein β-subunit-3(Gβ3)	*Gnb-3*	6	Danciger et al., 1992a
cGMP- phosphodiesterase α-subunit	*Pdea*	18	Danciger et al., 1990c
cGMP-phosphodiesterase γ-subunit	*Pdeg*	11	Danciger et al., 1989c
Interphotoreceptor retinoid binding protein (IRBP)	*Rbp-3*	14	Danciger et al., 1990b
33-kDa protein (also called phosphoreceptor protein-1 or phosducin)	*Rpr-1 or Pdc*	1	Danciger et al., 1991
48-kDa protein (also called arrestin or S-antigen)	*Sag*	1	Danciger et al., 1989b
Calcium/calmodulin-dependent protein kinase II α-subunit	*Camk2a*	18	Danciger et al., 1992b

date clones for the *rd* gene, zr.408, gave a different hybridization signal on Northern blots containing the mRNA of young *rd/rd* retinas compared with the mRNA of age-matched, *rd*/+ control retinas (Bowes et al., 1989).

Characterization of the Candidate cDNA for the *rd* Gene

The zr.408 clone hybridized to a set of apparently slightly larger mRNA transcripts from *rd/rd* retina than from normal retina (3.6 vs 3.3 kb). At that time, it was thought that this difference in mRNA size could result from an altered polyadenylation site, alternative splicing or structural differences in the *rd* gene. As discussed below, the second and third possibilities have been shown to occur. In addition, differences between the *rd/rd* and +/+ mouse DNAs revealed by genomic Southern blot analysis, confirmed that structural differences indeed exist between the mutant and normal genes hybridized by zr.408. Tissue specificity of the candidate cDNA was determined by hybridization on slot blots to RNA from several mouse tissues. Only retinal RNA gave a strong signal, whereas brain, kidney, liver, lung and spleen RNAs were negative (Bowes et al., 1989).

The chromosomal location of the gene corresponding to the zr.408 cDNA was established by hybridization of the candidate cDNA with the DNA of hamster-mouse somatic cell hybrids. The zr.408 clone mapped, without any discordancy, to mouse chromosome 5, the same chromosome to which the *rd* locus had been assigned. Expression of the mRNA encoded by the putative *rd* cDNA was detectable on slot blots on postnatal day 1 in both normal and *rd/rd* retinas, and it continued to increase steadily during development. However, the concentration of zr.408 mRNA was always lower in the diseased than in the +/+ retina. By postnatal day 14, following the degeneration pattern of the *rd* retina, the *rd* zr.408 mRNA level decreased sharply and, as expected, it was no longer detectable at 31 days of age. Therefore, the expression of the candidate *rd* cDNA is abnormal in the *rd/rd* retina from postnatal day 1, several days before cyclic GMP levels become elevated in the photoreceptor cells. This is the earliest detected molecular lesion in the *rd* disease (Bowes et al., 1989).

Molecular Genetic Studies

In order to further confirm that zr.408 was a part of the *rd* gene, two different investigations were carried out. The aim of the first was to establish whether the gene corresponding to the zr.408 cDNA and the *rd* gene mapped to the same location on chromosome 5. The second study was to verify if there was cosegregation of the candidate *rd* gene with the expression of the *rd* disease, photoreceptor degeneration, in mouse crosses. Interspecies backcrosses were used for both projects, since the mice resulting from these crosses have been shown to give restriction fragment length polymorphisms (RFLPs) more frequently than those resulting from inbred crosses.

To establish the location of the gene corresponding to the zr.408 cDNA within chromosome 5, the RFLPs detected by hybridization to the zr.408 cDNA were first analysed in the DNAs from NFS/N and *Mus musculus* mice. These RFLPs were then identified in DNAs from 62 mice which were the backcrossed progeny of the cross (NFS/N x *Mus musculus)F1 x Mus musculus*. This allowed us to determine how many of the 62 progeny were homozygous or heterozygous for zr.408. The same type of analysis was performed for RFLPs that were hybridized by *Afp* and *Gus*, two other genes that lie on opposite sides of the *rd* gene on chromosome 5. From a comparison of the inheritance pattern of the three genes, we established the number of recombinants present, and from the rate of recombination, the distance in centiMorgans of each gene relative to the others (Danciger et al., 1990). In this way, zr.408 was placed between *Afp* and *Gus* at distances similar to those established for the *rd* gene (Chapman et al., 1975).

To determine whether the zr.408 cDNA cosegregated with the retinal degeneration trait, we classified 72 mice, the backcrossed progeny of the cross (C57BL/6J *rd/rd x Mus spretus)*Fl *x* C57BL/6J *rd/rd,* either as homozygous *(rd/rd,* affected with retinal degeneration) or heterozygous (*rd*/+, morphologically normal) by two different methods. In the first method, backcross progeny were scored by RFLP Southern blot analysis to find out if they were homozygous or heterozygous for zr.408 hybridizing fragments corresponding to the C57BL/6J strain. In the second method, we performed a histological examination of each of the retinas of the 72 progeny mice. When the results of both approaches were compared, zr.408 was found to cosegregate with the *rd* disease with no recombinants; that is, each of the assignments to *rd/rd* or *rd*/+ determined by Southern blot hybridization with zr.408 corresponded exactly to the histologic identification (Danciger et al., 1990).

The molecular genetic results from these two different studies supported very strongly the notion that zr.408 was a part of the *rd* gene. As described below, this has since been confirmed in studies of transgenic mice.

Identification of the Product of the Candidate *rd* Gene

The initial search for homologies between zr.408 and cDNAs encoding the α and β subunits of rod cyclic GMP-phosphodiesterase and the cone α^1 subunits of the enzyme showed strong cross-hybridization with the rod β subunit cDNA. Since zr.408 had been isolated from a cDNA library prepared with RNA from *rd*/+ retinas, this clone was used to screen normal mouse (+/+) retinal cDNA libraries. Sequencing of several of the isolated clones and of PCR-derived fragments, established that zr.408 is a truncated clone which corresponds to 2/3 of the total sequence of the β subunit of cyclic GMP-phosphodiesterase (from base 1,114 to base 2,706). There is only one nucleotide changed in the zr.408 sequence, which may be due to a polymorphic difference in the mouse strains that were used for the generation of the libraries (Bowes et al., 1990).

In order to further confirm that zr.408 is a part of the β subunit of cyclic GMP-phosphodiesterase, a comparison of the patterns of hybridization of poly(A)+ RNA and genomic DNA from *rd/rd, rd*/+ and +/+ mouse tissues with zr.408 cDNA and with several murine and bovine cDNA probes of the β subunit of cyclic GMP-phosphodiesterase was carried out. The same transcripts as well as RFLPs were observed on the Northern and Southern blots with all different cDNA probes tested. These observations, along with the mapping of zr.408 to the *rd* region on mouse chromosome 5, supported the conclusion that the normal counterpart of the *rd* gene encodes the β subunit of cyclic GMP-phosphodiesterase (Bowes et al., 1990).

Conclusive Demonstration that the β Subunit of Cyclic GMP-Phosphodiesterase is the Site of the *rd* Mutation

In order to obtain definitive proof that the gene encoding the β subunit of cyclic GMP-phosphodiesterase is responsible for the retinal degeneration of the *rd* mouse, its normal cDNA was expressed in transgenic *rd* mice. The construct injected into one-cell stage mouse embryos consisted of the complete cDNA of the gene ligated on its 5' end to the 5' flanking region of the mouse rod opsin gene (to target expression of the transgene specifically to the photoreceptor cells) and on its 3' end to the simian virus 40 polyadenylation signal. Two lines of homozygous *rd* animals on different genetic backgrounds expressed the transgene. One of these lines showed complete rescue of the photoreceptors whereas the other only exhibited partial rescue of the visual cells, with only those cells in the superior temporal region of the retina appearing normal. The photoreceptors of the inferior nasal region had completely degenerated and those from the central retina showed an intermediate level of degeneration, with their inner segments intact but their outer segments either disorganized or absent. Both lines of transgenic mice had integrated a single copy of the transgene (Lem et al., 1992).

The cyclic GMP-phosphodiesterase activity in the retinas of the two different lines of transgenic mice and in those of their non-transgenic, age-matched *rd/rd* siblings was compared, in order to correlate enzyme expression with photoreceptor rescue. The retinas of mice that had complete rescue of the photoreceptors had enzyme activity similar to those of normal (+/+) animals of the same age, whereas the retinas of the non-transgenic *rd/rd* siblings had minimal or no activity. The retinas of the animals that showed a gradient in expression of the transgene had higher cyclic GMP-phosphodiesterase activity in the superior than in the inferior hemisphere. These results indicated that the enzyme activity correlates with the number of rescued photoreceptors present in the retina and, furthermore, that the transgene was capable of restoring the activity of cyclic GMP-phosphodiesterase in *rd* retina (Lem et al., 1992). The study of the transgenic mice allowed us unequivocally to conclude that the retinal degeneration phenotype in the homozygous *rd* mouse arises from a defect in the cyclic GMP-phosphodiesterase β subunit gene.

The *rd* mutation is an old one that is present in a wide variety of mouse inbred strains. The cyclic GMP-phosphodiesterase β-subunit gene is quite complex. It has 22 exons varying in length from 48 bp to several hundred base pairs. A nonsense mutation in the seventh exon (codon 347), which is predicted to give rise to a truncated gene product, has been found (Pittler and Baehr, 1991). This mutation potentially eliminates over one-half of the peptide, including the putative catalytic domain. We have recently reported the presence in the *rd* retina of cyclic GMP-phosphodiesterase β-subunit transcripts that are incorrectly spliced (Bowes et al., 1992). Aberrant processing and expression of these transcripts could result from the integration of an 8.5 kb murine leukemia provirus (*Xmv-28*) into the first intron (Bowes et al., 1992; 1993). A 100% correlation was found between the presence of *Xmv-28* at this site and the *rd* mutation in 16 mouse strains, while none of 36 non-*rd* strains contained *Xmv-28* (Bowes et al., 1993). The *Xmv-28/rd* allele has also been found in strains derived from wild mice caught recently in Asia, USA and Britain, which raises questions about the evolutionary significance of such a widespread and apparently ancient mutation. As mentioned before, the levels of β-subunit mRNA are greatly reduced in *rd* mice as well as abnormal in size. It is therefore unclear whether the aberrant transcription or the stop codon or both give rise to the *rd* phenotype. We are currently investigating the mechanisms that lead from cyclic GMP-phosphodiesterase deficiency to death of *rd* photoreceptors.

A variety of other animal models of retinal degeneration have been identified, one of which also appears to have a mutation in the β subunit of rod cyclic GMP phosphodiesterase (Farber et al., 1992). The Irish setter dog suffers from an early onset rod-cone photoreceptor degeneration during the period of outer segment elongation. The retinas of these dogs are deficient in cyclic GMP-phosphodiesterase activity and have greatly elevated levels of cyclic GMP that precede morphological signs of degeneration, similar to *rd/rd* mice (Aguirre et al., 1978). While the photoreceptor cells of the diseased retinas are viable, the mRNAs for opsin, transducin α_1 and β_1, arrestin, phosducin and the α and γ subunits of cGMP-phosphodiesterase have normal transcript sizes and levels. In contrast, the cyclic GMP-phosphodiesterase β-subunit mRNA is present in levels lower than normal from early life (Farber et al., 1992). Since this abnormality is observed prior to any signs of arrested photoreceptor development and prior to the described biochemical defect (Aguirre et al., 1982), our results indicate a specific involvement of cyclic GMP-phosphodiesterase β-subunit in the Irish setter disease. Recently, Suber et al. (1993) showed that this is associated with a nonsense mutation in the canine homologue of the *rd* gene which is predicted to eliminate 49 amino acid residues from the C terminal end of the β subunit of cyclic GMP-phosphodiesterase, including an isoprenylation addition site. This isoprenylation site of the normal β-subunit may facilitate its interaction either with the disk membranes or with other protein subunits of the enzyme.

Finally, several mutations have recently been reported in the human homologue of the *rd* gene in patients with autosomal recessive retinitis pigmentosa (ARRP). McLaughlin et al. (1993) studied 7 of the 22 exons of the human cyclic GMP-phosphodiesterase β-subunit gene in 99 unrelated patients. They found 4 mutations that were not observed in 100 control individuals without ARRP and that co-segregated with the disease in families. Two of these were nonsense mutations (Gln 298X and Arg 513X), one was a missense mutation (His 557 Tyr) and one was a one bp deletion (Pro 496). Although further work is required to establish the frequency of these mutations in ARRP, these exciting findings provide further validation of the *rd* mouse as a model for human retinal degeneration.

THE *rds* MOUSE

In the late 1970s, a new retinal degeneration was described in the 020/A inbred mouse strain which was found to be due to a single defective gene inherited as an autosomal recessive (van Nie et al., 1978). The gene was localized distal to and near the H-2 locus in the middle region of mouse chromosome 17 (Demant et al., 1979). Due to the relatively slow rate of degeneration of the retinal photoreceptors in the 020/A mice when compared with the rate of degeneration of these cells in homozygous *rd* mice, the mutation was named *retinal degeneration slow* or *rds* (van Nie and Demant, 1978).

Degeneration in homozygous *rds* mice is characterized by absence of the outer segments of rod photoreceptors and the gradual and complete loss of the rest of these visual cells over time. The rod outer segments, which normally begin to develop during the first week of life, simply fail to form in the *rds* mouse even though the other retinal layers appear to develop normally. In place of the rod outer segment, a bulb of plasma membrane without any disk structures accumulates on top of the cilium. The "outer segmentless" photoreceptor cells then gradually degenerate over a period of 7 months to a year until they are no longer found in the retina (for reviews of the *rds* phenotype, see Sanyal et al., 1985; Hawkins et al., 1985; Cohen, 1983). Although the *rds* mutation was first decribed as a recessive, *rds*/+ heterozygotes show rod outer segments that are reduced in length and contain irregular, swollen and vacuolated disks. The photoreceptors also undergo very slow degeneration, so that the *rds* mutation is semi-dominant.

Identification of the Defective *rds* Gene

Over a period of several years, many immunocytochemical studies were performed which compared the amount and localization of several photoreceptor proteins in *rds* and normal retina. The proteins investigated included the interphotoreceptor retinoid binding protein (IRBP), opsin, arrestin, transducin, actin, and sulfated proteoglycans (van Veen et al., 1988; Carter-Dawson and Burroughs, 1989; Schalken et al., 1990; Navon et al., 1987; Jansen et al., 1990; Chaitin et al., 1988; Tawara and Hollyfield, 1990; Cantera et al., 1990). Although each of these studies contributed to our understanding of the mechanism of photoreceptor degeneration, none was able to specifically point to the defective gene in the *rds* retina. In fact, the *rds* gene was identified by a completely different approach, again utilizing the techniques of molecular biology (Travis et al., 1989).

Retinal cDNA was made from adult *rd/rd* mice which, as described earlier, do not have any photoreceptors. This pool of photoreceptorless cDNAs was used to subtract non-photoreceptor messages from cDNAs isolated from normal retina. The remaining cDNAs from normal retina, highly enriched for photoreceptor cDNAs, were cloned and screened with single stranded cDNAs from developing normal *ds/rds* retinas. Twelve clones were selected which seemed to give different hybridization signals with mutant and normal probes. Northern blot analysis and chromosomal localization by means of somatic cell hybrids revealed a single cDNA clone that hybridized more weakly with *rds/rds* retinal mRNA (compared to normal retinal mRNA) that is increased in size and mapped to mouse chromosome 17. Further study of the *rds* mutation demonstrated that the gene was defective due to disruption of the coding region by insertion of a 2.7 kb mouse repetitive element at nucleotide position 899 (the entire coding sequence is 1038 bp, coding for 346 amino acids; Travis et al., 1989).

The normal product of the *rds* gene is a photoreceptor disk protein called rds/peripherin to distinguish it from an unrelated neurofilament protein called peripherin (Connell et al., 1991; Travis et al., 1992a; Arikawa et al., 1992). The bovine homologue of *rds*/peripherin had been identified independently of the *rds* study (Connell and Molday, 1990). Rds/peripherin is a 39 kDa glycoprotein of unknown function, localized by immunocytochemistry to the rims of photoreceptor disks (Molday et al., 1987) which is predicted to have four membrane-spanning α-helices and is expressed exclusively in the rod and cone photoreceptor disks. The *rds* mutation would be expected to truncate the protein product by 116 amino acids with loss of the glycosylated region of the intradiscal D2 loop, thought to be important in interaction with other proteins, and of the fourth membrane-spanning segment and the cytoplasmic C3 segment (Travis et al., 1989).

Complete rescue of the mutant *rds* phenotype has recently been achieved in transgenic mice that had incorporated normal copies of the rds/peripherin cDNA (Travis et al., 1992b). The transgene contained upstream regulatory sequences from the mouse opsin gene fused to a normal rds/peripherin mini-gene with the SV40 early splice and polyadenylation signals. The construct was microinjected into fertilized mouse eggs from *rds*/+ mutants. Three transgenic lines were obtained and crossed with homozygous *rds* mutants to observe the effects of the transgene in a homozygous *rds/rds* background. Two high rds/peripherin expressing lines showed retinas that were indistinguishable from the wild-type, with normal outer segment morphology and no photoreceptor degeneration. An intermediate phenotype was seen in a low expressing line, with about 10% of wild-type rds/peripherin levels, which showed whorls of disorganized disks in the subretinal space, representing dysplastic outer segments. Non-transgenic littermates showed the typical *rds/rds* mutant phenotype. These results indicate that the semi-dominant *rds* phenotype probably results from haplo-insufficiency – lack of sufficient normal gene product even in the heterozygous state – rather than a deleterious effect of the mutant product *per se* (dominant negative mutation). This would be consistent with the *rds* mutation being a null allele, with little or no expressed product, as predicted from the mutation analysis.

The introduction of upstream promoter sequences from the mouse opsin gene into transgenic mice coupled to a lacZ reporter gene had previously been shown to drive rod-specific transgene expression with onset shortly after the appearance of outer segment structures (around the first postnatal week) (Lem et al., 1991; Zack et al., 1991). The transgenic rescue of *rds* mutants would be expected to reverse rod but not cone photoreceptor degeneration, since *rds*/peripherin is expressed in both cell types and the opsin promoter is rod-specific. However, only about 3% of photoreceptors in mice are cones (Carter-Dawson et al., 1978) and it was not possible to confirm this prediction (Travis et al., 1992b).

The Human Counterpart of the Mouse *rds* Gene Causes Human Retinitis Pigmentosa and Macular Degenerations

The name retinitis pigmentosa (RP) represents a heterogeneous group of diseases that have in common a progressive retinal degeneration, the presence of pigment deposits in the retina and night blindness. The incidence in the US is about 1 in 3,500 people and the inheritance pattern is quite heterogeneous. Autosomal recessive, autosomal dominant, X-linked and simplex (no family history of disease) are all common patterns of inheritance of the various forms of RP. Furthermore, there are many clinical variations within each inheritance pattern.

Recently, the cause of a number of cases of autosomal dominant RP (ADRP) has been assigned to defects in the rhodopsin gene. The focus on rhodopsin started with its localization to the long arm of human chromosome 3 (3q) (Sparkes et al., 1986; Nathans et al., 1986). This was followed a few years later by the discovery of linkage of ADRP to the same region of chromosome 3q in a large Irish family (McWilliams et al., 1989) and then linkage to rhodopsin itself in the same pedigree (Farrar et al., 1990). From there, a number of point mutations were found in the rhodopsin gene that cosegregated perfectly with the ADRP cases in particular families (Dryja et al., 1990a,b; Keen et al., 1991; Gal et al., 1991; Inglehearn et al., 1991; Farrar et al., 1991a). However, in other families, genetic analysis revealed that the disease gene was specifically not linked to chromosome 3q (Farrar et al., 1990; Inglehearn et al., 1990; Blanton et al., 1990). In fact, it has been estimated that only about 20-30% of the cases of ADRP are due to defects in the rhodopsin gene (Kajiwara et al., 1991).

The *rds* gene story follows a similar line of discovery as that of the rhodopsin gene. After the cloning of the mouse rds/peripherin gene, its human homologue (RDS) was mapped to chromosomal region 6p12 (Travis et al., 1991). Then, linkage of ADRP to chromosome 6p and specifically to the RDS/peripherin locus was demonstrated in a large family (Farrar et al., 1991b; Jordan et al., 1992). Finally, as in the case of rhodopsin, mutations in the peripherin gene that segregated with RP were found in particular families (Kajiwara et al., 1991; Farrar et al., 1991c). The mutations were either single base changes or 3 base deletions and were discovered by single strand conformational polymorphism analysis. Further analysis of the RDS/peripherin gene in human retinal disease states has revealed further mutations associated with a surprising degree of clinical heterogeneity. These mutations have been found in autosomal dominant families with typical retinitis pigmentosa, with various types of macular dystrophy, with either retinitis pigmentosa or macular dystrophy in different members of the same kindred, and in a family with retinitis punctata albescens.

The main difference between the course of scientific discovery of the rhodopsin mutations and of the RDS/peripherin mutations in man was the direct involvement of the *rds* animal model of retinal degeneration in the latter case. Human mapping and pedigree linkage analysis put rhodopsin and ADRP together. Since rhodopsin is the most abundant protein in rod photoreceptors, it was a logical choice for the site of a gene defect. Interest in the human peripherin gene as a candidate for RP mutations occurred after it was demonstrated that its murine homologue was responsible for the *rds* disease, one of several mouse models of RP. This is a straightforward example of a case where the human homologue of a gene responsible for an inherited disease in an animal is responsible for a similar human disease. As discussed above, the human counterpart of the *rd* gene also now appears to cause retinal degeneration in man. It may well be that, when the molecular basis of the other remaining mouse retinal degeneration mutants becomes known, further human disease genes will be identified and explained.

MAPPING OF OTHER GENES EXPRESSED IN RETINAL TISSUES

The gene for rhodopsin *(Rho)* was localized to distal mouse chromosome 6 in the vicinity of the microophthalmia (*mi*) locus (Elliott et al., 1990), an inherited murine defect which results in a smaller eye. Another gene expressed in the retina has recently been mapped to the same chromosomal region, a third G protein β-subunit, *Gnb-3* (Danciger et al., 1992a). Further studies will need to be carried out with both *Rho* and *Gnb-3* now that they are potential candidates for *mi* based on their chromosomal location.

The gene for the α-subunit of calcium/calmodulin-dependent protein kinase II *(Camk2a)* has also recently been mapped to mouse chromosome 18 in the vicinity of the bouncy *(bc)* neurological mutant (Danciger et al., 1992b). The *Camk2a* gene product is present in brain, retina and other tissues. Since the only published information about the *bc* mutant is related to its neurological symptoms (bouncy gait and tremor) or its breeding capabilities (Lane et al., 1981; Green, 1989), it is possible that pathology involving other tissues (such as retina) may occur. Therefore, by virtue of its chromosomal location and expression in brain and other tissues, the *Camk2a* gene is a potential candidate for the *bc* mutation.

In the course of mapping the transducin β-subunit gene *(Gnb-1)* to mouse chromosome 4, we localized two other homologous sequences to mouse chromosomes 5 and 8 (Danciger et al., 1990a). Chromosomal mapping was done by Southern blot analysis of Chinese hamster-mouse somatic cell hybrid DNAs that were probed with the λ TB112 cDNA for the *Gnb-1* gene (provided by Dr. Melvin I. Simon [Sugimoto et al., 1985; Fong et al., 1986]; Figures 4 and 5 and Table 2). Linkage analysis with the marker genes *Afp* and *Gus* (Table 2) was used to localize the chromosome 5 sequence (*Gnb-2)* to a region that was homologous to the region of human chromosome 7 where a second G protein β-subunit gene (GNB2) had been mapped (Blatt et al., 1988; Hillyard et al., 1992; Lyon and Kirby, 1992; Searle et al., 1987). Linkage data was collected primarily by means of allelic RFLP analysis of Southern blots (Figure 6 and Table 3). The *Gnb-2* gene was in the vicinity of the *rd* locus, but more than 10 cM distal to it (Hillyard et al., 1992; Lyon and Kirby, 1992). Therefore, fine mapping demonstrated that *Gnb-2* is not the *rd* gene - which was later confirmed by the identification of the *rd* site as the cyclic GMP-phosphodiesterase β-subunit gene - avoiding the need for further effort on the *Gnb-2* gene.

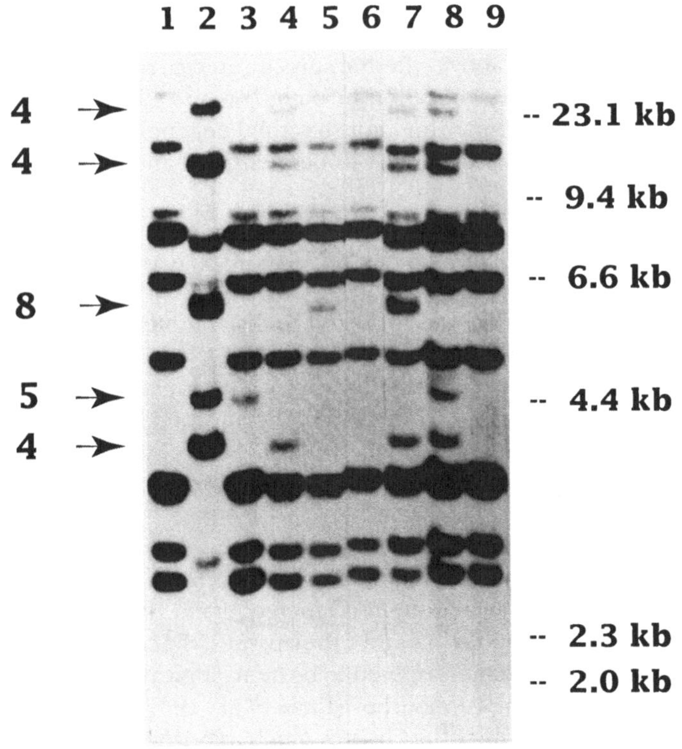

Figure 4 Autoradiogram of a Southern blot of hamster-mouse somatic cell hybrid DNAs digested with *Hind*III and hybridized with the λ TBl12 cDNA probe; each lane has 6 μg of DNA. Lane 1, Chinese hamster cell DNA; lane 2, NFS/N mouse liver DNA; lanes 3–9, representative somatic cell hybrid DNAs. The five arrows point to the >23.1-, 15.0-, 5.9-, 4.5- and 3.9-kb mouse bands used to score for the presence of homologous sequences in the hybrids. The numbers on the right (in kb) mark the positions of λ DNA fragments created by digestion with *Hind*III. The numbers on the left indicate chromosome assignments. DNA electrophoresis was carried out in 1.2% agarose gels and the separated fragments were transferred to nylon membranes by the technique of Southern (1975). Hybridization with labeled probe was carried out as described previously (Danciger et al., 1990a).

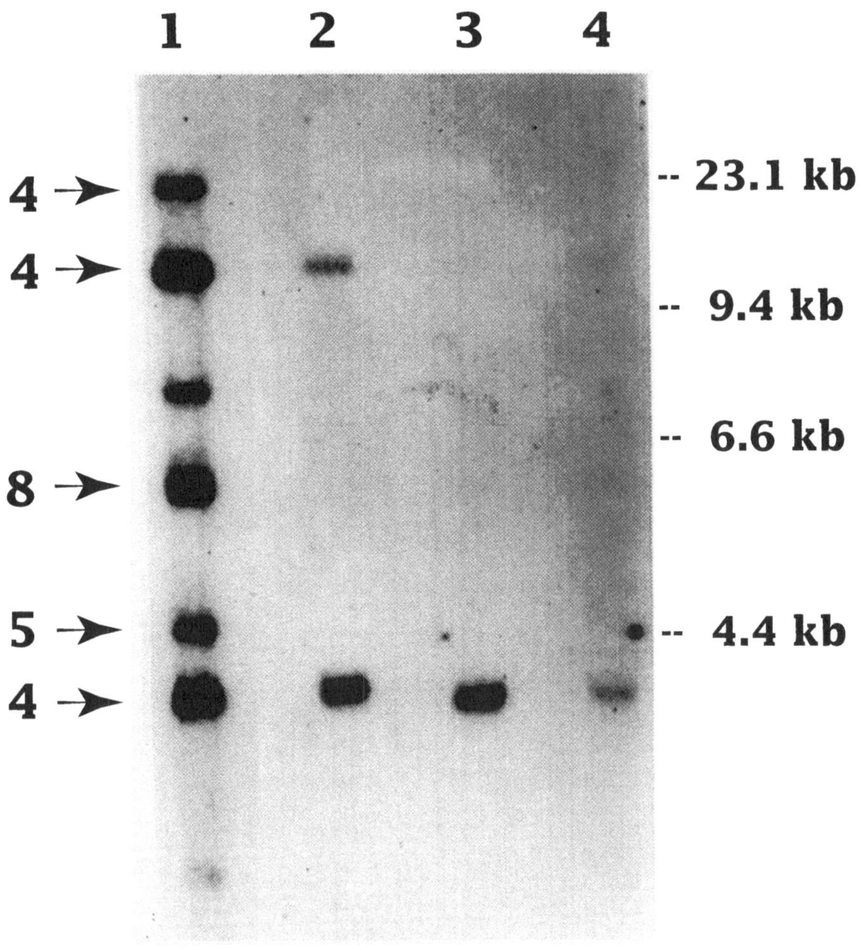

Figure 5 Autoradiogram of a Southern blot of DNAs from the mouse digested with *Hind*III, hybridized with the λ TB112 cDNA probe, and washed at increasing stringency from left to right; each lane was on a separate strip of nylon membrane and had 6 μg of DNA. The numbers on the left indicate chromosome assignments. The numbers on the right (in kb) mark the positions of λ DNA fragments created by digestion with *Hind*III. In the higher stringency washes only the fragments that map to mouse chromosome 4 resisted being separated from the probe. This indicates that the chromosome 4 site is the most homologous to the λ TB112 probe and suggests therefore, the site of the *Gnb-1* gene. This assignment is further corroborated by the fact that *Gnb-1 is* localized to a region of mouse chromosome 4 that is homologous to human chromosome 1, the chromosome to which the human homologue of *Gnb-1* (GNB1) maps (Blatt et al., 1988; Danciger et al., 1990a; Hillyard et al., 1992; Lyon and Kirby, 1992).

Table 2 Correlation between specific mouse chromosomes and restriction fragments hybridizing to the λTB112 cDNA (*Gnb-1, Gnb-2, Gnb1rs-1,* in Chinese hamster-mouse somatic cell hybrids[a]).

	Gnb-1	*Gnb-2*	*Gnb1rs-1*
Mouse Chromosome	*% Discordant*[b]	*%Discordant*[b]	*%Discordant*[b]
1	33	50	38
2	46	58	39
3	38	38	33
4	**0**	36	24
5	33	0	29
6	42	62	35
7	50	77	65
8	20	40	**4**[c]
9	23	35	23
10	27	15	19
11	33	11	22
12	42	74	47
13	38	58	46
14	32	32	24
15	59	88	71
16	30	26	39
17	46	58	39
18	29	38	29
19	32	52	24
X	36	44	36

[a] The chromosome content of the somatic cell hybrids was determined by trypsin-Giemsa banding followed by staining with Hoechst 33258, or by typing for specific marker loci.

[b] The number of somatic cell hybrids that had the particular mouse chromosome but did not show the mouse hybridizing sequence of the gene, plus the number of hybrids that did not have the chromosome but did show the hybridizing sequence of the gene, divided by the total number of hybrids evaluated for that particular chromosome.

[c] The 4%discordancy was due to one hybrid which contained Chr.8, but did not show the 5.9-kb scoring mouse fragment. The discrepancy can be explained by the fact that this hybrid line had a low percentage of cells containing Chr.8 (low copy number). Further, the Chr.8 assignment was corroborated by multilocus linkage analysis of backcross progeny (Danciger et al., 1990a).

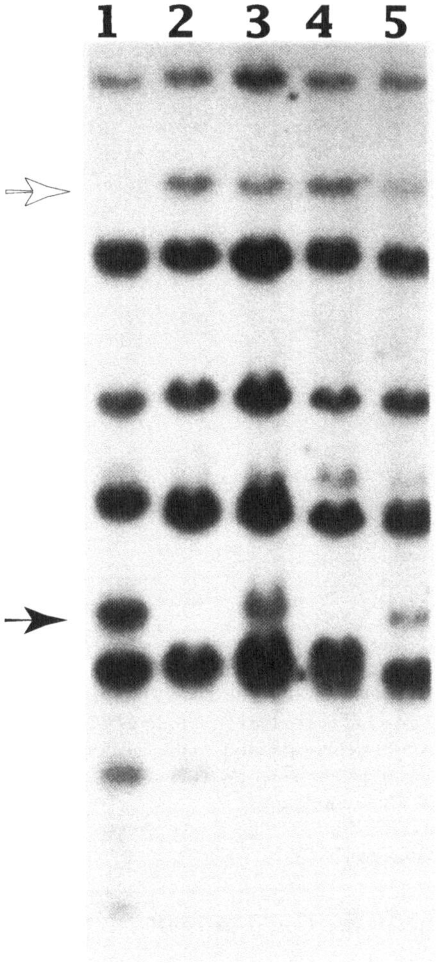

Figure 6 Autoradiogram of a Southern blot of test cross progeny DNAs from the cross (NFS/N x *M. m. musculus)*F1 *x M. m. musculus.* DNAs were digested with *Hind*III and hybridized with λ TB112 cDNA; each lane has 10 μg of DNA. Lane 1, parental NFS/N DNA; lanes 2-5, representative backcross progeny. The arrows point to a 14 kb *M. m. musculus* fragment (*M. m. m.)* present in all test cross progeny and a 4.5 kb fragment representing the unique NFS/N allele (for *Gnb-2*) present only in heterozygotes. DNAs were extracted from mouse livers, cleaved with *Hind*III, run on 0.4% agarose gels, transferred to nylon membranes and hybridized to labeled probe as described by Hoggan et al. (1988).

Table 3 Segregation of a 4.5-kb *Hind*III hybridizing fragment *(Gnb-2)* with alleles of *Afp* and *Gus*[a] in the cross (NFS/N × *M.m. musculus*) F1 × *M. m. musculus.*

LOCI			No. of progeny
Afp	*Gnb-2*	*Gus*	
parentals			
+[b]	+	+	21
–	–	–	23
single recombinants			
+	–	–	9
–	+	+	6
+	+	–	3
–	–	+	0
double recombinants			
–	+	–	1

% Recombination
(Afp, Gnb-2) = 16/63 = 25 ± 5.5[c]
(Gnb-2, Gus) = 4/63 = 6± 3.1

[a] *Afp* and *Gus* alleles were scored by Southern blot hybridization with appropriate cDNA probes. DNA was digested with *Eco*RI for *Afp*, and *Hind*III for *Gus*. A unique 6.0-kb NFS/N restriction fragment hybridized with a 960-bp cDNA probe released from the pBR322-AFP1 plasmid (provided by Dr. Shirley Tilghman, Princeton University [D'Eustachio et al., 1981]). A unique 3.7-kb NFS/N restriction fragment hybridized with a 1.45-kb cDNA probe corresponding to the *Gus* gene, released from the pGA1 plasmid (provided by Dr. Gordon Watson, UC Berkeley [Watson et al., 1985]).
[b] The symbol "+" refers to progeny that are heterozygous.
[c] Percent recombination and standard errors were calculated according to Green (1981); all recombinant fractions are significant to the 0.05 level.

Molecular Genetic Studies in the Mouse Could Increase our Understanding of Human Diseases

The human genome can be viewed as a rearrangement of chromosomal pieces of the mouse genome, or vice versa. For example, human chromosome 1 has large regions of homology with mouse chromosomes 1, 3 and 4, and human chromosome 5 has large regions of homology with mouse chromosomes 11, 13 and 18 (Searle et al., 1987; Lyon and Kirby, 1992; Hillyard et al., 1992). The rearrangements probably represent a history of translocations and other alterations in the two genomes as they diverged from a common ancestor millions of years ago. In fact, all of the genes that have been mapped by our laboratory either fell within, or extended in a distal or proximal direction, these mouse/human homologous chromosomal regions. Accordingly, nearly any localization of a murine gene can suggest the location of its human homologue based on homologies between man and mouse.

Once a gene has been localized in the mouse genome, it is an easy matter to check the chromosomal maps to see if the homologous region in the human contains any mutations. For example, North Carolina macular dystrophy has recently been localized to human chromosomal region 6q by linkage analysis of a single large family (Small et al., 1992). Human 6q has homology with small regions of mouse chromosomes 4, 9 and 17 and with a larger region of mouse chromosome 10. Thus, any murine retinal gene mapping to these regions becomes a potential candidate for this disease. There are other inherited retinal diseases that have been linked to regions of the human genome and to which the same logic can be applied: Usher syndrome type II (Lewis et al., 1990; Kimberling et al., 1990); several forms of X-linked retinitis pigmentosa (Musarella et al., 1991; 1989; Ott et al., 1990; Levy et al., 1989) and the RP1 form of autosomal dominant retinitis pigmentosa (Blanton et al., 1991).

In summary, the advances in the fields of molecular biology and molecular genetics that have occurred in the last few years, together with the availability of animal models of retinal degeneration – either inherited or created by the integration of mutant genes into the mouse genome – have opened up tremendous possibilities for the understanding of genetic diseases affecting the retina. We are encouraged by the recent findings and hope that in the near future not only the causes will be unravelled but also ways to prevent or cure the devastating disorders that lead to blindness.

ACKNOWLEDGEMENTS

We wish to thank Dr. Cathy Bowes for her continued contribution to the study of the *rd* mouse and Dr. John Flannery for providing us with the figures of the normal and *rd* mouse retinas. This work was supported by the National Institutes of Health grants EY02651 and EY08285 and Core grant EY0331; a Center grant from the National Retinitis Pigmentosa Foundation; and a grant from the George Gund Foundation. D.B.F. is the recipient of a Research to Prevent Blindness Senior Scientific Investigator Award.

REFERENCES

Aguirre, G. D., Farber, D., Lolley, R., Fletcher, R. T. and Chader, G. J. (1978). Rod-cone dysplasia in Irish setters: a defect in cyclic GMP metabolism in visual cells. *Science* **201**, 1133–1134.

Aguirre, G., Farber, D., Lolley, R., O'Brien, P., Alligood, R., Fletcher, T. and Chader, G. (1982). Retinal degenerations in the dog. III. Abnormal cyclic nucleotide metabolism in rod-cone dysplasia. *Experimental Eye Research* **35**, 625–642.

Arikawa, K., Molday, L. L., Molday, R. S. and Williams D.S. (1992). Localization of peripherin/rds in the disk membranes of cone and rod photoreceptors: Relationship to disk membrane morphogenesis and retinal degeneration. *The Journal of Cell Biology* **16**, 659–667.

Baehr, W., Devlin M. J., Applebury, M. L. (1979). Isolation and characterization of cGMP– phosphodiesterase from bovine rod outer segments. *Journal of Biological Chemistry* **254**, 11669–11677.

Blanks, J. C., Adinolfi, A. M. and Lolley, R. N. (1974). Photoreceptor degeneration and synaptogenesis in retinal-degenerative (*rd*) mice. *Journal of Comparative Neurology* **156**, 95–106.

Blanton, S. H., Heckenlively, J. R., Cottingham, A. W., Friedman, J., Sadler, L., Wagner, M., Friedman, L. and Daiger, S. P. (1991). Linkage mapping of autosomal dominant retinitis pigmentosa (RP1) to the pericentric region of human chromosome 8. *Genomics* **11**, 857–869.

Blanton, S. H., Cottingham, A. W., Giesenschlag, N., Heckenlively, J.R., Humphries, P. and Daiger, S. P. (1990). Further evidence of exclusion of linkage between type 2 autosomal dominant retinitis pigmentosa (ADRP) and D3S47 on 3q. *Genomics* 8, 179–181.

Blatt, C., Eversole-Cire, P., Cohn, V. H., Zollman, S., Fournier, R. E. K., Mohandas, L. T., Nesbitt, M., Lugo, T., Jones, D. T., Reed, R. R., Weiner, L. P., Sparkes, R. S. and Simon, M. I. (1988). Chromosomal localization of genes encoding guanine nucleotide-binding protein subunits in mouse and human. *Proceedings of the National Academy of Science (USA)* **85**, 7642–7646.

Bowes, C. and Farber, D. B. (1987). mRNAs coding for proteins of the cGMP cascade in the degenerative retina of the *rd* mouse. *Experimental Eye Research* **45**, 467–480.

Bowes, C., van Veen, T. and Farber, D. B. (1988). Opsin, G-protein and 48-kDa protein in normal and *rd* mouse retinas: developmental expression of mRNAs and proteins and light/dark cycling of mRNAs. *Experimental Eye Research* **47**, 369–390.

Bowes, C., Danciger, M., Kozak, C. A. and Farber, D. B. (1989). Isolation of a candidate cDNA for the gene causing retinal degeneration in the *rd* mouse. *Proceedings of the National Academy of Science (USA)* **86**, 9722–9726.

Bowes, C., Li, T., Danciger, M., Baxter, L. C., Applebury, M. L. and Farber, D. B. (1990). Retinal degeneration in the *rd* mouse is caused by a defect in the β subunit of rod cGMP-phosphodiesterase. *Nature* **347**, 677–680.

Bowes, C., Li, T., Frankel, W., Danciger, M., Coffin, J., Applebury, M. L. and Farber, D. B. (1992). Analysis of *Xmv-28* proviral integration within the *rd* β PDE gene. *Investigative Ophthalmology and Visual Science* **33**, 945.

Bowes, C., Li, T., Frankel, W. N., Danciger, M., Coffin, J. M., Applebury, M. L. and Farber, D. B. (1993). Localization of a retroviral element within the *rd* gene coding for the β subunit of cGMP phosphodiesterase. *Proceedings of the National Academy of Science (USA)* **90**, 2955–2959.

Caley, D. W., Johnson, C. and Liebelt, R. A. (1972). The postnatal development of the retina in the normal and rodless CBA mouse: a light and electron microscopic study. *American Journal of Anatomy* **133**, 179–212.

Cantera, R., von Schantz, M., Chader, G. J., Ehinger, B., Sanyal, S. and van Veen, T. (1990). Postnatal development of photoreceptor-specific proteins in mice with hereditary retinal degeneration. *Experimental Biology* **48**, 305–312.

Caravaggio, L. L. and Bonting, S. L. (1963). The rhodopsin cycle in the developing vertebrate retina. *Experimental Eye Research* **2**, 12–19.

Carter-Dawson, L. D., LaVail, M.M. and Sidman, R. L. (1978). Differential effect of the *rd* mutation on rods and cones in the mouse retina. *Investigative Ophthalmology and Visual Science* **17**, 489–498.

Carter-Dawson, L. D. and Burroughs, M. (1989). Interphotoreceptor retinoid-binding protein in the postnatal developing *rds* mutant mouse retina: EM immunocytochemical localization. *Experimental Eye Research* **49**, 829–841.

Chaitin, M. H., Carlsen, R.B. and Samara, G. J. (1988). Immunogold localization of actin in developing photoreceptor cilia of normal and *rds* mutant mice. *Experimental Eye Research* **47**, 437–446.

Chang, B., Heckenlively, J. R., Hawes, N. L. and Roderick, T. L. (1993). New mouse primary retinal degeneration (*rd-3*). *Genomics* **16**, 45–49.

Chapman, V. M., Noell, W. K. and Adler, D. (1975). In *Mouse Newsletter* **53**, 61.

Cohen, A. I. (1983). Some cytological and initial biochemical observations on photoreceptors in retinas of *rds* mice. *Investigative Ophthalmology and Visual Science* **24**, 832–843.

Connell, G. and Molday, R. S. (1990). Molecular cloning, primary structure, and orientation of the vertebrate photoreceptor cell protein peripherin in the rod outer segment disk membrane. *Biochemistry* **29**, 4691–4698.

Connell, G., Bascom, R., Molday, L., Reid, D., McInnes, R. R. and Molday, R. S. (1991). Photoreceptor peripherin is the normal product of the gene responsible for retinal degeneration in the *rds* mouse. *Proceedings of the National Academy of Science (USA)* **88**, 723–726.

D'Eustachio, P., Ingram, R. S., Tilghman, S.M. and Ruddle, F. H. (1981). Murine α-foetoprotein and albumin: II. Evolutionary linked proteins encoded on the same mouse chromosome. *Somatic Cell Genetics* **7**, 289.

Danciger, M., Kozak, C. A., and Farber, D. B. (1989a). The gene for the α-subunit of retinal rod transducin is on mouse chromosome 9. *Genomics* **4**, 215–217.

Danciger, M., Kozak, C. A., Tsuda, M., Shinohara, T. and Farber, D. B. (1989b). The gene for retinal S-antigen maps to the centromeric portion of mouse chromosome 1 near *Idh-1*. *Genomics* **5**, 378–381.

Danciger, M., Tuteja, N., Kozak, C. A., and Farber, D. B. (1989c). The gene for the γ-subunit of retinal cGMP-phosphodiesterase is on mouse chromosome 11. *Experimental Eye Research* **48**, 303–308.

Danciger, M., Farber, D. B., Peyser, M. and Kozak, C. A. (1990a). The gene for the β-subunit of retinal transducin *(Gnb-1)* maps to distal mouse chromosome 4, and related sequences map to mouse chromosomes 5 and 8. *Genomics* **6**, 428–435.

Danciger, M., Kozak, C. A., Nickerson, J., Redmond, T.M. and Farber, D. B. (1990b). Localization of the gene for interphotoreceptor retinoid-binding protein to mouse chromosome 14 near *Np-1*. *Genomics* **8**, 727–731.

Danciger, M., Kozak, C. A., Li, T., Applebury, M. L. and Farber, D. B. (1990c). Genetic mapping demonstrates that the α-subunit of retinal cGMP–phosphodiesterase is not the site of the *rd* mutation. *Experimental Eye Research* **51**, 185–189.

Danciger, M., Bowes, C., Kozak, C. A., LaVail, M. M. and Farber, D. B. (1990d). Fine mapping of a putative *rd* cDNA and its cosegregation with *rd* expression. *Investigative Ophthalmology and Visual Science* **31**, 1427–1432.

Danciger, M., Kozak, C. A., Abe, T., Shinohara, T. and Farber, D. B. (1991). The gene for retinal rod 33–kDa protein is on mouse chromosome 1, near Lamb2. *Cytogenetics and Cell Genetics* **56**, 202–205.

Danciger, M., Chakraborti, A., Farber, D. B., Kozak, C. A. (1992a). Localization of the gene for a third G protein β-subunit to mouse chromosome 6 near *Raf-1*. *Genomics* **12**, 688–692.

Danciger, M., Kozak, C. A., Adamson, C. M. and Farber, D. B. (1992b). Chromosomal localization of the murine genes for the α- and β-subunits of calcium/calmodulin dependent protein kinase II. *Mammalian Genome* **3**,122–125.

Demant, P., Ivanyi, D. and van Nie, R. (1979). The map position of the *rds* gene on the 17th chromosome of the mouse. *Tissue Antigens* **13**, 53–55.

Dryja, T. P., McGee, T. L., Hahn, L. B., Cowley, G. S., Olsson, J. E., Reichel, E., Sandberg, M. A. and Berson, E. (1990). Mutations within the rhodopsin gene in patients with autosomal dominant retinitis pigmentosa. *The New England Journal of Medicine* **323**, 1302–1307.

Dryja, T. P., McGee, T. L., Reichel, E., Hahn, L. B., Cowley, G. S., Yandel, D. N., Sandberg, M. A. and Berson, E. (1990). A point mutation of the rhodopsin gene in one form of retinitis pigmentosa. *Nature* **343**, 364–366.

Elliott, R. W., Sparkes, R. S., Mohandas, T., Grant, S. G. and McGinnis, J. F. (1990). Localization of the rhodopsin gene to the distal half of mouse chromosome 6. *Genomics* **6**, 635–644.

Farber, D. B. (1989). Cyclic nucleotides in normal retinal function and disease. In *Extracellular and Intracellular Messengers in the Vertebrate Retina*, edited by D. Redburn and H. Pasantes-Morales, pp. 13–24. New York: Alan R. Liss, Inc.

Farber, D. B. and Lolley, R. N. (1974). Cyclic guanosine monophosphate: elevation in degenerating photoreceptor cells of C3H mouse retina. *Science* **186**, 449–451.

Farber, D. B. and Lolley, R. N. (1976). Enzymatic basis for cyclic GMP accumulation in degenerative photoreceptor cells of mouse retina. *Journal of Cyclic Nucleotide Research* **2**, 139–148.

Farber, D. B. and Lolley, R. N. (1977). Light-induced reduction in cyclic GMP of retinal photoreceptor cells *in vivo:* abnormalities in the degenerative diseases of RCS rats and *rd* mice. *Journal of Neurochemistry* **28**, 1089–1095.

Farber, D. B. and Shuster, T. A. (1986). Cyclic nucleotides in retinal function and degeneration. In *The retina: A model for cell biology studies*, Part 1, edited by R. Adler and D. B. Farber, pp. 239–296. Orlando, Florida: Academic Press.

Farber, D. B., Park, S. and Yamashita, C. (1988). Cyclic GMP–phosphodiesterase of *rd* retina: biosynthesis and content. *Experimental Eye Research* **46**, 363–374.

Farber, D. B., Danciger, J. S. and Aguirre, G. (1992). The β subunit of cyclic GMP phosphodiesterase mRNA is deficient in canine rod-cone dysplasia 1. *Neuron* **9**, 349–356.

Farrar, G. J., McWilliam, P., Bradley, D. G., Kenna, P., Lawler, M., Sharp, E. M., Humphries, M. M., Eiberg, H., Conneally, P. M., Trofatter, J. M. and Humphries, P. (1990). Autosomal dominant retinitis pigmentosa: linkage to rhodopsin and evidence for genetic heterogeneity. *Genomics* **8**, 35–40.

Farrar, G. J., Kenna, P., Redmond, R., Shiels, D., McWilliam, P., Humphries, M. M., Sharp, E. M., Jordan, S. A., Kumar-Singh, R., and Humphries, P. (1991a). Autosomal dominant retinitis pigmentosa: A mutation in codon 178 of the rhodopsin gene in 2 ADRP families of Celtic origin. *Genomics* **11**, 1170–1171.

Farrar, G. J., Jordan, S. A., Kenna, P., Humphries, M. M., Kumar-Singh, R., McWilliam, P., Allamand, V., Sharp, E. M., and Humphries, P. (1991b). Autosomal dominant retinitis pigmentosa: Localization of a disease gene (RP6) to the short arm of chromosome 6. *Genomics* **11**, 870–874.

Farrar, G. J., Kenna, P., Jordan, S. A., Kumar-Singh, R., Humphries, M. M., Sharp, E. M., Shiels, D. and Humphries, P. (1991c). A three-base-pair deletion in the peripherin-rds gene in one form of retinitis pigmentosa. *Nature* **354**, 478–480.

Fong, H. K. W., Hurley, J. B., Hopkins, R. S., Miake-Lye, R., Johnson, M. S., Doolittle, R. F. and Simon, M. I. (1986). Repetitive segmental structure of the transducin β subunit: homology with the CDC4 gene and identification of related mRNAs. *Proceedings of the National Academy of Science (USA)* **83**, 2162–2166.

Fung, B. K.-K., Young, J. H., Yamane, H. K., Griswold-Prenner, I. (1990). Subunit stoichiometry of retinal rod cGMP phosphodiesterase. *Biochemistry* **29**, 2657–2664.

Gal, A., Artlich, A., Ludwig, M., Niemeyer, G., Olek, K., Schwinger, E. and Schinzel, A. (1991). Pro-347-Arg mutation of the rhodopsin gene in autosomal dominant retinitis pigmentosa. *Genomics* **11**, 468–470.

Grafstein, B., Murray, M. and Ingoglia, N. A. (1972). Protein synthesis and axonal transport in retinal ganglion cells of mice lacking visual receptors. *Brain Research* **44**, 37–48.

Green, E. L. (1981). *Genetics and Probability in Animal Breeding Experiments*, pp. 77–113. NewYork: MacMillan.

Green, M. C. (1989). Catalog of mutant genes and polymorphic loci. In M. F. Lyon and A. G. Searle (eds): *Genetic Variants and Strains of the laboratory Mouse*, 2nd edn, pp. 12–403. New York: Oxford University Press.

Hawkins, R. K., Jansen, H. G. and Sanyal, S. (1985). Development and degeneration of retina in *rds* mutant mice: photoreceptor abnormalities in the heterozygotes. *Experimental Eye Research* **41**, 701–720.

Heckenlively, J. R., Chang, B., Hawes, N. L. and Roderick, T. H. (1992). Two new mouse retinal primary degenerations. *Investigative Ophthalmology and Visual Science* **33**, 1063.

Hillyard, A. L., Doolittle, D., Davisson, M. T. and Roderick, T. H . (1992). Locus map of mouse. *Mouse Genome* **90**, 8–21.

Inglehearn, C. F., Jay, M., Leister, D. H., Bashir, R., Jay, B., Bird, A. C., Wright, A. F., Evans, H. J., Papiha, S. S. and Bhattacharya, S. S. (1990). No evidence for linkage between late onset autosomal dominant retinitis pigmentosa chromosome 3 locus D3S47 (C17): Evidence for genetic heterogeneity. *Genomics* **6**, 168–173.

Inglehearn, C. F., Bashir, R., Leister, D. H., Jay, M., Bird, A. C. and Bhattacharya, S. S. (1991). A 3bp deletion in the rhodopsin gene in a family with autosomal dominant retinitis pigmentosa. *American Journal of Human Genetics* **48**, 26–30.

Jansen, H. G., Aguirre, G. D., vanVeen, T. and Sanyal, S. (1990). Development and degeneration of retina in *rds* mutant mice: ultraimmunohistochemical localization of S-antigen. *Current Eye Research* **9**, 903–911.

Jordan, S. A., Farrar, G. J., Kumar-Singh, R., Kenna, P., Humphries, M. M., Allamand, V., Sharp, E. M. and Humphries, P. (1992). Autosomal dominant retinitis pigmentosa (adRP; RP6): Cosegregation of RP6 and the peripherin-rds locus in a late-onset family of Irish origin. *American Journal of Human Genetics* **50**, 634–639.

Kajiwara, K., Hahn, L. B., Mukai, S., Travis, G. H., Berson, E. L. and Dryja, T. P. (1991). Mutations in the human retinal degeneration slow gene in autosomal dominant retinitis pigmentosa. *Nature* **354**, 480–483.

Keen, T. J., Inglehearn, C. F., Lester, D. H., Bashir, M. J., Bird, A. C. and Bhattacharya, S. S. (1991). Autosomal dominant retinitis pigmentosa: Four new mutations in rhodopsin, one of them in the retinal attachment site. *Genomics* **11**, 199–205.

Kimberling, W. J., Weston, M. D., Moller, C., Davenport, S. L. H., Shugart, Y. Y., Priluck, I. A., Martini, A., Milani, M. and Smith, R. J. (1990). Localization of Usher Syndrome type II to chromosome 1q. *Genomics* **7**, 245–249.

Lane, P. W., Searle, A. G., Beechey, C. V. and Eicher, E. M. (1981). Chromosome 18 of the house mouse. *Journal of Heredity* **72**, 409–412.

LaVail, M. M. and Sidman, R. L. (1974). Retinal degeneration in the mouse. *Archives of Ophthalmology* **91**, 394–400.

Lee, R. H., Navon, S. E., Brown, B. M., Fung, B. K.-K. and Lolley, R. N. (1988). Characterization of a phosphodiesterase-immunoreactive polypeptide from rod photoreceptors of developing *rd* mouse retinas. *Investigative Ophthalmology and Visual Science* **29**, 1021–1027.

Lem, J., Applebury, M. L., Falk, J. D., Flannery, J. G. and Simon, M. I. (1991). Tissue-specific and developmental regulation of rod opsin chimeric genes in transgenic mice. *Neuron* **6**, 201–210.

Lem, J., Flannery, J. G., Li, T., Applebury, M., Farber, D. B. and Simon, M. (1992). Retinal degeneration is rescued in transgenic *rd* mice by expression of the cGMP phosphodiesterase β subunit. *Proceedings of the National Academy of Science (USA)* **89**, 4422–4426.

Levy, E. R., Powell, J. F., Buckle, V. J., Hsu, Y. P., Breakefield, X. O. and Craig, I. W. (1989). Localization of human monoamine oxidase–A gene to Xpl1.23-11.4 by in situ hybridization: implications for Norrie disease. *Genomics* **5**, 368-370.

Lewis, R. A., Otterud, B., Stauffer, D., Lalouel, J.-M. and Leppert, M. (1990). Mapping recessive ophthalmic diseases: Linkage of the locus for Usher Syndrome type II to a DNA marker on chromosome lq. *Genomics* **7**, 250–256.

Lolley, R. N., Farber, D. B. and Hollyfield, J. G. (1977). Cyclic GMP accumulation causes degeneration of photoreceptor cells: simulation of an inherited disease. *Science* **196**, 664–666.

Lolley, R. N. and Farber, D. B. (1980). Cyclic GMP metabolic defects in inherited disorders of *rd* mice and RCS rats. In *Neurochemistry of the Retina,* edited by N. G. Bazan and R.N. Lolley, pp. 427–441. New York, Oxford: Pergamon Press.

Lyon, M. and Kirby, M. C. (1990). Mouse chromosome atlas. *Mouse Genome* **90**, 22–44.

McLaughlin, M. E., Sandberg, M. A., Berson, E. L. and Dryja T. P. (1993). Recessive mutations in the gene encoding the β subunit of rod phosphodiesterase in patients with retinitis pigmentosa. *Nature Genetics* **4**, 130–134.

McWilliam, P., Farrar, G. J., Kenna, P., Bradley, D. G., Humphries, M. M., Sharp, E. M., McConnell, D. J., Lawler, M., Shiels, D., Ryan, C., Stevens, P., Daiger, S. P. and Humphries, P. (1989). Autosomal dominant retinitis pigmentosa (ADRP): Localization of an ADRP gene to the long arm of chromosome 3. *Genomics* **5**, 619–622.

Molday, R. S., Hicks, D., and Molday, L. (1987). Peripherin. A rim-specific membrane protein of rod outer segment discs. *Investigative Ophthalmology and Visual Science* **28**, 50–61.

Musarella, M. A., Weleber, R. G., Murphey, W. H., Young, R. S. L., Anson-Cartwright, L., Mets, M., Kraft, S. P., Polemeno, R., Litt, M. and Worton, R. G. (1989). Assignment of the gene for complete X-linked Congenital Stationary Night Blindness (CSNB1) to Xpll.3. *Genomics* **5**, 727–737.

Musarella, M. A., Anson-Cartwright, C. L., McDowell, C., Burghes, A. M., Coulson, S. E., Worton, R. G. and Rommens, J. M. (1991). Physical mapping at a potential X-linked Retinitis Pigmentosa locus (RP3) by pulsed-field gel electrophoresis. *Genomics* **11**, 263–272.

Nathans, J., Piantanida, T. P., Eddy, R. L., Shows, T. B. Hogness, D. S. (1986). Molecular genetics of inherited variation in human color vision. *Science* **232**, 203–210.

Navon, S. E., Lee, R. H., Lolley, R. N. and Fung, B. K.-K. (1987). Immunological determination of transducin content in retinas exhibiting inherited degeneration. *Experimental Eye Research* **44**, 115–125.

Nöell, W. K. (1965). Aspects of experimental and hereditary retinal degeneration. In *Biochemistry of the Retina,* edited by C. N. Graymore, pp. 51–72. New York: Academic Press.

Ott, J., Bhattacharya, S., Chen, J. D., Denton, M. J., Donald, J., Dubay, C., Farrar, G. J. et al. (1990). Localizing multiple X chromosome-linked retinitis pigmentosa loci using multilocus homogeneity tests. *Proceedings of the National Academy of Science (USA)* **87**, 701–704.

Palczewski, K., Farber, D. B. and Hargrave, P. A. (1991). Elevated level of protein phosphatase 2A activity in retinas of *rd* mice. *Experimental Eye Research* **53**, 101–105.

Pittler, S. J. and Baehr, W. (1991). Identification of a nonsense mutation in the rod photoreceptor cGMP-PDE β subunit gene of the *rd* mouse. *Proceedings of the National Academy of Science (USA)* **88**, 8322–8326.

Robb, R. M. (1974). Histochemical evidence of cyclic nucleotide phosphodiesterase in photoreceptor outer segments. *Investigative Ophthalmology* **13**, 740–747.

Sanyal, S. and Bal, A. K. (1973). Comparative light and electron microscopic study of retinal histogenesis in normal and *rd* mutant mice. *Zeitschrift fur Anatomie und Entwicklungsgeschichte* **142**, 219–238.

Sanyal, S., Chader, G. and Aguirre, G. (1985). Expression of retinal degeneration slow (*rds*) gene in the retina of the mouse. In *Retinal Degeneration: Experimental and Clinical Studies,* edited by M. M. LaVail, J. G. Hollyfield and R. E. Anderson, pp. 239–256. New York: Alan R. Liss, Inc.

Schalken, J. J., Janssen, J. J. M., Sanyal, S., Hawkins, R. K. and de Grip, W. J. (1990). Development and degeneration in *rds* mutant mice: immunoassay of the rod visual pigment rhodopsin. *Biochimica et Biophysica Acta* **1033**, 103–109.

Searle, A. G., Peters, J., Lyon, M. F., Evans, E. P., Edwards, J. H. and Buckle, V. J. (1987). Chromosome maps of man and mouse, III. *Genomics* **1**, 3–18.

Shuster, T. A. and Farber, D. B. (1984). Phosphorylation in sealed rod outer segments: effects of cyclic nucleotides. *Biochemistry* **23**, 515–521.

Shuster, T. A. and Farber, D. B. (1986). Rhodopsin phosphorylation in developing normal and degenerative mouse retinas. *Investigative Ophthalmology and Visual Science* **27**, 264–268.

Sidman, R. L. and Green, M. C. (1965). Retinal degeneration in the mouse. *Journal of Heredity* **56**, 23–29.

Small, K. W., Weber, J. L., Roses, A., Lennon, F., Vance, J. M. and Pericak-Vance, M. A. (1992). North Carolina macular dystrophy is assigned to chromosome 6. *Genomics* **13**, 681–685.

Sonohara, O. and Shiose, Y. (1968). Electron microscopic study of the visual cell of inherited retinal dystrophic mice. *Folia Ophthalmologica Japonica* **19**, 77–86.

Southern, E. M. (1975). Detection of specific sequences among DNA fragments separated by gel electrophoresis. *Journal of Molecular Biology* **98**, 503–517.

Sparkes, R. S., Klisak, I., Kaufman, D., Mohandas, T., Tobin, A. J. and McGinnis, J. F. (1986). Assignment of the rhodopsin gene to human chromosome 3, region 3q21–3q24 by hybridization studies. *Current Eye Research* **5**, 797–798.

Suber, M. L., Pittler, S. J., Qin, N., Wright, G.C., Holcomb, V., Lee, R. H., Craft, C. M., Lolley, R. N., Baehr, W. and Hurwitz, R. L. (1993). Irish setter dogs affected with rod-cone dysplasia contain a nonsense mutation in the rod cyclic GMP phosphodiesterase β subunit gene. *Proceedings of the National Academy of Science (USA)* **90**, 3968–3972.

Sugimoto, K., Nukada, T., Tanabe, T., Takahashi, H., Noda, M., Minamino, N., Kangawa, K., Matsuo, H., Hirose, T., Inayama, S. and Numa, S. (1985). Primary structure of the bovine transducin β-subunit deduced from the cDNA sequence. *FEBS Letters* **191**, 235–240.

Tawara, A., and Hollyfield, J. G. (1990). Proteoglycans in the mouse interphotoreceptor matrix. III. Changes during photoreceptor development and degeneration in the *rds* mutant. *Experimental Eye Research* **51**, 301–315.

Travis, G. H., Brennan, M. B., Danielson, P. E., Kozak, C. A. and Sutcliffe, J. G. (1989). Identification of a photoreceptor-specific mRNA encoded by the gene responsible for retinal degeneration slow (*rds*). *Nature* **338**, 70–73.

Travis, G. H., Christerson, L., Danielson, P. E., Klisak, I., Sparkes, R. S., Hahn, L. B., Dryja, T. P. and Sutcliffe, J. G. (1991). The human retinal degeneration slow (*rds*) gene: chromosome assignment and structure of the mRNA. *Genomics* **10**, 733–739.

Travis, G. H., Sutcliffe, J. G. and Bok, D. (1992a).The retinal degeneration slow (*rds*) gene product is a photoreceptor disc membrane-associated glycoprotein. *Neuron* **6**, 61–70.

Travis, G. H., Groshan, K. R., Lloyd, M. and Bok, D. (1992b). Complete rescue of photoreceptor dysplasia and degeneration in transgenic *retinal degeneration slow (rds)* mice. *Neuron* **9**, 113–119.

Tuteja, N. and Farber, D. B. (1988). γ-subunit of mouse retinal cyclic GMP-phosphodiesterase: cDNA and corresponding amino acid sequences. *FEBS Letters* **232**, 182–186.

Tuteja, N., Tuteja, R. and Farber, D. B. (1989). Cloning and sequencing of the γ-subunit of retinal cyclic GMP-phosphodiesterase from *rd* mouse. *Experimental Eye Research* **48**, 863–872.

van Nie, R., Ivanyi, D. and Demant, P. (1978). A new H-2 linked mutation, *rds* causing retinal degeneration in the mouse. *Tissue Antigens* **12**, 106–108.

van Veen, T., Ekstrom, P., Wiggert, B., Lee, L., Hirose, Y., Sanyal, S. and Chader, G. J. (1988). A developmental study of interphotoreceptor retinoid-binding protein (IRBP) in single and double homozygous *rd* and *rds* mutant mouse retinae. *Experimental Eye Research* **47**, 291–305.

Watson, G., Felder, M., Rabinow, L., Moore, K., LaBarca, C., Tietze, C., Vander Molen, G., Bracey, L., Brabant, M., Cai, J. and Paigen, K. (1985). Properties of rat and mouse β-glucuronidase mRNA and cDNA, including evidence for sequence polymorphism and genetic regulation of mRNA levels. *Gene* **36**, 15–25.

Zack, D. J., Bennett, J., Wang, Y., Davenport, C., Klaunberg, B., Gearhart, J. and Nathans, J. (1991) Unusual topography of bovine rhodopsin promoter-*lacZ* fusion gene expression in transgenic mouse retinas. *Neuron* **6**, 187–199.

6. AUTOSOMAL DOMINANT RETINITIS PIGMENTOSA

PETER HUMPHRIES, PAUL KENNA and G. JANE FARRAR

Department of Genetics, Trinity College, Dublin 2, Ireland

SUMMARY

Retinitis pigmentosa (RP) is the collective term for a group of inherited diseases of the eye involving a degeneration of the retina, the light sensitive membrane carrying the rod and cone photoreceptors. As these cells die off, symptoms of night blindness develop, usually followed by a gradual constriction of the patient's field of vision. In addition, pigmentary deposits may build up in the retina as a result of degeneration of the neural epithelial cells, which has given rise to the name retinitis pigmentosa. RP is a prevalent genetic disorder, affecting approximately one in every four thousand persons (probably about one and a half million people throughout the world). It can be inherited in an X-linked, an autosomal dominant or a recessive mode. Approximately fifty percent of cases have no family history and are presumed to be the result of autosomal recessive disease, reduced penetrance, new X-linked or autosomal dominant mutations or phenocopies. It is in the autosomal dominant form of the disease that much recent progress has been made in our understanding of the molecular pathology of RP. Here, the genetic aetiology has been shown to have its origins in defects either of the rhodopsin or rds/peripherin proteins, both of which are components of the rod photoreceptor outer segment disc membranes. Recently rhodopsin mutations have also been identified in patients with an autosomal recessive form RP. Genetic linkage studies have identified the location of further loci responsible for adRP on chromosomes 8, 7p, 7q and 19q indicating that further genes remain to be identified. An intriguing picture of the molecular basis for the degenerative processes involved in RP is now emerging.

INTRODUCTION

The group of degenerative conditions of the eye termed retinitis pigmentosa (RP) represents the most prevalent retinopathy (disease of the retina) showing clear-cut Mendelian inheritance. The disease is at present incurable, appears to be present in all of the world's populations and currently afflicts up to 1.5 million people. Pathological changes associated with the disease include the progressive death of rod photoreceptor cells. Of the two types of photoreceptor within the retina, rods and cones, the rods are by far the most light-sensitive, containing the visual pigment rhodopsin. Rhodopsin is too sensitive to function in normal daylight, but enables the eye to perceive light in dimly lit conditions (at night, for example). As the rod cells die off, night blindness, or nyctalopia, develops. In some cases, the progression of the disease is rapid, night vision problems developing within the first decade of life. In other instances, such symptoms may not manifest until the fifth decade or beyond. Variability of clinical expression is a classical feature in some genetic types of RP, particularly the dominant form. The death of the rod photoreceptors is followed by more extensive pathological changes in the retina. Cone cells, responsible for the perception of colour, begin to die off and there is a gradual constriction of the patient's visual fields (Figure 1).

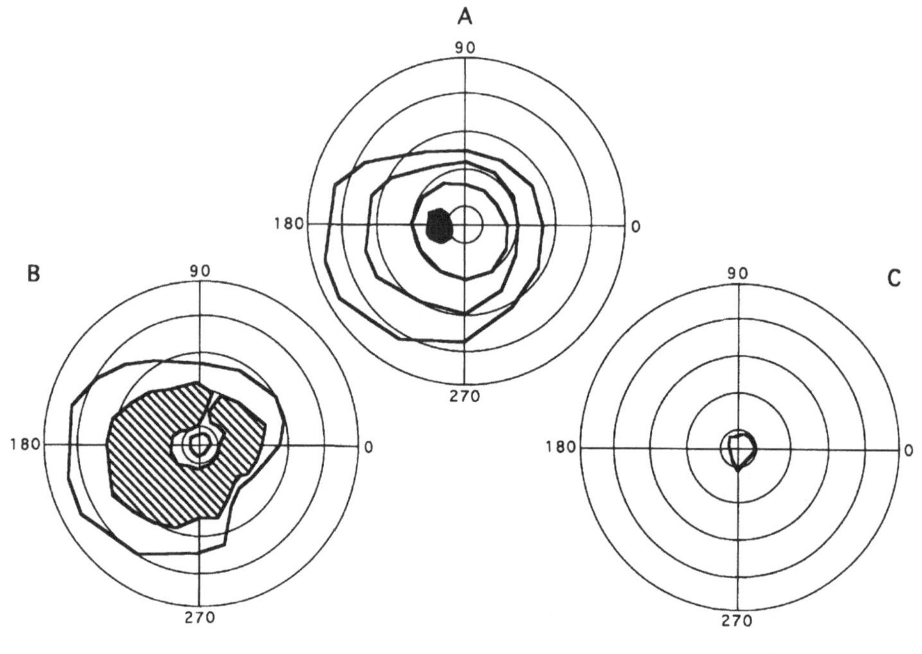

Figure 1 Diagrams of the visual fields of the left eye from normal and RP patients.
(A) shows the visual field of a normal individual, using targets of decreasing size. The black area is the extent of the physiological blind spot, caused by the optic nerve head.
(B) shows the visual field of a RP patient with moderately advanced retinopathy. The hatched area is the extent of a non-seeing area (scotoma) in the retinal mid-periphery to the largest sized target. Note also the loss of peripheral field in the lower right quadrant.
(C) shows the grossly constricted visual field, to the largest sized target of an RP patient with advanced retinopathy. This is the stage of 'tunnel vision'.

The retina visibly thins and the blood vessels supplying the retinal tissue begin to attenuate. Gradually, black pigmentary deposits build up in the neural retina, the result of damage to the photoreceptor and pigment epithelial cells, which leads to migration of pigment laden cells into the outer and inner retinal layers. Many patients eventually lose all of their sight. Pathological changes in the retinal fundus typical of RP are illustrated Figure 2.

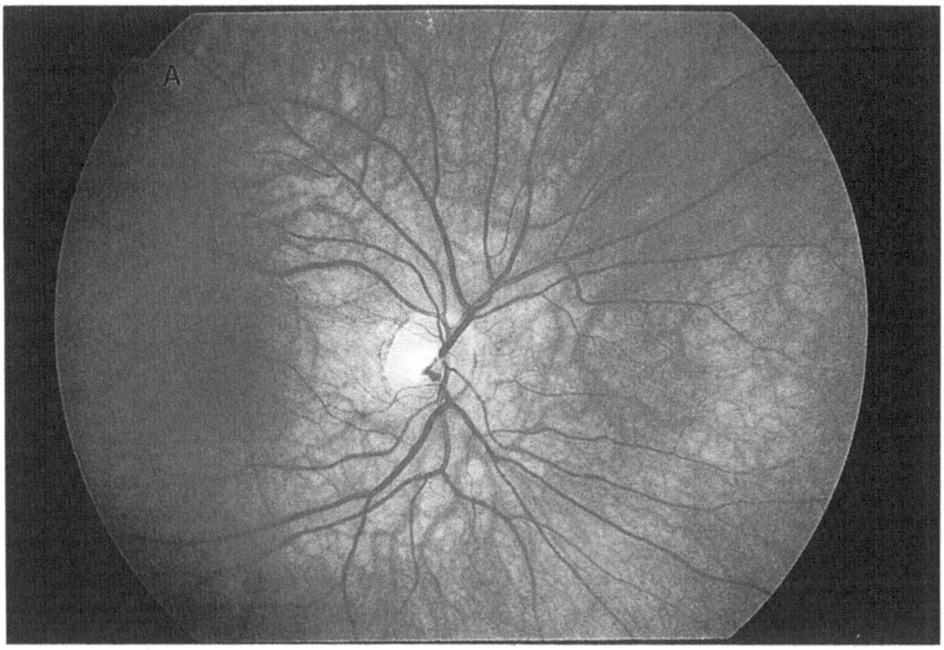

Figure 2 (A): show a normal optic disc and retinal vessels.

While the symptoms of RP clearly originate in the death of rod cells, nothing was known, until very recently, of the molecular events associated with this degenerative process. Rod cells carry a sophisticated battery of proteins forming components of the visual transduction cycle, the process through which light is absorbed and converted into neuronal signals (Chabre and Deterre, 1989; Chabre and Vuong, 1992). The disease could, in some way, be associated with defects in this cycle. Alternatively, defects could lie in structural components of the retinal cells or even in cells of the underlying pigment epithelium. Possibly, the problem could be one of vascular supply. A gene, or genes associated with the maintenance of the fine network of capillaries suffusing the retina could be defective, resulting in gradual cell death through ineffectual blood supply. With no clues, the search for the defective protein, or proteins, might seem virtually impossible. Fortunately, a powerful technique is available to enable the localization of defective genes, provided that sufficiently large families can be located in which the disease is

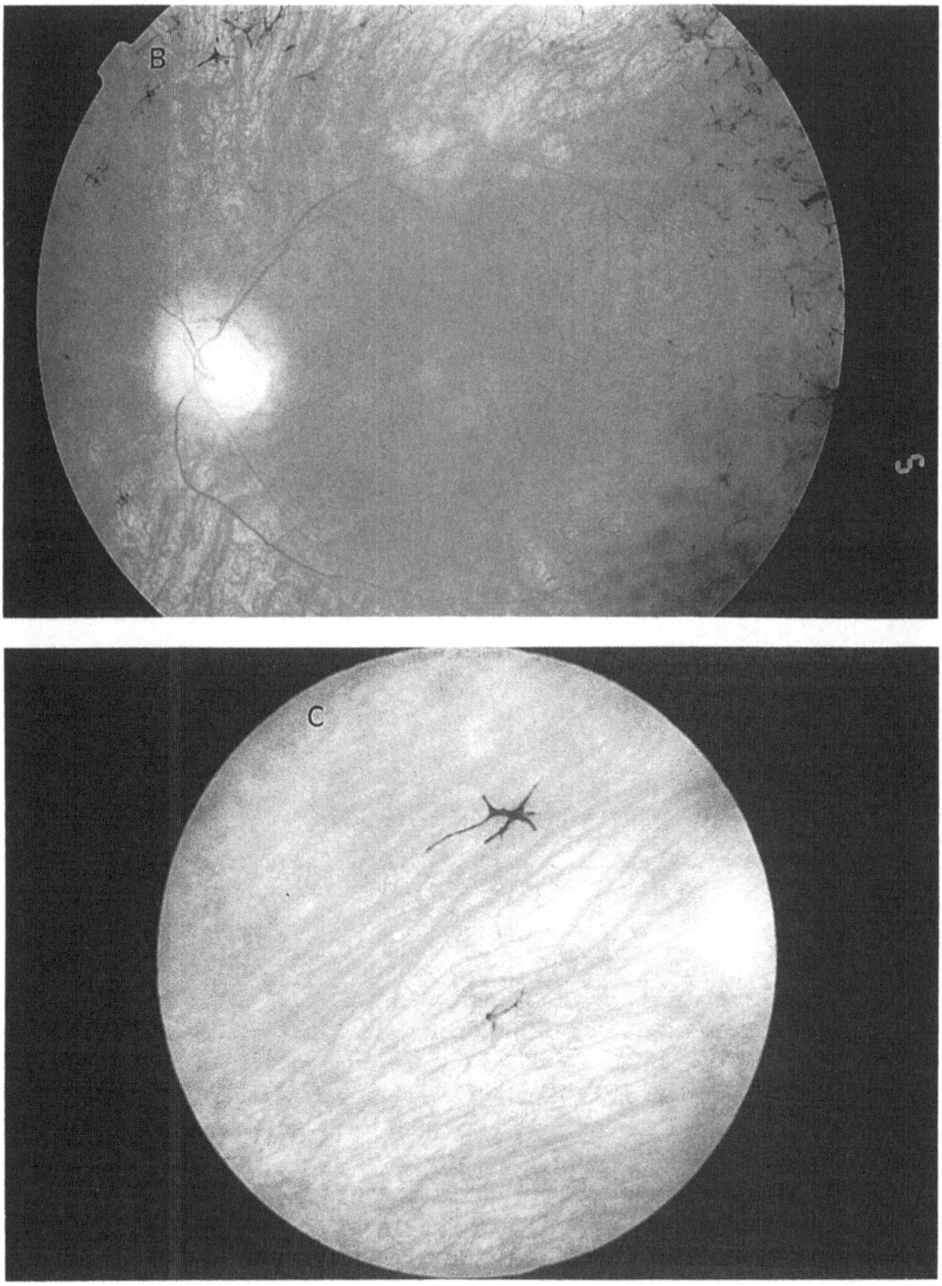

Figure 2 (B): shows the optic disc pallor and attenuation of retinal vessels in the fundus of an RP patient. Typical intra-retinal bone spicule pigmentary deposits are visible around the edges of the photograph.
(C): shows large bone spicule pigment deposit in the mid-peripheral retina of an RP patient. See Colour Plate I at the back.

segregating. The technique is called linkage analysis (Chapter 1). Genetic markers representing alleles (alternative forms) of a series of given chromosomal loci are tested, essentially randomly, in order to determine whether co-segregation can be observed within families between marker and disease locus. The establishment of "linkage" to a given marker is the first step in the isolation of the causative gene. Once the chromosomal location of the gene is established, the process of "positional cloning" (for review see Collins, 1992) can be deployed in order to isolate the gene from cloned DNA covering the region in question.

Two prerequisites for linkage analysis are the availability both of families with the disease and of suitable genetic markers. The first linkage of a genetic marker to a disease locus was established as early as the mid 1950s by Jan Mohr and colleagues (Mohr, 1951). These workers found that the dominant marker Se, now known to reside on chromosome 19 and governing the ability to secrete the A and B blood group substances in bodily fluids and saliva, showed genetic linkage to the myotonic dystrophy locus in families segregating this disease. Unfortunately however, the technique of genetic linkage analysis was available before the emergence of suitable marker technology. At the time of the Jan Mohr linkage, a mere handful of genetic markers, comprising polymorphic serum proteins and blood groups, had been characterized. While there were many hereditary conditions with large families segregating the disease, an insufficient number of markers were available to make the routine deployment of linkage analysis feasible. The technique had to await the development of a new generation of markers, which came with the establishment of RFLP (restriction fragment length polymorphism) technology in the late 1970s. Here, polymorphism at the level of the DNA sequence, in the form of variation in the size of restriction enzyme cleavage fragments at given loci, may be detected using the technique of "Southern blotting" (Chapter 1).

Armed with a growing number of markers of the RFLP type, members of our research team began, in the mid 1980s, to analyse DNA from a large family in Ireland who suffered from an autosomal dominant form of RP. Many markers were analysed (Bradley et al., 1989; Daiger et al., 1989; Farrar et al., 1989), and approximately 60% of the genome was covered in the search for a disease locus. Eventually, the marker C17 (D3S42) was shown to co-segregate with the disease locus with zero percent recombination (McWilliam et al., 1989). A portion of the family, illustrating the segregation of the C17 RFLP with the disease phenotype, is shown in Figure 3A. In all cases, allele A co-segregated with the disease, resulting in a lod score of 14.9. A lod score of 3 is taken to indicate an approximately 95% probability of linkage(Chapter 1). The observed lod score of 14.9 thus produced a very strong indication of linkage, and enabled the mapping of the disease locus to an interval of a few millions of base pairs on either side of the marker (Farrar et al., 1990).

C17 is located on the long arm of chromosome 3 and a number of additional markers were very rapidly run over the region. One of these markers happened to be derived from the recently mapped rhodopsin gene. In a multilocus analysis (Farrar et al., 1990), in which data from a number of mapped markers from the

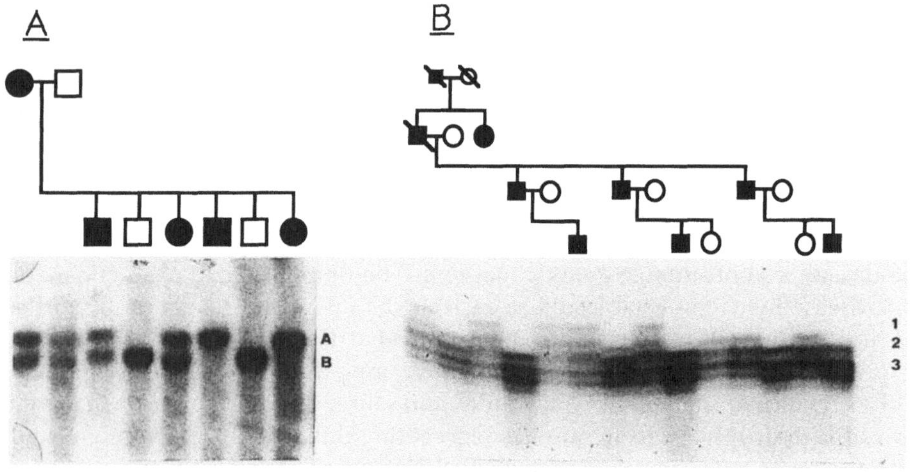

Figure 3 (A) shows a portion of an adRP family in which the disease phenotype segregates with allele A of the C17 RFLP.
(B) shows a portion of another adRP family in which the disease phenotype segregates with allele 3 of the rds poly T/A polymorphism.

long arm of chromosome 3 were simultaneously analysed, the rhodopsin marker was shown to map to the disease locus with zero recombination and a lod score of approximately 20. These results very strongly suggested that a mutation of the rhodopsin gene might be the underlying defect which caused RP in the Irish family.

MUTATIONS OF THE RHODOPSIN GENE IN ADRP

An intensive search for mutations within the rhodopsin gene in patients suffering from autosomal dominant RP (adRP) was mounted as a result of the above observations. A mutation was rapidly found in an American patient (Dryja et al., 1990). This was a single amino acid substitution (a C→A transversion) at codon 23 of the protein, resulting in the removal of proline and its substitution by histidine. This mutation has now been found in up to 15% of all dominant cases of RP in the United States. Interestingly, the mutation is either totally absent within the population of Europe, or present at a very low frequency (Farrar et al., 1990).

Using a variety of new techniques, the rapid screening of DNA for disease-related mutations is now possible. A combination of such methodologies, including single strand conformational polymorphism (SSCP), direct sequencing, heteroduplex analysis and allele-specific oligonucleotide analysis have been used in screening the rhodopsin gene from patients with RP. Using the first technique, SSCP analysis (Orita et al., 1989), small sectors of the gene (up to a few hundreds of base pairs) are amplified from genomic DNA using the polymerase chain reaction. The amplified DNA is denatured and subjected to electrophoresis in a neutral, non-denaturing acrylamide gel. Because the two single strands of DNA may assume different three dimensional conformations as a result of intra-strand pairing, they often separate in the gel, sometimes by considerable distances. Even the presence of a single base change in one of the two homologous DNA molecules is often sufficient to produce a radical mobility shift, sufficient to produce four bands (two originating from each double strand) instead of two. Evidence for the presence of a mutation within the rhodopsin gene, using this technique, is illustrated in Figure 4A. The DNA may then be subjected to direct sequence analysis in order to confirm the presence of a mutation. Again, sequencing is performed on amplified genomic DNA. Hence DNA from both homologous loci are amplified and sequenced at once (Figure 4B).

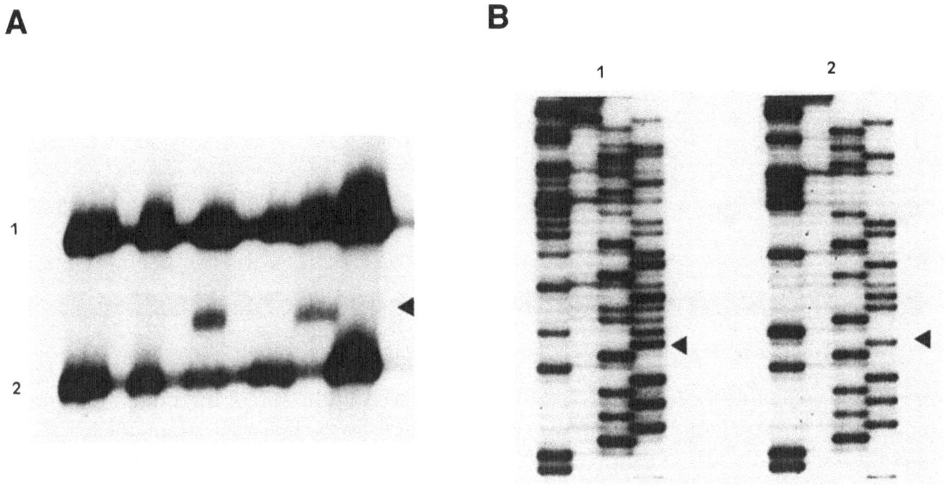

Figure 4 (A) shows results from an SSCPE analysis. Bands 1 and 2 represent the two denatured DNA strands from exon 3 of the rhodopsin gene. A third band is present in two of the lanes, indicating the presence of a sequence difference between the alleles of the DNA amplified from these individuals (highlighted by an arrow).
(B) shows direct sequence from PCR amplified DNA products from exon 1 of the rds gene. Sample 1 shows sequence from an individual heterozygous for a 3bp deletion, sample 2 shows the normal sequence from the same region. The site of the deletion at codons 118/119 is indicated by an arrow.

Point mutations are identified as double bands at the same position in the sequencing gel. The presence of a mutation may then be confirmed by the use of allele-specific oligonucleotides (ASOs). Short primers specific either for the normal or mutant DNA are synthesised and allowed to hybridise to DNA from members of the pedigree. DNA is usually immobilised on filters in the form of "dot blots". Normal DNA hybridises only to the normal ASO, while mutant DNA (because it carries both normal and mutant alleles) will hybridise to both normal and mutant ASOs. An example of the detection of a peripherin/RDS in mutation segregating in DNA from an RP pedigree using the ASO technique is illustrated in Figure 5.

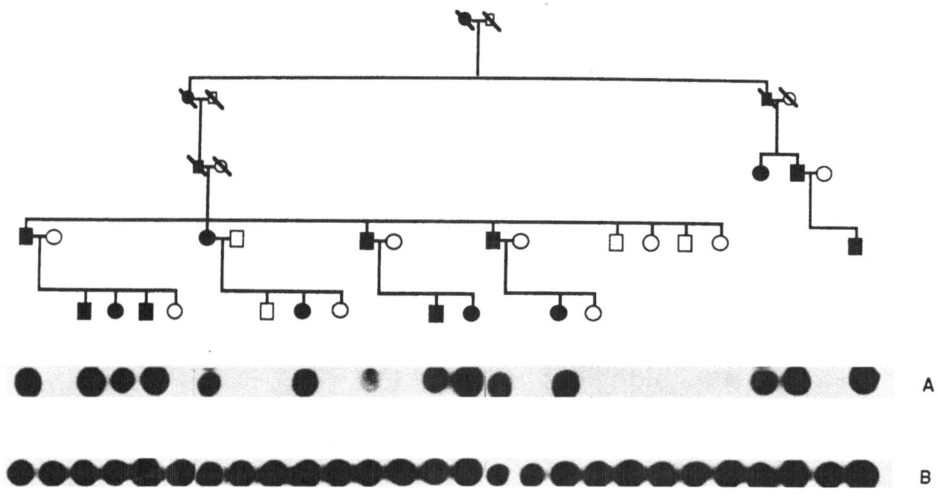

Figure 5 Segregation of the disease phenotype and a mutation at codon 212 of the rds gene in a family with adRP is shown. DNA was amplified around the site of the mutation from all members of the family and was then hybridised to both the normal and mutant allele specific oligonucleotides (ASO). Only affected individuals hybridised to the mutant ASO (A), both normal and affected individuals hybridised to the normal ASO (B)

Using a combination of these, and similar techniques, at least 60 mutations within the rhodopsin gene have been identified in patients with dominant RP over the last three years (reviewed by Humphries et al., 1992). The positions of these mutations within the protein are illustrated in Figure 6. In order to gain some insight into how such mutations might cause RP, an understanding of the function of rhodopsin within the visual transduction cycle is required (Chabre and Deterre, 1989; Chabre and Vuong, 1992)

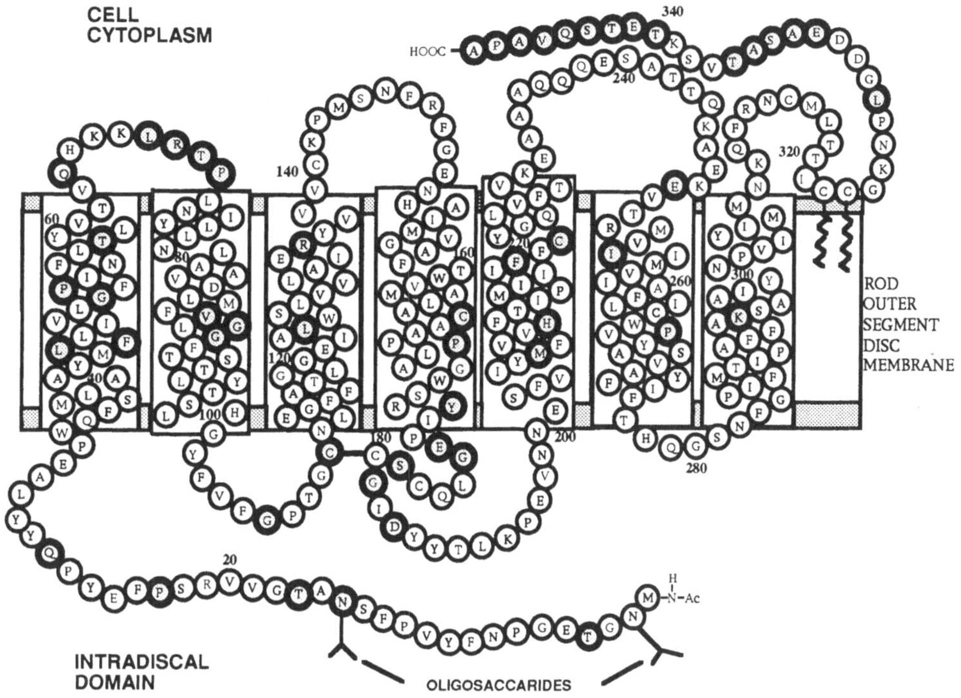

Figure 6 A diagramatic representation of the rhodopsin molecule indicating the positions at which mutations in adRP and arRP patients have been observed. These residues are indicated by thicker filled circles.

Rod photoreceptors consist of an outer and an inner segment (Figure 7). Rhodopsin synthesis takes place in the inner segment, and the protein is then transported to the outer segment discs. The discs are flattened structures stacked on top of each other rather like thin wafers. Each rod cell contains many such discs. The rhodopsin protein becomes incorporated into the membranes of the discs as illustrated in Figure 7, and it is here that light is absorbed. The amino terminus of the protein is situated in the lumen of the disc, while the carboxy terminus lies within the cell cytoplasm. The protein has seven transmembrane domains. Within the 7th transmembrane domain, at the lysine residue of position 296, a molecule of the chromophore 11-cis retinal is bound. Absorption of a single photon of light is sufficient to result in the isomerization of the chromophore to the all-trans form. This activates rhodopsin as a result of a change in its configuration, and in this form it may react with the second component of the cycle, a G-protein called transducin (Figure 8). Transducin consists of three components, alpha, beta and gamma subunits. As a result of its interaction with activated rhodopsin, the beta and gamma subunits dissociate and GDP, on the alpha subunit of the enzyme, isreplaced by GTP. This activated alpha subunit of transducin then activates,

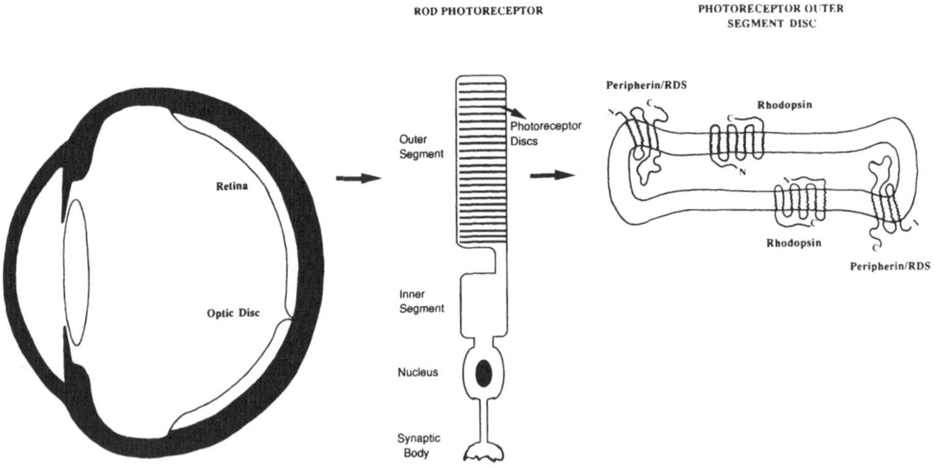

Figure 7 In this diagram, both the rhodopsin and rds proteins are seen embedded in the photoreceptor outer segment disc membrane. Both proteins twist in and out of the lipid bilayer having 7 and 4 transmembrane segments respectively.

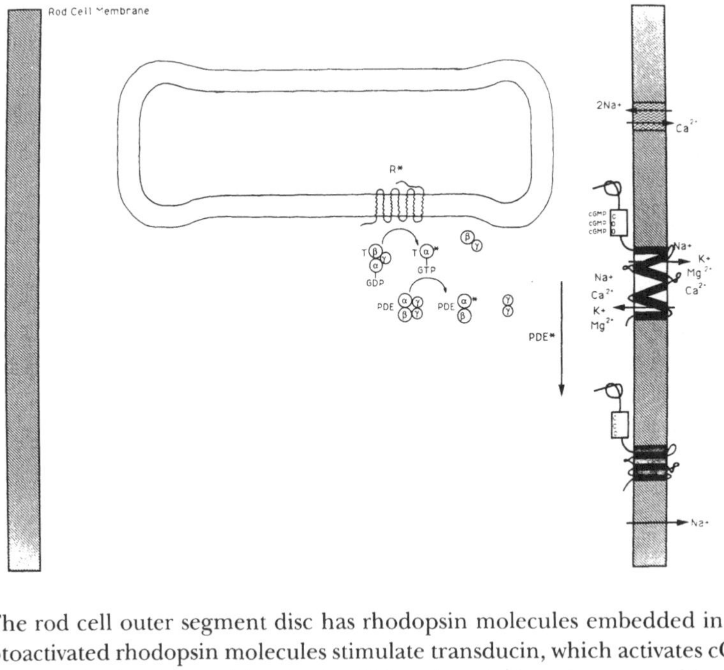

Figure 8 The rod cell outer segment disc has rhodopsin molecules embedded in the membrane. Photoactivated rhodopsin molecules stimulate transducin, which activates cGMP phosphodiesterase and leads to lowered levels of cGMP in the cytoplasm hence resulting in closure of cGMP-gated ion channels. The resulting lowered calcium levels leads to the stimulation of recoverin and guanylate cyclase activities and results in cGMP production and reopening of the channels.

in turn, the enzyme cyclic GMP phosphodiesterase (PDE). The latter protein consists of four subunits, alpha, beta and two gammas. Activation is achieved as a result of the removal from the protein of the two inhibitory gamma subunits. Activated PDE is able to rapidly hydrolyse cyclic GMP, which can bind to a transmembrane protein of the rod cell called the cyclic GMP-gated channel protein. While cyclic GMP is bound to this protein, the channel in the membrane of the rod cell is open, across which monovalent and divalent cations may diffuse. Cleavage of cyclic GMP lowers the intracellular concentration available for binding to the channel protein, which results in closure of the channel. This results in a rapid hyperpolarization of the cell, reducing neurotransmitter release and initiating a signal which is transmitted to the brain. In order to complete the cycle, cyclic GMP must be regenerated and rhodopsin deactivated. The former occurs as a result of continued removal of calcium ions from the rod cell via a sodium/calcium exchanger in the cell membrane. The lowering of calcium ion concentration results in the activation of the enzyme recoverin, which in turn stimulates guanylate cyclase. The latter enzyme regenerates cyclic GMP. Cyclic GMP then rebinds to the channel protein resulting in opening of the ion gate. Meantime rhodopsin is deactivated by the combined activities of two proteins, namely arrestin (S-antigen) and rhodopsin kinase. The cycle is represented diagrammatically in Figure 8.

Mutations within the rhodopsin gene in some way initiate a pathological process which ultimately results in the death of the rod cell. Most of the mutations so far detected in cases of adRP are single base pair substitutions (for a detailed list see Humphries et al., 1992). These have been found within the carboxy and amino terminus domains and also in each of the seven transmembrane domains of the molecule (Figure 6). Such mutations may have a number of effects. Firstly, they may interfere with the targeting and/or stability of the protein (e.g. thr-17-met, or pro-43-his/leu). Secondly, they may interfere with the folding or integration of the protein into the bilayer of the disc membrane (e.g. pro-53-arg or leu-125-arg). Thirdly, they may prevent binding of the chromophore 11-cis retinal (e.g. lys-296-glu). Fourthly, they may prevent rhodopsin from interacting with, or activating, transducin (e.g. arg-135-leu/trp). Finally, some mutations toward the extreme carboxy terminus of the molecule (e.g. thr-340-frame shift) may prevent phosphorylation of the molecule by rhodopsin kinase or its interaction with arrestin (S-antigen). Sung et al. (1991) succeeded in expressing 13 different mutant rhodopsins responsible for adRP by transfection into cultured human embryonic kidney cells. Three mutants resembled wild-type rhodopsin in their yield of pigment and their ability to bind 11-cis retinal and release all-trans retinal in response to light (Phe-45-Leu, Gln-344-ter, Pro-347-Leu). Two mutants produced photolabile pigments at greatly reduced yield compared with wild-type (Val-87-Asp, Gly-89-Asp). The remaining eight mutants failed to produce any detectable photopigments on incubation with 11-cis retinal (Thr-17-Met, Pro-23-His, Thr-58-Arg, Gly-106-Trp, Arg-135-Trp, Tyr-178-Cys, Asp-190-Gly). The last two groups were found to be inefficiently transported to the plasma membrane, remaining primarily in the endoplasmic reticulum (Sung et al., 1991). This suggests the possibility of a mechanism by which the endoplasmic reticulum retains and degrades incorrectly assembled or unstable proteins, which may be harmful to the photoreceptor. In addition to its role in absorbing light, rhodopsin is also a component of the outer disc membranes. Instability of the protein or incorrect folding into the membrane could interfere with the delicate molecular architecture of the outer segment discs, eventually resulting in rod cell destruction. Inhibition of the transduction cycle, a probable manifestation of some of the mutations described, might have similar effects.

TRANSGENIC MICE WITH THE ADRP PRO-23-HIS MUTATION

Olsson et al. (1992) have constructed a transgenic mouse containing the Pro-23-His mutation that is common in North America. The entire rhodopsin gene, including 4-5 kb of upstream and 6-8 kb of downstream regulatory sequences, was isolated from a patient with adRP who was heterozygous for this mutation. The normal and mutant genes were microinjected into the male pronuclei of single cell mouse embryos and five lines of transgenic mice were studied from the resultant progeny (two carrying the wild-type and four the mutant allele). All three mutant lines developed photoreceptor degeneration, the degree of degeneration correlating with the extent of mutant allele expression. The retinas of affected mice showed attenuated vessels and geographic patches of retinal pigment epithelial hyper- or hypopigmentation, contrasting with the dendritic clumps of pigment found in the human disorder. A monoclonal antibody specific to the human rod opsin detected expression of the mutant opsin in outer segments and failed to find accumulations in the endoplasmic reticulum. This contrasts with the findings of Sung et al. (1991) who showed a failure of the Pro-23-His opsin to be incorporated into plasma membranes in vitro and retention in the endoplasmic reticulum. Olsson et al. (1992) unexpectedly found abnormally located mutant opsin at the synaptic end of rods (outer plexiform layer), raising the possibility of disturbed intracellular transport.

Surprisingly, one of the two transgenics containing the wild-type allele also developed a very similar retinal degeneration. The ratio of human to mouse rod opsin mRNA expressed in this line was 5:1 compared to the unaffected wild-type transgenic which expressed equal amounts of human and mouse transcript. The degeneration was thought therefore to be related to overexpression of the (human) opsin gene.

These experiments show firstly that this mutation confers a degenerative phenotype, even in the presence of two normal mouse genes, which excludes the possibility that it is a null allele which gives rise to adRP as a result of a deficiency of normal gene product. Secondly, some proposed explanations for the degeneration, such as failure of RPE cells to phagocytose rod outer segments or inability of the mutant protein to be incorporated into outer segment discs, appeared less likely in the light of these findings. Finally, the finding that overexpression even of the normal human gene product causes a severe retinal degeneration perhaps emphasises how finely tuned the retina is for maintaining normal structure and function.

A DIVERSE MOLECULAR PATHOLOGY IN ADRP

Rhodopsin mutations account for only about 25% of all dominant cases of RP. Following the establishment of the initial rhodopsin linkage, we undertook additional linkage work in a second adRP pedigree of Irish origin which showed exclusion of a locus from the long arm of chromosome 3 (Farrar et al., 1990). In this family the disease was of much later onset than that in the first pedigree. RP is clinically divisible into two major forms, so called types 1 and 2, the former being the more severe (Chapter 3). The disease is also classifiable as either a diffuse or a regional retinopathy, depending on the nature and extent of pathological changes in the retina. Diffuse and regional forms of the disease usually correlate with type 1 and 2 forms respectively. Members of the second Irish family had a regional (type 2) form of RP with night blindness and field loss not usually manifesting until the fourth decade. By the time of this second linkage study, a new generation of genetic markers had become available. Such markers are called "microsatellites" (Weber and May, 1989; Litt and Lutty 1989). Human DNA contains many (one hundred thousand or more) short blocks of dimer repetitive sequence of the type, $d(CA)_n$, interspersed throughout the genome. The numbers of CA repeats at any given locus usually varies on each homologous chromosome. Oligonucleotides representing the sequence of DNA immediately flanking such repeats can be synthesized and used as primers in the polymerase chain reaction to amplify the segment of the genome spanning the block of repetitive DNA. Amplified DNA may then be analysed by electrophoresis in acrylamide gels, similar to the type used for DNA sequencing. Variation in the length of the repetitive block, even if this is as small as a single CA unit, is readily detectable. Microsatellites have two major advantages over conventional RFLPs. Firstly, they can be used without the need for enzymatic digestion or Southern blotting, in which lengthy gel separations, DNA transfer to membranes, hybridisations and autoradiography are required. Secondly, microsatellites are generally of much higher information content than conventional markers. This means that a proportionately greater number of genetic crosses are informative for linkage in any particular pedigree.

Continued linkage studies, based largely on the use of such markers, resulted in the localisation of a causative gene (Farrar et al., 1991; Jordan et al., 1991; Kumar-Singh et al., 1991). The gene resided close to the HLA locus on the short arm of chromosome 6. Interestingly, the gene for an additional protein component of the rod outer segment discs, peripherin/RDS, had recently been mapped to the same region (Travis et al., 1991). Co-segregation of the disease locus with the microsatellite marker at the 3' end of the peripherin/RDS gene is illustrated in Figure 3B. This was of special significance, since the rds/peripherin protein had previously been implicated in the cause of the naturally occurring retinopathy of mice called retinal degeneration slow (rds) (Chapter 5). The protein takes its name from this rodent disease which, in many respects, is similar to RP in man. The rds/peripherin protein thus became a prime candidate for human adRP and a mutation within the gene was soon found in a patient from an Irish adRP pedigree (Farrar et al., 1991). The mutation was a trinucleotide deletion of codons 118/119, removing one of a pair of conserved cysteine residues.

The human peripherin/RDS gene is highly conserved in mice, cattle, rats and humans and contains three exons, coding for a predicted 193, 83 and 70 amino acid residues respectively. The protein product is a 39 kDa glycoprotein but much less is known of its structure and function compared with rhodopsin. It is not believed to be a component of the visual transduction cycle but, like rhodopsin, it resides within the membranes of outer segment discs (Figure 7). Unlike rhodopsin, it is present in rod and cone photoreceptor outer segments (Arikawa et al., 1992). To date, about thirteen mutations within the peripherin/RDS gene have been reported in cases of adRP (Farrar et al., 1992; Kajiwara et al., 1992; S. Bhattacharya, A. Gal, personal communications). The original Irish pedigree in which linkage was established contains an amino acid substitution at codon 212 (Farrar et al., 1992). In addition, a number of "silent" polymorphic amino acid substitutions have been detected which are not associated with a disease phenotype (Jordan et al., 1992). Three of these polymorphisms cluster toward the carboxyl terminus of the molecule. (Interestingly, these "silent" polymorphic variations occur at residues hitherto considered to be conserved among mammalian rds/peripherin proteins). Current models suggest that the rds/peripherin protein has four membane-spanning α-helical domains with both the carboxyl and amino acid terminals facing the cytoplasmic side of the disc membrance (Figure 9). The protein is believed to have two intradiscal loops, one much larger than the other. It is possible that the protein may dimerise via at least one disulphide bridge, and it may have a role in maintaining disc structure.

It is too early to say what proportion of adRP is caused by mutations within the RDS gene. The figure may well be similar to the frequency of rhodopsin-associated RP. However, a number of mutations in the peripherin/RDS gene have recently been identified in families segregating dominantly inherited disorders other than RP. Kajiwara et al.(1993) screened over 350 patients with a variety of hereditary retinal disorders for peripherin/RDS gene mutations using SSCP analysis. Exons 1 and 2 together with some flanking DNA (intron or 5' untranslated) were screened, omitting exon 3 which contains frequent polymorphisms in normal individuals. Only one patient, with autosomal dominant retinitis punctata albescens, was found to have a mutation. The clinical phenotype consisted of macular hyperpigmentation and atrophy associated with widespread yellow-white flecks under the retina.

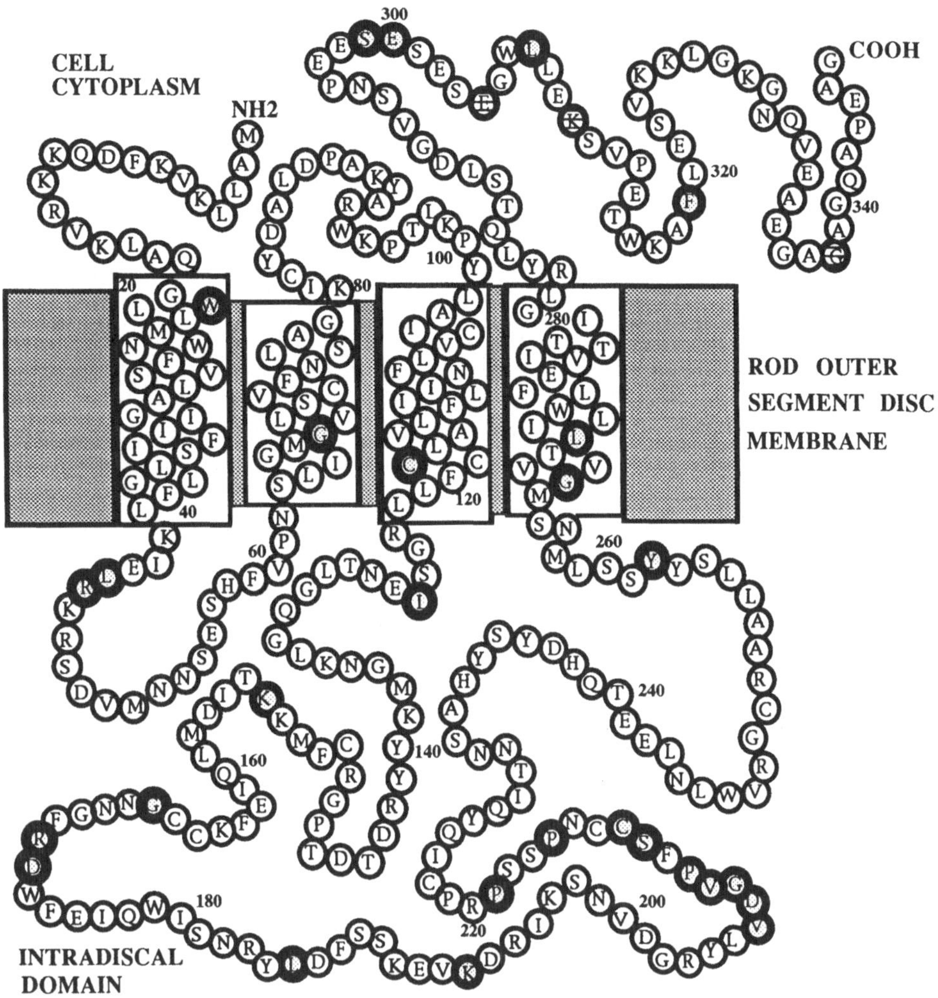

Figure 9 A diagramatic representation of the rds protein indicating the amino acids in which mutations have been identified in adRP patients. These residues are indicated by thicker filled circles. Nucleotide changes which are silent or which alter amino acids but are not associated with the adRP phenotype are indicated by circles filled with stripes.

Affected individuals had progressive loss of central vision, abnormal dark adaptation and abnormal rod and cone ERG responses, together indicating a generalised retinal degeneration. A 2-base pair deletion was found in codon 25 of the peripherin/RDS gene, representing a null allele since this would lead to a severely truncated protein product, similar to the murine rds mutation. Wells et al. (1993) reported three families with autosomal dominant macular degenerations associated with missense mutations at codon 172. They also reported finding a nonsense mutation at codon 258, producing a premature stop codon, in a family with autosomal dominant adult vitelliform macular dystrophy. This may also be a null allele, similar to that in the patient with retinitis punctata albescens, although the phenotypes are rather different. Finally, Nichols et al. (1993) described a family with autosomal dominant butterfly-shaped pigment dystrophy of the fovea associated with a missense mutation in peripherin/RDS codon 167, which would substitute an aspartate for a highly conserved glycine residue. The mutation segregated with the disease (lod score = 4 at a recombination fraction of zero) and was not present in 100 control chromosomes. The disorder is characterized by accumulations of yellowish or pigmented material at the level of the retinal pigment epithelium (RPE), central visual loss apparently related to the extent of accompanying RPE atrophy, normal dark adaptation thresholds and electroretinograms. Hence, like rhodopsin, mutant rds/peripherin proteins show a wide range of phenotypic expression. 'Indeed, most recently families with digenic adRP have been identified, in which only patients who inherit mutations in both the rds/peripherin gene and the gene encoding the rod outer membrane (ROM) protein are affected (T. Dryja, The Association for Research in Vision and Ophthalmology Meeting, 1994). Such families segregating with digenic adRP show autosomal dominant inheritance with reduced penetrance'.

Additional RP families have been identified in which neither rhodopsin nor rds/peripherin is involved. One such family forms the large adRP pedigree UCLA1 from the United States. Here, the causative gene has been localised by genetic linkage analysis to the pericentric region of chromosome 8 (Blanton et al., 1991). However, no potential candidate gene has, as yet, been identified. Most recently, two further genes have been identified in large adRP families on opposite arms of chromosome 7. Inglehearn et al. (1993) found linkage to a gene on the short arm of chromosome 7 (7p) in a large English adRP kindred, close to the marker D7S460 ($Z = 5.60$ at $\theta = 0$). Another large kindred was found by Jordan et al. (1993) to have a disease gene linked to markers on the long arm of chromosome 7, close to the marker D7S480 ($Z = 7.20$ at $\theta = 0$). Most recently an adRP gene has been localised to the long arm of chromosome 19 close to the markers D19S180 and D19S214 (Al-Maghteh et al., 1994). In addition, at least one other adRP family currently under study in our own laboratory shows no evidence for linkage on chromosomes 3, 6, 7 or 8, indicating the presence of at least one more, as yet unlocalised, adRP gene (Kumar-Singh et al., 1992, 1993; Jordan et al., 1992).

Thus in the dominant form of RP, a minimum of six genes are responsible for the cause of the disease in different families. Possibly about one third of all dominant cases can be accounted for by known mutations in the rhodopsin and peripherin/RDS genes. It is of interest that both of these genes code for proteins which are located within the membranes of the outer segment discs. Whether all adRP proteins are located within these structures remains to be determined.

CONCLUSIONS

In this review we have limited our coverage to the dominant form of RP, in which two of the first proteins responsible for RP were identified. Perhaps the most surprising finding in adRP has been the very extensive genetic heterogeneity. There now appear to be a minimum of six loci responsible for this disorder, on chromosomes 3, 6, 7p, 7q, 8 and at least one other as yet unidentified site. Different mutations in the rhodopsin gene on chromosome 3 give rise to very different phenotypes, from a very early onset and severe degeneration in some types of adRP to the clinically normal carriers of the truncated molecule associated with autosomal recessive disease. In the case of the peripherin/RDS gene on chromosome 6, the phenotypic variation is proving even more diverse. Different mutations are associated either with predominantly macular or peripheral retinopathy. Families with macular dystrophy, adult vitelliform macular dystrophy, butterfly-shaped pigment dystrophy, retinitis pigmentosa and retinitis punctata albescens have all been associated with RDS/peripherin mutations. In some cases, more than one phenotype has been found within the same family. The clinical heterogeneity is as striking as the genetic heterogeneity.

Developments are rapidly taking place in other forms of retinitis pigmentosa and in related conditions. The locations of two X-linked RP genes have been known for some time (Ott et al., 1990). In autosomal recessive forms of the disease, genes for Usher syndromes type 1 and 2 (RP associated with deafness) have also been localised by genetic linkage analysis (Kimberling et al., 1990) (Chapter 15). Moreover, a mutation within the rhodopsin gene and several mutations within the β subunit of cyclic GMP phosphodiesterase have recently been identified in families displaying autosomal recessive RP (Rosenfeld et al., 1992; McLaughlin et al., 1993). Our own studies of a large family of Dutch origin indicate that there is an additional autosomal recessive locus on 1q close to the marker F13B and the gene encoding the retinal protein phosducin (Bleeker-Wagemakers et al., 1992; van Soest et al., 1994). In addition, genes responsible for two other dominantly heritable forms of macular degeneration – North Carolina macular dystrophy and Best's disease – have been localised on chromosomes 6 and 11 respectively (Small et al., 1992; Stone et al., 1992) (Chapter 15). The availability of additional pedigrees suitable for genetic linkage analysis, coupled with the now well developed techniques for positional cloning and rapid mutational screening, will ensure that the many of these genes will be characterised within the next few years.

REFERENCES

Arikawa, K., Molday, L. L., Molday, R. S. and Williams, D. S. (1992) Localization of peripherin/rds in the disk membranes of cone and rod photoreceptors: relationship to disk membrane morphogenesis and retinal degeneration. *J. Cell Biol.* **116**, 659–667.

Al-Maghtheh, M., Inglehearn, C. F., Keen, J. T., Evans, K., Moore, A. T., Jau, M., Bird, A. C. and Bhattacharya, S. S. (1994). Identification of a sixth locus for autosomal dominant retinitis pigmentosa on chromosome 19.

Blanton, S. H., Heckenlively, J. R., Cottingham, A. W., Freidman, J., Sadler, L. A., Wagner, M., Freidman, L. H., and Daiger, S. P. (1991). Linkage mapping of autosomal dominant retinitis pigmentosa to the pericentric region of chromosome 8. *Genomics* **11**, 857–869.

Bleeker-Wagemakers, L., Li, Y., Bergen, A., van den Born 1, Kumar-Singh R., Humphries, P., Gal, A., Sandkjuil, S. A.., Echart, K., and Farrar, G. J. (1992). Evidence for non allelic genetic heterogeneity in autosomal recessive retinitis pigmentosa. *Genomics* (in press).

Bradley, D. G, Farrar, G. J., Sharp, E. M., Kenna, P., Humphries, M. M., McConnell, D. J., Daiger, S. P., McWilliam, P. and Humphries, P. (1989). Autosomal dominant retinitis pigmentosa: Exclusion of the gene from the short arm of chromosome 1 including the region around the Rhesus locus. *Amer. J. Hum. Genet.* **44**, 570–576.

Chabre, M. and Deterre, P. (1989). Molecular mechanisms of visual transduction. *Eur. J. Biochem.* **179**, 255–266.

Chabre, M. and Vuong, T. M. (1992). Kinetics and energetics of the rhodopsin–transducin-cGMP phosphodiesterase cascade of visual transduction. *Biochemica. et. Biophysica. Acta.* **1101**, 260–263.

Collins, F. S (1992) Positional Cloning. *Nature Genetics* **1**, 3–10.

Connell, G., Molday, L. L., Reid D., and Molday, R. S. (1991). Molecular structure and properties of peripherin/rds the normal product of the gene responsible for retinal degeneration in the rds mouse. In *Retinal Degenerations*, pp. 467–477, CRC Press, eds. J. G. Hollyfield, R. E. Anderson and M. M. LaVail.

Daiger, S. P., Humphries, M.M., Giesenschlag, N., Sharp, E. M., McWilliam, P., Farrar, G. J., Bradley, D., Kenna, P., McConnell, D. J., Sparkes, R. S., Spence, M. A., Heckenlively, J. R. and Humphries, P. (1989). Linkage analysis of human chromosome 4: exclusion of autosomal dominant retinitis pigmentosa (adRP) and detection of new linkage groups. *Cytogenet. Cell Genet.* **50**, 181–187.

Dryja, T. P., McGee, T. L., Reichel, E., Hohn, L. B., Cowley, G. S., Yandell, D. N., Sandberg, M. A. and Berson, E. L. (1990). A point mutation in the rhodopsin gene in one form of retinitis pigmentosa. *Nature* **343**, 364–366.

Farrar, G. J., McWilliam, P., Sharp, E. M., Kenna, P., Bradley, D. G., Humphries, M. M., McConnell, D. J. and Humphries, P. (1989). Autosomal dominant retinitis pigmentosa: exclusion of the gene from extensive regions of chromosomes 6,13, 20 and 21. *Genomics* **5**, 612–618.

Farrar, G. J., McWilliam, P., Bradley, D. G., Kenna, P., Lawler, M., Sharp, E. M., Humphries, M. M., Eigberg, H., Conneally, P. M., Trofatter, J. A., and Humphries, P. (1990). Autosomal dominant retinitis pigmentosa: linkage to rhodopsin and evidence for genetic heterogeneity. *Genomics* **8**, 35–40.

Farrar, G. J., Kenna P., Redmond, R., McWilliam, P., Bradley, D. G., Humphries, M. M., Sharp, E. M., Inglehearn, C. F., Bashir, R., Jay, M., Whatty, A., Ludwig, M., Schinzel, A., Samanns, C., Gal, A., Bhattacharya, S. and Humphries, P. (1990). Autosomal dominant retinitis pigmentosa: absence of the rhodopsin Proline->Histidine substitution (codon 23) in pedigrees from Europe. *Amer. J. Hum. Genet.* **47**, 941–945.

Farrar, G. J., Jordan, S. A., Kenna, P., Humphries, M. M., Kumar-Singh R., McWilliam, P., Allamand, V., Sharp, E. M. and Humphries, P. (1991). Autosomal dominant retinitis pigmentosa: localization of a disease gene (RP6) to the short arm of chromosome 6. *Genomics* **11**, 870–874.

Farrar, G. J., Kenna, P., Jordan, S. A., Kumar-Singh R., Humphries, M. M., Sharp, E. M., Sheils, D. and Humphries, P. (1991). A three base deletion in the peripherin/RDS gene in one form of retinitis pigmentosa. *Nature* **354**, 478–480.

Humphries, P., Kenna, P. and Farrar, G. J. (1992). On the molecular genetics of retinitis pigmentosa. *Science* **256**, 804–808.

Inglehearn, C. F., Carter, S. A., Keen, T. J., Lindsey, J., Stephenson, A., Bashir, R., Al-Maghtheh, M., Moore, A. T., Jay, M., Bird, A. C. and Bhattacharya, S. S. (1993). A new locus for adRP on 7p. *Nature Genetics* **4**, 51–53.

Jordan, S. A., Farrar, G. J., Kenna, P., Kumar-Singh R., Humphries, M. M., Allamand, V., Sharp, E. M. and Humphries, P. (1992). Autosomal dominant retinitis pigmentosa (adRP; RP6): co-segregation of RP6 and the peripherin-rds locus in a late-onset family of Irish origin. *Amer. J. Hum. Genet.* **50**, 634–639.

Jordan, S. A., Farrar, G. J., Kenna, P. and Humphries, P. (1992). Polymorphic variation within 'conserved' sequences at the 3' end of the human RDS gene which results in amino acid substitutions. *Human Mutation* **1**, 240–247.

Jordan, S. A., del Rio, T., Soriano, N., Garcia-Sandoval, B., Kenna, P., Ayuso, C., Benitez, J. and Humphries, P. (1992). Autosomal dominant Retinitis Pigmentosa (adRP): exclusion of a gene from three mapped loci provides evidence for the existence of a fourth locus. *Hum. Mol. Genet.* **1**, 411–415.

Jordan, S. A., Farrar, G. J., Kenna, P., Humphries, M. M., Sheils, D., Kumar-Singh R., Sharp, E. M., Benitez, J., Carmen, A. and Humphries, P. (1993). Localisation of an adRP gene to 7q. *Nature Genetics* **4**, 54–58.

Kajiwara, K., Hahn, L. B., Mukai, S., Travis, G. H., Berson, E. L. and Dryja, T. P. (1991). Mutations in the human retinal degeneration slow gene in autosomal dominant retinitis pigmentosa. *Nature* **354**, 480–483.

Kimberling, W. J., Weston, M. D., Moller, C., Davenport, S. L. H., Shugart, Y. Y., Priluck, I. A., Martini, A., Milani, M., Smith, R. J. (1990). Localisation of Usher syndrome type ll to a DNA marker on chromosome 1q. *Genomics* **7**, 245–249.

Kumar-Singh R., Jordan, S. A., Farrar, G. J., and Humphries, P. (1991). Poly (T/A) repeat polymorphism at the human retinal degeneration slow (RDS) locus. *Nucleic Acid Research* **19**, 5800.

Kumar-Singh R., Kenna, P., Farrar, G. J., Wang, K. and Humphries, P. (1992). Evidence for further genetic heterogeneity in autosomal dominant retinitis pigmentosa. *Genomics* **15**, 212–215.

Kumar-Singh R., Farrar, G. J., Mansergh, F., Kenna, P., Bhattacharya, S. S., Gal, S. S. and Humphries, P. (1993). Diverse clinical heterogeneity in autosomal dominant retinitis pigmentosa is mirrored by extensive genetic heterogeneity — evidence for a sixth locus (RP8). *Human Molecular Genetics* (in press).

Litt, M. and Lutty, J. A. (1989). A hypervariable microsatellite revealed by in vitro amplification of a dinucleotide repeat within the cardiac muscle actin gene. *Amer. J. Hum. Genet.* **44**,397–401.

McLaughlin, M. E., Berson, E. L. and Dryja, T. P. (1993). Mutations in the subunit of phosphodiesterase in patients with autosomal recessive retinitis pigmentosa. *Investigative Ophthalmology and Visual Science* **34**, 1148.

McWilliam, P., Farrar, G. J., Kenna, P., Bradley, D. G., Humphries, M. M. , Sharp, E. M., McConnell, D. J., Lawler, M., Shiels, D., Ryan, C., Stevens, K., Daiger, S. P. and Humphries, P. (1989). Autosomal dominant retinitis pigmentosa (ADRP): Localisation of an ADRP gene to the long arm of chromosome 3. *Genomics* **5**, 619–622.

Mohr, J. (1951) Acta Pathol Microbiol Scand 28: 207; ibid. **29**, 339 TITLES.

Orita, M., Suzuki, Y., Sekiy, A. T. and Hyashi, K. (1989). Rapid and sensitive detection of point mutations and DNA polymorphisms using the polymerase chain reaction. *Genomics* **5**, 874–879.

Ott, J., Bhattacharya, S., Chen, J. D., Denton, M. J., Donald, J., Dubay, C., Farrar, G. J, Fishman, G. A., Frey, D., Gal, A., Humphries, P., Jay B., Jay, M., Litt, M., Machler, M., Musarella, M., Neugebauer, M., Nussbaum, R. L., Terwilliger, J. D., Weleber, R. G., Wirth, B., Wong, F., Worton, R. G., and Wright, A. F. (1990). Localisation of multiple retinitis pigmentosa loci on the X chromosome. *Proc. Natl. Acad. Sci.* **87**, 701–704.

Rosenfeld, P. J., Cowley, G. S., McGee, T. L., Sandberg, M. A., Berson, E. L. and Dryja, T. P. (1992). A null mutation within the rhodopsin gene causes rod photoreceptor dysfunction and autosomal recessive retinitis pigmentosa. *Nature Genetics* **1**, 209–213.

Small, K. W., Weber, J. L, Roses, A., Lennon, F., Vance, J. M., and Pericak-Vance, M. A. (1992). North Carolina Macular Dystrophy is assigned to chromosome 6. *Genomics* **13**, 681–685.

Stone, E. M., Nichols, B. E., Streb, L. M., Kimura, A. E. and Sheffield, V.C. (1992). Genetic linkage of vitelliform macular degeneration (Best's disease) to chromosome 11q13. *Nature Genetics* **1**, 246–250.

Sung, C.-H., Schneider, B. G., Agarwal, N., Papermaster, D. S. and Nathans, J. (1991). Functional heterogeneity of mutant rhodopsins responsible for autosomal dominant retinitis pigmentosa. *Proc. Natl. Acad. Sci. USA* **88**, 8840–8844.

Travis, G. H, Christerson, L., Danielson, P. E., Klisak, 1, Sparkes, R. S., Hahn, L., Dryja, T. P. and Sutcliffe, J. G. (1991). The human retinal degeneration slow (RDS) gene: Chromosome assignment and structure of the mRNA. *Genomics* **10**, 733–739.

Travis, G. H., Sutcliffe, J. C. and and Bok, D. (1991). The retinal degeneration slow (rds) gene product is a photoreceptor disc membrane-associated glycoprotein. *Neuron* **6**, 61–70.

van Soest, S., van den Born, L. I., Gal, A., Farrar, G. J., Bleeker-Wagemakers L., Westerveld, A., Humphries, P., Sandkuijl, L. A. and Bergen, A. (1994). Assignment of a gene for autosomal recessive retinitis pigmentosa (RP12) to chromosome 1q31–q32.1 in an inbred and genetically heterogeneous disease population. *Genomics* (in press).

Weber, J. L. and May, P. E. (1989). Abundant class of human DNA polymorphisms which can be typed using the polymerase chain reaction. *Amer. J. Hum. Genet.169172* **44**, 388–396.

7. THE MOLECULAR GENETIC APPROACH TO MACULAR DEGENERATION

EDWIN M. STONE[1], and VAL C. SHEFFIELD[2]

Departments of Ophthalmology[1] and Pediatrics[2], the University of Iowa College of Medicine, Iowa City, Iowa 52242, USA

INTRODUCTION

A number of epidemiologic studies have shown that diseases affecting the retinal macula cause a significant proportion of severe visual loss in older individuals (Sorsby, 1966; Kahn and Moorhead, 1973; Kahn et al., 1977; Leibowitz et al., 1980; Ganley and Roberts, 1983; Klein and Klein, 1982; Martinez et al., 1982; Ghafour et al., 1983; Sommer et al., 1991). For practical reasons, such studies usually consider macular disease in patients over about 50 years of age to be a single clinical entity now known as age-related macular degeneration (AMD). Unfortunately, the creation of a single epidemiologic "pigeon hole" for older patients with macular disease, and the assignment of a single term to refer to it have tended to obscure two important facts:

1) AMD is not a single disease at the clinical, cellular, or biochemical level; and
2) the predisposition to AMD is inherited as an autosomal dominant trait in a significant percentage of affected patients.

In this chapter, we will review the published data that support these two assertions and put forth our view of their importance to the design of studies aimed at elucidating the pathogenesis of – and developing a better treatment for – the diseases that make up AMD. We will then discuss the ways in which molecular genetics can be used to exploit the heritable nature of these disorders to identify the precise molecular mechanisms involved in their pathogenesis and provide some examples of recent progress that has been made with this approach.

DEFINITION OF AMD

Before one can address the heritability or pathogenetic heterogeneity of AMD, one must decide what AMD is. Unfortunately, a universal definition of age-related macular degeneration is elusive, and the "working" definitions of clinicians, pathologists, epidemiologists and molecular biologists differ so widely that it can be difficult for each group to interpret and apply the findings of the others. Most clinicians use the term age-related macular degeneration to refer to patients over the age of 50 with some degree of visual loss associated with an ophthalmoscopic abnormality of the macular retinal pigment epithelium (Kahn et al., 1977; Klein and Klein, 1982; Hyman et al., 1983). In most such patients, drusen are present, and in 5–10% choroidal neovascularization eventually develops (Leibowitz et al., 1980, Hyman et al., 1983). Even when signs of a specific macular disease are present (old central serous retinopathy, the presumed ocular histoplasmosis syndrome, angioid streaks, etc.) the diagnosis of "age-related" macular degeneration is often applied to patients over 50, whereas the more specific diagnoses are more likely to be used in patients under 50.

HETEROGENEITY OF AMD

However pragmatic the foregoing definition is for the clinic, it poses some difficulties for investigators trying to work out the pathophysiology of the diseases that make up AMD. First, it is apparent from the designs of the two population-based epidemiologic studies (Leibowitz et al., 1980, Ganley and Roberts, 1983) that a variety of different pathophysiologic entities were grouped under this single heading. The fraction of AMD caused by each distinct macular disorder cannot be deduced from such studies. This is important because if one **assumes** that AMD is a single pathophysiological entity, and designs experiments to try to identify a single disease mechanism, one is likely to fail if the study population actually consists of patients with clinically similar but biochemically different disorders.

For example, suppose that 5% of AMD is caused by a genetically determined abnormality of a photoreceptor protein that increases the eye's vulnerability to ultraviolet light. If two groups of AMD patients with different light exposure histories are studied, the apparent effect of UV light on the entire "high exposure" group might be insignificant even though a very significant effect is present in the 5% of the population with high UV vulnerability. An example from the literature in which the study participants' specific diagnosis dramatically affected the apparent effectiveness of a treatment for macular disease is the juxtafoveal arm of the macular photocoagulation study (Figure 1). In this study, patients who had laser treatment for juxtafoveal choroidal neovascular membranes secondary to the presumed ocular histoplasmosis syndrome (POHS) had a 31% rate of membrane persistence or recurrence over five years (Macular Photocoagulation Study Group, 1989) while patients with juxtafoveal membranes secondary to non-POHS macular degeneration had a five-year rate of persistence or recurrence of 79% (Macular Photocoagulation Study Group, 1990).

Over the years, a number of different structural and functional abnormalities, physical exposures and dietary deficiencies have been proposed to play a role in the pathophysiology of macular degeneration, including:

1) abnormality of the neurosensory retina (Gartner and Henkind, 1981; Bird and Marshall, 1982; Katz et al., 1986);

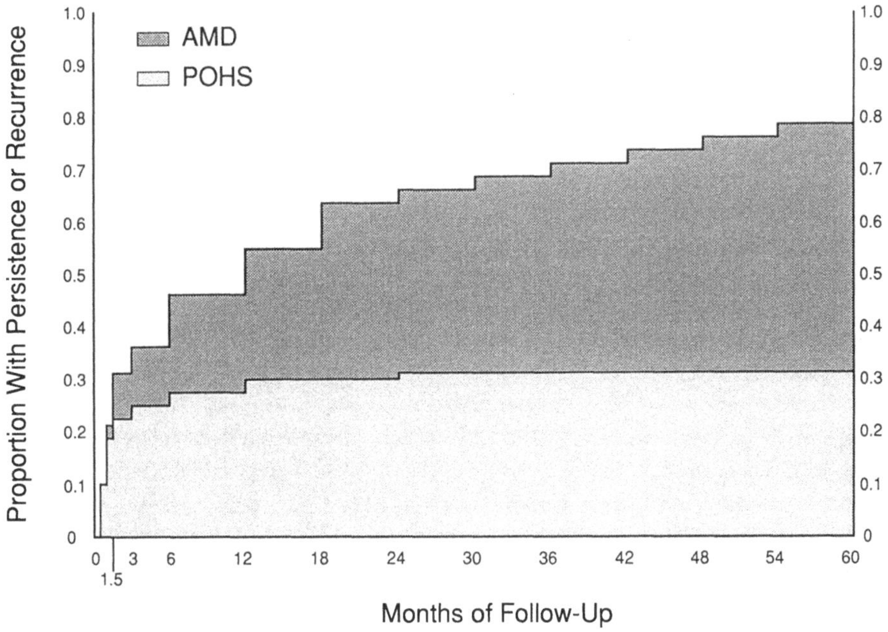

Figure 1 Comparison of membrane recurrence following laser treatment of juxtafoveal choroidal neovascular membranes secondary to the presumed ocular histoplasmosis syndrome (POHS) and age related macular degeneration. The fraction of treated patients suffering a persistence (less than 3 months after treatment) or recurrence of a choroidal neovascular membrane following laser treatment is given on the Y axis while months following treatment is given on the X axis. Within five years of treatment, the total fraction of krypton-treated POHS patients who had a persistent or recurrent membrane was 31% while 79% of AMD patients showed persistence or recurrence of their neovascularization during this same interval. This figure was drawn using data from two independent reports of the Macular Photocoagulation Study Group (see text).

2) abnormality of the retinal pigment epithelium associated with a co-factor deficiency (Newsome et al., 1988), or an "inborn error of metabolism" (Deutman and Jansen, 1970);
3) abnormality of Bruch's membrane associated with a faulty primary structure (e.g. pseudoxanthoma elasticum and Alport syndrome) or by deposition of an abnormal substance from an RPE or vascular source (Bird, 1986; Newsome et al., 1987);
4) abnormality of the choriocapillaris (Friedman et al., 1963; Tso, 1985);
5) immune system activation including development of anti-retinal antibodies (Gurne et al., 1991) and attraction of macrophages (Penfold et al., 1986);
6) abnormal scleral rigidity (Friedman et al., 1989);

7) mitochondrial dysfunction (Kuriyama et al., 1984; Modi et al., 1991);
8) light toxicity (Young, 1988; Taylor et al., 1992; Gottsch et al., 1993); and
9) various nutritional deficiencies (Eye Disease Case-Control Study Group, 1993).

Each of these hypotheses has some theoretical merit and some experimental support; and, in our opinion, there is currently little reason to completely exclude any of them from a significant role in a subset of AMD patients. One or more of these mechanisms could have a "natural" age-related component and more than one process could be active in any given patient. Thus, the demonstration of an age-related increase in disease prevalence or severity does not in and of itself support or preclude the existence of any of the previously listed mechanisms, nor does it preclude the existence of various types of genetic predisposition to the disease.

Other authors have previously expressed the opinion that AMD is not a single disease but is a common clinical manifestation of a variety of biochemical disorders (Delaney and Oates, 1982; Elman and Fine, 1989; Bird, 1991). If multiple mechanisms are involved in significant subsets of AMD, one would expect to find some clinical and histopathologic evidence for heterogeneity in populations of AMD patients. Such evidence exists but is often obscured by the varying definitions of different authors. For example, Bressler et al. (1988) broadly define AMD as "features of drusen and geographic atrophy of the retinal pigment epithelium as well as retinal changes associated with choroidal neovascularization in individuals over 50 years of age". In his atlas, Gass (1987) lists over 20 entities known to be associated with choroidal neovascularization ranging from inflammatory diseases, to drusen. If one were to take the latter half of the definition of Bressler and co-workers at face value, patients over the age of 50 with any of the different disorders listed by Gass as complicated by choroidal neovascularization could be considered to have AMD.

Even within the relatively narrow subcategory of "drusen" there is quite a bit of heterogeneity. "Typical" drusen can range from tiny hyaline bodies only a few RPE cells in diameter to relatively large detachments of the retinal pigment epithelium. In addition, there is an ophthalmoscopically and angiographically distinct form known as "cuticular" or "basal laminar" drusen (Gass et al., 1985). In contrast to the wide variation in the clinical appearance of drusen in the AMD population as a whole, it has been our experience that within families with AMD, the affected relatives are often distinctly similar to one another ophthalmoscopically (Figure 2).

There is also some histopathologic evidence that the susceptibility to drusen development may not be uniform within the population. Feeney-Burns and Ellersieck (1985) studied the ultrastructure of Bruch's membrane in 68 eyes spanning 10 decades of life. The authors assigned each specimen to one of six grades of ultrastructural abnormality and showed that younger patients on average had lower grades of abnormality than older patients (Figure 3). Although the authors did

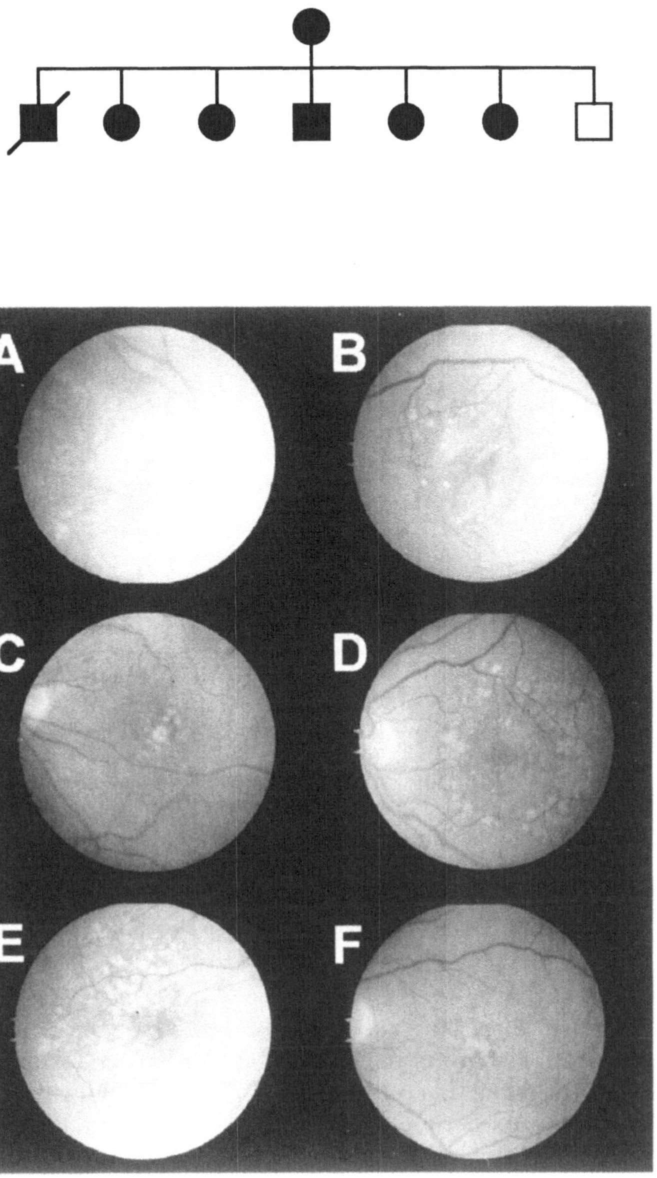

Figure 2 Ophthalmoscopic similarity of macular disease in relatives affected with familial macular degeneration. The pedigree is shown at the top. Closed symbols indicate individuals with macular degeneration. Fundus photographs of the mother (A) and five affected children (B-F) are shown in decreasing order of age. The drusen are soft, confluent and limited to the central macula. The peripheral fundus is normal. Three patients (A, B and D) have some geographic atrophy in the center of the affected areas. The view of the fundus in patient A is limited by cataract. See Colour Plate II at the back.

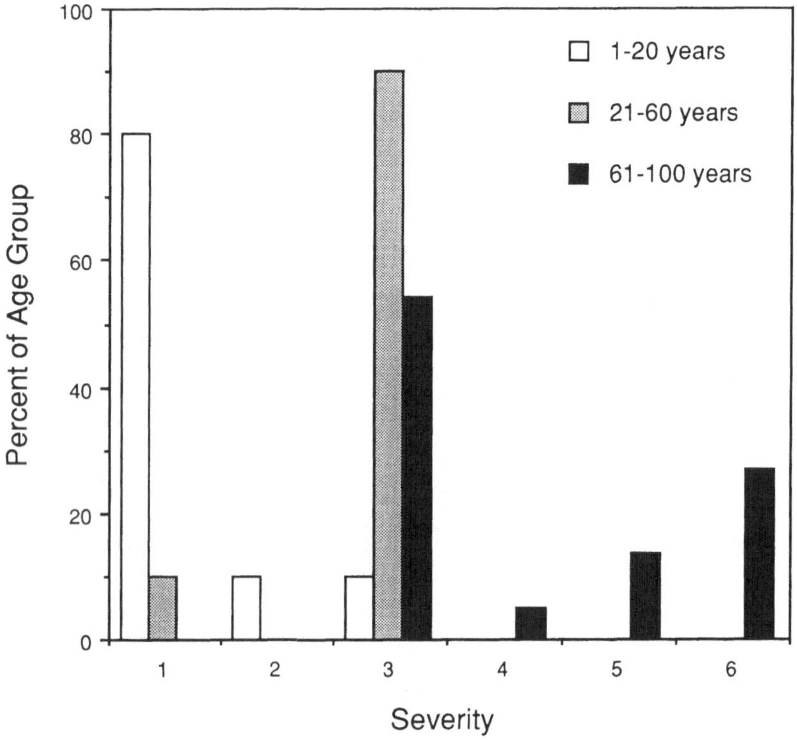

Figure 3 Prevalence of ultrastructural abnormalities in Bruch's membrane in patients of different ages. The figure is drawn using data reported by Feeney-Burns and Ellersieck (1985). The Y axis gives the fraction of patients in each age group assigned to each of six grades of ultrastructural severity that are distributed along the X axis. White bars indicate patients 1–20 years of age; grey bars, 21–60 years of age; and, black bars 61-100 years of age. Note that 10% of the youngest patients have the same degree of abnormality (grade 3) as 54% of the oldest patients.

not discuss the possibility of varying susceptibility to drusen formation within the population, it is interesting that one patient (10%) from their youngest group had the same ultrastructural severity as 20 patients (54%) from their oldest group. Possible explanations for this observation are that Feeney-Burns' and Ellersieck's "age-related sequence of changes" is 1) a common manifestation of a variety of different biochemical processes with different ages of clinical onset; or 2) true age-related phenomena are superimposed on a background of variable genetically-determined susceptibility.

Additional insight into this possibility was provided by Coffey and Brownstein (1986) who counted drusen in serial sections of 23 pairs of postmortem eyes. The unusual and important feature of this study is that their data were not averaged –

they were reported individually. Two of the 23 patients had an extraordinarily large number of drusen which the authors speculated might be due to a genetic predisposition.

HERITABILITY OF AMD

An important requirement for a molecular approach to any human disease is the existence of families affected with a heritable form. The way in which such families are used for the study of disease will be discussed more fully below, but at this juncture it might be reasonable to ask whether there is any evidence that a subset of AMD is heritable.

There are a number of dystrophies that affect the macula whose heritability has been undisputed for decades. Some of these are clinically similar enough to AMD that distinction from the latter is often difficult in affected patients over the age of fifty, especially if the family history is not known. These include Best's disease (Best, 1905), the pattern dystrophies (Marmor and Byers, 1977; Hseih et al. 1977; Watzke et al., 1982; DeJong and Delleman, 1982), Sorsby's macular dystrophy (Sorsby et al., 1949), and dominant drusen (Pearce, 1968; Deutman and Jansen, 1970). The familial nature of the latter entity was first described nearly 100 years ago (Doyne, 1899), and at least one author feels that it is clinically indistinguishable from typical AMD (Gass, 1973). Best's, Sorsby's and the pattern dystrophies also have some AMD-like features including an accumulation of abnormal material at the level of Bruch's membrane and the RPE as well as the eventual development of choroidal neovascularization or geographic atrophy in a subset of patients. North Carolina Macular Dystrophy can also mimic AMD ophthalmoscopically but differs from the other dystrophies in that it is usually not progressive (Small et al., 1991). The specific genetic defects that cause each of these diseases are of great interest because the affected genes may be involved in a sizeable proportion of "typical AMD". At the very least, identification of a gene that causes any macular dystrophy has the potential to greatly improve our understanding of the cascade of molecular events that occurs in AMD.

If mutations in macular dystrophy genes play an important role in AMD, one would expect to find evidence of heritability in a significant proportion of "typical AMD" patients. However, it is important to realize that even if all AMD has a primarily genetic basis, this would not necessarily be clinically obvious. Since the disease does not usually become symptomatic until the seventh or eighth decade, the parents of affected patients are not usually living and their siblings may also be deceased or otherwise unavailable for examination. Also, since only a minority of affected patients experience severe visual loss, it is impossible by history alone to determine whether a sibling's 20/50 vision is due to AMD or some other cause; and conversely, whether a parent with "excellent vision" actually had numerous asymptomatic drusen. Certainly if one assumes that AMD is a natural aging phenomenon and does not even question patients about their family, one is unlikely to discover evidence to support a genetic basis for the disease. To restate this important point: failing to look for evidence of heritability is not equivalent to carefully looking for it without success.

For example, a recently published study examined AMD risk factors in over 400 patients and 600 controls (Eye Disease Case Control Study Group, 1992). The subjects were questioned about cigarette use, alcohol and coffee consumption, sunlight exposure, marital status, physical activity, the presence of diabetes or cardiovascular disease, prior hysterectomy, and the use of estrogen or birth control pills. No questions regarding AMD in family members were asked and it is thus not surprising that these authors did not even mention the possibility of genetic risk factors, much less find any evidence to support them.

In contrast, when evidence for genetic factors has been looked for, it has been found. In 1973, Gass reported the results of his study of 200 patients with macular degeneration who had been followed for an average of four years. He stated: "although it was my initial impression that few patients with drusen and disciform detachment had a positive family history, more careful questioning of the patients and investigation of the few relatives available during this study has revealed a significant incidence of familial involvement." Specifically, he obtained a positive family history of central visual loss in 38 of the 200 patients (19%) and was able to confirm the diagnosis of AMD in the relatives of 10 of these patients by personal examination. In a case-control study by Hyman et al. (1983) the presence of AMD in parents and siblings of the study subjects was investigated via questionnaires distributed to the subjects, their siblings and their siblings' eye examiners. Inclusion of the latter group was revealing because the eye examiners of the case siblings reported the presence of AMD almost twice as often as the case siblings themselves, suggesting that patients are often not told that they have macular degeneration unless they have experienced visual loss. According to the eye examiners' responses, 19.9% of the siblings of AMD cases had AMD themselves in comparison to 7.9% of the siblings of controls. The age and sex distribution of the case siblings were the same as those of the controls. When the parental and sibling data were combined by defining a positive family history as either a parent or a sibling with a history of the disease, the data were similar with 21.6% of the cases having a positive family history as opposed to 8.6% of the controls.

There are a few reports of identical twins with AMD complicated by choroidal neovascularization (Melrose et al., 1985; Meyers and Zachary, 1988; Dosso and Bovet, 1992). In all three of these reports, the second twin developed symptoms of choroidal neovascularization within 16 months of the first. The twins reported by Meyers and Zachary belonged to a sibship of 13, and four of the other 11 siblings were also affected.

At least three studies have suggested that AMD is significantly more common in white patients than in blacks (Chumbley, 1977; Gregor and Joffe, 1978; Sommer et al., 1991). Although it is possible that increased pigmentation is directly responsible for the protective effect, it is also possible that AMD-predisposing genes are less common in the black population.

Unfortunately, neither the Framingham study (Kahn et al., 1977; Leibowitz et al., 1980) nor the National Health and Nutrition Examination Survey (Ganley and Roberts, 1983) collected data regarding a family history of AMD. However, these studies were already underway when Gass's 1973 paper was published and were complete before the report of Hyman et al. in 1983. It is less understandable why more recent studies (Weiter et al., 1985; Bressler et al., 1989; Klein et al., 1991; Eye Disease Case Control Study Group, 1992) have chosen to ignore the possibility of genetic predisposition.

In any case, there seems to be sufficient evidence in the literature to support the assumption that a significant fraction of AMD is heritable and therefore amenable to molecular genetic attack. Moreover, the evidence that AMD is pathophysiologically heterogeneous is also compelling and we believe that some method of reliably dividing patients according to disease mechanism is essential for new therapies to be rationally designed, tested and administered.

MOLECULAR APPROACHES TO DISEASE GENE IDENTIFICATION

Identification of genes that cause inherited diseases has been accomplished with a variety of approaches. The first genetic disease genes were identified based on a knowledge of the specific cellular proteins involved in the pathogenesis of the diseases. Examples of this are the globin genes involved in the thalassemias and sickle cell anemia (Ingram, 1956; Weatherall and Clegg, 1981; Bunn and Forget, 1986). For many genetic diseases, including hereditary eye diseases, information concerning proteins involved in the pathogenesis of the disease is lacking or has failed to aid in the identification of the disease-causing locus. For such disorders, the use of genetic linkage analysis with polymorphic DNA probes to identify the genetic location of the gene has proven valuable.

Figure 4 provides an overview of a stepwise genetic approach to macular disease. The ideal starting material is a large family affected with an autosomal dominant disease which can be used for linkage analysis. Linkage analysis does not require any specific hypothesis about the pathogenesis of the disease. It requires only the availability of a sufficient number of correctly diagnosed individuals to allow a statistically significant relationship between marker alleles and the disease phenotype to be demonstrated. Linkage data are expressed as lod scores which are derived and interpreted as discussed in Chapter 1.

The first macular disease to be successfully linked to a genetic marker was an atypical vitelliform dystrophy linked to the marker GPT-1 (Ferrell et al., 1983) which is now known to lie on the long arm of chromosome 8 (Rocha et al., 1988). More recently, Small and co-workers (1992) mapped North Carolina Macular Dystrophy (Figure 5) to chromosome 6. Two microsatellite markers located in chromosomal region 6q13-q21 (MFD171, MFD97) were found to be linked to the disease locus with lod scores of 8.4 and 13.1 at recombination fractions of 0.004 and 0.017 respectively. A patient with a cone dystrophy and mental retardation as-

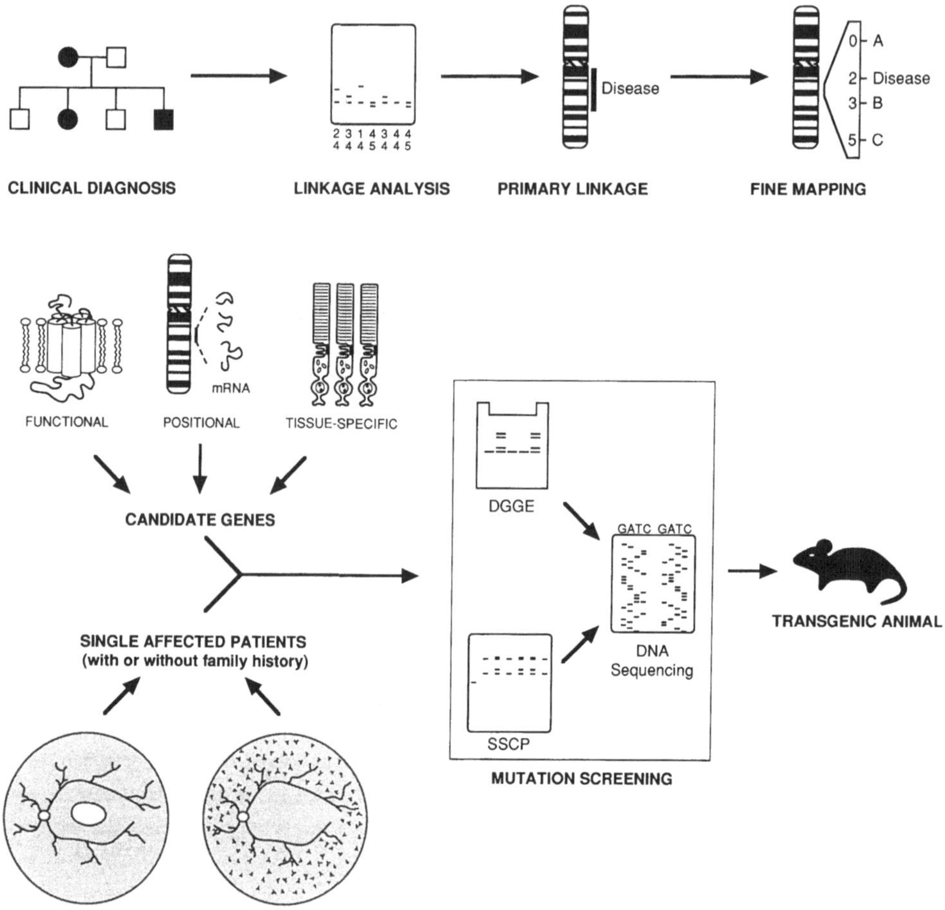

Figure 4 Diagrammatic representation of a stepwise molecular genetic approach to inherited diseases. The upper panel depicts a family-oriented approach. Starting with a large family affected with an inherited disease, linkage analysis is used to determine the chromosomal location of the disease-causing gene. The ability to refine the locus with fine mapping is dependent upon family size as well as the density of markers in the area of primary linkage. The lower panel depicts a candidate gene approach to inherited diseases. Candidate genes are of three types: 1) genes encoding proteins whose function can be theoretically associated with the disease phenotype (e.g. rhodopsin and retinitis pigmentosa); 2) genes known to map at a locus linked to the disease phenotype; and, 3) genes whose proteins are primarily expressed only in the tissue affected by the disease. DNA samples from members of families affected with an inherited disease (as in the top panel) or single affected patients without a family history are used for screening these candidate genes for disease-causing mutations via denaturing gradient gel electrophoresis (DGGE) or single-strand conformational polymorphism (SSCP) analysis. DNA sequencing is used to determine the exact nucleotide changes in each mutation detected. Transgenic animals containing the mutated gene can be created and used to study the pathophysiology of the disease and can potentially be used to test experimental therapies as well.

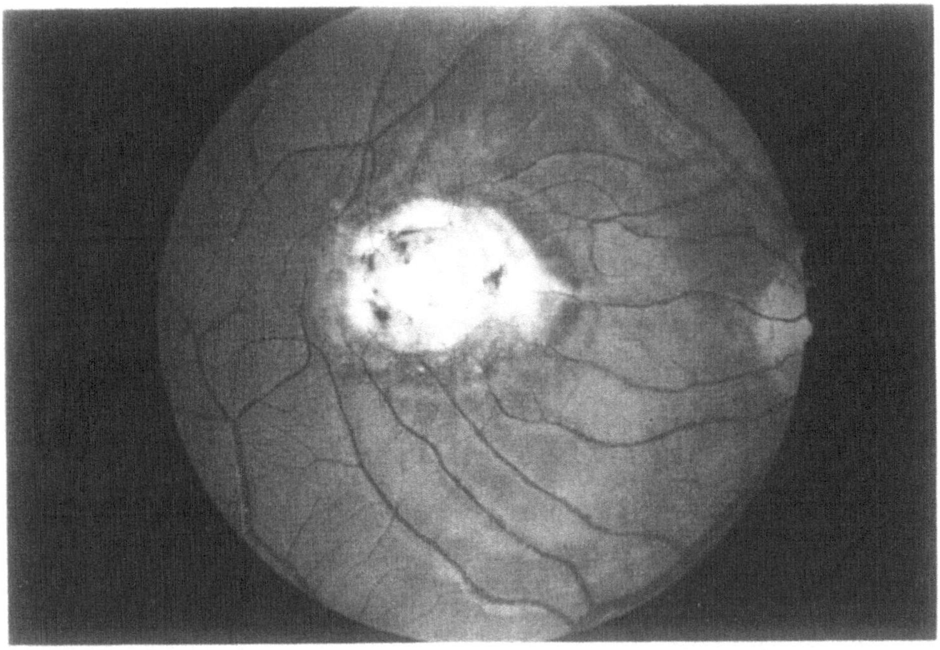

Figure 5 North Carolina macular dystrophy. Typical features of the disease are the central crater with a shiny base surrounded by a gliotic rim. See Colour Plate III at the back.

sociated with a translocation breakpoint in the same region has been reported by Tranebjaerg et al. (1986). These rare chromosomal rearrangements are potentially important for the task of identifying the gene directly. Also in 1992, we (Stone et al., 1992) and another group (Forsman et al., 1992) independently mapped Best's Vitelliform Dystrophy (Figure 6) to the long arm of chromosome 11.

This disorder has been studied histopathologically and found to be associated with lipofuscin-like deposits within and beneath the retinal pigment epithelium. Our group mapped a large five-generation family with this early-onset autosomal dominant form of vitelliform macular dystrophy to the 11q13 region, between markers *INT2* and *D11S871* (Z_{max} = 9.3 at 3.9 cM from marker *D11S871*) (Stone et al., 1992). The other group mapped the Best gene in a large Swedish family to the same chromosomal region (11q13) with a lod score of 15.12 at a recombination fraction of 0.01 for the closest marker, *INT2* (Forsman et al., 1992). A possible candidate gene is the ROM-1 gene (Bascom et al., 1992) which maps to same chromosomal region and shows similarity to the perpherin/RDS gene (see below). The gene product, rom-1, is a membrane-associated protein of rod outer segments that is of similar size and amino acid sequence as rds/peripherin.

Other macular diseases that should be amenable to linkage analysis include dominant drusen (both typical and cuticular), Sorsby's fundus dystrophy, various pattern dystrophies, and Stargardt's disease.

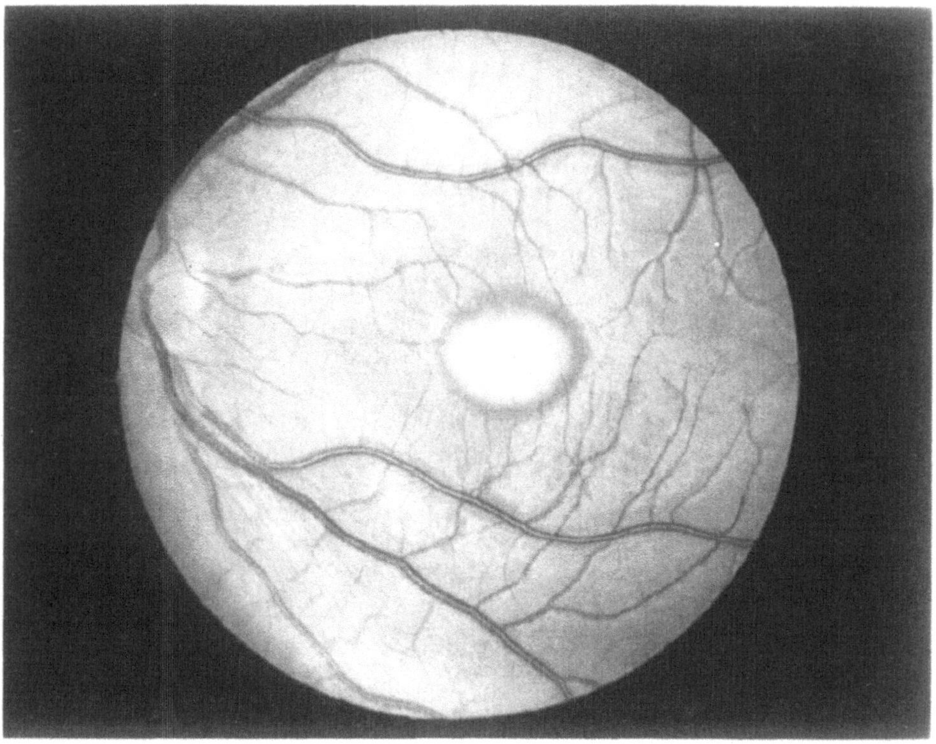

Figure 6 Best's vitelliform dystrophy. A classic "egg yolk" deposit of lipofuscin-like material is present beneath the retinal pigment epithelium of the central macula. See Colour Plate IV at the back.

Once a disease is mapped to a region, one may find that a candidate gene has been previously mapped to the same location. This was the case for autosomal dominant retinitis pigmentosa in which the rhodopsin gene had been localized to the long arm of chromosome 3 (Sparkes et al., 1986) before linkage analysis suggested that a disease-causing mutation might exist at the same locus (McWilliam et al., 1989 and Chapter 6).

An alternative method used to identify disease-causing genes is the "candidate gene" approach. Genes that are likely candidates for playing a role in a given genetic disease are selected and tested for their involvement in the disease either by mutation analysis or by linkage studies that use polymorphisms previously identified within the gene sequence. A candidate gene can be selected based on its function, its chromosomal location, or the tissue in which it is primarily expressed. Successful identification of a gene involved in human retinal disease was recently accomplished using the candidate gene approach. Specifically, the murine RDS gene was shown to be mutated in animals with retinal degeneration (Travis et al., 1989) and this finding made the human homologue (Travis et al., 1991) a good candidate gene for retinitis pigmentosa. This was confirmed when mutations in the human RDS gene were identified in patients with autosomal dominant retinitis pigmentosa (Farrar et al., 1991; Kajiwara et al., 1991; Wells et al., 1993). Using the same reasoning, we recently identified two different mutations in the RDS gene (Nichols et al., 1993a and 1993b) in families affected with an autosomal dominant pattern dystrophy (Figures 7 and 8) resembling Deutman's butterfly dystrophy of the fovea (Deutman et al., 1970). The first was a G to A substitution at codon 167 that substitutes an aspartate for a highly conserved glycine residue. The mutation segregated with the disease in eleven affected and four unaffected (>55 yrs) family members ($Z_{max} = 4.0$ at $\theta = 0$). The glycine^{167} residue is also conserved in the peripherin-related protein rom-1, which may associate non-covalently with peripherin in rod photoreceptors (Bascom et al., 1992). The phenotype in this disorder involves accumulation of yellow material within and beneath the retinal pigment epithelium in the perifoveal region. The parallels with the *rds/+* mouse are interesting, since the latter shows dysplastic outer segments which appear to be unstable and to shed at an increased rate. The underlying pigment epithelium contains large and abnormal phagosomes which may result from impaired digestion of the abnormal outer segments (Sanyal and Hawkins, 1988). It is possible that accumulation of such phagosomes in the macular pigment epithelium gives rise

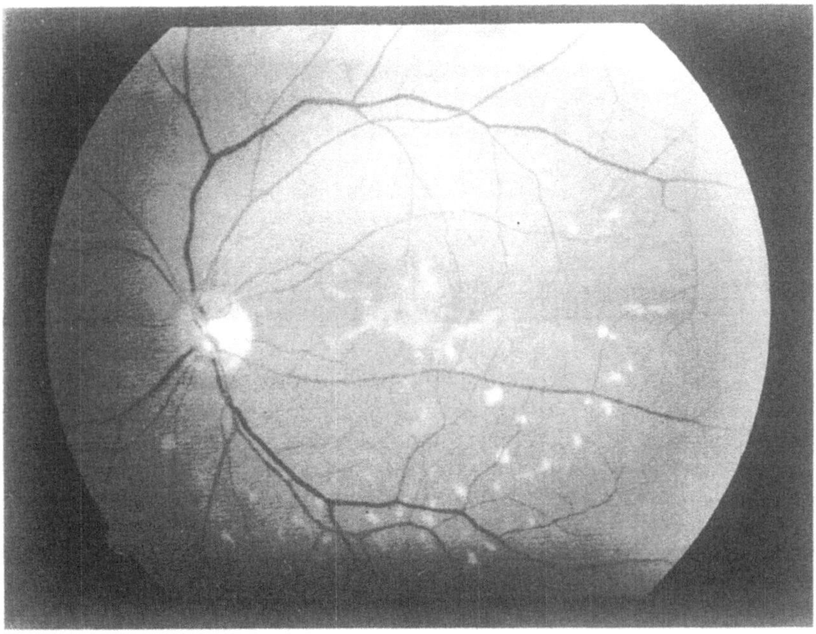

Figure 7 Butterfly dystrophy. There is a yellowish subfoveal lesion with five radiating arms. In addition, round pigment deposits that are larger than typical drusen are present just posterior to the major vascular arcades. From Nichols et al., *Nature Genetics* (1993); used with permission. See Colour Plate V at the back.

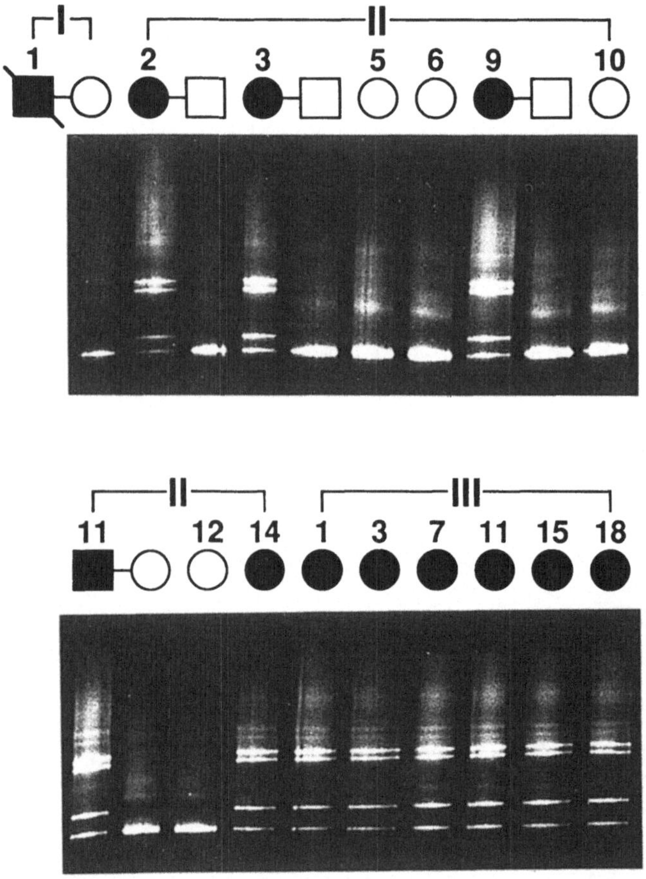

Figure 8 Denaturing gradient gel electrophoresis of PCR products containing RDS exon 1 sequences. Clinically affected individuals are indicated by closed symbols while clinically unaffected individuals are indicated by open symbols. Spouses are indicated by symbols without numbers which are connected to an affected patient's symbol by a horizontal line. Each gel lane contains a sample from the individual whose pedigree symbol is directly above. Affected individuals have multiple bands on the gel (both homoduplexes and heteroduplexes) while normal individuals have only one band corresponding to normal homoduplex molecules. From Nichols, et al., *Nature Genetics* (1993); used with permission.

to the deposits of yellowish material seen in the butterfly lesions (Nichols et al., 1993a). The second mutation seen in a family with butterfly-shaped pigment dystrophy of the fovea was caused by a 2 base pair deletion in the RDS gene overlapping codons 299 and 300 that results in a translational frameshift. This mutation causes 56 incorrect amino acids to be translated before a stop codon is encountered 131 nucleotides past the end of normal translation.

Kajiwara et al. (1993) screened over 350 patients with a wide variety of retinopathies for mutations in the peripherin/RDS gene using single strand conformation polymorphism (SSCP) analysis and found one mutation in a family with retinitis punctata albescens. This condition is a progressive disorder of the central and peripheral retina associated with subnormal rod and cone ERGs and subretinal white or yellow punctate deposits. These accumulations may represent the same material as found in the family of Nichols et al. (1993a). The mutation gives rise to a premature stop codon and is therefore probably a null allele which truncates the rds peripherin protein to 42 amino acids instead of the normal 346. Similarly, Wells et al. (1993) screened RDS/peripherin exons for mutations in 100 patients with RP and 13 with various macular dystrophies. They found a 3 base pair deletion associated with loss of a cysteine at codon 118 or 119 that segregated with typical autosomal dominant RP in one family. In three families with autosomal dominant macular dystrophy, they found missense mutations at codon 172 of the RDS gene, associated with a predicted substitution of glutamine for arginine in one family and a tryptophan substitution in the other two. Finally, a family with adult vitelliform macular dystrophy was found to have a nonsense mutation at codon 258 (Tyr 258 stop) which would lead to a truncated protein product, similar to that in the *rds/+* heterozygous mouse (Chapters 5, 6). In this case, the macula shows a fried egg appearance again with yellow deposits at the level of the retinal pigment epithelium rather than in the photoreceptor layer. One member of the family with butterfly-shaped pigment dystrophy of the fovea described by Nichols et al. (1993a) also showed bilateral vitelliform lesions reminiscent of Best's disease while others showed either punctate or patterned lesions. The single affected patient with the Tyr 258 stop mutation was only aware of a relatively mild central vision loss. Her father had a similar disorder but no other affected relatives were available to confirm segregation of the mutation with the illness. However, the severe nature of the mutation seems to leave little doubt that it is causally related to the disorder.

This surprising finding that mutations in the same gene can cause phenotypes as widely differing as butterfly dystrophy and retinitis pigmentosa, serves to illustrate two important points. First, current clinical and electrophysiological techniques are not capable of separating diseases into molecularly similar groups. That is, mutations in the rhodopsin gene on chromosome 3 cause a phenotype that is clinically and electrophysiologically much more similar to RDS-associated retinitis pigmentosa than the pattern dystrophies caused by mutations elsewhere in the RDS gene. Second, clinical, psychophysical, and electrophysiologic methods are not capable of predicting the primary site of expression of the disease causing gene. That is, prior to the discovery of the vitelliform and pattern dystrophy associated RDS mutations, most retinologists would have predicted the molecular defect in these dystrophies to exist at the level of the retinal pigment epithelium or perhaps Bruch's membrane. In this case, molecular biology was capable of demonstrating that the primary expression of the genetic defect is actually in the neurosensory retina.

Some caution should be introduced at this point: not all sequence changes in candidate genes are disease-causing mutations. When hundreds of patients are screened for changes in almost any gene sequence, some variation is likely to be found. A change is likely to be a disease causing mutation if it has the following characteristics.

1) It results in an altered amino acid (this is especially convincing if the mutation results in a stop codon, a frameshift, or a dramatic change in charge, size or hydrophobicity of a single amino acid).
2) It is not found in normal individuals to any measurable extent. For example, if screening fifty normal controls (100 chromosomes) does not reveal a similar sequence change, it is less likely to be a clinically silent polymorphism.
3) The mutation segregates with the disease phenotype in a statistically significant fashion (Figure 8).

The latter criterion is important because if a heterozygous polymorphism is detected in an affected patient the probability that a second affected sibling or parent will harbor the same polymorphism just by chance is 50%; the probability that a third would have the polymorphism is 25%, and so on. Thus, pedigrees containing only one or a few affected patients cannot by themselves prove the participation of a given candidate gene in the disease. It is remotely possible for a DNA sequence change to satisfy all three of the above criteria without actually causing any disease. For example, suppose the RDS sequence change illustrated in Figure 8 was actually a silent polymorphism, and that the true disease-causing mutation was in another gene a centiMorgan or two from RDS. If the polymorphisms were in phase with the disease-causing mutation, one would get exactly the same result as shown in Figure 8, even although the RDS gene was not really involved in the disease. Of course, the likelihood of this hypothetical situation occurring (especially in two different families) is so low that statistically significant segregation of an amino-acid-changing mutation is usually accepted as tentative proof of the mutation's involvement in the disease. In fact, once a given phenotype has been significantly associated with a mutation in a given gene, it is probably sufficient to satisfy only the first two criteria. Each new family with a given phenotype that harbors a different mutation in the same candidate gene lends additional support to the hypothesis that the mutations are the primary cause of the disease.

The ultimate proof that a given mutation actually causes the disease is the creation of the disease in a transgenic animal by the introduction of a specific mutation initially identified in an affected human. The existence of such animals also provides a platform for studying a disease in ways that are impossible in humans. For example, no human eyes with butterfly dystrophy have ever been studied histopathologically, and if such specimens do eventually become available they will likely be from an elderly patient with advanced atrophic disease. The existence of a transgenic animal model would allow the morphological features of the disease to be worked out in great detail, including the temporal changes from birth to advanced disease. Ultrastructural, biochemical, and cell biological studies using transgenic animals would have a much greater chance of elucidating the molecular sequence of events that cause the clinically evident disease than would similar studies in the human population. In addition, once the disease mechanism was well worked out, physical, dietary or pharmacologic interventions could be devised and tested much more readily in a transgenic animal model than in the human population.

Figure 4 demonstrates two other important features of the molecular genetic approach to human disease; the dependence at nearly every level on the availability of actual human patients affected with various disorders and the need for skilled clinicians to correctly diagnose these patients and carefully define their phenotypes. It is the clinician who initially discovers a family that is suitable for linkage analysis. More importantly, if such an analysis is to be successful, each patient must be diagnosed with a high degree of accuracy. For example, it could be very damaging to a linkage study to incorrectly diagnose a patient with the presumed ocular histoplasmosis syndrome as "affected" in a family with pattern dystrophy. In a fully informative sibship of 11 sibs and 2 living parents, such a misdiagnosis could decrease the lod score from 3 to 1.3. If the nearest genetic marker was 10 centiMorgans from the disease gene, a single misdiagnosis could result in a maximal lod score of only 0.5 which might not even result in additional intensive searching in that area.

At the candidate gene level, clinicians can participate by contributing patients with various phenotypes for mutation searches. When such mutations are found, clinical ophthalmologists can attempt to correlate specific features of the disease with individual mutations. Such correlations have the potential to make molecular diagnosis a very useful clinical tool. For example, if certain mutations are found to be associated with a high risk of exudative macular degeneration while others tend to be associated with slowly progressive geographic atrophy, such information could be used to counsel patients as well as to help plan follow-up visits so that the greatest amount of scrutiny could be focused upon the patients at greatest risk.

Two recent advances have greatly facilitated both linkage analysis and the candidate gene approach. The development of highly polymorphic DNA markers known as microsatellites (Weber and May, 1989; Litt and Luty, 1989; Weber, 1990) has greatly simplified genetic linkage studies. Microsatellite markers are superior to restriction fragment length polymorphisms (RFLPs) because they are more informative and they can be assayed using the polymerase chain reaction (Chapter 1). The second advance has been the development of methods for screening single-base sequence substitutions. These methods include denaturing gradient gel electrophoresis (DGGE) (Fischer and Lerman, 1980; Meyers et al., 1987; Sheffield et al., 1989), single-strand conformational gel analysis (SSCP) (Orita et al., 1989), chemical cleavage of mismatch (Cotton et al., 1988), and heteroduplex analysis (Keen et al., 1991). As a group, they allow rapid screening of a candidate gene for mutations in multiple individuals. These techniques decrease the amount of labor-intensive DNA sequencing that needs to be performed and thereby increase the number of affected individuals and controls that can be screened.

THERAPY FOR AMD

Everyone involved in the study of age related macular degeneration is hoping that more effective therapy can be developed. As outlined at the beginning of this chapter, a variety of pathogenic mechanisms are likely to be involved in the large clinical entity known as AMD and it is unrealistic to expect that any given treatment will be effective for all of them. It is essential to develop methods to reliably subdivide patients into pathophysiologically similar groups. Within such groups (e.g. within a family whose affected members all carry the same dominant mutation) it is common to find that some patients do relatively well clinically while others do poorly. This suggests that dietary, physical or additional genetic factors are capable of modulating the effect of the disease gene. Such modulators have the potential to be the basis of effective therapy. That is, the pathological process does not have to be totally overcome in a given type of macular degeneration to make a tremendous difference in these patients. If the rate of progression of the disease can be altered by only 10 or 15%, it would make a significant number of affected patients asymptomatic for their entire lives.

Molecular genetics can facilitate the development of effective therapy in the following ways.

1) When patients can be divided into pathophysiological groups based on an objective molecular diagnosis, "risk factor" studies will be more likely to be able to identify agents that accelerate or retard the disease process.
2) The identification of disease causing genes will suggest specific agents that might be investigated. For example, if a disease is found to result from a mutation in a zinc-requiring enzyme of the RPE, zinc augmentation might be a reasonable approach to therapy in these patients.
3) Transgenic animals will allow experimental therapies to be investigated rapidly and in a controlled fashion.
4) The ability to identify disease-causing mutations in presymptomatic individuals will allow specific therapy to be administered over a longer period of time. For example, if light exposure plays a role in a subset of AMD, sunglasses would be more likely to make an impact if they were prescribed for children with the genetic predisposition than if their use was delayed until the clinical signs of macular degeneration were evident.

CONCLUSION

The human genetic map is being refined to the centiMorgan level with highly polymorphic genetic markers and many genes are being characterized and placed on this map. The increasing availability of good markers and the increasing resolution of the map greatly facilitates the mapping of disease-causing mutations as well as the isolation of the affected genes themselves. In the past two years alone, the genes for two types of macular disease have been mapped and specific mutations involved in a third have been identified.

A decade ago, the prospect of understanding any form of macular degeneration at the molecular level seemed remote. However, recent progress on the clinical, genetic and molecular fronts makes it likely that such understanding will be forthcoming in the near future and makes us hopeful that more effective therapy will follow closely on its heels.

ACKNOWLEGEMENTS

Preparation of this chapter was supported in part by the National Retinitis Pigmentosa Foundation Fighting Blindness; the George Gund Foundation; the C. S. O'Brien Center for Macular Diseases; the Forrest Lattner Foundation; Public Health Service Research grants EY08426 (EMS), P50HG00835 (VCS) and HG00457 (VCS), and an unrestricted grant from Research to Prevent Blindness, New York, N.Y.

REFERENCES

Bascom, R. A., Manara, S., Collins, L., Molday, R. S., Kalnins, V. I. and McInnes, R. R. (1992). Cloning of the cDNA for a novel photoreceptor membrane protein (rom-1) identifies a disk rim protein family implicated in human retinopathies. *Neuron* **8**, 1171–1184.

Best, F. (1905). Uber eine hereditare macula-affektion: Bietrage zur vererbungslehre. *Z Augenheilkd* **13**, 199–212.

Bird, A. C. (1991). Pathogenesis of retinal pigment epithelial detachment in the elderly; the relevance of Bruch's membrane change. *Eye* **5**, 1–12.

Bird, A. C. and Marshall, J. (1982). Retinal receptors disorders without known metabolic abnormalities. *Pathobiology of Ocular Disease: A Dynamic Approach Part B,* Alec Garner and Gordon Klintworth, eds., Marcel Dekker, New York.

Bird, A. C. and Marshall, J. (1986). Retinal pigment epithelial detachments in the elderly. *Transactions of the Ophthalmological Society UK* **105**, 674–682.

Bressler, N. M., Bressler, S. B. and Fine, S. L. (1988). Age-related macular degeneration. *Survey of Ophthalmology* **32**, 375–413.

Bressler, N. M., Bressler, S. B., West, S. K., Fine, S. L. and Taylor, H. R. (1989). The grading and prevalence of macular degeneration in Chesapeake Bay watermen. *Archives of Ophthalmology* **107**, 847–852.

Bunn, H. F. and Forget, B. G. (1986). Hemogloblin: molecular, genetic and clinical aspects. Saunders, Philadelphia.

Chumbley, L. C. (1977). Impressions of eye diseases among Rhodesian Blacks in Mashonaland. *SA Medical Journal* **52**, 316–318.

Coffey, A. J. H. and Browstein, S. (1986). The prevalence of macular drusen in postmortem eyes. *American Journal of Ophthalmology* **102**, 164–171.

Cotton, R. G. H., Rodriquez, N. R. and Campbell, R. D. (1988). Reactivity of cystosine and thymine in single base pair mismatches with hydroxylamine and osmium tetroxide and its applications to the study of mutations. *Proceedings of the National Academy of Sciences USA* **85**, 4397–4401.

De Jong, P. T. V. M. and Delleman, J. W. (1982). Pigment epithelial pattern dystrophy: Four different manifestations in a family. *Archives of Ophthalmology* **100**, 1416–1421.

Delaney, W. V. and Oates, R. P. (1982). Senile macular degeneration: A preliminary study. *Annals of Ophthalmology* **14**, 21–24.

Deutman, A. F. and Jansen, L. M. A. A. (1970). Dominantly inherited drusen of Bruch's membrane. *British Journal of Ophthalmology* **54**, 373–382.

Deutman, A. F., van Blommestein, J. D. A., Henkes, H. E., Waardenburg, P. J. and Solleveld-van Driest, E. (1970). Butterfly-shaped pigment dystrophy of the fovea. *Archives of Ophthalmology* **83**, 558–569.

Dosso, A. A. and Bovet, J. (1992). Monozygotic twin brothers with age-related macular degeneration. *Ophthalmologica* **205**, 24–28.

Doyne, R. W. (1899). Peculiar condition of choroiditis occurring in several members of the same family. *Transactions of the Ophthalmological Societies of the United Kingdom* **19**, 71.

Elman, M. J. and Fine, S. L. (1989). Exudative age-related macular degeneration. in *Retina, Volume Two* , Schachat, A. P., Murphy, R. P., Patz, A., and Ryan, S., eds., C. V. Mosby, St. Louis.

Eye Disease Case-Control Study Group. (1992). Risks factors for neovascular age-related macular degeneration. *Archives of Ophthalmology* **110**, 1701–1708.

Eye Disease Case-Control Study Group. (1993). Antioxidant status and neovascular age-related macular degeneration. *Archives of Ophthalmology* **111**, 104–109.

Farrar, J. F., Kenna, P., Jordan, S. A., Kumar-Singh, R., Humphries, M. M., Sharp, E.M., Sheils, D. M. and Humphries, P. (1991). A three-base-pair deletion in the peripherin-RDS gene in one form of retinitis pigmentosa. *Nature* **354**, 478–480.

Feeney-Burns, L. and Ellersieck, M. R. (1985). Age-related changes in the ultrastructure of Bruch's membrane. *American Journal of Ophthalmology* **100**, 686–697.

Ferrell, R. E., Mintz-Hittner, H. and Antoszyk, J. H. (1983). Linkage of atypical vitelliform macular dystrophy (VMD-1) to the soluble glutamate pyruvate transaminase (GPT1) locus. *American Journal of Human Genetics* **35**, 78–84.

Fischer, S. G. and Lerman, L. S. (1980). Separation of random fragments of DNA according to properties of their sequences. *Proceedings of the National Academy of Sciences USA* **77**, 4420–4424.

Forsman, K., Graff, C., Nordstrom, S., Johansson, K., Westermark, E., Lundgren, E., Gustavson, K. H., Wadelius, C. and Holmgren, G. (1992). The gene for Best's macular dystrophy is located in 11q13 in a Swedish family. *Clinical Genetics* **42**, 156–159.

Friedman, E., Ivry, M., Ebert, E., Glynn, R., Gragoudas, E. and Seddon, J. (1989). Increased scleral rigidity and age-related macular degeneration. *Ophthalmology* **96**, 104–108.

Friedman, E., Smith, T. R. and Kuwabara, T. (1963). Senile choroidal vascular patterns and drusen. *Archives of Ophthalmology* **69**, 220–230.

Ganley, J. P. and Roberts, J. (1983). Eye conditions and related need for medical care. Vital and Health Statistics, Series 11, no. 228, DHHS Publication no (PHS) 83–1678. U.S. Government Printing Office, Washington, DC.

Gartner, S. and Henkind, P. (1981). Aging and degeneration of the human macula. 1. Outer nuclear layer and photoreceptors. *British Journal of Ophthalmology* **65**, 23–28.

Gass, J. D. M. (1973). Drusen and disciform macular detachment and degeneration. *Archives of Ophthalmology* **90**, 206–217.

Gass, J. D. M., (1987). *Stereoscopic Atlas of Macular Diseases.* C. V. Mosby Co., St. Louis.

Gass, J. D. M., Jallow, S. and Davis, B. (1985). Adult vitelliform macular detachment occurring in patients with basal laminar drusen. *American Journal of Ophthalmology* **99**, 445–459.

Ghafour, I. M., Allan, D. and Foulds, W.S. (1983). Common causes of blindness and visual handicap in the west of Scotland. *British Journal of Ophthalmology* **67**, 209–213.

Gottsch, J. D., Bynoe, L. A., Harlan, J. B., Rencs, E. V. and Green, W. R. (1993). Light-induced deposits in Bruch's membrane of protoporphyric mice. *Archives of Ophthalmology* **111**, 126–129.

Gregor, Z. and Joffe, L. (1978). Senile macular changes in the black African. *British Journal of Ophthalmology* **62**, 547–550.

Gurne, D. H., Tso, M. O., Edward, D. P. and Ripps, H. (1991). Antiretinal antibodies in serum of patients with age-related macular degeneration. *Ophthalmology* **98**, 602–607.

Hsieh, R. C., Fine B. S. and Lyons, J. S. (1977). Patterned dystrophies of the retinal pigment epithelium. *Archives of Ophthalmology* **95**, 429–435 .

Hyman, L. G., Lilienfeld, A. M., Ferris, F. L. and Fine, S. L. (1983). Senile macular degeneration: A case-control study. *American Journal of Epidemiology* **118**, 213–227.

Ingram, U. M. (1956). A special chemical difference between the globins of normal human and sickle-cell anemia hemoglobin. *Nature* **178**, 792–794.

Kahn, H. A. and Moorhead, H. B. (1973). Statistics on blindness in the model reporting area 1969–70. *United States Department of Health, Education, and Welfare Publication No. (NIH) 73–427.* U.S. Government Printing Office, Washington, DC.

Kahn, H. A., Leibowitz, H. M., Ganley, J. P., Kini, M. M., Colton, T., Nickerson, R. S. and Dawber, T. R. (1977). The Framingham eye study I. Outline and major prevalence findings. *American Journal of Epidemiology* **106**, 17–32.

Kajiwara, K., Hahn, L. B., Mukai, S., Travis, G. H., Berson, E. L. and Dryja, T. P. (1991). Mutations in the human retinal degeneration slow gene in autosomal dominant retinitis pigmentosa. *Nature* **354**, 480–483.

Kajiwara, K., Sandberg, M. A., Berson, E. L. and Dryja, T. P. (1993). A *null* mutation in the human peripherin/RDS gene in a family with autosomal dominant retinitis punctata albescens. *Nature Genetics* **3**, 208–212.

Katz, M. L., Drea, C. M., Eldred, G. E., Hess, H. H. and Robison Jr., W. G. (1986). Influence of early photoreceptor degeneration on lipofuscin in the retinal pigment epithelium. *Experimental Eye Research* **43**, 561–573.

Keen, J., Lester, D., Inglehearn, C., Curtis, A. and Bhattacharya, S. (1991). Rapid detection of single base mismatches as heteroduplexes on hydrolink gels. *Trends in Genetics* **7**, 5.

Klein, B. E. and Klein, R. (1982). Cataracts and macular degeneration in older Americans. *Archives of Ophthalmology* **100**, 571–573.

Klein, R., Klein, B. E., Linton, K. L. and De Mets, D. L. (1991). The Beaver Dam eye study: Visual acuity. *Ophthalmology* **98**, 1310–1315.

Kuriyama, M., Umezaki, H., Fukuda, Y., Osame, M., Koike, K., Tateishi, J. and Igata, A. (1984). Mitochondrial encephalomyopathy with lactate-pyruvate elevation and brain infarctions. *Neurology* **34**, 72–77.

Leibowitz, H., Krueger, D. E., Maunder, L. R., Milton, R. C., Kini, M. M., Kahn, H. A., Nickerson, R. J., Pool, J., Colton, T. L., Ganley, J. P., Loevenstein, J. I., and Dauber, T. R. (1980). The Framingham Eye Study Monograph; an ophthalmological and epidemiological study of cataract, glaucoma, diabetic retinopathy, macular degeneration and visual acuity in a general population of 2631 adults, 1973–75. *Survey of Ophthalmology* **24**, (Suppl.) 335–610.

Litt, M. and Luty, J. A. (1989). A hypervariable microsatellite revealed by in vitro amplification of a dinucleotide repeat within the cardiac muscle actin gene. *American Journal of Human Genetics* **44**, 397 401.

Macular Photocoagulation Study Group (1989). Persistent and recurrent neovascularization after krypton laser photocoagulation for neovascular lesions of ocular histoplasmosis. *Archives of Ophthalmology* **107**, 344–352.

Macular Photocoagulation Study Group (1990). Persistent and recurrent neovascularization after krypton laser photocoagulation for neovascular lesions of age-related macular degeneration. *Archives of Ophthalmology* **108**, 825–831.

Marmor, M. F. and Byers, B. (1977). Pattern dystrophy of the pigment epithelium. *American Journal of Ophthalmology* **84**, 32–44.

Martinez, G. S., Campbell, A. J., Reinken, J. and Allan, B. C. (1982). Prevalence of ocular disease in a population study of subjects 65 years old and older. *American Journal of Ophthalmology* **94**, 181–189.

McWilliam, P., Farrar, G. J., Kenna, P., Bradley, D. G., Humphries, M. M., Sharp, E. M., McConnell, D. J., Lawler, M., Sheils, D., Ryan, C., Stevens, K., Daiger, S. D. and Humphries, P. (1989). Autosomal dominant retinitis pigmentosa (ADRP): localization of an ADRP gene to the long arm of chromosome 3. *Genomics* **5**, 619–622.

Melrose, M. A., Magargal, L. E. and Lucier, A. C. (1985). Identical twins with subretinal neovascularization complicating senile macular degeneration. *Ophthalmic Surgery* **16**, 648–651.

Meyers, S.M. and Zachary, A. A. (1988). Monozygotic twins with age-related macular degeneration. *Archives of Ophthalmology* **106**, 651–653.

Modi, G., Heckman, J. M. and Saffer, D. (1991). Vitelliform macular degeneration associated with mitochondrial myopathy. *British Journal of Ophthalmology* **76**, 58–60.

Myers, R. M., Maniatis, T. and Lerman, L. S. (1987). Detection and localization of single base changes by denaturing gradient gel electrophoresis. *Methods in Enzymology* **155**, 501–527.

Newsome, D. A., Hewitt, A. T., Huh, W., Robey, P. G. and Hassell, J. R. (1987). Detection of specific extracellular matrix molecules in drusen, Bruch's membrane and ciliary body. *American Journal of Ophthalmology* **104**, 373–381.

Newsome, D. A., Swartz, M., Leone, N. C., Elston, R. C. and Miller, E. (1988). Oral zinc in macular degeneration. *Archives of Ophthalmology* **106**, 192–198.

Nichols, B. E., Sheffield, V. C., Vandenburgh, K., Drack, A. V., Kimura, A. E. and Stone, E. M., (1993a). Butterfly-shaped pigment dystrophy of the fovea is caused by a point mutation in codon 167 of the RDS gene. *Nature Genetics.* **3**, 202–207.

Nichols, B. E., Drack, A. V., Vandenburgh, K. Kimura, A. E., Sheffield, V. C. and Stone, E. M. (1993b). A 2 base pair deletion in the RDS gene associated with Butterfly-shaped pigment dystrophy of the fovea. *Human Molecular Genetics*, in press.

Orita, M., Iwahara, H., Kanazawa, H., Hayashi, K. and Sekiya, T. (1989). Detection of polymorphisms of human DNA by gel electrophoresis as single-strand conformation polymorphisms. *Proceedings of the National Academy of Sciences USA* **86**, 2766–2770.

Pearce, W. G. (1968). Doyne's honeycomb retinal degeneration. Clinical and genetic features. *British Journal of Ophthalmology* **52**, 73–78.

Penfold, P. L., Killingsworth, M. C. and Sarks, S. H. (1986). Senile macular degeneration: The involvement of giant cells in atrophy of the retinal pigment epithelium. *Investigative Ophthalmology & Visual Science* **27**, 364–371.

Rocha, J., Amorim, A., Almeida, V. M., Oliveira, J. P., Leao, M., Tavares, M. C., Pereira, M. S. and Vidal-Pinheiro, L. (1988). Gene dosage evidence for the regional assignment of GPT (glutamate-pyruvate transaminase; E.C.2.6.1.2) locus to 8q24.2—8qter. *Human Genetics* **80**, 299–300.

Sanyal, S. and Hawkins, R. K. (1988). Development and degeneration of retina in *rds* mutant mice: altered disc shedding pattern in the albino heterozygotes and its relation to light exposure. *Vision Research* **28**, 1171–1178.

Sheffield, V. C., Cox, D. R. and Myers, R. M. (1989). Attachment of a 40-base-pair G+C sequence (GC-clamp) to genomic DNA fragments by the polymerase chain reaction results in improved detection of single-base changes. *Proceedings of the National Academy of Sciences USA* **86**, 232–236.

Small, K. W., Killian, J. and McLean W. C. (1991). North Carolina's dominant progressive foveal dystrophy: how progressive is it? *British Journal of Ophthalmology* **75**, 401–406.

Small, K. W., Weber, J. L., Roses, A., Lennon, F., Vance, J. M. and Pericak-Vance, M. A. (1992). North Carolina macular dystrophy is assigned to chromosome 6. *Genomics* **13**, 681–685.

Sommer, A., Tielsch, J. M., Katz, J., Quigley, H. A., Gottsch, J. D., Javitt, J. C., Martone, J. F., Royall, R. M., Witt, K. A. and Ezrine, S. (1991). Racial differences in the cause-specific prevalence of blindness in East Baltimore. *The New England Journal of Medicine* **325**, 1412–1417.

Sorsby, A. (1966). Reports on Public Health and Medical subjects. Her Majesty's Stationery Office, London.

Sorsby, A., Joll Mason, M. E. and Gardener, N. (1949). A fundus dystrophy with unusual features. *British Journal Ophthalmology* **33**, 67–97.

Sparkes, R. S., Klisak, I., Kaufman, D., Mohandas, T., Tobin, A. J. and McGinnis, J. F. (1986). Assignment of the rhodopsin gene to human chromosome three, region 3q21-3q24 by in situ hybridization studies. *Current Eye Research* **5**, 797–798.

Stone, E. M., Nichols, B. E., Streb, L. M., Kimura, A. E. and Sheffield, V. C. (1992). Genetic linkage of vitelliform macular degeneration (Best's disease) to chromosome 11q13. *Nature Genetics* **1**, 246–250.

Taylor, H. R., West, S., Munoz, B., Rosenthal, F. S., Bressler, S. B. and Bressler, N. M. (1992). The long-term effects of visible light on the eye. *Archives of Ophthalmology* **110**, 99–104.

Tranebjaerg, L., Sjo, O. and Warburg, M. (1986). Retinal cone dysfunction and mental retardation associated with a de novo balanced translocation 1;6 (q44D7). *Ophthalmic Paediatrics and Genetics* **7**, 167–173.

Travis, G. H., Brennan, M. B., Danielson, P. E., Kozak, C. A. and Sutcliffe, J. G. (1989). Identification of photoreceptor-specific mRNA encoded by the gene responsible for retinal degeneration slow (rds). *Nature* **338**, 70–73.

Travis, G. H., Christerson, L., Danielson, P. E., Klisak, I., Sparkes, R. S., Hahn, L. B., Dryja, T. P. and Sutcliffe, J. G. (1991). The human retinal degeneration slow (RDS) gene: chromosome assignment and structure of the mRNA. *Genomics* **10**, 733–739.

Tso, M. O. (1985). Pathogenetic factors of aging macular degeneration. *Ophthalmology* **92**, 628–635.

Watzke, R. C., Folk, J. C. and Lang, R. M. (1982). Pattern dystrophy of the retinal pigment epithelium. *Ophthalmology* **89**, 1400–1406.

Weatherall, D. J. and Clegg, J. B. (1981). The thalassaemia syndrome. Blackwell, Oxford.

Weber, J. L. (1990). Informativeness of human $(dC\text{-}dA)_n$ o $(dG\text{-}dT)_n$ polymorphisms. *Genomics* **7**, 524–530.

Weber, J. L. and May, P. E. (1989). Abundant class of human DNA polymorphisms which can be typed using the polymerase chain reaction. *American Journal of Human Genetics* **44**, 388–396.

Weiter, J. J., Delori, F. C., Wing, G. L. and Fitch, K. A. (1985). Relationship of senile macular degeneration to ocular pigmentation. *American Journal of Ophthalmology* **99**, 185–187.

Wells, J., Wroblewski, J., Keen, J., Inglehearn, C., Jubb, C., Eckstein, A., Jay, M., Arden, G., Bhattacharya, S. S., Fitzke, F. and Bird, A. C. (1993). Mutations in the human retinal degeneration slow (*RDS*) gene can cause either retinitis pigmentosa or macular dystrophy. *Nature Genetics* **3**, 213–218.

Young, R. W. (1988). Solar radiation and age-related macular degeneration. *Survey of Ophthalmology* **32**, 252–269.

8. MOLECULAR GENETICS OF RETINOBLASTOMA

JOHN K. COWELL

ICRF Oncology Group Institute of Child Health
30 Guilford Street, London WC1N 1EH, UK

INTRODUCTION

It appears that many human solid tumours result from the inactivation of critical genes which are responsible for the normal development of the particular tissue in question. Because these genes, through their normal function, ensure histiogenesis and prevent tumorigenesis, they have been called 'tumour suppressor genes' or 'recessive oncogenes'. The first such gene to be isolated was that responsible for the development of the children's eye cancer, retinoblastoma. The study of this gene and the application of this knowledge to the clinical management of the disease has led, and continues to lead, the way in our understanding of molecular events which result in tumour development. The details about the isolation and characterisation of this gene is the subject of this chapter.

RETINOBLASTOMA GENETICS

As the name implies, Rb is a tumour of retinal cells and, with only rare exceptions, affects children under the age of 5 years. Individuals can present with Rb at birth, demonstrating that the tumours have been growing since early fetal life. This view is supported by the histopathology of the tumour which demonstrates a relatively undifferentiated, embryonic-like organisation, implying an arrest in development of a retinal precursor cell. Thus, pools of cells are frozen in a state in which further genetic changes can occur, giving rise to the full tumour phenotype. The exact identity of these precursor cells, however, remains unknown.

Approximately 10% of patients will have a prior family history, the rest being apparently sporadic. Since the new mutation rate is relatively high (Vogel, 1979) many of these apparently sporadic cases will carry 'new' germ line mutations. In the familial form, the tumour phenotype segregates as an autosomal dominant trait (Figure 1). This means that inheritance of a single mutant gene is apparently sufficient to result in tumorigenesis. In fact, pedigree analysis (Figure 2) shows that, in 10% of cases, individuals who inherit the mutant gene do not develop a tumour – so called "incomplete penetrance" – so it is clearly only a predisposition to tumorigenesis that is inherited and other genetic events must happen. Thus, at the cellular level, Rb gene mutations act in a recessive manner. In fact, this must be the case since not all retinal cells in predisposed individuals develop into tumours.

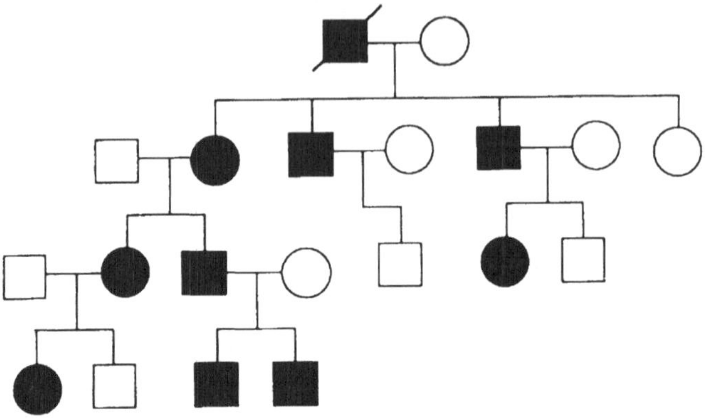

Figure 1 Family pedigree showing the segregation of the Rb tumour phenotype. All individuals are affected bilaterally as indicated by the black symbols;
■ = male ●= female

In a consideration of how many additional events are required, Knudson analysed the incidence and age distribution of tumour development to produce the now classic mathematical theory 'the two-hit hypothesis' (Knudson, 1971). In hereditary cases, the first mutation is present in all cells of the body and only one additional mutation is required for tumour development. Note that this event only affects the developing retina suggesting a highly specific role for this gene in retinal development (but see later). Since only one additional mutation is required in these cells – the chances of which are high – hereditary Rb is characterised by the presence of multiple tumours in both eyes. For this reason all bilaterally affected individuals must be considered carriers of Rb gene mutations which accounts for 40–50% of all patients. In truly sporadic cases, both mutations must occur in the same gene, in the same cell, during an early stage of development and the possibility of this occurring by chance is relatively small. Sporadic cases, therefore, are generally characterised by the presence of unilateral, unifocal tumours. However, we know empirically that approximately 10–15% of families have unilaterally affected individuals and, therefore, some unilaterally affected, sporadic cases will carry a predisposing mutation although it is difficult to identify which ones. It might be expected, for example, that early onset unilateral cases would potentially be mutant gene carriers. In our experience, this group probably represents less than 5%. In some families, apparently unaffected individuals have been seen to have retinal scars which resemble successfully treated tumours. These have been described as benign tumours – retinomas (Gallie et al., 1982) – or as regressed tumours. Occasionally, several affected children can be born to unaffected parents with no prior family history. One possibility in these cases is that an unusual insertional translocation is segregating in the family (see later) and they should certainly be referred for cytogenetic analysis. It is also possible, however,

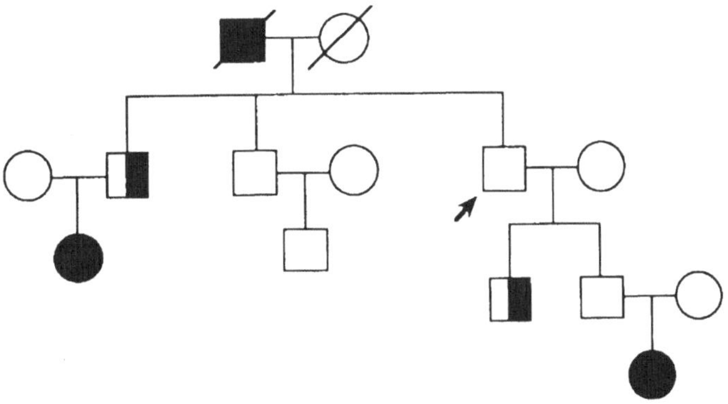

Figure 2 Family pedigree showing incomplete penetrance of the Rb predisposition gene. The male, indicated by the arrow, has inherited the predisposition gene from his affected father and passed it on to his affected son, although he is not affected himself. His second son is also an unaffected gene carrier since his daughter is also affected. Note that two individuals in this family are only unilaterally affected (half-filled symbols). The affected twins have bilateral disease (black symbols). For further details see text.

that one of the parents is a tissue mosaic (see Ribeiro et al., 1988 for review) carrying the mutation in the germline but not in their own retinal cells.

If tumours are detected early, they are usually more easily treated than those presenting later, although exactly where the tumour arises in the eye is important in this regard. Treatment of small tumours usually involves cryosurgery, photocoagulation or radiation therapy, whereas larger tumours usually require enucleation. Tumours left to develop in the eye will eventually metastasise, often down the optic nerve, and prognosis in these cases is very poor indeed. Since early diagnosis offers a better prognosis, all 'at risk' patients are screened regularly during the first years of life. In practice, this involves all relatives of Rb patients since, because of the possibility of incomplete penetrance, tumour formation is, for most patients, the only unequivocal means of identifying mutant gene carriers. Clearly a system to identify those patients with germ line mutations would make the clinical management of this disease more efficient (see below).

FINDING THE RB GENE

Our understanding of the finer details of the genetics of Rb came with the cloning of the predisposition gene, RBl. The circumstances leading to the actual cloning of RB1 resulted from painstaking analysis of many patients over many years.

Thirty years ago Stallard and colleagues (Stallard, 1962) showed that the association between Rb and phenotypic abnormalities such as mental retardation, distinctive dysmorphic features and, sometimes, abnormal gonadal development were due to the presence of constitutional chromosome deletions in these individuals. The deletions involved one of the D-group of chromosomes (Lele et al., 1963), later shown to be chromosome 13 (Yunis and Ramsay, 1978). In all cases, region 13q14 was involved (Figure 3), indicating the site of the RBl gene. Since the esterase-D gene (ESD) was also shown to be located in 13q14 (Sparkes et al., 1980), deletion carriers could be detected by a quantitative assay for this enzyme, which formed the basis for population studies (Cowell et al., 1986). Deletion carriers, who have 50% of normal enzyme levels, constitute approximately 3% of all Rb patients (Cowell et al., 1989). Close genetic linkage between Rb and ESD was reported by Sparkes et al. (1983), demonstrating the unequivocal location of the hereditary form of the disease and, to date, there have been no reported recombination events between the two loci (Cowell et al., 1987b). The exact orientation of the two genes was established by the analysis of a patient with a constitutional 13q14-q31chromosome deletion and normal ESD levels (Cowell et al., 1987a). Using somatic cell hybrids the ESD gene was shown to remain on the deletion chromosome but the RBl gene did not. This placed the RBl gene distal to ESD (Mitchell and Cowell, 1988), probably in 13q14.3.

Sporadic tumours from individuals heterozygous for polymorphic ESD alleles (allowing each of the parental chromosomes to be distinguished) often lost one allele (Godbout et al., 1983). This 'loss of heterozygosity' was also demonstrated using polymorphic DNA probes (Cavenee et al., 1983, Dryja et al., 1984). Chromosome analysis of these tumours showed that two copies of 13 were still present. The interpretation of this observation was that an acquired mutation in retinal precursor cells was duplicated at some stage and the normal homologue was then lost. In this way, the cell becomes homozygous for the initial loss of function mutation in RBl and the cells do not have a functional protein. The mechanisms by which this loss of heterozygosity (LOH) occurs were most frequently due to chromosome non-disjunction and mitotic recombination (Cavenee et al., 1983). Cavenee et al., (1985) showed that, in a tumour from a patient with heredutary Rb, the chromosome retained in the tumour was that transmitted by the affected parent. It appears that up to 70% of tumours experience this LOH (Cavenee et al., 1983). These kind of analyses allowed the origin of the parental mutation to be determined (Dryja et al., 1989). In sporadic cases there was no differential susceptibility to somatic mutation between the homologous copies of the gene. However, for new germline mutations, the heritable mutation arose on the paternally derived chromosome. These findings and those of Zhu et al. (1989) argue against genomic imprinting being important in Rb tumorigenesis but point to new mutational events arising predominantly during spermatogenesis. It has not, however, been possible to attribute these to a paternal age effect (Matsunaga et al., 1990).

CYTOGENETIC ANALYSIS OF RB TUMOURS

The mechanisms leading to LOH can clearly occur without structural chromosome rearrangement. It was hardly surprising, therefore, that chromosome 13

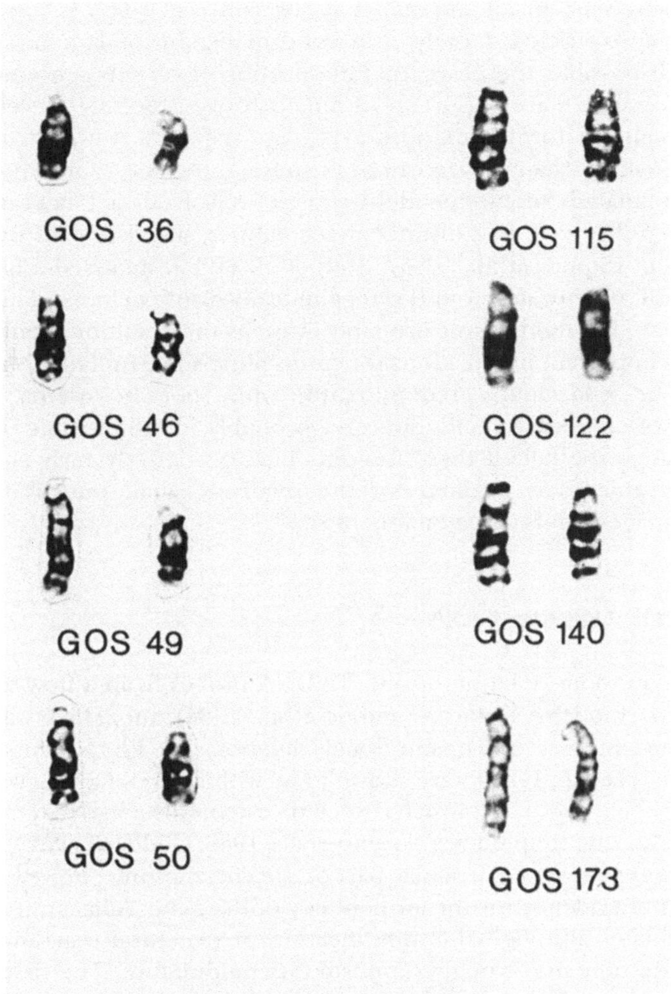

Figure 3 Chromosome 13 pairs from patients with constitutional deletions. The normal chromosome is shown on the left in all cases. Deletions from the four different patients GOS 36, 46, 49 and 50 carry large deletions, whereas the other four patients only show small deletions involving region 13q14.

abnormalities were found infrequently in tumour cells, although several interesting observations have come out of these cytogenetic analyses (see Cowell and Hogg, 1992, for review). The most consistent finding was the presence of (usually two copies of) an isochromosome 6p (iso-6p) and trisomy for all or part of the long arm of chromosome 1 (1q+), in 45% and 44% of tumours respectively. Chromosome abnormalities involving chromosome 13, usually resulting in monosomy 13, was only found in 20% of cases. Abnormalities involving lq are the most commonly seen in all tumour cells. By contrast iso-6p is less frequently observed, being restricted, largely, to Rb and malignant melanomas (Becher et al., 1983). It is possible, therefore, that duplication of certain genes on the short arm of chromosome 6 are important in tumour progression. However, since this analysis is usually performed on advanced stage tumours, it is difficult to determine whether these changes are causal in tumorigenesis or consequences of it. Chromosome analysis of independent tumour foci from a bilaterally affected patient showed that each had distinct abnormalities, suggesting an independent origin for them (Squire et al., 1985). Tien et al. (1989) analysed a large, apparently unilateral, tumour and found cytogenetically distinct clones. This was interpreted to mean that the tumour probably arose as the result of fusion of several foci. This has important implications for counselling since unilateral tumours are thought to be associated, predominantly, with sporadic events. Multifocal tumours, however, even in only one eye, probably identifies that patient as a hereditary case, especially if these tumours have a relatively early age of onset. Whether this analysis is justified for the relatively small return in terms of improved management is questionable.

ISOLATION OF THE RB1 GENE

Following the random isolation of only 12 DNA probes from a flow-sorted chromosome 13-specific DNA library (Lalande et al., 1984) one, H3-8, was shown to be within the smallest of constitutional deletions in Rb patients. Adjacent sequences (Dryja et al., 1986) were shown to be within a gene which was soon isolated (Friend et al., 1986) and which detected structurally abnormal mRNA in Rb tumours with varying frequencies (Friend et al., 1986; Goddard et al., 1988). Simply isolating a gene from a particular part of the chromosome, however, is not, in itself, conclusive evidence for the authenticity of the gene. Adjacent genes would also be candidates although the identification of structural rearrangements of this gene in Rb tumours strongly supported its candidature. The tissue distribution of expression of RB1, however, was slightly surprising, being present in relatively high levels in all tissues examined (Friend et al., 1986). This was unexpected since the hypothesis was that this gene controls important aspects of the developing fetal retina. The demonstration of predisposing mutations involving RB1 provided more convincing evidence for its authenticity. In patients showing constitutional, predisposing reciprocal translocations the breakpoints on chromosome 13 always interrupted the RB1 gene (Higgins et al., 1989; Mitchell and Cowell, 1989). Thus, adjacent genes could not be involved.

A few cases have been reported where the translocation partner chromosome is the X (Ejima et al., 1982; Hida et al., 1980; Nichols et al., 1980). Because random X-inactivation occurs in females, when the derivative chromosome is inactivated in retinal precursor cells the position of the breakpoint on 13 is not important, since the whole chromosome experiences 'genetic silencing', thereby constituting the first hit. In two families in our series, with t(1:13) (q22:q14) and t(13:20) (q14:p12) translocations, the patient with Rb inherited the rearrangement from a parent who was not affected (B. Gibbons, pers. comm). A similar situation was discussed by Dryja et al. (1984) for 13q14 deletions, in that deletion carriers often only have unilateral, unifocal tumours. In our survey (Cowell et al., 1989) half of the 16 cases reported had unilateral tumours. There are also reports of deletion patients who have never developed tumours (Cowell et al., 1988; Fukushima et al., 1987). The age of onset of tumours from patients with chromosome 13 abnormalities also appears to be later than those with germinal mutations (Ejima et al., 1988). The explanation for the low penetrance in deletion patients is not clear although one proposal is that these deletions expose lethal mutations which are deleterious to the rapidly growing tumour cells (Dryja et al., 1984). This does not explain the low penetrance of the reciprocal translocation carriers, unless large deletions are associated with the rearrangement. In one case of a t(1:13) rearrangement (Mitchell and Cowell, 1989), although a deletion was associated with the translocation, it was maximally 8 kb long and confined to the RB1 gene. Keith et al. (1985) described a patient with a 13q14 deletion with a single tumour in one eye and a retinoma in the other, both of which are considered to be 'mild' forms of the disease.

MUTATIONS IN RB TUMOURS

Only 20% of tumours showed structural abnormalities of RB1 and very few predisposing mutations involve chromosome translocations. Clearly, the majority of mutations were more subtle. The nature of these mutations in tumours has been demonstrated in a variety of ways. Dunn and colleagues (Dunn et al., 1989) analysed RNA from tumour cells, although this is not always possible for reasons discussed by Cowell and Hogg (1992). Nonetheless a variety of different mutations in tumours and cell lines were reported. The RB1 cDNA is 4.7 kb long (Friend et al., 1986) and consists of 27 exons clustered into three groups each separated by two very large introns (Figure 4). The 27 exons represent the fragmented coding region of the gene as they appear on the chromosome which are subsequently joined together in the final mRNA. In the absence of mRNA, each individual exon can be sequenced from DNA by first amplifying the exon and flanking intron regions using the polymerase chain reaction (PCR). Direct sequencing of these PCR products would identify mutations by comparison with the normal sequence. This approach has been successful (Yandell et al., 1989) and was improved using the single strand conformation polymorphism (SSCP) technique to prescreen the PCR amplified exons before sequencing, which identified those DNA molecules most likely to carry mutations (Hogg et al., 1992).

In summmary, the available data shows that there is no apparent 'hot-spot' for mutations within RB1. The majority of mutations are insertions, deletions or single base pair substitutions which result in the production of premature stop codons. These mutations would be predicted to result in structurally grossly

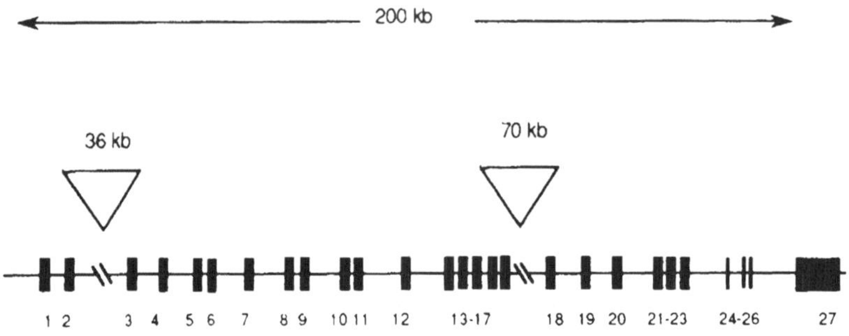

Figure 4 Diagrammatic representation of the RBI gene. The 27 individual exons, represented by the vertical bars, are clustered along a 200 kb region of the chromosome and separated by two relatively large (36 kb and 70 kb long) introns.

abnormal proteins, which is consistent with antibody analysis where Rb tumours mostly had no detectable Rb protein (Horowitz et al., 1990). In approximately 10% of tumours, the mutations affect the correct processing of the mRNA (splicing) by damaging sequences essential for this process. Missense mutations, simply substituting one amino acid for another, appear to be less common. In our own survey of hereditary Rb patients, one example was found in exon 20, which was associated with a 'low penetrance' phenotype (Onadim et al., 1992). It is tempting to speculate that the substitution of a single amino acid only compromises the function of the protein and, unless the second mutation in the tumour precursor cell causes loss of RB1 function, duplication of the 'weak' mutation allows a sufficiently functional Rb protein to be produced, so preventing tumorigenesis. This is consistent with our observation in this particular family since many of the mutant gene carriers were either unaffected or had regressed tumours (Onadim et al., 1991). Sakai et al. (1991) also investigated low penetrance families and found mutations in recognition sequences for different transcription factors – the promoter – which lie at the beginning of the gene and which controls the production of RB1 mRNA. Again the suggestion is that, as a result, a quantitative decrease in transcription occurs rather than complete inactivity. Sufficient pRB is produced, however, and any phenotypic consequences are mild. Whether single amino acid changes will generally be found in patients with mild phenotypes is still not clear.

THE FUNCTION OF THE RB1 GENE

The fact that Rb tumours are relatively undifferentiated, histopathologically, suggests an arrest in development of a retinal precursor cell at an early stage. By inference it might appear, therefore, that RB1 controls the transition from this immature precursor cell to a photoreceptor or other neuroepithelial cell in the retina. Either directly, or indirectly, RB1 must therefore control this differentiation process although analysis of the structure of the gene does not reveal any 'tell-tale' motives implicating it as a regulator of transcription. The first clues to the function of pRB came from the demonstration that it could bind to proteins from certain DNA tumour viruses which dominantly transform normal cells into cancer cells. After invading cells, DNA tumour viruses produce an early set of proteins which trigger the normally quiescent cell into division. This is essential for successful virus propagation. Thus, the E1A protein from adenovirus (Whyte et al., 1988), Large-T-antigen (LT) from SV40 (DeCaprio et al., 1988) and E7 protein from human papilloma virus (Dyson et al., 1989), all form complexes with pRB. All of the viral transforming proteins share conserved regions which are necessary for their transforming function. Mutations in these conserved regions, which prevent cellular transformation, also prevent binding to pRB (Whyte et al., 1989). pRB can be phosphorylated at many positions. In resting cells pRB is unphosphorylated but, as the cell moves into S-phase of the cell cycle, pRB becomes phosphorylated until the end of mitosis where it is dephosphorylated again (Mihara et al., 1989). This led to the suggestion that the unphosphorylated form of pRB promotes cell quiescence, i.e. prevents cell growth.

By selectively deleting parts of the RB1 gene, two non-contiguous regions were identified which were necessary for E1a and LT binding (Hu et al., 1990; Huang et al., 1990; Kaelin et al., 1990). These protein binding domains comprised amino acids 393-572 and 646-772 respectively, which have collectively been called the RB 'pocket'. It appears, therefore, that by sequestering pRB from the cell during G1, the viral transforming proteins allow the cell to enter S-phase. The association of the viral early genes with pRB is almost certainly an *in vitro* phenomenon, since it is unlikely that fetal retinal cells have been infected with these viruses. Rather, this model system points to associations of pRB with other naturally occurring proteins. In fact, it has emerged that pRB participates in the establishment of protein complexes which associate and dissociate during the cell cycle (Figure 5). Many of these processes are under tight control and the whole system appears to be regulated by the biochemical modification of the participants in these complexes and their availability to join the complex. The actual mechanisms, real and implied, have been discussed in many recent reviews and the minutiae of these models will not be discussed here (Horowitz, 1993). In essence, pRB can bind to many cell cycle regulating proteins and regulate their function. The net result is that if pRB is missing, several genes are left active which should not be so and the cell continues to divide instead of going into quiescence.

pRB has also been shown to complex with a developmentally regulated cellular transcription factor, DRTF-1 (Bagchi et al., 1991; Bandara et al., 1991; Partridge and La Thangue, 1991). The suggestion is that, by binding with a sequence-specific transcription factor, pRB can control transcription of target genes. Since DRTF is developmentally regulated, and its expression is tissue dependent, this association suggests a role for pRB in differentiation (Bandara and La Thangue, 1991; Partridge and La Thangue, 1991).

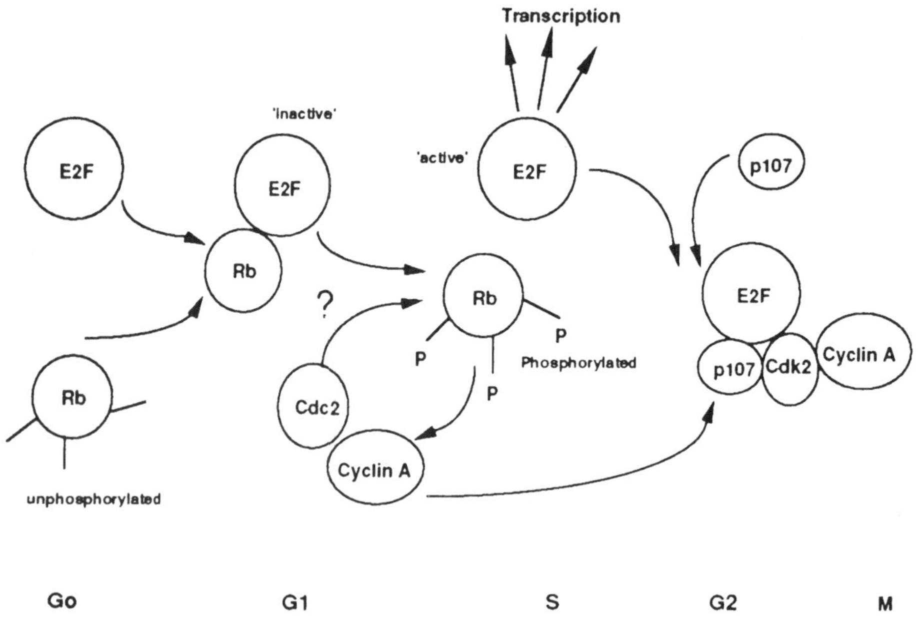

Figure 5 Schematic representation of the function of the Rb protein. The unphosphorylated protein present in G0 binds a transcription factor, E2F, at the transition to G1, inactivating it. The Rb protein is then phosphorylated, possibly mediated by an association with cdc2 kinase and cyclin A and E2F is released, for a brief period, to activate other genes which are presumably related to cell division. After this brief period of freedom, E2F is again captured by the Rb-related protein, p107, which then forms a protein complex repressing E2F activity. Exactly what happens between G2 and G0 is unclear.

RB1 MUTATIONS IN OTHER TUMOURS

It has been known for some time that mutant RB1 gene carriers also develop second non-ocular tumours; the conclusion always being that RB1 also controls the normal development of these tissues. Patients carrying constitutional RB gene mutations are at significantly higher risk for the development of second, non-ocular tumours later in life (Draper et al., 1986).These are usually osteosarcomas and soft tissue sarcomas. Both of these tumours were shown to lose heterozygosity for markers on chromosome 13 (Dryja et al., 1986; Friend et al., 1987). The same classes of tumours also show frequent structural and transcriptional abnormalities of RB1 suggesting it plays a role in establishing the malignant phenotype in these cells. The risk of second tumours is enhanced within the irradiated tissues following radiation treatment. This risk is greater for patients with germline mutations in RB1. This raises important issues about whether to treat these patients with radiation although the risks must be offset with the benefits.

In addition to those tumours known to be associated with RB1 gene mutation carriers, structural abnormalities were also found in breast cancer (Lee et al., 1988; T'Ang et al., 1988) and small cell lung carcinoma (Harbour et al., 1988) cells at high frequency. A series of other tumours showing less frequent involvement was presented by Horowitz et al. (1990). It is likely, however, that somatic RB1 mutations in these other tissues only contribute to tumour progression since, in many cases, the frequency of tumours with mutations is still relatively low.

One interesting tumour associated with hereditary Rb is pineoblastoma. Because the pineal is thought to be a vestigial photoreceptor organ, patients with retinal and pineal tumours have been described as having 'trilateral' retinoblastoma (Bader et al., 1982; Holladay et al., 1991). The chances of two such rare tumours occurring coincidentally is negligible and suggests that the RB1 gene also contributes to the development of pineoblastoma.

CLINICAL APPLICATIONS OF RB1 ISOLATION

The standard of metaphase chromosome analysis is sufficiently good these days in most routine labs that patients carrying the 13q14 deletions will be readily identifiable. Because of the cost and effort in setting up ESD quantitation studies for those rare cases with small deletions, we consider it no longer necessary to perform ESD quantitation on all Rb patients. In the majority of cases, because most of these have associated congenital abnormalities typical of $13q^-$ patients, cytogeneticists will be alerted to the possibility of a deletion. *In situ* hybridisation, using cosmids for RB1, to metaphase chromosome from Rb patients could easily identify small deletions where conventional G banding is equivocal. In this case hybridisation would not occur to one (the deletion) chromosome. For the clinical management of Rb, it is essential to be able to predict accurately who is a mutant gene carrier and who is not. Linkage analysis using the ESD protein polymorphism (Cowell et al., 1987b) has also been superseded by DNA probe analysis. Using a series of five unique DNA sequences, derived by Wiggs et al., (1988) from within the RB1 gene, the majority of individuals with a family history of Rb can be successfully analysed. These DNA probes all recognise restriction fragment length polymorphisms (RFLPs) with varying frequencies and, to date, no recombination has been reported between any of these probes and the Rb phenotype (Onadim et al., 1990; Scheffer et al., 1989; Wiggs et al., 1988). This analysis has since (Figure 6) been extended to prenatal and perinatal screening (Mitchell et al., 1988; Onadim et al., 1992). In our own experience, approximately 85% of families would qualify for prenatal screening using these 5 probes alone. The development of other highly informative polymorphisms, based on a single base pair polymorphism (Yandell and Dryja, 1989) and a small variable number tandem repeat (Figure 7) in the intron adjacent to exon 20 (McGee et al., 1989), now means that virtually all gene carriers in families with a history of Rb can be identified (Onadim et al., 1992b). Family pedigree analysis, however,

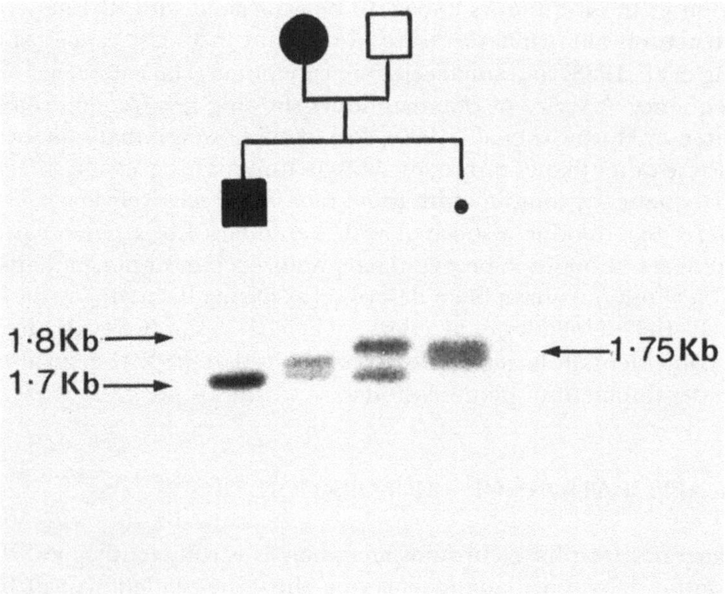

Figure 6 Prenatal screening using the polymorphic DNA probe RS2. The pedigree of a family requesting prenatal screening is shown above. The transmitting parent is heterozygous for the 1.75/1.7 kb alleles. The unaffected father is heterozygous 1.7/1.8 kb. Since the affected first-born child is homozygous for the 1.7 kb allele, the chromosome from the mother carrying the 1.7 kb allele also carries the mutant Rb gene. The chorionic villus sample from the second pregnancy was shown to inherit the 1.75 kb allele from the mother and hence the fetus has not inherited the mutant gene.

has shown considerable heterogeneity in the Rb phenotype which has significant implications for clinical management. Using intragenic polymorphic DNA probes, we have shown that it is possible to demonstrate that individuals with regressed tumours in Rb families carry the mutant allele (Figure 8) and have identified unaffected gene carriers whose children are at high risk of tumour development (Onadim et al., 1991). The prenatal screening option for these families is to have chorionic villus sampling after 10 weeks of pregnancy. In most cases the results are available after 2–5 days. We have recently published our experience with prenatal screening using DNA probes (Onadim et al., 1992b) where all patients have been followed for at least 18 months and so far all predictions have been correct.

RISK ASSESSMENT

There have been many estimates of the relative risks for the development of Rb in the children and siblings of affected individuals. Many of these calculations are reviewed by Vogel (1979) but the figures do vary from study to study and almost certainly reflect referral bias. In a review of the UK population of Rb patients involving over 900 cases over a 25 year period, Draper and colleagues (Draper et al., 1992) have re-evaluated the figures. Clearly, where there is a family history the chances of inheriting the mutant gene is 50% but, because only 90% of gene carriers actually develop a tumour, the actual risk is nearer 45%. It was noted some time ago (Cowell et al., 1986) that bilaterally affected mutant gene carriers generally have bilaterally affected children and that unilaterally affected gene carriers have a greater chance of having unilaterally affected children, or sometimes unaffected gene carriers (Onadim et al., 1991). The risk of tumorigenesis for the children of unilaterally affected mutant gene carriers is therefore lower, estimated by Draper and colleagues at around 30%. In the UK series, where there is no family history of Rb the probability of a sibling being affected is 1.1% if the proband is bilaterally affected and 0.6% if the proband is unilaterally affected. The difference, however, is not statistically significant. Finally, using the method of maximum likelihood estimation the probability of a unilateral case with no family history being a gene carrier was calculated to be 1.7% giving their children a risk of approximately 0.8%.

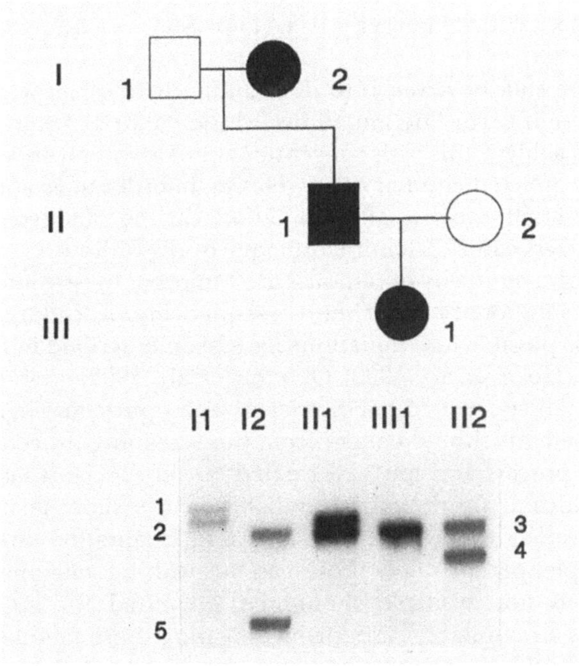

Figure 7 Pedigree analysis using the Rb1.20 polymorphism. The affected grandparent I.2 is heterozygous (3/5) and passes the 3-allele to her son who is affected. He in turn passes this same allele to his daughter who is also affected.

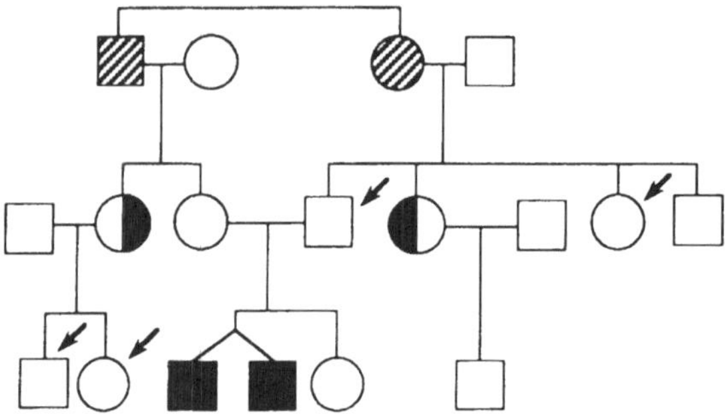

Figure 8 Example of a family pedigree showing mild expression of the RB phenotype and a missense mutation in exon 20. DNA sequence analysis shows the mutant gene is present in the grandparents (hatched symbols) who both had regressed tumours as well as in the unaffected individuals indicated by the arrows. Affected individuals are shown with solid symbols. Individuals with unilateral Rb are represented by half-filled symbols, those with bilateral Rb by solid symbols.

IDENTIFICATION OF CAUSATIVE MUTATIONS

The big challenge still, however, is to determine which of those apparently sporadic cases represent germ line mutations, whose children are at risk of tumour development. To achieve this with conventional linkage analysis, the "at risk" parent must have an affected child or other relative, in order to establish the linkage phase (Chapter 1). If the causative mutation can be characterised, however, mutant gene carriers can be identified unequivocally without extensive pedigree analysis. Since only a minority of tumours are removed, these mutations must be identified in constitutional cells. The PCR-sequencing procedure described earlier has made this possible and mutations have been described in individuals predisposed to Rb (Hogg et al., 1992; Onadim et al., 1992a; Yandell and Dryja, 1989). Although all concerned paediatricians and ophthalmologists would consider that their patients should benefit from this screening procedure, it is a very time consuming process and must be carried out in specialist laboratories. It is unlikely that sufficient funds will be available in the short term to screen all patients in a cost effective way and some logical rationalisation, therefore, should occur. For example, perhaps only those patients with a family history for whom linkage analysis is not available should be submitted for mutation analysis although there is a reasonable case for investigating those families with unusual family pedigrees to confirm the gene carrier status of unaffected family members.

All apparently sporadic, bilaterally affected cases must be considered carriers of germline mutations and should probably have mutational analysis performed. In our series, priorities are given to those individuals approaching childbearing age although, ideally, all individuals should be analysed. Unilaterally affected individuals, for whom the risk of being carriers of gene mutations is low should possibly be assigned a lower priority given limited resources although, if their tumours had an early age of onset, there is perhaps a more urgent need to carry out the study. Since the clinical screening programme for Rb patients with an affected relative is well established, it is usually the unilateral cases which present late and experience enucleation. The importance of examining the tumour cells from these patients cannot be overemphasised. Since approximately 70% of these tumours will be homozygous for the causative mutation it is important to study the RB1 gene. If a mutation is identified that is not present in constitutional cells from the same individual, then the patient can be classified as truly sporadic. It remains to be seen whether the nature of the mutations being identified in gene carriers will allow us to predict outcome, prognosis and susceptibility to second tumours. A review of the types of mutation which cause Rb shows a variety of different types. In some cases, the mutations coincidentally create or destroy a restriction enzyme site. In these cases, subsequent analysis of family members merely requires digestion of the PCR product with the appropriate enzyme, and mutant gene carriers will have the abnormal restriction pattern (Onadim et al., 1992a). In constitutional cells the mutation will always be heterozygous which makes the analysis of sequencing gels more difficult. The fact that several exons can be analysed simultaneously makes the procedure less labour intensive. PCR analysis can also be carried out on formalin-fixedparaffin-embedded tissue sections (Onadim and Cowell, 1991), which means archival material can also be used to identify causative mutations. In practice, it is not urgent to identify the mutation in sporadic cases, since those individuals will not be considering children for 16–20 years although they may be at increased risk of developing second tumours.

The availability of prenatal screening clearly raises ethical issues. At present, screening is only offered to parents who are considering terminating the pregnancy if the fetus carries the mutation. Otherwise the test can be performed perinatally. The justification for considering termination is that, although 90% curable, Rb is a lethal tumour if it escapes the confines of the eye. Even if the patient survives the tumour there may be side effects of the treatment and no guarantee that the sight will be preserved. In addition, survivors of Rb are at risk of developing second tumours later in life for which there is no effective screening programme. Although first trimester screening is preferable to amniocentesis, there may be psychological effects on parents who may change their minds about termination. Given these considerations, there is a possibility of performing preimplantation screening in some cases, especially where the nature of the mutation is known. It should soon be possible to study mutations using PCR in single cells which could be derived from *in vitro* fertilised embryos. Only mutation-free embryos would be reimplanted although the success of this procedure remains to be evaluated.

REFERENCES

Bader J. L., Meadows A. T., Zimmerman L. E., Rorke L. B., Voute P. A., Champion L. A. et al. (1982) Bilateral retinoblastoma with ectopic intracranial retinoblastoma: trilateral retinoblastoma. *Cancer Genet.Cytogenet.*, **5**:203–213.

Bagchi S., Weinmann R. Raychaudhuri P. (1991) The retinoblastoma protein copurifies with E2F-I, and E1A-regulated inhibitor of the transcription factor E2F. *Cell,* **65**:106–31072.

Bandara L. R., Adamczewski J. P., Hunt T. La Thangue N. B. (1991) Cyclin A and the retinoblastoma gene product complex with a common transcription factor. *Nature,* **352**:249–251.

Bandara L. R. and La Thangue N. B. (1991) Adenovirus E1a prevents the retinoblastoma gene product from complexing with a cellular transcription factor. *Nature,* **351**:494–497.

Becher R., Gibas Z. Sandberg A. A. (1983) Chromosome 6 in malignant melanoma. *Cancer Genet.Cytogenet.,* **9**:173–175.

Cavenee W., Dryja T. P., Phillips R. A., Benedict W. F., Godbout R., Gallie B. L. et al. (1983) Expression of recessive alleles by chromosomal mechanisms in retinoblastoma. *Nature,* **305**:779–784.

Cavenee W. K., Hansen M. F., Nordenskjold M., Kock E., Maumenee I., Squire J. A. et al. (1985) Genetic origin of mutations predisposing to retinoblastoma. *Science,* **228**:501–503

Cowell J. K. and Hogg A. (1992) The genetics and cytogenetics of retinoblastoma. *Cancer Genet.Cytogenet.,* **64:** 1–11.

Cowell J. K., Hungerford J., Rutland P. Jay M. (1987a) A chromosomal breakpoint which separates the esterase-D and retinoblastoma predisposition loci in a patient with del(13)(q14-q31). *Br.J.Cancer,* **55**:661–664.

Cowell J. K., Hungerford J., Rutland P. Jay M. (1989) Genetic and cytogenetic analysis of patients showing reduc ed esterase-D levels and mental retardation from a survey of 500 individuals with retinoblastoma. *Ophthal.Paed.Genet.,* **10**:117–127.

Cowell J. K., Jay M., Rutland P. Hungerford J. (1987b) An assessment of the usefulness of electrophoretic varients of esterase-D in antenatal diagnosis of retinoblastoma in the United Kingdom. *Br.J.Cancer,* **55**:661–664.

Cowell J. K., Rutland P., Hungerford J. Jay M. (1988) Deletion of chromosome region 13q14 is transmissible and does not always predispose to retinoblastoma. *Hum.Genet.,* **80**:43–45.

Cowell J. K., Rutland P., Jay M. Hungerford J. (1986) Deletions of the esterase-D locus from a survey of 200 retinoblastoma patients. *Hum.Genet.,* **72**:164–167.

DeCaprio J. A., Ludlow J. W., Figge J., Shew J.-Y., Huang C.-M., Lee W.-H. et al. (1988) SV40 large tumour antigen forms a specific complex with the product of the retinoblastoma susceptibility gene. *Cell,* **58**:1085–1095.

Draper G. J., Sanders B. M., Brownbill P. A. Hawkins M. M. (1992) Patterns of risk of hereditary retinoblastoma and applications to genetic counselling. *Br.J.Cancer,* **66**:2112 1 9

Draper G. J., Sanders B. M. Kingston J. E. (1986) Second primary neoplasms in patients with retinoblastoma. *BrJ.Cancer.* **53**:661–671.

Dryja T. P., Cavenee W. K., White R., Rapaport J. M., Peterson R., Albert D. M. et al. (1984) Homozygosity of chromosome 13 in retinoblastoma. *New Eng.J.Med.,* 310:550553.

Dryja T. P., Mukai S., Petersen R., Rapaport J. M., Walton D. Yandell D. W. (1989) Parental origin of mutations of the retinoblastoma gene. *Nature,* **339**:556–558.

Dryja T. P., Rapaport J. M., Epstein J., Goorin A. M., Weichselbaum R., Koufos A. et al. (1986a) Chromosome 13 homozygosity in osteosarcoma without retinoblastoma. *Am.J.Hum.Genet.,* **38**:59–66.

Dryja T. P., Rapaport J. M., Joyce J. M. Petersen R. A. (1986b) Molecular detection of deletions involving band q14 of chromosome 13 in retinoblastomas. *Proc.Natl.Acad.Sci.USA,* **83**:7391–7394.

Dunn J. M., Phillips R. A., Zhu X., Becker A. Gallie B. L. (1989) Mutations in the RBl gene and their effects on transcription. *Mol.Cell Biol.*, **9**:4596–4604.
Dyson N., Howley P. M., Munger K. Harlow E. (1989) The human papilloma virus-16 E7 oncoprotein is able to bind to the retinoblastoma gene product. *Science*, **243**:934–936.
Ejima Y., Sasaki M. S., Kaneko A., Tanooka H., Hara Y., Hilda T. et al. (1982) Possible inactivation of part of chromosome 13 due to 13q Xp translocation associated with retinoblastoma. *Clin.Genet.*, **21**:357–361.
Ejima Y., Sasaki M. S., Kaneko A. Tanooka H., (1988) Types, rates, origin and expressivity of chromosome mutations involving 13q14 in retinoblastoma patients. *Hum.Genet.*, **79**:118–123.
Friend S. H., Bernards R., Rogeli S., Weinberg R. A., Rapaport J. M., Albert D. M. et al. (1986) A human DNA segment with properties of the gene that predisposes to retinoblastoma and osteosarcoma. *Nature*, **323**:643–646.
Friend S. H., Horowitz J. M., Gerber M. R. Wang X.-F. (1987) Deletions of a DNA sequence in retinoblastoma and mesenchymal tumors: organization of the sequence and its encoded protein. *Proc.Natl.Acad.Sci.USA*, **84**:9059–9063.
Fukushima Y., Kuroki Y., Ito T., Kondo I. Nishigaki I. (1987) Familial retinoblastoma (mother and son) with 13q14 deletion. *Hum.Genet.*, **77**:104–107.
Gallie B. L., Ellsworth R. M., Abramson D. H. Phillips R. A. (1982) Retinoma: spontaneous regression of retinoblastoma or benign manifestation of mutation? *Br.J.Cancer*, **45**:513–521.
Godbout R., Dryja T. P., Squire J., Gallie B. L. Phillips R. A. (1983) Somatic inactivation of genes on chromosome 13 is a common event in retinoblastoma. *Nature*, **304**:451–453.
Goddard A. D., Balakier H., Canton M., Dunn J., Squire J., Reyes E. et al. (1988) Infrequent genomic rearrangement and normal expression of the putative RBl gene in retinoblastoma tumours. *Mol.Cell Biol.*, **8**:2082–2088.
Harbour J. W., Lai S.-L., Whang-Peng J., Gazdar A. F., Minna J. Kaye F. J. (1988) Abnormalities in structure and expression of the human retinoblastoma gene in SCLC. *Science*, **241**:353–357.
Hida T., Kinoshita Y., Matsumoto R., Suzuki N. Tanaka H. (1980) Bilateral retinoblastoma with 13qXp translocation. *J.Paediat.Ophthalmol.Strab.*, **17**:144–146.
Higgins M. J., Hansen M. F., Cavenee W. K. Lalande M. (1989) Molecular detection of chromosomal translocations that disrupt the putative retinoblastoma susceptibility locus. *Mol.Cell Biol.*, **9**:1–5.
Hogg A., Onadim Z., Baird P. N. Cowell J. K. (1992) Detection of heterozygous mutations in the RBl gene in retinoblastoma patients using single strand conformation polymorphism (SSCP) analysis and PCR sequencing. *Oncogene*, **7**:1444–1451.
Holladay D. A., Montebello J. F. Redmond K. P. (1991) Clinical presentation, treatment and outcome of trilateral retinoblastoma. *Cancer*, **67**:710–715.
Horowitz, J.M. (1993) Regulation of transcription by the retinoblastoma protein. *Genes, Chromosomes and Cancer*, **6**:124–131.
Horowitz J. M., Park S.-H., Yandell D. W. Weinberg R. A. (1990) Frequent inactivation of the retinoblastoma anti-oncogene is restricted to subset of human tumour cells. *Proc.Natl. Acad.Sci.*, **87**:101–107.
Hu Q., Dyson N. Harlow E. (1990) The regions of the retinoblastoma protein needed for binding to adenovirus E1A or SV40 large T antigen are comrnon sites for mutation. *EMBO Journal*, **9**:1147–1566.
Huang H. S., Wang N.-P., Tseng B. Y., Lee W.-H. Lee E. H.-H. P. (1990) Two distinct and frequently mutated regions of the retinoblastoma protein are required for binding to SV40 T antigen. *EMBO J.*, **9**(6):1815–1822.
Kaelin W. G., Ewen M. E. Livingston D. M. (1990) Definition of the minimal simian virus large T antigen and adenovirus E1A-binding domain in the retinoblastoma gene product. *Mol.Cell.Biol.*, **10**:3761–3769.

Keith C. G. and Webb G. C. (1985) Retinoblastoma and retinoma occurring in a child with a translocation and deletion of the long arm of chromosome 13. *Arch. Ophthalmol.*, **103**:941–944.

Knudson A. G. (1971) Mutation and cancer: statistical study of retinoblastoma. *Proc.Natl. Acad.Sci.USA*, **68**:820–823.

Lalande M., Dryja T. P., Schreck R. R., Shipley J., Flint A. Latt S. A. (1984) Isolation of human chromosome 13-specific DNA sequences cloned from flow sorted chromosomes and potentially linked to the retinoblastoma locus. *Cancer Genet.Cytogenet.*, **13**:283–295.

Lee E. Y.-H. P., To H., Shew J.-H., Bookstein R., Scully P. Lee W.-H. (1988) Inactivation of the retinoblastoma susceptibility gene in human breast cancers. *Science*, **241**:218–221.

Lele K. P., Penrose L. S. Stallard H. B. (1963) Chromosome deletion in a case of retinoblastoma. *Ann.Hum.Genet.Lond.* **27**:171–174.

Matsunaga E., Minoda K. Sasaki M. S. (1990) Parental age and seasonal variation in the births of children with sporadic retinoblastoma: a mutation-epidemiologic study. *Hum.Genet.*, **84**:155–158.

McGee T. L., Yandell D. W. Dryja T. P. (1989) Structure and partial genomic sequence of the human retinoblastoma susceptibility gene. *Gene*, **80**:119–128.

Mihara K., Cao X-R, Yen A., Chandler S., Driscoll B., Murphree A. L. et al. (1989) Cell cycle-dependent regulation of phosphorylation of the human retinoblastoma gene product. *Science*, **246**:1300–1303.

Mitchell C. D. and Cowell J. K. (1988) Molecular evidence that the esterase-D gene lies proximal to the retinoblastoma susceptibility locus in chromosome region 13q14. *Hum.Genet.*, **81**:57–60.

Mitchell C. D. and Cowell J. K. (1989) Predisposition to retinoblastoma due to a translocation within the 4.7R locus. *Oncogene*, **4**:253–257.

Mitchell C. D., Nicolaides K., Kingston J., Hungerford J., Jay M. Cowell J. K. (1988) Prenatal exclusion of hereditary retinoblastoma. *Lancet*, **1**:826.

Nichols W. W., Miller R. C., Sobel M. Hoffman E. (1980) Further observations on a 13qXp translocation associated with retinoblastoma. *Am.J.Ophthalmol.*, **89**(5):621–627.

Onadim Z. and Cowell J. K. (1991) Application of PCR amplification from paraffin embedded tissue sections to linkage analysis in familial retinoblastoma. *J.Med.Genet.*, **28**:312–316.

Onadim Z., Mitchell C. D., Rutland P. C., Buckle B. G., Jay M., Hungerford J. L. et al. (1990) Application of intragenic DNA probes in prenatal screening for retinoblastoma gene carriers in the United Kingdom. *Arch.Dis.Child.*, **65**:651–656.

Onadim Z., Hykin P. G., Hungerford J. L. Cowell J. K. (1991) Genetic counselling in retinoblastoma: importance of ocular fundus examination of first degree relatives and linkage analysis. *Br.J.Ophthalmol.*, **75**:147–150.

Onadim Z., Hogg A., Baird P. N. Cowell J. K. (1992a) Oncogenic point mutations in exon 20 of the RB1 gene in families showing incomplete penetrance and mild expression of the retinoblastoma phenotype. *Proc.Natl.Acad.Sci.USA*, **89**:6177–6181.

Onadim Z., Hungerford J. Cowell J. K. (1992b) Follow-up of retinoblastoma patients having prenatal and perinatal predictions for mutant gene carrier status using intragenic polymorphic probes from the RB1 gene. *Br.J.Cancer*, **65**:711–716.

Partridge J. F. and La Thangue N. B. (1991) A developmentally regulated and tissuedependent transcription factor complexes with the retinoblastoma gene product. *EMBO*, **10**(12):3819–3827.

Ribeiro M. C. M., Andrade J. A., Erwenne C. M. Brunoni D. (1988) Bilateral retinoblastoma associated with 13q-mosaicism: possible manifestation of a germinal mutation. *Cancer Genet. Cytogenet.*, **32**:169–175.

Sakai T., Ohtani N., McGee T. L., Robbins P. D. Dryja T. P. (1991) Oncogenic germ-line mutations in Spl and ATF sites in the human retinoblastoma gene. *Nature*, **353**:83–86.

Scheffer H., Meerman G. J., Kruise Y. C. M., van den Berg A. H. M., Penning D. P., Tan K. E. W. P. et al. (1989) Linkage analysis of families with hereditary retinoblastoma: non-penetrance of mutation, revealed by combined use of markers within and flanking the RB1 gene. *Am.J.Hum.Genet.*, **45**:252–260.

Sparkes R. S., Murphree A. L., Lingua R. W., Sparkes M. C., Field L. L., Funderbuck S. et al. (1983) Gene for hereditary retinoblastoma assigned to human chromosome 13 by linkage to esterase-D. *Science*, **217**:971–973.

Sparkes R. S., Sparkes M. C., Wilson M. G., Towner J. W., Benedict W., Murphree A. L. et al. (1980) Regional assignment of genes for human esterase-D and retinoblastoma to chromosome band 13q14. *Science*, **208**:1042–1044.

Squire J., Gallie B. L. Phillips R. A. (1985) A detailed analysis of chromosomal changes in heritable and non-heritable retinoblastoma. *Hum.Genet.*, **70**:291–301.

Stallard H. B. (1962) The conservative treatment of retinoblastoma. *Trans.Ophthalmol.Soc.U.K.*, **83**:473–535.

T'Ang A., Varley J. M., Chakraborty S., Murphree A. L. Fung Y.-K. T. (1988) Structural rearrangements of the retinoblastoma gene in human breast carcinoma. *Science*, **242**:263–242.

Tien H.-F., Chuang S.-M., Chen M. S., Lee F.-Y. Hou P.-K. (1989) Cytogenetic evidence of multifocal origin of a unilateral retinoblastoma. *Cancer Genet.Cytogenet.*, **42**:203–208.

Vogel W. (1979) The genetics of retinoblastoma. *Hum. Genet.*, **52**:1–54.

Whyte P., Buchkovitch K., Horowitz J. M., Friend S. H., Raybuck M., Weinberg R. A. et al. (1988) Association between an oncogene and an anti-oncogene: the adenovirus E1A proteins bind to the retinoblastoma gene product. *Nature*, **334**:124–129.

Whyte P., Williamsson N. M. Harlow E. (1989) Cellular targets for transformation by the adenovirus E1A proteins. *Cell*, **56**:6775.

Wiggs J., Nordenskjeld M., Yandell D., Rapaport J., Grondin V., Janson M. et al. (1988) Prediction of the risk of hereditary retinoblastoma using DNA polymorphisms within the retinoblastoma gene. *New Eng.J.Med.*, **318**:151–157.

Yandell D. W., Campbell T. A., Dayton S. H., Petersen R., Walton D., Little J. B. et al. (1989) Oncogenic point mutations in the human retinoblastoma gene: their application to genetic counselling. *New Eng.J.Med.*, **321**:1689–1695.

Yandell D. W. and Dryja T. P. (1989) Detection of DNA sequence polymorphisms by enzymatic amplification and direct genomic sequencing. *Am J.Hum.Genet.*, **45**:547–SSS.

Yunis J. J. and Ramsay N. (1978) Retinoblastoma and subband deletion of chromosome 13. *Am.J.Dis.Child.*, **132**:16 1– 163.

Zhu X., Dunn J. M., Phillips R. A., Goddard A. D., Paton K. E., Becker A. et al. (1989) Preferential germline mutation of the paternal allele in retinoblastoma. *Nature*, **340**:312–313.

9. COLOR VISION DEFECTS

JAY NEITZ and MAUREEN NEITZ

Medical College of Wisconsin, Department of Cellular Biology and Anatomy, and Department of Ophthalmology, 8701 Watertown Plank Road, Milwaukee, WI 53226, USA

SUMMARY

In this chapter, the light sensitive visual pigments that underlie normal and defective color vision, and the genes that encode them are reviewed. New understanding of these has developed quickly in the last few years since the techniques of molecular biology have been brought to bear on long pondered problems. Decades before molecular methods were available, a theory to explain the biological basis of normal and defective color vision was widely accepted. A favored version of that theory held that everyone with normal color vision has, in common, three types of cone photoreceptor, each containing a different type of visual pigment. The common color defects were proposed to occur when one type of cone pigment was lost, or when the photopigment in one of the cone types was replaced by an abnormal (or anomalous) pigment, different in absorption spectrum from any of the three stereotyped normal pigments. However, over the years, results have accumulated that are difficult to fit within the framework of the classic view. Thus, an alternative theory is outlined. Its essentials are as follows.

(1) Humans can have many genes on the X-chromosome that encode cone photopigments (as many as ten are not uncommon).
(2) Spectrally distinct subtypes of pigments occur within each of the two classes of X-encoded cone pigments. Both normal and color defective observers draw from the same pool of pigments.
(3) People with normal color vision all have at least one pigment from each of the two X-encoded pigment classes, but they can have more than one subtype from either class and, thus, can have more than three spectrally distinct cone types.
(4) All of the common color vision deficiencies can be explained by a mutational pathway in which genes that produce pigments required for normal color vision are deleted by unequal crossing over between normal pigment gene arrays.

This is not offered as the only possible model, but here it serves as a framework around which the facts of color vision phenotypes and genotypes can be organized and re-evaluated.

INTRODUCTION

Color vision capacities are characterized by the abilities to distinguish lights that differ in their wavelength compositions. Although human color capacities are extensive, there are many differences in wavelength composition to which we are all quite blind. Deficiencies in our ability to tell spectrally different lights apart are exploited, for example, in color television that employs mixtures of just three color primaries, red, green, and blue to synthesize a wide range of colors. The wavelength composition of the light from a banana seen on television is quite different from a real banana, but we are nearly blind to the physical difference. Inabilities to see the physical differences between lights that differ in wavelength content have proved convenient for classifying different forms of color vision. For example, given a test light of any wavelength composition, humans with normal color vision can make an identical color match using just three appropriately chosen color primaries, e.g., red, green, and blue – either by mixing the right amounts of the three primaries, or by mixing two primaries and adding the third to the test light. This three-dimensionality or trichromacy is the hallmark of normal human color vision. Other forms of color vision (or color vision deficiency) can be similarly classified by the number of primaries required in color matches. An individual who is completely color-blind, classified in this way, is a monochromat. A monochromat can exactly match the appearance of any light, regardless of its wavelength content, with a single fixed primary light by simply adjusting the intensity of the comparison primary. The intermediate category, dichomats, need only two primaries. This form of color vision seems to be a standard among other species (Jacobs, 1981). For example, among mammals, trichromacy has been found to occur only among primates. Many other familiar mammals, e.g., dogs, cats, and squirrels, are dichromats. Dichromacy is also fairly common among humans. In Caucasian populations about 1 person out of every 100 is a dichromat.

The terms monochromacy, dichromacy and trichromacy strictly refer to the number of primary colors required in color matches, however, these terms are often (mis-) used to refer to the number of spectrally different cone pigments an individual has, or is believed to have. While it is true that three pigments are the minimum required for trichromatic color vision and at least two are needed for dichromacy, there may be hazards in equating visual capacities with the number of cone pigment types. It is possible, for example, to imagine a system with dozens of spectrally different receptors that still has monochromatic (color blind) vision, since color vision requires that the receptors be "wired" in a way that yields information about spectral differences. Or, more to the point in this review, the simplest (and favorite) theory of human color vision holds that we have three types of receptors each tuned to a different region of the visible spectrum. Trichromacy is a fact of normal human color vision. No doubt, we have at least three types of receptors. But, we could have more than three and still be trichromats if the signals from multiple receptors were channeled into three outputs at any stage in the visual system. As will be seen, recent evidence suggests that many people may, indeed, have more than three spectrally different cone pigments.

Classes of Cone Pigments

Human color vision is based on light sensitive pigments that absorb maximally in different regions of the visible spectrum (Dartnall, Bowmaker, & Mollon, 1983; Schnapf, Kraft, & Baylor, 1987; Oprian, Asenjo, Lee, & Pelletier, 1991; Merbs & Nathans, 1992a). Each pigment molecule is composed of an apoprotein component, opsin, linked to the light absorbing chromophore *11-cis* retinal. Structural differences between the opsins govern the absorption properties of the pigments. Pigments can be classed according to the spectral region of maximal absorption. Not everyone agrees on the appropriate terms for these different pigment classes. The terms red, green and blue are sometimes used, however, there has been a move to avoid them since they can be misleading if they are taken to mean that the different pigment classes are most sensitive to the regions of the spectrum seen as red, green, and blue, which is not true. For example, the "red" pigments have peak absorption near 560 nm, a wavelength that appears greenish-yellow, not red, and the blue pigment peaks somewhere below 430 nm, a spectral region that appears violet. The sensation of "redness" comes about because of the way output from the receptors is processed by the nervous system, not because humans have a receptor that is most sensitive to the spectral region that appears red. Another common misunderstanding is that people who are missing a "red" pigment can see green and blue but not red, while those missing a "green" pigment see red and blue but not green. This is not an adequate conception. In people with normal color vision the ability to distinguish lights with wavelengths longer than about 550 nm, e.g., red, orange, yellow, and yellowish-green comes from comparison, by the nervous system, of outputs from the "red" and "green" cones. The absence of either cone type results in dichromatic color vision in which all middle-to-long wavelengths are virtually indistinguishable. Confusion can be minimized if color terms are avoided when referring to pigments. Human pigments can be classed into three categories: long-wavelength sensitive (pigments that peak near 560 nm), middle-wavelength sensitive (pigments that peak near 530 nm), and short-wavelength sensitive (pigments that peak below 430 nm) and conveniently abbreviated (e.g., following Mollon, 1991) long-wave, middle-wave, and short-wave. These terms will be used here.

Forms of congenital color blindness

(1) Monochromacy

Among humans, complete color blindness (monochromacy) is extremely rare, affecting about 1 in 100,000 of the general population. Within this group are three types of congenital monochromacy (reviewed by Nathans et al., 1989;

McKusick, 1992). Individuals with **congenital rod monochromacy** (congenital achromatopsia) lack all cone-mediated responses but have normal rhodopsin and rod function. Cones are present but functionally defective. These individuals function poorly in bright light and see better in dim light. In some cases, the central cones mediate a light response with the action spectrum of rhodopsin, suggesting that rhodopsin may be present in rods and cones (typical rod monochromacy). In others, typical cone pigments are present, suggesting that the defect lies beyond the point of light absorption (atypical rod monochromacy). Both conditions are non-progressive and are inherited as autosomal recessive traits.

Blue-cone monochromats only have functional rods and short-wavelength cones. This is an X-linked condition most commonly associated with one of two classes of gene rearrangement, either (1) the reduction, presumably by unequal homologous recombination, to a single X-linked pigment gene (as in dichromacy) plus a deleterious mutation in the remaining functional gene, or (2) the deletion of the locus-control region (LCR). The LCR is located about 4 kb upstream of the transcription initiation site and is necessary for the correct transcription of the X-linked pigment genes (Nathans et al., 1989; Wang et al., 1992). Both long- and middle-wave genes are presumed to become inactive as a result of LCR deletion. Some patients who are reduced to a single X-linked pigment gene lose the function of the remaining gene as a result of a point mutation changing a cysteine to an arginine at amino acid position 203, a highly conserved site among G-protein coupled transmembrane receptors (Nathans et al., 1989). This residue is apparently involved in forming a disulfide linkage that is essential for the assembly of a functional visual pigment (Karnik and Khorana, 1990).

In all three types of monochromacy, affected individuals have poor visual acuity, cannot distinguish colors and often show nystagmus and photophobia. Some blue cone monochromats develop a slowly progressive central retinal degeneration.

(2) Dichromacy

The most severe forms of congenital color blindness that are commonly encountered, are the dichromacies. As there are three classes of human cone pigments, there are three different types of dichromacy. Protanopes have a single middle- and a single short-wave pigment but no long-wave pigments. Deuteranopes have a single long- and a short-wave pigment but lack middle-wave pigment. Tritanopes are missing short-wave receptors but have the other two classes. **Tritanopia** is a rare condition that is inherited as an autosomal dominant disorder. It is much less common than the X-linked forms of color blindness but in some populations the incidence is as high as 1 in 500 (van Heel et al., 1980). Recently, three different mutations have been identified within the gene encoding the short-wave pigment in individuals with tritanopia, which segregated with the disorder and were absent from general population controls (Weitz et al., 1992a; Weitz, Went & Nathans, 1992b). In each case, a non-conservative amino acid substitution was predicted (Gly-79-Arg, Ser-214-Pro, Pro-264-Ser). One of these residues (Pro-264) is highly conserved both in visual pigments and in G-protein coupled receptors. **Protanopia** and **deuteranopia** each occur at a frequency of about 1% in Caucasian males.

(3) Trichromacy

Less severe forms of color vision deficiency are the **anomalous trichromacies**. Anomalous trichromats, like normal trichromats, require three primaries in color matches. However, they generally have poorer color discrimination ability than normal. The terms for the anomalous trichromacies: protanomaly, deuteranomaly, and tritanomaly, parallel those for the dichromatic types. **Deuteranomaly** is the most common form of color vision deficiency affecting about 5% of males in Caucasian populations. The frequency of **protanomaly** is about 1% of males. **Tritanomaly** is a rare condition.

There is considerable variation in the ability to discriminate colors within each class of X-linked anomalous trichromacy. Some protanomals and deuteranomals have red/green color vision that is nearly as poor as a dichromat's. Others have much more mild color deficiency; the color discrimination ability of some approaches normal. Perceptual tests that involve color matching were useful for broad classification of color vision types. Color matching tests are also used in the diagnosis of color anomalies. A dramatic difference between the mild anomalous trichromacies and normal can be seen in behavior in a color matching task that is referred to as the Rayleigh color match (after Lord Rayleigh, 1881, who introduced the use of this color match for diagnosing red-green color vision variations). A person is given just two primary lights, a red and a green and is asked to mix them in a proportion that will exactly match the appearance of a monochromatic yellow comparison light. Only two primary lights are required in this color match because the short-wave cones are very insensitive to the lights used in this test. The observer needs only to satisfy the requirement that the middle- and long-wave pigments absorb photons at the same rate when stimulated by the red/green mixture as they do when presented with the monochromatic yellow light. Similar to normals, mildly anomalous trichromats can find a fairly narrow range of red/green mixtures that identically match the yellow comparison. However, they choose a ratio of red to green light that is very different from normal. Mildly protanomalous observers require considerably more red light and mildly deuteranomalous observers, considerably more green light in the mixture than observers with normal color vision. For example, mildly deuteranomalous observers typically match the monochromatic yellow light with a red/green ratio that is 4 to 5 times lower than normal. People with more extreme color anomalies, because of their poor color discrimination ability, will accept a wide range of red/green ratios as indistinguishable from the yellow standard, often including the normal red/green ratio and the ratio of the corresponding mild anomaly.

Color matches occur at the level of the pigments. In the color match described above, the middle- and long-wave pigments each absorb some proportion of incident photons from the monochromatic yellow light. The mixture ratio of red and green light can be adjusted to produce the same proportion of quantal absorption in these pigments as the yellow light does. The important consequence of the fact that the match occurs at the level of pigment absorption is that once the lights are set to match in their effects on the pigments, the two lights cannot be made distinguishable by any trick of the nervous system, or by changing either the ratio of middle-to-long wave cones or the distribution of pigments among the cone receptors. For example, a pigment match made by a person who has long-wave pigment in one set of cones and middle-wave pigment in a separate set of cones, would be the same as a (hypothetical) person who has one cone type with long-wave pigment and a second cone type containing a mixture of middle- and long-wave pigments.

Individual differences in the density of spectrally selective ocular filters, such as the yellow pigments in the lens and macula, do produce slight individual differences in the ratio of red-to-green light reaching the pigments, and these can influence the color match. Also, differences in the effective optical density of the visual pigments within the cones can produce slight differences in the absorption spectra of the receptors. Other than the slight influences of these factors, however, two people can have different color matches only if they have complements of pigments that differ in absorption spectra. The fact that anomalous trichromats require extremely abnormal ratios of red-to-green light in color matches indicates that they have very different complements of pigment spectral sensitivities than normals.

The visual pigments that underlie the two X-linked anomalous trichromacies have, at least to a first approximation, been characterized (e.g. DeMarco, Pokorny, & Smith, 1992; Pokorny & Smith, 1977; Piantanida, 1976; Pokorny, Smith, & Katz, 1973). Protanomalous observers, like protanopes, have no long-wave pigments but they maintain two slightly different pigments that peak in the middle wavelengths. These two pigments are sufficiently different in their spectral properties to support some color discrimination in the middle-to-long wavelengths. The absence of long-wave pigments in protanomalous observers confers a loss in sensitivity to very long-wavelength (red) lights. In color matching, protanomalous observers compensate for their insensitivity to red light by adding a much higher proportion of the red primary to the mixture. In contrast, deuteranomalous observers lack any measurable contribution to their vision from middle-wave pigments, but they maintain two types of long-wave pigments. The spectra of the two long-wave pigments expressed in deuteranomaly are similar but their difference supports some degree of red-green color vision that is variable across individuals.

Until recently it was believed that anomalous trichromacy was caused by either the replacement of the gene that encodes normal long-wave pigment (in protanomaly) or the gene that encodes normal middle-wave pigment (in deuteranomaly) with an abnormal gene. The abnormal gene (previously believed to be allelic with the normal gene) was thought to produce an anomalous pigment with a spectrum shifted from that of normals. Thus, for example, deuteranomalous observers were conceived as having a mutant (anomalous) middle-wave pigment and a normal long-wave pigment. Similarly, protanomalous observers were conceived as having a normal middle-wave pigment and a mutant (anomalous) long-wave pigment. This is now thought to be incorrect (Nathans et al., 1986a). Current understanding suggests, to us, that a much more useful conception is that anomalous trichromats, like their dichromatic counterparts (deuteranopes and protanopes), are each missing all members of one class of cone pigment. Deuteranomalous observers have no middle-wave pigments, however, unlike deuteranopes, they maintain two slightly different long-wave pigment types. Protanomalous observers have no long-wave pigments but they maintain two slightly different middle-wave pigment types.

Variation in normal color vision and in normal pigments

Dichromacy and anomalous trichromacy are phenotypic color vision variants that occur with relatively high frequency in the human population. People with these color vision deficiencies have different complements of cone pigments than normal. In addition to the large color vision differences between normal and color defective vision, most people with normal color vision have experienced disagreements with others about colors. Are some of these normal individual differences in color perception caused by individual differences in cone pigments? Recent evidence indicates that they are.

Individual differences in normal color vision have been long recognized. However, only recently has there been renewed interest in these normal variations and their cause. Fifteen years ago, Alpern and colleagues measured individual differences in color defective observers that they attributed to variation in the cone pigments (Alpern & Moeller, 1977; Alpern & Pugh, 1977; Alpern & Wake, 1977). They concluded that the long-wave pigments of different deuteranopes are variable in their spectral position as are the middle-wave pigments of protanopes. They reasoned that if the dichromacies are reduced forms of normal vision, then the middle- and long-wave pigments of normals must vary as well. They emphasized that a finding of individual differences in normal pigments would dictate major alterations in theories of the basis of color deficiency. However, the idea that normal pigments vary in spectral sensitivity has been controversial over the past decade and a half and the possibility of variation in normal pigments has continued to be ignored in the formulation of favored theories to explain color defects.

While all the controversies have yet to be settled, results have now accumulated to provide strong evidence that the spectra of normal cone pigments do vary across individuals. The picture that emerges is that among color normals there may be discrete forms of both the long-wave pigments and the middle-wave pigments. Most common are, perhaps, two major alternative forms of the long- and two major alternative forms of the middle-wave pigment. The idea of discrete variation in normal human pigments was suggested earlier by Waaler (1967) from analysis of normal color matches, by Eisner & MacLeod (1981) who used psychophysical methods to isolate responses from normal middle- and long-wave pigments and by Dartnall, et al. (1983), whose microspectrophotometric study of the spectra from individual human photoreceptors suggested two types of middle- and two types of long-wave receptors among people with normal color vision.

The largest number of recent studies that provide evidence for variation in normal pigments have examined normal individual differences in color matching. The Rayleigh color match (described earlier) where a mixture of red and green light is adjusted to match the appearance of a monochromatic yellow light, has long served to classify abnormal vision. Since the red/green ratio set in the match is directly related to the absorption spectra of the pigments, this test can be a very sensitive indicator of individual differences in normal middle- and long-wave pigments as well. If the cone pigments of normal observers vary in spectral position, the distribution of normal color matches will reflect that variability. Recent studies specifically designed to examine such normal individual differences agree that there is considerable variability in normal color matches, and the range of differences is similar across studies (Neitz & Jacobs, 1986; Elsner & Burns, 1987; Jordan & Mollon, 1988; Webster & MacLeod, 1988; Lutze, Cox, Smith, & Pokorny, 1990; Neitz & Jacobs, 1990; Lindsey, Winderickx, Sanocki, Teller, Deeb, & Motulsky, 1992; Shevell & He, 1992; Winderickx, Lindsey, Sanocki, Teller, Motulsky, & Deeb, 1992b; Piantanida & Gille, 1992). These differences in normal color vision can be striking. Normal observers at one extreme of the distribution require a red/green ratio that is more than twice as high as those at the other extreme. This means that a red/green mixture that a normal person from one end of the distribution sees as exactly matching the monochromatic yellow is seen as a very conspicuous red-orange mismatch to a normal person that falls at the other extreme.

Agreement about the issue of discretely different pigments has been a different matter. If there are discrete forms of the normal middle- and long-wave pigments, and if the steps in spectral peak are large enough, then different groups of individuals with distinct pigment complements should be evident in the distribution of normal color matches. Evidence for such modality in the distribution of color matches is claimed by some studies (e.g. Neitz & Jacobs, 1986; Neitz & Jacobs, 1990; Elsner & Burns, 1987; Lindsey, et al., 1992; Winderickx, et al., 1992b; Piantanida & Gille, 1992) but clear evidence of modality is lacking, or at least, is not as evident in other studies (e.g. Jordan & Mollon, 1988; Lutze, et al., 1990; Webster, 1992).

In the past year, the idea that a small number of discretely different types of pigments occur among observers with normal color vision has gained considerable impetus from studies that reveal a strong correlation between individual differences in color vision and a polymorphism at a single amino acid position in the middle and long-wave pigments (Neitz, Neitz, & Jacobs, 1991c; Lindsey, et al., 1992; Winderickx, et al., 1992b; Merbs & Nathans, 1992a; Neitz, Neitz, & Jacobs, 1993). These are discussed in detail in the section on spectral tuning, below.

PIONEERING STUDIES OF THE MOLECULAR GENETICS OF COLOR VISION

Nathans et al. (1986a; Nathans, Thomas, & Hogness, 1986b) were the first to isolate and characterize genes that encode cone visual pigments. They described the goal of their studies: "to test the hypothesis that human color vision is mediated by a family of rhodopsin-like molecules, the apoproteins of which are encoded by the corresponding members of a family of genes... [and the] related hypothesis that the common inherited variations in human color vision (color blindness) are due to mutations in the members of this gene family." A long history of research on color vision and color blindness made clear predictions as to what they would find. Fundamental to theories of color vision is the idea that normal human vision is mediated by three pigments, each presumed to be encoded by a single gene. Color vision deficiencies that involve the short-wave pigment are inherited in an autosomal fashion whereas red-green color vision deficiencies are X-chromosome linked. Thus, it was expected that two cone pigment genes, one encoding a long-wave, and the other a middle-wave pigment would be found on the X-chromosome, and a third, encoding the short-wave pigment, would be found on an autosome.

Nathans et al. (1986b) isolated clones containing fragments of at least three pigment genes that map to the X-chromosome from a genomic DNA library of a male (J. Nathans) who has normal color vision. They concluded that the cloned fragments, from this one person, corresponded to one gene encoding the long-wave pigment and two copies of a gene encoding the middle-wave pigment. They also isolated three X-linked pigment genes from a cDNA library made from several dozen human eyes obtained at autopsy. They noted that these X-linked genes were polymorphic. All the genes were slightly different in nucleotide sequence.

The finding of more than two X-linked cone pigment genes from one normal male was quite unexpected. Subsequently, Southern hybridization analysis of the X-linked pigment genes from a sample of males with normal color vision suggested that the presence of "extra" pigment genes was not uncommon, and that the number of pigment genes was variable across individuals. Nathans and colleagues concluded that everyone had a single copy of the long-wave gene, but individuals could have either one, two, or three copies of the middle-wave gene. Two copies of the middle-wave gene were proposed to be the most common arrangement.

In a second study (Nathans, et al., 1986a), the genes of 25 subjects with congenital color defects were examined using Southern hybridization analysis. It was long believed that X-linked color vision deficiencies are caused by alterations in genes that encode middle- and long-wave pigments and, indeed, in every case, the gene arrangements of the color blind subjects, as inferred from Southern analysis, were different from any observed in the Nathans et al. (1986b) sample of normals. The results and conclusions can be summarized as follows.

1) **Deuteranopia.** Six of nine deuteranopes examined had a single long-wave gene and no middle-wave genes. The three other deuteranopes had multiple

X-linked pigment genes but none of those genes corresponded, in structure, to the prototypical middle-wave genes, i.e., none had a restriction fragment, from near the 3' end, identified as that derived from a normal middle-wave gene. It was concluded that the color defect in this sample of deuteranopes was associated with the absence of all genes that encode middle-wave pigments.

2) **Protanopia.** The genes from six protanopes were examined. Two had a single X-linked pigment gene. Three others had multiple genes but all their genes had a restriction fragment, near the 3' end, corresponding to Nathans' own middle-wave gene. The color defect in these five observers seemed associated with the absence of all genes that could encode long-wave pigment. Inexplicably, the sixth protanope had all the fragments corresponding to an intact long-wave gene. He also had more middle-wave genes (perhaps six) than had been attributed to any normal observer.

3) **Protanomaly.** Neither of the two protanomalous observers had any genes with 3' fragments identified as being derived from long-wave genes. It appeared, however, that compared to Nathans' own normal genes, each had one gene whose 5' sequences were derived from a long-wave gene and 3' sequences derived from a middle-wave gene. It seemed logical that these 5'long- 3'middle-wave fusion genes would encode a pigment with a spectrum intermediate to the two normal pigments. It was concluded that while the protanomalous observers were lacking any normal long-wave genes, they had normal middle-wave genes and, in addition, fusion genes to encode the anomalous pigments thought to characterize protanomaly.

4) **Deuteranomaly.** The results from deuteranomalous subjects were surprising. They continue to be the greatest challenge to our understanding of the genetic mechanisms that underlie color vision. At the same time, solving the conundrums of deuteranomaly may be a key to understanding the relationship between pigment genes and color vision. The reason for color deficiency can be understood easily in the case of only one deuteranomalous subject in Nathans' study. This one deuteranomalous observer has several genes but none with a 3' fragment identified as being derived from a middle-wave gene. He, thus, has no intact genes that could encode normal middle-wave pigment. The other seven deuteranomalous observers all had fragments corresponding to the identified complete normal long- and complete normal middle-wave genes. In addition, from the stoichiometries of long- and middle-wave gene fragments, it appeared that all the deuteranomalous subjects had one or more fusion genes with a 5' end derived from a normal middle- and 3' end derived from a normal long-wave gene. The fusion genes could encode a pigment with a slightly different spectrum from the pigment encoded by the normal long-wave gene and, thus, explain the anomalous pigments that were believed to characterize deuteranomaly. But, how could the presence of normal middle-wave genes be reconciled with the deuteranomalous phenotype? Nathans et al. (1986a) offered the explanation that the fusion genes "would be ex-

pressed in the green cones together with the normal green pigment gene or genes. This arrangement therefore confers a G' R+ [mild] or G" R+ [extreme deuteranomalous] phenotype." This cannot, however, explain the deuteranomalous phenotypes. As described earlier, the X-linked pigments that underlie deuteranomaly are similar in spectrum. Close enough, in fact, that they can both be classed as long-wave pigments. Mixing both long- and middle-wave pigments in the same middle-wave cones would not cause subjects to set the abnormal Rayleigh matches that characterize deuteranomaly. As discussed earlier, this is because color matches occur at the level of the pigments and not the cones. Individuals who expressed both normal middle- and deuteranomalous pigments in the same middle-wave cones would set normal or near normal pigment matches. From a study that investigated this issue in more detail (Neitz, Neitz, & Jacobs, 1991b) it was concluded that deuteranomalous subjects do not express any detectable amount of middle-wave pigment even though they often appear, by Southern analysis, to have multiple normal middle-wave genes.

With the exception of an adequate explanation for the behavior of deuteranomalous observers, the interpretations of Nathans and colleagues seem to offer simple explanations of the genetic basis for red-green color blindness. One greatly simplifying aspect of their theory is that some of the color-blind genotypes are explained as arising from unequal but homologous recombination of normal gene arrangements. This can be simply diagrammed as in the example of Figure 1.

The importance of the work of Nathans and colleagues cannot be underestimated. They established the primary structures of the pigment genes and their chromosomal locations. They suggested that the X-linked pigment genes are in a tandem array, and demonstrated that most people have more pigment genes than previously believed. They also demonstrated differences between normal gene arrangements and those of color blind observers and illustrated how some of the gene arrangement might come about by homologous recombination.

Nathans et al. (1986a,b) came to some new conclusions about the genetic mechanisms of color vision: 1) many X-chromosomes of normal observers were proposed to have an extra copy (or two) of the middle-wave gene, 2) the abnormal pigments of anomalous trichromats were proposed to result from recombination between the long- and middle-wave genes, and 3) many deuteranomalous observers were proposed to have both normal middle- and long-wave genes in addition to middle- /long-wave fusion genes. With these exceptions, their results were interpreted as generally confirming conventional theories. Botstein (1986), for example, interpreted the molecular genetic results as, at last, confirming the Young–Helmholtz theory of color vision in which the existence of three types of light absorbing particles (one for each primary color) are proposed to explain why normal human vision is trichromatic (Young, 1802). Furthermore, inheritance of mutant pigments with abnormal spectra, different from any expressed in normal observers, was interpreted as the cause of anomalous trichromacy.

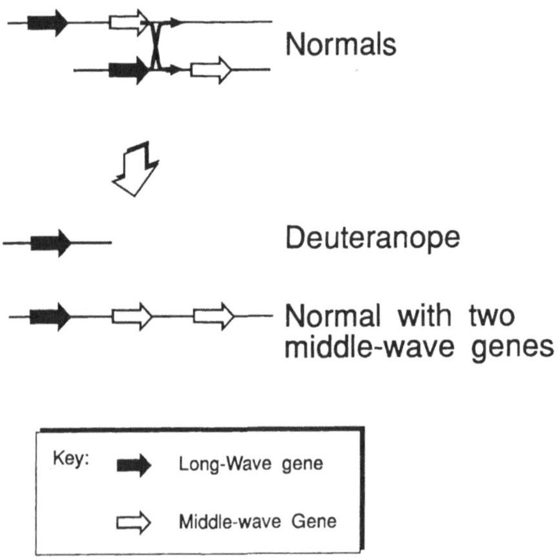

Figure 1 Unequal crossing over between X-linked pigment gene arrays can explain polymorphism in human pigment gene arrangements. In this example, an intergenic recombination between two gene arrays that confer normal color vision produces one X-chromosome with a single long- and no middle-wave genes and another X-chromosome with one long- and two middle-wave genes. A male who inherits an X-chromosome with only a single pigment gene will have dichromatic color vision. One who inherits one long- and two middle-wave genes will have normal trichromatic color vision.

Now, seven years later, the body of experimental results that relate to the molecular biology of color vision has grown. It may be time to reconsider the theories put forth in 1986. A new theory may better explain the experimental results – a theory in which all the ideas in the preceding paragraph would be held as substantially incorrect. As will be elucidated in subsequent sections, the evidence suggests that it is likely that people with normal color vision have, on average, more X-linked pigment genes than suggested by Nathans and colleagues. Furthermore, many observers with normal color vision may have multiple long- as well as multiple middle-wave genes. The facts are consistent with an alternative theory in which the fusion genes neither produce abnormal pigments, nor do they cause color defects. That is, fusion genes and the variant pigments they produce may be found and expressed in individuals with normal color vision. Finally, results from molecular biology may refute, rather than confirm, the Young-Helmholtz theory as the explanation for human trichromacy.

MUTATIONAL PATHWAYS: FROM NORMAL TO DEFECTIVE GENOTYPES

The discoveries that the middle- and long-wave pigment genes are highly homologous, lie closely spaced in a tandem array, and are variable in number suggested a mutational pathway to gene arrangements that confer color defects that is simple and compelling – the arrangement of the pigment genes is one that would promote a high frequency of unequal but homologous recombination, a mechanism that could increment and decrement gene number and produce fusion genes, all seemingly, just as required to explain the gene arrangements that underlie color defects. This idea is illustrated in Figure 1; an intergenic exchange between two normal arrays transfers a middle-wave gene from one chromosome to the other, giving rise to one chromosome with only a single long- and no middle-wave pigment genes. The other chromosome that results from this crossover has one long- and two middle-wave genes. In a stroke, both the occurrence of deuteranopes with only a long-wave gene and normal trichromats with extra middle-wave genes are explained.

Difficulties arise, however, when one tries to explain the gene arrangements of other color defective observers by similar one-step mutational pathways. For example, several of the genotypes inferred for deuteranomalous subjects cannot result from just one crossover between normal arrays (see Figure 2). Fusion genes can result from crossovers between normal gene arrays but no more than one fusion gene can be produced in each crossover. Yet, frequently, deuteranomalous subjects seem to have multiple fusion genes. A crossover between two normal arrays is illustrated at the top of Figure 2. The product of the crossover that is assembled with a 5'middle- 3' long-wave fusion gene also has the complete long-wave gene from one parental chromosome and the complete middle-wave gene from the other. To get two fusion genes by crossing over, an arrangement that already has one fusion gene (Figure 2) must undergo a second recombination. The same type of recombination must occur three separate times for three fusion genes. The probability of a recombination occurring twice is expected to be the square of the probability of one occurrence and the likelihood of accumulating three fusion genes by this mechanism is expected to be equal to the cube of the probability of a single occurrence. The frequency of deuteranomalous phenotypes should equal the sum of the frequencies of each of the different genotypes. Satisfying these requirements yields that, in the population, 95.2% of deuteranomalous subjects should have a single fusion gene, 4.5% should have two fusion genes, and only 0.2% should have three fusion genes.

Table 1 shows computations of the number of fusion genes for each of the eight deuteranomalous subjects examined by Nathans et al. (1986a). The calculations suggest that two subjects may have two fusion genes and two more may have three fusion genes each. This is wildly different from what would be expected if the inheritance of a single fusion gene was sufficient to cause deuteranomaly. If 95% of deuteranomalous individuals in the population have only a single fusion gene, then the probability of drawing four subjects (out of eight) with multiple fusion

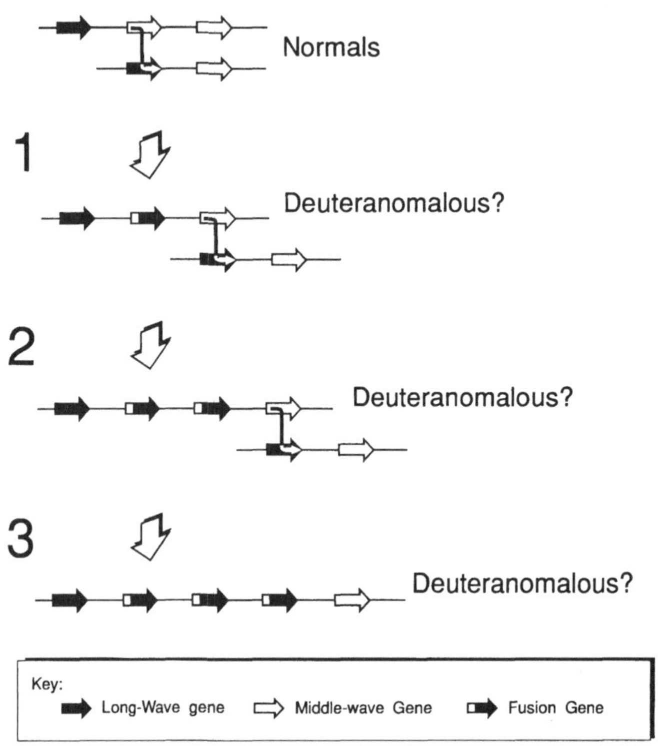

Figure 2 Mutational pathways to deuteranomaly are not known. Two gene arrangements that confer normal color vision (shown at the top) could crossover to produce an X-linked pigment gene array that has a 5'middle- 3' long-wave fusion gene in addition to normal middle- and normal long-wave genes. This gene arrangement has been proposed to confer a deuteranomalous phenotype. If all observers with normal color vision have only a single long-wave gene as has been proposed by Nathans et al. (1986b), then to produce a gene array with two fusion genes requires a gene arrangement with one fusion gene to undergo a second crossover (illustrated as crossover number 2). For an array to have three fusion genes requires three independent crossover events. If the presence of a fusion gene could cause deuteranomaly then deuteranomalous observers with one fusion gene would be expected to substantially out number those that have accumulated multiple fusion genes. Instead, deuteranomalous observers with multiple fusion genes appear to be common, suggesting that the inheritance of a fusion gene does not cause color defects.

genes and two of those with three fusion genes is less than one in a million. Earlier, it was noted that a mysterious feature of deuteranomaly is the presence of seemingly normal middle-wave genes. Now we have an added curiosity. Deuteranomalous subjects seem to have more fusion genes than would be expected from a model in which the inheritance of a fusion gene causes deuteranomaly.

Table 1. The number of fusion genes in deuteranomalous observers are calculated from the results of Nathans et al. (1986a). The total number of X-linked pigment genes in each subject was determined from his A_g/A_r fragment ratio. The ratio of band intensities was translated to a gene fragment ratio using the intensity ratios suggested to correspond to a 1:1 gene ratio (A_r:A_g = 0.97; D_r:D_g = 1.24). Gene number was taken to equal $A_g/A_r + 1$; these values were rounded to the nearest whole number. The number of genes that contain fragment D_r was calculated using the equation: Genes with D_r = Total number of X-linked pigment genes / (D_g/D_r +1). These were rounded to the nearest whole number. The number of fusion genes equals 1 less than the number of genes with fragment D_r because one gene with D_r is the normal long-wave gene. These calculations assume that all subjects have only one copy of fragment A_r. If some subjects have more than one copy of A_r, the number of fusion genes would be higher.

Subject	*Number of X-linked pigment genes*	D_g/D_r	*Number of long-wave fusion genes*
34	5	0.27	3
35	3	0	2
36	6	0.57	3
37	6	1.61	1
38	5	0.84	2
39	3	0.32	1
40	3	0.45	1
41	4	1.32	1

Nathans et al. (1986a) examined the genes of twenty-five color deficient males. We have organized the gene arrangements that can be inferred from their results into 10 different categories. These are illustrated in the left-hand column of Figure 3. The popular idea is that color defective arrangements can be explained by simple, one-step recombinations between normal arrays that result in gene deletions or produce a fusion gene. Arrangement **number 1** of Figure 3, can be produced by a crossover in which both products of the recombination fit perfectly with this idea (as was shown in Figure 1); one product confers deuteranopia and the other has two middle-wave genes just as found to occur among normals. The other nine color-blind gene arrangements are more difficult to fit with this scheme. The protanope arrangements, like **numbers 2 and 3** and the protanomalous arrangements, like **number 4** of Figure 3 can arise by a simple crossover. However, two distinct gene arrangements arise from every crossover. When a protan arrangement is produced, another array with an intact normal long-wave gene from one parent, all the normal middle-wave genes from the other parent, and a 5'middle- 3'long fusion gene is also produced. The 5'middle- 3' long fusion gene is expected to encode a pigment with a long-wave spectrum. Protanope gene arrangements are missing sequences that encode a long-wave pigment, but in their production, another chromosome effectively gains the long-wave gene (as a fusion gene) that the protanope lost.

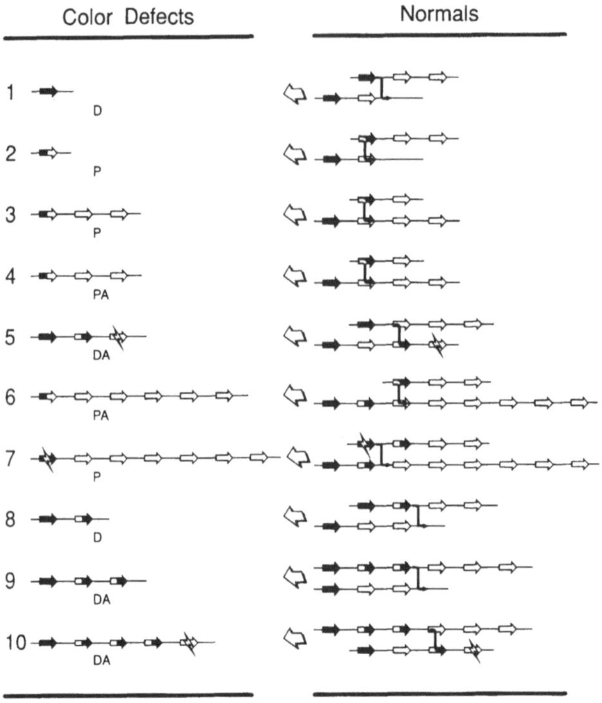

Figure 3 A new theory of the photopigment gene arrangements that underlie normal color vision allows a very simple explanation for how color defects arise. The inferred gene arrangements of the 25 color defective observers studied by Nathanset al. (1986a) can be classed into 10 categories shown in the left-hand column (abbreviations: D, deuteranope; P protanope; PA, protanomalous; DA deuteranomalous). In some cases, the results from Southern analysis indicate the presence of an intact gene whose expression is not evident in the subjects' behavior. For example, the color vision of deuteranomalous observers is that predicted by the expression of two slightly different long-wave pigments but no middle-wave pigments, however, deuteranomalous subjects often have all the gene fragments corresponding to intact middle-wave genes. A "lightning bolt" symbol drawn through a gene indicates lack of expression or function. The gene arrangements of color defective subjects are often complicated, having many genes, apparently unexpressed genes, and multiple fusion genes. These suggest a previously unrealized complexity in the normal gene arrangements from which they arose. The right-hand column shows hypothetical gene arrangements that could confer normal color vision. Each pair of normal gene arrangements could undergo a single crossover to yield the color-defective gene arrangement shown directly to its left. In this new theory, normal observers are proposed to carry defective genes sometimes but in each case they have at least one functional middle- and one functional long-wave pigment gene. Some normal observers are shown to have fusion genes. The expression of these genes in normal observers is proposed to contribute to the observed variation in normal color vision but does not interfere with normal color vision. Previous theories required protanomaly and deuteranomaly to arise by different mutational pathways, often with many steps. In contrast, the crossovers shown here all give rise to color defects by the same one-step mechanism. They produce gene arrangements that lack either functional middle-wave genes (deutan defects) or long-wave genes (protan defects).

It seems that either one of two alternatives must be true with regard to the gene arrangements that are produced as counterparts to the protan gene arrays. One is that the inheritance of a 5'middle- 3'long-wave fusion gene usually prevents the expression of the otherwise normal middle-wave genes and, thus, deutan gene arrangements are produced as the complement to protan ones. This would explain why deuteranomalous subjects, like those illustrated by arrangement **number 5** of Figure 3, can often have normal middle-wave genes. The alternative is that the gene arrays that are produced in the same crossovers that give rise to protan defects can confer normal color vision. This would allow the possibility that normal observers can have, and may express fusion genes. The question of whether or not color normal observers can have and express fusion genes is central to understanding the molecular genetics of color vision.

The remaining five categories of gene arrangements illustrated in Figure 3 are all bewildering because of their complexity relative to the simple gene arrangements that have been proposed to underlie normal color vision. None of them can be explained by a simple single crossover between two of the proposed common normal gene arrays.

Number 6, a protanomalous gene arrangement, apparently has at least six genes. Protanomalous arrangements are proposed to arise from deletion of the one normal long-wave gene. To result from a single crossover, an arrangement like number 6 would have to arise from a normal arrangement with at least seven genes. The suggested counts of the number of pigment genes in normals would indicate that the occurrence of normals with seven genes is exceptionally rare. Such a protanomalous arrangement cannot have arisen by a single crossover from any of the proposed common normal gene arrays.

Number 7, a protanope, also has many more genes (perhaps seven) than are proposed for color normal observers. In addition, he appears, by Southern analysis, to have a normal long-wave gene. A likely explanation of his protanopia, is that his long-wave gene has a mutation that renders the pigment produced non-functional, perhaps similar to the mutations found to occur in blue-cone monochromats (Nathans et al., 1989) (as are discussed earlier). In Figure 3, genes which do not appear to contribute to the phenotype have a "lighting bolt" symbol drawn through them. A deleterious mutation may explain the loss of long-wave pigment function. But, the extremely high gene numbers found in protan subjects relative to those proposed for normals are not what would be predicted from a single crossover between common normal arrays.

Some deuteranopes (**number 8**) have only two genes, a normal long-wave gene and a 5'middle- 3'long-wave fusion gene. Middle-wave genes cannot be deleted in the same crossover that produced the fusion gene. Nathans and colleagues noted that either two independent crossovers or a gene conversion are required to explain these arrangements.

Number 9, a deuteranomalous arrangement with multiple fusion genes, and no middle-wave genes can only be explained by some combination of crossovers, or crossovers and gene, conversions. This is another example of an arrangement that must be proposed to arise by a mutational pathway with many steps.

As discussed at the begining of this section, many deuteranomalous observers do not have one fusion gene, which is expected to be by far the most common arrangement, but multiple fusion genes (**Number 10**).

The idea that color defects can be explained by a one-step recombination is an attractive one. If this explanation is correct, then nearly all the observed color-blind arrangements should be explainable by a one-step path. This is not the case. Instead, the arrangements associated with color deficiencies often have too many normal genes, too many fusion genes, or they lack genes that would be expected from one-step mutational pathways. Of course, more complicated arrangements should occasionally occur among color deficient observers, but they would be tremendously outnumbered by simpler arrangements.

There is a simple solution. If color-deficient phenotypes arise via one-step recombination of normal gene arrangements, then the model for what constitutes a normal array must be revised. That is, the complexity of the color-blind gene arrangements provides a unique window through which we can glimpse a previously unrealized complexity in the normal arrangements from which they were derived. In developing their model of normal gene arrangements, Nathans and colleagues made two assumptions: 1) everyone with normal color vision has a single long-wave gene, and 2) fusion genes cause color defects and they, thus, do not typically occur in normal arrays. However, the evidence may favor a radically different view of the genes that underlie normal color vision; one that does not rely on these assumptions and that allows all color defecient phenotypes to be explained as arising from one-step recombinations.

A new concept of normal genes makes abnormal gene arrangements easily explainable

The right hand column of Figure 3 illustrates a revised model of what constitutes normal arrays that enormously simplifies the explanation of how the color deficient genotypes are produced. In this new model, all the gene arrangements in the right hand column of Figure 3 are proposed to confer normal color vision and overall these might typify the arrangements of people with normal color vision. These were constructed to obey the following rules. 1) To confer normal color vision an individual must have at least one functional long- and one functional middle-wave pigment gene. 2) Color normal individuals can have ten or more X-linked pigment genes in total, but to fall within the ratios of middle- to long-wave genes suggested from Southern analysis, the ratio of middle- to long-wave genes is typically no larger than four to one. 3) Normal arrays can have multiple long-wave genes but the additional long-wave genes are always illustrated as 5'middle- 3'long-wave fusion genes to emphasize that they include, and/or are preceded by 3' sequences that have been believed, in normal subjects, to be exclusive to middle-wave genes. 4) Normal subjects can have genes with defects that interfere with their expression or function; such mutations would be expected by genetic drift if people usually have additional genes to provide the function.

In Figure 3 a single crossover is illustrated for each pair of chromosomes in the right-hand column. Each crossover produces the color-blind genotype shown to its left. All common forms of color-blindness can arise by the same mechanism – an unequal homologous recombination that deletes a gene (or genes) required for normal color vision. Quite simply, if normal arrangements are as proposed in this new model, illustrated on the right-hand side of Figure 3, then the color-blind gene arrangements could:

1) all arise from mutation of normal arrays,
2) all come about by the same one-step pathway, and
3) all be reduced forms of normal arrangements.

Notably absent from this scheme are the complications that deuteranomalous and protanomalous arrangements come about by different mechanisms, that color-blindness can be caused by the addition of fusion genes to normal arrangements, and that many color-blind arrangements need to be explained by improbable mutational pathways from normal arrangements.

Results from Southern analysis re-examined

Genetic mechanisms that give rise to color defects could be quite simple, if: 1) people with normal color vision can often have multiple long-wave genes, 2) color normal people often have fusion genes, and 3) the average number of pigment genes per X-chromosome of color normal observers is substantially larger than previously imagined. The current view of normal gene arrangements comes from the analysis of pigment gene fragments by Southern hybridization (Nathans, et al., 1986b; Vollrath, Nathans, & Davis, 1988; Drummond-Borg, Deeb, & Motulsky, 1989; Deeb et al., 1992). An examination of those results lends insight into whether or not they may have been misinterpreted. For example, is it likely that a normal array with multiple long-wave genes and with fusion genes might have been interpreted as having only a single long-wave gene, and no fusion genes?

Figure 4 shows the restriction maps for middle- and long-wave genes. Four restriction fragments, labeled A, B, C, and D have been used in Southern analysis to distinguish middle- from long-wave genes. The A fragment from the long-wave gene (A_r) is larger than the A fragment from the middle-wave gene (A_g) because of an insert (of about 1.9 kb, in the first intron) about one third of the way into the gene. This same insert accounts for the size difference between the C_g and C_r fragments. The fragment, B_r, is smaller than B_g because it is produced by cleavage at a Bam HI site that occurs in unique sequences flanking the 5' end of the array. These unique sequences are not part of the tandemly repeated sequences

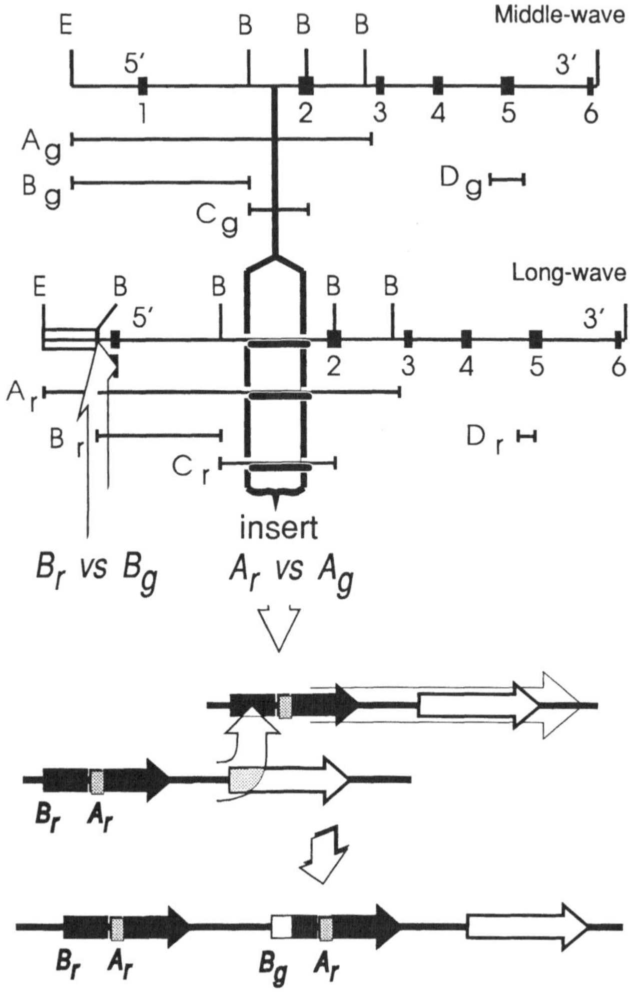

Figure 4 (Top) Restriction maps of middle- and long-wave genes showing fragments that have been used to distinguish the two genes in Southern analysis. An insert in the long-wave gene is responsible for the size difference between fragments A_r and A_g and between C_r and C_g. The fragment B_r is foreshortened compared to the corresponding fragment from the middle-wave gene, B_g, because of a Bam HI site (B) that occurs upstream of the first gene, outside of the repeated photopigment gene array. The B_r fragment occurs only once in each gene cluster and cannot be duplicated or deleted by homologous recombination.

(Bottom) Because the insert that distinguishes fragments A_r and C_r from A_g and C_g occurs within the gene, a crossover anywhere in about the first third of the gene will duplicate the long-wave gene including fragments A_r, C_r, and D_r. The duplicated long-wave gene would have fragment B_g. It may also have exon 1 derived from a middle-wave gene but that exon is identical between middle- and long-wave genes. Thus, the duplicated long-wave gene and the long-wave gene from which it was derived will have identical coding sequences.

that constitute the array. Thus, a corresponding Bam HI site is not found flanking the genes other than the first in the array.

The sequences flanking the first gene in the array are not homologous with any that are within the array. Thus, the fragment B_r can neither be deleted or duplicated by homologous recombination. On the other hand, the insert that distinguishes A_r from A_g and C_r from C_g will be duplicated by any crossover that occurs in the first one-third of the gene. Nathans et al. (1986a,b) proposed that observers with normal color vision have a single long-wave gene. They explain that because of its position at the 5' edge of the array, the long-wave gene can never be completely duplicated or deleted by homologous exchange. The lower half of Figure 4 shows that, while this is true, the sequences that distinguish A_r, C_r, and D_r will all be duplicated by any crossover in the first third of the gene. The fusion gene that results may have exon 1 derived from a middle-wave gene, but Nathans et al. (1986b) found exon 1 to be identical between middle- and long-wave genes. Thus, long-wave genes can be duplicated. The long-wave gene duplicated by any crossover in the first third of the gene will have a coding sequence that is identical to its parent long-wave gene. If the likelihood of a crossover in any homologous region is related to the length of that region, then about one-third of all crossovers between middle- and long-wave genes should duplicate the coding sequence of the long-wave gene in its entirety. People with multiple long-wave genes are expected from the same logic that explains multiple middle-wave genes, i.e., the homology and arrangement of the pigment genes promotes a high frequency of homologous recombination.

Figure 5 shows the effect multiple long-wave genes would be expected to have on the results from Southern analysis. Three gene arrays are shown with one, two, and three long-wave genes respectively. Even though they differ in gene number, all would have A_g/A_r fragment ratios of 2.0. They are similarly identical in their C_g/C_r and D_g/D_r ratios of 2.0. The only indicator of the fact that they are different is the B_g/B_r ratio, which increases with gene number. Previously, the A_g/A_r ratio was used to calibrate the absolute number of pigment genes (Nathans, et al., 1986a; Nathans, et al., 1986b; Vollrath, et al., 1988; Drummond-Borg, et al., 1989). Thus, all three arrangements of Figure 5 would be interpreted as having one long-wave gene and two middle-wave genes. However, because A_r is expected to be duplicated (perhaps frequently) by recombination; A_g/A_r cannot be taken as a reliable indicator of gene number.

The B_g/B_r ratio, on the other hand, is predicted to be a reliable indicator of differences in gene number. Suppose that the observers with normal color vision examined previously by Southern analysis vary in their number of long-wave genes and that the total number of genes people have is often much higher than is indicated by their A_g/A_r ratio. Since the B_g/B_r ratio is an indicator of gene number, then there are two predictions about the B fragment ratio. First, B_g/B_r should be greater in magnitude than the A or C ratios reflecting that the average gene number is higher than A_g/A_r would predict. Second, the B fragment ratio should be more variable than the A or C fragment ratios reflecting variability in the number

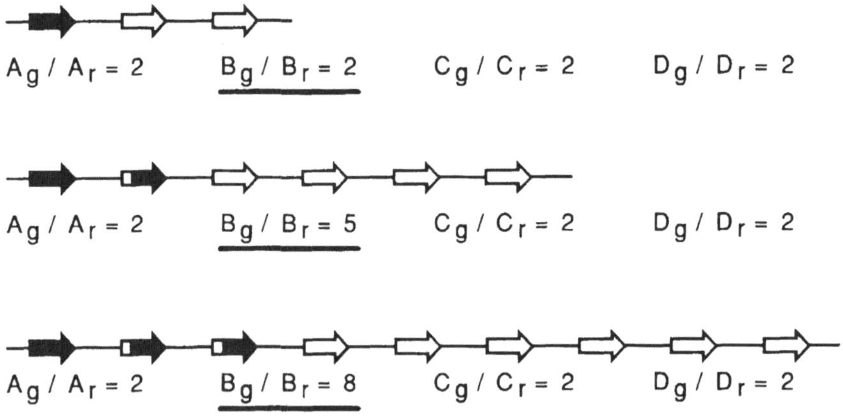

Figure 5 The presence of multiple long-wave genes in males with normal color vision would not be evident from the ratios of A, C, and D gene fragments visualized in Southern analysis. The gene arrangements shown are dramatically different from one another, yet, they have identical ratios of A_g/A_r, C_g/C_r, and D_g/D_r. In Southern analysis, only the B_g/B_r ratio would reveal the differences among these arrays. Previous interpretations of Southern blot data have emphasized the A_g/A_r fragment ratio in the categorization of normal gene arrays. That practice would lead to the incorrect conclusion that the three gene arrays shown here are identical to one another. Each would be interpreted as having one long-wave gene, two middle-wave genes and no fusion genes.

of long-wave genes within groups that are classified by A_g/A_r ratio. This is exactly what is found. For example, Nathans et al. (1986a) found that for people with a 1:1 A_g/A_r ratio (mean $A_g/A_r = 0.97$) the other fragments had the following-means and standard deviations: B_g/B_r, 2.00 ± 0.49;C_g/C_r, 0.88 ± 0.18; D_g/D_r, 1.24 ± 0.30.Just as predicted if some people have multiple long-wave genes and, thus, their higher B_g/B_r ratios are included in the sample, the mean B_g/B_r ratio is high, more than twice the A or C ratios. As predicted by the idea that the sample is inhomogeneous in gene number, the B ratio is highly variable across subjects. The B_g/B_r ratio has a standard deviation 2.7 times greater than the C_g/C_r ratio. If everyone had a single long-wave gene, B_g/B_r would be expected to be always equal to A_g/A_r and C_g/C_r both in magnitude and in degree of variability. Neither of these are borne out by experimental results.

Two specific examples can serve to illustrate, in detail, how nicely a new interpretation of normal gene arrangements can explain color defective genotypes and can explain results from Southern analysis of normal subjects as well. Consider two cases from Nathans et al. (1986a), subject #27, a protanope, and subject #40, a deuteranomalous observer. Figure 6 shows gene arrangements that could yield the fragment stoichiometries observed for each subject. Both are problems for earlier theories because the arrangements cannot come about by any simple pathway from normal arrangements. Also shown for each subject are theoretical normal arrangements that could crossover just once to yield the color defective genotype. These differ from the normal genotypes proposed previously in that they have more genes (as many as eight) and they all have one or more 5'middle- 3'long-wave fusion genes. Could such arrangements really be typical of color normal observers? Drummond-Borg et al. (1989) examined the pigment genes of 134 Caucasian males using Southern analysis. Subjects were classified according to their A_g/A_r ratios. The results for the males who were considered to have normal gene arrangements with nominal A_g/A_r ratios of 3:1 (about 20% of their subjects) are illustrated in the lower half of Figure 6. The observed mean for each of the A, B, and D fragment ratios is shown together with bars that extend two standard deviations on each side of the mean. The C ratio is omitted because the information it yields about the genes is redundant with the A ratio. Overlaid are the fragment ratios that correspond to the two hypothetical normal arrangements illustrated (in the upper half of Figure 6) to give rise to the protanope (squares) and the two proposed to give rise to the deuteranomal (circles) (middle panel, Figure 6). Even though the arrangements are very different from those previously thought to underlie normal color vision, the stoichiometries are consistent with those considered normal. This suggests that the normal gene arrays examined previously probably included such arrangements – e.g., with large numbers of genes, multiple long-wave genes and fusion genes. It would seem that normal arrangements similar to those illustrated in Figure 6 were interpreted as having only a single long-wave gene and no fusion genes because the interpretation emphasized the results for the A fragments.

Each recombination illustrated in Figure 6 simply deletes genes required for normal color vision but the process alters the fragment stoichiometries substantially, such that those of the color defective observers fall outside normal limits. This is illustrated in the bottom panel of Figure 6. The protanope inherits only one long-wave gene (a non-functional one). Even though his X-chromosome has the same number of pigment genes (seven) as one of the normal arrangements from which it is proposed to arise, the reduction to a single long-wave gene increases his A_g/A_r and D_g/D_r ratios to the high value considered rare for normals. In the case of the production of the deuteranomalous arrangement, previous interpretations would be oblivious of the fact that the two normal arrays shown as giving rise to the deuteranomalous genotype have fusion genes, but the crossover imparts a deuteranomalous gene arrangement in which the presence of fusion genes becomes obvious because of the extremely low D_g/D_r ratio (even though no additional fusion genes were produced).

Pigment gene fragment stoichiometries of color-deficient observers almost always differ from those of typical normal observers. However, there are exceptions; Deeb et al., (1992) found four of 129 normal observers in which the presence of fusion genes was obvious from the fragment stoichiometries. It was previously believed that the abnormal stoichiometries were the result of fusion genes present in color defects but nearly always absent in normal subjects (Nathans, et al.,

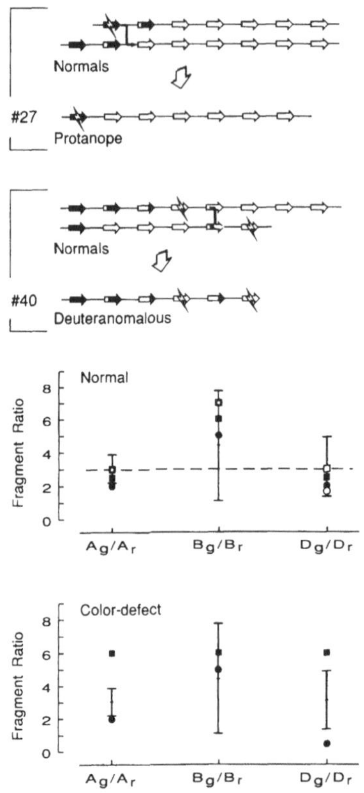

Figure 6 Two specific examples of how a new theory of normal gene arrangements can account for results from color defective and color normal subjects. The results from Southern analysis for two subjects examined by Nathans et al. (1986a) are as follows: subject #27 (a protanope), $A_g/A_r = 6.08$, $B_g/B_r = 5.40$, $C_g/C_r = 3.77$, $D_g/D_r = 5.87$; subject #40 (deuteranomalous), $A_g/A_r = 2.31$, $B_g/B_r = 5.36$, $C_g/C_r = 1.42$, $D_g/D_r = 0.56$. Those fragment stoichiometries suggest the gene arrangements shown (marked #27 and #40). The order of the genes was arbitrarily chosen. Above each color-defective gene array, a pair of hypothetical arrays proposed to confer normal color vision is shown. The pairs of normal gene clusters could each undergo a single crossover to yield the observed color defective genotypes. Each normal array illustrated has more pigment genes than previously proposed for normals and has multiple genes that encode long-wave pigments. Three of the normal arrays have defective genes in addition to both normal middle- and long-wave genes. Plotted in the lower half of the figure are the ranges of fragment stoichiometries (bars, drawn as the mean±2 SD) determined for genes, considered to confer normal color vision, from a sample of males that had nominal A_g/A_r fragment ratios of 3:1 (from Drummond-Borg, et al., 1989). Superimposed on those ranges are the fragment ratios for the four hypothetical normal gene arrangements shown in the upper half of the figure (circles, parents to the deuteranomalous gene arrangement; squares, parents to the protanope gene arrangement). Even though the hypothetical normal gene arrangements are much different than those previously proposed for normal subjects they yield fragment stoichiometries that are consistent with those previously considered normal. Fragment stoichiometries for the two color defective gene arrangements are plotted in the bottom-most panel (circles, deuteranomalous; squares, protanope). They fall outside the normal limits.

1986a). It can be seen that the presence of (even multiple) fusion genes in an array is not sufficient to impart stoichiometries that fall outside normal limits. In fact, it appears that the presence of fusion genes in a large subset of normal subjects is required to account for the observed range of normal stoichiometries. Thus, for example, in the deuteranomalous observer, it is the small number of middle-wave genes relative to the number of long-wave and fusion genes that accounts for his abnormal stoichiometries. This occurs because of the deletion of normal middle-wave genes not because of the production of fusion genes. This would suggest that the fusion genes do not cause the color deficiency; the loss of normal middle-wave genes does.

The theory that is presented here allows all color defective arrangements to arise from normal arrays by one-step deletions of genes required for normal color vision. Nothing could be simpler. More than that, however, the theory is attractive because it can account for facts earlier theories would have difficulties explaining. In previous sections we discussed that earlier theories have difficulties explaining, 1) why a deuteranomalous observer could seemingly inherit all the normal middle-wave genes from a normal array yet not express any of them or 2) why so many color-blind observers have more complex gene arrangements than can be explained by the simple addition or deletion of a gene from normal arrays. Now we see that earlier theories would have difficulty explaining a third observation, why the Bg/Br ratio is both so high and so variable compared to the other fragment ratios. All these can be explained by a new interpretation of the normal gene arrangements.

Earlier theories posed a deleterious role for the fusion genes, their presence was believed to cause color anomaly. To the contrary, in this new scheme, the fusion genes have a beneficial role. For example, if fusion genes that encode slightly different pigments were not present in normal observers, then the loss of functional middle-wave pigments would always cause deuteranopia. Instead, many people are protected by the presence of fusion genes, and inherit the often much milder color deficiency deuteranomaly when they suffer the loss of functional middle-wave genes.

Defective genes

We suggest that defective genes are common among color defective and color normal populations. By defective we mean that individual genes have altered coding or regulatory sequences, as might come about by point mutations or deletions, that prevent expression or function of expressed product. Such defects might be expected to be especially common in the X-linked pigment genes because often people have genes to spare (perhaps, sometimes as many as seven or eight more than are required to support trichromatic color vision). Thus, defects might be acquired with little consequence to vision. We propose that deuteranomalous individuals with middle-wave genes may be people who inherit all their middle-wave genes as defective genes. They could arise from normal gene arrangements with defective genes by unfortunate recombinations (as illustrated in Figure 6) that leave no functional middle-wave genes. Recent evidence suggests that defects in individual genes are a common cause of other visual defects. It is, thus, plausible that defective genes underlie deuteranomaly as well. Three different point mutations within the gene encoding the short-wave pigment have been found to be associated with tritanopia (Weitz et al., 1992a; Weitz, Went, & Nathans, 1992b). Point mutations within X-linked pigment genes have also been found associated with some cases of blue cone monochromacy (Nathans, et al., 1989). An example of defective genes in red-green color blindness has also been found. Winderickx et al. (1992c) examined the middle-wave pigment genes of a deutan observer who had an apparently normal gene array, i.e., by Southern analysis he had long- and middle-wave genes and no fusion genes. From these results he would be expected to have normal color vision. Analysis of his middle-wave genes revealed sequences encoding the same cysteine to arginine substitution observed in blue cone monochromats. These mutations probably explain the lack of middle-wave pigment function in this observer. The same alteration was detected in one middle-wave gene of a mildly deuteranomalous observer. It was also detected in a middle-wave gene of one out of 52 normal subjects demonstrating that defective genes can occur in observers with normal color vision.

The possibility that defective cone pigment genes may be fairly common in observers with normal color vision may have important clinical implications in light of the discovery that photopigment gene defects cause retinal degeneration. At least 32 mutations in the rhodopsin gene have been found associated with autosomal dominant retinitis pigmentosa (e.g., Dryja et al., 1990; Sung et al., 1991a; Sung et al., 1991b). Cone pigment gene defects may similarly cause cone-specific degenerations. Some blue cone monochromats show retinal degeneration (Nathans, et al., 1989). Cone degeneration has also been observed in one family where functional middle-wave genes were present but there was a deletion in the long-wave gene (Reichel, Bruce, Sandberg, & Berson, 1989). A possibility that seems worth considering is that while in most cases the inheritance of a defective cone pigment gene in an array of otherwise normal genes may have no serious consequences for vision in young eyes, some cone pigment gene defects may cause late-onset retinal (macular) degenerations.

Another recent study relates to the issue of defective pigment genes and their expression. The expression of photopigment genes was investigated in an examination of cone pigment mRNA transcripts from human retinas (Winderickx, Battisti, Motulsky, & Deeb, 1992a). A common silent polymorphism (A vs. C at the third position of codon 283) occurs in exon 5 of the middle-wave gene. Genomic DNA derived from blood samples and cDNAs from retinas were examined from 13 males. Both the C and the A versions of the middle-wave genes were detected in the genomic DNA from 10 of the males. Middle-wave cDNAs were detected in samples from all eyes, however, middle-wave cDNAs with C at the third position of codon 283 were not detected in the samples from any of those 10 males. The expression of the "C allele" was detected in only one subject, a male who had no "A alleles" of the gene. One possible explanation for this result is that the C serves to mark middle-wave genes that have another mutation that interferes with their expression. If this is true, middle-wave genes with such defects occur at a high frequency in the normal population, as we propose. An alternative explanation (preferred by Winderickx, et al., 1992a) is that only the first two genes are expressed from the pigment gene arrays of all observers. There is good evidence for a locus-control region (LCR) about 4 kb upstream of the transcription initiation site (Nathans, et al., 1989; Wang et al., 1992). An interaction between the LCR and a pigment gene is proposed to be required to switch on transcription. Winderickx et al. (1992a) suggest that the LCR allows transcription of only the two genes closest to the LCR; more downstream genes are supposed not to interact with the LCR and are, thus, not transcribed. If so, it is curious that in individuals with multiple middle-wave genes, the expression of only the A allele was observed. Considering the high frequency of recombination between opsin genes, one might expect that either allele could occur in the second position, and across individuals both alleles would be found to be expressed.

There is precedent from other systems for the idea that more proximally located genes have a competitive advantage to interact with the LCR. It is imaginable that the probability of a particular pigment gene being expressed in a cone is proportional to its proximity to the LCR, i.e., the most 5' gene would be expressed in the highest number of cones, the second gene expressed in fewer cones than the first and so on, until the most distal genes of large arrays are expressed in very few cones. We have suggested that the number of genes per array might often be larger than has been suggested from previous interpretations of results from Southern analysis. It could be that the A alleles of the middle-wave genes usually out number the C alleles in individual arrays and the C alleles may tend to occur at more distal positions relative to the LCR, where they are expressed at low levels. This might explain why Winderickx et al. (1992a) detected the expression of only the A version in their subjects who carry both alleles. Such a proposal is attractive, and it does not conflict with a theory in which more than two X-linked pigment genes can be expressed in an individual eye. Even the expression of a gene at very low levels is expected to have important consequences for vision in some cases. There are about 5 million cones in each eye. If a gene were expressed at a level of only 5%, it might not be detected in experiments like those of Winderickx et al. (1992a) but a person would still have a quarter-million cones with that expressed pigment. It is virtually certain that a person would not suffer extreme deuteranomaly if he had that number of normal middle-wave cones. Extremely biased cone ratios still seem to support normal color vision. For example, Pokorny, Smith & Baron (1988) reported the case of a female heterozygous carrier of a protan color defect who, they concluded, has an extremely small number of long-wave cones; the subject has normal color vision. The possibility that levels of expression depend on gene position (Winderickx, et al., 1992a; Deeb, et al., 1992) is interesting and extremely important, but it needs to be tested further before the consequences for genotype-phenotype relationships in color vision are understood.

SPECTRAL TUNING AND NORMAL COLOR VISION VARIATIONS

Red-green color vision depends on having at least two pigments with different spectral sensitivities in the middle-to-long wavelengths. Understanding the genetic basis of normal color vision and the cause of color defects depends critically on knowing which amino acid differences between the cone pigment opsins are responsible for giving the middle-to-long wave pigments different spectra. A comparative study of the genes that encode eight pigments with various spectra, two from humans, and three each from two species of South American monkey, indicated that amino acid substitutions at just three amino acid positions could account for the spectral difference between human long- and middle-wave pigments (Neitz, et al., 1991c). The positions of these substitutions in the opsin are illustrated in Figure 7. Two amino acid differences between long- and middle-wave pigments, at positions 277 and 285 are encoded by nucleotides in exon 5 of the pigment genes. A substitution of Tyr for Phe at position 277 produces a spectral shift of about 10 nm, and a substitution of Thr for Ala at 285 produces a shift of about 15 nm. Together, these two substitutions shift pigments from ones that can be categorized as middle-wave to those that can be categorized as long-wave, confirming an earlier conclusion that substitutions in exon 5 are principally responsible for the difference between middle and long-wave pigment classes (Neitz, Neitz, & Jacobs, 1989). Thus, if all common human X-linked pigments are categorized into two classes, middle- or long-wave sensitive, any gene that has exon 5 from a long-wave gene will encode a long-wave pigment. Any gene that has exon 5 derived from a middle-wave gene will encode a middle-wave pigment. A third amino acid substitution, of Ser for Ala at position 180, encoded by a nucleotide substitution in exon 3 of the genes is responsible for a smaller spectral shift of about 4-6 nm. This is one of the substitutions that appears to be responsible for spectral differences within the classes of middle- and long-wave pigments. Subtypes of long-wave pigments and sub-types of middle-wave pigments that have differences at position 180 could arise from point mutations or from crossovers between the middle- and long-wave genes occuring downstream of exon 3 but upstream of exon 5.

The finding that substitutions at the three positions, 180, 277, and 285 are responsible for spectral differences among the pigments has been confirmed (Chan, Lee, & Sakmar, 1992; Ibbotson et al., 1992; Merbs & Nathans, 1992a; Merbs & Nathans, 1992b; Williams et al., 1992). However, some studies suggested that these three are not the only substitutions involved in tuning the spectra of middle to long-wave pigments. Specifically, several results (Ibbotson, et al., 1992; Williams, et al., 1992; Winderickx, et al., 1992b) hinted that one or both of two closely separated substitutions, at positions 230 and 233 encoded in exon 4 might also be involved. Recently, Merbs and Nathans (1992b) have measured the absorption spectra of hybrid pigments expressed in vitro. Pigments encoded by genes with long-wave exon 4 were found to have spectra that differ by about 2–4 nm from those encoded by genes with middle-wave exon 4. Further, they found that substi-

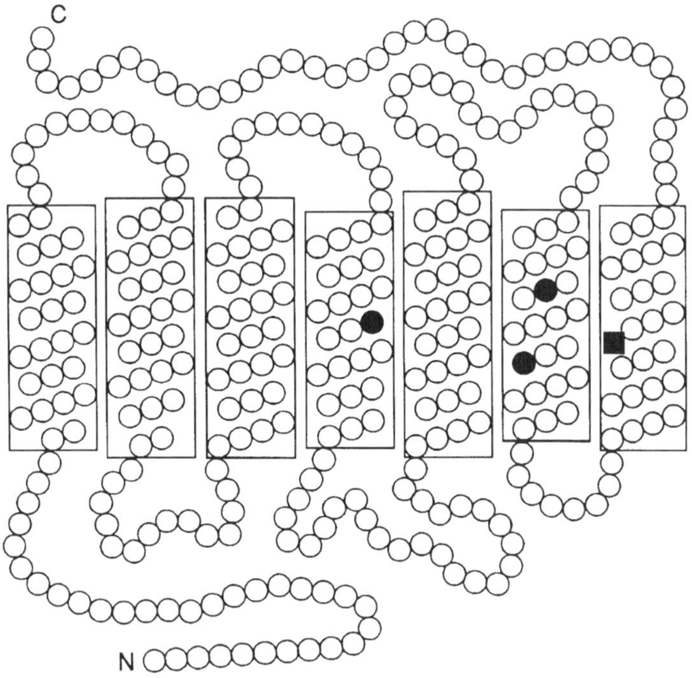

Figure 7 Amino-acid substitutions at three sites can account for most of the differences in spectral sensitivity among middle-to-long wavelength sensitive visual pigments. The pigment molecules consist of seven membrane-spanning helices. The filled square in the seventh helix indicates the lysine residue that binds the chromophore, 11-cis-retinal. Filled circles indicate the positions of three amino acid residues that participate in controlling the pigment spectrum. The identities of the two residues in helix six determine whether a pigment will fall into the middle- or the long-wave class. Both residues are encoded in exon 5 of the gene. Substitutions at the site in helix four (position 180) produce a smaller spectral shift that accounts for spectral differences between pigment subtypes within each of the middle- and long-wave classes. In addition to the three residues indicated, substitutions at other sites encoded by exons 2 and 4 of the genes are associated with small spectral shifts that are likely to contribute to spectral variation within the middle- and long-wave classes.

tutions in exon 2 were also associated with small shifts in the spectral peak of the long-wave pigments.

The small spectral shifts produced by nucleotide substitutions in exons 2, 3 and 4 may seem hardly significant, but together they may occasionally be responsible for important differences between subtypes within each of the classes of middle- and long-wave pigments. For example, we would class as long-wave pigments, all pigments produced by genes with a long-wave exon 5, however, a gene with middle-wave exons 2, 3, and 4, but a long-wave exon 5 is predicted to have a spectrum as much as 11–12 nm shorter than the longest long-wave pigment that occurs in humans (Merbs & Nathans, 1992b). It is an attractive idea that spectral differences within each of the classes of middle- and long-wave pigments may be responsible for the differences between mild and extreme forms of color anomaly. For example, a deuteranomalous observer who inherited two well separated long-wave pigments might be expected to have better color vision, i.e., a milder form of deuteranomaly, than one who draws a set of long-wave pigments that are very similar in spectra. However, there are several lines of evidence that would seem to contradict this as the complete explanation (e.g., Hurvich, 1972; Nagy, 1982; Pokorny & Smith, 1982). They suggest that factors other than the spectral separation of the pigments contribute to differences in discrimination ability among anomalous observers.

As we pointed out earlier, and as shown in Figure 8, there is considerable variability in the Rayleigh color matches of people classed as having normal color vision. It has now been demonstrated that substitutions at a single amino acid position, residue 180, are largely responsible for those individual differences. Two observations suggested to us (Neitz, et al., 1991c) that substitutions at position 180 produce normal color vision variations, 1) substitutions of Ser for Ala at position 180 of the X-linked pigments produce a shift in spectrum (Neitz, et al., 1991c; Neitz, Neitz, & Jacobs, 1991d), and 2) the long-wave pigment cDNAs (presumably isolated from normal observers) analyzed by Nathans et al. (1986b) differed from Nathans' own genomic long-wave gene at position 180, suggesting that the amino acid at this position was polymorphic in the color normal population. We tested the hypothesis that spectral variations in both the middle- and long-wave pigments produced by Ser/Ala substitutions at position 180 can explain normal color vision variations by comparing individual differences in Rayleigh matches with individual differences in the relative proportion of X-linked pigment genes that encode Ala180 vs. Ser180. There was a high correspondence between the two measures indicating that individual differences in the proportion of X-linked pigment genes that encode Ala180 vs. Ser180 are, alone, sufficient to predict 89% of the variance in the observed normal color matches (see Figure 9) (Neitz, et al., 1993). In a similar experiment, Winderickx et al. (1992b) used two different primers in the polymerase chain reaction (PCR) to specifically amplify two subsets of DNA fragments containing the codon for amino acid 180 from the X-linked pigment genes in each individual. They used single strand conformation polymorphism (SSCP) analysis to examine individual differences and found that the migration patterns of the DNAs that they believed were from long-wave genes correlated with individual differences in Rayleigh matches. Even though Winderickx et al. (1992b) emphasized a conclusion that polymorphism at position 180 of the long-wave genes influences the Rayleigh matches of color normal males, they also found that the fragments that they believed were from middle-wave genes were polymorphic at position 180; while those gene from 84% of individuals were concluded to have only Ala, 16% were concluded to have genes with Ser among at least some of the putative middle-wave genes. Psychophysical results (e.g., Webster & MacLeod, 1988; Neitz & Jacobs, 1990) suggest that individual differences in both middle- and long-wave genes underlie normal color vision variation. Although the exact frequencies still need to be clarified, it appears that there are two subtypes of middle- and two subtypes of long-wave pigments with respect to the identity of the amino acid, Ser or Ala, at position 180 in the population with normal color vision.

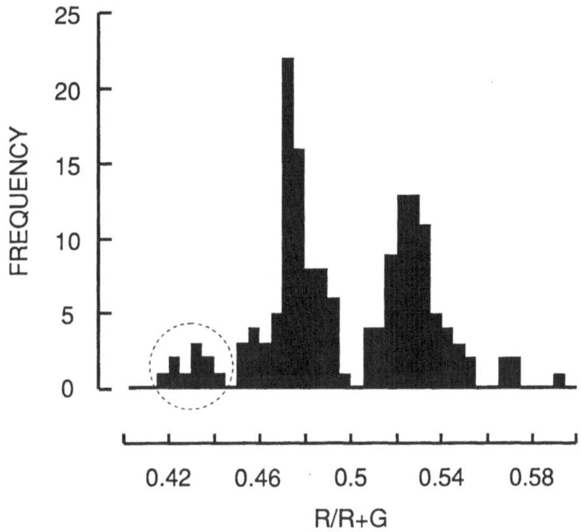

Figure 8 Individual differences among males with normal color vision. Complements of middle-and long-wave pigments vary among people with normal color vision. These are manifest as individual differences in a color matching task in which subjects are asked to adjust the proportion of red light in a mixture of red and green so that it exactly matches the appearance of a monochromatic standard light. Illustrated is the distribution of Rayleigh color matches for males with normal color vision compiled from Neitz and Jacobs (1986; 1990). In these studies most males fell into one of two large groups. However, a few subjects fell into smaller groups that flank the major modes. About 8% of males fell into a group (circled) who require less red light in a red/green mixture than the other normal observers. Results from experiments designed to distinguish individual differences in color vision produced by shifts in the spectral peaks of the middle- and long-wave pigments from those produced by other factors suggest that individuals in the small group are different from the larger groups in their complements of pigments as are the two large groups different from one another in their pigments (Neitz & Jacobs, 1990).

As discussed earlier, one of the most engaging questions raised by the finding that most people have extra X-linked pigment genes is: what are the extra genes doing? Are they expressed; and if so is each gene expressed in its own subset of cones? The investigations of polymorphism in normal color vision lend insight into these questions. One idea is that only the first two genes in the array are expressed (Deeb, et al., 1992; Winderickx, et al., 1992a). If the favored view of normal gene arrangements was true and almost everyone with normal color vision has only a single long-wave gene, the first in the array, then the second gene could always be a middle-wave gene in normals. If only the first two genes were expressed, then everyone with a normal gene arrangement would express one long- and one middle-wave gene. This would fit well with the traditional view of normal color vision in which everyone is supposed to express only three different cone pigments. Furthermore, this mechanism could account for the fact that deuteranomalous observers do not express the normal middle-wave genes they are often believed to carry. The idea is that, in deuteranomaly, the addition of a 5'middle- 3'long-wave fusion gene in the second position of the array displaces the normal middle-wave genes to more 3' positions where they are not expressed (Deeb, et al., 1992; Winderickx, et al., 1992a). Conversely, however, we have proposed here that many color normal subjects have multiple long-wave genes. In Figure 3, the extra long-wave genes are illustrated as occuring in the second position of the array to explain how crossovers between normal gene arrangements can produce color defects. This would seem to preclude the possibility that only the first two genes are expressed.

Normal individual differences lend insight into this issue as follows. We find that the normal color matches of most males fall into one of two main groups as shown in Figure 8 (Neitz & Jacobs, 1986; Neitz & Jacobs, 1990). However, there is a small group of males, circled in Figure 8, who require less red light in the Rayleigh mixture than those in either main group. The differences in color vision between individuals in the two main groups correlate well with amino acid differences at position 180 (Neitz, et al., 1993; Winderickx, et al., 1992b). We find that the more extreme color matches are also predicted by the proportion of X-linked genes with Ala vs. Ser at position 180 (Neitz, et al., 1993). It has been suggested (Merbs & Nathans, 1992a; Merbs & Nathans, 1992b; Winderickx, et al., 1992b) that the two large main groups of observers differ because individuals in the large group who require less red light in the mixture, express a long-wave gene with Ser180 and middle-wave genes with Ala180 while those in the other main group have Ala at position 180 of both middle- and long-wave pigments. Since the most extreme color matches also correlate with the proportion of genes encoding Ala180 vs. Ser180, in order to be consistent with the idea that only two genes are expressed, people who fall into the extreme group (circled in Figure 8) must have Ser at position 180 of both their expressed middle- and long-wave genes. Winderickx et al. (1992b) suggest that although about 60% of the long-wave genes encode Ser at position 180, only a small percentage of the middle-wave genes have Ser180. They used nucleotide differences in exon 2 to identify middle- from long-wave genes. It is unlikely that those changes would reliably distinguish middle- from long-wave genes, but nonetheless, it does appear, from their experiment, that Ala is much more common than Ser at position 180 among the middle-wave pigment genes. From this, under the hypothesis that everyone has a single long-wave gene and only the first two genes are expressed, the common gene arrangements that should give rise to the extreme color matches (circled in Figure 8) can be predicted. They are illustrated in Figure 9, the first two genes must have Ser at 180 but the other, unexpressed genes are likely to have Ala. It can be seen that if people are allowed to have unexpressed genes with Ala180, the proportion of genes that encode Ala vs. Ser at 180 should vary widely for this extreme phenotype depending on how many unexpressed middle-wave genes a person has. Indeed, there should be very little correspondence between genotype and phenotype for this group. To the contrary the correlation is very high. Our experiment suggests that all these extreme subjects have numerous genes (perhaps, between 4 and 10) of which the vast proportion have Ser at position 180. The chance of obtaining this result if only the first two genes were expressed is vanishingly small. Conversely, however, if all the functional genes are expressed, then the only way a person could fall into this extreme group is if the preponderance of his genes have Ser180, just as is observed.

If many or all the functional pigment genes are expressed, then because of the polymorphism in the normal pigments it seems likely that many people would individually have and express genes for both subtypes of middle- and/or both subtypes of long-wave pigments. That is, contrary to the long favored view that only three pigment types underlie normal color vision, many people could have four or five spectrally different pigments. This is consistent with the distribution of color matches among color normal males (Neitz, et al., 1993). Another recent result suggests that the expression of additional pigments is also evident in the behavior of individual males. A fundamental property of color vision based on three cone types is that color matches made with the unadapted eye will persist after selective color adaptation (Brindley, 1970; Jameson & Hurvich, 1972). To the contrary, in an experiment designed to test the idea that normal males have more than three spectrally different cone types, it was found that some normal males change their color matches after adaptation to red light, as predicted if they have more than two different types of long-wave cones (Neitz, Neitz, & Jacobs, 1991a).

No anomalous pigments in color anomaly

Protanomalous observers can be thought of as having two subtypes of middle-wave pigments and deuteranomalous observers, two subtypes of long-wave pigments. Merbs and Nathans (1992b), recently pointed out that a deuteranomalous observer could inherit two normal long-wave genes that differ at position 180 (one normal and one hybrid gene with only exon 1 from the middle-wave gene). In such a case, neither deuteranomalous pigment is an abnormal one, instead he has variants of the long-wave pigments found in normal trichromats, consistent with the model for anomalous trichromacy proposed years ago by Alpern and colleagues (Alpern & Moeller, 1977; Alpern & Pugh, 1977; Alpern & Wake, 1977). Similarly, a protanomalous observer could inherit two normal middle-wave genes that differ at position 180 and, thus, not have an anomalous pigment.

We would take the idea that anomalous observers express normal variants of the middle and long-wave pigments several steps further. The results of Merbs and Nathans (1992b) suggest that substitutions in exons 2 and 4 also influence the pig-

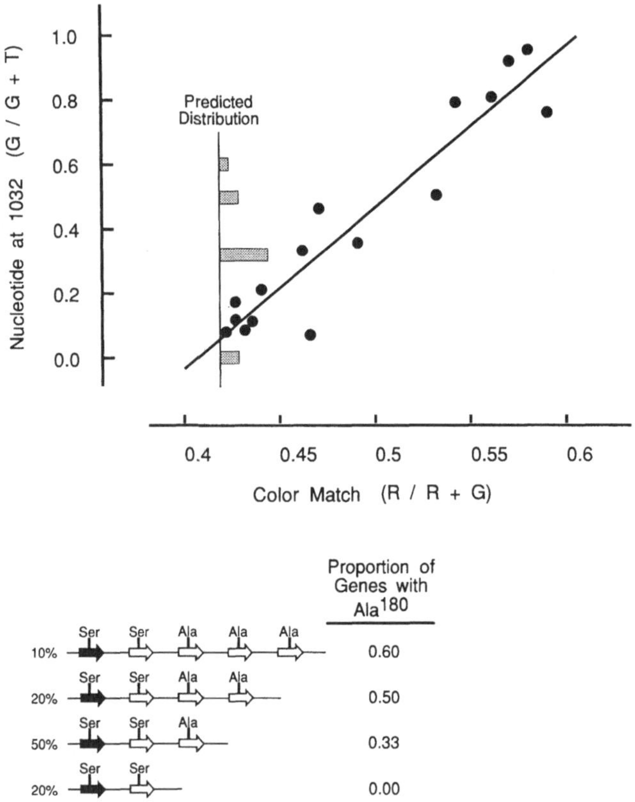

Figure 9 Individual differences at a single amino acid position (180) of the middle- and long-wave pigments can account for individual differences in normal red-green color vision. A G/T polymorphism at nucleotide position 1032 of the genes produces a Ser/Ala 180 polymorphism among the X-encoded pigments. Filled circles plot the proportion of red light in a mixture of red and green light that each subject required to match the appearance of a monochromatic standard light versus the relative proportion of X-chromosome pigment genes that have G at position 1032. The six subjects with the smallest (R/R+G) settings in the color match would fall into the extreme group that is circled in Figure 8. Each has a low proportion of genes with G at position 1032 among his total X-chromosome pigment genes, i.e., they all have a preponderance of genes encoding Ser 180. Compare this result with the prediction from the hypothesis that males with normal color vision each express only the two 5'-most genes in their array, one long- and one middle-wave gene. If only two of the genes were expressed, to explain their color matches, these extreme subjects must have Ser180 for both the expressed middle- and long-wave pigments, i.e., they must express the more long wavelength sensitive subtype of each pigment. Because of a preponderance of genes that encode Ala180 in the population, the unexpressed genes are expected to most often specify Ala180. The most common gene arrangements predicted for these extreme subjects are illustrated at the bottom of the figure. This distribution predicted by the hypothesis that only two genes are expressed, is superimposed in histogram form, on the observed results in the upper graph. The descrepency between the prediction and the observed results suggests that most or all of the functional X-chromosome genes are expressed and contribute to the observed behavior.

ment spectra. Although they recognize position 180 as a polymorphic site in the normal pigments, long-wave genes that include the active substitutions in exons 2 or 4 of the middle-wave genes would have to be fusion genes. If normal observers do not express fusion genes, then only a subset of the pigments that occur in anomalous observers would correspond to normal pigment variants, others, produced by genes with fusions between exons 2 and 5 would be exclusively expressed in anomalous trichromats. To the contrary, here, it is suggested that fusion genes are characteristic of normal gene arrangements. We further suggest that normal observers can express their fusion genes along with their other middle- and long-wave pigment genes. The expression of fusion genes in normal subjects merely contributes to the variability in normal color vision. Thus, normal observers can have and express any of the pigments that occur in anomalous trichromacy and vice versa. In the traditional sense, anomalous pigments have been conceived as pigments that differ in spectrum from those that ever occur in normal eyes. If the picture of the pigments and genes of normal color vision presented here is the correct one, then anomalous pigments do not exist. As originally envisioned by Alpern and colleagues, anomalous trichromats draw from exactly the same set of long- and middle-wave pigments as do normal trichromats. Thus, there are no pigments that are unique to color anomaly – i.e., no anomalous pigments. The difference between normal and anomalous trichromacy is that for X-linked pigments, normal subjects draw a set that includes one or more members from each of the middle- and long-wave classes; deuteranomalous observers have functional pigments from only the long-wave class; protanomalous observers have pigments from only the middle-wave class.

In summary, the traditional theory held that the replacement of a normal pigment with an anomalous one caused color anomaly. This thinking has carried over in more recent explanations of anomaly, i.e., that the fusion genes cause anomalous trichromacy. We hold that the fusion genes do not, in any way, cause color anomaly. Quite the opposite, the normal variant pigments, including those produced by fusion genes, serve the advantageous role, of protecting some observers who loose the genes for (or function of) one class of pigments, from becoming dichromats. They can inherit anomalous trichromacy instead. The advantage of anomalous trichromacy compared to dichromacy may be quite significant for individuals who inherit the mildest forms of anomaly. This advantage may explain why polymorphisms within the middle- and long-wave pigment classes have been maintained in the normal human population.

Our view of the relationship between genes, cone pigments, and color vision is a simple one (illustrated in Figure 10). The cone pigments are encoded by a gene for the short-wave pigment on chromosome 7 and an array of genes on the X-chromosome. The number of X-chromosome pigment genes is variable, with a range across individuals, of one to ten (or more). Genes for middle-wave pigments are more numerous than genes for long-wave pigments, but people with multiple long-wave genes are not unusual. The genes that encode middle-wave and long-wave pigments are highly polymorphic. An enormous variety of pigment genes

SUMMARY — A THEORY OF THE CONES, CONE PIGMENTS AND CONE PIGMENT GENES OF HUMAN COLOR VISION

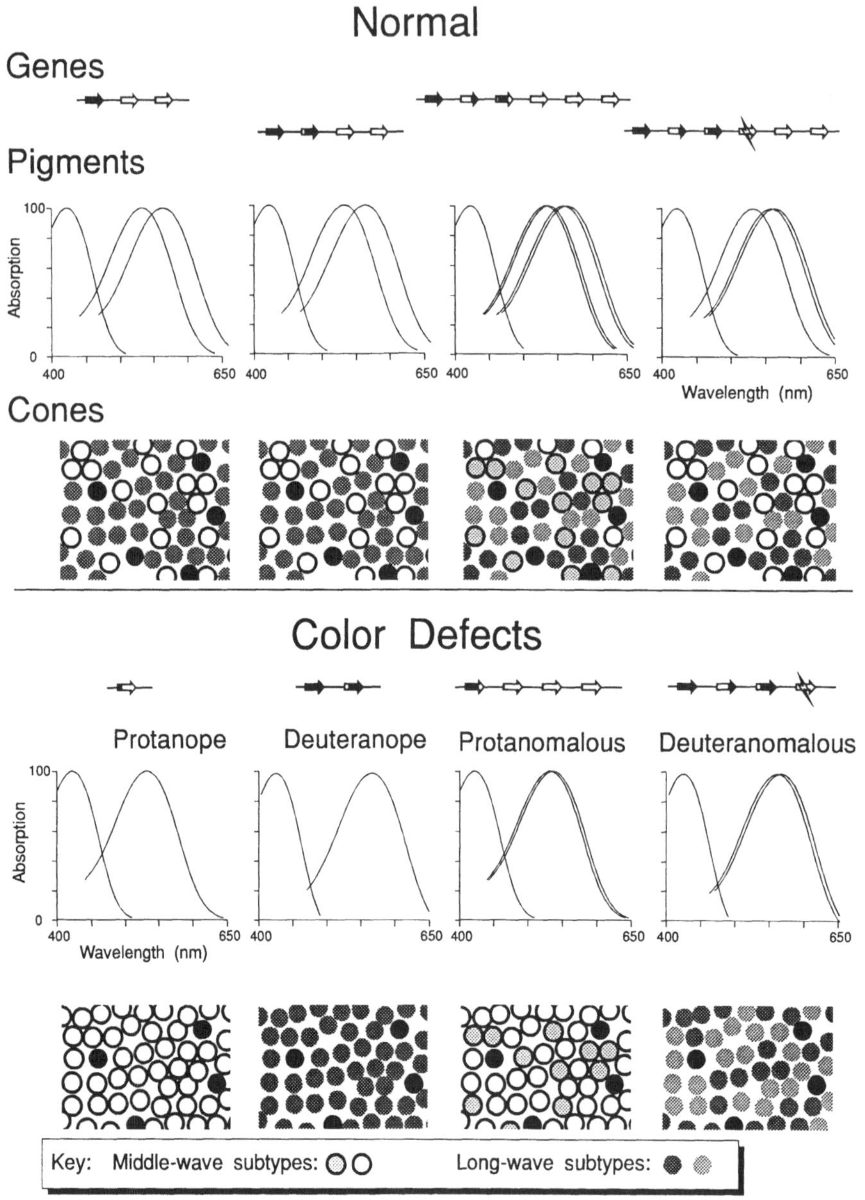

Figure 10 Summary of a new theory of the genes, pigments, and cone types that underlie normal and defective human color vision. Among observers with normal color vision there is an enormous variety of different pigment gene arrangements. Some are relatively simple as illus-

with different structures exist because of point mutations and because of frequent recombinations that have occurred between the long- and middle-wave genes over the span of evolutionary time. Some of these differences among the genes are responsible for spectrally different subtypes of middle- and long-wave pigments in the population, and within individuals. Perhaps, two subtypes of long-wave and two subtypes of middle-wave are most common. Deleterious mutations in the pigment genes or their regulatory sequences that prevent expression, or interfere with normal pigment function are not uncommon. These may have accumulated particularly among the X-linked genes because most people have many more genes than are necessary to serve normal function. Although some of these may ultimately lead to retinal disorder, one or more gene defects can be tolerated without serious consequences to vision through the reproductive years of life. With the exception of genes with defects that specifically affect expression, many or all of the X-linked pigment genes are expressed (although the level of expression may vary). Only one pigment gene is expressed in any one cone so that each gene is expressed in its own subpopulation of cones. Thus, contrary to the long held idea that normal trichromats have only three cone types, individual males and females with normal color vision can, sometimes, have four or more spectrally different cone types.

To have normal color vision, a person must express a short-wave pigment and at least one pigment subtype from both the middle-wave class and the long-wave class. All inherited color vision deficiencies are caused by the loss of pigments required for normal color vision. The most common color deficiencies can be ex-

trated by the example in the upper left. Many are more complicated and the gene complements can include fusion genes and genes with defects, as illustrated by the more right-hand upper gene arrays. Most or all of the functional genes are expressed. The pigments produced by the X-chromosome genes fall into two classes, middle- and long-wave, but each of these classes is composed of subtypes. Perhaps, two major subtypes of middle-wave and two major subtypes of long-wave pigment are most common. Examples of absorption spectra of the expressed pigments are shown directly below each example gene array. Each includes a short wave-pigment (encoded by a separate gene on chromosome 7) and at least one pigment from each of the middle- and long-wave classes. However, normal individual males can express multiple middle and/or multiple long-wave pigment subtypes. Below the spectral curves, small patches of the cone mosaic that would result from each gene arrangement, are diagrammed. Each cone is illustrated as a circle. Differences in fill and outline illustrate cones with spectrally different pigments (short-wave cones, solid black; middle-wave cone types, dark outline; long-wave cone types; no outline; subtypes have different fills). Each gene is expressed in its own subpopulation of cone, thus, individuals can have as many as four or five spectrally different cone types. The situation in the color defects is simple. Each of the four illustrated is reduced in functional genes, in pigments, and in cone types from the normal arrangement shown above it. All have a short-wave pigment. The protanope and deuteranope are each reduced to a single type of X-encoded pigment. The protanomalous observer is reduced to only normal middle-wave subtypes, the deuteranomalous observer to only normal long-wave subtypes.

plained by the loss of normal pigment genes. All such losses can occur by one-step recombinations in which normal genes are deleted from gene complements that underlie normal color vision. Anomalous trichromats with red-green deficiencies maintain two or more of either the middle-wave subtypes or the long-wave pigment subtypes that are expressed in normal color vision. Dichromats with red-green defects have X-linked genes that produce only a single type of cone pigment.

We offer this model as an alternative to conventional theories of the basis of normal and defective color vision. It has been useful for us in providing a framework for organizing and re-examining what is known about the biological mechanisms that underlie normal and defective color vision. We believe it is the simplest model that can explain the facts as we understand them. This model will, no doubt, be controversial. If it stimulates further research directed toward understanding the intriguing mysteries of color vision, it will have served its purpose.

ACKNOWLEDGEMENTS

Supported by NIH grants EY09303, EY09620 and a RPB career development award to M. N.

REFERENCES

Alpern, M., and Moeller, J. (1977). The red and green cone visual pigments of deuteranomalous trichromacy. *Journal of Physiology*, **266**, 647–675.

Alpern, M., and Pugh, E. N. (1977). Variation in the action spectrum of erythrolabe among deuteranopes. *Journal of Physiology*, **266**, 613–646.

Alpern, M., and Wake, T. (1977). Cone pigments in human deutan color defects. *Journal of Physiology*, **266**, 595–612.

Botstein, D. (1986). The molecular biology of color vision. *Science*, **232**, 142–143.

Brindley, G. S. (1970). *Physiology of the Retina and Visual Pathway*. London: Camelot.

Chan, T., Lee, M., and Sakmar, T. P. (1992). Introduction of hydroxyl-bearing amino acids causes bathochromic spectral shifts in rhodopsin: amino acid substitutions responsible for red-green color pigment spectral tuning. *Journal of Biological Chemistry*, **267**, 9478–9480.

Dartnall, H. J. A., Bowmaker, J. K., and Mollon, J. D. (1983). Human visual pigments: microspectrophotometric results from the eye of seven persons. *Proceedings of the Royal Society of London* B, **220**, 115–130.

Deeb, S. S., Lindsey, D. T., Hibiya, Y., Sanocki, E., Winderickx, J., Teller, D. Y., and Motulsky, A. G. (1992). Genotype-phenotype relationships in human red/green color-vision defects: molecular and psychophysical studies. *American Journal of Human Genetics*, **51**, 687–700.

DeMarco, P., Pokorny, J., and Smith, V. C. (1992). Full spectrum cone sensitivity functions for X-chromosome-linked anomalous trichromats. *Journal of the Optical Society of America* A, **9**, 1465–1476.

Drummond-Borg, M., Deeb, S. S., and Motulsky, A. G. (1989). Molecular patterns of X-chromosome-linked color genes among 134 men of European ancestry. *Proceedings of the National Academy of Sciences U.S.A.*, **86**, 983–987.

Dryja, T. P., McGee, T. L., Reichel, E., Hahn, L. B., Cowley, G. S., Yandell, D. W., Sandberg, M. A., and Berson, E. L. (1990). A point mutation of the rhodopsin gene in one form of retinitis pigmentosa. *Nature*, **343**, 364–366.

Eisner, A., and MacLeod, D. I. A. (1981). Flicker photometric study of chromatic adaptation: selective suppression of cone inputs by colored backgrounds. *Journal of the Optical Society of America*, **71**, 705–718.

Elsner, A., and Burns, S. A. (1987). Classes of color normal observers. *Journal of the Optical Society of America* A, **4**, 123.

Hurvich, L. M. (1972). Color vision deficiencies. In D. Jameson and L. M. Hurvich (Eds.), *Handbook of Sensory Physiology*, Vol. VII/4: Visual Psychophysics (pp. 582–624). Berlin: Springer-Verlag.

Ibbotson, R. E., Hunt, D. M., Bowmaker, J. K., and Mollon, J. D. (1992). Sequence divergence and copy number of the middle- and long-wave photopigment genes in Old World monkeys. *Proceedings of the Royal Society of London* B, **247**, 145–154.

Jacobs, G. H. (1981). *Comparative Color Vision*. New York: Academic Press.

Jameson, D., and Hurvich, L. H. (1972). Color adaptation: sensitivity, contrast, after-images. In D. Jameson and L. H. Hurvich (Eds.), *Handbook of Sensory Physiology*, Vol. VII/4: Visual Psychophysics (pp.568–581). New York: Springer-Verlag.

Jordan, G., and Mollon, J. D. (1988). Two kinds of men? *Investigative Ophthalmology and Visual Science* (Suppl.), **29**, 164.

Karnik, S. S., and Khorana, H. G. (1990). Assembly of functional rhodopsin requires a disulfide bond between cysteine residues 110 and 187. *Journal of Biological Chemistry*, **265**, 17520–17524.

Lindsey, D., T., Winderickx, J., Sanocki, E., Teller, D., Deeb, S. S., and Motulsky, A. G. (1992). Individual differences in Rayleigh matches are related to differences in L cone pigment structure. Advances in Color Vision: *Journal of the Optical Society of America*, Technical Digest Series, **4**, 17–19.

Lutze, M., Cox, N. J., Smith, V. C., and Pokorny, J. (1990). Genetic studies of variation in Rayleigh and photometric matches in normal trichromats. *Vision Research*, **30**, 149–162.

McKusick, V. A. (1992) *Mendelian Inheritance in Man*. 10th edn., Baltimore, Johns Hopkins University Press.

Merbs, S. L., and Nathans, J. (1992a). Absorption spectra of human cone pigments. *Nature*, **356**, 433–435.

Merbs, S. L., and Nathans, J. (1992b). Absorption spectra of the hybrid pigments responsible for anomalous color vision. *Science*, **258**, 464–466.

Mollon, J. (1991). Hue and the heptahelicals. *Nature*, **351**, 696–697.

Nagy, A. L. (1982). Homogeneity of large-field color matches in congenital red-green color deficients. *Journal of the Optical Society*, **72**, 571–577.

Nathans, J., Davenport, C. M., Maumenee, I. H., Lewis, R. A., Hejtmancik, J. F., Litt, M., Lovrien, E., Weleber, R., Bachynski, B., Zwas, F., Klingaman, R., and Fishman, G. (1989). Molecular genetics of blue cone monochromacy. *Science*, **245**, 831–838.

Nathans, J., Piantanida, T. P., Eddy, R. L., Shows, T. B., and Hogness, D. S. (1986a). Molecular genetics of inherited variation in human color vision. *Science*, **232**, 203–210.

Nathans, J., Thomas, D., and Hogness, D. S. (1986b). Molecular genetics of human color vision: the genes encoding blue, green, and red pigments. *Science*, **232**, 193–202.

Neitz, J., and Jacobs, G. H. (1986). Polymorphism of the long-wavelength cone in normal human color vision. *Nature*, **323**, 623–625.

Neitz, J., and Jacobs, G. H. (1990). Polymorphism in normal human color vision and its mechanism. *Vision Research*, **30**, 620–636.

Neitz, J., Neitz, J., and Jacobs, G. H. (1989). Analysis of fusion gene and encoded photopigment of color-blind humans. *Nature*, **342**, 679–682.

Neitz, J., Neitz, J., and Jacobs, G. H. (1991a). Matches from normal trichromatic males suggest that some eyes contain more than three cone types. *Investigative Ophthalmology and Visual Science* (Suppl.), **32**, 1092.

Neitz, M., Neitz, J., and Jacobs, G. H. (1991b). Relationship between cone pigments and genes in deuteranomalous subjects. In *Color Vision Deficiencies X*, edited by B. Drum, J.D. Moreland and A. Serra, pp. 397–403, Dordrecht, Kluwer Academic Publishers.

Neitz, M., Neitz, J., and Jacobs, G. H. (1991c). Spectral tuning of pigments underlying red-green color vision. *Science*, **252**, 971–974.

Neitz, M., Neitz, J., and Jacobs, G. H. (1991d). Genes and encoded M-cone pigments from two types of protanope. *Journal of the Optical Society of America*, Technical Digest Series, **17**, 23.

Neitz, J., Neitz, M., and Jacobs, G. H. (1993). More than three different cone pigments among people with normal color vision. *Vision Research*, **33**, 117–122.

Oprian, D. D., Asenjo, A. B., Lee, N., and Pelletier, S. L. (1991). Design, chemical synthesis, and expression of genes for the three human color vision pigments. *Biochemistry*, **30**, 11367–11372.

Piantanida, T. P. (1976). Polymorphism of human color vision. *American Journal of Optometry and Physiological Optics*, **53**, 647–657.

Piantanida, T. P., and Gille, J. (1992). Methodology-specific Rayleigh-match distributions. *Vision Research*, **32**, 2375–2377.

Pokorny, J., and Smith, V. C. (1977). Evaluation of a single pigment shift model of anomalous trichromacy. *Journal of the Optical Society of America*, **67**, 1196–1209.

Pokorny, J., and Smith, V. C. (1982). New observations concerning red-green color defects. *Color Research and Application*, **7**, 159–164.

Pokorny, J., Smith, V. C., and Baron, R. (1988). How many cones do you need for normal color perception? *Investigative Ophthalmology and Visual Science* (Suppl.), **29**, 298.

Pokorny, J., Smith, V. C., and Katz, I. (1973). Derivation of the photopigment absorption spectra in anomalous trichromats. *Journal of the Optical Society of America*, **63**, 232–237.

Reichel, E., Bruce, A. M., Sandberg, M. A., and Berson, E. L. (1989). An electroretinographic and molecular genetic study of X-linked cone degeneration. *American Journal of Ophthalmology*, **108**, 540–547.

Schnapf, J. L., Kraft, T. W., and Baylor, D. A. (1987). Spectral sensitivity of human cone photoreceptors. *Nature*, **325**, 439–441.

Shevell, S. K., and He, J. C. (1992). Dual Rayleigh-like color matches sensitive to individual differences in cone photopigments. Advances in Color Vision: *Journal of the Optical Society of America*, Technical Digest Series, **4**, 20–22.

Strutt, J. W. (Lord Rayleigh) (1881). Experiments in color. *Nature*, **25**, 64–66.

Sung, C. H., Davenport, C. M., Hennessey, J. C., Maumenee, I. R., Jacobson, S. G., Heckenlively, J. R., Nowakowski, R., and Fishman, G. (1991a). Rhodopsin mutations in autosomal dominant retinitis pigmenotosa. *Proceedings of the National Academy of Sciences of the United States of America*, **88**, 6481–6485.

Sung, C. H., Schneider, B. G., Agarwal, N., Papermaster, D. S., and Nathans, J. (1991b). Functional heterogeneity of mutant rhodopsins responsible for autosomal dominant retinitis pigmentosa. *Proceedings of the National Academy of Sciences of the United States of America*, **88**, 8840–8844.

van Heel, L., Went, L. N. and van Norren, D. (1980) Frequency of tritan disturbances in a population study. In: Verreist G. (ed.) *Color Vision Deficiencies*, V. Hilger, Bristol, pp. 256–260.

Vollrath, D., Nathans, J., and Davis, R. W. (1988). Tandom array of human visual pigment genes at Xq28. *Science*, **240**, 1669–1672.

Waaler, G. H. M. (1967). Heredity of two types of normal color vision. *Nature*, **215**, 406.

Wang, Y., Macke, J. P., Merbs, S. L., Zack, D. J., Klaunberg, B., Bennett, J., Gearhart, J., and Nathans, J. (1992). A locus control region adjacent to the human red and green visual pigment genes. *Neuron*, **9**, 429–440.

Webster, M. A. (1992). Reanalysis of λ_{max} variations in the Stiles-Birch 10° color- matching functions. *Journal of the Optical Society of America* A, **9**, 1419–1421.

Webster, M. A., and MacLeod, D. I. A. (1988). Factors underlying individual differences in the color matches of normal observers. *Journal of the Optical Society of America* A, **5**, 1722–1735.

Weitz, C. J. Miyake, Y., Shinzato, K., Montag, E., Zrenner, E., Went, L. N., and Nathans, J. (1992a). Human tritanopia assiciated with two amino acid substitutions in the blue sensitive opsin. *American Journal of Human Genetics*, **50**, 498–507.

Weitz, C. J., Went, L. N., and Nathans, J. (1992b). Human tritanopia associated with a third amino acid subsitution in the blue sensitive visual pigment. *American Journal of Human Genetics*, **51**, 444–446.

Williams, A. J., Hunt, D. M., Bowmaker, J. K., and Mollon, J. D. (1992). The polymorphic photopigments of the marmoset: spectral tuning and genetic basis. *EMBO Journal*, **11**, 2039–2045.

Winderickx, J. Battisti, L., Motulsky, A. G., and Deeb, S. S. (1992a). Selective expression of human X chromsome-linked green opsin genes. *Proceedings of the National Academy of Sciences of the United States of America*, **89**, 9710–9714.

Winderickx, J., Lindsey, D. T., Sanocki, E., Teller, D. Y., Motulsky, A. G., and Deeb, S. S. (1992b). Polymorphism in red photopigment underlies variation in color matching, *Nature*, **356**, 431–433.

Winderickx, J., Sanocke, E., Lindsey, D. T., Teller, D. Y., and Moutulsky, A. G. (1992c). Defective color vision associated with a missense mutation in the human green visual pigment gene. *Nature Genetics*, **1**, 251–256.

Yong, T (1802). On the theory of light and colors. *Philosophical Transactions of the Royal Society of London*, **92**, 12–48.

10. X-LINKED RETINITIS PIGMENTOSA

M. A. ALDRED[1], M. JAY[2] and A. F. WRIGHT[1]

[1] *MRC Human Genetics Unit, Western General Hospital, Crewe Road, Edinburgh EH4 2XU, UK and*

[2] *Department of Clinical Ophthalmology, Moorfields Eye Hospital, City Road, London EC1V 2PD, UK*

SUMMARY

X-linked retinitis pigmentosa (xlRP) is one of the most severe forms of this group of retinal dystrophies. It has been found to be genetically heterogeneous, with two loci (RP2 and RP3) on the short arm of the X-chromosome. The RP3 gene has been localised by genetic linkage analysis and more finely mapped by analysis of deletion patients. The RP2 locus is less well defined by linkage analysis alone and the lack of associated cytogenetic abnormalities is likely to make the identification of this gene more problematic. Heterogeneity is a source of difficulty in genetic counselling and gene mapping studies, since many families cannot be unambiguously assigned to one or the other locus.

INTRODUCTION

Genetic disorders are an important cause of blindness in Western countries. A study in Newfoundland by Green et al. (1986) showed that 24% of all blindness had an established monogenic cause, while about 40% of blindness in children is genetically determined (Fraser and Friedmann, 1967). Tapeto-retinal dystrophies form the most common genetic cause of blindness (Warburg, 1990), with 31% of all genetic blindness being due to retinitis pigmentosa or rod-cone dystrophy and 22% due to cone, cone-rod or macular dystrophy (Green and Johnson, 1983).

Retinitis pigmentosa (RP) is a group of progressive retinal degenerations characterised by night blindness, pigmentary retinopathy and visual field loss. Deterioration of central vision commonly occurs in the later stages of the disease and RP is one of the major causes of blindness in middle age in the UK. There is wide variation in clinical expression but typically, an initial ring scotoma is found in the mid-periphery and this slowly enlarges both towards the macula and peripherally (Pagon, 1988). The scotopic electroretinogram (ERG) is abnormal in the early stages of the disease and soon becomes unrecordable (Heckenlively, 1990). Dark adaptation is also abnormal, with raised thresholds (Deutman, 1977) and night blindness is one of the earliest symptoms of the disease. Ophthalmoscopy reveals depigmentation and atrophy of the retinal pigment epithelium (RPE) and migration of pigment laden cells into the retina. Pigment often adheres to the narrowed retinal vessels giving a characteristic "bone-spicule" appearance, but it may also be diffusely scattered. The retinal blood vessels, particularly the arterioles, gradually constrict as the disease progresses (Heckenlively, 1988).

RP may be inherited as an autosomal dominant, autosomal recessive or X-linked recessive trait. In general, the different modes of inheritance cannot be distinguished clinically, although carriers of the X-linked disorder may show a characteristic fundus appearance (see below). The majority of cases with bilateral, symmetrical disease probably have a genetic basis (see Jay, 1982). However, 40-50% of patients have no family history of RP ("simplex" RP) (Fishman, 1978; Jay, 1982; Bundey and Crews, 1984b; Kaplan et al., 1990).

X-LINKED RP

X-linked RP (xlRP) is one of the most severe forms of RP. In the first decade of life, affected males commonly have night blindness, and early visual field loss may be apparent in the child tripping over things or appearing clumsy. If there is a family history of RP, then the significance of these symptoms may be recognised and a diagnosis made as early as age four years (Heckenlively, 1988). In the teens, degeneration of the RPE becomes apparent, visual field loss is more marked and visual acuity may deteriorate. The patient starts to experience problems in bright light as well as in the dark. In the third decade, the loss in visual acuity progresses and visual fields may be reduced to 10–15°. Secondary cataracts may develop. By the age of forty, visual acuity is often reduced to counting fingers or the perception of light and searching nystagmus may develop.

RP patients tend to be myopic and those with high myopia have the added problem of myopic degeneration. This is particularly true of X-linked RP. Sieving and Fishman (1978) found that 75% of all RP patients (268 eyes) and 90% of xlRP patients (41 eyes) were myopic, compared to 12% of the general population. xlRP patients had, on average, higher myopia (mean −5.51D, SD 4.04D) than those with non-X-linked RP (mean −1.20D, SD 2.79D). There was a suggestion that the X-linked group might be divided into two subgroups with low or high myopia respectively, but the sample size was small. RP patients were also found to have a much higher frequency of astigmatism, with 74% having cylindrical errors of >0.5D compared to 19% of the general population. There was no significant difference between the X-linked and non-xlRP groups with regard to astigmatic correction.

In females who are heterozygous for xlRP, random inactivation of one or other of the two X chromosomes in each cell during early embryogenesis (Lyon, 1961) causes the phenotype to range from asymptomatic to severely affected, depending on the proportion of retinal cells with an active disease-bearing X-chromosome (Bird, 1975; Fishman et al., 1986). Many carriers over the age of 20–30 years have one or more of the following features: peripheral retinal degeneration with pigment deposition, ERG abnormalities or a tapetal reflex in the macular region. The latter is a golden metallic sheen which is seen in the fundus of some, but not all, carriers (Fishman et al., 1986) and occasionally also in young hemizygous males (Heckenlively, 1988; van Osch et al., 1990). However, its detection can be subjective and the reflex diminishes with age, so it is not always a helpful indicator of carrier status (Bird, 1975). Carriers may have clinical signs at any age but tend not to become symptomatic until middle to late life (Bird, 1975). Severely affected carriers may be night blind, have reduced visual field and sometimes loss of visual acuity. Like affected males, myopia and astigmatism are common in heterozygous females and carriers may show myopic degeneration of the fundus (Fishman et al., 1986).

Fishman et al. (1986) examined 33 obligate carriers (i.e. females with an affected father, son, or both) from 22 families, ranging from 13 to 83 years of age. The carrier state was detectable by the presence of pigment and/or an abnormal ERG in 28 patients (85%), although 3 of these (aged 28–45) showed only myopic changes of the fundus and the abnormal ERG may have been due to high myopia rather than RP. A fourth patient (aged 31) who showed myopic degeneration as the only fundus abnormality refused ERG. The remaining four patients showed a borderline normal phenotype. Three had normal ERG and the only fundus abnormality was a tapetal reflex. One of these patients was only 18 years old, but the other two were both aged 49. The remaining patient had a completely normal fundus appearance and was only abnormal in one of five ERG tests, at age 16 years. Details of this study are summarised in Table 1.

Table 1 - Phenotypes of 33 obligate xlRP carriers (summarised from Fishman et al. (1986))

	RP-type changes (27 patients)		*No RP-type changes (6 patients)*	
Fundi	Peripheral pigment (16)	Tapetal reflex only (11)	Myopia only(4)	Normal (2)
ERG (5 tests)	Abnormal (14) Refused (2)	Abnormal (8) 5/5 Normal (3)	Abnormal (3) Refused (1)	Abnormal (1) 4/5 Normal (1)

Therefore, 10–20% of obligate carriers show a normal or near-normal phenotype when assessed in specialist centres (Fishman et al., 1986; Friedrich et al., 1985), and the percentage may be considerably higher outside such centres. This has important implications for carrier detection in females who are at 50% risk of being heterozygous for xlRP. It cannot be assumed that those who have no detectable carrier phenotype do not carry the gene, and while retinal changes may become apparent in later life, the question of genetic status is most important to those of child-bearing age.

PREVALENCE

The overall prevalence of RP is about 1 in 4000 of the general population (Boughman et al., 1980; Bundey and Crews, 1984a). Estimates of the proportions of the different genetic forms vary between countries and with the method of ascertainment. The high proportion of simplex cases is an additional complication.

In the UK, X-linked inheritance accounts for about 15% of all RP families (Jay, 1982; Bundey and Crews, 1984b). xlRP seems to be more common in the UK than the USA, where it accounts for only about 8% of all families with RP (Fishman, 1978). A higher proportion of xlRP is found in genetic clinics (>20% of families), due to ascertainment bias, and compares with 11% found in an electrodiagnostic clinic (Bird, 1975; Jay, 1982).

GENETIC MAPPING OF XLRP GENES

xlRP was first localised by Bhattacharya et al. (1984), who reported linkage of xlRP in five kindreds to DXS7, a marker localised in Xp11.4-p11.3 (Figure 1). However, ensuing reports were in conflict over the location of xlRP relative to DXS7. Friedrich et al. (1985) and Wright et al. (1987) found that xlRP was most likely to be proximal to DXS7, while others favoured a more distal location between DXS7 and DXS84, tightly linked to OTC (Nussbaum et al., 1985; Denton et al., 1988; Musarella et al., 1988, 1989b). This distal localisation was supported by two patients who had deletions of Xp21 associated with RP (Francke et al., 1985; de Saint-Basile et al., 1988). Wright et al. (1987) suggested that xlRP is genetically heterogeneous and Chen et al. (1989) found evidence of genetic heterogeneity in nine xlRP families. Meitinger et al. (1989) reported linkage of xlRP to DXS255, a probe 15-20cM proximal to DXS7. Four of their five families showed no recombinants between DXS255 and xlRP in a total of 17 meioses, but the other family contained two such recombinants in 12 meioses. One of these also recombined with DXS7, placing xlRP distal to DXS7 (Figure 1) in that family. Several authors suggested that xlRP with tapetal reflex in one or more carriers was a distinct genetic form mapping to the distal location (Nussbaum et al., 1985; Denton et al., 1988; Curtis and Blank, 1989; Musarella et al., 1989b), although one out of 14 carrier females in a family with the proximal locus has been reported to show a tapetal reflex (Friedrich et al., 1985).

Multipoint linkage analysis of 62 families with xlRP gave overwhelming evidence for the existence of two xlRP loci, with odds of 6.4×10^9:1 (Ott et al., 1990). One locus, termed RP2, was mapped to Xp11.3-p11.2. The most likely location of RP2 was 2cM proximal to DXS14, but the confidence limits were wide, extending to 3cM proximal to DXS7. The second locus, RP3, was mapped to Xp21.1-p11.4 at a position 1cM distal to OTC (Figure 1). There was also evidence that the propor-

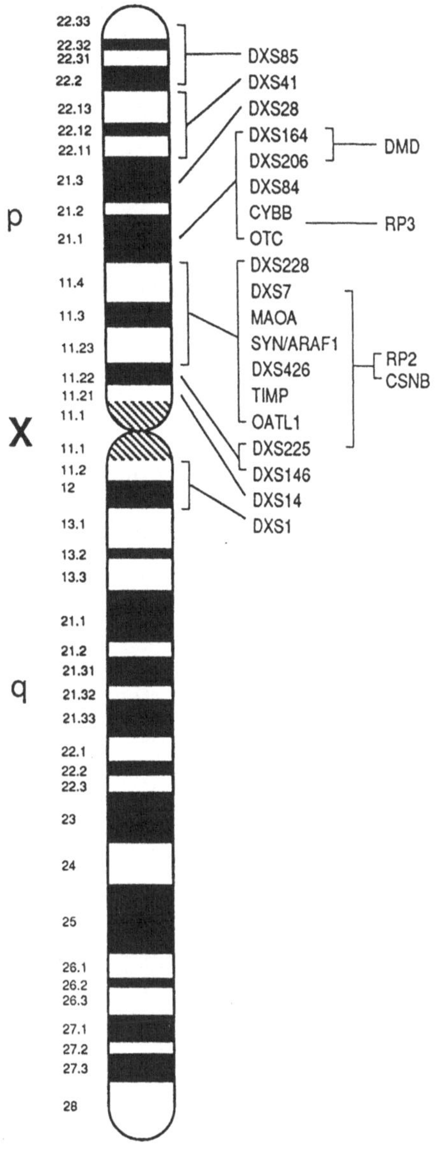

Figure 1 Location of genes and marker loci in Xp. The figure shows the order and location of marker loci used in linkage analysis of X-linked RP and the localisations of the disease genes relative to these markers.

tion of RP2 and RP3 families varied between investigators, explaining the previously discrepant results from different groups.

Musarella et al. (1988) had previously proposed another xlRP locus located between DXS28 and the DMD locus. Ott et al. (1990) found some evidence for a third locus, located in the same region, with odds of 293:1, while Musarella et al. (1990) found weaker evidence from an analysis of 20 xlRP families. However, patients deleted for this region have not been reported to have RP (Clarke et al., 1986; Francke et al., 1987; reviewed by Wright, 1990), which argues against an xlRP gene in this location.

The following sections summarise our current knowledge with regard to the localisations of RP3 and RP2.

RP3

The region containing RP3 has been defined both by linkage analysis and by identification of patients with deletions of the X-chromosome. Linkage analysis indicates that RP3 is proximal to DXS84 and distal to OTC (Nussbaum et al., 1985; Denton et al., 1988; Musarella et al., 1988 and 1989b), a genetic distance of about 6cM. This localisation is supported by two patients that were identified with RP and an associated deletion. Patient BB (Francke et al., 1985), a male with a visible deletion in Xp21, suffered from Duchenne muscular dystrophy (DMD), chronic granulomatous disease (CGD), McLeod phenotype, RP and mental retardation. Another male, SB, was reported with CGD, RP and McLeod phenotype (de Saint-Basile et al., 1988). No visible cytogenetic abnormality was detectable in SB, but both the CGD locus (CYBB) and DXS140 were deleted from his DNA. There was evidence of X-linked transmission, as his mother had diminished neutrophil cytochrome b content, weak expression of Kell antigens and some bilateral retinal pigmentation, indicating that she is a carrier of all three traits.

Other patients with deletions overlapping those of BB and SB but who do not have RP have been reported. Patient NF, a nine year old male, was reported with DMD and CGD (Baehner et al., 1986), while patient OM, a twelve year old male, had CGD and McLeod phenotype but not RP (Bertelson et al., 1988). Both patients had deletions that included and extended distally from CYBB, indicating that RP3 is proximal to CYBB (Figure 2). Thus physical and genetic linkage data indicate that the region containing the RP3 gene is flanked distally by DXS84 and CYBB, and proximally by OTC and the proximal BB deletion breakpoint.

As a first step towards isolating the RP3 gene, the sequences at the proximal BB deletion breakpoint have been cloned (Bruns et al., 1988; Musarella et al., 1991) and used as a marker for long-range physical mapping and chromosome walking. A detailed physical map spanning the region assumed to contain the McLeod locus (XK) and RP3 has been constructed (Musarella et al., 1991; Ho et al., 1992) and five CpG islands have been identified that could mark candidate genes for XK and RP3. One of these islands is located 50 kb distal to the BB proximal breakpoint and an adjacent DNA fragment showed cross-species homology (McDowell et al., 1990). This conserved fragment was used to screen retinal and retinal pigment epithelial cDNA libraries. A 2.1 kb cDNA clone was identified. It is expressed in retina and brain, and shows homology to the mouse t complex sterility locus (McDowell et al., 1990). However, extensive examination of the gene in RP3 patients has failed to identify any mutation that would indicate that this is the RP3 gene (Roux et al., 1992).

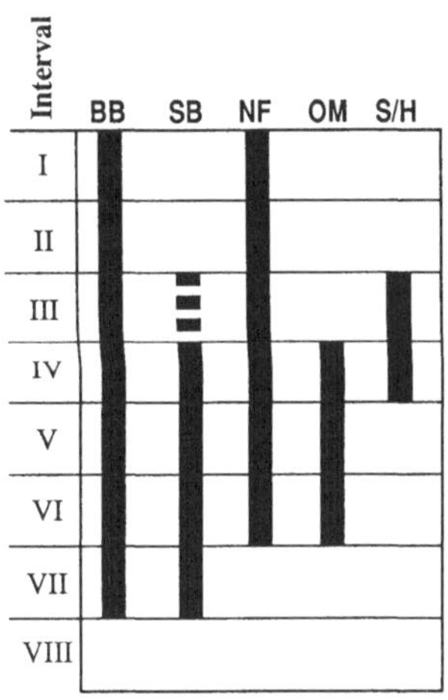

LOCUS	BB	SB	NF	OM	S/H
DXS164	–	NT	–	NT	+
DXS142	–	NT	–	NT	+
DXS84	–	+	–	+	+
DXS141	–	NT	–	+	–
CGD cDNA	–	–	–	–	+
DXS140	–	–	–	–	+
OTC	+	+	NT	NT	NT
DXS7	+	+	NT	NT	NT

Figure 2 Deletion analysis and RP3. Schematic diagram showing the extent of deletions in 5 patients, BB (Francke et al., 1985; Kunkel et al., 1985), SB (de Saint-Basile et al., 1988), NF (Royer-Pokora et al., 1986), OM and S/H (Bertelson et al., 1988) (S/H are first cousins with McLeod phenotype and have identical deletions). Intervals I to VIII contain the following loci: Duchenne muscular dystrophy (DMD) (I), DXS84 (II), DXS141 (III), McLeod phenotype (XK) (IV), chronic granulomatous disease (CGD) (V), DXS140 (VI), RP3 (VII) and OTC and DXS7 (VIII). NF and OM do not have RP, indicating that the RP3 gene is proximal to CGD, while deletions in OM and S/H define the smallest region of overlap that must contain the McLeod locus, distal to CGD. The table shows the presence or absence of various markers in DNA from these patients, summarised from the references detailed above. NT = not tested.

The RP3 gene may be associated with one of the other CpG islands, but there could be other candidate genes between CYBB and the BB breakpoint that do not have CpG islands. It is also possible that part of the RP3 gene is situated proximal to the BB breakpoint and that BB was only deleted for the terminal part of it. Alternatively, the deletion in that patient may have caused a position effect, such that a change in the chromatin structure influenced the expression of the RP3 gene, even though it lay outside the deletion. However, the patient BB suffered from a very severe form of RP, suggesting that a complete loss of gene function was more likely than reduced expression due to a chromosomal position effect.

RP2

The RP2 locus has been mapped by linkage analysis alone, since no associated cytogenetic abnormalities have been reported. As mentioned above, Ott et al. (1990) localised RP2 to a broad region extending from DXS7 to the centromere. More recent data have refined the localisation of RP2 to a position proximal to DXS7 and distal to DXS255 (Friedrich et al., 1992; Aldred et al., submitted). However, this still represents a region of 13–18cM (Mahtani et al., 1991). Coleman et al. (1990) suggested that RP2 is localised distal to DXS426 on the basis of a recombinant in one family. However, that family has since been shown also to be linked to markers in the RP3 region (L. Hardwick, personal communication) and therefore is not unambiguously RP2-type.

One of the main problems with the genetic mapping of RP2 is the difficulty in identifying a homogeneous subset of families to study. Bayesian probabilities of RP2 versus RP3 can be calculated for individual families (Wright et al., 1991; van Dorp et al., 1992), but the posterior probabilities are very sensitive to the assumed locations of these loci, which in the case of RP2 is not accurately known.

The most robust method of classifying families is to inspect the haplotypes and identify recombinant meioses. Those families that have recombination events localising the gene relative to OTC and/or DXS7 can be unambiguously classified, as illustrated in Figure 3. However, only a very small number of families can be classified in this manner. A recent multi-point analysis of seven such RP2 families that includes both new and previously published data indicates that the most likely localisation for RP2 is proximal to DXS7 and distal to DXS255, with the maximum lod score at the TIMP locus (Aldred et al., submitted). This compares well with the data from a heterogeneity analysis of 37 predominantly British xlRP families, in which the most likely location for RP2 was 3.5 cM distal to TIMP (Teague et al., 1994).

There is still some scope for refining the localisation of RP2 further by linkage analysis, but given the classification difficulties outlined above, the resolution of the genetic localisation is unlikely to be better than 5 cM. Further mapping studies will increasingly rely on physical mapping techniques, such as the development of a YAC contig spanning the region. Candidate genes could be identified by searching for CpG islands or for sequences with cross-species homology. However, detailed physical mapping of a 5 Megabase (Mb) region is a major undertaking. An alternative approach is to identify expressed sequences from the region of interest. Human-specific cDNAs could be cloned from a YAC introduced into a rodent cell line (Eliceiri et al., 1991), or alternatively from a human-rodent somatic cell hybrid by subtractive hybridisation or by cDNA synthesis from unprocessed heterogeneous nuclear RNA (Sive and St John, 1988; Liu et al., 1989; Corbo et al., 1990).

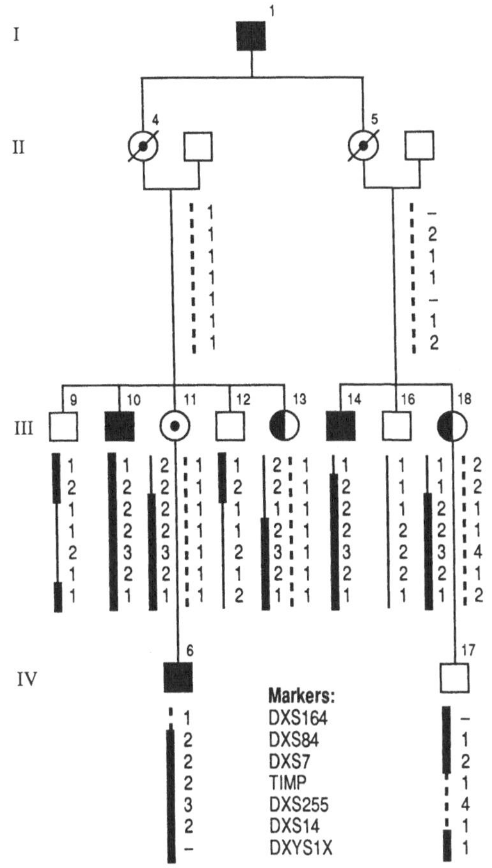

Figure 3 Distinguishing between RP2 and RP3 types of xlRP by haplotype analysis. Sections of two xlRP families illustrate the use of haplotype analysis to determine which locus is segregating. In (a), unaffected male IV_{17} has the affected haplotype at DXS7 and more distal markers, indicating that this is an RP2 family. This is supported by another recombinant in individual III_{13}, a female with phenotypic evidence of the carrier state. A graph of the multipoint linkage data from this family confirms a broad peak in the RP2 region (Wright et al., 1991).

The search would be greatly facilitated by the identification of a cytogenetic abnormality associated with RP2, which might reduce the region of search to less than 1 Mb. However, none have been identified. Aldred et al. (1994) have described a family in which X-linked RP co-segregates with mental retardation and in which linkage analysis indicates that it may be RP2-type. This raises the possibility that there is a sub-microscopic deletion involving the RP2 gene in this family. If this proves to be the case, the deletion could significantly reduce the region of search for the RP2 gene, as with RP3.

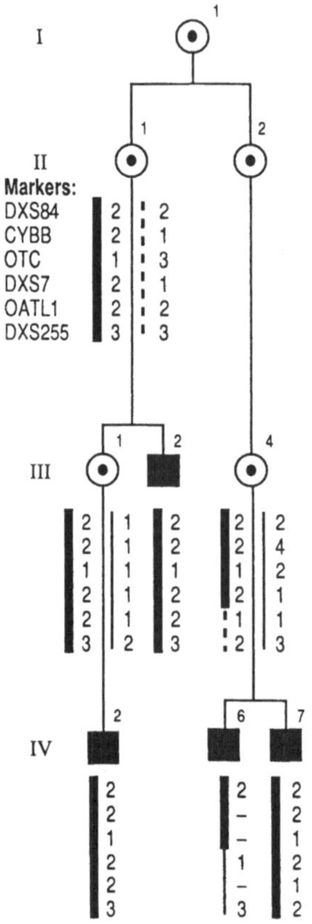

Figure 3 (b) shows the reverse situation, in which affected male IV_6 is recombinant with DXS7 and DXS255, indicating that the disease locus is distal to DXS7, consistent with an RP3 localisation. It should be noted that the majority of families (86% in this laboratory) do not show recombinants that allow such unambiguous classification. While probabilities can be calculated, assignment difficulties remain one of the major problems in research and counselling in xlRP.

PROBLEMS CAUSED BY HETEROGENEITY

A number of problems arise from genetic heterogeneity in RP and in the X-linked form in particular. The different Mendelian forms cannot be distinguished on clinical grounds alone and the pedigree structure often fails to provide unequivocal evidence of the mode of inheritance. Linkage analysis can sometimes resolve this and establish whether the disease is X-linked or autosomal dominant, but genetic heterogeneity within the X-linked group is a further complication. Many X-linked families cannot be unambiguously assigned to one or the other locus as they lack the recombinants necessary to make a clear distinction. The difficulty that this raises in mapping studies has been discussed above, but it is also a major problem in genetic counselling. In some cases, this can be circumvented by examining polymorphic markers spanning the entire RP3/RP2 interval (Bergen et al., 1991). Provided that there is no recombination in this interval, which would be true in about 70% of cases (Aldred et al., submitted), prenatal diagnosis, presymptomatic tests and carrier identification may be possible without determining which locus is involved.

GENOTYPE–PHENOTYPE CORRELATIONS IN XLRP

There are no unequivocal phenotypic differences between RP2 and RP3 that allow them to be distinguished on clinical grounds alone. However, clinical studies in families that can be distinguished genetically have led to some tentative correlations between genotype and phenotype. A tapetal reflex, when reliably observed in xlRP carriers, seems to correlate with an RP3 localisation (Nussbaum et al., 1985; Denton et al., 1988; Curtis and Blank, 1989; Musarella et al., 1989b), but the absence of a tapetal reflex does not exclude this locus. Wright et al. (1991) reported that the presence of moderate to high myopia is a characteristic feature of the RP2-type disease in the affected males of one large kindred, but it was not clear whether this was related to the primary disease process or was secondary to severity or early visual dysfunction. Kaplan et al. (1992) have found that if xlRP families were classified according to the onset symptom in affected males, families showing night blindness as the initial symptom at about age 10 years were linked to the RP3 region, while families with myopia presenting at the age of about 3 years showed linkage to RP2. Further clinical studies are required to confirm this proposed distinction, but the confounding effect of allelic differences within each genetic type may make it impossible to clarify the issue until the genes are isolated.

WHAT ARE THE RELATIVE PROPORTIONS OF THE RP3 AND RP2-TYPE OF XLRP?

Since it is difficult to distinguish the two xlRP loci, both clinically and genetically, the answer to this interesting and important question remains somewhat elusive. In the multicentre analysis of Ott et al. (1990), 75% of families were RP3-type with confidence limits extending from 45% to 90%. However, the relative proportions varied between different investigators. The highest proportion of RP2 was found in families contributed from this laboratory, whereas Musarella et al. (1988 and 1990) found no evidence of the RP2 locus amongst 20 families. This raised the possibility that RP2 might be more common in Britain than elsewhere.

To investigate this further, published families that included linkage data for two or more markers between DXS84 and DXS255 inclusive were reviewed. Where possible, they were tentatively classified as RP2 or RP3-type on the basis of haplotypes and lod scores. As detailed above, families showing recombination with OTC and/or DXS7 can be unambiguously classified. In addition, for this study, families that showed one or more recombinants with DXS84 but were linked to the RP2 region (TIMP-DXS255) were classified as possible RP2 families and vice versa. Those families showing no recombination, or recombination in both regions, and those with recombinants that only involved more proximal or distal markers were not classified. Families which were neither RP2 nor RP3 (four from this laboratory and one reported by Musarella et al. (1988 and 1990)) were excluded.

The results (Table 2) show that overall, 50 of the 83 families (60%) can be classified. Comparing Britain with the combined data from the rest of Europe and Scandinavia, the proportions of classified families are very similar, as are the relative proportions of RP2 and RP3 amongst the classified families. Thus there is no evidence that RP2 is more common in Britain than the rest of Europe on the basis of this approximate classification.

The combined British and European data suggest that 38% of the 34 classified families are RP2-type. This is almost twice the proportion of RP2 found in Australia (22%) and nearly three times that in America (14%). RP2 may therefore be more common in Europe than elsewhere, but the number of classified families from America and Australia is small (n = 16), so it is not clear whether this represents a true difference.

Determining the relative proportions of the two genes and whether there is variation between countries is an important issue with implications for genetic counselling. Overall, 32% of families in this study were RP2-type, agreeing with Teague et al. (1994) who found 30% RP2 in a heterogeneity analysis of 37 predominantly British families. However, given the difficulties in categorising families and the small sample sizes, the question of ethnic differences may not be resolved until one of the genes is cloned.

RP2, XL-CSNB AND MYOPIA

It is interesting that X-linked congenital stationary night blindness (XL-CSNB) maps to the same region as RP2 (Gal et al., 1989; Bech-Hansen et al., 1990, 1992; Musarella et al., 1989a, 1992; Aldred et al., 1992). Both XL-CSNB and RP2 are associated with moderate to high myopia and night blindness, raising the possibility that the two conditions could be allelic (Wright, 1990). Perhaps some mutations could cause rod cell death and progressive retinal degeneration, as seen in RP, while other mutations in the same gene might produce a stationary disorder with non-functional rod cells but no photoreceptor death.

Table 2 Proportions of RP2 and RP3 in published xlRP data

Nationality	*Total no. classified*	*Total no. unclassified*	*No. RP2 families*	*No. RP3 families*	*Reference*
Danish	1	0	1	0	Friedrich et al. (1985)
Danish	1	0	1	0	Friedrich et al. (1992)
Swedish	3	0	2	1	Dahl et al. (1991)
German	1	0	0	1	Meitinger et al. (1989)
French	5	4	2	3	Kaplan et al. (1992)
Dutch	1	0	0	1	Wirth et al. (1988)
Dutch	6	8	2	4	Bergen et al. (in press)
Dutch	1	0	0	1	van Dorp et al. (1992)
Italian	1	0	0	1	footnote 3
Total Europe excl. Britain	20	12 (38%)[1]	8 (40%)[2]	12 (60%)	
British	13	8	5	8	Wright et al. (1987,1991); Aldred et al. (submitted);&[3]
British	1	0	0	1	Curtis and Blank (1989)
Total Britain	14	8 (36%)	5 (36%)	9 (64%)	
Mixed origin	5	13	0	5	Musarella et al. (1988, 1989b, 1990)
Latin-American	1	0	0	1	Nussbaum et al. (1985); Musarella et al. (1989b)
American	1	0	1	0	Litt et al. (1987)
Total America	7	13 (65%)	1 (14%)	6 (86%)	
Australian[4]	9	0 (0%)	2 (22%)	7 (78%)	Denton et al. (1988); Wirth et al. (1988); Chen et al. (1989)
Combined Totals	50	33 (40%)	16 (32%)	34 (68%)	

[1]Expressed as a percentage of total families.
[2]Expressed as a percentage of all classified families.
[3]And additional unpublished data from this laboratory.
[4]All British origin.

It has been suggested that there are closely linked genes for CSNB and myopia on the X-chromosome and that hyperopic XL-CSNB patients result from crossing-over between these genes (Khouri et al., 1988). However, although X-linked myopia has been documented (Bartsocas and Kastrantas, 1981), a linkage study of a family with two cousins affected by XL-CSNB, one myopic and one hyperopic, showed no evidence of exchange of flanking markers (Dry et al., 1993). In both CSNB and RP, myopia may be a pleiotropic effect of the mutation, perhaps influencing the growth of the eye pre- and/or post-natally. Since ocular refraction is a multifactorial trait, the variation in refraction observed in RP and CSNB could be explained by modulation of the disease phenotype by the genetic background. Alternatively, myopia might be secondary to the defective vision. It has been shown that neonatal lid fusion in monkeys causes the eye to enlarge and become myopic (Wiesel and Raviola, 1977). It is therefore possible that the early visual defect in CSNB or RP causes myopia. This might explain why Miyake et al. (1986) found high myopia to be associated with a total lack of rod function in CSNB, while those with some residual function tended to have smaller refractive errors, and why patients with non-X-linked RP have lower myopia that those with the more severe X-linked disease (Sieving and Fishman, 1978). However, the question of allelic relationships between these disorders will not be answered definitively until the genes are isolated.

CONCLUSION

Following the identification of genetic heterogeneity in X-linked retinitis pigmentosa, genetic mapping studies have also revealed extensive genetic heterogeneity in autosomal forms of the disease. Considerable progress has been made towards the identification of the RP3 gene, located in the Xp21.1 region, due largely to the availability of patients with deletions in this region. The RP3 type appears to be the most common form of xlRP, affecting approximately two-thirds of families that can be classified, although there may be ethnic differences. The RP2 gene was localised by linkage analysis to Xp11.4-p11.2 and has been shown to lie in the DXS7-DXS255 interval. It has been more difficult to refine this localisation because of genetic heterogeneity and the absence of chromosomal rearrangements affecting this locus.

ACKNOWLEDGEMENTS

We are grateful to the British Retinitis Pigmentosa Society, the National Retinitis Pigmentosa Foundation and George Gund Foundation for financial support.

REFERENCES

Aldred M. A., Dry K. L., Sharp D. M., Van Dorp D. B., Brown J, Hardwick L. J., Lester D. H., Pryde F. E,, Teague P. W., Jay M, Bird A. C., Jay B, Wright A. F. (1992). Linkage analysis in X-linked congenital stationary night blindness. *Genomics* **14**, 99–104.

Aldred M. A., Dry K. L., Knight-Jones E. B., Hardwick L. J., Teague P. W., Lester D. H., Brown J., Spowart G., Carothers A.D., Raeburn J. A., Bird A. C., Fielder A. R., Wright A. F. (1994). Genetic analysis of a kindred with x-linked mental handicap and retinitis pigmentosa. *Am. J. Hum. Genet.* (in press).

Aldred M. A., Bergen A. A. B., Jay M, Teague P. W., Bhattacharya S. S., Bird A. C. And Wright A. F. (Submitted). Multipoint linkage analysis of the RP2-type of X-linked retinitis pigmentosa.

Aldred M. A., Hardwick L. J., Dry K. L., Bundey S. and Wright A. F. (submitted). Prenatal diagnosis of X-linked retinitis pigmentosa.

Baehner R. L., Kunkel L. M., Monaco A. P., Haines J. L., Conneally P. M., Palmer C., Heerma N., Orkin S. H. (1986). DNA Linkage analysis of X-chromosome-linked chronic granulomatous disease. *Proc. Natl. Acad. Sci.* **83**, 3398–3401.

Bartsocas C. S., Kastrantas A. D. (1981). X-linked form of myopia. *Hum. Hered.* **31**, 199–200.

Bech-Hansen N. T., Field L. L., Schramm A. M., Reedyk M., Craig I. W., Fraser N. J., Pearce W. G. (1990). A locus for X-linked congenital stationary night blindness is located on the proximal portion of the short arm of the X chromosome. *Hum. Genet.* **84**, 406–408.

Bech-Hansen N. T., Moore B. J., Pearce W. G. (1992). Mapping of locus for X-linked congenital stationary night blindness (CSNB1) proximal to DXS7. *Genomics* **12**, 409–411.

Bergen A. A.B., Platje Ejm, Craig I., Bakker E., Bleeker-Wagemakers E. M., Van Ommen G. J. B. (1991). Carrier detection in X-linked retinitis pigmentosa by multipoint DNA analysis. Problems due to genetic heterogeneity. *Ophthalmic Paediatrics And Genetics* **12**, 99–103.

Bergen A. A.B., Van Den Born L. I., Schuurman E. J. M., Diergaarde P. J., Pinckers A. J. L.G., Bakker E., Van Ommen G. J. B., Bleeker-wagemakers E. M. (in Press). *DNA recombination analysis and clinical evaluation of genetically heterogeneous X-linked retinitis pigmentosa: X1RP family typing.*

Bertelson C. J., Pogo A.O., Chaudhuri A, Marsh W. L., Redman C. M., Banerjee D., Symmans W. A., Simon T., Frey D., Kunkel L. M. (1988). Localisation of the Mcleod locus (XK) Within Band Xp21 by deletion analysis. *Am. J. Hum. Genet.* **42**, 703–711.

Bhattacharya S. S., Wright A. F., Clayton J. F., Price W. H., Phillips C. I., Mckeown C. M. E., Jay M., Bird A. C., Pearson P. L., Southern E.M., Evans H. J. (1984). Close genetic linkage between X-linked retinitis pigmentosa and a restriction fragment length polymorphism identified by recombinant DNA probe L1.28. *Nature* **309**, 253–255.

Bird A (1975). X-linked retinitis pigmentosa. *Brit. J. Ophthalmol.* **59**: 177–199.

Boughman J. A., Conneally P. M., Nance W. E. (1980). Population genetic studies of retinitis pigmentosa. *Am. J. Hum. Genet.* **32**, 223–235.

Bruns G. A. P., Posner D., Reichel E., Koenig M., Berson E. L., Kunkel L. M. (1988). A probe for the Xp21 retinitis pigmentosa (RP3) locus. *Am. J. Hum. Genet.* (Suppl.) **43**, A178.

Bundey S., Crews S. J. (1984a). A study of retinitis pigmentosa in the city of Birmingham. I: Prevalence. *J. Med. Genet.* **21**, 417–420.

Bundey S., Crews S. J. (1984b). A study of retinitis pigmentosa in the city of Birmingham. II: Clinical and genetic heterogeneity. *J. Med. Genet.* **21**, 421–428.

Chen J. D., Halliday F., Keith G., Sheffield L., Dickinson P., Gray R., Constable I., Denton M. (1989). Linkage heterogeneity between X-linked retinitis pigmentosa and a map of 10 RFLP Loci. *Am. J. Hum. Genet.* **45**, 401–411.

Clarke A., Roberts S. H., Thomas N. S. T, Whitfield A., Williams J., Harper P. S. (1986). Duchenne muscular dystrophy with adrenal insufficiency and glycerol kinase deficiency: high resolution cytogenetic analysis with molecular, biochemical and clinical studies. *J. Med. Genet.* **23**, 501–508.

Coleman M. P., Bhattacharya S. S., Lindsay S., Wright A. F., Jay M., Litt M., Craig I. W., Davies K. E. (1990). Localization of the microsatellite probe DXS 426 between DXS 7 and DXS 255 On Xp and linkage to X-linked retinitis pigmentosa. *Am. J. Hum. Genet.* **47**, 935–940.

Corbo L., Maley J. A., Nelson D. L., Caskey C. T. (1990). Direct cloning of human transcripts with HNrNA from hybrid cell lines. *Science* **249**, 652–655.

Curtis D., Blank C. E. (1989). Recombination between DXS7, DXS84 and a rare form of x-linked retinitis pigmentosa (Mck-30320). *Hum. Genet.* **81**, 200–202.

Dahl N., Sundvall M., Pettersson U., Andréasson S., Anvret M., Kugelberg U., Hagbyhn-Gericke A., Goonewardena P. (1991). Genetic mapping of loci for X-linked retinitis pigmentosa. *Clinical Genetics* **40**, 435–440.

De Saint-Basile G., Bohler M. C. Fischer A., Cartron J, Dufier J. L., Griscelli C., Orkin S. H. (1988). Xp21 DNA Microdeletion in a patient with chronic granulomatous disease, retinitis pigmentosa, and mcleod phenotype. *Hum. Genet.* **80**, 85–89.

Denton M. J., Chen J. D., Serravalle S., Colley P., Halliday F. B., Donald J. (1988). Analysis of linkage relationships of x-linked retinitis pigmentosa with the following Xp loci: L1.28, OTC, 754, XJ-1.1, PERT87, and C7. *Hum. Genet.* **78**, 60–64.

Deutman A. F., (1977). Rod-cone dystrophy: primary, hereditary pigmentary retinopathy, retinitis pigmentosa. In: *Hereditary Retinal and Choroidal Diseases.* Volume II. Clinical Characteristics (AE Krill, ed.) pp479–576, Harper & Row Publishers Inc., Hagerstown, USA.

Dry K. L., Van Dorp D. B., Aldred M. A., Brown J., Hardwick L. J., Wright A. F. (1993). Linkage analysis in a family with complete type congenital stationary night blindness with and without myopia. *Clin. Genet.* **43**, 250–254.

Eliceiri B., Labella T., Hagino Y., Srivastava A., Schlessinger D., Pilia G., Palmieri G., D'urso M. (1991). Stable integration and expression in mouse cells of yeast artificial chromosomes harbouring human genes. *Proc. Natl. Acad. Sci.* **88**, 2179–2183.

Fishman G. A. (1978). Retinitis Pigmentosa. genetic percentages. *Arch. Ophthalmol.* **96**, 822–826.

Fishman G. A., Weinberg A. B., Mcmahon T. T. (1986). X-linked recessive retinitis pigmentosa. Clinical characteristics of carriers. *Arch. Ophthalmol.* **104**, 1329–1335.

Francke U., Ochs H. D., De Martinville B., Giacalone J., Lindgren V., Distéche C., Pagon R. A., Hofker M. H., Van Ommen G. J. B, Pearson P. L. , Wedgwood R. J. (1985). Minor Xp21 chromosome deletion in a male associated with expression of Duchenne muscular dystrophy, chronic granulomatous disease, retinitis pigmentosa and McLeod syndrome. *Am. J. Hum. Genet.* **37**, 250–267.

Francke U., Harper J. F., Darras B. T., Cowan J. M., McCabe E. R. B., Kohlschütter A., Seltzer W.K., Saito F., Goto J., Harpey J. P., Wise J. E. (1987). Congenital adrenal hypoplasia, myopathy, and glycerol kinase deficiency: molecular genetic evidence for deletions. *Am. J. Hum. Genet.* **40**, 212–227.

Fraser G. R., Friedmann A. I. (1967). *The causes of blindness in childhood.* Johns Hopkins Press, Baltimore, USA.

Friedrich U., Warburg M., Wieacker P., Wienker T. F., Gal A., Ropers H. H. (1985). X-linked retinitis pigmentosa: linkage with the centromere and a cloned DNA sequence from the proximal short arm of the X chromosome. *Hum. Genet.* **71**, 93–99.

Friedrich U., Warburg M., Kruse T. A., Andréasson S. (1992). X-linked retinitis pigmentosa: New map studies of xlRP2, and a possible human centromere effect. *Hum. Genet.* **88**, 683–687.

Gal A., Schinzel A., Orth U., Fraser N. A., Mollica F., Craig I. W., Kruse T., Mächler M., Neugebauer M., Bleeker-Wagemakers L. M. (1989). Gene of X-chromosomal Congenital stationary night blindness is closely linked to DXS 7 on Xp. *Hum. Genet.* **81**, 315–318.

Green J. S., Johnson G. J. (1983). Hereditary diseases as causes of blindness in Newfoundland: preliminary report. *Can. J. Ophthalmol.* **18**, 281–284.

Green J. S., Bear J. C., Johnson G. J. (1986). The burden of genetically determined eye disease. *Brit. J. Ophthalmol.* **70**, 696–699.

Heckenlively J. R.(1988). *Retinitis Pigmentosa.* J. B. Lippincott Co., Philadelphia, USA.

Heckenlively J. R. (1990). Hereditary retinal and choroidal degenerations. In: *Principles And Practice of Medical Genetics,* 2nd Edition, Volume 1 (AEH Emery, D. L. Rimoin, eds) pp683–703, Churchill Livingstone, Edinburgh, Uk.

Ho M. F., Monaco A. P., Blonden L. A. J., Van Ommen G. J. B., Affara N. A., Ferguson-Smith M. A., Lehrach H. (1992). Fine mapping of the McLeod locus (XK) to a 150–380ÿkb region in Xp21. *Am. J. Hum. Genet.* **50**, 317–330.

Jay M. (1982). On The heredity of retinitis pigmentosa. *Br. J. Ophthalmol.* **66**, 405–416.

Kaplan J., Bonneau D., Frézal J., Munnich A., Dufier J. l. (1990). Clinical and genetic heterogeneity in retinitis pigmentosa. *Hum. Genet.* **85**, 635–642.

Kaplan J., Pelet A., Martin C., Delrieu O., Aymé S., Bonneau D., Briard M. L., Hanauer A., Larget-piet L., Lefrançois P., Michel-Awad A., Plauchu H., Dufier J. L., Frézal J., Munnich A. (1992). Phenotype-genotype correlations in X-linked retinitis pigmentosa. *J. Med. Genet.* **29**, 615–623.

Khouri G., Mets M. B., Smith V. C., Wendell M., Pass A. S. (1988). X-linked congenital stationary night blindness. review and report of a family with hyperopia. *Arch. Ophthalmol.* **106**, 1417–1422.

Kunkel L. M., Monaco A. P., Middlesworth W., Ochs H. D., Latt S. A. (1985). Specific cloning of DNA fragments absent from the DNA of a male patient with an X-chromosome deletion. *Proc. Natl. Acad. Sci.* **82**, 4778–4782.

Litt M., Weleber R. G., Lovrien E.W., Litt R. H., Payne N., Murphey W., Dubay C., Willard H. F. (1987). A Family study of X-linked retinitis pigmentosa. *Cytogenet. Cell Genet.* **46**, 651.

Liu P., Legerski R., Siciliano M. J. (1989). Isolation of human transcribed sequences from human-rodent somatic cell hybrids. *Science* **246**, 813–815.

Lyon M. F. (1961). Gene action in the X-chromosome of the mouse (Mus Musculus L.). *Nature* **190**, 372–373.

Mahtani M. M., Lafreniére R. G., Kruse T. A., Willard H. F. (1991). An 18-locus linkage map of the pericentromeric region of the human X-chromosome: genetic framework for mapping X-linked disorders. *Genomics* **10**, 849–857.

McDowell C., Burghes A. H., Anson-Cartwright L., Coulson S., Worton R. G., Musarella M. A. (1990). X-linked retinitis pigmentosa (xlRP): Mapping of the gene to Xp21, pulsed field electrophoresis (PFGE) of the region and cloning strategies. *Am. J. Hum. Genet.* (Suppl.) **47**, A256.

Meitinger T., Fraser N. A., Lorenz B., Zrenner E., Murken J., Craig I. W. (1989). Linkage of X-linked retinitis pigmentosa to the hypervariable DNA marker M27 β (DXS255). *Hum. Genet.* **81**: 283–286.

Miyake Y., Yagasaki K., Horiguchi M., Kawase Y., Kanda T. (1986). Congenital stationary night blindness with negative electroretinogram. *Arch. Ophthalmol.* **104**, 1013–1020.

Musarella M. A., Burghes A., Anson-Cartwright L., Mahtani M. M., Argonza R., Tsui L. -C., Worton R. (1988). Localisation of the gene for X-linked recessive type of retinitis pigmentosa (xl-RP) to Xp21 by linkage analysis. *Am. J. Hum. Genet.* **43**, 484–494.

Musarella M. A., Weleber R. G., Murphey W.H., Young R. S. L., Anson-Cartwright L., Mets M., Kraft S. P., Polemeno R., Litt M., Worton R. G. (1989a). Assignment of the gene for complete X-linked congenital stationary night blindness (CSNB1) to Xp11.3. *Genomics* **5**, 727–737.

Musarella M. A., Anson-Cartwright L., Burghes A., Worton R. G., Lesko J. G., Nussbaum R. L. (1989b). Linkage analysis of a large Latin-American family with X-linked retinitis pigmentosa and metallic sheen in the heterozygote carrier. *Genomics* **4**, 601–605.

Musarella M. A., Anson-Cartwright L., Leal S. M., Gilbert L. D., Worton R. G., Fishman G. A., Ott J. (1990). Multipoint linkage analysis and heterogeneity testing in 20 X-linked retinitis pigmentosa families. *Genomics* **8**, 286–296.

Musarella M. A., Anson-Cartwright C. L., McDowell C., Burghes A. H. M., Coulson S. E., Worton R. G., Rommens J. M. (1991). Physical mapping at a potential X-linked retinitis pigmentosa locus (RP3) by pulsed field gel electrophoresis. *Genomics* **11**, 263–272.

Musarella M. A,, Kirchgessner C., Trofatter J., Miyakcy, Degennaro L. (1992). Assignment of the gene for incomplete congenital stationary night blindness (CSNB2) to proximal Xp. *Invest. Ophthalmol. Vis. Sci.* **33**, 792.

Nussbaum R. L., Lewis R. A., Lesko J. G., Ferrell R. (1985). Mapping X-linked ophthalmic dieases: ii. linkage relationship of X-linked retinitis pigmentosa to X chromosomal short arm markers. *Hum. Genet.* **70**, 45–50.

Ott J., Bhattacharya S., Chen J. D., Denton M. J., Donald J., Dubay C., Farrar G. J, Fishman G. S. A., Frey D., Gal A., Humphries P., Jay B., Jay M., Litt M., Mächler M., Musarella M., Neugebauer M., Nussbaum R. I., Terwilliger J. D., Weleber R. G., Wirth B., Wong F., Worton R. G., Wright A. F. (1990). Localizing multiple X chromosome-linked retinitis pigmentosa loci using multilocus homogeneity tests. *Proc. Natl. Acad. Sci.* **87**, 701–704.

Pagon R. A. (1988). Retinitis pigmentosa. *Surv. Ophthalmol.* **33**, 137–177.

Roux A. F., Yuan C., Musarella M. A. (1992). Identification of a (CA) sequence polymorphism in a gene cloned in the vicinity of the RP3 locus. *Am. J. Hum. Genet.* (Suppl.) **51**, A372.

Royer-Pokora B., Kunkel L. M., Monaco A. P., Goff S. C., Newburger P. E., Baehner R. L., Cole F. S., Curnutte J. T., Orkin S. H. (1986). Cloning the gene for an inherited human disorder – chronic granulomatous disease – on the basis of its chromosomal location. *Nature* **322**, 32–38.

Sieving P. A., Fishman G. A.(1978). Refractive errors of retinitis pigmentosa patients. *Brit. J. Ophthalmol.* **62**, 163–167.

Sive H. L., St John T. (1988). A simple subtractive hybridisation technique employing photoactivatable biotin and phenol extraction. *Nucl. Acids Res.* **16**, 10937.

Teague P. W., Aldred M. A., Jay M., Bundey S., Dempster M., Harrison C., Carothers A. D., Hardwick L. J., Strain L., Brock D. J. H., Van Dorp D. B., Carbonara A., Jay B., Bird A. C., Bhattacharya S. S., Evans H. J., Wright A. F. (1994). Heterogeneity analysis in 40 X-linked retinitis pigmentosa families. *Am. J. Hum. Genet.* (in press).

Van Dorp D. B., Wright A. F., Carothers A. D., Bleeker-Wagemakers E. M. (1992). A Family With RP3 type of X-linked retinitis pigmentosa: an association with ciliary abnormalities. *Hum. Genet.* **88**, 331–334.

Van Osch L., Van Schooneveld M., Bleeker-Wagemakers E. M.(1990). Golden tapetal reflex in male patients with X-linked retinitis pigmentosa. *Ophthalmic Paediatrics and Genetics* **4**, 287–291.

Warburg M. (1990). Congenital blindness. In: *Principles and practice of medical genetics*, 2nd Edition, Volume 1 (AEH Emery, DL Rimoin, eds) pp631–636, Churchill Livingstone, Edinburgh, UK.

Wiesel T. N., Raviola E. (1977). Myopia and eye enlargement after neonatal lid fusion in monkeys. *Nature* **266**, 66–68.

Wirth B., Denton M. J., Chen J. D., Neugebauer M., Halliday F. B., Van Schooneveld M., Donald J., Bleeker-wagemakers E., Pearson P. L., Gal A. (1988). Two different genes for X-linked retinitis pigmentosa. *Genomics* **2**, 263–266.

Wright A. F. (1990). Towards the identification of genes in X-linked retinitis pigmentosa. *Prog. Retina Res.* **9**, 1–31.

Wright A. F., Bhattacharya S. S., Clayton J. F., Dempster M., Tippett P., Mckeown C. M. e., Jay M., Jay B., Bird A. C. (1987). Linkage Relationships between X-linked retinitis pigmentosa and nine short-arm markers: exclusion of the disease locus from XP21 and localisation to between DXS7 and DXS14. *Am. J. Hum. Genet.* **41**, 635–644.

Wright A. F., Bhattacharya S. S., Aldred M. A., Jay M., Carothers A. D., Thomas N. S. T., Bird A. C., Jay B., Evans H. J. (1991). Genetic localisation of the RP2 type of X-linked retinitis pigmentosa in a large kindred. *J. Med. Genet.* **28**, 453–457.

11. CONGENITAL STATIONARY NIGHT BLINDNESS: A CRITICAL REVIEW FOR MOLECULAR APPROACHES

ELISE HÉON *and* MARIA A. MUSARELLA

The Hospital for Sick Children, Research Institute, Departments of Ophthalmology and Genetics, 555 University Avenue, Toronto, Ontario M5G lX8, Canada

SUMMARY

In this chapter, the group of disorders referred to as congenital stationary night blindness (CSNB) is reviewed. The term CSNB will be used to describe a subgroup of inherited disorders characterized by non-progressive night blindness, normal retinal appearance and Mendelian inheritance. The understanding of CSNB requires knowledge of tests that measure the function of the retina in various conditions of light- and dark-adapted states. The principles of electrophysiological testing are reviewed. The reader is also introduced to the classifications of CSNB based on these tests. The chapter discusses two animal models for CSNB, linkage studies done on CSNB families, and current developments in the relationship of CSNB to Åland Island eye disease and Duchenne muscular dystrophy.

INTRODUCTION

Night blindness, or nyctalopia, is a symptom of a wide group of retinal disorders. It can be acquired or inherited, and if inherited, can be either progressive or non-progressive (stationary). Unfortunately, no uniform classification has been developed for the group of disorders known as stationary night blindness, and the nomenclature is often confusing. For instance, the French literature uses "hemeralopie" (Franceschetti, Babel & François, 1963) to describe night blindness (Larousse & Auge, 1974), whereas in English, hemeralopia refers to day blindness. This chapter focuses on a group of non-progressive inherited retinal disorders, present at birth, in which the retina has a normal appearance. This group is referred to as congenital stationary night blindness (CSNB).

Classifications for CSNB are provided that are useful in understanding the various aspects of these diseases. We also review the historical background, including the exceptionally well-documented pedigrees that have given CSNB a special place in human genetics. The fundamentals of electrodiagnostic tests are discussed, since they are crucial not only to the diagnosis but also to an understanding of the pathogenesis of CSNB. Finally, we provide a summary of the present state of knowledge of molecular studies carried out to map the genes for CSNB. Although some work has been done on the autosomal dominant CSNB (Yijian et al., 1991), most of the linkage studies have been conducted on the X-linked form (Volker-Dieben & Went, 1975; Musarella et al., 1989; Bech-Hansen et al., 1990; Schwartz & Rosenberg, 1991; Aldred et al., 1992; Bech-Hansen, Moore & Pearce, 1992; Musarella et al., 1992). The results suggest a single gene may be responsible for several clinical entities on the X-chromosome.

CLASSIFICATION OF INHERITED STATIONARY NYCTALOPIA BASED ON FUNDUS APPEARANCE

We define any non-progressive night blindness that is inherited and present at birth as inherited stationary nyctalopia. This category is further divided into two groups depending on the appearance of the retinal eyegrounds (also known as the fundus). The first group includes patients with a normal fundus, who are classified as having congenital stationary night blindness (CSNB), a term initially used by Nettleship (1912) and later by Carr (Carr & Siegel, 1964; Carr, 1974) to designate a congenital non-progressive nyctalopia with normal or abnormal fundus. To avoid confusion, in this chapter CSNB will only be used to define a group of inherited retinal disorders characterized by non-progressive night blindness present at birth, normal daylight vision and visual fields, normal appearance of the fundus, and specific results on electroretinographic (ERG) and psychophysical testings. CSNB may be inherited through one of the three modes of Mendelian inheritance (Table 1).

Table 1 Inherited stationary night blindness. Classification by fundus appearance

Normal fundus	*Abnormal fundus*
Congenital stationary night blindness	Oguchi's disease (autosomal dominant)
• Autosomal dominant	Fundus albipunctatus (autosomal recessive, autosomal dominant uncommon)
• Autosomal recessive	
• X-linked recessive	Flecked retina of Kandori (autosomal recessive?)

The second group, inherited stationary nyctalopia associated with abnormal fundus appearance (Carr, 1974; Vaghefi et al., 1978; Jimenez-Sierra, Ogden & Van Boemel, 1989), includes three specific disorders that will be briefly discussed for the sake of completeness: Oguchi's disease, fundus albipunctatus and flecked retina of Kandori. In Oguchi's disease (Oguchi, 1907), the fundus is characterized by a peculiar greenish-golden discoloration. After prolonged dark adaptation (2-3 hrs), this metallic sheen disappears, leaving a perfectly normal-looking retina. This change in the retina is called the "Mizuo phenomenon" (Mizuo, 1913). If the fundus shows flecks, a diagnosis of either fundus albipunctatus (small flecks) (Mooren, 1882; Lauber, 1910) or flecked retina of Kandori (large irregular flecks)

(Kandori, 1959; Kandori, Setogawa & Tamai, 1966; Kandori et al., 1972) should be considered.

INHERITANCE OF CSNB: HISTORICAL BACKGROUND

Three Mendelian forms of inheritance of CSNB have been described: autosomal dominant (Cuniers, 1838), autosomal recessive (Gassler, 1925) and X-linked recessive (Donders, 1855). The pedigrees for CSNB include some of the most extensive and well-documented in human genetics. CSNB has supplied us with one of the most complete genealogical records of an autosomal dominant disorder. R. Cuniers, a Belgian ophthalmologist (Cuniers, 1838), was the first to report this famous pedigree of the oldest family known to be affected with CSNB. It has been traced back to Jean Nougaret, a butcher, who lived in the latter part of the 17th century in the small French village of Vendemian. The pedigree was discovered only in the early 1830s, when a fifth-generation descendant of Nougaret claimed exemption from the French army because of poor night vision. The examining surgeon found that the young man could read by candlelight, thought he was malingering, found him fit for service and enrolled him in the army for 7 years. It was not until years later, when the man was re-examined, that he conveyed the familial nature of his night blindness and was finally exempted from military duty (Snyder, 1963; Krill, 1977).

In 1907, Nettleship published an updated pedigree of the Nougaret family and confirmed the stationary nature of their night blindness (Nettleship, 1907). This study encompassed nine generations and included 2121 persons, of whom 135 were affected (Carr, 1974). An attempt was made in 1949 to update this pedigree further (DeJean & Gassenc, 1949), but the two world wars and changing lifestyles had led to the dissolution of the original genetic isolate. Today this form of CSNB (dominantly inherited) is referred to as the "Nougaret variety" (Carr, 1974). Many other autosomal dominant CSNB pedigrees have been described (Stievenart, 1847; Sedan, 1885; Hudson, 1903; Bessonnet, 1904; Sinclair, 1905; Snell, 1907; Bordley, 1908; Carroll & Haig, 1952; Latte, Manunta & Pisano, 1954; Rambusch, 1954; Noble, Carr & Siegel, 1990; Yijian et al., 1991).

The first X-linked form of CSNB was reported by Donders (1855); the most famous pedigree of this form is that of Pfluger-Amman-Kleiner (1898). The first autosomal recessive form of CSNB was reported by Gassler in 1925.

CLINICAL MANIFESTATIONS

Because of the paucity of symptoms and ocular findings, CSNB is probably one of the most frequently misdiagnosed group of ocular disorders. The most common complaint is decreased night vision, but because night blindness is present at birth, patients are often unaware of their handicap until later in life. Since the most frequent sign is nystagmus (jerky eye movements), patients are often initially diagnosed as having ocular albinism or congenital nystagmus. It may be difficult to differentiate these conditions from CSNB without proper electrodiagnostic evaluation.

The phenotypic distinction of each pattern of inheritance is not strict and overlaps do occur. However, both visual acuity and refraction bear some relationship to inheritance. Visual function varies from normal acuity (20/20) to legal blindness (20/200). In autosomal dominant CSNB, impaired night vision is often the main defect (Cuniers, 1838). Visual acuity is usually within the normal range (20/20–20/40) with a mild refractive error (−0.50 D – +0.50 D) if any. Myopia is rarely seen. In the autosomal recessive phenotype, vision is usually moderately impaired (about 20/80 or better), and the myopia tends to be mild to moderate.

In males affected with X-linked CSNB, visual loss can show intra- and inter-familial variability (Pearce, Reedyk & Coupland, 1990). In these cases vision is usually abnormal, ranging from 20/30 to 20/200. Visual acuity of 20/20 appears to be rare in this subgroup. Although refractive error is usually of high myopia, it may range from mild hyperopia (+0.50) to severe myopia (up to −20 D) (Donders, 1855; Carroll & Haig, 1952; François, Verriest & De Rouck, 1965; Carr, 1974; Krill, 1977; Haim, 1985; Khouri et al., 1988; Fishman, 1990; Noble, Carr & Siegel, 1990; Pearce, Reedyk & Coupland, 1990).

Nystagmus is a common finding in patients born with subnormal vision (less than 20/60) (Krill, 1977). In most cases of CSNB the nystagmus is horizontal and may be fine or coarse. Fine rotary nystagmus has also been described (Merin et al., 1970). Strabismus is often present, but there is no predominance of esotropia or exotropia (Nettleship, 1912; Merin et al., 1970; Carr, 1974; Krill, 1977; Merin, 1991). Anisometropia (a difference in refractive error greater than 1.5 D between eyes) and astigmatism are usually mild (1.0–2.0 D). Amblyopia, which is not specific, may also be seen (Merin et al., 1970; Krill, 1977; Khouri et al., 1988; Pearce, Reedyk & Coupland, 1990).

Findings of the fundus examination are generally normal. However, when myopia is more than 4.0 D, the retina may show characteristic changes of myopia, such as a tigroid fundus and thinning of the retina, but blood vessel caliber is always normal and no pigment clumping is seen. Cases have been reported of the optic disc being tilted, pale or dysplastic (Mintz Hittner, Borda & Justice, 1981; Heckenlively, Martin & Rosenbaum, 1983). Carr (1966b) also describes an occasional increase in granularity of the macula, which can alter the normal foveal reflex. Barricks et al. (1977) described paradoxical pupillary responses, in which the pupil constricts in the dark, in three boys with CSNB of unspecified inheritance (Barricks, Flynn & Kushner, 1977). Since this phenomenon is not seen in older affected males (> 55 y.o.), it is considered a useful diagnostic finding in young children when present (Khouri et al., 1988).

Vitreous fluorophotometry, a useful tool in the evaluation of the blood-retinal barrier, was performed on three types of night blindness: Oguchi's disease, X-linked CSNB and fundus albipunctatus. All eyes showed normal values, strongly suggesting that the blood-retinal barrier is intact (Miyake et al., 1983).

Colour vision is usually unaffected in CSNB, but occasionally a mild tritan defect can be seen (Carroll & Haig, 1952; Armington & Schwab, 1954; Franceschetti, Babel & François, 1963; Miyake et al., 1986). Red-green deficiency has been reported, regardless of the pattern of inheritance (Merin et al., 1970; Ponte, Lodato & Lauricella, 1974; Miyake et al., 1986), and is not felt to be related specifically to CSNB but rather to an unrelated red-green colour gene defect located at Xq 28 (McKusick, 1990a).

Visual fields are usually normal when tested under lighted condition (photopic). However, if patients are tested in dim illumination, visual field constriction may be noted (Carr, 1974; Krill, 1977; Miyake et al., 1986).

Physical anomalies are usually not reported in patients affected with CSNB; however, Maeder (1946) reported a family of autosomal dominant CSNB in four generations where plantar syndactyly (fusion of two toes) was a regular finding. Unfortunately, no long-term follow-up of this family is available.

ELECTROPHYSIOLOGIC EVALUATION OF CSNB

The diagnosis of retinal diseases can be substantially aided by the use of electrophysiologic tests. Electrodiagnostic procedures used to study the retina will be briefly described as a preparation for detailed discussion of their importance in the diagnosis of CSNB.

Basics of Electroretinography

The electroretinogram (ERG) is a recording of a flash-evoked electrical potential that arises in the retina. The most effective stimulus is delivered by a full-field homogeneous illumination known as a Ganzfeld system. The electrical response is recorded from the cornea with a double-electrode contact lens, introduced by Riggs in 1941, and represents an objective measurement of total mass response of the retina. Localized diseases therefore may not show any changes on ERG testing. The output of this response is amplified and displayed on an oscilloscope. The normal ERG is made up of several components that can be separated, depending on the stimulus presented, by varying the intensity and/or wavelength as well as the frequency of the stimulus and the state of the retinal adaptation between light and dark (Fishman, 1985). These differences form the basis of tests designed to activate rods only (scotopic), both rods and cones (mesopic), or cones only (photopic) (Fishman, 1985; Ogden, 1989) (Figure 1). The ability to separate rods and cones permits many retinal disorders to be differentiated. Rod-specific responses can be elicited after a retina has been dark-adapted for 30 to 40 minutes with a single flash of blue light (usually less than 470 nm) or a very low intensity white light. A bright white flash produces responses from both cones

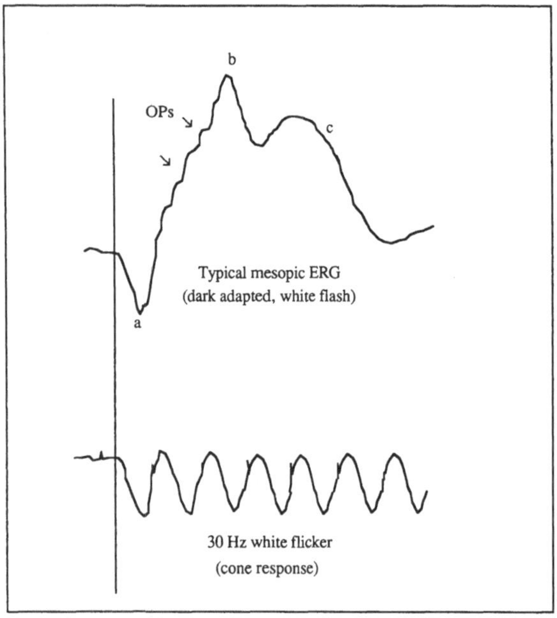

Figure 1 Normal ERG tracings.

and rods, as does a single red flash on a dark-adapted retina. Pure cone responses are obtained with either a red or white light flickering at a frequency of 30 cycles per second (Hz), or from a light-adapted retina following a single flash of yellow-red or blue-green stimulus (Fishman, 1985).

The three main components of the ERG (Einthoven & Jolly, 1908; Jimenez-Sierra, Ogden & Van Boemel, 1989) are the a-wave, b-wave and c-wave (Figure 1). The a-wave, the initial cornea-negative component, is generated by the photoreceptor layer. It is followed by a cornea-positive b-wave that arises from the inner nuclear layer (bipolar, Müller and amacrine cells). The c-wave, produced by the retinal pigment epithelial (RPE) cells, is a cornea-positive response that is not routinely used clinically (Fishman, 1985; Jimenez-Sierra, Ogden & Van Boemel, 1989). The ganglion cell layer does not contribute to the ERG.

The most frequent measurements taken with the ERG are the a- and b-wave amplitudes and the b-wave implicit time (time from flash to peak of b-wave). In CSNB, the b-wave is most severely affected, and in some forms of the condition, its amplitude can be reduced so drastically that it becomes negative. Several other components of the ERG are clinically useful. For example, the amplitudes of the oscillatory potentials (OPs) (Ogden, 1973), wavelets on the ascending limb of the b-wave, can be affected by diseases that are due to neurotransmission defects in the vicinity of the inner plexiform layer (Lachapelle, Little & Polomeno, 1983; Miyake, Yagasaki & Horiguchi, 1987c).

Electroretinography and CSNB

The ERG plays a key diagnostic role in CSNB. Characteristically, in any of the inherited stationary night blindness disorders, the scotopic or rod ERG is predominantly affected (Ponte, Lodato & Lauricella, 1974). However, in CSNB the dysfunction is not limited to the scotopic system (Goodman & Bornschein, 1957). The cone photoreceptors also show a variable degree of involvement (Auerbach, Godel & Rowe, 1969; Kawasaki, Tusuchida & Jacobson, 1971; Krill & Martin, 1971; Lachapelle, Little & Polomeno, 1983). Two patterns of ERG have been described for CSNB: the Schubert-Bornschein and the Riggs types (Table 2 and Figure 2).

Table 2 Congenital stationary night blindness. Classification by electroretinography (ERG)

Schubert–Bornschein ERG	*Riggs ERG*
Scotopic a-wave: normal to subnormal	Scotopic a-wave: normal to subnormal
Scotopic b-wave: reduced to absent, smaller than scotopic a-wave	Scotopic b-wave: near normal to reduced in amplitude, larger than scotopic a-wave
$b/a < 1$	$b/a > 1$
i. complete (no rod adaptation)	
ii. incomplete (rod adaptation present)	

The first authentic case of CSNB studied electrophysiologically was that of Schubert and Bornschein (Schubert & Bornschein, 1952; Bornschein & Vukovich, 1953) (Figure 3). The Schubert–Bornschein ERG is a negative type in which the amplitude of the a-wave (normal or slightly subnormal) is larger than that of a reduced b-wave (Vukovich, 1952; Bornschein & Vukovich, 1953). The Riggs-type ERG is characterized by a reduced scotopic b-wave that remains larger in amplitude than the a-wave (which is either normal or subnormal) (Riggs, 1954; Fishman, 1990). Although most cases of the Riggs type have frequently been associated with autosomal dominant CSNB and those of the Schubert–Bornschein type (negative b-wave ERG) described for all forms of CSNB inheritance, Auerbach et al. (1969) illustrated that the division into groups derived from the ERG patterns had no relation to heredity (Auerbach, Godel & Rowe, 1969). Miyake et al. (1986) subdivided the patients with Schubert–Bornschein ERG into two categories, complete and incomplete, based on the presence or absence of rod function on ERG and psychophysical dark adaptation (Miyake et al., 1986) (Table 3). The complete form of CSNB lacks rod function entirely, has no demonstrable scotopic rod-mediated ERG b-wave, and produces night blindness. Visual acuity is moderately poor and psychophysical dark adaptation of rods is absent. Patients affected by this condition are known to have moderate to severe myopia (mean ± SD; -7.9 ± 8.1 D), and oscillatory potentials are markedly reduced (Hill, Arbel & Berson, 1974; Heckenlively, Martin & Rosenbaum, 1983; Lachapelle, Little & Polom-

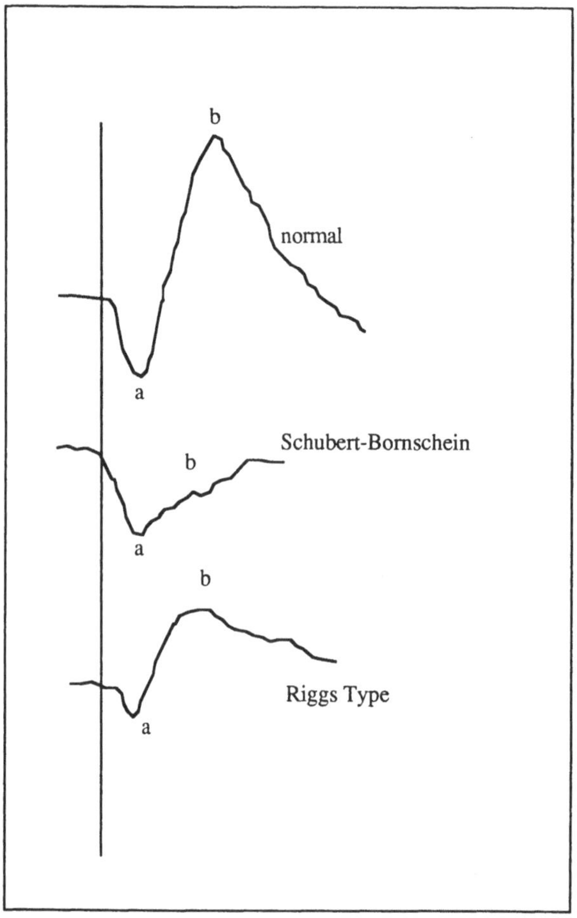

Figure 2 Schubert–Bornschein ERG versus Riggs-type ERG.

eno, 1983). Although this phenotype is most frequently seen with X-linked inheritance, it has also been reported in autosomal dominant and autosomal recessive pedigrees (Noble, Carr & Siegel, 1990; Yijian et al., 1991).

Patients with incomplete CSNB exhibit measurable, albeit reduced, scotopic b-wave amplitudes, decreased night vision sensitivity and less severe myopia than those with the complete form (mean ± 2SD; −0.8 ± 10.8 D) (Heckenlively, Martin & Rosenbaum, 1983; Miyake et al., 1987a; Miyake, Yagasaki & Horogushi, 1987b). Visual acuity may be normal or near normal. Oscillatory potentials are more easily recorded than in complete CSNB (Lachapelle, Little & Polomeno, 1983; Miyake, Yagasaki & Horiguchi, 1987c; Miyake & Kawase, 1984). This phenotype has also been described in the three Mendelian patterns of inheritance. Initially, the incomplete CSNB was thought to be a variant of the complete form, distinguished mainly by the severity of functional disturbance. However, Miyake et al. (1986) fur-

Table 3 Subclassification of Schubert–Bornschein ERG

	Complete	*Incomplete*
Dark Adaptation	absent	present
ERG		
• Bright flash	negative	negative
• Oscillatory potentials	absent	present
• 30 Hz flicker	normal	poor

ther characterized both subgroups and provided evidence that the complete and the incomplete types of CSNB are different clinical entities. Others have debated whether these types are phenotypically and genotypically distinct (Khouri et al., 1988; Pearce, Reedyk & Coupland, 1990). In considering these questions, we believe it is important to recognize that the ERG test stimulus used by Miyake et al. (1986) is unique. Their use of a rectangular rather than a flash-evoked stimulus resulted in different ERG patterns, making comparison with other groups difficult. This may contribute to their distinction between complete and incomplete CSNB. As we note later, linkage studies of Miyake's incomplete families do not support separation into two disorders.

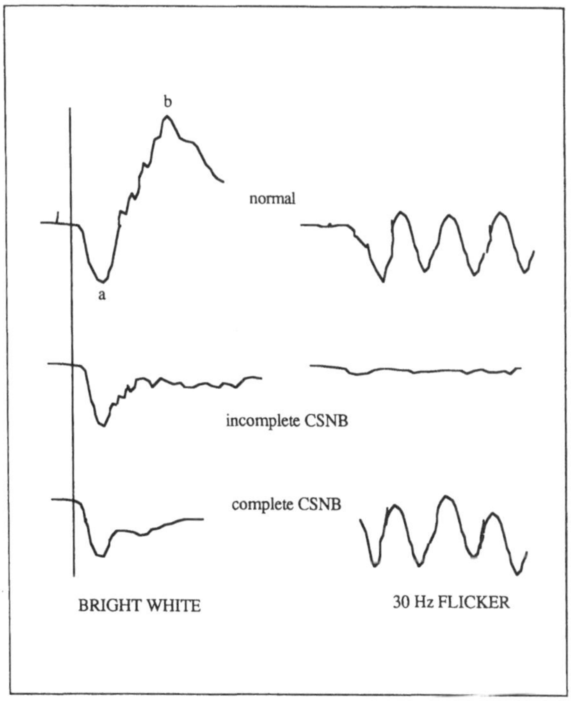

Figure 3 Electroretinographic comparison of the Schubert–Bornstein complete and incomplete CSNB.

The photopic (cone) system is also affected in CSNB. Patients with negative ERGs have been shown to have reduced amplitudes of their photopic (cone) ERGs and an altered flicker response (Krill & Martin, 1971; Hill, Arbel & Berson, 1974; Lachapelle, Little & Polomeno, 1983; Miyake et al., 1986, 1987a). These changes are more severe in incomplete CSNB, in which cone amplitudes are markedly reduced along with the 30 Hz flicker cone responses (Miyake et al., 1986, 1987a). In complete CSNB, the photopic system is essentially normal (Figure 3).

When Miyake (1987) further studied patients with incomplete CSNB, he was able to demonstrate a peculiar phenomenon of the ERG waves after the eye was dark-adapted and stimulated continuously for 12 to 15 min with a 30 Hz flicker. The ERG showed an exaggerated increase in amplitude and a characteristic change in ERG wave shape, a "separation phenomenon" not seen in normal subjects, complete CSNB, or any other kind of retinal dystrophy examined. This change appears unique to incomplete CSNB (Miyake et al., 1987a) and may result from a type of rod-cone interaction specific to this type (Granit, 1938; Hood, 1972a,b). Although there has been controversy (Krill & Martin, 1971; Hill, Arbel & Berson, 1974) about the normality of the implicit time of the scotopic ERG in CSNB, it has been reported to be normal in most cases.

In complete CSNB, the associated decreased amplitude b-wave has been interpreted as reflecting a focal destruction of retinal elements (Berson, Gouras & Hoff, 1969; Berson & Howard, 1971; Krill & Martin, 1971). This has led some investigators to postulate a decrease in the cone population or at least an impairment in the photopic (cone) system (Hill, Arbel & Berson, 1974). However, affected patients do not report visual anomalies that could be explained on the basis of a decrease in the cone population. For example, photopic visual fields and colour vision are usually within the normal range (Armington & Schwab, 1954; Krill, 1977). Furthermore, decreased vision is not a major complaint in patients with CSNB (Krill, 1977).

By studying on- and off-responses under photopic conditions, Miyake, Yagasaki & Horiguchi (1987c) were able to determine the site of impairment of the photopic visual pathway. The rapid off-response represents the primary component of the b-wave, and the late off-response reflects the decay of the late receptor potential of the cones (Kawasaki, Tusuchida & Jacobson, 1971). The positive on-response (b-wave) was significantly smaller than normal in both complete and incomplete CSNB. The rapid off-response was normal in the complete type but significantly smaller than normal in the incomplete type. These results indicate that, in the complete type, the photopic visual disturbance is in the middle retinal layer, and the cone function is essentially intact. In the incomplete type, however, the visual disturbance is in the cone itself.

Although the negative ERG is characteristic of CSNB, it can be seen in other disorders. The most important differential diagnoses of the negative ERG to be considered are listed in Table 4 (Miyake, Yagasaki & Horogushi, 1987b; Weleber et al.,

1989). Of these, special attention will be given to Åland Island eye disease (ÅIED) and Duchenne muscular dystrophy (DMD) in a later section.

Table 4 Differential diagnosis of "Negative ERGs"

1) Stationary
   1.1 complete congenital stationary night blindness
   1.2 incomplete congenital stationary night blindness
   1.3 Oguchi's disease
2) Retinal dystrophies
   2.1 Early retinitis pigmentosa
   2.2 Infantile Refsum's disease
   2.3 Goldman-Favre vitreoretinopathy
   2.4 Juvenile retinoschisis
   2.5 Progressive myopia
3) Vascular disorders
   3.1 Ischemic central vein occlusion
   3.2 Ischemic central artery occlusion
4) Retinal toxicity
   4.1 Quinine
   4.2 Vincristine
5) Paraneoplastic melanoma
6) Familial optic nerve atrophy
7) Duchenne muscular dystrophy
8) Åland Island eve disease

Modified from Weleber, RG; Pillers, D -A; Powell, B R, Hanna, CE, Magenis, R E, Buist, N R M (Weleber, 1989)

Electrophysiology of the X-Linked Carrier

Most carriers of X-linked CSNB are asymptomatic. However, because of random inactivation of the X chromosome (lyonisation), they may show signs and symptoms of the disease. Some may show evidence of nystagmus and/or ERG changes (Pearce, Reedyk & Coupland, 1990), but the majority of these female carriers are not identified by routine ocular examination (Miyake, Yagasaki & Horogushi, 1987b; Bech-Hansen et al., 1990). Subtle changes have been documented by ERG testing. Miyake and Kawase (1984) recorded statistically smaller oscillatory potentials (OPs) and amplitudes in carriers of X-linked CSNB than in non-carrier women. The implicit time of the OPs, and the amplitudes of the a- and b-waves, were within normal limits. Young et al. (1989) established that the OPs are optimally recorded in the X-linked carriers when the dark-adapted eye is stimulated by a blue flash.

Recently, Ruttum, Lewandowski and Bateman (1992) reported the clinical findings and the results of ERG testing and dark adaptometry in four symptomatic carrier females from a five-generation family with X-linked CSNB. The scotopic ERGs showed a negative b-wave consistent with the Schubert–Bornschein type of CSNB. The poor rod-mediated dark adaptations found in some symptomatic carriers are thought to be a consequence of lyonization.

SPECIAL CONSIDERATIONS WITH ÅIED AND DMD

Although many of the disorders causing a "negative ERG" can be recognized by an adequate evaluation, a "gray zone" does exist; this must be kept in mind when evaluating these cases. We will focus our attention initially on ÅIED, a disorder marked by controversy. It has been thought to be related to incomplete CSNB and later to DMD because of a unique case with a deletion of Xp21, complex glycerol kinase deficiency, DMD, and a negative ERG.

The family in whom ÅIED was first described (Forsius & Eriksson, 1964) originated in the Åland Islands in the Gulf of Bothnia, and was affected with an X-linked ocular disorder thought to be a form of ocular albinism, OA2 (Forsius & Eriksson, 1964; McKusick, 1990b). OA2 is characterized by hypopigmentation of the fundus, hypoplasia of the fovea, impaired vision, nystagmus, progressive myopia, astigmatism and dyschromatopsia. In 1967, Waardenburg, Eriksson and Forsius (1969) reviewed the features present in 10 affected males and 2 carriers from the original pedigree of Forsius and Eriksson. Visual acuity was variable in affected males, ranging from 20/45 to 20/400. Myopia ranged from –1.50 D to − 19 D, with regular astigmatism between 1.25 D and 7.0 D. Except for a mild increase of axial myopia in older patients, OA2 was known to be a non-progressive disorder (Waardenburg, Eriksson & Forsius, 1969; McKusick, 1990b). The irides transilluminated to a variable degree in half of the patients, but further studies showed that, unlike what is seen in Nettleship-Falls ocular albinism, hypopigmentation varied in different areas of the fundus and was mostly limited to the posterior pole (Witkop, Quevedo & Fitzpatrick, 1983). Female carriers demonstrated mild colour vision defects and fine nystagmus, but did not demonstrate the characteristic mosaic fundus pattern of Nettleship-Falls ocular albinism (Forsius & Eriksson, 1964). The work of O'Donnell et al. (1980) showed that, unlike Nettleship-Falls albinism, carriers and affected OA2 patients did not show macromelanosomes on skin biopsy (Witkop, Quevedo & Fitzpatrick, 1983). In addition, van Dorp et al. (1985) demonstrated that optic nerve fibre misrouting does not occur in OA2, as it does in all types of albinism (van Dorp et al., 1985). This finding strongly suggested that OA2 was not an albinotic disorder. Consequently, the disease symbol was changed from OA2 to ÅIED at the Tenth International Human Gene Mapping Workshop in 1989 (Rosenberg, Schwartz & Simonsen, 1990).

Confusion arose when Weleber et al. (1989) reported a child, Oregon J R, with a deletion at Xp21 and glycerol kinase deficiency, congenital adrenal hypoplasia, and ÅIED (Weleber et al., 1989). The ERG responses of this child were similar to those for ÅIED and identical to those reported for incomplete CSNB, as described earlier. It was suggested that ÅIED and incomplete CSNB might be the same disease, thus initially localizing this form of CSNB to Xp21 (Waardenburg, 1970; Weleber et al., 1989). Further molecular studies of this patient localized the deletion to the subbands Xp21.3-p21.2, between DXS67 and DMD (Miyake, Yagasaki & Horogushi, 1987b; Pillers et al., 1989a, 1989b, 1990a, 1990b; Weleber et al., 1989) (Figure 4). In 1991, linkage studies were reported on two families with ÅIED (Alitalo et al., 1991; Schwartz & Rosenberg, 1991). Both studies positioned ÅIED on the proximal part of the short arm of the X chromosome. Although these results did not support the work of Pillers et al., their data suggested that ÅIED and complete CSNB may be allelic and that the ocular findings in the patient described by Weleber were probably part of the syndrome resulting from the deletion at Xp21, rather than ÅIED. To clarify the matter, the ocular disorder of this patient was named "Oregon eye disease", and the origin of the negative ERG of Oregon JR was suggested to lie in the Xp21.3-p21.2 region, the location of the glycerol kinase deficiency and DMD genes.

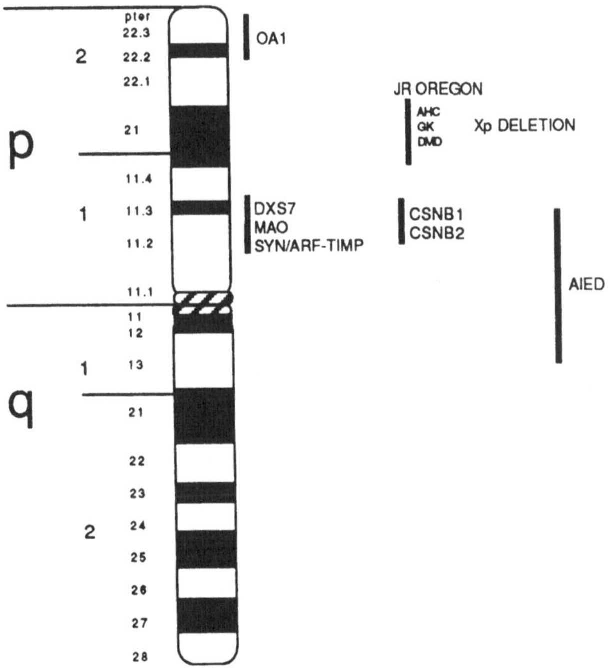

Figure 4 Localization of X-linked CSNB.

Since then, the gene for DMD has been isolated and its gene product, dystrophin, has been identified. Antibodies to dystrophin were used to determine whether it was expressed in the retina, and dystrophin was identified in the outer plexiform layer, the putative site for CSNB (Bulman et al., 1992). In our experience, most young DMD patients tested by ERG were found to have a significantly reduced b-wave amplitude, suggesting that dystrophin may play a role in normal retinal neurotransmission (Musarella et al., unpublished data). Further studies to determine the role of dystrophin in the retina may contribute to a better understanding of the X-linked forms of CSNB.

PSYCHOPHYSICAL EVALUATION OF CSNB

Basics of Dark Adaptation

Because dark adaptometry measures the ability of the photoreceptors to adapt over time in the dark, it is mainly indicated for patients suffering from nyctalopia. In this test the patient is initially dark-adapted for 2 minutes, then light-adapted for 10 minutes to bleach the photoreceptors. The background light is then extinguished and the patient is asked to identify the test target, which has adjustable intensities (Jimenez-Sierra, Ogden & Van Boemel, 1989; Fishman, 1990). The ability to dark-adapt is then measured over a 30-minute period with the dark adaptometer. A normal curve has a bipartite shape; the first 10-minute segment represents the cones' adaptation, and the rest of the curves represent the rods' adaptation (Figure 5).

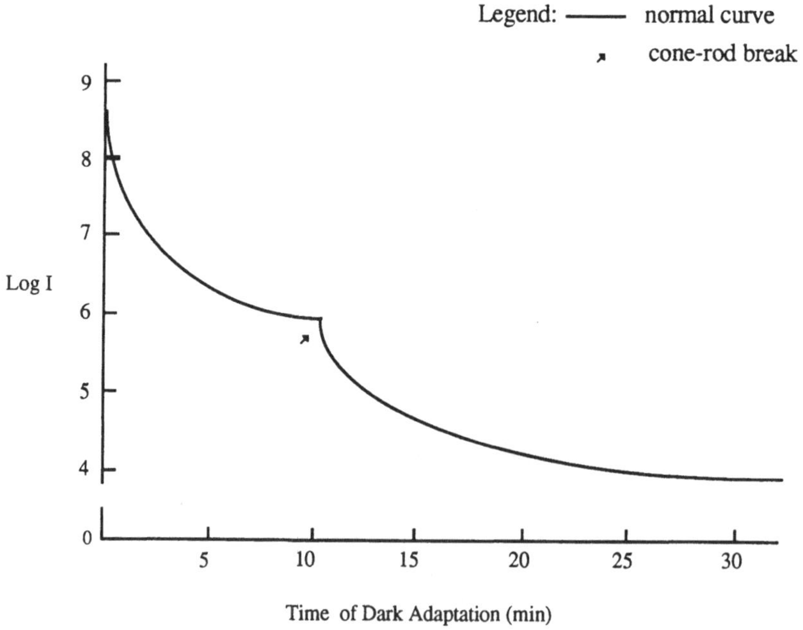

Figure 5 Normal dark adaptation curve.

Dark Adaptation and CSNB

The initial psychophysical studies of CSNB performed by Dieter (1929) and later by Garabedian and Meunier (1942) identified two types of dark adaptation curves in CSNB patients. Further studies by Auerbach, Godel and Rowe (1969) and by Miyake et al. (1986,1987b) substantiated the existence of these two curves, which today are correlated with the presence or absence of rod function (Miyake, Yagasaki & Horogushi, 1987b). The curve may have a bipartite shape if some rod adaptation is present or a monophasic shape with no cone–rod break if it is absent.

The dark-adaptation curves of the Riggs and incomplete Schubert–Bornschein types of ERG have a similar shape (Miyake et al., 1986). Since some rod adaptation is still present, a bipartite component is preserved, although the overall adaptation threshold is elevated (Miyake, Yagasaki & Horogushi, 1987b). In the complete Schubert–Bornschein ERG, where there is no rod adaptation, the dark-adaptation curve is monophasic and the residual cones threshold is elevated (Miyake, Yagasaki & Horogushi, 1987b) (Figure 6).

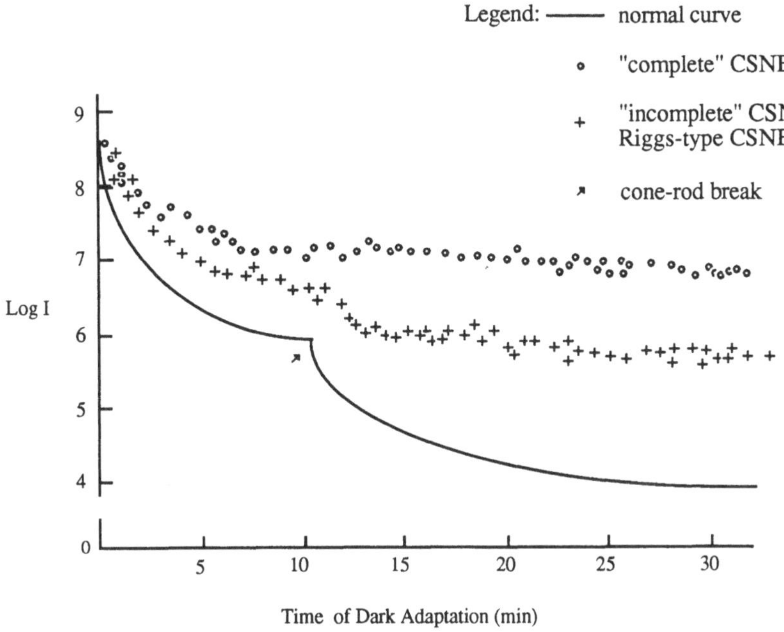

Figure 6 Dark adaptation curves in CSNB.

CLINICAL EVOLUTION OF CSNB

By definition CSNB is a stationary disorder, and long-term follow-up of the well-documented pedigrees have shown lack of progression (Franceschetti, Babel &

François, 1963; Carr, 1974; Krill, 1977). However, there appears to be a subgroup of retinitis pigmentosa (RP), a form of night blindness, marked by extremely slow progression. This *forme fruste* of retinitis pigmentosa should not be confused with CSNB (Maeder, 1946; Babel, 1963; Franceschetti, Babel & François, 1963; Auerbach, Godel & Rowe, 1969). Because many tapetoretinal degenerations may show normal retinal findings, usually in the early stages, it is essential to repeat retinal examinations periodically. If progression is documented, the diagnosis of CSNB must be re-evaluated and a tapetoretinal degeneration considered (Franceschetti, Babel & François, 1963). The ERG is useful in distinguishing these disorders. However, cases of patients with initial findings consistent with CSNB (Zorn, 1920; Ammann, 1946; François, 1961; Auerbach, Godel & Rowe, 1969), including the ERG characteristics, have been reported, but in these cases the condition developed very slowly, like a progressive retinitis pigmentosa. Sieving et al. (1992) reported a family that fell between the definitions for CSNB and RP, since they demonstrated ERG criteria for CSNB and a progressive course consistent with RP. In most family members, this autosomal dominant pedigree has no detectable rod vision, normal cone function and normal Goldmann visual fields. Older members have bone-spicules, a sign usually seen in retinitis pigmentosa. Since all family members tested had normal (or near normal) cone function, they are believed to have a congenital night blindness, probably progressive (a type of rod dystrophy) rather than retinitis pigmentosa (which would imply a progressive rod-cone dysfunction). Molecular studies of this family will be discussed later.

HISTOPATHOLOGY

In 1957, Goodman and Borschein suggested that the near absence of scotopic function was evidence of a pathological abnormality in the distal segments of the rod photoreceptors. This theory, however, has never been supported by histopathological findings; reports by Babel (1963) and Vaghefi et al. (1978) demonstrated relatively well-preserved inner segments of the photoreceptor cells. Babel also noted that numerous cone nuclei were displaced anterior to the external limiting membrane. No mention was made of the outer segment. The internal retina, retinal pigment epithelium, and the choroid were normal.

PATHOGENESIS

Fundus reflectometry is a laboratory test for evaluation of rhodopsin concentration and kinetics. It was introduced by Ripps and Weale (1969) to study patients with night blindness (Carr & Ripps, 1967; Ripps et al., 1984). The test depends on the ability of the retina to absorb light. When light traverses the retina, it is partially absorbed by the pigment epithelium and the visual pigment. With reflectometry, light of different spectral wave lengths is transmitted into the eye and measured as it exits. Comparison of measurements after bleaching to those after dark-adaptation indicates the amount of rhodopsin present (Ripps & Weale, 1969).

The inability of nyctalopic patients to see at low levels of illumination is often attributed to a defect involving the biosynthesis of the visual pigment, rhodopsin. However, a photochemical defect was ruled out when the concentration, spectral position and kinetics of regeneration were found to be normal regardless of the pattern of CSNB inheritance (Ripps & Weale, 1969). In view of normal photochemistry, analysis of the ERG suggests that both the a-wave of the ERG and the light rise in the standing potential originate proximal to the outer segment of the receptors. This suggests that the origin of the defect in CSNB is one of neural transmission.

ANIMAL MODELS

Two animal models have been used in the study of CSNB. In the first, Witzel et al. (1978) studied electroretinographically nyctalopic Appaloosa horses. Under photopic and scotopic conditions, ERG changes were similar to those in humans with the Schubert–Bornschein type of CSNB. The photopic abnormalities consisted of reduced b-wave and slower-than-normal implicit time. The dark-adapted ERGs showed a simple negative potential, the b-wave being non-recordable. Furthermore, the presence of a normal a-wave in the dark-adapted eye suggested that the rod photoreceptors were responding normally and that the rod-RPE relationship was not disturbed. Absence of the scotopic b-wave localizes the functional defect to the outer plexiform layer and/or inner nuclear layers. Histological and ultrastructural studies showed no abnormality.

The second animal model studied was the pearl mutant mouse (Balkema, Mangini & Pinto, 1983; Pinto et al., 1985). The mouse is known to have no gross malformations, decreased sensitivity in dark-adapted conditions, an abnormal ERG a-wave, absence of photoreceptor degeneration, and normal rhodopsin concentration. The pearl is an autosomal recessive mutation in the mouse that has been mapped to mouse chromosome 13. The mouse is characterized by hypopigmentation of the RPE and reduced ipsilateral retinofugal projections, as observed in albinos. This phenotypic characterization of the pearl mouse raises questions about whether it should be considered an animal model of CSNB.

MAPPING OF THE GENES FOR CSNB

X-linked CSNB is distinguished from other hereditary types by its association with high myopia (Gal et al., 1989). It has been suggested that this association is due to close linkage of these two traits on the X-chromosome, but no data are available to support this observation. Moreover, several families with X-linked CSNB do not show an association with myopia (Cutler, 1895; Newman, 1913; Khouri et al., 1988) a finding that could be explained by a crossing-over event between the two genes.

Initial linkage studies performed by White et al. in 1940 excluded linkage to the loci of colour blindness (Xqter). Subsequently, Volker-Dieben and Went (1975) failed to show linkage to the XG locus (Xpter). More recently, linkage studies involving the locus for X-linked CSNB have been conducted by three independent groups. Gal et al. (1989) studied three large European families affected with X-linked CSNB and mapped this gene to locus DXS7 at Xp11.3. Although no recombinations (θ) were detected between the disease locus and DXS7, the LOD score was 2.02. Musarella and coworkers (1989) assigned the gene for complete CSNB (CSNB1) to the Xp11.3 region, very close to the DXS7 locus (LOD 7.35 at θ = 0.00), by performing linkage analysis on eight families with X-linked CSNB (seven with the complete type and one with the incomplete). Recombinations were not detected between the disease locus and DXS7 but were present distal and proximal to this locus. These eight families were subjected to heterogeneity testing but heterogeneity was not found to be significant. Recently, the linkage analysis of these families was extended to include synapsin, monoamine oxidase (MAO) and the synaptophorin/Araf 1 gene loci. Recombinations of CSNB1 with all of these gene loci excluded them as the candidate gene for X-linked CSNB (Musarella, personal communication). Analysis of informative crossovers places the gene for CSNB proximal to MAO and distal to TIMP. This localization was confirmed by linkage studies done by Bech-Hansen and colleagues (1990), and more recently by Aldred et al. (1992). Both groups found no distinction between the complete and the incomplete forms of CSNB in their families.

The question of whether incomplete CSNB (CSNB2) is localized to Xp21 or to Xp11.3 was recently addressed (Musarella et al., 1992). In two of the families previously studied by Miyake, tight linkage was found between CSNB2 and markers located at Xp11.3-p11.2. These data supported the localization of CSNB2 to the same region as CSNB1 and ÅIED. Interestingly, this locus has two other X-linked eye disorders closely linked to it, X-linked retinitis pigmentosa (RP2) (Bhattacharya et al., 1984) and Norrie's disease (Bleeker-Wagemakers et al., 1985). The reason for this clustering remains to be elucidated. Further testing with more polymorphic markers in families with CSNB1, ÅIED and CSNB2 will be needed to determine if these are allelic or are tightly clustered on the short arm of the X-chromosome.

Few studies have been done on the autosomal dominant forms of CSNB. In a large Chinese family with typical dominant CSNB, close linkage between autosomal dominant CSNB and markers of PGM1 (chromosome 1), ABO (chromosome 9) and MN (chromosome 4) was excluded (Yijian et al., 1991). Sieving et al. (1992) carried out mutational analysis on rhodopsin in an autosomal dominant congenital night blindness pedigree discussed earlier and found a single base substitution in codon 90 of rhodopsin, resulting in a change in amino acid from glycine to asparagine.

CONCLUSION

Although there are distinct clinical entities referred to as CSNB, little is known about the molecular defects or even the defective neurotransmitters that are responsible for any of their forms. The localization of putative genes for the autosomal forms of CSNB remains unknown. The X-linked type of CSNB is located at Xp11.3 in a region near monoamine oxidase A and B, synapsin and synaptophorin. However, linkage studies done with these markers have excluded these neurotransmitters as putative sites for this form of CSNB. At this point, little else remains at this location as possible candidate genes for X-linked CSNB. The identification of dystrophin in the outer nuclear layer of normal retinas may indirectly provide clues to this enigmatic group of disorders.

ACKNOWLEDGEMENTS

This work was supported by the MacLaughlin Foundation (E. H.) and the R P Foundation of Canada (MAM). This chapter was prepared with the assistance of Medical Publications, The Hospital for Sick Children, Toronto.

REFERENCES

Aldred, M. A., Dry, K. L., Sharp, D. M., Van Dorp, D. B., Brown, J., Hardwick, L. J., Lester, D. H., Pryde, F. E., Teague, P. W., Jay, M., Bird, A. C., Wright, A. F. (1992). Linkage analysis in X-linked congenital stationary night blindness. *Genomics* **14**, 99–104.

Alitalo, T., Kruse, T. A., Forsius, H., Eriksson, A. W. and de la Chapelle, A. (1991) Localization of the Åland Island eye disease locus to the pericentromeric region of the X chromosome by linkage analysis. *American Journal of Human Genetics* **48**, 31–38.

Ammann, F. (1946) Zur Verebung der Hemeralopia herediteria und tapeto-retinalen Degeneration. *Ophthalmologica* **112**, 78–87.

Armington, J. C. and Schwab, G. J. (1954) Electroretinogram in nyctalopia. *Archives of Ophthalmology* **52**, 725–733.

Auerbach, E., Godel, V. and Rowe, H. (1969) An electrophysiological and psychophysical study of two forms of congenital night blindness. *Investigative Ophthalmology* **8**, 332–345.

Babel, J. (1963) Constatations histologiques dans l'amaurose infantile de Leber et dans diverses formes d'héméralopie. [Histological findings in Leber's infantile amaurosis and in different forms of hemeralopia.] *Ophthalmologica (Basel)* **145**, 399–402.

Balkema, G. W., Mangini, N. J. and Pinto, L. H. (1983) Discrete visual defects in pearl mutant mice. *Science* **219**, 1085–1087.

Barricks, M. E., Flynn, J. T. and Kushner, B. J. (1977) Paradoxical pupillary responses in congenital stationary night blindness. *Archives of Ophthalmlogy* **95**, 1800–1804.

Bech-Hansen, N. T., Moore, B. J. and Pearce, W. G. (1992) Mapping of locus for X-linked congenital stationary night blindness (CSNB1) proximal to DXS7. *Genomics* **12**, 409–411.

Bech-Hansen, N. T., Field, L. L., Schramm, A. M., Reedyk, M., Craig, I. W., Fraser, N. J., et al. (1990) A locus for X-linked congenital stationary night blindness is located on the proximal portion of the short arm of the X chromosome. *Human Genetics* **84**, 406–408.

Berson, E. L., Gouras, P. and Hoff, M. (1969) Temporal aspects of the electroretinogram. *Archives of Ophthalmology* **81**, 207–214.

Berson, E. L. and Howard, J. (1971) Temporal aspect of the electroretinogram in sector retinitis pigmentosa. *Archives of Ophthalmology* **86**, 653–665.

Bessonnet, J. (1904) Héméralopie héréditaire sans lesions rétiniénnes. *Annales Medico-Chirurgicales* **28**, 505-506.

Bhattacharya, S. S., Wright, A. F., Clayton, J. F., Price, W. H., Phillips, C. I., McKeown, C. M. E., et al. (1984) Close genetic linkage between X-linked retinitis pigmentosa and a restriction fragment length polymorphism identified by recombinant DNA probe L1.28. *Nature* **309**, 253–255.

Bleeker-Wagemakers, L. M., Friedrich, U., Gal, A., Wienker, T. F., Warburg, M. and Ropers, H.-H. (1985) Close linkage between Norrie disease, a cloned DNA sequence from the proximal short arm, and the centromere of the X chromosome. *Human Genetics* **71**, 211–214.

Bordley, J. (1908) A family of hemeralopes. *Bulletin of Johns Hopkins Hospital* **19**, 278.

Bornschein, H. and Vukovich, V. (1953) Das Elektroretinogramm bei Mangelhemeralopia. *Albrecht von Graefes Archiv fur Ophtalmologie vereinigt met Archiv fur Augenheilkunde (Berlin)* **153**, 484–487.

Bulman, D. E., Pillers, DeA. M., Weleber, R. G., Powell, B. R., Becker, L. E. and Ray, P. N. (1992) Do mutations in the Duchenne muscular dystrophy gene cause Oregon eye disease? *Investigative Ophthalmology and Visual Science* **33**, 1396 (abstract 3521).

Carr, R. E., Ripps, H., Siegel, I. M. and Weale, R. A. (1966b) Visual functions in congenital night blindness. *Investigative Ophthalmogy* **5**, 508–514.

Carr, R. E. and Siegel, I. M. (1964) Electrophysiologic aspect of several retinal diseases. *American Journal of Ophthalmology* **58**, 95–107.

Carr, R. E. (1974) Congenital stationary nightblindness. *Transactions of the American Ophthalmological Society* **72**, 448–487.

Carr, R. E. and Ripps, H. (1967) Rhodopsin kinetics and rod adaptation in Oguchi's disease. *Investigative Ophthalmology* **6**, 426–436.

Carroll, F. D. and Haig, C. (1953) Congenital stationary night blindness without ophthalmoscopic or other abnormalities. *Transactions of the American Ophthalmological Society* **50**, 193–209.

Cuniers, F. (1838) Histoire d'une hemeralopie hereditaire depuis deux siecles dans une famille de la commune de Vendemian, pres de Montpellier. *Annales d'Oculistiques 1,* 383–395.

Cutler, C. (1895) Ueber angeborene nachtblindheit und pigmentdegeneration. *Klinisches Monatsblatter fur Augenheilkunde* **30**, 92–116.

DeJean, Ch. and Gassenc, R. (1949) Note sur la généalogie de la famille Nougaret, de Vendemian. *Bulletin des Societes d'Ophtalmologie de France (Paris)* **1**, 96–99.

Dieter, W. (1929) Untersuchungen zur Duplizitatstheorie; die angeborene familiar-erblicke, stationare (idiopathische) Hemeralopie. *Pfluger's Archiv fur die gesamte Physiologie des Menschen und der Tiere (Berlin)* **222**, 381–394.

Donders, F. (1855) Torpeur de la rétine congénitale héréditaire. *Annales d'Oculistiques* **34**, 270–272.

Einthoven, W. and Jolly, W. (1908) The form and magnitude of electrical response of the eye to stimulation by light at various intensities. *Quarterly Journal of Experimental Physiology* **1**, 373–416.

Fishman, G. A. (1985) Basic principles of clinical electroretinography. *Retina* **5**, 123–126.

Fishman, G. A. (1990) Stationary night-blinding disorders. In *Ophthalmology Monographs 2. Electrophysiologic Testing in Disorders of the Retina, Optic Nerve, and Visual Pathway,* edited by G. A. Fishman and S. Sokol, pp. 39–43. San Franscisco: American Academy of Ophthalmology.

Forsius, H. and Eriksson, A. W. (1964) Ein neues Augensyndrom mit x-chromosomal Transmission. Eine Sippe mit Fundusalbinismus, Foveahypoplasie, Nystagmus, Myopie, Astigmatismus und Dyschromatopsie. *Klinische Monatsblaetter fur Augenheilkunde (Stuttgart)* **144**, 447–457.

Franceschetti, A., François, J., and Babel, J. (1963) Anomalies fonctionnelles. In *Les h_r_do-dégénérescences chorio-rétiniennes (dégénérescences tapéto-rétiniennes), tome II,* edited by Masson and Cie, pp. 1205–1250. Paris: Librairies de l'Académie de Médecine.

François, J. (1961) *Heredity in Ophthalmology.* St Louis: CV Mosby Company.

François, J., Verriest, G. and De Rouck, A. (1965) A new pedigree of idiopathic congenital night-blindness. Transmitted as a dominant hereditary trait. *American Journal of Ophthalmology* **59**, 621–625.

Gal, A., Schinzel, A., Orth, U., Fraser, N. A., Mollica, F., Craig, I. W., et al. (1989) Gene of X-chromosomal congenital stationary night blindness is closely linked to DXS7 on Xp. *Human Genetics* **81**, 315–318.

Garabedian, M. D. and Meunier, P. (1942) L'adaptometrie. Methode clinique pour le depistage des carences nutritives, les avitaminoses A en particulier. *Ophthalmologica* **104**, 65–85.

Gassler, V. (1925) Uebereine bis jetzt nicht bekannte rezessive Verknupfung von hochgradiger myopie mit angeborener hemeralopie. *Thesis,* Art Institut Orell Fussii, Zurich.

Goodman, G. and Bornschein, H. (1957) Comparative electroretinographic studies in congenital night blindness and total color blindness. *Archives of Ophthalmology* **58**, 174–182.

Granit, R. (1938) Processes of adaptation in vertebrate retina in light of recent photochemical and electrophysiological research. *Documenta ophthalmologica* **1**, 7–77.

Haim, M. (1986) Congenital stationary night blindness. *Acta Ophthalmologica (Copenh)* **64**, 192-198.

Heckenlively, J. R., Martin, D. A. and Rosenbaum, A. L. (1983) Loss of electroretinographic oscillatory potentials, optic atrophy, and dysplasia in congenital stationary night blindness. *American Journal of Ophthalmology* **96**, 526–534.

Hill, D. A., Arbel, K. F. and Berson, E. L. (1974) Cone electroretinograms in congenital nyctalopia with myopia. *American Journal of Ophthalmology* **78**, 127–136.

Hittner, H. M., Borda, R. P. and Justice, J., Jr. (1981) X-linked recessive congenital stationary night blindness, myopia, and tilted discs. *Journal of Pediatric Ophthalmology and Strabismus* **18**, 15–20.

Hood, D. C. (1972a) Adaptational changes in the cone system of the isolated frog retina. *Vision Research* **12**, 875–888.

Hood, D. C. (1972b) Suppression of the frog's cone system in the dark. *Vision Research* 12, 889–907

Hudson, A. (1903) Notes on a case of congenital night blindness. *Ophthalmoscope* **1**, 89–90.

Jimenez-Sierra, J. M., and Ogden, T. E. (1989) Abnormalities of photoreceptors, rod system disorders (congenital stationary night blindness [CSNB]). In *Inherited Retinal Diseases: A Diagnostic Guide,* edited by J. M. Jimenez-Sierra, T. E. Ogden, and G. B. Van Boemel, pp. 175–187. St. Louis: CV Mosby Company.

Kandori, F. (1959) Very rare cases of congenital non-progressive night blindness witih fleck retina. *Japanese Journal of Ophthalmology* **13**, 384–386.

Kandori, F., Setogawa, T. and Tamai, A. (1966) Electroretinographical studies on "fleck retina with congenital nonprogressive night blindness". *Yonago Acta Medica* **10**, 98–108.

Kandori, F., Tamai, A., Kurimoto, S. and Fukunaga, K. (1972) Fleck retina. *American Journal of Ophthalmology* **73**, 673–685.

Kawasaki, K., Tsuchida, Y. and Jacobson, J. H. (1971) Positive and negative deflections in the off response of the electroretinogram in man. *American Journal of Ophthalmology* **72**, 367–375.

Khouri, G., Mets, M. B., Smith, V. C., Wendell, M. and Pass, A. S. (1988) X-linked congenital stationary night blindness. Review and report of a family with hyperopia. *Archives of Ophthalmology* **106**, 1417–1422.

Krill, A. E. (1977) Congenital stationary night blindness. In *Hereditary Retinal and Choroidal Diseases, Vol 2,* pp. 391–420. Hagerstown: Harper and Row.

Krill, A. E. and Martin, D. (1971) Photopic abnormalities in congenital stationary night blindness. *Investigative Ophthalmology and Visual Science* **10**, 625–636.

Lachapelle, P., Little. J. M. and Polomeno, R. C. (1983) The photopic electroretinogram in congenital stationary night blindness with myopia. *Investigative Ophthalmology & Visual Science* **24**, 442–450.

Larousse, P. (1974) *Petit Larousse Illustré*, edited by P. Larousse, p. 532. Paris: Librairie Larousse.

Latte, B., Manunta, G. and Pisano, E. (1954) Comportamento dell'elettroretinogramma in alcuni soggetti affetti di emeralopia ereditaria congenita stazionaria. *Bollettino della Societa Italina di Biologica Sperimentale* **30**, 1046–1049.

Lauber, H. (1910) Die sogenannte Retinitis punctata albescens. *Klinische Monatsblatter fur Augenheilkunde* **48**, 133–148.

Maeder, G. (1946) Hemeralopie et syndactylie familiales. *Ophthalmologica* **111**,278–284.

McKusick, V. A. (199Oa) 300600 Albinism, ocular, type 2 [OA2; Forsius-Eriksson type ocular albinism; Åland Island eye disease; AIED]. In *Mendelian Inheritance in Man. Catalogs of Autosomal Dominant, Autosomal Recessive, and X-linked Phenotypes, 9th edition,* pp. 1555–1556. Baltimore: The Johns Hopkins University Press.

McKusick, V. A. (1990b) 310500 Night blindness, congenital stationary, with myopia [NBM1; CSNB1; hemeralopia-myopia; myopia-night blindness; nyctalopia, included]. In *Mendelian Inheritance in Man. Catalogs of Autosomal Domonant, Autosomal Recessive, and X-linked Phenotypes, 9th edition,* pp. 1692–1693. Baltimore: The Johns Hopkins University Press.

Merin, S. (1991) Night vision disorders. In *Inherited Eye Diseases: Diagnosis and Clinical Management,* edited by S. Merin, pp. 345–362. New York: Marcel Dekker. Inc.

Merin, S., Rowe, H., Auerbach, E. and Landau, J. (1970) Syndrome of congenital high myopia with nyctalopia. Report of findings in 25 families. *American Journal of Ophthalmology* **70**, 541–547.

Miyake, Y., Gto, S., Ando, F. and Ichikawa, H. (1983) Vitreous fluorophotometry in congenital stationary nightblindness. *Archives of Ophthalmology* **101**, 574–576.

Miyake, Y., Horiguchi, M., Ota, I. and Shiroyama, N. (1987a) Characteristic ERG flicker anomaly in incomplete congenital stationary night blindness. *Investigative Ophthalmology & Visual Science* **28**, 1816–1823.

Miyake, Y. and Kawase, Y. (1984) Reduced amplitude of oscillatory potentials in female carriers of X-linked recessive congenital stationary night blindness. *American Journal of Ophthalmology* **98**, 208–215.

Miyake, Y., Yagasaki, K. and Horiguchi, M. (1987b) A rod-cone dysfunction syndrome with separate clinical entity: incomplete-type congenital stationary night blindness. In *Degenerative Retinal Disorders: Clinical and Laboratory Investigations,* edited by J. G. Hollyfield, R. E. Anderson, and M. M. LaVail, pp. 137–145. New York: Alan R. Liss.

Miyake, Y., Yagasaki, K. and Horiguchi, M. (1987c) On- and off-response in photopic electroretinogram in complete and incomplete types of congenital stationary night blindness. *Japanese Journal of Ophthalmology* **31**, 81–87.

Miyake, Y., Yagasaki, K., Horiguchi, M., Kawase, Y. and Kanda, T. (1986) Congenital stationary night blindness with negative electroretinogram. A new classification. *Archives of Ophthalmology* **104**, 1013-1020.

Mizuo, G. (1913) On a new discovery in the dark adaptation on Oguchi's disease. *Acta Societatis Ophtalmologicae Japonicae* **17**, 1148–1150.

Mooren, A. (1882) Funf Lustren Ophthalmology. Wirksamkeit. *Weisbaden* **8**, 216–218.

Musarella, M. A., Weleber, R. G., Murphey, W. H., Young, R. S., Anson-Cartwright, L., Mets, M., et al. (1989) Assignment of the gene for complete X-linked congenital stationary night blindness (CSNB1) to Xp11.3. *Genomics* **5**, 727–737.

Musarella, M. A., Kirshgessner, C., Trofatter, J., Miyake, Y., and DeGennaro, L. (1992) Assignment of the gene for incomplete congenital stationary night blindness (CSNB2) to proximal Xp. *Investigative Ophthalmology and Visual Science* **33**, 792 (abstract 502–8).

Nettleship, E. (1907) A history of congenital stationary night blindness in nine consecutive generations. *Transactions of the Ophthalmological Societies of the United Kingdom* **27**, 269-293.

Nettleship, E. (1912) A pedigree of congenital night-blindness with myopia. *Transactions, Ophthalmological Society of the United Kingdom* **32**, 21–45.

Newman, H. (1913) Five generations of congenital stationary night blindness in an American family. *Journal of Genetics* **3**, 25–38.

Noble, K. G., Carr, R. E. and Siegel, I. M. (1990) Autosomal dominant congenital stationary night blindness and normal fundus with an electronegative electroretinogram. *American Journal of Ophthalmology* **109**, 44–48.

O'Donnell, F. E., Green, W. R., McKusick, V. A., Forsius, H. and Eriksson, A. W. (1980) Forsius-Eriksson syndrome: its relation to the Nettleship-Falls X-linked ocular albinism. *Clinical Genetics* **17**, 403–408.

Ogden, T. E. (1973) The oscillatory waves of the primate electroretinogram. *Vision Research* **13**, 1059–1074.

Ogden, T. (1989) Clinical electrophysiology. In *Retina,* edited by S. Ryan, pp. 285–296. St. Louis: CV Mosby Company.

Oguchi, C. (1907) Ueber einen fall von eigenartiger hemeralopie. *Nippon Gankakai Zasshi* **11**, 123–134.

Pearce, W. G., Reedyk, M. and Coupland, S. G. (1990) Variable expressivity in X-linked congenital stationary night blindness. *Canadian Journal of Ophthalmology* **25**, 3–10.

Pillers, D. -A., Powell, B. P., Seltzer, W. K., Ray, P. N., Tremblay, F., La Roche, G. R., et al. (1989a) Åland Island eye disease localized by molecular genetic analysis to Xp21.3-p21.2 in a patient with Duchenne muscular dystrophy, glycerol kinase deficiency, congenital adrenal hypoplasia, and a chromosomal deletion. *Cytogenetics and Cell Genetics* **51**,1059–1060 (abstract).

Pillers, D. -A., Towbin, J. A., Chamberlain, J. S., Wu, D., Ranier, J., Powell, B. R., et al. (1989b) Åland Island eye disease (MIM 30060) localized between DXS67 and Duchenne muscular dystrophy locus (DMD) in the Xp21.3-21.2 region (Abstract). *American Journal of Human Genetics* **45**(suppl), A157

Pillers, D. -A. M., Towbin, J. A., Chamberlain, J. S., Wu, D., Ranier, J., Powell, B. R., et al. (1990b) Deletion mapping of Åland Island eye disease to Xp21 between DXS67 (B24) and Duchenne muscular dystrophy. *American Journal of Human Genetics* **47**, 795–801.

Pillers, D. -A., Weleber, R. G., Powell, B. R., Hanna, C. E., Magenis, R. E. and Buist, N. R. M. (1990a) Åland Island eye disease (Forsius-Eriksson ocular albinism) and an Xp21 deletion in a patient with Duchenne muscular dystrophy, glycerol kinase deficiency, and congenital adrenal hypoplasia. *American Journal of Medical Genetics* **36**, 23–28.

Pinto, L. H., Williams, M. A., Suzuki, H., Magini, N. J., Balkema, G. W., Jr. and Vanable, J. W., Jr. (1985) A hypothesis for the mechanism for subnormal retinal sensitivity in the pearl mouse model for human congenital night blindness. In *Retinal Degeneration: Experimental and Clinical Studies,* edited by M. M. LaVail, J. G. Hollyfield, and R. E. Anderson, pp. 257–267. New York: Alan R. Liss.

Ponte, F., Lodato, G. and Lauricella, M. (1974) Electrophysiological tests in congenital stationary night-blindness with regard to the different types of inheritance. *Documenta Ophthalmologica Proceedings series* **4**, 161–167.

Rambusch, J. (1954) Die Erblichkeitsverhaltnisse der kongenitalen Nachtblindheit. *Cbl Prakt Augenheilk* 34, 52.

Riggs, L. A. (1941) Continuous and reproducible records of the electrical activity of the human retina. *Proceedings of the Society of Experimental Biology and Medicine* **48**, 204–207.

Riggs, L. A. (1954) Electroretinography in cases of night blindness. *American Journal of Ophthalmology* **38**, 70–78.

Ripps, H., Carr, R. E., Siegel, I. M. and Greenstein, V. C. (1984) Functional abnormalities in vincristine-induced night blindness. *Investigative Ophthalmology & Visual Science* **25**, 787–794.

Ripps, H. and Weale, R. A. (1969) Rhodopsin regeneration in man. *Nature* **222**,775–777.

Rosenberg, T., Schwartz, M. and Simonsen, S. E. (1990) Åland eye disease (Forsius-Eriksson-Miyake syndrome) with probability established in a Danish family. *Acta Ophthalmologica (Copenh)* **68**, 281–291.

Ruttum, M. S., Lewandowski, M. F. and Bateman, J. B. (1992) Affected females in X-linked congenital stationary night blindness. *Ophthalmology 99,* 747–752

Schubert, G. and Bornschein, H. (1952) Beitrag zur Analyse des menschlichen Elektroretinogramms. *Ophthalmologica* **123**, 396–412.

Schwartz, M. and Rosenberg, T. (1991) Åland eye disease: linkage data. *Genomics* **10**, 327–332.

Sedan, L. (1885) Une famille d'hemeralope. *Recherche d'Ophtalmologie* **7**, 675–678.

Sieving, P. A., Richards, J. E., Bingham, E. L. and Naarendorp, F. (1992) Dominant congenital complete nyctalopia and Gly90Asp rhodopsin mutation. *Investigative Ophthalmology and Visual Science* **33**, 1397 (abstract 3530).

Sinclair, W. (1905) Hereditary congenital night blindness without visible ophthalmoscopic changes. *Ophthalmic Revue* **24**, 255–258.

Snell, S. (1907) Retinitis pigmentosa occuring in several members of a family. *Transactions of the Ophthalmólogical Society of the United Kingdom* **27**, 217–219.

Snyder, C. (1963) Jean Nougaret, the butcher from Provence, and his family. *Archives of Ophthalmology* **69**, 676–678.

Stievenart, P. (1847) Note sur l'hemeralopie hereditaire. *Annales d'Occulistiques* **18**, 163–164.

Vaghefi, H. A., Greeen, W. R., Kelley, J. S., Sloan, L. L., Hoover, R. E. and Patz, A. (1978) Correlation of clinicopathologic findings in a patient. Congenital night blindness, branch retinal vein occlusion, cilioretinal artery, drusen of the optic nerve head, and intraretinal pigmented lesion. *Archives of Ophthalmology* **96**, 2097–2104.

van Dorp, D. B., Eriksson, A. W., Delleman, J. W., van Vliet, A. G. M., Collewijn, H., van Balen, A. T., et al. (1985) Åland eye disease: no albino misrouting. *Clinical Genetics* **28**, 526–531.

Volker-Dieben, H. J. and Went, L. N. (1975) Ophthalmologic and genetic study of a family with nyctalopia and myopia. *Ophthalmologica* **171**, 358–359.

Vukovich, V. (1952) Das ERG des Achromaten. *Ophthalmologica* **124**, 354–359.

Waardenburg, P. J. (1970) Some notes on publications of Professor Arnold Sorsby and on Åland eye disease (Forsius-Eriksson syndrome). *Journal of Medical Genetics* **7**, 194–199.

Waardenburg, P. J., Eriksson, A. and Forsius, H. (1969) Åland eye disease (syndroma Forsius-Eriksson). *Progress in Neuro-Ophthalmogy* **2**, 336–339.

Weleber, R. G., Pillers, D. -A. M., Powell, B. R., Hanna, C. E., Magenis, R. . and Buist, N. R. M. (1989) Åland Island eye disease (Forsius-Eriksson syndrome) associated with contiguous deletion syndrome at Xp21. Similarity to incomplete congenital stationary night blindness. *Archives of Ophthalmology* **107**, 1170–1179.

White, T. (1940) Linkage and crossing-over in the human sex chromosomes. *Journal of Genetics* **40**, 403–437

Witkop, C. J. Jr., Quevedo, W. C., Jr. and Fitzpatrick, T. B. (1983) Albinism and other disorders of pigment metabolism. In *The Metabolic Basis of Inherited Disease, 5th edition,* edited by J. B. Stanbury, J. B. Wyngaarden, D. S. Fredrickson, J. L. Goldstein and M. S. Brown, pp. 301–346. New York: McGraw-Hill Book Co.

Witzel, D. A., Smith, E. L., Wilson, R. D. and Aguirre, G. D. (1978) Congenital stationary night blindness: an animal model. *Investigative Ophthalmology & Visual Science* **17**, 788–795.

Yijian, F., Chengren, L., Anren, L., Jiumu, Z. and Yongzhi, H. (1991) Linkage analysis of autosomal dominant congenital stationary night blindness in a large chinese pedigree. In *Degenerative Retinopathies: Advances in Clinical and Genetic Research. Proceedings of the Sixth World Congress of the International Retinitis Pigmentosa Association (IRPA), Dublin, Ireland, July 20–22,1990,* edited by P. Humphries, S. Bhattacharya and A. Bird, pp. 49–54. Boca Raton: CRC Press.

Young, R. S. L., Chaparro, A., Price, J. and Walters, J. (1989) Oscillatory potentials of X-linked carriers of congenital stationary night blindness. *Investigative Ophthalmology & Visual Science* **30**, 806–812.

Zorn, B. (1920) Ueber familiare atypische pigmentdegeneration der netzhaut (totale aderhautatrophie). *Graefes Archives Ophthalmology* **101**, 13.

12. CHOROIDEREMIA

FRANS P. M. CREMERS and HANS-HILGER ROPERS

Department of Human Genetics, University Hospital Nijmegen
P.O. Box 9101, 6500 HB Nijmegen, The Netherlands

SUMMARY

Choroideremia (CHM) is an X-linked progressive degeneration of the retinal pigment epithelium (RPE), choroid and retina. Microscopically visible deletions associated with contiguous gene syndromes have assigned the CHM gene to Xq21. Deletions in patients with classical choroideremia were instrumental in refining this assignment and in identifying DNA markers in the vicinity of the CHM gene locus. Using evolutionary conserved sequences from the critical region, CHM cDNAs were isolated which encode an ubiquitously expressed mRNA of approximately 6 kb. Thus far, 4.7 kb of cDNA containing a 1.3 kb open reading frame (ORF) as well as 9 corresponding exons of the structural gene have been cloned and characterized.

In the Western European population 14% of patients with CHM have deletions encompassing part of the CHM gene and 21% have subtle mutations. Invariably, these mutations give rise to truncated proteins. Missense mutations have not been observed. A closely related gene, CHML, was isolated and mapped to lq42-qter in the vicinity of the gene for Usher syndrome type 2 (USH2). Whether CHML has a role in USH2 is not yet known.

There is recent evidence that the CHM gene codes for a component of a geranylgeranyl transferase (GGT). Six tryptic peptides of component A of rat Rab GGT show almost complete identity with the mouse CHM homologue. Rat Rab GGT is a multisubunit enzyme that prenylates Rab3A, RablA and possibly other small GTP-binding proteins. In addition, the CHM and CHML proteins show weak homology to bovine Rab3A GDI, a protein that inhibits the exchange of bound GDP by GTP on Rab3A. Since Rab3A and Rab GGT are expressed in a variety of tissues, their specific role in the aetiology of CHM remains to be elucidated.

CLINICAL ASPECTS

Choroideremia was first described by Mauthner in 1872 who coined the name of this disease because of presumed congenital absence of the choroid. Progressive degeneration of the choroid, retinal pigment epithelium (RPE) and retina was established only much later (Bedell, 1937; Schutzbach, 1938; Friedman, 1940;

McCulloch and McCulloch, 1948). Choroideremia is an X-linked disorder with full manifestation in males and minor clinical signs in female carriers (Waardenburg, 1942; Goedbloed, 1942). The incidence of CHM is estimated as 1 in 100,000 (Heckenlively et al., 1988).

The clinical picture and rate of progression can vary between affected individuals from different families, but also within the same family. In affected males, fine pigment mottling in the midperiphery is the first sign of the disorder. Gradually, there is atrophy of the RPE which is accompanied or directly followed by atrophy of the capillaries of the choroid. These initial fundus changes cause a diffuse dysfunction of the photoreceptors, and as a result, most patients experience night blindness since early childhood (Bedell, 1937; Francois, 1968). At a later stage, the outer retinal layers and large choroidal vessels become atrophic, first in the midperiphery, and then centrally towards the macula. Consequently, vision deteriorates and often results in so-called "tunnel vision". Other patients develop ring scotomas (Kärnä, 1986).

Female carriers are mostly asymptomatic, but their fundi show characteristic pigment changes in the midperiphery closely resembling the fine mottling observed in the initial stage of the disease in males. Occasionally, female carriers have serious visual impairment (Fraser and Friedman, 1968; Harris and Miller, 1968) which can be ascribed to skewed X-inactivation. Severe manifestation of the disease in females can also result from disruption of the CHM gene by X-autosome translocations (Kaplan et al., 1989; Siu et al., 1990). In these females, X-chromosome inactivation is non-random: the normal X is preferentially inactivated while both translocation fragments remain active.

PHYSICAL MAPPING OF THE CHM LOCUS

Contiguous Gene Deletion Syndromes

Several males have been described who show cytogenetically detectable deletions of (parts of) the Xq21 band which are associated with complex phenotypes (Tabor et al., 1983; Rosenberg et al., 1986,1987; Schwartz et al., 1986; Hodgson et al., 1987; Nussbaum et al., 1987; Wells et al., 1991; Reardon et al., 1992). These phenotypes almost invariably include CHM, congenital mixed deafness, mental retardation (MR), and in a few cases cleft lip and palate (CLP), obesity, urinary tract abnormalities, hypertelorism, macrocephaly, and myopia. In order to position the CHM locus more accurately, these deletions were characterized in great detail employing Southern analysis (Schwartz et al., 1986, 1988; Hodgson et al., 1987; Cremers et al., 1988, 1989a; Merry et al., 1990; Yang et al., 1990; Bach et al., 1992). Based on these results, the Xq21 region was subdivided into a number of intervals. As depicted in Figure 1, the smallest region of deletion overlap thought to contain the CHM gene encompasses interval 3 which is defined by the loci DXS233 (pJL68), ZNF6 (CMPX1), DXS165 (p1bD5), and DXS95 (pXG7c). The

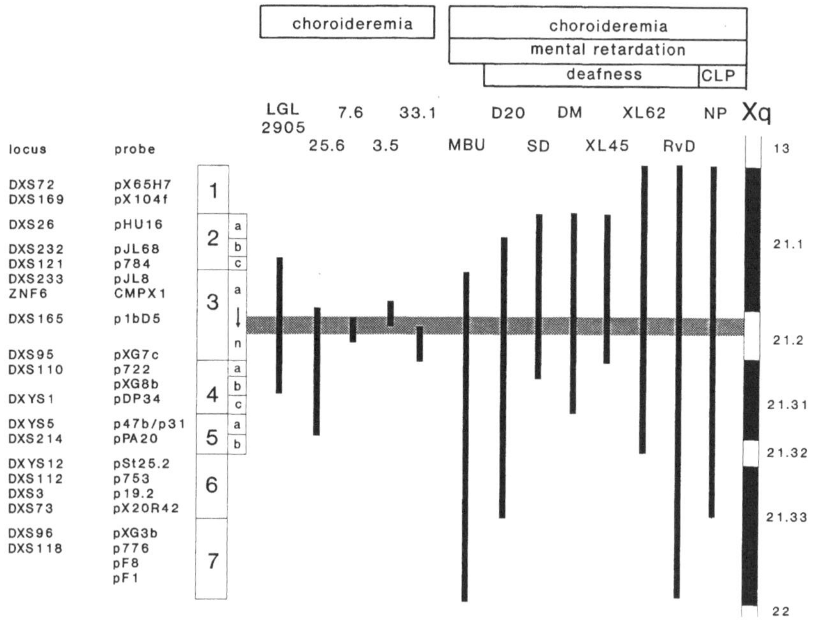

Figure 1 Deletion map of the Xq21 region. The DNA probes used in Southern analysis are indicated at the left. Deleted chromosomal segments are indicated by vertical bars. The shaded horizontal bar represents the critical region for the CHM locus. The probe order for intervals 6 and 7 is arbitrary; intervals and banding pattern are not drawn to scale. CLP = cleft lip and palate.

onlY discrepancy in this assignment is the absence of clinical signs of CHM in patient RvD, who, at the age of 15, does not show the characteristic features of CHM despite the fact that interval 3 is completely absent (J. J. P. van de Kamp and E. M. Bleeker-Wagemakers, personal communication). All other patients show CHM or fundus abnormalities reflecting an earlier stage of CHM because of their young age.

Chromosomal Abnormalities Associated with Classical CHM

In an attempt to identify deletions in patients with classical CHM, the DNA of 80 male CHM patients was investigated for the absence of three loci from interval 3, i.e. DXS233, DXS165, and DXS95. Five males were identified with deletions spanning one or more of these DNA markers (Figure 1; Cremers et al., 1987, 1990b). In four of them, i.e. patients LGL2905, 25.6, 7.6, and 3.5, the DNA marker plbD5 (DXS165) was deleted, suggesting a location near or in the CHM gene. Additional DNA probes were isolated in the vicinity of the DXS165 locus employing the following methods.

1. Chromosomal walking and jumping

Employing the DNA marker plbD5 as a probe on a human genomic cosmid library a 36 kb genomic DNA fragment was isolated. Invariably, single copy probes located proximal to DXS165 (probe p1bD5) were found to be present in the DNA of one patient (7.6 in Figure 2), whereas DXS165 itself was deleted. Thus, the critical region harboring the CHM gene could be positioned distal to DXS165. Employing this marker as a probe, two clones, i.e. pJ11 and pJ15 (Figure 2), were isolated (Cremers et al., 1989b) from a human chromosomal jumping library (Collins et al., 1987). To generate starting probes for further jumps towards the telomere, probe pJ15 was subsequently used to isolate two overlapping clones spanning a total of 23 kb. Rescreening the jumping library with two single copy clones yielded clones pJ59 and pJ60, the former of which could be positioned proximal to pJ11 (Cremers et al., 1989b) while the latter was mapped distal to another deletion (patient 3.5 in Figure 2). Thus, this jump provided the distal demarcation of the segment that deletions 3.5 and 7.6 have in common.

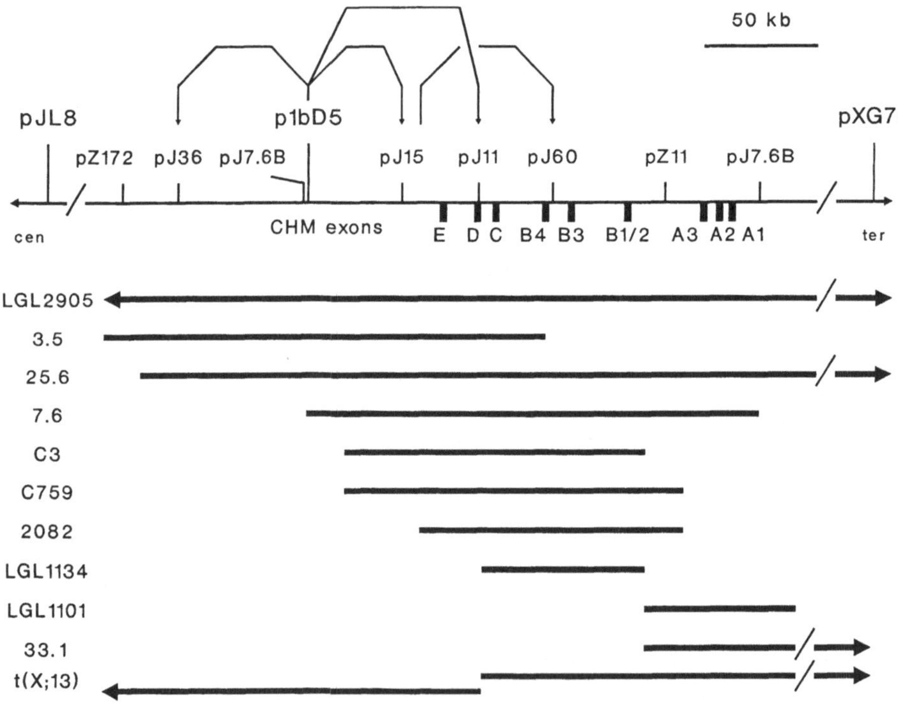

Figure 2 Deletion map of the CHM locus at Xq21.2. Deleted segments are given as horizontal bars. The location of the X-chromosomal breakpoint in a female with an X;13 translocation and choroideremia is shown as a broken bar at the bottom. The exact physical distances between the anonymous probes is not known, except for pJ7.6A/p1bD5 and pJ15/pJ11/pJ60 which have been linked by genome walking (Cremers et al., 1989b, 1990c).

2. Preparative field inversion gel electrophoresis (FIGE)

Employing FIGE analysis of control DNA, the DNA probes demarcating the critical region for CHM, i.e. p1bD5 and pJ60, were shown to hybridize to the same 625kb SfiI fragment indicating that at least part of the CHM gene was located on this fragment (Van de Pol et al., 1990). High molecular weight DNA from a cell line that contained only the human X chromosome in a rodent background was digested with SfiI and subjected to FIGE. DNA was extracted from agarose slices containing the 625 kb SfiI fragment, digested with EcoRI and subcloned into the lambdaZAP vector (Van de Pol et al., 1990). From this library, 7 clones were isolated which derived from the correct 625 kb SfiI fragment. Two of these probes, i.e. pZ172 and pZ11, are depicted in Figure 2. pZ172 was located centromeric to plbD5 and pZ11 was positioned telomeric to pJ60.

3. Deletion breakpoint cloning

As indicated above, the centromeric breakpoint of the deletion in patient 7.6 was located proximal to p1bD5. Conventional Southern analysis using a centromeric junction probe showed that the deletion junction was situated on a 10.5 kb EcoRI fragment (Cremers et al., 1990a). This fragment was gel-purified from genomic DNA of patient 7.6 and cloned into a lambda vector. Using the centromeric junction clone, several clones were isolated, one of which was analysed in more detail. The deletion junction was located on a 0.9 kb HindIII fragment which in normal genomic DNA detected the expected 7.5 kb EcoRI fragment (pJ7.6A) and a 8.5 kb EcoRI fragment (pJ7.6B) spanning the distal deletion breakpoint. pJ7.6B is located distal to the jumpclones and to pZ11 (Figure 2), and is located on a different SfiI fragment (Cremers et al., 1990a).

Employing the new probes, six additional deletions were found in patients with CHM (Figure 2) which enabled us to locate part of the CHM gene to a 50 kb region flanked by pJ11 and pZ11. Recently, two balanced translocations were found in females showing choroideremia (Kaplan et al., 1989; Siu et al., 1990). In both translocations, the breakpoints on the X-chromosome were situated in band Xq21.2, while the autosomal breakpoints were at 7p14 and 13p12, respectively. Apart from CHM, both females showed primary amenorrhoea. The X;13 translocation breakpoint could be positioned between pJ60 and pJ11, in the critical region for the CHM gene (Figure 2; Cremers et al., 1989b). Merry et al. (1990) isolated a jumping clone in the vicinity of pJ11 which also detected the X;13 translocation breakpoint. Southern blotting has indicated that the X;7 translocation breakpoint is located more distally (C. Philippe et al., unpublished data).

ISOLATION OF THE CHM GENE

Cloning of CHM cDNAs

Employing two jumpclones from the critical region for CHM, i.e. pJ11 and pJ60 (Figure 2), human genomic DNA phage clones were isolated, which together spanned a region of approximately 45 kb. Single and low copy probes were subsequently tested for evolutionary conservation by hybridization to Southern blots containing DNAs from a variety of vertebrates. Using one of the probes that detected homologous sequences in e.g. mouse and chicken, eight overlapping cDNAs were isolated from a adult human retinal cDNA library (Cremers et al., 1990c). Rescreening of the adult retina, a fetal retina, and a fetal brain cDNA library with DNA probes from the 5' and 3' ends of the consensus cDNA resulted in the isolation of a number of different clones, together spanning 4.7 kb of the cDNA including a poly(A) tail indicative of the 3' end. An overview of the most relevant CHM cDNA clones is given in Figure 3; the sequence of the ORF is given in Figure 4. The consensus cDNA contains an ORF of 1257 bps encoding a polypeptide of 419 amino acids. At the 5' end, no in-frame stopcodons have been found indicating that the gene might be longer.

Merry et al. (1992) independently isolated a CHM cDNA clone from an adult human retina library which contains a sequence similar to the above cDNA, T120, except for the presence of a 190 bp inversion flanked by short inverted repeats at the 5' end. The correct orientation of the 5' sequence could be deduced from the sequences of the closely related mouse CHM and human CHML cDNAs (Cremers et al., 1992).

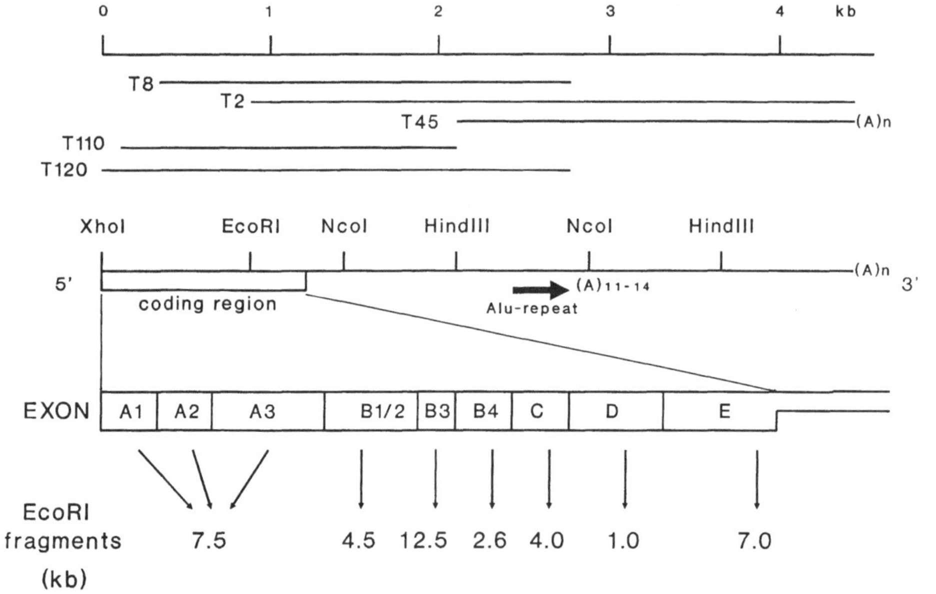

Figure 3 cDNA and exon map of the choroideremia gene. Schematic representation of the overlapping cDNA clones corresponding to the CHM gene with the restriction map of the consensus cDNA sequence shown in the middle. The coding region is indicated by an open box. cDNA clones T2 and T8 were isolated from a human retina cDNA library (Cremers et al., 1990c), clones T45 and T110 from a human fetal brain library, T120 from a fetal retina cDNA library. Clone T45 contains a poly(A) sequence indicative of the 3' end of the corresponding mRNA. The open reading frame encodes a predicted polypeptide of 419 amino acids.

```
<        .         .         .         .         .         .         .         .         .               Exon
CTGCTGTATTCTCGAGGATTACTAATTGATCTTCTAATCAAATCTAATGTTAGTCGATATGCAGAGTTTAAAAATATTACCAGGATTCTT    90     A1
 L  L  Y  S  R  G  L  L  I  D  L  L  I  K  S  N  V  S  R  Y  A  E  F  K  N  I  T  R  I  L     30
                          ><
GCATTTCGAGAAGGACGAGTGGAACAGGTTCCGTGTTCCAGAGCAGATGTCTTTAATAGCAAACAACTTACTATGGTAGAAAAGCGAATG   180     A2
 A  F  R  E  G  R  V  E  Q  V  P  C  S  R  A  D  V  F  N  S  K  Q  L  T  M  V  E  K  R  M     60
                                                         ><
CTAATGAAATTTCTTACATTTTGTATGGAATATGAGAAATATCCTGATGAATATAAAGGATATGAAGAGATCACATTTTATGAATATTTA   270
 L  M  K  F  L  T  F  C  M  E  Y  E  K  Y  P  D  E  Y  K  G  Y  E  E  I  T  F  Y  E  Y  L     90

AAGACTCAAAAATTAACCCCCAACCTCCAATATATTGTCATGCATTCAATTGCAATGACATCAGAGACAGCCAGCAGCACCATAGATGGT   360     A3
 K  T  Q  K  L  T  P  N  L  Q  Y  I  V  M  H  S  I  A  M  T  S  E  T  A  S  S  T  I  D  G    120

CTCAAAGCTACCAAAAACTTTCTTCACTGTCTTGGGCGGTATGGCAACACTCCATTTTTGTTTCCTTTATATGGCCAAGGAGAACTCCCC   450
 L  K  A  T  K  N  F  L  H  C  L  G  R  Y  G  N  T  P  F  L  F  P  L  Y  G  Q  G  E  L  P    150
             ><
CAGTGTTTCTGCAGGATGTGTGCTGTGTTTGGTGGAATTTATTGTCTTCGCCATTCAGTACAGTGCCTTGTAGTGGACAAAGAATCCAGA   540     B1/2
 Q  C  F  C  R  M  C  A  V  F  G  G  I  Y  C  L  R  H  S  V  Q  C  L  V  V  D  K  E  S  R    180

AAATGTAAAGCAATTATAGATCAGTTTGGTCAGAGAATAATCTCTGAGCATTTCCTCGTGGAGGACAGTTACTTTCCTGAGAACATGTGC   630
 K  C  K  A  I  I  D  Q  F  G  Q  R  I  I  S  E  H  F  L  V  E  D  S  Y  F  P  E  N  M  C    210
                ><                                                              ><
TCACGTGTGCAATACAGGCAGATCTCCAGGGCAGTGCTGATTACAGATAGATCTGTCCTAAAAACAGATTCAGATCAACAGATTTCCATT   720     B3
 S  R  V  Q  Y  R  Q  I  S  R  A  V  L  I  T  D  R  S  V  L  K  T  D  S  D  Q  Q  I  S  I    240
                                                                                       ><
TTGACAGTGCCAGCAGAGGAACCAGGAACTTTTGCTGTTCGGGTCATTGAGTTATGTTCTTCAACGATGACATGCATGAAAGGCACCTAT   810     B4
 L  T  V  P  A  E  E  P  G  T  F  A  V  R  V  I  E  L  C  S  S  T  M  T  C  M  K  G  T  Y    270

TTGGTTCATTTGACTTGCACATCTTCTAAAACAGCAAGAGAAGATTTAGAATCAGTTGTGCAGAAATTGTTTGTTCCATATACTGAAATG   900     C
 L  V  H  L  T  C  T  S  S  K  T  A  R  E  D  L  E  S  V  V  Q  K  L  F  V  P  Y  T  E  M    300
      ><
GAGATAGAAAATGAACAAGTAGAAAAGCCAAGAATTCTGTGGGCTCTTTACTTCAATATGAGAGATTCGTCAGACATCAGCAGGAGCTGT   990     D
 E  I  E  N  E  Q  V  E  K  P  R  I  L  W  A  L  Y  F  N  M  R  D  S  S  D  I  S  R  S  C    330
                                                                             ><
TATAATGATTTACCATCCAACGTTTATGTCTGCTCTGGCCCAGATTGTGGTTTAGGAAATGATAATGCAGTCAAACAGGCTGAAACACTT  1080
 Y  N  D  L  P  S  N  V  Y  V  C  S  G  P  D  C  G  L  G  N  D  N  A  V  K  Q  A  E  T  L    360

TTCCAGGAAATCTGCCCCAATGAAGATTTCTGTCCCCCTCCACCAAATCCTGAAGACATTATCCTTGATGGAGACAGTTTACAGCCAGAG  1170     E
 F  Q  E  I  C  P  N  E  D  F  C  P  P  P  P  N  P  E  D  I  I  L  D  G  D  S  L  Q  P  E    390

GCTTCAGAATCCAGTGCCATACCAGAGGCTAACTCGGAGACTTTCAAGGAAAGCACAAACCTTGGAAACCTAGAGGAGTCCTCTGAATAA  1260
 A  S  E  S  S  A  I  P  E  A  N  S  E  T  F  K  E  S  T  N  L  G  N  L  E  E  S  S  E  *    419
```

Figure 4 Sequence of the protein coding region of the CHM cDNA and of the predicted protein product (in one-letter amino acid code). Exon–exon boundaries are indicated above the nucleotide sequence; exons are given at the right.

Cloning of Part of the CHM Structural Gene

The genomic organization of the central portion and 3' end of the CHM gene could be inferred by comparing cDNA sequences and their genomic counterparts (Figures 2 and 3). The CHM gene spans a minimum of 12 exons, 9 of which have been subcloned into plasmid vectors from phage, cosmid and YAC clones spanning the CHM gene (van den Hurk et al., 1992; E.Bogerd et al., unpublished data). To ascertain the relationship between the cloned cDNA and the CHM gene, genomic DNAs of eight patients with CHM carrying submicroscopic deletions of the Xq21 band were reexamined with a single copy cDNA fragment which spans exons A3-B4 (Figure 3). In all patients, the deletions had removed at least part of the ORF (Figure 5). Also, by employing different parts of the cDNA as probes, we could locate the X-chromosomal breakpoint of the t(X;13) associated with CHM in a female, to intron sequences separating exons C and D (Figure 2; Cremers et al., 1990c). These findings provide convincing evidence that this cDNA is part of the CHM gene.

It is noteworthy that Merry et al. (1992) were unable to find deletions with the CHM cDNA probe in 34 unrelated patients. The marked difference between these results and those of Cremers et al. (1990b,c), who found 11 deletions in 80 patients with classical CHM, might be explained by founder effects resulting in a relatively low number of different CHM mutations in the American population.

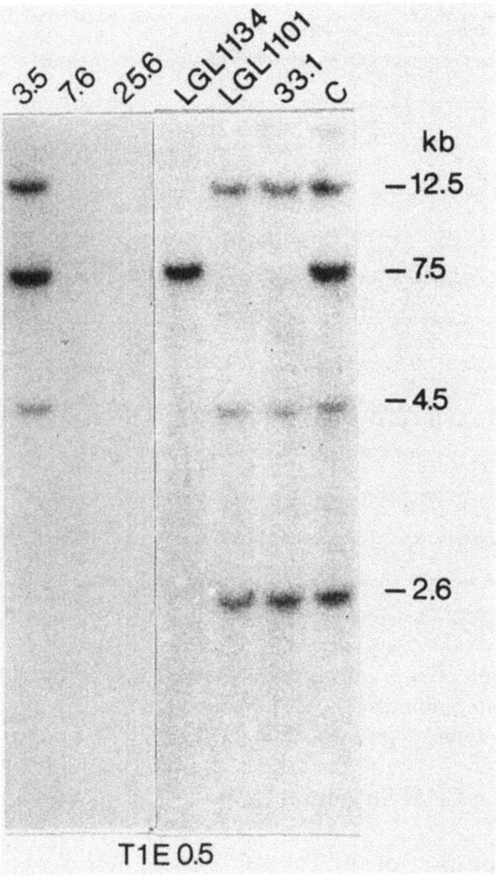

Figure 5 Southern blot analysis of EcoRI digested DNAs from several patients with classical CHM. The cDNA probe T1E0.5 encompasses exons A3, B1/2, B3, and B4 which are located on EcoRI fragments of 7.5 kb, 4.5 kb, 12.5 kb, and 2.6 kb, respectively (From Cremers et al., 1990c, by permission of *Nature*).

Expression Pattern of the CHM Gene

Northern blot analysis of RNA from several human cell lines and tissues using CHM cDNA clones revealed that the CHM gene is not only expressed in ocular tissues or ocular derived cell lines, but also in various cells of non-ocular origin, e.g. HeLa, lymphoblast cells and fibroblasts (Figure 6, Cremers et al., 1990c; Merry et al., 1992). Yet clinical symptoms are confined to the eye. This finding is not unprecedented, however. Gyrate atrophy (GA), a choroidal disease with clinical similarity to choroideremia, is also caused by a deficiency of an ubiquitously expressed protein, the enzyme ornithine aminotransferase (Valle and Simell, 1989). Apart from ocular symptoms resulting from the typical sharply demarcated, circular areas of chorioretinal degeneration, most patients with GA are asymptomatic. Proximal muscle weakness which is seen in less than 10% of the patients is the only extraocular manifestation of this disease (Valle and Simell, 1989).

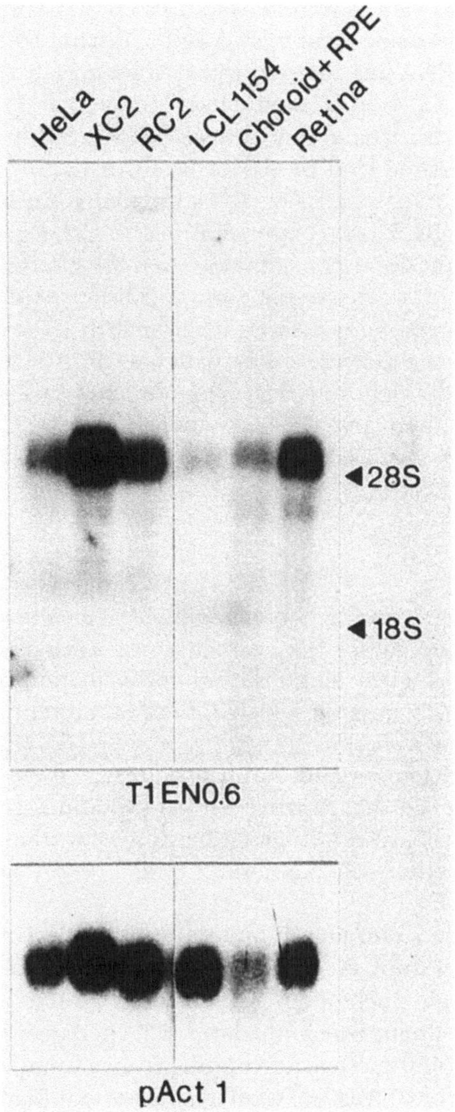

Figure 6 Northern blot analysis of RNA from several human cell lines and tissues using cDNA clone T1. **a.** Blot containing RNA from HeLa, two retinal cell lines (HER RC2 and HER XC2), an EBV-immortalized lymphoblastoid cell line (LCL1154), as well as human choroid/retinal pigment epithelium and retina screened with clone T1EN0.6. **b.** Hybridization of a hamster actin cDNA clone (pAct-1). The positions of the 18S and 28S ribosomal RNA bands are indicated (From Cremers et al., 1990c, by permission of *Nature*).

MUTATION SPECTRUM

Northern Analysis

As shown in Figure 6, the CHM gene is expressed at a low level in lymphoblasts. Northern blot analysis using total mRNA isolated from EBV-immortalized B-cells showed that CHM mRNA was absent, or present at greatly reduced levels, in 8 out of 11 patients (F.P.M.C., unpublished data). Merry et al. (1992) used the reverse transcriptase-polymerase chain reaction (RT-PCR) technique employing exon-primers and observed reduced or absent levels of CHM mRNA in 25 out of 34 patients. This kind of analysis, using RNA isolated from EBV immortalized cell lines has several pitfalls. Firstly, expression of the CHM gene in B-cells or B-cell derived cell lines might differ considerably from the expression of the CHM gene in ocular tissues of the respective patients. Secondly, the expression of the CHM gene varies among B cell lines derived from healthy controls thereby complicating the quantitation of mRNA levels in patients with CHM (Merry et al., 1992). Thirdly, the highly homologous CHML gene on chromosome 1, might interfere with both Northern blot and RT-PCR analysis of the CHM gene. The high sequence identity between CHM and CHML necessitates the use of very specific primers for RT-PCR.

SSCP Analysis

Forty-five CHM patients and an equal number of healthy individuals from the Western European population were screened for nucleotide sequence variation in exons B3-E employing the single strand conformation polymorphism (SSCP) technique. Significantly altered electrophoretic mobilities were observed in eleven patients but in none of the healthy individuals. Direct DNA sequencing revealed five small deletions and 3 substitutions in the exon sequences and 3 mutations altering splice sites (Figure 7). All mutations result in stopcodons at the site of mutation or just downstream because of a frameshift (Van den Hurk et al., 1992; Sankila et al., 1992; Schwartz et al., 1993). No missense mutations have yet been found.

The most frequently occurring mutation found so far is the CHM*SALLA mutation, where a splice donor site just downstream of exon C is inactivated through the insertion of a single nucleotide (Sankila et al., 1992). This mutation cosegregates with the disease phenotype in the large Salla pedigree comprising more than 60 living patients and 100 carriers descending from a founder couple born 13 generations ago (Sankila et al., 1987). Detailed analysis of mRNA isolated from a B cell line of one of the patients using RT-PCR indicated that the mutation results in two aberrant cDNA products both of which give rise to stopcodons.

An overview of the small mutations is given in Figure 7. It is of note that two patients (COP-E and 2086), carry the same mutation, i.e. a four base deletion in exon C. Patient COP-E is a Danish patient, whereas 2086 is an unrelated patient from Southern Germany. The deleted sequence (TGTT) is present in tandem which may indicate that this mutation is due to a replication slippage event. Also, there is an apparent clustering of mutations in exon B4; 4 mutations were observed in a stretch of 97 bps. Up to now, only 610 bps of the ORF have been investigated for small mutations. This segment corresponds to approximately half of the cloned CHM cDNA and to approximately one-third of the anticipated total length of the ORF. The mouse cDNA homologue of the CHM gene as well as a highly homologous gene from chromosome 1 show open reading frames which, at their 5' ends, are 173 and 236 amino acids longer respectively (see below). Hence, it is assumed that approximately 35% of the CHM ORF is still missing. Taking into account that, under the conditions used, the SSCP technique only reveals 80% of all mutations (Michaud et al., 1992) the observed percentage of small mutations among non-deletion CHM patients (11/45 = 24%) suggests a relatively even distribution of mutations throughout the ORF and the promoter of the CHM gene. So far, small mutations or larger rearrangements like deletions and translocations could be detected in approximately 35% of patients. Together with the results from mRNA analysis, these findings provide convincing evidence for the candidate gene being identical with the choroideremia gene.

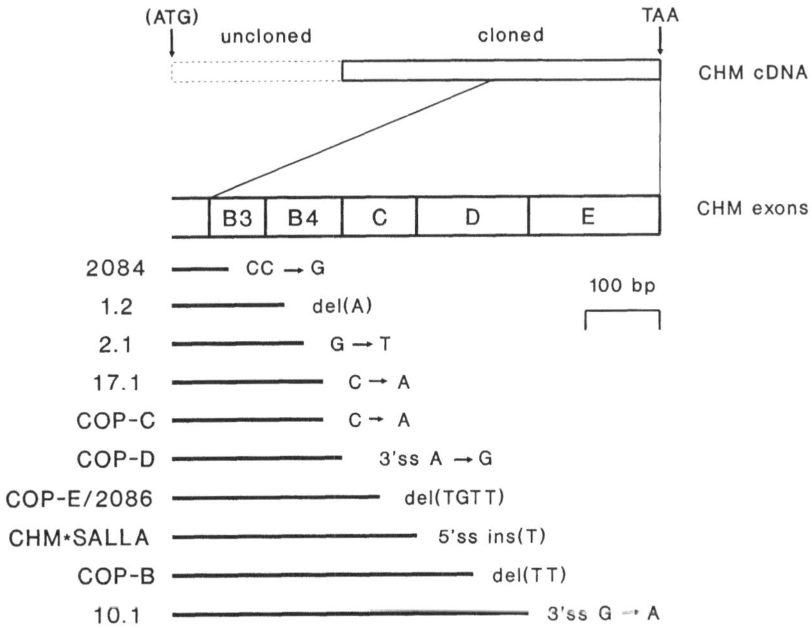

Figure 7 Nonsense mutations in the CHM gene. Upper half: gene segment that has been investigated for small mutations by employing the SSCP technique. Lower half: mutations observed and length of the resulting truncated proteins.

CLONING OF THE MURINE HOMOLOGUE AND A HUMAN AUTOSOMAL HOMOLOGUE FROM CHROMOSOME 1

Employing a cDNA clone from the 5' end of the CHM gene (T8; Figure 3), a cDNA corresponding to the murine CHM gene, mCHM, was isolated. The mCHM cDNA sequence is more complete than the consensus human CHM cDNA sequence; it contains 173 additional amino acids at its 5' end. Overall, mCHM and hCHM show 88% nucleotide and amino acid sequence identity, illustrating their high degree of evolutionary conservation. When conservative amino acid replacements are disregarded, mCHM displays 97% similarity to hCHM (Cremers et al., 1992; Figure 8).

```
hCHML   MADNLPTEFDVVIIGTGLPESILAAACSRSGQRVLHIDSRSYYGGNWASFSFSGLLSWLKEYQQNNDIGEESTVVWQDLIHETEEAITLR  90
mCHM                                                                              M--EQ-L-N----L-S  16

hCHML   KKDETIQHTEAFSYASQDMEDNVEEIGALQKNPSLGVSNTFTEVLDSALPEESQLSYFNSDEMPAKHTQKSDTEISLEVTDVEESVEKEK 180
mCHM    S--K----V-V-C-----LHKD---A-------ASVM-AQA--AAEA-EAA-ATEA  AEAAEA-EAACLPTA-E--STRSC-LPA-QSQ 104

hCHML   YCGDKTCMHT    VSDKDGDKDESKST         VEDKADEPIRNRITYSQIVKEGRRFNIDLVSKLLYSQGLLIDLLIKSDVSRYV 257
mCHM    CM-PESSPQVNDAE-GEKETQS-AKS--EQSSEILPK-Q-NTET-KK--V------I----------------R----------N----A 194
hCHM                                                                         ----R----------N----A  21

hCHML   EFKNVTRILAFREGKVEQVPCSRADVFNSKELTMVEKRMLMKFLTFCLEYEQHPDEYQAFRQCSFSEYLKTKKLTPNLQHFVLHSIAMTS 347
mCHM    ----I---------T---------------Q----------------V---D--G--K-YEETT-------Q-------Y---------- 284
hCHM    ----I---------R---------------Q----------------M---KY----KGYEEIT-Y-----Q-------YI-M------- 111

hCHML   ESSCTTIDGLNATKNFLQCLGRFGNTPFLFPLYGQGEIPQGFCRMCAVFGGIYCLRHKVQCFVVDKESGRCKAIIDHFGQRINAKYFIVE 437
mCHM    -TTSS-V---K---K-------Y--------------L--C----------------S---L------RK----V-Q-----IS-H-VI- 374
hCHM    -TASS-----K------H----Y--------------L--C----------------S---L------RK------Q-----ISEH-L-- 201

hCHML   DSYLSEETCSNVQYKQISRAVLITDQSILKTDLDQQTSILIVPPAEPGACAVRVTELCSSTMTCMKDTYLVHLTCSSSKTAREDLESVVK 527
mCHM    ------N---G---R----------G-V--P-S---V---T--AE-S-SF----I-----------G--------M----------R--Q 464
hCHM    ---FP-NM--R---R----------R-V----S---I---T--AE---TF----I-----------G--------T-------------Q 291

hCHML   KLFTPYTETEINEEELTKPRLLWALYFNMRDSSGISRSSYNGLPSNVYVCSGPDCGLGNEHAVKQAETLFQEIFPTEEFCPPPPNPEDII 617
mCHM    --------I-AEN-QVE---I------------D---DC--D-------------N---DN--Q---IV--K-C-N-D---A-------- 554
hCHM    ---V----M--EN-QVE---I------------D----C--D-----------------DN------------C-N-D------------ 381

hCHML   FDGDDKQPEAPGTNNVVMAKLESSEESKNLESPEKHLQN 656
mCHM    L---SS-Q-VSESSVIPETNS-TPK--TV-GDS-EPSE  592
hCHM    L---SL----SESSAIPE-NS-TFK--T--GNL-ESSE  419
```

Figure 8 Alignment of the human CHML amino acid sequence with mouse CHM and human CHM sequences. Dashes indicate identical amino acid residues. The numbers at the right correspond to the amino acid residues in the respective sequences (Adapted from Cremers et al., 1992, by permission of *Hum. Mol. Genet.*).

In an attempt to isolate sequences from the 5' end of the hCHM cDNA, a mCHM cDNA clone was used as a probe to screen a human fetal brain cDNA library. Among six phage recombinants identified, four were partly overlapping but different from the hCHM cDNA consensus (Cremers et al., 1992). The consensus cDNA of these new clones contains an ORF encoding 656 amino acids and is designated hCHML which stands for human choroideremia-like. The nucleotide sequence and the deduced amino acid sequence of hCHML was compared with hCHM sequences. Amino acid residues 237-655 of hCHML show 76% identity and 95% similarity to hCHM; the nucleotide sequences are 80% identical (Figure 8). Northern analysis has shown that the CHML gene is expressed in the same tissues as the CHM gene, although at different and generally lower levels (F.P.M.C., unpublished data). One of the hCHML cDNA clones was hybridized to a panel of 32 human-mouse or human-hamster cell lines containing various different human chromosomes. These studies showed that the hCHML gene resides on chromosome 1. In addition, hCHML subclones were hybridized to Southern blots with DNAs from three somatic cell hybrids carrying defined segments of chromosome 1. In this way, hCHML could be assigned to 1q42-qter (Cremers et al., 1992; F.P.M.C., unpublished data). Recently, the gene for another form of retina degeneration, Usher syndrome type 2 (USH2), has been localized in approximately the same chromosomal interval by linkage analysis (Kimberling et al., 1990; Lewis et al., 1990). Usher syndrome consists of retinitis pigmentosa with hearing loss and is inherited as an autosomal recessive condition (Chapter xx). The chromosomal region containing the USH2 locus and hCHML is estimated to comprise no more than 2% of the human genome. This co-localization, together with the clinical similarities between CHM and USH2, renders hCHML a promising candidate gene for this disorder. More accurate physical mapping of hCHML and closely linked flanking markers of USH2, as well as hCHML segregation studies in USH2 families, are needed to substantiate this hypothesis. Final proof for the identity of hCHML and USH2 may come from the identification of mutations in the hCHML gene in patients with the USH2 phenotype. It is of note, however, that so far, screening of 10 patients with USH2 has failed to detect gross rearrangements or point mutations in the hCHML gene (T. J. R. van de Pol et al., unpublished data).

THE CHM PROTEIN MAY BE A COMPONENT OF A GERANYLGERANYL TRANSFERASE

Many membrane-associated eukaryotic proteins contain a covalently bound farnesyl or geranylgeranyl group attached in thioether linkage to cysteine residues near the COOH termini (Farnsworth et al., 1990; Rilling et al., 1990). This modification facilitates proper binding to specific membranes and is essential for the activity of the relevant proteins. At least three different protein prenyltransferases have been identified in cytosolic extracts of animal tissues. One of these, a Rab geranylgeranyl transferase (designated Rab GGT) isolated from rat brain, consists of two components, A and B, of 95 kD and 90 kD respectively. The latter of these is composed of two subunits (Seabra et al., 1992a,b). Six tryptic peptides from component A show almost complete identity to the mouse homologue of the CHM protein, strongly suggesting that component A is the rat homologue of the CHM protein.

This important finding sheds new light on an earlier observation that the CHM and CHML polypeptides show weak but significant homology to Rab3A (or smg p25A) GDP dissociation inhibitor (GDI), a bovine brain protein which inhibits the exchange of bound GDP by GTP on the small ras-like GTP-binding protein Rab3A (Matsui et al., 1990; Fodor et al., 1991). CHML shows two regions of approximately 50 amino acid residues which are 50% identical to bovine Rab3A GDI (Cremers et al., 1992). Because both Rab3A GDI and Rab GGT interact with Rab3A, their sequence homology may indicate that these proteins have similar binding domains for this small GTP-binding protein.

Assuming that CHM and Rab GGT are identical, there are several remaining questions related to the pathogenesis of choroideremia, as follows.

1. Both CHM (Rab GGT) and Rab3A are expressed in a variety of tissues, including brain. Why then, is the defect limited to the choroid and retina?
2. What is the role of the CHML gene product? Does it geranylgeranylate the same or another class of small GTP binding proteins? In view of its high degree of similarity with CHM, is it possible that CHML expression modulates the CHM phenotype? Is there a CHML homologue in mice and rats?
3. Rab3A is thought to regulate vesicle transport in brain synapses (Fischer von Mollard et al., 1990; Mizoguchi et al., 1990). Could Rab3A or a related geranylgeranylated GTP binding protein be involved in membrane turnover in rod outer segments? In accordance with this hypothesis, the fundus abnormalities observed in early stages of the disease suggest that lesions first occur in or near the retinal pigment epithelium (Waardenburg, 1942; Pameijer et al., 1960; McCulloch, 1969; Hammerstein et al., 1979; Hammerstein and Bohm, 1985).

Future studies will concentrate on the identification of Rab GGT/CHM geranylgeranylated small GTP binding proteins in the retina, RPE, and choroid. The subcellular expression of these molecules as well as the CHM gene product will be studied by immunohistochemical techniques. Also, a better understanding of the pathological processes in choroideremia may be achieved by the construction of a mouse model via insertional inactivation of the mouse CHM gene. Eventually, complete understanding of the pathological processes underlying choroideremia may provide a basis for rational therapy.

REFERENCES

Bach, I., Cremers, F. P. M., Robinson, D., Thomas, N., and Ropers, H-H. (1992) Physical fine mapping of genes underlying X-linked deafness and non fra(X)-X-linked mental retardation at Xq21. *Hum Genet,* **89**, 620–624.

Bedell, A. J. (1937) Choroideremia. *Arch Ophthalmol,* **17**, 444–467.

Collins, F. S., Drumm, M. L., Cole, J. L., Lockwood, W. K., Van de Woude, G. F., and Iannuzzi, M. C. (1987) Construction of a general human chromosome jumping library, with application to cystic fibrosis. *Science,* **235**, 1046–1049.

Cremers, F. P. M., Brunsmann, F., Van de Pol, T. J. R., Pawlowitzki, I. H., Paulsen, K., Wieringa, B., and Ropers, H-H. (1987) Deletion of the DXS165 locus in patients with classical choroideremia. *Clin Genet,* **32**, 421–423.

Cremers, F. P. M., Van de Pol, T. J. R., Wieringa, B., Hofker, M. H., Pearson, P. L., Pfeiffer, R. A., Mikkelsen, M., Tabor, A., and Ropers, H.-H. (1988) Molecular analysis of male-viable deletions and duplications allows ordering of 52 DNA probes on proximal Xq. *Am J Hum Genet,* **43**, 452–461.

Cremers, F. P. M., Van de Pol, T. J. R., Diergaarde, P. J., Wieringa, B., Nussbaum, R. L., Schwarz,

M., and Ropers, H-H. (1989a) Physical fine mapping of the choroideremia locus using Xq21 deletions associated with complex syndromes. *Genomics*, **4**, 41–46.

Cremers, F. P. M., Van de Pol, T. J. R., Wieringa, B., Collins, F. S. , Sankila, E-M., Siu, V. M., Flintoff, W. F., Brunsmann, F., Blonden, L. A. J., and Ropers, H-H. (1989b) Chromosomal jumping from the DXS165 locus allows molecular characterization of four microdeletions and a de novo chromosome X/13 translocation associated with choroideremia. *Proc Natl Acad Sci USA*, **86**, 7510–7514.

Cremers, F. P. M., Brunsmann, F., Berger, W., Van Kerkhoff, E. P. M. , Van de Pol, T. J. R., Wieringa, B., Pawlowitzki, I. H., and Ropers, H-H. (1990a) Cloning of the breakpoints of a deletion associated with choroideremia. *Hum Genet*, **86**, 61–64.

Cremers, F. P. M., Sankila, E-M., Brunsmann, F., Jay, M., Jay, B., Wright, A., Pinckers, A. J. L. G., Schwartz, M., Van de Pol, T. J. R. , Wieringa, B., De la Chapelle, A., Pawlowitzki, I. H., and Ropers, H.-H. (1990b) Deletions in patients with classical choroideremia vary in size from 45kb to several Megabases. *Am J Hum Genet*, **47**, 622–628.

Cremers, F. P. M., Van de Pol, T. J. R., Kerkhoff, E. P. M., Wieringa, B., and Ropers, H-H. (1990c) Cloning of a gene that is rearranged in patients with choroideraemia. *Nature*, **347**, 674–677.

Cremers, F. P. M., Molloy, C. M., Van de Pol, T. J. R., Van den Hurk, J. A. J. M., Bach, I., Geurts van Kessel, A. H. M., and Ropers, H-H. (1992) An autosomal homologue of the choroideremia gene colocalizes with the Usher syndrome type II locus on the distal part of chromosome 1q. *Hum Mol Genet*, **1**, 71–75.

Farnsworth, C. C., Gelb, M. H., and Glomset, J. A. (1990) Indentification of geranylgeranyl-modified proteins in HeLa cells. *Science*, **247**, 320–322.

Fischer von Mollard, G., Sudhof, T. C., and Jahn, R. (1991) A small GTP-binding protein dissociates from synaptic vesicles during exocytosis. *Nature*, **349**, 79–81.

Fodor, E., Lee, R. T., and O'Donnell, J. J. (1991) Analysis of choroideraemia gene. *Nature*, **351**, 614.

François, J. (1968) Choroideremia (Progressive Chorioretinal Degeneration). *Int Ophthalmol Clin*, **8**, 949–964.

Fraser, G. R. F. and Friedmann, A. I. (1968) Choroideremia in a female. *Brit Med J*, 2, 732–734

Friedman, B. (1940) Choroideremia. *Arch Ophthalmol*, **23**, 1285–1287.

Goedbloed, J. (1942) Mode of inheritance in chorioderemia. *Ophthalmologica*, **104**, 308–315.

Hammerstein, W., Bischof, G., and Leide, E. (1979) Chorioideremie im Kindesalter. *Klin Monatsbl Auqenh*, **174**, 599–604.

Hammerstein, W. and Böhm, T. (1985) *Fluoreszenzanqiographische Verlaufsbeobachtunqen bei Chorioideremie* 300–303. Stuttgart: Ophthalmologische Genetik.

Harris, G. S. and Miller, J. R. (1968) Choroideremia. Visual defects in a heterozyqote. *Arch Ophthalmol*, **80**, 423–429,

Heckenlively, J. R. (1988) *Retinitis Piqmentosa* Philadelphia: Lippincott.

Hodgson, S. V., Robertson, M. E., Fear, C. N., Goodship, J., Malcolm, S., Jay, B., Bobrow, M., and Pembrey, M. E. (1987) Prenatal diagnosis of X-linked choroideremia with mental retardation, associated with a cytologically detectable X-chromosome deletion. *Hum Genet*, **75**, 286–290.

Kaplan, J., Gilgenkrantz, S., Dufier, J. L., and Frezal, J. (1989) Choroideremia and ovarian dysgenesis associated with an X;7 de novo balanced translocation. Human Gene Mapping 10: Tenth International Workshop on Human Gene Mapping. *Cytoqenet Cell Genet*, **51**, 1022.

Kärnä, J. (1986) Choroideremia: a clinical and genetic study of 84 Finnish patients and 126 female carriers. *Acta Ophthalmol*, **176**(suppl), 1–68.

Kimberling, W. J., Weston, M. D., Moller, C., Davenport, S. L. H., Shugart, Y. Y., Priluck, I. A., Martini, A., Milani, M., and Smith, R. J. (1990) Localization of Usher syndrome type II to chromosome 1q. *Genomics*, **7**, 245–249.

Lewis, R. A., Otterud, B., Stauffer, D., Lalouel, J.-M., and Leppert, M. (1990) Mapping recessive ophthalmic diseases: linkage of the locus for Usher syndrome type II to a DNA marker on chromosome 1q. *Genomics*, **7**, 250–256.

Matsui, Y., Kikuchi, A., Araki, S., Hata, Y., Kondo, J., Teranishi, Y., and Takai, Y. (1990) Molecular cloning and characterization of a novel type of regulatory protein (GDI) for smg p25A, a ras p21-like GTP-binding protein. *Mol Cell Biol*, **10**, 4116–4122.

Mauthner, H. (1872) Ein fall von choroideremia. *Ber Naturw-med Ver Innsbruck*, **2**, 191.

McCulloch, C. and McCulloch, R. J. P. (1948) A hereditary and clinical study of choroideremia. *Trans Am Acad Ophthalmol Otolaryngol*, **52**, 160–190.

McCulloch, J. C. (1969) Choroideremia: a clinical and pathologic review. *Trans Am Ophthalmol Soc*, **67**, 142–195.

Merry, D. E., Lesko, J. G., Sosnoski, D. M., Lewis, R. A., Lubinsky, M., Trask, B., and Van den Engh, G.,et al (1989) Choroideremia and deafness with stapes fixation: a contiguous gene deletion syndrome in Xq21. *Am J Hum Genet*, **45**, 530–540.

Merry, D. E., Lesko, J. G., Siu, V. M., Flintoff, W. F., Collins, F. S., Lewis, R. A., and Nussbaum, R. L. (1990) DXS165 detects a translocation breakpoint in a woman with choroideremia and a de novo X;13 translocation. *Genomics*, **6**, 609–615.

Merry, D. E., Janne, P. A., Landers, J. E., Lewis, R. A., and Nussbaum, R. L. (1992) Isolation of a candidate gene for choroideremia. *Proc Natl Acad Sci USA*, **89,** 2135–2139.

Michaud, J., Brody, L. C., Steel, G., Fontaine, G., Martin, L. S., Valle, D., and Mitchell, G. (1992) Strand-separating conformational polymorphism analysis: efficacy of detection of point mutations in the human ornithine -aminotransferase gene. *Genomics*, **13**, 389–394.

Mizoguchi, A., Kim, S., Ueda, T., Kikuchi, A., Yorifuji, H., Hirokawa, N., and Takai, Y. (1990) Localization and subcellular distribution of smg p25A, a ras p21-like GTP-binding protein, in rat brain. *J Biol Chem*, **265**, 11872–11879.

Nussbaum, R. L., Lesko, J. G., Lewis, R. A., Ledbetter, S. A., and Ledbetter, D. H. (1987) Isolation of anonymous DNA sequences from within a submicroscopic X chromosomal deletion in a patient with choroideremia, deafness, and mental retardation. *Proc Natl Acad Sci USA*, **84**, 6521–6525.

Pameyer, J. K., Waardenburg, P. J., and Henkes, H. E. (1960) Choroideremia. *Brit J Ophthalmol*, **44**, 724–738

Reardon, W., Robers, S., Phelps, P. D., Thomas, N. S., Beck, L., Isaac, R., and Hughes, H. E. (1992) Phenotypic evidence for a common pathogenesis in X-linked deafness pedogrees and in Xq13-q21 deletion related deafness. *Am J Med Genet*, **44**, 513–517.

Rilling, H. C., Breunger, E., Epstein, W. W., and Crain, P. F. (1990) Prenylated proteins: the structure of the isoprenoid group. *Science*, **247**, 318–320.

Rosenberg, T., Schwartz, M., Niebuhr, E., Yang, H-M., Sardemann, H., Andersen, O., and Lundsteen, C. (1986) Choroideremia in interstitial deletion of the X chromosome. *Ophthalmic Paediatr Genet*, **7**, 205–210.

Rosenberg, T., Niebuhr, E., Yang, H-M., Parving, A., and Schwartz, M. (1987) Choroideremia, congenital deafness and mental retardation in a family with an X chromosomal deletion. *Ophthalmic Paediatr Genet*, **8**, 139–143.

Sankila, E.-M., De la Chapelle, A., Kärnä, J., Forsius, H., Frants, R., and Eriksson, A. (1987) Choroideremia: close linkage to DXYS1 and DXYS12 demonstrated by segregation analysis and historical-genealogical evidence. *Clin Genet*, **31**, 315–322.

Sankila, E.-M., Tolvanen, R., Van den Hurk, J. A. J. M., Cremers, F. P. M., and De la Chapelle, A. (1992) Aberrant splicing of the CHM gene is a significant cause of choroideremia. *Nat Genet*, **1**, 109–113.

Schutzbach, M. (1938) Über erbliche Aderhaut-Netzhaut- erkrankung. *Albrecht von Graefes Arch Ophthalmol*, **138**, 315–331.

Schwartz, M., Rosenberg, T., Niebuhr, E., Lundsteen, C., Sardemann, H., Andersen, O., Yang, H-M., and Lamm, L. U. (1986) Choroideremia; further evidence for assignment of the locus to Xql3-Xq21. *Hum Genet*, **74**, 449–452.

Schwartz, M., Yang, H-M., Niebuhr, E., Rosenberg, T., and Page, D.C. (1988) Regional localization of polymorphic DNA loci on the proximal long arm of the X chromosome using deletions associated with choroideremia. *Hum Genet*, **78**, 156–160.

Schwartz, M., Rosenberg, T., Van den Hurk, J. A. J. M., Van de Pol, T. J. R., and Cremers, F. P. M. (1993) Identification of mutations in Danish choroideremia families. *Hum Mutation*, **2,** 43–47.

Seabra, M. C., Brown, M. S., Slaughter, C. A., Sudhof, T. C., and Goldstein, J. L. (1992a) Purification of component a of rab geranylgeranyl transferase: apparent identity with the choroideremia gene product. *Cell*, **70**, 1049–1057.

Seabra, M. C., Goldstein, J. L., Sudhof, T. C., and Brown, M. S. (1992b) Rab Geranylgeranyl Transferase: A Multisubunit Enzyme That Prenylates GTP-Binding Proteins Terminating in Cys-X-Cys or CysCys. *J Biol Chem*, **267**, 14497–14503.

Siu, V. M., Gonder, J. R., Jung, J. H., Sergovich, F. R., and Flintoff, W. F. (1990) Choroideremia associated with an X-autosomal translocation. *Hum Genet*, **84**, 459–464.

Tabor, A., Andersen, O., Lundsteen, C., Niebuhr, E., and Sardemann, H. (1983) Interstitial deletion in the "critical region" of the long arm of the X chromosome in a mentally retarded boy and his normal mother. *Hum Genet*, **64**, 196–199.

Valle, D. and Simell, O. (1989) The hyperornithinemias. In *The Metabolic Basis Of Inherited Disease I*, C. R. Scriver, A. L. Beaudet, W. S. Sly, and D. Valle, eds. (pp. 599–627.New York: McGraw-hill, Inc).

Van de Pol, T. J. R., Cremers, F. P. M., Brohet, R. M., Wieringa, B., and Ropers, H.-H. (1990) Derivation of clones from the choroideremia locus by preparative field inversion gel electrophoresis. *Nucleic Acids Res*, **18**, 725–731.

Van den Hurk, J. A. J. M., Van de Pol, T. J. R., Molloy, C. M., Brunsmann, F., Ruther, K., Zrenner, E., Pinckers, A. J. L. G., Pawlowitzki, I. H., Bleeker-Wagemakers, E., Wieringa, B., Ropers, H-H., and Cremers, F. P. M. (1992) Detection and characterization of point mutations in the choroideremia candidate gene by PCR-SSCP analysis and direct DNA sequencing. *Am J Hum Genet*, **50**, 1195–1202.

Waardenburg, P. J. (1942) Chorioideremie als Erbmerkmal. *Acta Ophthalmol*, **20**, 235–274

Yang, H.-M., Lund, T., Niebuhr, E., N_rby, S., Schwartz, M., and Shen, L. (1990) A deletion panel of the long arm of the X chromosome: subregional localization of 22 DNA probes. *Hum Genet*, **85**, 25–30.

13. NORRIE DISEASE

A. A. B. BERGEN[1], W. BERGER[2], Z.-Y. CHEN[3], P. J. DIERGAARDE[1], E. M. BLEEKER-WAGEMAKERS[1], E. M. BATTINELLI[3], M. WARBURG[4], H.-H. ROPERS[2], I. CRAIG[3]

[1]*The Netherlands Ophthalmic Research Institute, P.O. Box 12141, NL-1100 AC Amsterdam, The Netherlands*

[2]*University Hospital Nijmegen. Department of Human Genetics, P.O. Box 9101, 6500 HB Nijmegen, The Netherlands*

[3]*University of Oxford. Department of Biochemistry. Genetics laboratory. South Park Road, Oxford OX 13QU, UK*

[4]*Gentofte Hospital, Department of Ophthalmology, Division of Paediatric Ophthalmology and Handicaps, 40 Sognevej, DK-2820 Gentofte, Denmark*

CLINICAL FEATURES

Norrie disease (NDP) was delineated by Warburg (Warburg, 1961; 1966) who subsequently found that patients with this disorder had been mentioned by the Danish ophthalmologist to the Royal Institute for the Blind, Gordon Norrie (Norrie, 1927). As Norrie was the first to recognise the genetic origin of the disorder, Warburg decided to name the disease after him. The condition has been observed in all ethnic groups. Since the disease is X-linked it is manifest in boys only. Two affected females have been observed (I. Craig, personal communication). It is probably the most common disorder with bilateral retrolental vascular masses.

Ophthalmological findings

The main pathology of NDP is congenital or very early bilateral blindness due to opacities behind the lens presenting in eyes of normal size. The patients are born at full term and need no oxygen treatment. Within the first month of life, a white or yellow vascular membrane is seen behind the lens, and by ultrasonography this is recognised as a totally detached retina. In rare cases, the detachment appears in the second eye only after some months and a progressive cystic schisis is then observed in the posterior pole of this eye (Figure 1). The cyst soon changes into a total detachment. Most affected boys are blind from the detachment, but some perception of light may remain for a couple of years. In the beginning the lens is clear, and elongated ciliary processes are seen behind it (Figure 2). The affected eyes have shallow anterior chambers, and may in rare cases develop glaucoma. After a few years the lens becomes cataractous, the cornea becomes opaque, and the eyes begin to shrink (Figure 3). At that stage microphthalmos may be erroneously suspected (Warburg, 1966; 1975). Treatment with vitrectomy or peeling the detached retinal membranes has been tried but was in no case successful (Liberfarb et al., 1985; Enyedi et al., 1991).

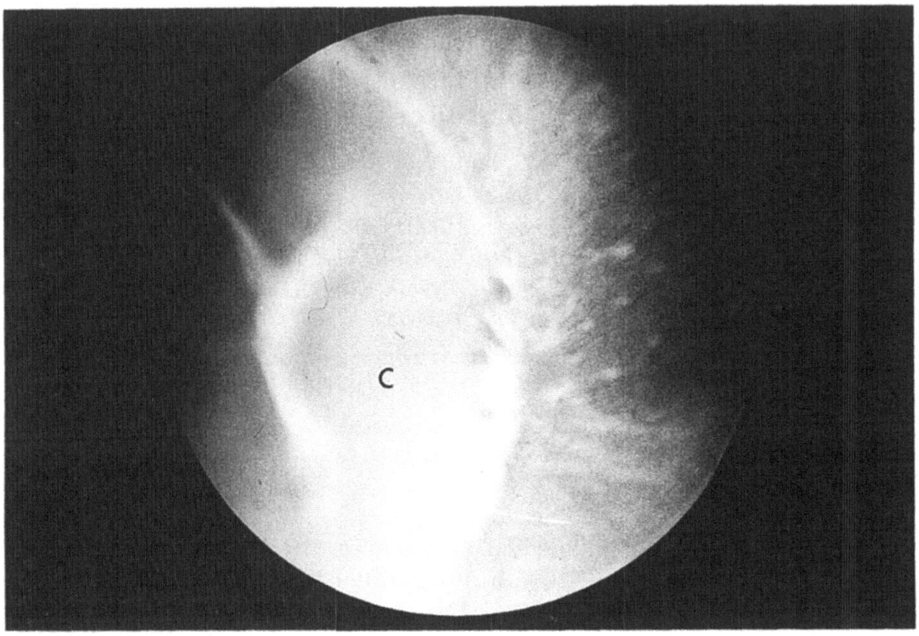

Figure 1 Delayed retinal detachment in NDP. Progressive cystic schisis in the posterior pole in the less affected second eye of a Norrie disease patient (C).

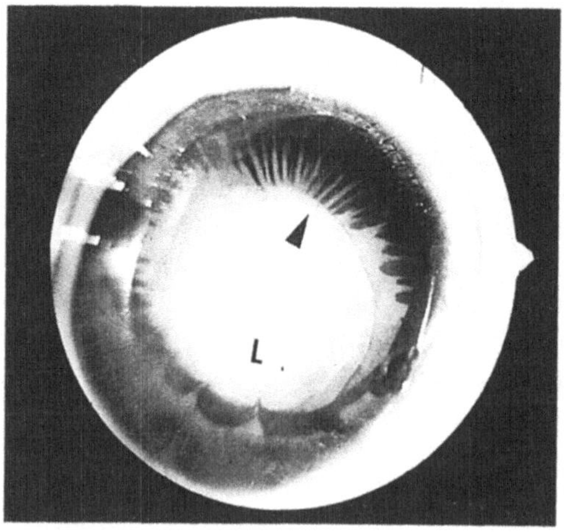

Figure 2 Elongated ciliary processes (arrow), which can be seen behind a clear lens (L).

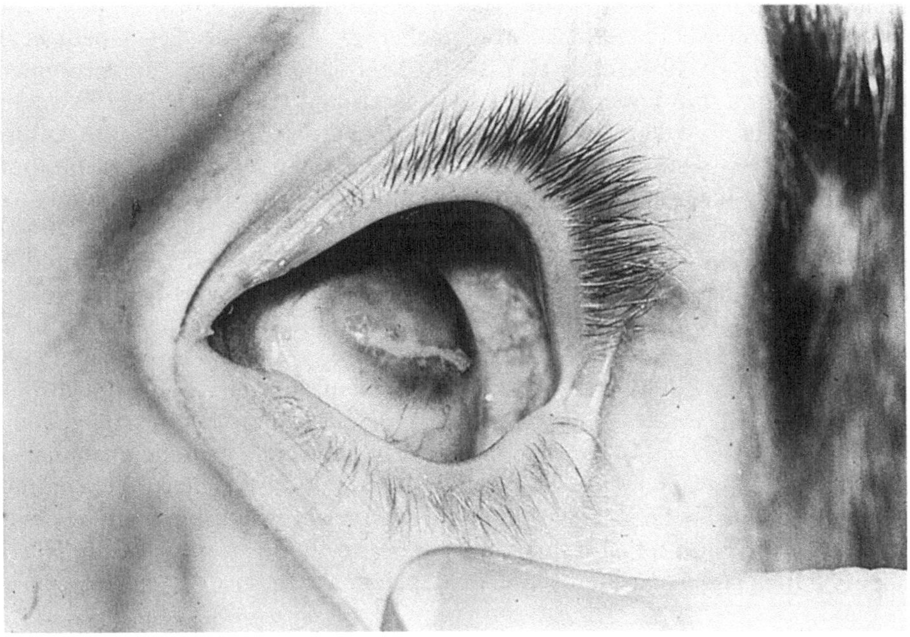

Figure 3 Phthisis of the globe in NDP. The shrinking of the eye, and opacity of the cornea car be seen in this NDP patient.

Histology

Sometimes the retrolental masses are mistaken for retinoblastoma and the eye is enucleated. Histological examination shows that the detached retina contains malformative rosettes, many large retinal vessels, haemorrhage, gliosis, and optic atrophy. Hyperplasia of the retinal pigment epithelium was also observed (Warburg, 1966; Andersen and Warburg, 1961; Liberfarb et al., 1985; Blodi and Hunter, 1969). In ultramicroscopical sections (Liberfarb et al., 1985) the inner and outer neuroblastic cells are undifferentiated.

Systemic signs

The patients are usually healthy and have a normal life span. Many are very intelligent; in most cases procreation is normal. During their lifetime, about one third of the patients develop progressive neurosensory hearing impairment. This may begin in childhood or as late as in the 6th decade. One quarter of those affected show dementia or autistic signs which may develop at any time between the age of 2 and 60 years. In some cases the psychiatric impairment may be a complication of the combined visual and auditory communicative disorder, while in others the blind men and boys show autistic signs even when hearing is still normal. In a few families, the affected blind males had systemic malformations, seizures, and severe mental retardation (Bleeker-Wagemakers et al., 1988; De la Chapelle et al., 1985; Donnai et al., 1988; Zhu et al., 1989). Microcephaly, large ears, prominent nasal bridge, and thin upper lips have been described (Donnai et al., 1988). One patient (Zhu et al., 1989) was autistic and self abusive, he had growth failure, abnormal sexual maturation, hypotension, paroxysmal irregularity of respiration, flushing, atonic seizures, myotonic contractions and hyper-reflexia without spastic paresis. He was unable to talk and walk. A Dutch patient (Bleeker–Wagemakers et al., 1988) had cutaneous hyperkeratosis and microcephaly, cryptorchidism and penis hypoplasia. Brain stem audiography showed sensory hearing loss. A Danish patient recently reported was mildly mentally retarded and lived in an open institution. His only extraocular malformations were micropenis and cryptorchidism. He was childish and had very little confidence in the people around him or in himself. In these atypical patients, deletions or other rearrangements involving the NDP locus were demonstrated by molecular genetic examination (Berger et al., 1992a; Chen et al., 1992b). It is conceivable that their complex phenotype is due to the involvement of other genes which have not yet been identified. An X/autosomal translocation was also found in some of the Norrie patients (Ohba and Yamashita, 1986; McMahan et al., 1991).

Carrier phenotype

In general, female carriers are completely healthy. Ophthalmological and audiological examinations are usually normal (Warburg, 1966; 1975; Parving and Schwartz, 1991) and both carriers of deletions and of other mutations have had children. Two female carriers manifesting signs of the disease have been observed. For one of these cases evidence has been found supporting a modest inactivation bias (70:30) in favour of expression of the defective allele (I. Craig, personal communication).

Pathogenesis

Ultrastructural studies (Enyedi et al., 1991) showed that the inner and outer neuroblastic layers of the retina were undifferentiated. This may explain the lack of coaption of the retinal pigment epithelium and the neuroretina, leading to congenital detachment. Most of the clinical signs are secondary, namely the persistence of the primary vitreous, the elongation of the ciliary processes, the cataract and the shrinkage with resulting corneal opacities. The progressive hearing impairment and the frequent autistic signs may be interpreted as a progressive neurological disorder of unknown character.

Differential diagnosis

Diagnosis is simple if there are other affected males in the family. When only a single person is affected, the diagnosis can only be tentatively established. The differential diagnosis depends on the time when the patient is first observed. In the neonate, it includes retinoblastoma, retinopathy of prematurity, persistent hyperplastic primary vitreous (PHPV), oculo-palato-cerebral dwarfism (Frydman et al., 1985), the osteoporosis-pseudoglioma syndrome (Saraux and Frézal, 1967), X-linked primary retinal dysplasia (Godel and Goodman, 1981) which is probably identical with X-linked familial exudative vitreo-retinopathy, trisomy 13, and incontinentia pigmenti. In this respect, it is of particular interest to note that the locus for X-linked exudative vitreo-retinopathy is localized to the same Xp region as NDP, and, recently, a mutation in the NDP gene has been found in affected members of an X-linked exudative vitreo-retinopathy family (Chen et al., in press).

The patients with NDP are usually born at term and need no oxygen, which excludes retinopathy of prematurity. Their disorder is bilateral while PHPV is unilateral, except in the autosomal recessive oculo-palato-cerebral dwarfism characterised by microcephaly, hypotonia, bilateral leukocornia, and cleft lip-palate. X-linked retinal dysplasia and X-linked falciform folds manifest as a falciform retinal fold, and carriers of X-linked retinal dysplasia may show peripheral folds; in NDP the affecteds are male – not females as in incontinentia pigmenti. In children, the differential diagnosis comprises infections, X-linked congenital cataract, X-linked microphthalmos and autosomal dominant familial exudative vitreoretinopathy (FEVR). Cataract and small eyes are not congenital in NDP, and FEVR is rare in children, in addition to which one of the relatives may show the retinal vascular anomalies of FEVR (von Nouhuys, 1991). In adults, the differential diagnosis comprises xerophthalmia and microphthalmos or phthisis due to a different aetiology.

Clearly, molecular genetic examination will soon facilitate the differential diagnosis.

Penetrance and Incidence

More than 300 male patients have been described worldwide, (Warburg, 1961; 1966; 1975; Blodi et al., 1969; Nance et al., 1969; Lomickova and Raska, 1969; Moreira–Filho and Neustein, 1979; Bleeker–Wagemakers 1981; Johnston et al., 1982; Bleeker–Wagemakers et al., 1985; Liberfarb et al., 1985; Phillips et al., 1986; Hill et al., 1987; Kivlin et al., 1987; Esakowitz et al., 1988; Gal et al., 1988; Harenda de Silva et al., 1988; Curtis et al., 1989; Nadol et al., 1990). Penetrance was complete, but expression was variable except for congenital blindness which is always present. The incidence is not known. Phillips (Phillips et al., 1986) calculated a mutation rate of 3.9×10^{-6} in the Scottish population, but ascertainment was probably not complete.

Table 1 Lod scores between NDP and DXS7

Author	z^{max}	Θ^{max}
Gal et al. (1985a)	3.50	0.00
Bleeker-Wagemakers (1985)	3.81	0.00
Warburg et al. (1985)	4.10	0.00
Kivlin et al. (1987)	5.42	0.00
Hill et al. (1986)	5.60	0.00
Ngo et al. (1988)	7.58	0.05
Curtis et al. (1989)	7.99	0.05

CHROMOSOMAL ASSIGNMENT OF THE NDP LOCUS

Linkage

As early as 1965, Warburg employed the classical X chromosomal Xg blood group marker in an attempt to establish linkage with NDP (Warburg, 1965). No linkage could be found between NDP and the Xg blood group system. The same results were reported within a Brazilian Negro sibship (Moreira–Filho and Neustein, 1979). A loose and unreliable linkage was found between NDP and the glucose-6-phosphate dehydrogenase marker in an American Negro family (Nance et al., 1969). In Dutch and Danish families, Gal et al. (1985a) noted close linkage of NDP with an anonymous RFLP marker, L1.28 (DXS7) (Bakker et al., 1983) from the Xp11.4-p11.3 region. After this report, extensive linkage studies were performed on Norrie families from all over the world by several research groups. The results from these studies clearly assigned NDP to the proximal Xp near other regionally assigned eye disease loci (Figure 4). The accumulated LOD scores are summarized in Table 1. A recombination between NDP and DXS7 was reported in only one family (Ngo et al., 1988). Subsequent analysis of the same family with an RFLP detected by a human ornithine–δ–aminotransferase (OAT1) cDNA probe (Ngo et al., 1989, Lafreniere et al., 1991b), a polymorphic CA-repeat at the 5'-end of the MAOB gene (Sims et al., 1992) and the DXS426 CA-repeat (Lindsay et al., 1992) suggested the genetic order: Xpter-(DXS7,MAOB)-NDP-(DXS426,OATL)-Xcen (Figure 5).

Cytogenetics

The first physical confirmation for the localization of NDP in band Xp11.4 on the short arm of the X chromosome came from a paper by Ohba and Yamashita in which they described a female infant NDP patient carrying a balanced X;10 translocation (Ohba and Yamashita, 1986). The expression of the mutated gene in this female could be explained by the fact that in an X-autosome translocation the translocated X chromosome is preferentially activated (Therman et al., 1974). More recently, a second X-chromosomal aberration associated with NDP

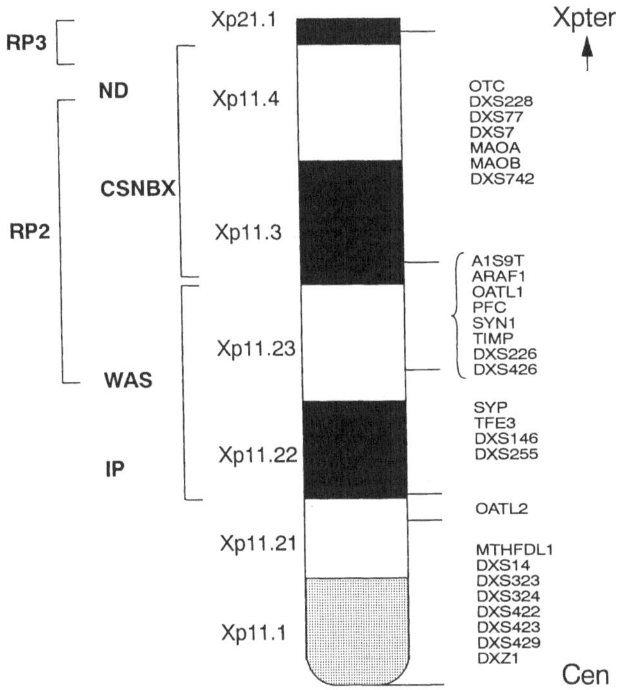

Figure 4 Genetic disorders and anonymous DNA probes assigned to the proximal Xp. Disorders on proximal Xp: RP2, RP3; Retinitis Pigmentosa types 2 and 3. NDP; Norrie disease. CSNB1; Congenital Stationary Night Blindness type 1. WAS; Wiskott-Aldrich Syndrome. IP; Incontinentia Pigmenti. The DNA probes on the right were arranged in intervals demarcated by chromosomal breakpoints (Diergaarde et al., 1989; Lafrenière et al., 1991).

was published in which a familial pericentric inversion (X)(p11.4-q22) was found in males affected with NDP. Female carriers of this inversion showed no clinical symptoms of NDP (McMahan et al., 1991).

Regional assignment of DNA probes to the proximal Xp

Important progress towards the mapping and isolation of the NDP gene came from the cloning and ordering of genes and anonymous DNA probes to defined X-chromosomal intervals. One of the first markers, L1.28 (DXS7) (Bakker et al., 1983), was assigned to the proximal Xp by linkage analysis in Duchenne Muscular Dystrophy families (Wieacker et al., 1984). Subsequently, a large number of markers and genes have been localized to the proximal Xp region using linkage analysis, deletion mapping, *in situ* hybridization and by mapping probes in human hamster cell hybrids containing parts of the human X chromosome (Lafreniere et al., 1991; Davies et al., 1991) (Figure 4).

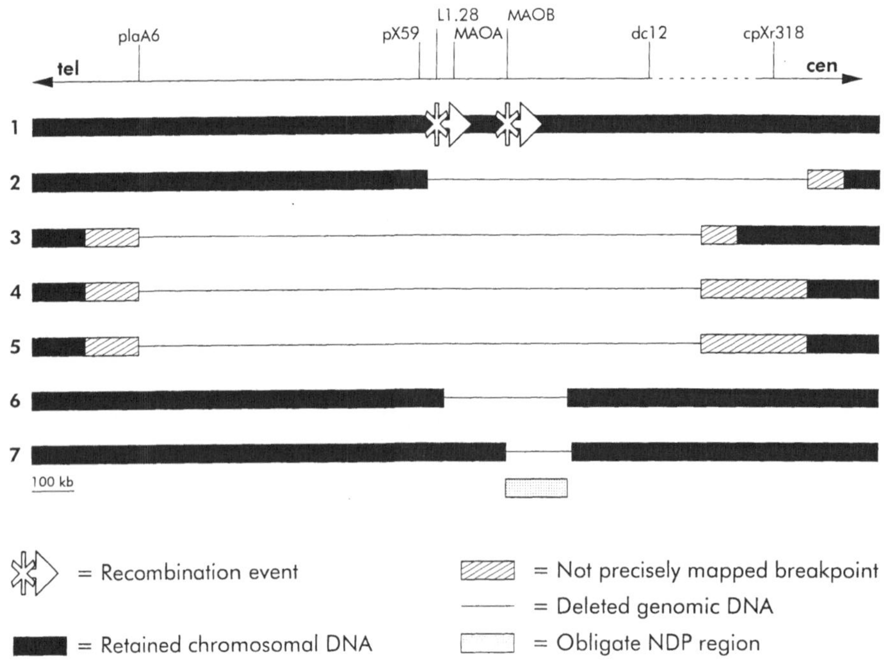

Figure 5 Genetic and physical fine mapping of the NDP locus. Top line: Relative order of anonymous DNA markers spanning the NDP region. Broken line: physical distance unknown. Line 1: recombination events between closely linked markers and NDP. (Ngo et al., 1988; Sims et al., 1991). Deletions: 2; Gal et al., 1986, 3; de la Chapelle et al., 1985, 4; Forrest et al., 1987, 5; Zhu et al., 1989, 6; Sims et al., 1992, 7; Bergen et al., 1992.

PHYSICAL FINE-MAPPING OF THE NDP GENE.

Deletion studies

In 1985, two research groups simultaneously reported physical evidence for close proximity of the NDP gene to the DXS7 locus (Gal et al., 1985b; 1986; de la Chapelle et al., 1985). In each case the DXS7 locus was found to be deleted, in a Dutch and in a Finnish male with atypical features of NDP respectively. Co-deletion of the Norrie gene together with the DXS7 locus could explain the NDP phenotype of these patients. The physical length of the deletions was not known, but as no changes could be found in high resolution chromosome banding experiments, the deletion in the Dutch patient could not be much larger than about 3 million bp (Gal et al., 1986).

Subsequently, two additional submicroscopic interstitial deletions encompassing the DXS7 and the NDP locus were described (Forrest et al., 1987; Donnai et al., 1988; Zhu et al., 1989). Further molecular analysis revealed that these four NDP-associated deletions included several regionally assigned anonymous DNA markers and the human monoamine oxidase A and B loci (MAOA, MAOB) (Bach et al., 1988; Hsu et al., 1988; Lan et al., 1989) (Figure 5).

Since the MAOA and MAOB genes were absent in the DNA of these four patients, and because of the role of biogenic amines in neuronal development (Lauder, 1985; Lam et al., 1981) and retinal neurotransmission (Osborne, 1981), they were initially considered to be implicated in NDP. However, enzyme, metabolite and DNA studies in several classic NDP patients did not reveal any MAOA or MAOB related abnormalities, and these genes were excluded as candidate genes for NDP (Sims et al., 1989).

The discovery in 1991 of submicroscopic interstitial NDP-deletions which did not include DXS7 (Sims et al., 1992), MAOA or part of the MAOB loci (Bergen et al., 1993), assigned the NDP locus unambiguously proximal to the MAOB locus. By this time, the physical order Xpter-(DXS228,DXS77)-DXS7-(MAOA, MAOB)-NDP-DXS742-Xcen could be established on the basis of different NDP deletion intervals (Figure 5).

Long range restriction mapping

After the discovery of the first submicroscopic Xp-deletions associated with NDP, information on the relative gene marker order in the proximity of NDP came from yet another source: long range restriction analysis.

Diergaarde et al., (1989) assigned the probes p1Aa6 (DXS228), pX59 (DXS77), L1.28 (DXS7) and MAOA/B to a region of 1500 kb, as they all hybridized to an identical BssHII restriction fragment of this size. In the same study, the distal breakpoint of the Dutch deletion (Gal et al., 1985b) was mapped within a 400 kb SalI fragment shared by pX59 (DXS77) and L1.28 (DXS7), but which could not be identified with MAOA/B. Furthermore, with different combinations of single and double digests using a set of restriction enzymes, the physical order DXS228-DXS77-DXS7-(MAOA/B, NDP) was suggested (Diergaarde et al., 1989). Refinement and extension of this map came from physical mapping experiments using YACs corresponding to the L1.28 and MAOA/MAOB loci (Chen et al., 1992a, Sims et al., 1992), and from genomic pulsed-field gel electrophoresis (PFG) studies with a novel probe, cpXr318 (DXS742), which is localized just proximal to the Norrie disease locus (Bergen et al., 1993). The resulting order of markers and physical distances between them were in general agreement with the data obtained from deletion studies (Figure 5).

THE CLONING AND CHARACTERISATION OF THE CANDIDATE NDP GENE

Isolation of the candidate cDNA

By 1992, linkage analysis, deletion studies and long range PFG restriction mapping reduced the obligate NDP gene region to approximately 150 kb proximal to the MAOB locus. This mapping information paved the way for the positonal cloning of the gene implicated in NDP.

Several research groups set out to survey the obligate NDP region in detail. Of particular importance was the availability of a large YAC of 640 kb corresponding to the DXS7 locus (Larin et al., 1991) and which encompassed at least part of the NDP locus (Chen et al., 1992a; Sims et al., 1992). The group of Berger et al., (1992a) cloned 350 kb spanning the obligate NDP region into a set of 14 overlapping cosmids. In parallel, Chen et al. (1992b) were able to isolate the region as a 160 kb BsshII-SstII YAC fragment. Subsequently, these cosmids as well as the entire 160 kb YAC-fragment were employed as probes to look for genomic abberations in classic NDP patients (non-deleted for established markers), and to isolate homologous cDNAs from retina-specific (both fetal and adult) and brain-specific libraries.

Using a cosmid, G8, from the NDP region Berger et al. identified at least three new microdeletions smaller than 20 kb associated with ND. Five retinal and brain cDNAs homologous to G8 were shown to be evolutionarily conserved and to be specifically expressed in retinal tissue (Berger et al., 1992a). These overlapping cDNAs formed a sequence of 1.9 kb including a 3' poly A tail. The sequence contains two overlapping reading frames of 474 and 417 bp, respectively, of which probably only the first encodes a protein (Berger et al., 1992a).

Simultaneously, Chen et al. (1992b) used the 160 kb YAC fragment to isolate four retina specific cDNAs which were partially or completely deleted in the previously described NDP microdeletion patients. One of these clones, FR2, also detected very small deletions in five out of 20 unrelated classical NDP patients surveyed. Northern analysis with FR2 as a probe revealed specific expression in brain of two RNA transcripts of about 1.7 and 2 kb. Analysis of the sequence obtained from the longest cDNA clone homologous to FR2 provided strong evidence for an open reading frame of 399 bp (Chen et al., 1992b).

The structure of the NDP gene

Berger et al. (1992b) and Meindl et al. (1992) established the intron-exon structure of the Norrie disease candidate gene by sequencing. The genomic size of the gene is 28 kb in length, and it contains 3 exons of approximately 201–221, 381 and 1246–1249 bp separated by introns of approximately 16 and 9 kb (Chen et al., 1993) (Figures 6 and 7). In genomic DNA, the 5' region of the gene is interrupted by the first intron with the result that the first exon is entirely untrans-

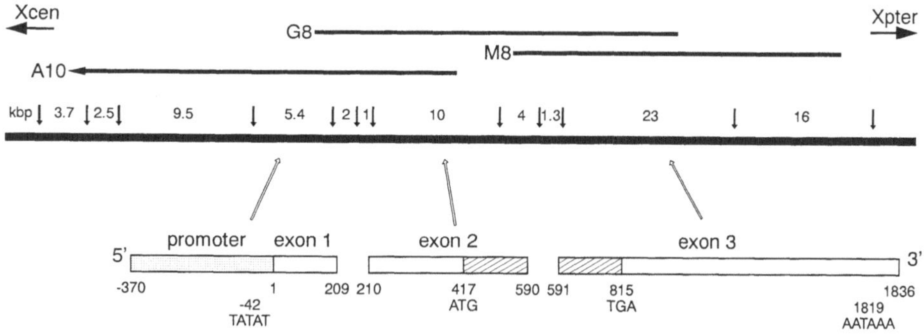

Figure 6 The exon-intron-structure of the Norrie disease gene as published elsewhere (Berger et al., 1992c; Meindl et al., 1992). The cosmid clones used for sequence analysis are given by the codes M8, G8 and A10. EcoRI restriction sites are indicated by arrows and fragment sizes in kb. Open arrows indicate exon containing genomic fragments. The protein coding part of the exons is given as hatched boxes, the promoter region as dotted, and the 3' and 5' flanking regions as open boxes. Positions of the TATA box, the start and stop codon as well as the polyadenylation signal correspond to sequences published earlier (Berger et al., 1992a; Meindl et al., 1992).

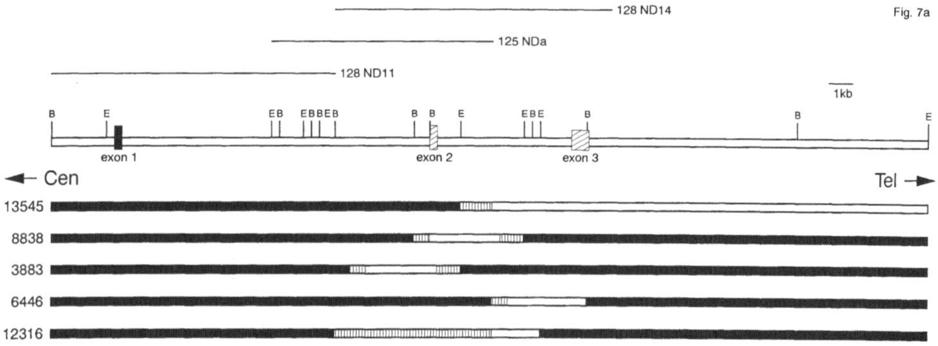

Figure 7 (a) Genomic restriction map of the NDP gene and the locations of the three exons according to Chen et al. (1992c; and unpublished results). The three overlapping lambda phage clones containing the NDP gene are depicted above the map. Below the restriction map, the regions deleted in the microdeletion patients are indicated (open bars deleted; filled bars: present; hatched bars: precise endpoints not determined). In (b) the genomic structure according to Chen et al. (1992b, c; and unpublished results) of the NDP gene is shown. The nucleotide sequences of the regions flanking the three exons (bold and underlined) are shown. Potential TATA-boxes are underlined. The start of the first exons has been taken as coincident with the longest cDNA clone (FR7).

AAATTCAGAT	ATTCATTGGC	CTCTTATTAG	TTCCATAATA	CCATTAAAAA
AGAAAGAAAG	AAAGAAACTT	CCTCGCCCTT	GTTCTCGTAC	GCTGTTCCCA
TCGTAAGATG	CTCCGTGGAA	GGGAGCCGAG	CGGTGGGCAG	AGGCTGAGTC
CCCGATAACG	AGCGCCTCAC	ATTTCCGTGG	CATTCCCATT	TGCTAGTGCG
CTGCTGCGGC	CGCACGCCTG	ATTGGATATA	TGACTGCAAT	GGCACTTTTC
CATTTGACAT	TCTTC EXON	1 (201bp)	TAAG	GTAAGCAGGA
AAACAAGCGC	TGCCGGCTTG	---16kb---	ACCCAATTCG	GTTACGTTGT
TGCCAGAACA	ACATGTTTAA	TCTTTAACAT	GGGTTCAAAC	TATTCTTGGC
CCTAGGAACA	TGGACTTCAG	CAATTAAAGT	CAACATGTGC	TTCCATTAAA
CCATTGTGTC	CACCTCCAAA	TGGTTATAAA	TATCTAGCAT	AAGCTATGGG
AGTTGGGGTG	GAATGGATGA	CAGCCTTTGC	TAATGACCCT	CTAGAAACCA
ATATTCTCCT	CTCAAAATAA	CATGGAAAAA	TTCTACTTAA	TACCTCCTGG
CTTCCATTAG	TGGTTCTGGG	TAAATAATTC	TGGGGAAAGT	AATTTCTGTT
TTCATTCCAG	CTGT EXON	2 (380 bp)	AAAG	GTAAGACCAA
GGTCTCTGTG	AGGAGAGCAT	--- 8kb---	CCCACAGATA	ACCTCCTTTT
CTTTCCCCAG	ATGG EXON	3 (1257 bp)	TC	AAATTTGACT
GTTAGTATTT				

Figure 7 (b)

lated. The ATG start codon lies within exon 2, and the stop codon as well as the polyadenylation signal are located in exon 3 (Figures 6 and 7). The open reading frame of 399 bp encoding a protein of 133 amino acids, is completely contained within exons 2 and 3. Within the first intron, a canonical TATAAAT sequence was identified 185 bp 5' to the start of the second exon. This structural organisation would allow an alternative start of transcription, suggesting that expression of the NDP gene may be regulated by the activity of two independent promotor regions (Chen et al., 1993).

Mutation analysis and diagnostic implications

Convincing evidence for the involvement of the isolated candidate gene in the etiology of Norrie disease has come from the detection of point mutations in patients. Berger et al., (1992b) described the identification of small sequence alterations in 10 out of 12 non-deletion patients with classical NDP. Furthermore, mutation analysis established the diagnosis in 2 out of 5 patients, in whom the clinical diagnosis of NDP was suspected. Similar results were reported by Meindl et al., (1992) and Chen et al., (1993). Most of the mutations described so far give rise to an amino acid substitution (7 out of 15). In 6 patients, mutation events predict a truncated protein. Only one mutation was found to affect a splice site, while one other mutation extends the predicted polypeptide by 127 amino acids. It is of interest to note that 6 out of these mutations alter or generate a cysteine codon which may indicate that cysteine residues are particularly important for the function of the protein.

Furthermore, these studies prove that mutation screening is a powerful diagnostic tool in clinically uncertain, sporadic cases. Comparison between the type of mutation and the clinical phenotype has failed to reveal any obvious correlation (Berger et al., 1992b). Apart from the typical ophthalmological signs, both mental changes and progressive deafness were seen in several patients with different point mutations.

Expression of the NDP gene

Northern analysis of mRNA derived from human glioma, glioblastoma, and neural cells reveals strong hybridisation to a transcript of 1.8 kb. Also, other transcripts with different sizes have been observed from Northern analyses (I. Craig, personal communication; Chen et al., 1992b; Berger et al., 1992a).

Using PCR primers designed for specific amplification of cDNA (RT-PCR), expression is found in fetal eye, fetal and adult brain, fetal lung and adult muscle. No expression is found in fetal liver, adult liver cells or cultured liver cells (Chen et al., 1993). Thus, the expression of the NDP gene is not only confined to the eye or brain. Analysis of the tissue expression pattern may help to understand the disease manifestation in typical Norrie patients.

The function of the predicted protein

An extensive DNA and protein sequence comparison between the human NDP gene and related genes from the NBRF and Swiss protein database revealed significant homology with C-terminal cysteine-rich domains shared by a family of mucin-like proteins and von Willebrand factor (Meindl et al., 1992; Chen et al., 1993). Many mucins are secreted proteins and some are integral membrane proteins anchored at a mucin-like, glycosylated domain. No glycosylation sites are identified in the NDP gene product. However, an N-terminus hydrophobic region of the NDP product contains features of a signal sequence containing a basic region, a hydrophobic "core" and a more polar region (Chen et al., 1993). A weak homology between this N-terminal domain and the rat-horsefall protein supports the secretory status of the NDP gene product (Meindl et al., 1992). Prediction of signal sequence cleavage sites based on the method of G. von Heijne (1986) defined a signal peptide of 24 amino acid residues with a 0.744 probability (Berger et al., unpublished data). The same region also exhibits homology with two immediate early genes; the phorbol ester-repressive and v-src-inducible Cyr61 gene identified in murine fibroblast (Meindl et al., 1992; Chen et al., 1993).

Based on the observed homologies, it can be postulated that the NDP gene is involved in a pathway that regulates neural cell differentiation and proliferation. Also, because of the potential of cysteine rich regions to form disulphide bonds, it can be postulated that the function of the protein may concern interaction with other growth factors important in neurological development.

Outlook

The precise function of the NDP gene remains to be established. Direct information on the role, tissue specificity and localization of the protein can come from studies using specific antibodies raised against the *in vitro* synthesized gene product (Marston, 1986). The isolation of the mouse gene and the construction of transgenic mice or other animal models should provide a more detailed insight into the function of the Norrie disease gene.

At present, direct mutational analysis for coding changes in the second and the third exons is possible, while SSCP (Orita et al., 1989) will assist in rapid screening for additional lesions outside these regions.

The occurrence of small DNA deletions in 20% of the NDP patients and the identification of point mutations in 73% of the cases offers the possibility of reliable DNA diagnosis in 93% of all cases (Berger et al., unpublished results).

Finally, it is encouraging to note that the small size and the secretory status of the Norrie gene translation product raises perhaps the possibility of therapeutic approaches in some instances.

ACKNOWLEDGEMENTS

We thank T. Put for technical assistance in preparing this manuscript.

REFERENCES

Andersen S. R., Warburg M. (1961) Congenital bilateral pseudotumor of the retina with X-chromosomal recessive inheritance. *Archives of Ophthalmology* **66**, 614.

Bach A.W. L., Lan N. C., Johnson D. L., Abell C. W., Bembenek M. E., Kwan S-W. et al. (1988) *cDNA cloning of human liver monoamine oxidase A and B: molecular basis of differences in enzymatic properties.* Proceedings of the National Academy of Sciences of the USA **85**, 4934–4938.

Bakker E., Wieacker P., Beverstock G. C., Pearson P. L. (1983) Recombinant DNA techniques for mapping the human X chromosome. *Clinical Genetics* **23**, 225

Bergen A. A. B., Wapenaar M. C., Schuurman E. J. M., Diergaarde P. J., Lehrach H., Monaco A. P., Bakker E., Bleeker-Wagemakers E. M., van Ommen G. J. B. (1993) Detection of a new Norrie disease deletion interval with a novel DNA probe isolated by differential ALU PCR fingerprint cloning. *Cytogenetics Cell Genetics* **62**, 231–235.

Berger W., Meindl A., van de Pol T. J. R., Cremers F. P. M., Ropers H. H., Döerner C, Monaco A., Bergen A. A. B., Lebo R., Warburg M., Zergollern L., Lorenz B., Gal A., Bleeker-Wagemakers E. M., Meitinger T. (1992a) Isolation of a candidate gene for Norrie disease by positional cloning. *Nature Genetics* **1**, 199–203.

Berger W., van de Pol D., Warburg M., Gal A., Bleeker-Wagemakers E. M., de Silva H., Meindl A., Meitinger T., Cremers F., Ropers H. -H. (1992b) Mutations in the candidate gene for Norrie disease. *Human Molecular Genetics* **1**, 461–465.

Bleeker-Wagemakers E. M. (1981) *On the causes of blindness in the mentally retarded.* Thesis, Bartimeus Foundation, Box 87, Doorn, the Netherlands, 82–88.

Bleeker-Wagemakers E. M., Friedrich U., Gal A., Wienker T. F., Warburg M., Ropers H-H. (1985) Close linkage between Norrie disease, a cloned DNA sequence from the proximal short arm, and the centromere of the X chromosome. *Human Genetics* **71**, 211–214.

Bleeker-Wagemakers E. M., Zweije-Hofman I., and Gal A. (1988) Norrie disease as part of a complex syndrome explained by a submicroscopic deletion of the X chromosome. *Ophthalmic Paediatrics and Genetics* **9**, 3: 137–142.

Blodi F. C., Hunter W. S. (1969) Norrie disease in North America. *Documenta Ophthalmology* **26**, 434–450.

Chapelle A. de la, Sankila E-M., Lindlöf M., Aula P., Norio R. (1985) Norrie disease caused by a gene deletion allowing carrier detection and prenatal diagnosis. *Clinical Genetics* **28**, 317–320.

Chen Z.-Y., Sims K. B., Coleman M., Donnai D., Monaco A., Breakefield X. O., Davies K. E., Craig I. W. (1992a) Characterization of a YAC containing part or all of the Norrie disease locus. *Human Molecular Genetics* **1**, 3: 161–164.

Chen Z.-Y., Hendriks R. W., Jobling M. A., Powell J. F., Breakefield X. O., Sims K. B., Craig I. (1992b) Isolation and characterisation of a candidate gene for Norrie disease. *Nature Genetics* **1**, 204–208

Chen Z.-Y., Battinelli B., Hendriks R. W., Powell J. F., Middleton P. H., Sims K., Breakefield X. O., Craig I. W. (1993) Norrie disease gene: characterisation of deletions and possible function. *Genomics,* **16**, 533–535.

Chen Z. -Y., Battinelli. E. M., Fielder, A., Bundey, S., Sims, K., Breakefield, X. O. and Craig, I. W. *Nature Genetics* (in press).

Curtis D., Blank C. E., Parsons M. A., Hughes H. N. (1989) Carrier detection and prenatal diagnosis in Norrie disease. *Prenatal Diagnosis* **9**, 735–740.

Davies K. E., Mandel J-L., Monaco A. P., Nussbaum R. L., Willard H. F. (1991) Report of the committee on the genetic constitution of the X chromosome. Eleventh International Workshop on Human Gene Mapping. *Cytogenetics and Cell Genetics* **58**, 853–966.

Diergaarde P. J., Wieringa B., Bleeker-Wagemakers E. M., Sims K. B., Breakefield X. O., Ropers H. -H. (1989) Physical fine-mapping of a deletion spanning the Norrie gene. *Human Genetics* **84**, 22–26.

Donnai D., Mountford R. C., Read A. P. (1988) Norrie disease resulting from a gene deletion: clinical features and DNA studies. *Journal of Medical Genetics* **25**, 73–78.

Esakowitz L., Clark C., Haites N., Kelly K., Johnston A. W. (1988) A genetic linkage study of a family with Norrie disease. *Eye* **2**, 443–447.

Enyedi E., Juan D. E., Gaitan A. (1991) An ultrastructural study of Norrie Disease. American *Journal of Opthalmology* **111**, 439–445.

Forrest S. M., Smith T. J., Kenwrick S. J., Dorknis H. R., King A. W., Read A. P. et al. (1987) Genetic and physical mapping of markers localized to Xp11.0-Xp11.23 on the human X chromosome. (9th International Workshop on Human Gene Mapping). *Cytogenetics and Cell Genetics* **46**, 615.

Frydman M., Kauschansky A., Leshem I., Savir H. (1985) Oculo-palato-cerebral dwarfism: a new syndrome. *Clinical Genetics* **27**, 414–41.

Gal A., Stolzenberger C., Wienker T., Wieacker P., Ropers HH., Friedrich U. et al. (1985a) Norrie disease: close linkage with genetic markers from the proximal short arm of the X chromosome. *Clinical Genetics* **27**, 282–283.

Gal A., Stolzenberger C., Wienker T, Wieacker P., Ropers H. H. (1985b) Localization of the gene for Norrie disease by linkage to the DXS7 locus (Abstract). Eight International Workshop on Human Gene Mapping. *Cytogenetics and Cell Genetics* **40**, 633.

Gal A., Wieringa B., Smeets D. F. C. M., Bleeker-Wagemakers L., Ropers H. -H. (1986) Submicroscopic interstitial deletion of the X chromosome explains a complex genetic syndrome dominated by Norrie disease. *Cytogenetics and Cell Genetics* **42**, 219–224.

Gal A., Ulhaas S., Glaser D., Grimm T. (1988) Prenatal exclusion of Norrie disease with flanking DNA markers. *American Journal of Medical Genetics* **31**, 449–453.

Godel V., Goodman R. M. (1981) X-linked recessive primary retinal dysplasia: clinical findings in affected males and carrier females. *Clinical Genetics* **20**, 260–266.

Harendar de Silva D. G., de Silva D. B. (1988) Norrie disease in an Asian family. *British Journal of Ophthalmology* **72**, 62–64.

Heijne, von G. (1986) A new method for predicting signal sequence cleavage sites. *Nucleic Acids Research* **14**, 4683–4690.

Hill D. F., Chapman C. J. Gardner R. J. M. (1987) The use of a DNA marker for carrier diagnosis in an X-linked disorder: Norrie disease. *New Zealand Medical Journal* **100**, 166–168.

Hsu Y-P., Weyler W., Chen S., Sims K. B., Rinehart W. B., Utterback M. et al. (1988) Structural features of human monoamine oxidase A elucidated from cDNA and peptide features. *Journal of Neurochemistry* **51**, 1321–1324.

Johnston S. S., Hanna J. E., Nevin N. C., Bryars J. H. (1982) Norrie disease. *Birth Defects* **18**, 729–738.

Kivlin J. D., Sanborn G. E., Wright E., Cannon L., Carey J. (1987) Brief clinical report: further linkage data on Norrie disease. American *Journal of Medical Genetics* **26**, 733–736.

Lafrenière R. G., Brown C. J., Powers V. E., Carrel L., Davies K. E., Barker D. F. et al. (1991) Physical mapping of 60 DNA markers in the p21.1 q21.3 region of the human X chromosome. *Genomics* **11**, 352–363.

Lafrenière R. G., Geraghty M. T., Valle D., Shows T. B., Willard H. F. (1991b) Ornithine aminotransferase-related sequences map to two nonadjacent intervals on the human X chromosome short arm. *Genomics* **10**, 276–279.

Lam D. M. -K., Fung S. -C., Kong Y., Y. -C. (1981) Postnatal development of dopaminergic neurons in the rabbit retina. *Journal Neuroscience* **1**, 1117–1132.

Lan N. C., Heinzmann C., Gal A., Klisak I., Orth U., Lai E. et al. (1989) Human monoamine oxidase A and B genes map to Xp11.23 and are deleted in a patient with Norrie disease. *Genomics* **4**, 552–559.

Larin Z., Monaco A. P., Lehrach H. (1991) *Yeast artificial chromosome libraries containing large inserts from mouse and human DNA.* Proceedings of the National Academy of Sciences of the USA **88**, 4123–4127.

Lauder J. M. (1985) Roles for neurotransmitters in development: possible interaction with drugs during fetal and neonatal periods. *Prog Clin Biol Res* **163**, 375–380.

Liberfarb R. M., Eavey R. D., DeLong G. R., Albert D. M., Dieckert J. P., Hirose T. (1985) Norrie disease: a study of two families. *Ophthalmology* **92**, 1445–1451.

Lindsay S., Thiselton D. L., Bateman J. B., Ngo J. T., Sparkes R. S., Coleman M., Davies K. E. and Bhattacharya SS (1992) Localisation of the gene for Norrie disease to between DXS7 and DXS426 on Xp. *Human Genetics* **88**: 349–350.

Lomickova H., Raska B. (1969) La maladie de Norrie. *Médecine et Hygiène* **27**, 1168–1169.

Marston F. A. O. (1986) The purification of eukaryotic polypeptides synthesized in Escherichia coli. *Biochemical Journal* **240**, 1–12.

McMahan M. R., Weaver R. G. Jr, Kerr J., Thomas I. T., Rao N., Pettenari M. J. (1991) Inversion (X)(p11.4;q22) associated with Norrie disease in a 3 generation family. *American Journal of Human Genetics* **49**, 270a.

Meindl A., Berger W., Meitinger T., van de Pol D., Achatz H., Dörner C., Haassemann M., Hellebrand H., Gal A., Cremers F., Ropers H. -H. (1992) Norrie disease caused by mutations in extracellular protein resembling C-terminal globular domain of mucins. *Nature Genetics* **2**, 139–143.

Moreira-Filho C. A., Neustein I. (1979) A presumptive new variant of Norrie disease. Journal of *Medical Genetics* **16**, 125–128.

Nadol J. B. Jr, Eavey R. D., Liberfarb R. M., Merchant S. N., Williams R., Climenhager D. et al. (1990) Histopathology of the ears, eyes, and brain in Norrie's disease (Oculoacousticocerebral Degeneration). *American Journal of Otolaryngology* **11**, 112–124.

Nance W. E., Hara S., Hansen A., Eliott J., Lewis M., Chown B. (1969) Genetic linkage studies in a Negro kindred with Norrie disease. *American Journal of Human Genetics* **21**, 423–429.

Ngo J. T., Spence M. A., Cortessis V., Sparkes R. S., Bateman J. B. (1988) Recombinational event between NDP and DXS7 loci. *Clinical Genetics* **34**, 43–47.

Ngo J. T., Bateman J. B., Cortessis V., Sparkes R. S., Mohandas T., Inana G. et al. (1989) Norrie disease: linkage analysis using a 4.2 kb RFLP detected by a human ornithine aminotransferase cDNA probe. *Genomics* **4**, 539–545.

Norrie G. (1927) Causes of blindness in children. *Acta Ophthalmology* (Kbh) **5**, 357–386.

Ohba N., Yamashita T. (1986) Primary vitreoretinal dysplasia resembling Norrie disease in a female: association with X autosome chromosomal translocation. *British Journal of Ophthalmology* **70**, 64–71.

Orita M., Iwahana H., Kanazawa H., Hayashi K., Sekiya T. (1989) *Detection of polymorphisms of human DNA by gel-electrophoresis as single strand conformation polymorphisms.* Proceedings of the National Academy of Sciences USA **86**, 2766–2770.

Osborne, N. N. (1981) Noradrenaline, a transmitter candidate in the retina. *Journal of Neurochemistry* **36**, 43–47.

Parving A., Schwartz M. (1991) Audiometric tests in gene carriers of Norrie disease. *International Journal of Pediatric Otorhinolaryngology* **21**, 103–111.

Phillips C. I., Newton M., Duvall J., Holloway S., Levy A. M. (1986) Probable Norrie disease due to mutation. *British Journal of Ophthalmology* **70**, 305–313.

Saraux H., Frézal J., Roy C., Aron J. J., Hayat B., Lamy M. (1969) Pseudogliome et fragilité osseuse héréditaire a transmission autosomal recessive. *Annale Ocul.* **200**, 1241–1252

Sims K. B., Ozelius L., Corey T., Rinehart W. B., Liberfarb R., Haines J., Chen W. J., Norio R., Sankila E., de la Chapelle A., Murphy D. L., Gusella J., Breakefield X.O. (1989) The Norrie disease gene is distinct from the monoamine oxidase genes. *American Journal of Human Genetics* **45**: 424–434.

Sims K. B., Lebo R. V., Benson G., Shalish C., Shuback D., Chen Z. -Y., Bruns G., Craig I.W., Golbus M. S., Breakefield X. O. (1992) The Norrie disease gene maps to a 150 kb region on chromosome Xp11.3. *Human Molecular Genetics* **1**, 2: 83–89.

Therman E., Patau K. (1974) Abnormal X chromosomes in man: origin, behaviour and effects. *Humangenetik* **25**, 1–16.

Von Nouhuys C. E. (1991) Signs, complications, and platelet aggregation in familial exudative vitreoretinopathy. *American Journal of Ophthalmology* **11**, 34–41.

Warburg M. (1961) Norrie's disease. Atrophia bulborum hereditarium. *Acta Ophthalmology* **39**, 757–772.

Warburg M., Hauge M., Snager R. (1965) Norrie's disease and the Xg blood group system: Linkage data. *Acta Genetics* (Basel) **15**, 103.

Warburg M. (1966) Norrie's disease: a congenital progressive oculo-acoustico-cerebral degeneration. *Acta Ophthalmology* (Copenh) **89** [Suppl],1–147.
Warburg M. (1975) Norrie's disease. Differential diagnosis and treatment. *Acta Ophthalmology* (Copenh) **53**, 217–236.
Wieacker H. F., Davies K., Cooke H. J., Pearson P. L., Williamson R., Bhattacharya S., and Ropers H. H. (1984) Toward a complete linkage map of the human X chromosome: regional assignment of 16 cloned single copy DNA sequences employing a panel of somatic cell hybrids. *American Journal of Human Genetics* **36**: 265–276.
Zhu D. P., Antonarakis S. E., Schmeckpeper B. J., Diergaarde P. J., Greb A. E., Maumenee I. H. (1989) Microdeletion in the X-chromosome and prenatal diagnosis in a family with Norrie's disease. *American Journal of Medical Genetics* **33**, 485–488.

14. X-LINKED JUVENILE RETINOSCHISIS

ALBERT DE LA CHAPELLE, TIINA ALITALO and HENRIK FORSIUS

Department of Medical Genetics, University of Helsinki, and Folkhalsan Institute of Genetics, Helsinki, Finland

INTRODUCTION

Retinoschisis is a descriptive term indicating a splitting of the retina into two layers and is often subdivided into **congenital** and **acquired** forms.

The differential diagnosis of congenital retinoschisis includes a diverse group of disorders. Gorlin and Knobloch (1972) distinguished 6 different genetic syndromes in which juvenile retinal detachment occurs, including juvenile retinoschisis. In most textbooks, the lists of conditions to be considered in patients with retinal detachment are long. Two main developments now make further subdivisions of the retinoschisis syndromes easier than before. First, improved clinical methods allow distinct syndromes to be delineated. Second, the genetic mapping of several congenital eye disorders provides a means of genetic classification.

The subject of this review is X-linked juvenile retinoschisis (MIM 312700, McKusick, 1990), a disease that belongs to the vitreoretinal dystrophies. We shall use the designation RS adopted by the Human Gene Mapping Workshops (McAlpine et al., 1991). According to this classification *RS* indicates the gene and RS the disease phenotype; for simplicity this distinction is not always necessary. While the first reports suggesting that RS may reside in the distal part of the short arm of the X chromosome appeared over 20 years ago (Eriksson et al., 1969; Race and Sanger, 1975) its more precise localization has been accomplished in the last 5-10 years. This has paved the way for detailed linkage studies, carrier detection, and pre- and postnatal diagnostics within families. Thus the concept of a disorder well defined both by clinical and genetic criteria is emerging.

EPIDEMIOLOGY

As reviewed by Alitalo (1990), the first description of RS probably appeared in 1898; however its birth as a distinct entity occurred as late as the middle of this century (Mann and McRae, 1938; Sorsby et al., 1951; Jager, 1953). Already in their textbook published in 1974 Franceschetti et al., listed 218 patients with RS described in the literature. Most patients were from Western European populations. Combined clinical and genealogical studies revealed a strikingly high prevalence in a small district in southwestern Finland where over 100 patients probably descended from a common ancestor (Vainio-Mattila et al., 1969). In 1973 Forsius et al. described 183 cases. Later, as more patients with RS were diagnosed in Finland (over 300 today) the concept emerged that RS is more prevalent in Finland than elsewhere in the world (Forsius and Eriksson, 1980; Norio, 1981).

The idea of an uneven geographic or ethnic distribution now needs to be re-examined. In the past, case reporting has dominated the literature on RS making it difficult to assess its frequency and distribution. However, it is obvious that RS occurs in most parts of the world and in different ethnic groups such as American Blacks (Constantaras et al., 1972; Ide and Wilson, 1973), Chinese (Zhang, 1984), Japanese (Kawano et al., 1981; Minato, 1991), Indonesians (Deutman, 1971) and East Europeans (Janaky and Benedek, 1990; Nizankowska and Wozny, 1991).

PREVALENCE

From the medical and sociological points of view, the **prevalence** (the number of living affected individuals per total number of individuals in the population) is an important figure. This is distinct from the *number of mutations*, that is, the number of *unrelated* affected individuals, families or pedigrees in a given population. An example of a discrepancy between the two figures comes from Finland, where the prevalence exceeds 1: 17,000 (over 300 patients in a population of 5 million). However, as determined by genealogy and haplotype analysis, as few as 2 mutations (one in the Pori district, one in the Oulu district) may account for the majority of all patients in Finland (Alitalo et al., 1991a). Corresponding data are not available from other countries. However, RS should no longer be regarded as rare. Already in 1953 Jager in the Netherlands had seen 25 patients. Kraushar (1972) described 40 patients seen in Boston, USA; Deutman (1977) working in the Netherlands reported having seen over 100 cases, and Verdaguer (1982) had seen 48 patients in Chile. More recent comprehensive reports from France describe 60 patients (Turut et al., 1989; Castier et al., 1990) and 22 patients from 7 unrelated families, (Kaplan et al., 1991), respectively. In a German series 52 patients were described (Kellner et al., 1990). In Michigan, USA, 56 affected members of 7 families were recently reported (Sieving et al., 1990). These data allow us to suggest that RS is much more common than previously thought, and that a geographically or ethnically uneven distribution of mutations worldwide does not necessarily exist. Very high local prevalences such as the one in the Pori region in Finland (Vainio-Mattila et al., 1969) and Canada (Ewing and Ives, 1969; Condon et al., 1986), obviously exist in stable populations. These are usually the result of a fairly recent (<20 generations) founder having by chance transmitted the mutation to numerous offspring.

Few reports address the question of the actual prevalence of RS. In an effort to determine the prevalence of the main hereditary retinal dystrophies Puech et al. (1991) studied the occurrence of these disorders in a defined population of 4 million people in northern France during 1972 to 1989. For RS the prevalence figure was 1 in 28,092. As a comparison, the corresponding figure for "typical" retinitis pigmentosa was 1 in 4,225 and for choroideremia 1 in 177,454. Even though the authors attempted to correct for incomplete ascertainment, the above figures should be minimum estimates.

In a similar study in central Sweden, Skoog et al. (1990) encountered 9 patients with RS among a total of 487 patients referred to an opthalmological centre for suspected or obvious hereditary degenerative or dystrophic eye disorders. While no estimates of prevalence were offered, the same series comprised 38 patients with retinitis pigmentosa and 2 patients with choroideremia, respectively. These data are compatible with the data from northern France and indicate that RS indeed accounts for a relatively large proportion of these disorders. Taken together with the Finnish prevalence figure of 1 in 17,000 (uncorrected for incomplete ascertainment) we tentatively conclude that the prevalence of RS is between 1 in 15,000 and 1 in 30,000, or even higher. At face value this means, for instance, that there are over 30,000 RS patients in Europe and over 300,000 in the world. The condition is certainly underdiagnosed for many reasons, including the mildness of symptoms in some cases, and the difficulty ophthalmologists have in making the diagnosis. In the study by Skoog et al. (1990) for instance, only 4 of the 9 RS patients had been correctly diagnosed by the referring ophthalmologists.

CLINICAL FEATURES

Pathogenesis

According to Arden et al. (1988), RS is to be considered as a disorder of retinal development rather than a dystrophic process. The primary defect may be in Muller cells, as suggested by electrodiagnostic findings. The first sign of the disease appears to be the abnormal development of the cortical vitreum that is manifest at birth and so must have already begun in the fetal state. Progression results in a slow and gradual deterioration of retinal structure and functions.

In 1968 Yanoff and collaborators performed a histopathological examination in a 4-year-old affected boy. A large cystic cavity was found between the equator and the ora serrata. This cavity was the result of the internal limiting membrane and part of the innermost retinal tissue having split away from the remainder of the sensory retina. Sections from the inferior calotte showed early schisis in the nerve fibre layer. Ultrastructural findings led Condon et al. (1986) to the hypothesis that defective Muller cells produce the filaments that accumulate extracellularly within the retina and secondarily lead to retinoschisis.

Anatomical features

In mild cases comprising 60% of all patients changes occur in the macular area only. In this site a wheel-like cystic formation about the size of the optic disc is seen (Figure 1). The changes at the macula are probably congenital but are difficult to differentiate from their surroundings before the age of about 2 years.

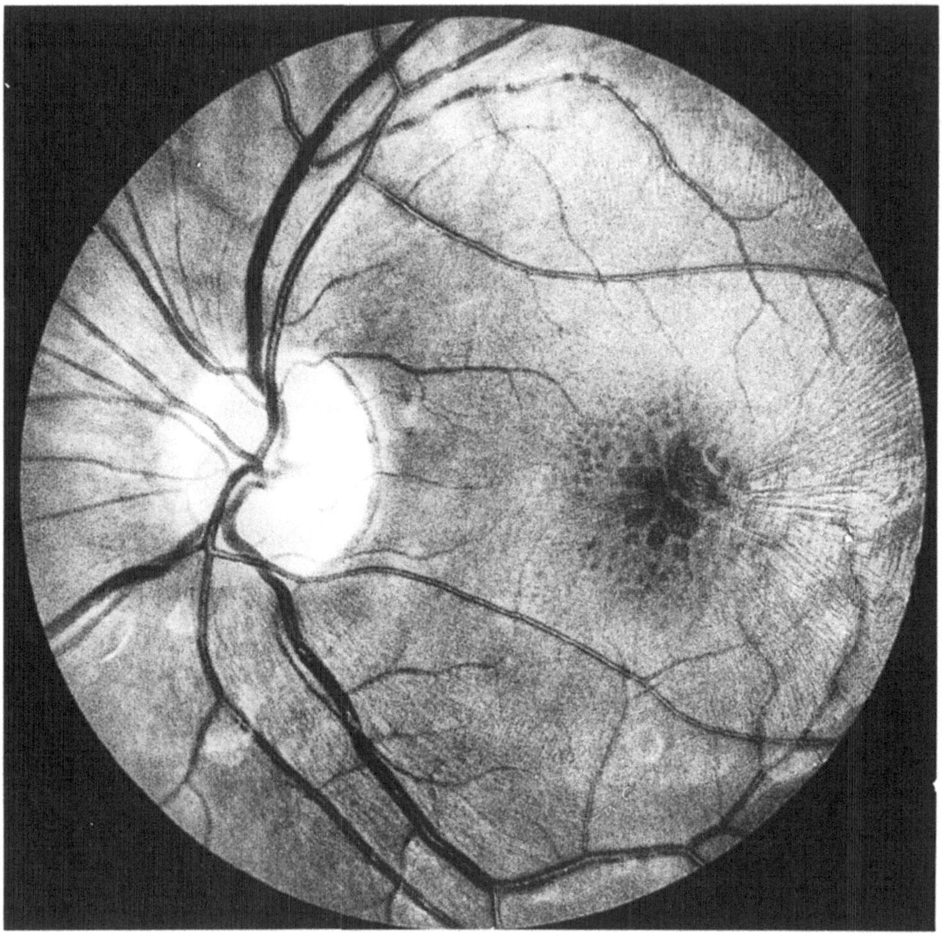

Figure 1 Retinoschisis in an 11-year-old boy showing typical cystic fovea.

A radiation of superficial retinal tissue is sometimes seen outside the cystic area in young subjects. Around the age of 40, the highly typical and readily distinguishable central schisis flattens out, atrophies and becomes more difficult to recognise. In aged patients, the pigment layer disappears so that the disease can no longer be differentiated from circumscribed macular degeneration (Figure 2). In addition to the macular changes, moderately severe cases show semitranslucent vitreous veils in the lower temporal periphery (Figure 3). These veils consist of detached portions of the superficial layer of the retina, and contain retinal blood vessels. Often a demarcation line is seen at the border of the schisis (Figure 4), although this may also occur in retinal detachment unassociated with schisis. The

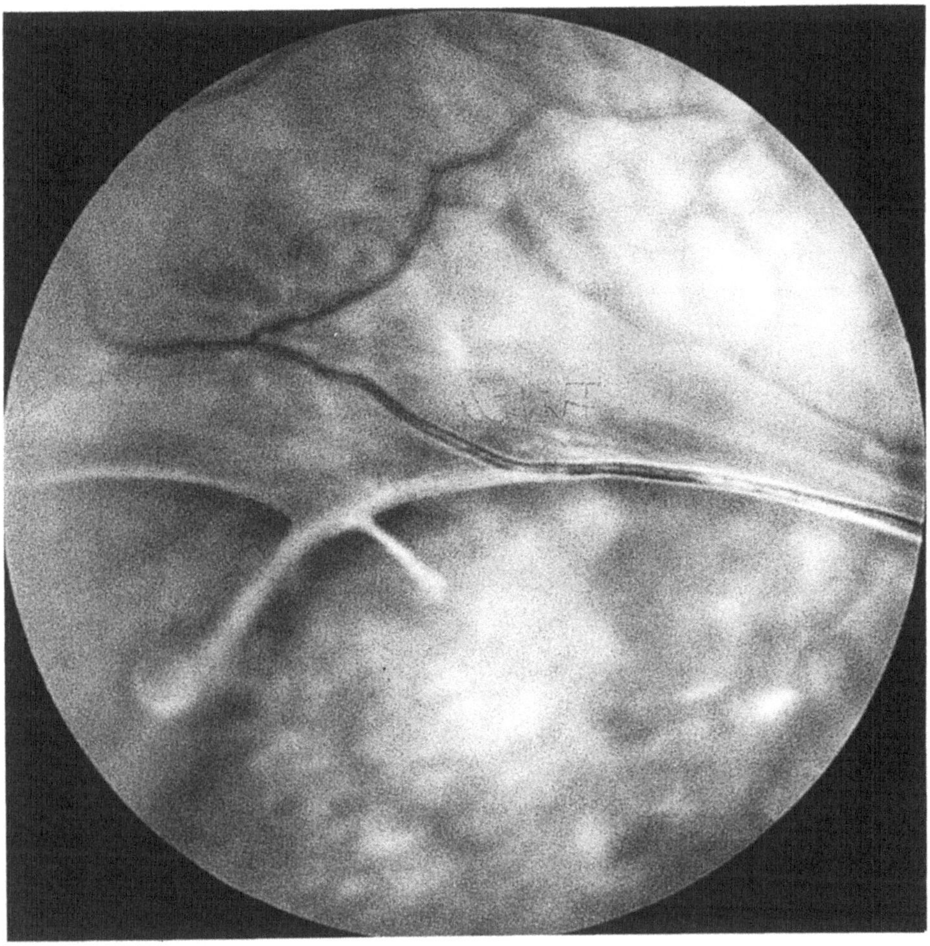

Figure 2 Retinoschisis in a 27-year-old man showing an elevated superficial leaf of the retina, a so-called vitreous veil.

veils can remain unchanged for decades. Parts of them may float freely in the degenerated vitreous body, but as a rule they appear elevated from the ocular fundus and form coherent sheets which seem to start from apparently normal retinal tissue close to the optic disc. In the periphery, the veil again has contact with the fundus. In young subjects up to a few years of age, the veil can form a large cyst in which holes gradually appear, and the cyst collapses. In adults, the holes often increase in size and finally only remnants of the veil around blood vessels may be seen. Sorsby et al. (1951) were the first to describe this phenomenon. In severe cases, the veil may form a roll below the optic disc. Increasing atrophy including degeneration of the pigment layer occurs in the attached part of the retina. Up to middle age, garland-like glistening formations nearly always

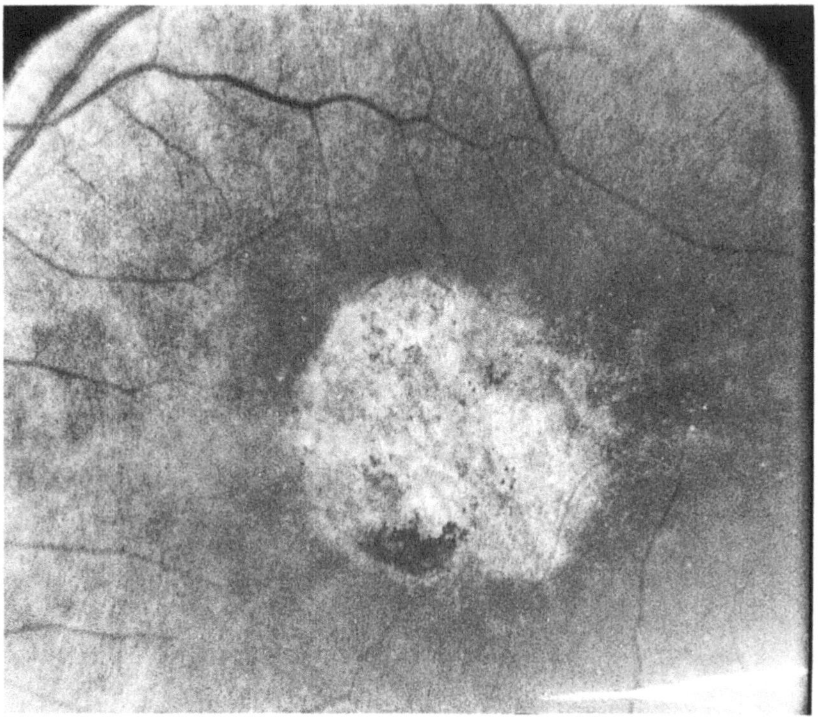

Figure 3 Retinoschisis in a 69-year-old man showing disappearance of the foveal cysts leading to a picture indistinguishable from senile macular degeneration.

occur in the periphery of the fundus, which otherwise appears normal. Drusen of the optic disc have been described, as well as whitish streaks along the blood vessels and pigmented areas; in extreme cases, the entire retina and choroid are disorganised. Patients with these features can be totally blind from birth.

Some investigators have reported follow-up observations of several years' duration beginning in patients at the age of a few months. Forsius et al. (1990) followed 40 patients for an average of 22 years. An initial high bullous detachment is followed by hole formation in the wall, after which the balloon-shaped detached retina flattens out. Conversely, even in affected individuals as old as 50–60 years, retinoschisis without holes has been observed by some investigators.

Visual acuity and refraction

In young subjects, visual acuity can be nearly normal. Typically, the disease is not detected until the first school year when a vision of 0.5 to 0.7 (20/40–20/30) is found. However, occasionally the acuity can be very low at this stage. Among our own 300 cases, two boys were nearly totally blind from birth.

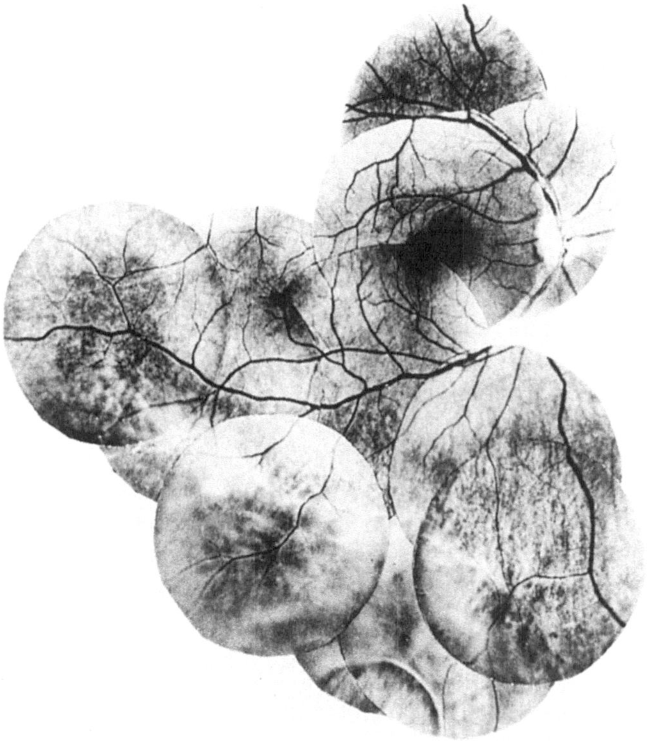

Figure 4 Retinoschisis in a 6-year-old boy showing both central and peripheral changes. A large hole is seen in the ablated retinal leaf at the bottom.

Up to 40–50 years of age the vision is stable but then gradually deteriorates, as shown in large patient series (Ewing and Ives, 1969; Forsius et al., 1973). The variability of this parameter is exemplified by patients in whom good visual acuity can be preserved up to 60 years of age (Kellner et al., 1990). The disease rarely results in total blindness.

Patients are usually hyperopic and in addition can be astigmatic, sometimes quite remarkably so. Myopia is rare.

Colour vision

Secondary dyschromatopsia is the rule in more advanced cases. Disturbances both in tritanopic and red–green vision have been demonstrated. The abnormalities have been attributed to dysfunction of the outer sensory retina occurring secondarily to changes in the inner sensory retina (Helve, 1972; Minato, 1991).

Electro-oculography, electroretinography, dark adaptation, fluorescein angiography

Except in advanced cases, electro-oculography usually shows a slight reduction of normal values according to Tanino et al. (1985).

On electroretinography, the a-wave is normal in light-adapted eyes but subnormal in dark-adapted eyes. Likewise the b-wave shows normal values in light-adapted eyes and somewhat subnormal values at lower light intensities (Kellner et al., 1990). In severely affected eyes the electroretinograms are abnormal as a result of dysfunction of the Muller cells.

The final threshold in dark adaptation is 1–2 log units above normal. (Miyake et al., 1981) The visual fields correspond to the area of retinal involvement .

According to Krause et al. (1970), fluorescein angiography is normal in young patients with an intact pigment layer and with changes only in the macular area. In old subjects, the pigment layer disappears and large choroidal vessels appear in the macula. In cases with both central and peripheral changes, epithelial damage involving pigment occur both centrally and peripherally of the vascular veil.

Complications

Bleeding in the retina and vitreous body have been reported in up to 25% of all cases. These often constitute the cause of the first consultation with an ophthalmologist. In large series studied as part of family investigations, the incidence is much lower, e.g. 4.7% according to Kellner et al. (1990). In our series of 300 mostly familial cases, the figure is less than 10%. Haemorrhages occur both in the cavity of the schisis and in the vitreous body and are probably caused by increased splitting of the ablated layer. In advanced cases, neovascularisation has been frequently observed in the periphery but can also occur in the optic nerve (Pearson and Jagger, 1989). Blood vessels can be seen in both the ablated layer and in the part of the retina that remains attached to the fundus. The bleeding is mostly quickly resorbed but sometimes requires vitreous surgery (Greven et al., 1990).

True, full thickness retinal detachment is reported at highly varying frequencies up to 20% depending on whether the series emanate from hospital reports or from studies on families. For instance, in family studies Deutman (1971) found retinal detachment to be rare and we have observed only 5 cases out of 300. In these cases a hole has appeared even in the inner retinal layer. Treatment is the same as for regular retinal detachments and the results are acceptable.

Therapy

No therapy to cure the degenerative condition is known. The rare true retinal detachments should be treated by conventional surgical techniques. Large intraocular bleeding may require vitrectomy. Peripheral schisis rarely progresses to the central area but may need photocoagulation. Prophylatic photocoagulations were frequently used earlier but are hardly beneficial as progression of an elevated leaf is rare. In the last decade it has been observed that prophylactic photocoagulation can itself cause detachment; it should be abandoned (Castier et al., 1990; Kellner et al., 1990). The indications for vitrectomy have been reviewed by Schulman et al. (1985).

Differential diagnosis

Retinoschisis is sometimes not noticed at all. It is often misdiagnosed as heredodegeneration of the macula, or as retinal periphlebitis. In the aged the central schisis is nearly impossible to distinguish from senile macular degeneration. Other types of similar retinoschisis exist. The most common one, senile retinoschisis, has a different age distribution and seldom poses differential diagnostic difficulties.

Central retinoschisis may have autosomal dominant inheritance (Deutman et al., 1976; Yassur et al., 1982). Autosomal recessive central retinoschisis has also been reported (Forsius et al., 1978; Lewis et al., 1977; Noble et al., 1978; Zografos and Faggioni, 1980). Several syndromes display retinoschisis as one component.

A combination of central and peripheral retinoschisis is seen in Wagner disease (Wagner, 1938), which has autosomal dominant heredity. This syndrome is further characterized by myopia, hemeralopia, extinguished electroretinogram, and cataract. A variant of Wagner disease was studied by Yamaguchi and Hara (1989). Other disorders sometimes posing diagnostic difficulties are congenital retinal folds, and Norrie disease (Godel and Goodman, 1981). Stickler disease may have the same symptoms and heredity but in addition joint abnormalities occur (Stickler et al., 1965). Pedigrees published by Yassur et al. (1982) (autosomal dominant RS) and Shimazaki and Masakazu (1987) probably represent diseases belonging to the same family as RS.

The Goldmann–Favre type of hereditary idiopathic retinoschisis is characterised by vitreous degeneration with liquefaction and the formation of preretinal strands and cords, peripheral and central retinoschisis, pigmentary and degenerative changes of the retina, hemeralopia with abolition of the electroretinogram, progressive decrease in visual function, lens opacities, and autosomal recessive heredity (Favre, 1958). A radiating stellate pattern in the macula may occur in this disorder (Nasr et al., 1990).

Penetrance and expressivity

The penetrance in males appears to be 100%. As described above, expression is quite variable. That this is in large part due to factors other than the type of mutation in the RS gene is shown by examples of highly different severity within pedigrees and even within sibships. We have seen brothers with phenotypes ranging from slight cystic macular changes to almost complete blindness.

Phenotypes of heterozygotes

Most authors hold that even careful ophthalmoscopy of the central and peripheral retina does not disclose changes in RS gene carriers that can be distinguished from those seen in normal women of the same age (e.g. Vainio–Mattila et al., 1969; Deutman, 1971). However, occasional reports of pathological foveal findings such as mottling and gliosis in the periphery have appeared (Ewing and Ives, 1969; Kaplan et al., 1991). In several carrier women, a consistent disturbance of retinal function was detected by Arden et al. (1988), who examined the relationship of cone flicker threshold to changing light intensities of the background at a single retinal site.

It may indeed be that some heterozygous women can display phenotypic effects of the gene mutation they carry. However, these changes are neither typical enough nor do they occur in a sufficient proportion of carriers so as to provide reliable carrier detection by conventional clinical investigation.

A small number of affected females have been described (Forsius et al., 1962; Uchino and Shimizu, 1976; Hamaguchi et al., 1989). These are women who are homozygous as a result of the mating of an affected male and a heterozygous female. As expected, their phenotype resembles that of affected males. The same was true of a 45,X female with RS (Hommura et al., 1982).

MOLECULAR STUDIES

Linkage studies in retinoschisis

The first attempts to map the RS gene were made by typing RS families for the Xg blood group. These linkage studies suggested that RS might be located at a measurable distance from the Xg locus (Eriksson et al., 1969; Ives et al., 1970; Race and Sanger, 1975; Boman et al., 1976; Forsius and Eriksson, 1980). Subsequently Wieacker et al. (1983) used a polymorphic DNA marker (RC8) to study two RS families which had also been typed for Xg. Their linkage results with RC8 (Zmax = 1.74 at max = 0.15) gave additional support for the localisation of the RS gene on the distal part of Xp, although no definite location could be ascertained. One of the two RS families was further analyzed with additional markers by Gal et al. (1985). These preliminary two-point linkage studies, reported in an abstract, suggested that RS is also linked to DXS43, DXS16, and DXS85.

All two-point linkage studies between RS and anonymous RFLPs published to date (with maximum lod score values >3) are in agreement that RS is closely linked to markers DXS16, DXS43, DXS207, DXS9, DXS274, and DXS41 on Xp22.1-p22.2 (Table 1). With the exception of one family in the initial study by Wieacker et al., (1983), the marker locus DXS9 has shown no recombinations with RS in the families studied (Alitalo et al., 1988; Gellert et al., 1988; Sieving et al., 1990). In addition, no recombinations have been found between RS and DXS197 (Zmax = 4.82 at θ max = 0.00; Alitalo et al., 1991a). The results of one study appear divergent, but all the lod scores reported are smaller than 3 (Gellert et al., 1988). Whether the families described by Wieacker et al. (1983) and Gellert et al. (1988) represent different eye disorders, genetic heterogeneity in RS, or indicate greater distances between RS and the markers DXS9, DXS16, DXS43, and DXS41 than is suggested by the other studies, remains to be resolved. Incidentally, when comparing the results of various two-point linkage studies, it is good to remember that although significant lod scores can be obtained between various pairs of markers by two-point linkage analysis, the confidence intervals tend to be wide (e.g. Alitalo et al., 1991b). Thus, while two-point linkage analysis is a good method for establishing linkage, the genetic distances obtained should be interpreted with caution. This applies especially when maximum lod score values are close to or smaller than 3 and the number of informative meioses is small.

Table 1 Two-point linkage data (^{Z}max at $^{\Theta}$max) between RS and six markers

DXS43 Z_{max}	θ_{max}	*Confidence interval*	*DXS207* Z_{max}	θ_{max}	*Confidence interval*	*DXS9* Z_{max}	θ_{max}	*Confidence interval*	*Reference**
13.87	0.02	0.00–0.06	5.25	0.05	0.01–0.15	3.61	0.00	0.00–0.10	1.
4.98	0.00	NA	2.93	0.00	NA	NA	NA	NA	2.
1.36	0.08	NA	NA	NA	NA	2.66	0.00	NA	3.
5.91	0.00	NA	4.17	0.00	NA	NA	NA	NA	4.
7.37	0.05	0.01–0.16	NA	NA	NA	4.93	0.00	0.00–0.12	5.

DXS16 Z_{max}	θ_{max}	*Confidence interval*	*DXS274* Z_{max}	θ_{max}	*Confidence interval*	*DXS41* Z_{max}	θ_{max}	*Confidence interval*	*Reference*
12.60	0.06	0.02–0.13	9.02	0.05	0.01–0.14	9.60	0.06	0.02–0.13	1.
1.41	0.00	NA	NA	NA	NA	4.09	0.00	NA	2.
0.07	0.40	NA	NA	NA	NA	0.00	0.50	NA	3.
2.42	0.00	NA	3.28	0.08	NA	1.16	0.18	NA	4.
7.81	0.07	0.02–0.18	NA	NA	NA	9.47	0.07	0.02–0.16	5.

*1. Alitalio et al., 1987, 1988, 1991a; 2. Dahl et al., 1988, 3. Gellert et al., 1988; 4. Kaplan et al., 1991; 5. Sieving et al., 1990
NA=not available

Multipoint linkage analysis provides a more effective and quantitative means of answering the question of the most probable location of the RS gene with respect to the markers. Two independent studies have localised the RS gene between the markers DXS43 and DXS41 (Alitalo et al., 1988; Sieving et al., 1990). The multipoint results presented in an abstract by van Schooneveld et al. (1989) are in agreement with this localisation. Only the study by Gellert et al. (1988) is in disagreement. Estimates for the distance between DXS43 and DXS41 vary from 10 cM to 15 cM (Page et al., 1990; Sieving et al., 1990; Alitalo et al., 1991a; Benham and Rowe, 1992). The genetic map around the RS locus has been further improved by multilocus linkage studies in an expanded database including 31 Finnish RS families and a set of 40 normal CEPH families (Figure 5; Alitalo et al., 1991b). The refined linkage map is in agreement with other linkage maps from the area with slight differences in distances between loci (Drayna and White, 1985; Donis-Keller et al., 1987; Brown et al., 1988; Econs et al., 1990; Page et al., 1990; Sieving et al., 1990; Thakker et al., 1990; Davies et al., 1991). With the help of multipoint linkage studies, the location of the RS gene has been refined to between the markers DXS274 and (DXS43,DXS207) with odds of $1.8x10^4$:1 favouring this

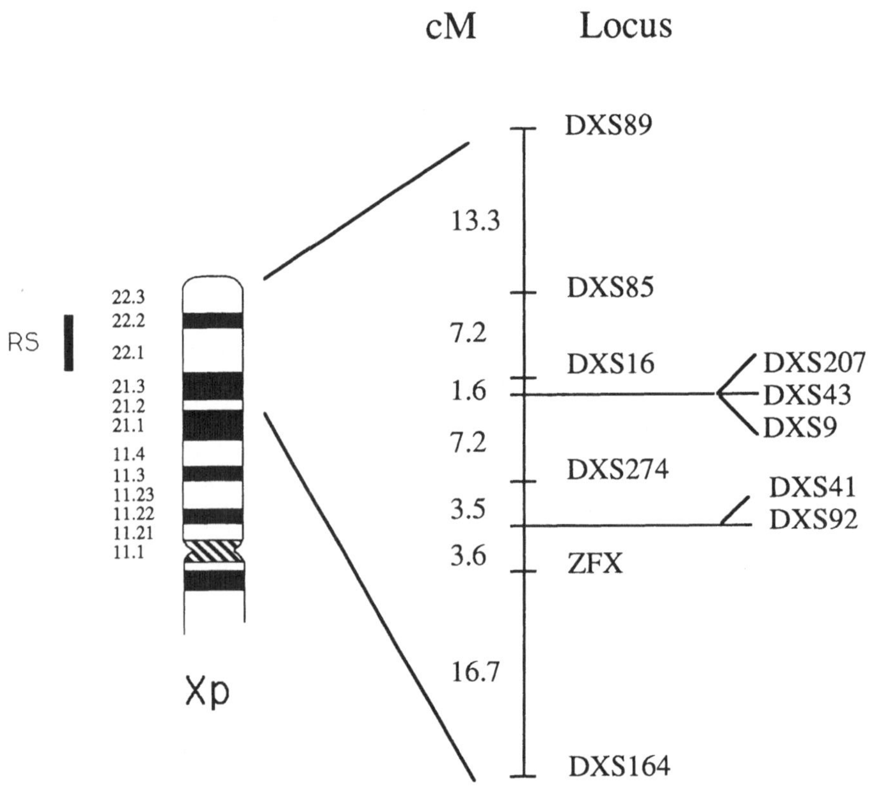

Figure 5 Estimated distances (cM) between markers on Xp. The distances are compiled from five separate seven-point analyses. The order of the markers DXS207, DXS43, and DXS9 with respect to each other, as well as the order of the markers DXS274 and DXS41, has not been established (data taken from Alitalo 1990; Alitalo et al., 1991b).

location over the second best location (Figure 6 and Table 2). It has been estimated that the genetic distance between the flanking markers is 7 cM and that the RS gene is located 2 cM proximal to (DXS43,DXS207). The exact order of DXS9 and DXS197 relative to RS or the markers DXS43 and DXS207 has not been determined due to lack of recombinations.

Diagnostic procedures

Based on all available linkage data, it is now possible to predict which members of RS families are carrier females and to diagnose the syndrome pre- and postnatally. Since close linkage (with Zmax > 3) between RS and markers in the Xp22.1-p22.2 region has been observed in all families reported, there is no strong evidence for genetic heterogeneity of RS. The closest distal markers DXS43 and DXS207 are about 2 cM from the RS gene, and the closest proximal

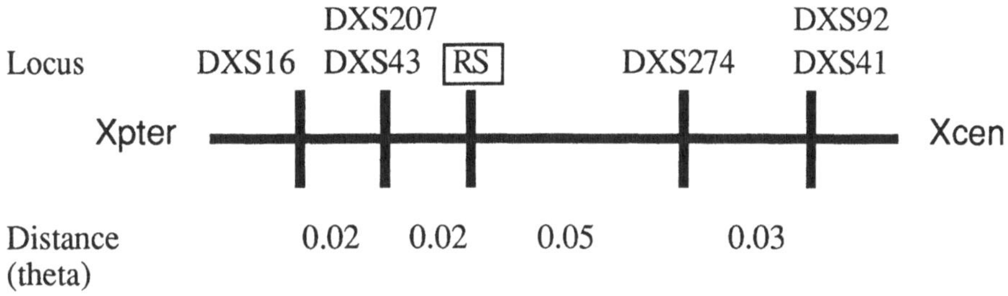

Figure 6 Map of chromosome Xp showing locations of six linked marker loci and the RS gene under the best-supported gene order (Table 2).

marker DXS274 is about 5 cM from the gene (Alitalo et al., 1991a). Because it is known that linked markers flanking the disease locus provide much more diagnostic reliability than linked but non-flanking markers, we have used these three flanking markers in carrier and prenatal diagnosis. In a few cases it has also been necessary to screen the families with the markers DXS16 and DXS41, because the diagnostic utility of linked RFLPs is sometimes limited by a lack of heterozygosity at the marker loci in female carriers. We have performed risk calculations using the risk-calculation option of the MLINK program from the LINKAGE package of computer programs (Lathrop et al., 1984). Microsatellite repeats isolated from cosmid clones corresponding to the marker loci DXS207 and DXS274 are now available, greatly improving the informativeness of these markers (Biancalana et al., 1992; Rowe et al., 1992) and thus reducing the need to use additional markers in diagnostics.

In diagnostic work, the population structure and available genealogical data should be taken into consideration. Finland provides an example. First, allele frequency studies (on 12 markers from Xp) in unrelated Finnish individuals have shown that the observed heterozygosity value for DXS207 deviates significantly from the value found in other populations, thus making DXS207 less informative in Finnish families than elsewhere (Alitalo et al., 1991a). Corresponding data on a new microsatellite repeat are not yet available. Second, most of the known RS patients come either from South West Finland and can be traced to one super-pedigree, or from North Central Finland and can be traced to a few pedigrees. Haplotype analysis with the markers DXS207, DXS43, DXS197, DXS274, and DXS41 have shown that patients from South West Finland have a haplotype association which differs from the haplotype association found in the patients from North Central Finland. Both haplotypes differ from the most frequently observed haplotype in normal chromosomes, which is shared by both groups (Alitalo et al., 1991a). These data can sometimes be useful in genetic counselling; moreover they favour the hypothesis that the mutations in the two groups arose independently.

We have performed 17 analyses to determine carriership and one for prenatal diagnosis. The results do not show any discrepancies with clinical and genealogical evidence regarding phenotype.

Pointers for future studies

With the aim of cloning and characterizing the RS gene, ongoing and future studies will first attempt to refine the genetic map of the region. A number of new markers have been mapped close to DXS43 and DXS41, although these markers have not yet been used in linkage studies of RS. Econs et al. (1990) mapped GLR close to DXS43 ($Zmax = 5.40$ at $\theta max = 0.00$), and markers DXS365 and DXS257 have been localised between the markers DXS43 and DXS41 (Davies et al., 1991; Browne et al., 1992; Econs et al., 1992). In addition, markers DXF22S4/S5 and 16d/E are located close to DXS43 (Davies et al., 1991; Rowe et al., 1992). New markers isolated from the Xp22.1-p22.2 region are presently being mapped (Benham and Rowe, 1992; Biancalana et al., 1992).

The next step is the physical mapping of RS, that is, the delineation of a contiguous stretch of DNA and the localisation of RS within it. This will require the building of contigs with the aid of yeast artificial chromosomes and cosmids. Physical mapping of the RS region is in progress (Alitalo et al., 1992) with the aim of ordering the markers DXS9, DXS197, DXS43, DXS207 and GLR with respect to each other, and allowing new markers closer to the RS gene to be isolated.

SUMMARY AND CONCLUSIONS

X-linked juvenile retinoschisis is caused by a mutation in the RS gene on the distal half of the short arm of the X chromosome. We estimate the prevalence to be approximately one in 15,000 to 30,000, that is, higher than previously estimated. While there are geographical regions with very high prevalences such as in Finland and Canada, the frequency of independent mutations may not be unequally distributed. The disease is a developmental defect of the cortical vitreus being manifest at birth. Its main feature is poor eyesight that occurs in all affected males (penetrance 100%). Expression is highly variable (from almost total blindness to a slight reduction in visual acuity) between and even within families and sibships. There is little worsening of eyesight up to 40–50 years of age whereafter deterioration occurs, probably as a result of the degeneration of several retinal structures and deterioration of related functions. Heterozygotes have normal vision and have no or minimal recognisable signs of disease.

The RS gene has been mapped by linkage analysis to an interval of less than 10 cM between cloned polymorphic markers. Definitive flanking markers are DXS207 and DXS43 on the distal, and DXS274, DXS41 and DXS92 on the proximal side. Until recently this region has been relatively poor in mapped markers and genes. Newly generated markers of high informativity in conjunction with yeast artificial chromosome- and cosmid-mediated building of contigs will eventually allow RS to be placed on a physical map of the region. This in turn will make it possible to clone and characterise the RS gene. The structure and function of the gene and the characterisation of its protein product and mutations are needed to provide leads regarding the pathogenesis of RS, and ultimately, even therapy and prevention.

ACKNOWLEDGMENTS

Our RS studies have been supported by grants from the Academy of Finland, the Sigrid Juselius Foundation, the Emil Aaltonen Foundation, the Jenny and Antti Wihuri Foundation and the Finnish Cultural Foundation. We thank the RS patients and families for their enthusiastic participation. Dr Tiina Alitalo is presently at the Department of Genetics, Case Western Reserve University, Cleveland, Ohio, USA.

REFERENCES

Alitalo, T. (1990) *Gene mapping of juvenile retinoschisis and Aland Island eye disease on the human X chromosome.* D. Phil. thesis, University of Helsinki, Helsinki. ISBN 951–45–5610-0.

Alitalo, T., Karna, J., Forsius, H. and de la Chapelle, A. (1987) X-linked retinoschisis is closely linked to DXS41 and DXS16 but not DXS85. *Clinical Genetics* **32**, 192–195.

Alitalo, T., Forsius, H., Karna, J. Frants, R. R., Eriksson, A.W., Wood. S., Kruse, T. A. and de la Chapelle, A. (1988) Linkage relationships and gene order around the locus for X-linked retinoschisis. *American Journal of Human Genetics* **43**, 476–483.

Alitalo, T., Kruse, T. A. and de la Chapelle, A. (199la) Refined localization of the gene causing X-linked juvenile retinoschisis. *Genomics 9*, 505–510.

Alitalo, T., Kruse, T. A., Ahrens, P., Albertsen, H. M., Eriksson, A. W. and de la Chapelle A. (199lb) Genetic mapping of 12 marker loci in the Xp22.3-p21.2 region. *Human Genetics* **86**, 599–603.

Alitalo, T., Francis, F., Baker, A., Kere, J., Lehrach H., de la Chapelle, A. and Willard, H. F. (1992) Identification of yeast artificial chromosomes in the region close to the retinoschisis locus: physical linkage of DXS43 and DXS197. *American Journal of Human Genetics* **51**, A233.

Arden, G. B., Gorin, M. B., Polkinghorne, P. J., Jay, M. and Bird, A. C. (1988) Detection of the carrier state of X-linked retinoschisis. *American Journal of Ophthalmology* **105**, 590–595.

Benham, F. and Rowe P. (1992) Use of Alu-PCR to characterize hybrids containing multiple fragments and to generate new Xp21.3-p22.2 markers. *Genomics* **12**, 368–376.

Biancalana, V., Weber, C., Hors-Cayla, M. C., Holvoet-Vermaut, L., Chery, M., Oudet, C. and Hanauer, A. (1992) Toward the development of a genetic map of the Xp22.1-p22.2 region. *Abstract of 3rd X Chromosome Workshop,* Amalfi, Italy, pp. 29–30.

Boman, H., Heilig, P., Kolder, H. E., Giblett, E. R. and Fialkow, P. J. (1976) Hereditary retinoschisis. Linkage studies in a family and considerations in genetic counselling. *Canadian Journal of Ophthalmology* **10**, 11–16.

Brown, C. J., Mahtani, M. M. and Willard, H. F. (1988) Genetic mapping of four DNA markers (DXS16, DXS43, DXS85, and DXS143) from the p22 region of the human X chromosome. *Human Genetics* **80**, 296–298.

Browne, D., Barker, D. and Litt, M. (1992) Dinucleotide repeat polymorphisms at the DXS365, DXS443 and DXS451 loci. *Human Molecular Genetics* **1**, 213.

Castier, P., Turut, P., Francois, P. and Woillez, J. -P. (1990) La photocoagulation dans le retinoschisis heredofamilial lie au sexe. *Bulletin des Societes d'Ophthalmologie de France* **90**, 149–152.

Condon, G. P., Brownstein, S., Wang, N. S., Kearns, A. F. and Ewing, C. C. (1986) Congenital hereditary (juvenile X-linked) retinoschisis. Histopathologic and ultrastructural findings in three eyes. *Archives of Ophthalmology* **104**, 576–583.

Constantaras, A. A., Dobbie, J. G., Choromokos, E. H. and Frenkel, M. (1972). Juvenile sex-linked recessive retinoschisis in a black family. *American Journal of Ophthalmology* **74**, 116–1178.

Dahl, N., Goonewardena, P., Chotai, J., Anvret, M. and Pettersson, U. (1988) DNA linkage analysis of X-linked retinoschisis. *Human Genetics* **78**, 228–232.

Davies, K. E., Mandel, J. -L., Monaco, A. P., Nussbaum, R. L. and Willard, H. F. (1991) Report of the committee on the genetic constitution of the X chromosome. *Cytogenetics and Cell Genetics* **58**, 853–966.

Deutman, A. F. (1971) The hereditary dystrophies of the posterior pole of the eye. Thomas, Springfield, Illinois, USA. p. 48.

Deutman, A. F., Pinckers, A. J. L. G., and de Klerk, A. L. (1976) Dominantly inherited cystoid macular edema. *American Journal of Ophthalmology* **82**, 540.

Deutman, A. F. (1977) Vitreoretinal dystrophies. In (A. E. Krill, ed.) *Krill's Hereditary Retinal and Choroidal Diseases* Vol. II. Harper & Row Publishers, Inc. p. 1043.

Donis-Keller, H., Green, P., Helms, C., Cartinhour, S., Weiffenbach, B., Stephens, K. et al. (1987) A genetic linkage map of the human genome. *Cell* **51**,319–337.

Drayna, D. and White, R. (1985) The genetic linkage map of the human X chromosome. *Science* **230**, 753–758.

Econs, M. J., Pericak-Vance, M. A., Betz, H., Bartlett, R. J., Speer, M. C. and Drezner M. K. (1990) The human glycine receptor: A new probe that is linked to the X-linked hypophosphatemic rickets gene. *Genomics* **7,** 439–441.

Econs, M. J., Barker, D. F., Speer, M. C., Pericak-Vance, M. A., Fain, P. R. and Drezner, M. K. (1992) Multilocus mapping of the X-linked hypophosphatemic rickets gene. *Journal of Clinical Endocrinology and Metabolism* **75**, 201–206.

Eriksson, A. W., Fellman, J., Vainio-Mattila, B. and Forsius, H. (1969) X-chromosomal retinoschisis (retinal detachment) in Finland. *Bulletin of the European Society of Human Genetics* **3,** 67–76.

Ewing, C. C. and Ives, E. J. (1969) Juvenile hereditary retinoschisis. *Transactions of the Ophthalmological Society of the UK* **89**, 29–39.

Favre, M. (1958) A propos de deux cas de degenerescence hyaloideoretinienne. *Ophthalmologica* **135**, 604–609.

Forsius, H., Eriksson, A. W. and Vainio-Mattila, B. (1962) Geschlechtsgebundene erbliche Retinoschisis in zwei Familien in Finnland. *Klinische Monatsblatter fur Augenheilkunde* **143,** 806–816.

Forsius, H., Krause, U., Helve, J., Vuopala V., Mustonen, E., Vainio-Mattila, B., Fellman, J. and Eriksson, A. W. (1973) Visual acuity in 183 cases of X-chromosomal retinoschisis. *Canadian Journal of Ophthalmology* **8**, 385–393.

Forsius, H., Eriksson, A. W., Damsten, M. and Alanko, H. (1978) Recessive retinoschisis. *Documenta Ophthalmologica Proceedings Series* **17**, 345–359.

Forsius, H. R. and Eriksson, A. W. (1980) Retinoschisis X-chromosomalis. In *Population structure and genetic disorders* pp. 673–676. Eds.: Eriksson, A. W., Forsius, H., Nevanlinna, H. R., Workman, P. L. & Norio, R. K. Academic Press, London.

Forsius, H. R., Eriksson, A. W. and Damsten M. (1990) Progression in juvenile X-chromosomal retinoschisis. *Acta Ophthalmologica Suppl* 95 **68**, 113–119.

Franceschetti, A., Francois, J. and Babel, J. (1974) Hereditary vitreoretinal degenerations. In *Chorioretinal heredodegenerations* Chapter VII: Cystic and vitreoretinal degenerations, Charles C. Thomas, Springfield, Illinois. 637–646.

Gal, A., Wienker, T., Davies, K., Pearson, P., Kunkel, L., Latt, S., Willard, H. and Ropers, H. (1985) Further linkage studies between retinoschisis and cloned DNA sequences from the distal Xp. *Cytogenetics and Cell Genetics* **40**, 635.

Gellert, G., Petersen, J., Krawczak, M. and Zoll, B. (1988) Linkage relationship between retinoschisis and four marker loci. *Human Genetics* **79**, 382–384.

Godel, V. and Goodman, R. M. (1981) X-linked recessive primary retinal dysplasia: clinical findings in affected males and carrier females. *Clinical Genetics* **20**, 260–266.

Gorlin, R. J. and Knobloch, W. H. (1972) Syndromes of genetic juvenile retinal detachment. *Zeitschrift fur Kinderheilkunde* **113**, 81–92.

Greven, C. M., Moreno, P. J. and Tasman, W. (1990) Unusual manifestations of X-linked retinoschisis. *Transactions of the American Ophthalmological Society* **88**, 211–228.

Hamaguchi, H., Wada, I., Takigawa, J., Yamada, H., Mori, K. and Uji, Y. (1989) A case of congenital retinoschisis in a 34-week gestation female infant (in Japanese). *Nippon Ganka Gakkai Zasshi* **93**, 575–580.

Helve, J. (1972) Colour vision in X-chromosomal juvenile retinoschisis. *Modern Problems in Ophthalmology* **11**, 122–129.

Hommura, S., Nakano, H. and Hamaguchi, H. (1982) A family of foveal retinoschisis including a female with Turner's syndrome. *Ophthalmology* **36**, 291–296.

Ide, C. H. and Wilson, R .J. (1973) Juvenile retinoschisis. *British Journal of Ophthalmology* **57**, 560–562.

Ives, E. J., Ewing, C. C. and Innes, R. (1970) X-linked juvenile retinoschisis and Xg linkage in five families. *American Journal of Human Genetics* **22**, 17A–18A.

Janaky, M. A. and Benedek, G. (1990) Elektrophysiologische Zeichen bei Makulakrankheiten. *Fortschritte der Ophthalmologie* **87**, 594–598.

Jager, G. M. (1953) A hereditary retinal disease. *Transactions of the Ophthalmological Society of the UK* **73**, 617–619.

Kaplan, J., Pelet, A., Hentati, H., Jeanpierre, M., Briard, M. L., Journel, H., Munnich, A. and Dufier, J. L. (1991) Contribution to carrier detection and genetic counselling in X linked retinoschisis. *Journal of Medical Genetics* **28**, 383–388.

Kawano, K., Tanaka, K., Murakami, F. and Ohba, N. (1981) Congenital hereditary retinoschisis: evolution at the initial stage. *Albrecht von Graefes Archiv fur Klinische und Experimentelle Ophthalmologie* **217**, 315–323.

Kellner, U., Brummer, S., Foerster, M. H. and Wessing, A. (1990) X-linked congenital retinoschisis. *Graefe's Archive for Clinical and Experimental Ophthalmology* **228**, 432–437.

Kraushar, M. F., Schepens, Ch. L., Kaplan, J. A. and Mackenzie F. H. (1972) Congenital retinoschisis. *Contemporary Ophthalmology*. The Williams & Wilkins Company, Baltimore, p. 265.

Krause, U., Vainio-Mattila, B., Eriksson, A. W. and Forsius, H. (1970) Fluorescein angiographic studies on X-chromosomal retinoschisis. *Acta Ophthalmologica* **48**, 794–807.

Lathrop, G. M., Lalouel, J. M., Julier, C. and Ott, J. (1984) Strategies for multipoint linkage analysis in humans. *Proceedings of the National Academy of Sciences, USA* **81**, 3443–3446.

Lewis, R. A., Lee, G. B. and Martonyi, C. L. (1977) Familial foveal retinoschisis. *Archives of Ophthalmology* **95**, 1190–1196.

Mann, I. and MacRae A. (1938) Congenital vascular veils in the vitreous. *British Journal of Ophthalmology* **22**, 1–10.

McKusick, V. A. (1990). MIM312700 Retinoschisis. In: *Mendelian Inheritance in Man*, 9th ed., p. 1714, Baltimore, Johns Hopkins University Press.

McAlpine, P. J., Shows, T. B., Boucheix, C., Huebner, M. and Anderson, W. A. (1991) The 1991 catalog of mapped genes and report of the nomenclature committee. *Cytogenetics and Cell Genetics* **58**, 5–102.

Minato, H. (1991) Color vision defects of macular diseases. *Nippon Ganka Gakkai Zasshi* **95**, 354–362.

Miyake, Y., Miyake, S. and Yanagida, K. (1981) X-chromosomal congenital retinoschisis. *Acta Societatis Ophthalmologicae Japonicae* **83**, 97–112.

Nasr, Y. G., Cherfan, G. M., Michels, R. G. and Wilkinson, C. P. (1990) Goldmann-Favre maculopathy. *Retina* **10**, 178–180

Nizankowska, H. and Wozny, J. (1991) Congenital retinal detachment with predominant macular changes. *Klinika Oczna* **93**, 178–179.

Noble, K. G., Carr, R. E. and Siegel, I. M. (1978) Familial foveal retinoschisis associated with a rod-cone dystrophy. *American Journal of Ophthalmology* **85**, 551–557.

Norio, R. (1981) The diseases of Finland and Scandinavia. In: *Biocultural aspects of disease.* Rotschild H (ed.), Academic Press, New York. pp 359–415.

Page, D. C., Disteche, C. M., Simpson, E. M., de la Chapelle, A., Andersson, M., Alitalo, T., Brown, L. G., Green, P. and Akots, G. (1990) Chromosomal localization of ZFX – a human gene that escapes X inactivation – and its murine homologs. *Genomics* **7**, 37–46.

Pearson, R. and Jagger, J. (1989) Sex linked juvenile retinoschisis with optic disc and peripheral retinal neovascularisation. *British Journal of Ophthalmology* **73**, 311–313.

Puech, B., Kostrubiec, B., Hache J. -C. and Francois, P. (1991) Epidemiologie et prevalence des principales dystrophies retiniennes hereditaires dans le Nord de la France. *Journal Francais d'Ophtalmologie* **14**, 153–164.

Race, R. R. and Sanger, R. (1975) *Blood groups in man,* 6th ed. Blackwell Scientific Publications, Oxford.

Rowe, P. S. N., Benham, F., Read, A., Mountford, R., Meier-Ewart, S., Lehrach, H., Davies, K. E. and O'Riordan, J. L. H. (1992) A new microsatellite marker for X-linked hypophosphataemic rickets. *Abstract of 3rd X Chromosome Workshop,* Amalfi, Italy, pp. 31–32.

Schulman, J., Peyman, G. A., Jednock, N. and Larson, B. (1985) Indications for vitrectomy in congenital retinoschisis. *British Journal of Ophthalmology* **69**, 482–486.

Shimazaki, J. and Masakazu, M. (1987) Familial retinoschisis in female patients. *Documenta Ophthalmologica* **65**, 393–400.

Sieving, A. P., Bingham, E. L., Roth, M. S., Young, M. R., Boehnke, M., Kuo, C. -Y. and Ginsburg, D. (1990) Linkage relationship of X-linked juvenile retinoschisis with Xp22.1-p22.3 probes. *American Journal of Human Genetics* **47**, 616–621.

Skoog, K. -O., Textorius, O. and Nilsson S. E. G. (1990) Evaluation of patients with obvious or suspected degenerative and dystrophic disorders of the retina, pigment epithelium and choroid. Experience from a Swedish referral center. *Acta Ophthalmologica* **68**, 131–138.

Sorsby, A., Klein, M., Hurndall, M. J. and Siggins, G. (1951) Unusual retinal detachment, possibly sex-linked. *British Journal of Ophthalmology* **35**, *1-*

Stickler, G. B., Belau, P. G., Farrell, F. J., Jones, J. D., Pugh, D. G. and Steinberg, A. G. (1965) Hereditary progressive arthro-ophthalmopathy. *Mayo Clinic Proceedings* **40**, 433–447.

Tanino, T., Katsumi, O. and Hirose, T. (1985) Electrophysiological similarities between two eyes with X-linked recessive retinoschisis. *American Journal of Ophthalmology* **60**, 149–161.

Thakker, R. V., Davies, K. E., Read, A. P., Tippett, P., Wooding, C., Flint, T. et al. (1990) Linkage analysis of two cloned DNA sequences, DXS197 and DXS207, in hypophosphatemic rickets families. *Genomics* **8**, 189–193.

Turut, P., Francois, P., Castier, P. and Milazzo, S. (1989) Analysis of results in the treatment of peripheral retinoschisis in sex-linked congenital retinoschisis. *Graefe's Archive for Clinical and Experimental Ophthalmology* **227**, 328–331.

Uchino, M. and Shimizu, H. (1976) A family with congenital retinoschisis including a female patient. *Japanese Journal of Clinical Ophthalmology* **30**,145–151.

Vainio-Mattila, B., Eriksson, A. W. and Forsius, H. (1969) X-chromosomal recessive retinoschisis in the region of Pori. An ophthalmo-genetical analysis of 103 cases. *Acta Ophthalmologica* **47**, 1135–1148.

van Schooneveld, M., Hogenkamp, Th., Orth, U., Neugebauer, M., Lisch, K., Bleeker-Wagemakers, E.M. and Gal, A. (1989) Linkage studies in X-linked juvenile retinoschisis. *Cytogenetics and Cell Genetics* **51**, 1096.

Verdaguer, J. T. (1982) Juvenile retinal detachment. *American Journal of Ophthalmology* **93**, 145–156.

Wagner, H. (1938) Ein bisher unbekanntes Erbleiden des Auges (Degeneratio hyaloideo-retinalis hereditaria), beobachtet im Kanton Zurich: *Klinische Monatsblatter fur Augenheilkunde* **100**, 840–857.

Wieacker, P., Wienker, T. F., Dallapiccola, B., Bender, K., Davies, K. E. and Ropers, H. H. (1983) Linkage relationships between retinoschisis, Xg, and a cloned DNA sequence from the distal short arm of the X chromosome. *Human Genetics* **64**, 143–145.

Yamaguchi, K. and Hara, S. (1989) Autosomal juvenile retinoschisis without foveal retinoschisis. *British Journal of Ophthalmology* **73**, 470–473.

Yanoff, M., Rahn, E. K. and Zimmerman, L. E. (1968) Histopathology of juvenile retinoschisis. *Archives of Ophthalmology* **79**, 49–53.

Yassur, Y., Nissenkorn, I., Ben-Sira, I., Kaffe, S. and Goodman, R. M. (1982) Autosomal dominant inheritance of retinoschisis. *American Journal of Ophthalmology* **94**, 338–343.

Zhang, Y. M. (1984) Study of families with congenital hereditary retinoschisis. *Chung Hua Yen Ko Tsa Chih* **20**, 174–175.

Zografos, L. and Faggioni, R. (1980) Hereditary macular retinoschisis. *Klinische Monatsblatter fur Augenheilkunde* **176**, 603–608.

15. CLINICAL AND GENETIC HETEROGENEITY OF USHER SYNDROME

W. J. KIMBERLING, M. WESTON AND C. MOLLER

Boys Town National Research Hospital, Omaha NE, 68131 and University of Linkoping, Linkoping, Sweden

INTRODUCTION

Usher Syndrome (US)[1] is an autosomal recessive disorder characterized by hearing loss and retinitis pigmentosa. Von Grafe (1858) was the first to note the simultaneous occurrence of deafness with retinitis pigmentosa. The heritable nature of Usher syndrome was recognized early by Liebreich (1861) but Usher (1914) was the first to emphasize its recessive pattern of inheritance and he is the one after whom the disorder has been named. The clinical heterogeneity of Usher syndrome was noted by several early researchers (Bell, 1933). Hallgren's 1959 monograph, however, was the first large study on Usher syndrome and he clearly pointed out that at least two distinct types of Usher syndrome existed. Only just recently has the underlying genetic heterogeneity been detected using linkage analysis and these findings indicate that the heterogeneity of Usher syndrome may be more extensive than first thought, in that five or more types may exist each with a different responsible locus. It is the purpose of this paper to give the reader an update on the progress of the identification of the genes involved and to speculate on the implication of these findings.

CLINICAL CONSIDERATIONS

Usher Syndrome Phenotypes

The clinical heterogeneity of Usher syndrome has been recognized for several years (Nuutila, 1970; Merin et al., 1974; Fishman, 1979) . Hallgren (1959) studied a large series of Usher syndrome patients in Sweden and brought attention to the fact that two different clinical types appeared to exist. The first, Usher type 1, was thought to be characterized by a profound hearing loss, vestibulocerebellar ataxia, and retinitis pigmentosa. The second, Usher type II, presents with a

---

[1]The nomenclature of Usher syndrome used is as follows: US refers to the syndrome and the types are indicated by USla, USlb, etc, and are used to refer to the clinical syndrome; USH refers to the corresponding locus, so that USHla is responsible for the clinical suptype USla.

Table 1 Genetic and Clinical Classification of Usher syndrome

	Gene	*Localization*	*Hearing*	*Vestibular*	*Visual*
US I	USH1A (rare) USH1B (common) USH1C (Acadian)	14q31-ter 11q13-14 11p13-14	Severe to profound hearing loss involving all frequencies (Deaf). Congenital.	Absent vestibular responses.	Retinitis pigmentosa with diagnosis usually in the preteenage years
US II	USH2A (common) USH2B (rare)	1q32-41 Unknown	Sloping moderate to profound hearing loss (Hard of hearing). Congenital.	Normal vestibular responses.	Diagnosis often made later, after puberty.
US III	USH3 (rare)	Unknown	Progressive hearing Loss. May have normal hearing at birth	Hypoactive to normal vestibular reflexes	Not well differentiated from types I and II

milder hearing loss, normal vestibular reflexes, and retinitis pigmentosa. In fact, type I patients are often described as deaf and going blind while type II patients are described as going blind with hearing impairment. The distinction between the two is dramatic and there is seldom reason for confusing the two types. The two phenotypes seldom, if ever, occur together in the same family. This observation suggested that different loci were involved, a fact which was established once the gene localizations were found. In fact, so far 5 different loci have been identified, 3 of which produce the type I phenotype and two which cause the type II symptoms. The genes and their corresponding phenotypes are listed in Table 1. The genes for Usher type IIb and type III remain unlocalized. Other rarer, and clinically unusual, types may exist but their relationship to types I and II is as yet unclear.

HEARING LOSS IN USHER SYNDROME

Audiologic evaluations are an important part of the clinical workup of the patient suspected to have Usher syndrome. The severity of involvement is highly correlated with genetic type. Typical audiograms for type I and II patients are presented in Figure 1 for the purposes of comparison. Three patients of each type are shown and all were drawn from known linked Usher type I (11q) and II (1q) families. Type I patients have a profound hearing loss, have poor speech, are seldom helped by hearing aids, and prefer communication through signing . Type II patients show a sloping loss which is milder in the low frequencies. Bilateral hearing aids frequently benefit the Usher type II patient; sometimes the patient, when fitted with bilateral hearing aids, hears and speaks well enough that the hearing impairment is often not noticeable. One characteristic of Usher syndrome has been the consistency of the audiometric profile within subtypes. The extent of the variability for both types is quite limited. Any deviations away from the expected sloping audiogram, when occuring without reasonable explanation, is reason to suspect a typical type I or II diagnosis. Furthermore, the hearing loss is symmetric and of the same approximate extent in both ears (Moller et al., 1989).

Progressive hearing loss has been observed in some families with Usher syndrome and may represent a type III, although its occurrence had been generally accepted as rare. (Gorlin, 1979). However, there are now several reports that suggest that progressive hearing loss may be more common than expected. For example, Merin et al. (1974), reported observing a slowly progressive hearing loss in 26% of their series of patients. Kumar et al. (1984) observed significant differences in speech discrimination between type II patients with similar hearing deficits and suggested that these differences could be due to progressive losses in some patients. Type III may be more frequent in Scandinavian countries since two of the reports showing a high frequency are in Finnish populations (Karjalainen et al., 1989; Nuutilla, 1970). Our observations, however, indicate that type III Usher syndrome with progressive hearing loss among Usher type II patients is not common in a large series of families from the USA and Sweden (Moller, 1989). It may be that the progressive loss seen by other investigors is due to the normal process of ageing and is not an effect of the gene. Another possibility is that the progression of hearing loss apparent in some type II patients is really a progressive loss in communication skills due to the gradual loss in visual abilities and which affect lipreading skills. Nonetheless, the possibility that type III actually exists and is due to allelic variation at one of the known Usher loci must be considered. In fact, Merin et al. (1974) postulated that type III was simply a variant of type II. Whether type III is linked to one of the known chromosomal locations of the 4 USH loci would answer this question but the possibility remains to be investigated.

There has been little discussion of the question of progressive hearing loss in Usher syndrome type I. It is generally assumed that a total or profound hearing deficit is present at birth. There have been, however, a few unpublished accounts of residual hearing detectable in infancy by auditory evoked brain stem responses (ABR) which disappear quickly. Although speech communication is generally not useful for type I patients, there is variability and one could postulate that this is due to the presence of hearing in some patients. So-called corner audiograms are frequently seen in type I patients; these may be due to either simply the sensation of vibration or, in some cases, to the presence of a very minimal residual hearing reflex presumably in the cochlear apex. It may be that, in type I Usher syndrome, the cochlea develops normally until about the 7–9th month of gestation and then starts on a rapidly progressive deterioration which is usually complete by birth. This hypothesis is attractive because it makes cochlear damage progressive just like

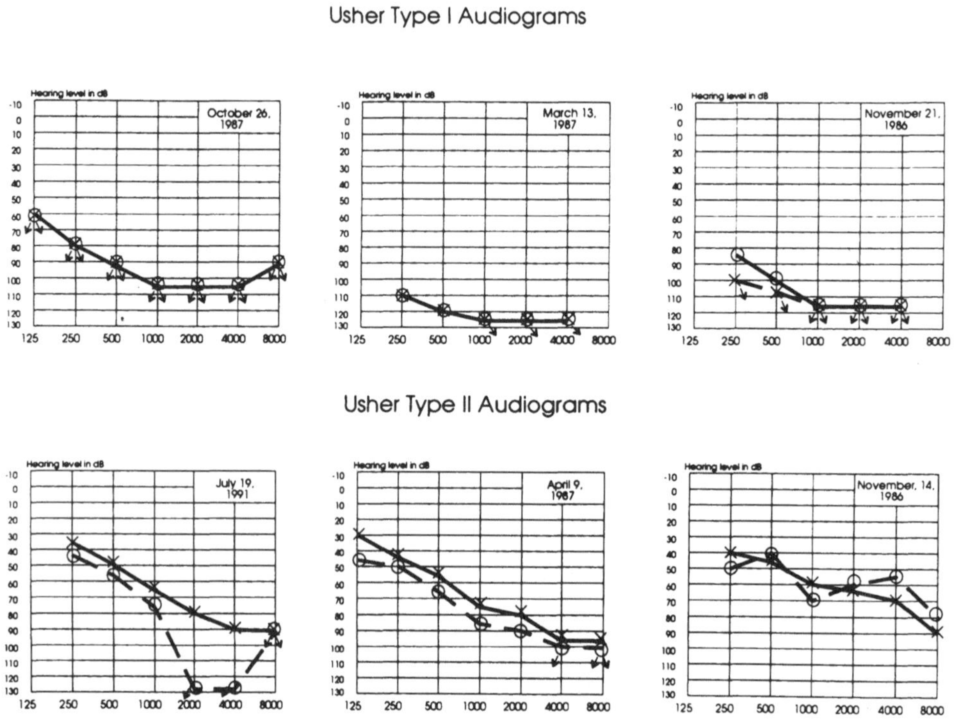

Figure 1 Audiograms of both Usher I and II patients. The top three panels show a profound hearing loss characteristic of type I; the arrows indicate that the hearing threshold is below detection. The bottom three show the sloping audiogram typically seen with type II.

the retinitis pigmentosa. In terms of a mechanism for the neurosensory damage, it is more palatable intellectually to have both systems act in an abiotrophic manner rather than having the hearing deficit appear developmental and the retinopathy being degenerative. Knowledge that a progressive component exists, for either types I or II, will be important in the evaluation of strategies to either restore or prevent the loss of hearing in Usher syndrome. Obviously, if the hearing loss is progressive, then we may have an opportunity to interfere with the deterioration so long as the time of the onset of the deterioration is known.

Auditory evoked emissions are consistent with the degree of hearing loss in both type I and II patients. Auditory evoked brain stem (ABR) responses are, as expected, absent in type I patients. In type II patients the results indicate that the hearing impairment is due solely to a peripheral lesion involving the cochlea (Moller et al., 1989).

VESTIBULAR DEFICIT

The second important distinguishing feature between types I and II Usher syndrome is the presence or absence of vestibular responses. The vestibular system in type I Usher patients is virtually non-functional. Occasionally, an affected type I individual will be seen with minimal vestibular responses, but this is extremely rare. This deficiency is manifest in the adult as a non-progressive ataxia in the form of a swaying and swinging gait. The ataxia is entirely of peripheral origin and there is no evidence to indicate an additional cerebellar component. The vestibular deficit appears to be congenital, or at least occurs at a very early age, since Usher I children are late to sit and walk. None of the patients (Moller et al., 1989) in one large well studied series walked before the age of 18 months. In fact, walking age is a good predictor of the US subtype, since few type II children are delayed in walking for as long.

The postural system in humans relies on input from three sources: vestibular, visual, and somatosensory. While the patients have compensated well for the vestibular deficit, postural stability is aggravated because of the progressive visual loss due to RP. The result is that Usher I patients are more posturally unstable than expected based upon experience with other patients with hearing or visual loss alone. Type II patients, however, are essentially normal with regard to the vestibular system. There have been some reports of mild vestibular abnormalities, such as hypo- or hyperactivity, but it is unclear whether these findings are within the range of normal. Whether there are vestibular deficits present in Usher II remains to be definitely established.

Sophisticated vestibular studies, including rotary chair, bithermal calorics, dynamic and posturography, are often necessary to determine if the "ataxia" of USH1 is central or vestibular in origin and to determine the degree to which the visual loss is affecting postural control. Visual impairment complicates, but does not invalidate, the otoneurologic examination. The RP affects the corneo-retinal potential which is the basis of electro-nystagmography (ENG); the gain of the response must sometimes be increased and the test performed in the dark since even dim light will decrease the signals. If the ataxia is not due to a labyrinthine defect, then the diagnosis of Usher syndrome type I should be questioned. The vestibular phenotype is highly correlated with the severity of hearing loss. The finding of a normal vestibular response with profound deafness or absent vestibular responses with the sloping type II audiometric profile should alert the researcher that the diagnosis is not certain. Such patients may represent new subtypes of Usher syndrome or could be due to as yet unrecognized allelic variation at one or the other known loci.

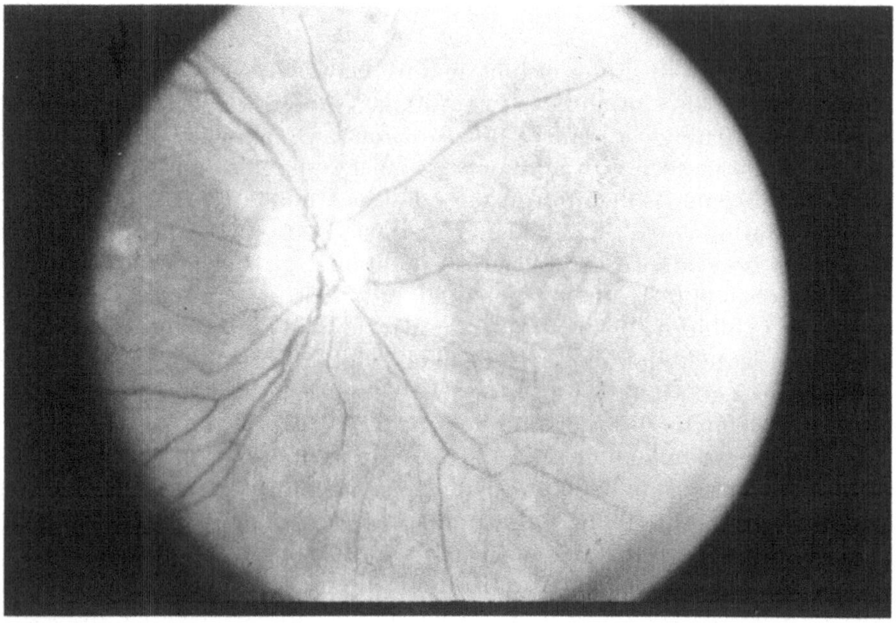

Figure 2 A

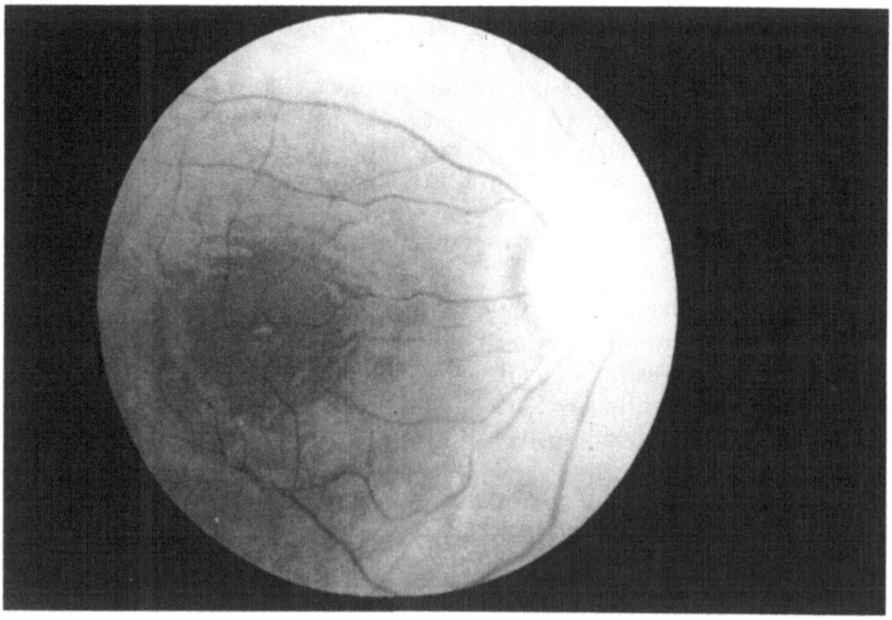

Figure 2 B

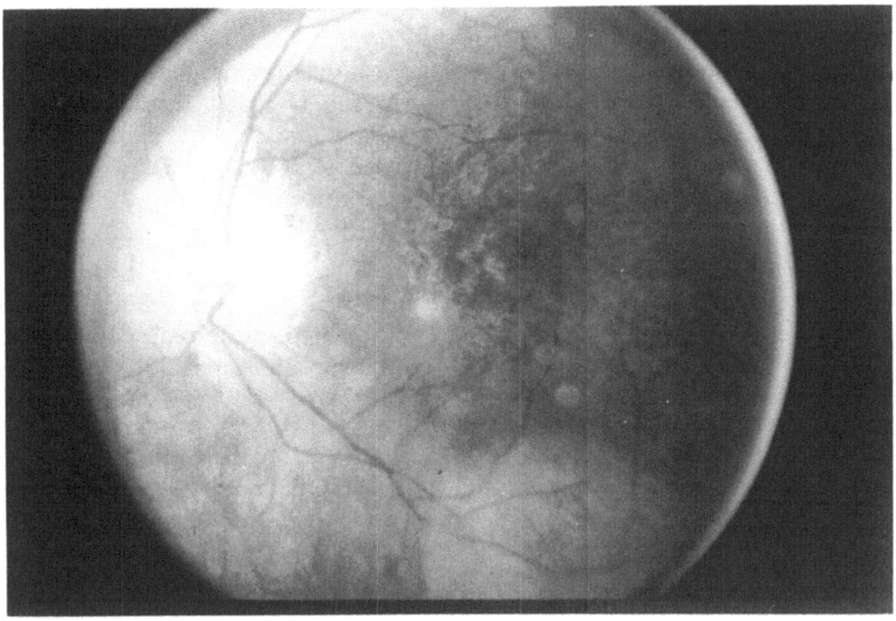

Figure 2 C

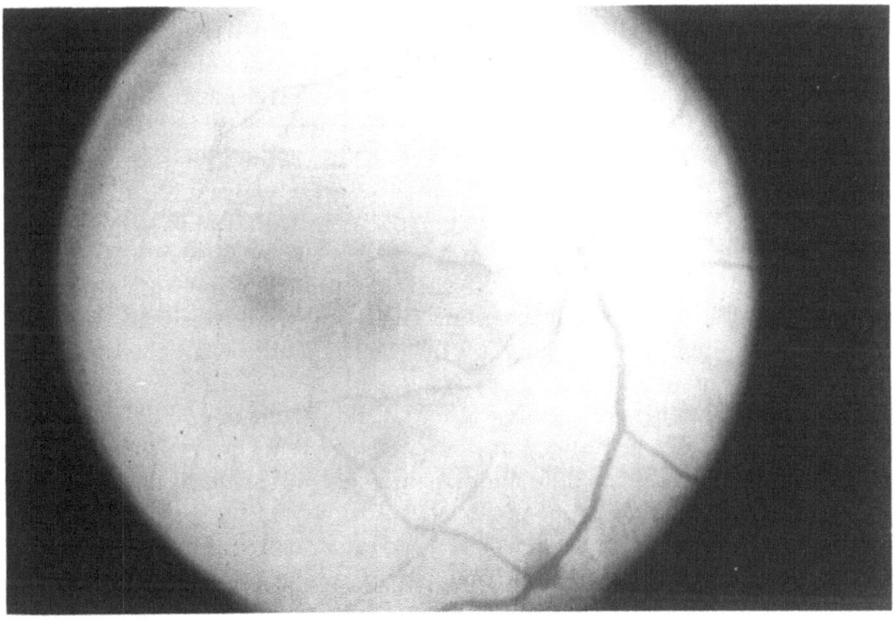

Figure 2 D

Figure 2 Photographs of the fundi of 4 independent patients with Usher syndrome showing the range of involvement. The first two panels are from type I patients and the other two are from type II patients. See Colour Plate VI at the back.

Visual Phenotype

Fundus changes are typical for patients with retinitis pigmentosa and currently there is no recognized pattern of fundus involvement that is correlated with any specific subtype. In some cases the pigmentary abnormalities are minimal and in others they are extensive. Bone spicule formation is common. A set of typical fundus photographs are illustrated in Figure 2 for both type I and II patients. However, there are preliminary indications of fundoscopic differences related to genetic subtype, at least for type II. One family with Usher type II unlinked to 1q32-41 was found to have a milder course of RP which had been described as RP *sine pigmento.* Some bone spicule formation was noted in the oldest affected sibling. On the other hand, RP *sine pigmento* has been found in both type I and type II persons and is believed to represent an early stage of the retinal deterioration common to all RP (Heckenlively, 1988).

The most frequent foveal change described by Fishman (1979) is that of an atrophic appearing lesion which was seen in 31% of the patients. An additional 11% showed a cystic-like foveal lesion associated with leakage from the perifoveolar arteries.

Several studies have shown that electroretinography (ERG) might be subnormal as early as 2 to 3 years of age, before abnormalities can be seen functionally or on fundoscopic examination (Merin, 1974; De Haas, 1970; Davenport, 1978). Both scotopic and photic responses are involved. Responses extinguish totally as the degeneration progresses. The visual evoked responses also show decreased amplitudes and longer latencies (Abraham, 1977). There is no general consensus about the severity of the RP as a function of subtype, but it has been suggested that type I may have a more severe visual involvement (Fishman, 1979). In type I patients, cone and rod responses are usually undetectable by the middle of the second decade and often in the first. Many type II patients have a recordable, but depressed, cone response (Fishman, 1979). In one type II family with four affected brothers, the gradual decline of the ERG over two decades has been documented. While electrophysiological studies provide the earliest and most reliable diagnosis of retinitis pigmentosa, they are not necessary in all cases. An absent or diminished ERG in at least one affected sibling is a necessary condition for the diagnosis of Usher Syndrome in a family.

Restricted visual fields are an integral part of the syndrome. Impairment of peripheral vision may not be noticed until the teenage years but is probably present early in life. Generally, visual field abnormalities are not associated with scotomas in type I patients. Several type II patients have shown a ring scotoma which may either be a characteristic of the specific type II genotype or could simply reflect the milder course of the disease for some type II patients. Searching nystagmus is uncommon in Usher syndrome and the retinal degeneration is uniformly progressive from the peripheral regions to the central.

Nightblindness is a concomitant to retinitis pigmentosa. Most patients have a history of nightblindness that dates back to childhood. The nightblindness is particularly evident in type I patients where disorientation is more severe in the evening and night because of the combined visual and vestibular impairment. The Usher I patient may appear especially clumsy in low light and bump into things and fall. This may lead to an earlier recognition of the RP which could be interpreted incorrectly as relating to differences in severity between types I and II.

Other ophthalmologic complications are cataracts and Coat's disease. Hallgren (1959) considered cataracts as part of the syndrome. While posterior subcapsular cataracts occur in about 30%–60% of Usher patients, the general consensus now is that the cataracts are a secondary result of the RP. This hypothesis is supported by the observation that cataracts occur at about the same frequency in all types of retinitis pigmentosa and are neither characteristic nor more frequent in Usher syndrome.

While total blindness in not common, the severity of the visual defect restricts the patient's mobility. For most Usher patients, significant visual loss is due to either posterior subcapsular lens opacities or the development of a foveal lesion. The foveal lesion can cause significant central vision loss. While the degree of hearing loss has been observed to correlate with the degree of peripheral field loss, there is no association of hearing impairment with central vision loss (Fishman et al., 1983; von Alberth et al., 1958). The discrepancy between auditory and visual involvement was also noted by other investigators (de Haas et al., 1970). These findings suggest, but do not prove, that clinical and genetic type may not be correlated with the presence or absence of foveal lesions and/or cataracts.

It should be recognized that the audiologic profile and vestibular responses, while the most reliable indicators of clinical type, are relatively non-specific and, except for progressive versus non-progressive hearing loss, hold little hope for further subclassification. Visual phenotype, on the other hand seems more variable and may be used to determine if any of the 5 genetic subtypes can be classified on the basis of ophthalmologic parameters. For example, is the fundoscopic appearance different between types? Specifically, does Usher syndrome IIb have a different ophthalmologic picture than IIa? Are cataracts or foveal lesions more frequent in one type or another? Because of the extensive genetic heterogeneity, answers to these questions may prove invaluable in *a priori* assignation of families to different subtypes and thus assist in the identification and cloning of the responsible genes.

Inter- and Intrafamilial Variability

Usher types I and II are distinct phenotypes and only rarely do the two occur in the same family. For that reason, it was hypothesized that the two were genetically distinct as well, a fact subsequently established by linkage analysis. There is little variability in expression within sibships for type I patients but greater intrafamilial variation in both hearing and visual involvement has been observed for type II (Fishman, 1979; Hallgren, 1959). The origin of the variation in hearing acuity of type II patients has not been rigorously investigated. Certainly, one might expect some of the variation to be due to the different loci known to be involved. On the other hand, some of the variation could be due to different mutant alleles at either of the type II loci.

Mental Retardation and/or Other CNS Dysfunction.

The existence of psychiatric disturbance and mental retardation has been the subject of discussion with regard to Usher syndrome. Hallgren originally noted a high frequency of mental deficiency (24%) in his series. Although Hallgren was a psychiatrist, this could have represented a bias of ascertainment or, more likely, a simple misinterpretation of the available records. Few of the patients were actually examined by Dr. Hallgren and much of the results depended upon an interpretation of the existing medical records. Furthermore, with Usher syndrome type I, the coexistence of both vestibular dysfunction (leading to a walking age of 18 months or greater) and speech delay could easily lead to a misdiagnosis of mental deficiency. Whatever the cause, the association has not been noted by most other investigators (Nuutila, 1970; Fishman et al. 1983; Moller et al., 1989) with the exception of Lindenov (1945). Merin et al. (1974) designated Usher patients with mental retardation (MR) as type IV. Several investigators have questioned the existence of type IV (Nuutila, 1970; Fishman et al., 1983; Moller et al., 1989) which has sometimes been referred to as Hallgren syndrome. Although the literature is confused and confusing with regard to the nosology of type IV/ Hallgren syndrome, there is little doubt that retardation is an uncommon aspect of all types of Usher syndrome. The occasional cases where it does occur may represent a contiguous syndrome where the Usher gene is deleted along with adjacent DNA and the MR is the result not of a defective Usher gene but of abnormality in neighboring genes. Also, there are a number of other disorders associated with hearing impairment and retinitis pigmentosa which do include MR (e.g. Laurence–Moon and Bardet–Biedl syndromes). Variation of the phenotype of these disorders could possibly lead to their inclusion in a series of supposedly Usher syndrome patients.

Various psychiatric disorders have been associated with Usher syndrome. This includes schizophrenia-like disorder, atypical psychosis, and recurrent depression (Usher 1914; Hallgren, 1959; Vernon, 1962; Nuutila, 1970; Karjalainen; 1983). Hallgren (1959) estimated the highest frequency of mental disturbance (>20%) and while subsequent studies did not confirm such a high prevalence, Usher syndrome patients are undoubtedly at a higher risk for psychiatric disturbance than the normal hearing, normal seeing population. Two different etiologies have been hypothesized as the cause of this association. One hypothesis is that it is part of the syndrome and reflects a pleiotropic effect of the Usher gene. The pleiotropic theory would have been more reasonable had the putative association between mental deficiency and Usher syndrome been upheld by subsequent studies. The second, and more likely, hypothesis is that the psychiatric dysfunction is due to stress and is a reaction to the loss of both of the two major senses. This premise is supported by the finding that patients who have gone blind or deaf have often been reported to experience temporary reactive psychosis. Furthermore, the schizophrenia-like symptoms seen in Usher syndrome are also temporary, a feature not seen in true schizophrenia.

However, there is now accumulating evidence that there may be a cerebral lesion associated with Usher syndrome. Bloom et al. (1983) found indications of cerebellar atrophy by CT scan on 6 of 12 Usher patients. Further evidence of CNS changes has been found by other investigators (Koizumi et al., 1988; Schaefer et al., 1992). The findings are relatively non-specific and indicate both cerebellar and cerebral anomalies. In one study, similar anomalies were NOT seen in deaf or RP controls (Schaeffer et al., 1992) suggesting that the brain anomalies are not secondary to the sensory deficit. The brain anomalies were seen in both type I and II patients although they appeared to be more frequent in type II, which is somewhat surprising in that type II has the milder sensory deficit. The cerebellar and cerebral anomalies have not yet been associated with any clinical neurologic problem.

Congenital Anomalies And Other Associated Defects

Congenital anomalies are not part of Usher syndrome. Occasionally, congenital anomalies have been noted in association with Usher syndrome. However, no clear pattern of dysmorphology has emerged. In fact, it is likely that whenever major or minor anomalies are found in conjunction with Usher syndrome, they occur either by chance or are the indication of a different syndrome. One might expect to find a few legitimate cases of Usher syndrome with anomalies and/or mental impairment as the result of a contiguous deletion of the Usher gene and its neighbors. Such a finding would be extremely useful in attempts to clone the Usher gene(s). Unfortunately, this has never been found.

THE GENETICS OF USHER SYNDROME

Pattern Of Inheritance

It has long been recognized that Usher syndrome is inherited as an autosomal recessive disorder with full penetrance. Boughman and Fishman (1983) showed that the segregation of Usher syndrome did not deviate from that expected for an autosomal recessive. A recent study by Shugart et al. (1991) showed that the segregation ratio is 25% for both type I and II and that the frequency of sporadic cases did not exceed that expected for recessive disorders of the same frequency. Furthermore, the penetrance was not significantly different from 1.0. Thus there is no evidence to suggest that any fraction of Usher syndrome is due to a nongenetic etiology.

X-linked inheritance was considered in one type II family (Davenport et al., 1978) because of four affected brothers, but this family was later shown to be linked to 1q32-41 markers. Boughman observed a slight excess of Usher syndrome males in a population study of retinitis pigmentosa in Louisiana. This observation was not made in earlier studies (Hallgren, 1959; Bell, 1933) nor in subsequent studies (Kimberling et al., 1989; Shugart et al., 1992).

Prevalence and Population Distribution

It has been estimated that Usher type I represents 90% of Usher syndrome in the United States, with type II accounting for nearly 10%, and other types accounting only about 1%. However, Fishman's group reported 71 out of 106 patients with type II (Piazza, 1986). A Norwegian study of 28 cases were distributed as: type I – 14 cases; type II – 10 cases; and other – 4 cases (Grondahl, 1986; Grondahl and Mjoen, 1986)). Our own series has almost 50% type II cases both from the USA and Sweden (unpublished data). We believe Usher type II is at least as common as type I. However, type II patients are not usually ascertained in schools for the deaf because their hearing loss is not severe enough to place them there. Schools and programs for the blind and ophthalmologic clinics have a relatively higher proportion of type II patients compared to ascertainment from schools for the deaf.

The frequency of Usher syndrome has been estimated at 3.0/100,000 in Scandinavia (Hallgren, 1959; Nuutila, 1970) and at 4.4/100,000 in the United States (Boughman et al., 1983). The prevalence of Usher syndrome among deaf individuals has been reported to range from 0.6 to 28% (Vernon, 1969). Conversely, the frequency of deafness in the retinitis pigmentosa population is estimated to range between 8.0 and 33.3%. Overall, there are about 16,000 deaf and blind people in the United States, of which more than half are believed to have Usher syndrome. Most of these will have Usher type I. Most Usher type II individuals, with a milder hearing loss, may not be noted in surveys which have focused on the profoundly deaf. Piazza et al. (1986) postulated that the frequency of type II is at least as great as that for type I.

Usher Type I

Localization of Usher Type I to Chromosome 14q

The Usher syndrome type I gene was first localized to chromosome 14q by a linkage with marker D14S13 in fifteen families (Kaplan et al., 1992). A linkage was suggested by the fact that for the total sample the lod score maximized at a recombination fraction (θ) of 0.20 with lod score (Z) of 0.83 (Chapter 1). However, the test for genetic heterogeneity was significant, assuming linkage, at the 5% level suggesting the possibility that two or more different genes might be involved. Ten of the 15 families showed linkage to the 14q markers and 8 of these came from the same region known as Poitu-Charentes in France but there was no evidence of consanguinity. When these families were pooled, the two point lod score went to 3.78 at $\theta = 0$ with D14S13. Based upon this location, Kaplan et al. (1992) hypothesized that the USH gene may be involved in an abnormality of ciliary structures of the retina, vestibule, and cochlea. The sensory cilium of the rods and cones separate the outer segment of stacked discs containing the visual pigment from the inner segment and it is reasonable to expect that defective cilia could lead to degeneration of the outer segment of the retina. However, vestibular and cochlear hair cells do not have true sensory cilium with the characteristic 9+0 microtubules. Vestibular hair cells have kinocilia with 9+2 microtubules and are missing the dynein arms found in motile cilia. During development, the cochlear hair cells possess a transitory kinocilium that may be responsible for organizing the development of the stereocilium characteristic of the cochlear hair cells. Thus, a defective kinocilium could lead to defective hair cells and abnormalities of hearing and balance. Because of the transitory nature of kinocilium development, it would be expected that any hearing or vestibular abnormalities would be congenital but the sensory cilia of the eye are required for normal retinal maintenance throughout life and a defect would be more likely to cause a progressive deterioration (Hunter et al., 1986).

Photoreceptor and sperm axoneme abnormalities had been previously observed in Usher syndrome (Hunter et al., 1986); nine of the 10 Usher patients studied had type II. In a subsequent report, Barrong et al. (1992) also observed abnormal connecting cilia in a type II patient. These abnormalities are not seen in the general population of X-linked and autosomal retinitis pigmentosa patients with normal hearing. Studies of axoneme structure of nasal cilia showed abnormalities associated with a group of RP patients including Usher syndrome but the results were not clearly broken down into diagnostic group so it is hard to draw any conclusions. Sperm motility and structure has also been found to be abnormal in Usher patients (Hunter et al., 1986; Nuutila, 1970). Kaplan et al. (1992) point out that the gene controlling heart position in the mouse has been mapped to a region homeologous to human 14q32 and was identified by a recessive situs inversus mutation, *iv*. The human homologue of the mouse *iv* gene should lie exactly between the human alpha-1 antitrypsin gene and the heavy chain immunoglobulin which is where the Usher gene was mapped. These are intriguing observations. There is confusion about which types of Usher syndrome actually have abnormal cilia and most cases reported have been type II and not clinically equivalent with the type I phenotype localized to 14q. A ciliary defect makes sense for type I Usher syndrome since both the vestibular and cochear functions are impaired. However, the reported ciliary abnormalities mostly occur in type II individuals. Further investigation into this phenomenon is warranted and, now that the different types can be differentiated from each other at the molecular level, the research can be directed towards defining the differences between types. It is unreasonable, although not impossible, to expect that all types of Usher syndrome share a ciliary defect. The expectation is that one or more types would have a different biologic basis. If this is found to be the case, then ciliary or axonemal structure could be an additional method to distinguish between types without prior reliance on linkage groups.

Other investigators have been unable to confirm the localization to 14q in families outside of the original French group (Keats et al., 1992). This analysis included a large series of families from the USA and Northern Europe, notably England and Sweden. This created some initial doubt about the linkage and raised the question that it may have occurred by statistical chance. However, the fact that 8 of the 10 linked families come from the same region in France supports the argument for linkage and for the existence of a locus on chromosome 14. Definitive evidence will come when those families are shown not to be linked to the other two regions now shown to contain the Usher gene. This locus has been given the designation of USH1a.

Linkage of Usher type I to chromosome 11

The discovery of the linkage of Usher syndrome type I to chromosome 11 was made almost simultaneously by two different investigators. The 11q linkage was made originally with the marker D11S527 in 19 families. The multipoint lod score reached a maximum of 4.75 at a position about 7 centiMorgans centromeric to D11S527. The gene localized here was designated USH1b to distinguish it from the one localized to 14q. Further analysis of the data indicated that the USH1 gene was flanked by the markers D11S527 and D11S35. However, the odds of placement of the USH1b gene to within this flanking region was not at the traditional 1000:1 level. The map of chromosome 11 is given in Figure 3 which shows the most recent evidence for localization of the Usher Ib gene to the 11q13-14 region. Recent results (unpublished data) indicate that the USH1b gene lies between the markers D11S527 and D11S416. This is a 2 to10 centiMorgan interval of DNA.

There was a slight possibility that the gene could be located between D11S527 and FGF3 or even more centromeric than FGF3. This raised the possibility that FGF3 could be the Usher Ib gene. Induced mutation in FGF3 produced abnormalities of both the ear and eye in mice (Mansour et al., 1992) but these abnormalities were primarily structural and not at all like those observed in Usher syndrome. Furthermore, FGF3 shows several crossovers with Usher type I thus reducing its attractiveness as a candidate gene. ROM1, a retinal membrane protein gene which is part of a photoreceptor specific protein family, is located proximal to FGF3. It show crossovers with Usher Ib and hence can also be excluded as a candidate.

The third localization of Usher syndrome was to chromosome 11p (Smith et al., 1992). This linkage has so far been found only in families of French Acadian descent. The French Acadians migrated to the bayous of Louisiana from Quebec over 100 years ago and had, until the second world war, remained genetically isolated. Presumably because of a founder effect, Usher syndrome has been found to be at a high frequency in that population. In fact, southern Louisiana may have the highest frequency of Usher syndrome in the USA, perhaps even the world. The

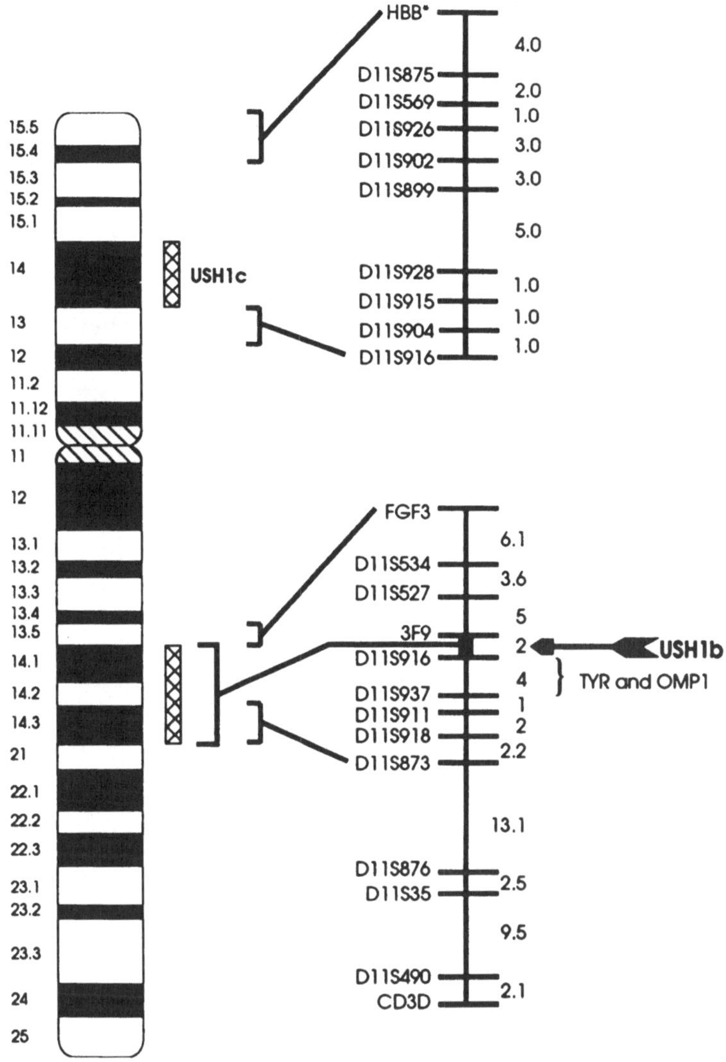

Figure 3 Map of chromosome 11 which shows the most recent evidence for localization of the Usher Ib gene to the 11q13-14 region.

Acadians were originally from the Normandy area in France and there is no evidence that they share ancestry with the families from Poitu-Charentes showing linkage to 14q. It was originally thought that the use of the Acadian isolate avoided the complications of heterogeneity and allowed the pooling of data across sibships without mixing types I and II. However, it was subsequently discovered that both types I and II occur in the population. Although type I appears to be the most prevalent and type II may have a frequency approximately equal to that in the non-Acadian USA population, the frequency distribution in this population needs to be established with full attention to the clinical differences between types.

The linkage found by Smith's group was between Usher syndrome and the markers D11S419 and D11S875. The lod score maximized at about 5.0 and indicated with a high degree of confidence that the USH1c gene is located between the markers D11S875 and Mfd132. Physically, this region corresponds to 11p13-p14.

It is interesting that the two regions 11p13-p14 and 11q13-q14 are paralogous and share an evolutionary ancestry (Lundin, 1993). Tyrosinase is on 11q and there is a tyrosinase-like (tyrosine hydroxylase) gene on 11p. Also, potassium channel genes are found in both regions. This raises the question whether USH1b and 1c are loci coding for related products. Given the paralogy of the two regions, this would not be surprising and means that once either of the two genes are cloned, then the other should be detectable because of probable homologous DNA segments.

The USH1b locus lies close to the region where the mouse mutation known as *shaker-1* (*sh-1*) is found. *Sh-1* causes deafness with vestibular disturbances. No retinitis pigmentosa has been observed (K. Steele, personal communication). Nonetheless, *sh-1* remains high on the list as a possible mouse homologue for Usher 1b. *Sh-1* is closely linked to the olfactory marker protein (OMP1). OMP1 and *sh-1* are not the same, since no mutations have been found in OMP1 which correspond to the defect seen in *shaker-1* mice. However, OMP1 was a attractive candidate for Usher syndrome given the evidence cited above about the possible abnormalities in nasal cilia seen in some Usher patients. The role of OMP1 is not known and it is tempting to speculate that it may have a function in nasal ciliary development. OMP1 has been shown to have a dinucleotide polymorphism and was found to have crossovers with Usher type Ib families, so that it is excluded as a candidate.

The fact that there are at least 3 different genes for Usher type I considerably complicates the problem of refining their localizations well enough to ensure gene identification and cloning. Usher type Ib is the more frequent type among people of presumed northern European ancestry. Types Ia and Ic have yet to be found outside of the populations in which they were originally described. However, as much as a 15% rate of heterogeneity could exist in the USA sample and still give a non-significant chi-squared for heterogeneity. Thus a considerable fraction of cases other than type 1b could exist in the population without our having the means to detect them reliably.

Future research will focus on two aspects. First, there is the question of possible clinical differences between the three types. Will ciliary abnormalities be found in one type and not in the other types? Are the ophthalmologic findings consistent across subtypes? Does one type have a more frequent association with the presence of foveal lesions and/or cataracts? Are there any significant differences in the ERG or progression of retinal degeneration? Are there any hearing or vestibular differences (some type I patients appear to benefit from hearing aids)? It is critical to define now the clinical differences between the subtypes of Usher type I in order to have ways of *a priori* dividing families used in fine gene mapping of USH1a, 1b, or 1c.

Further refinement of location will depend upon the observation of critical crossover events that will bisect and reduce the region of flanking markers. If a "critical crossover" is actually due to the inappropriate inclusion of a type Ic family in a type Ib cohort, then the assessment of location may be off by a considerable physical distance. If the investigation is limited to the Acadians, there is little chance of contamination by non-Ush1c families. However, until we know the true frequencies of the non-Ush1b families in the general population, the possibility of error will always be present for USH1b. Two ways to minimize this effect is first to develop clinical criteria that allow subclassification of the Usher type I families without resorting to molecular data. The second is to type all three sets of markers from 11q, 11p, and 14q and to include only those which show definite non-linkage to two of the three loci.

Usher Type II

Linkage of the Usher Type II Gene to 1q

Usher syndrome type II was localized to chromosome 1q32 (Kimberling, 1990; Lewis, 1990). The gene has subsequently been shown to be flanked by the markers D1S237 and D1S229. The most recent map of chromosome 1q is shown in Figure 4 along with the putative location of the Usher type II gene. There is some ambiguity in the correspondence between the CEPH collaborative study map and the second generation chromosome 1 map (Weissenbach et al., 1992), so that the exact position of the markers between D1S245 and D1S227 markers is provisional. D1S237 and D1S229 are definitely between D1S70 and D1S81 (unpublished data). The best estimate of the physical location of the Usher II gene is in 1q32-1q42. Localization of the Usher II gene is narrowed to about 4 centiMorgans. This region amounts to about 4 Megabases of DNA.

There are few genes in that region that represent true candidates. An autosomal homologue of the X-linked choroideremia gene, known as hCHML, has been localized to 1q. This gene shows homology with a bovine protein which regulates the GDP/GTP exchange of GTP binding protein smg 25A. It is expressed in the choroid/retinal pigment epithelium and several other tissues (Chapter 12). It is not known whether this gene is expressed in the cochlea. hCHML has been localized to 1q31-1qter and hence must be considered a logical candidate for Usher syndrome.

Two other genes in the region are transforming growth factor β2 (TGFB2) and the homeobox gene HB24 (HLX1). Both genes map between D1S81 and D1S70, which places them in close proximity to the Usher II gene. Both TGFB2 and HLX1 have been localized to 1q41. TGFB2 is a member of a polypeptide family which regulates growth and cellular differentiation; it is expressed in a wide variety of tissues. HLX1 has been found to be expressed in hematopoietic tissues and has been

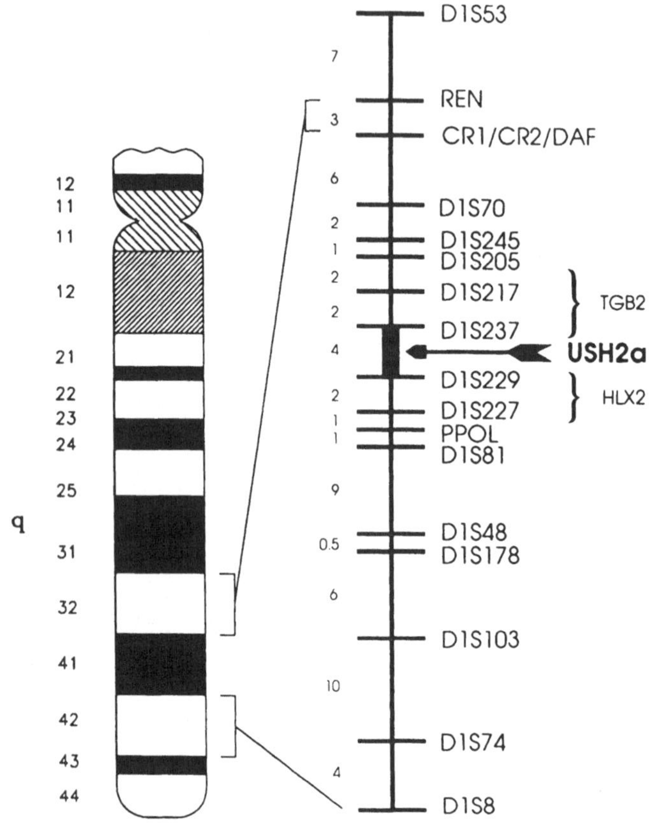

Figure 4 Map of chromosome 1 showing the probable location of the Usher II gene in the 1q32-1q42 region.

proposed as a regulator of hematopoietic differentiation. HLX1 expression was also found in the fetal brain. It seems unlikely that either of these genes are the Usher II gene but until they are localized more accurately or until their functions are better understood, they remain legitimate considerations.

The gene for poly (ADP-ribose) polymerase (PPOL) is localized in the region between D1S70 and D1S81. Its exact location relative to the Genethon (Weissenbach et al., 1992) marker set is not yet known but it seems likely to lie outside the D1S237-D1S229 segment. Nonetheless, there is some evidence that it might somehow be involved in Usher Syndrome. Lymphoid and fibroblast cells from patients with Usher syndrome have been implicated in being more sensitive to DNA-damaging agents (Robbins et al., 1984). One patient treated with cytotoxic agents was reported to have an accelerated course of retinitis pigmentosa which was hypothesized as being due to sensitivity of the retinal cells to the DNA damaging properties of the chemotherapeutic agent. If these speculations are true, it would imply that the Usher gene affects either DNA repair or synthesis which provides a clue towards the selection of candidate genes. Unfortunately, these studies do not well characterize the type of Usher syndrome. In fact, the case reported by Blanchet et al. (1992) had an atypical RP which was not diagnosed until the 6th decade and an optic atrophy not typical of Usher syndrome. There was no mention of vestibular responses. Clearly, studies of hypersensitivity to X-irradiation or similar agents need to be repeated with attention to the clinical and molecular subtype. PPOL (or ADPRT) activity is induced by single strand breaks in DNA and is required for cellular repair. It has been suggested as the gene for either Fanconi's anemia or xeroderma pigmentosa. A gene involved in the complementation of the defect in group A xeroderma pigmentosa cells has been localized to 1q42-1qter. The PPOL gene was localized 1q42 by *in situ* hybridization and it is interesting that a second hybridization peak was observed on 14q24, close to the localization of USH1a. One can only speculate about the possibility of contiguous deletions involving both the Usher II gene and PPOL giving rise to the observation of increase chromosome breakage in some type II patients.

A potential mouse model for Usher syndrome type II has recently been discovered by Chang et al. (1993). The RBF/DnJ albino mouse carries a recessive mutation, *rd-3*, which causes a retinal degeneration similar to retinitis pigmentosa. Electroretinograms and retinal appearance are both suggestive of human type RP. There is nothing particularly notable about the RP seen in Usher II patients which would suggest that Usher II and the *rd-3* mouse are the same. The RP phenotype is too non-specific. However, the *rd-3* gene maps to 10 centiMorgans distal to the alkaline phosphatase gene on mouse chromosome 1. This region shows considerable homology with human chromosome 1. The region where the *rd-3* mouse gene would lie is theoretically close to the position where the mouse homologue for the Usher II gene should lie. There was no report of whether these mice are hearing impaired.

Discovery of USH2 Heterogeneity

One family with Usher type II was observed to have at least 3 double crossovers in the 1q32 region between markers D1S70 and D1S81 if their USH2 gene were located in that segment. A formal heterogeneity analysis using Ott's HOMOG program was statistically significant when this family was included in an anlysis of a large sample of Usher type II families. The significance disappears when the family is removed. The affected family members had audiograms typical of Usher type II along with normal vestibular function. The ERG was diminished to absent. There was a history of nightblindness and visual fields were restricted. Thus, a diagnosis of Usher syndrome had been given to the family. However, it was noted that the retinitis pigmentosa was described as *sine pigmento* and one ophthalmologist had observed minute "white flecks" in the retina. In retrospect, the retinal findings are subtly different suggesting that there is a good chance that this type of family may be identified prior to molecular genetic testing.

The family did not show linkage to the 1q32 - q41 region containing the Usher type II gene. Nor does the Usher syndrome in the family show linkage to markers on chromosomes 11q, 11p, or 14. This establishes the fact that a second locus for Usher type II exists which is not located on 1q32-41 and that the gene is not an allele at any of the other 3 Usher I loci. This new gene has been given the designation USH2b, leaving USH2a to refer to the original 1q locus. The frequency of Usher IIb relative to IIa is not known, although it must be relatively less common inasmuch as it has been seen only once in a series of over 30 type II families.

CONCLUSIONS

Usher syndrome was first described as hearing impairment with retinitis pigmentosa. With time, investigators realized that at least two clinical types existed and that they could be consistently and reliably distinguished from each other. However, linkage analysis showed that a minimum of five different genes are involved, three for type I and two for type II. There are no clear clinical differences between the different subtypes within each type.

The next stage of Usher research is focused on identifying and cloning the responsible genes. Saturation mapping of the critical regions by chromosome walking might be necessary to accomplish this task. Saturation mapping of the chromosome region will eventually produce a series of "contigs" of overlapping clones that collectively span the region of interest, which hopefully will involve only 1 Megabase or less. Genetic heterogeneity will remain a problem until the genes are identified and sequenced. For example, a family with two children affected with a separate type of Usher syndrome would show non-double crossovers in the critical region 25% of the time that their markers are informative: for three affected children, this frequency would be 6.25%. Most Usher families are small with only 1 affected and in most of those cases it will be impossible to determine which subtype of Usher syndrome is present by linkage analysis. When a family has two or more affected members, the probability that linkage can rule out a subtype is related to the number of affected children. At the present time, the diagnosis of one subtype or the other by molecular testing must rely solely on exclusion of the alternate types. Clinical differences between the subtypes may exist and the possibility of being able to differentiate one type from another needs to be further investigated. Resolution of the question of clinical differences may help in the selection of families helpful in identifying the Usher genes.

Meanwhile, until the Usher genes are found, DNA markers close to the Usher gene can be used in genetic counseling. Given tight linkage and an informative family with one previously affected child, early diagnosis is possible for subsequent children. The markers showing tight linkage could also be used to determine which of the unaffected relatives are carriers. Unfortunately, effective screening of spouses who marry into a family will not be possible until the appropriate Usher gene is cloned.

Ultimately, the goal of genetic research is to develop a treatment for Usher syndrome, however, the underlying defect causing it is unknown. Most biochemical and physiologic studies have centered on the study of retinitis pigmentosa and the results have been disappointingly negative in terms of elucidating the basic defect. We would expect that the gene that causes Usher syndrome does so by an effect on a substrate that is essential for some component of inner ear function, balance, and visual ability. Determination of the function of the gene would further our understanding of the cellular and biochemical bases for these sensory abilities. An understanding of the underlying pathologic process occurring from gene to phenotype is an important step towards finding an effective method of remediation and/or prevention. Gene cloning and its subsequent characterization may be the only approach that could lead to the discovery of the basic defect of Usher syndrome.

Considering the fact that Usher syndrome is the major cause of deafness/blindness, it is important to pursue research into the cause(s) of Usher Syndrome in the hope that someday an effective remedy may be possible. About 1 out of every 20,000 individuals is affected with Usher syndrome. The burden caused by the loss of the two most vital human senses is tremendous. It isolates patients from the mainstream of society, resulting in severe psychological stresses and reduces their ability to act independently. The answers to many intriguing questions regarding Usher syndrome will not be available until the genes are cloned. An understanding of the etiologies of Usher syndrome will provide critical information about the underlying mechanisms of function and expression of the genes.

REFERENCES

Barrong, S. D., Chaitin, M. H., Fliesler, S. J., Possin, D. E., Jacobson, S. G., and Milam, A. H. (1992) Ultrastructure of connecting cilia in different forms of retinitis pigmentosa. *Arch. Ophthalmol.* **110**, 706–710.

Bell, J. (1933). Retinitis pigmentosa and allied diseases. In: *Treasury of Human Inheritance*, **Vol. 2**, Cambridge University Press, London.

Blanchet, P., Wellemeyer, M. L., and Burton, G. V. (1992) Case report: retinitis Pigmentosa following cytotoxic chemotherapy in Usher's syndrome. *Am J. Med Sci.* **303**, 319–320.

Bloom, T. D., Fishman, G. A., and Mafee, M. F. (1983) Usher's syndrome: CNS defects determined by computed tomography. Retina **3**, 108–113.

Boughman, J. A., and Fishman, G. A. (1983) A genetic analysis of retinitis pigmentosa. *Brit. J. Ophthalmol.* **67**, 449–454.

Boughman, J. A., Vernon, M., and Shaver, K. A. (1983) Usher syndrome: definition and estimate of prevalence from two high risk populations. J. Chron. Dis. **36**, 595–603.

Chang, B., Heckenlively, J. R., Hawes, N. L., and Roderick, T. H., (1993) New mouse primary retinal degeneration (*rd-3*). *Genomics* **16**, 45–49.

Davenport, S. L. H., and Omenn, G. S. (1977) The heterogeneity of Usher's syndrome. Publication no. 426, Amsterdam, Excerpta Medica Foundation, International Congress Series (abst 215), pp 87–88.

Davenport, S. L. H., O'Nuallain, S., Omenn, G. S., and Wilkus, R. J. (1978) Usher syndrome in four hard of hearing siblings. *Pediatrics* **62**, 578–583.

De Haas, E. B. H., Van Lith, G. H. M., Rijnders, J., Rumke, A. M. L., and Volmer, C. H. (1970) Usher's syndrome with special reference to heterozygous manifestations. *Doc. Ophthalmol.* **28**, 166–190.

Fishman, G. A. (1979) Usher's Syndrome: Visual loss and variations in clinical expressivity. *Perspect. Ophthalmol.* **3**, 99–103.

Fishman, G. A., Basquez, V., Fishman, M., and Berger, D. (1979) Visual loss and foveal lesions in Usher's syndrome. *Br. J. Ophthal.* **63**, 484–488.

Fishman, G. A., Kumar, A., Joseph, M E., Torok, N., and Anderson, RJ. (1983) Usher's syndrome: ophthalmic and neuro-otologic findings suggesting genetic heterogeneity. *Arch. Ophthalmol.* **110**, 1367–1374.

Gerber, S., Bonneau, D., Rozet, J. M., Delrieu, O., Briard, M. L., Dollfus, H., Bhazi, I., Dufier, J. L., Frezal, J., and Munnich, A. (1992) A gene for Usher syndrome type I (USH1A) maps to chromosome 14q. Genomics **14**, 979–987.

Gorlin, R. J., Tilsner, T. J., and Feinstein, S. (1979) Usher's syndrome type III. *Arch. Otolaryngol.* **105**, 353–354.

Grondahl, J. (1986) Tapeto-retinal degeneration in four Norwegian counties. I. Diagnostic evaluation of 89 probands. *Clin. Genet.* **29**, 1–16.

Grondahl, J., and Mjoen, S. (1986) Usher syndrome in four Norwegian counties. **30**, 14–28.

Hallgren, B. (1959). Retinitis Pigmentosa combined with congenital deafness; with vestibulo-cerebellar ataxia and neural abnormality in a proportion of cases. *Acta Psychiat. Scand.* **138**(Suppl.), 1–101.

Hunter, D. G., Fishman, G. A., Mehta, R. S., and Kretzer, F. L. (1986) Abnormal sperm and photoreceptor axonemes in Usher's Syndrome. *Arch. Ophthalmol.* **104**, 385–389.

Karjalainen, S., Pakarinen, L., Terasvirta, M., Kaariainen, H., and Vartiainen, E. (1989) Progressive hearing loss in Usher's syndrome. *Ann. Otol. Rhinol. Laryngol.* **38**, 863–866.

Keats B. J. B., Todorov A. A., Atwood L. D., Pelias, M. Z., Hejmancik, J. F., Kimberling, W. J., Leppert, M., Lewis, R. A., and Smith, R. J. H. (1992) Linkage studies of Usher syndrome type I: exclusion results from the Usher syndrome consortium. *Genomics* **14**, 707–714.

Koizumi, J., Ofuku, K., Sakuma, K., Shiraishi, H., Iio, M., and Nawano, S. (1988) CNS changes in Usher's syndrome with mental disorder: CT, MRI and PET findings. *J. Neurol. Neurosurg. Psychiatry* **51**, 987–990.

Kumar, A., Fishman, G. A., and Torok, N. (1984) Vestibular and auditory function in Usher's syndrome. *Ann. Otol. Rhinol. Laryngol.* **93**, 600–608.

Liebrich, R. (1861) Abkunft aus ehen unter blulsverwandten als grund von retinitis pigmentosa. *Dtsch. Klin.* **13**, 53–55.

Lindenov, H. (1945) The etiology of deaf-mutism with special reference to heredity. *Op. Ex. Domo. Biol. Hered. Hum. Univ. Hafnienses* **8**, 1–268.

Lundin, L. (1993) Evolution of the vertebrate genome as reflected in paralogous chromosomal regions in man and the house mouse. *Genomics* **16**, 1–19

Mansour, S. L., Thomas, K. R., and Capecchi, M. R. (1988) Disruption of the proto-oncogene *int-2* in mouse embryo-derived stem cells: a general strategy for targeting mutations to non-selectable genes. *Nature* **336**, 348–352.

Merin, S., Abraham, F.A., and Auerbach, E. (1974) Usher's and Hallgren's syndrome. *Acta Genet. Med. Gemellol. (Roma)* **23**, 49–55.

Moller, C. G., Kimberling, W. J., Davenport, S. L., Priluck, I., White, V., Biscone, Halterman K. et al. (1989) Usher syndrome: an otoneurologic study. *Laryngoscope* **99**, 73–79.

Nuutila, A. (1970) Dystrophia retinae pigmentosa-dysacusis syndrome (DRD): A study of the Usher or Hallgren syndrome. *J. Genet. Hum.* **18**, 57–58.

Piazza, L., Fishman, G. A., Farber, M., Derlacki, D., and Anderson, R. J. (1986) Visual acuity loss in patients with Usher's syndrome. *Arch. Ophthalmol.* **104**, 1336–1339.

Robbins, J. H., Scudiero, D. A., Otsuka, F., Tarone, R. E., Brumback, R. A., Wirtschafter, J. D., Polinsky, R. J., Barrett, S. F., Moshell, A. N., and Scarpinato, R G. (1984) Hypersensitivity to DNA-damaging agents in cultured cells from patients with Usher's syndrome and Duchenne muscular dystrophy. J. Neurol. Neurosurg. Psychiatry **47**, 391.

Schaefer, G. B., Thompson, J. N., Bodensteiner, J. B., McConnell, J. M., Kimberling, W. J., and Gay, C.T. (1992) The cerebellar vermis in autism and conditions with autistic behavior. *Am. J. Hum. Genet.* **51** (suppl), 417A.

Smith, R. J. H., Lee, E. C., Kimberling, W. J., Daiger, S. P., Pelias, M. Z., Keats, B. J. B., Jay, M., Bird, A., Reardon, W., Guest, M., Ayyagari, R., and Hejtmancik, F. (1992) Localization of two genes for Usher syndrome type I to chromosome 11. Genomics **14**, 995–1002.

Shugart, Y. Y., Kimberling, W. J., Pieke Dahl, S. A., and Kenyon, J. B. (1991) Segregation analysis of Usher Syndrome. 8th International Congress of Human Genetics, Washington D.C.

Usher, C. H. (1914) On the inheritance of retinitis pigmentosa with notes of cases. *Roy. Lond. Ophthalmol. Hosp. Rev.* **19**, 130–236.

Vernon, M., (1969) Usher's syndrome: deafness and progressive blindness. Clinical cases, prevention, theory and literature survey. *J. Chronic. Dis.* **22**, 131–151.

von Alberth, B., Balint, A., and Kosa, A. (1958) Audiometrische untersuchungen an kranken mit degeneratio pigmentosa retinae. *Klin. Monatsbl. Augenheilkd.* **132**, 797–806.

Von Graefe, A. (1858) Exceptionelles verhalten des gesichlsfeldes bei pigmententartung der netzhaut. *Von Graefe's Arch. Klin. Exp. Ophthalmol.* **4**, 250–253.

Weissenbach, J., Gyapay G., Dib C., et al. (1992) A second-generation linkage map of the human genome. *Nature* **359**, 794–801.

16. MARFAN SYNDROME

PETROS TSIPOURAS AND MICHAEL W. KILPATRICK

Department of Pediatrics, University of Connecticut Health Center, Farmington, CT 06030, USA

SUMMARY

In 1896 Marfan (Marfan, 1896) described the musculoskeletal manifestations (dolichostenomelia) characteristic of the condition that presently bears his name; some doubts remain to date as to whether the original patient was actually affected with the Marfan syndrome (MFS). The ocular abnormalities associated with the MFS were identified in the 1940s (Lutman and Neel, 1949) and the cardiovascular complications began to receive attention when studied by McKusick in the 1950s (McKusick, 1955). The diagnosis of MFS is clinical and based on well established criteria (Beighton et al., 1988). Clinical variability of expression within families but also between families is a hallmark of the disorder. The ætiology of MFS remained elusive until recently when several mutations in the fibrillin gene on chromosome 15 were identified.

CLINICAL MANIFESTATIONS AND NATURAL HISTORY

MFS is a systemic disorder of connective tissue which primarily involves the musculoskeletal, cardiovascular and ocular systems (Pyeritz and McKusick, 1979). MFS has been reported in different races and ethnic groups (McKusick, 1972). The prevalence of MFS has been estimated to be approximately 4–6 per 100,000 population in the United States (Pyeritz and McKusick, 1979).

Musculoskeletal Manifestations

Dolichostenomelia (long and narrow frame) is the characteristic skeletal abnormality observed in MFS. Tall stature, decreased upper to lower segment ratio, armspan in excess of height, dolichocephaly, and arachnodactyly of fingers and toes are some of the manifestations directly associated with dolichostenomelia (Pyeritz and McKusick, 1979) (Figure 1). In most individuals affected with MFS the distal bones of the extremities tend to be longer. Thus, a decreased ratio (approximate value of 0.85) of upper segment to lower segment is frequently found in adults with MFS. The face is long and narrow, the palate is high-arched,

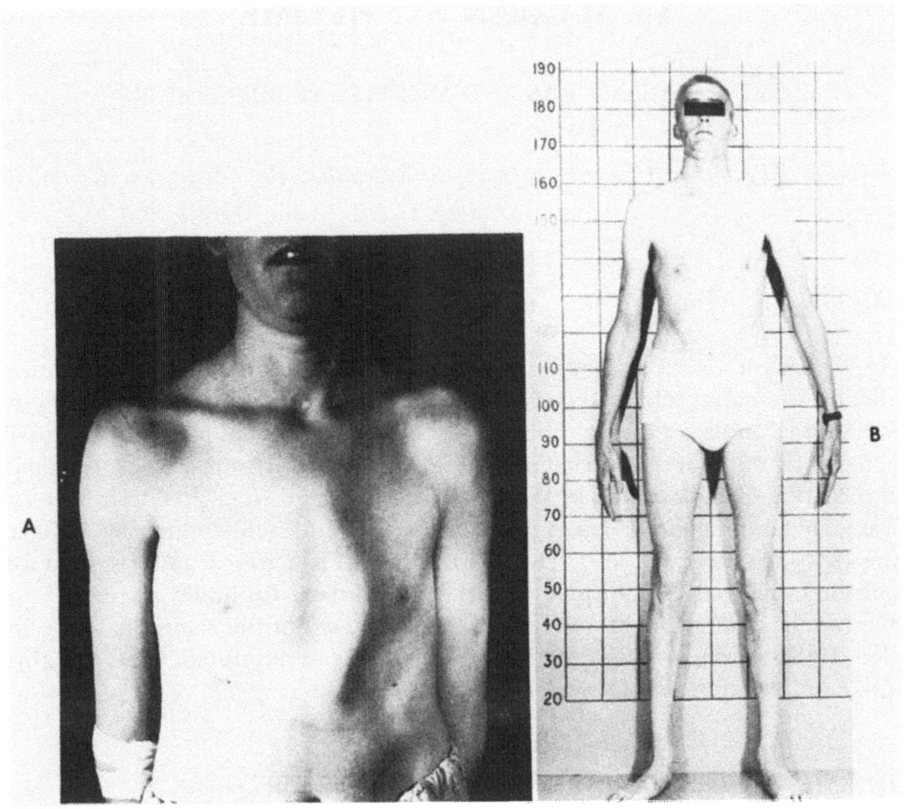

Figure 1 Individual affected with Marfan syndrome . Panel A: asymmetric pectus carinatum at the age of 14 years. Panel B: same patient at the age of 29 years.

prognathism and crowding of the maxillary and mandibular teeth are frequently observed. Scoliosis or kyphoscoliosis, whose age of onset is during childhood, occurs in 30 to 60% of individuals with MFS. The kyphoscoliosis is severe in the infantile form of MFS. Thoracic kyphosis is associated with pulmonary deficiencies including reduced vital capacity and residual volume (Magid et al., 1990). The spinal canal is often enlarged in depth or width due to the ectasia of the neural foramina and sacral erosion. This feature, which is termed "dural ectasia", has been observed by computed tomography and it is of considerable diagnostic value (Pyeritz et al., 1988). The thoracic deformities, pectus excavatum or pectus carinatum, present in MFS, result from excessive longitudinal growth of the ribs. The presence of a pectus deformity in a child makes it imperative to consider the diagnosis of MFS. The "asthenic" habitus frequently observed in individuals affected with MFS is due partly to muscular underdevelopment and also to a significant decrease of the subcutaneous fat. Some patients appear to have a facial expression reminiscent of a myotonic facies. Ligamentous laxity and generalized joint hypermobility are common in MFS. Pes planus, genu recurvatum, recurrent dislocation of hips and other joints result from the redundancy and weakness of joint capsules, ligaments, tendons and fasciae.

Cardiovascular Manifestations

Overall, in the majority of infants and children, manifestations related to multivalvular incompetence predominate. This contrasts with the majority of adults in whom the manifestations are primarily from the aorta.

Dissecting aneurysm of the aorta is the most serious and potentially fatal cardiac manifestation in MFS; it develops post-natally and reflects the underlying pathology of the tunica media (McKusick, 1955). Dilatation of the aorta usually precedes the development of the aneurysm and it is found in approximately 50% of children and 70 to 80% of adults. The dilatation is confined either to the area around the sinuses of Valsalva or it is generalized. The generalized aortic dilatation is usually associated with higher morbidity (Roman et al., 1988). Both types of aortic dilatation breed true in families, suggesting the presence of distinct phenotypic subtypes (Devereux, personal communication, 1992; Tsipouras, unpublished data). Not all affected individuals with MFS develop aortic dilatation and a small percentage of individuals with unequivocal MFS maintain normal aortic diameters into their sixties. The dilatation of the aorta could extend to the thoracic or abdominal segments and occasionally to medium size arterial branches such as the internal or external carotid arteries or even intracranial branches. Enlargement of the pulmonary artery is seen in MFS albeit at a much lower frequency compared to the aorta.

Aortic regurgitation is a common and usually progressive complication of MFS. It is relatively uncommon in childhood, but its frequency in adults has been estimated to be up to 70%. Aortic regurgitation occurs only rarely in individuals affected with MFS but having normal-sized aortic roots. A positive correlation between aortic root diameter and aortic regurgitation has been observed by Lima et al. (1985). The acceleration of aortic root enlargement described in some patients with aortic regurgitation reflects the causative relationship between the two phenomena.

The Marfan syndrome is the leading cause of dissecting aortic aneurysms in individuals under the age of 40 years. Although a dissecting aneurysm might be the first presenting manifestation in MFS, it is usually superimposed on underlying dilatation of the aortic root and frequently of the ascending aorta as well (Pyeritz et al., 1981). Most aortic dissections in MFS arise in the aortic root or proximal ascending aorta and extend distally through a variable portion of the aorta and its branches, but dissections can also begin in more distal aortic segments (Roberts and Honig, 1982; Crawford, 1983; Larson and Edwards, 1984).

Approximately 60 to 70% of persons with MFS exhibit mitral valve prolapse and in a much smaller fraction tricuspid valve prolapse is also observed (Pyeritz and McKusick, 1979). The mitral cusps and chordae tendinae may be redundant resulting in mitral regurgitation.

Ocular manifestations

Ectopia lentis, almost always bilateral, is a clinical hallmark of MFS (Lutman and Neel, 1949; Maumenee, 1981). Only 50 to 80% of individuals with MFS have this abnormality. The ciliary zonules, when visualized with a slit-lamp, appear redundant, attenuated and frequently broken. The lenses are normal in shape and size (Maumenee, 1981). The lens displacement is usually supero-nasal. The observation of iridodonesis indicates the presence of lens dislocation. Myopia is found in a large number of individuals affected with MFS (Maumenee, 1981). Spontaneous retinal detachment occurs with an unusually high incidence in MFS; it is also an important complication of the extirpation of the lens.

Miscellaneous manifestations

Spontaneous pneumothorax occurs in approximately 10% of individuals with MFS (Wood et al., 1984). Apical bullae, congenital pulmonary cysts and emphysema are some rarer pulmonary findings. Cutaneous striae distensae in the pectoral, deltoid, and lumbar areas are frequent. Inguinal, femoral and incisional hernias occur rather commonly in men with MFS. Finally, the voice of individuals with MFS is sometimes high pitched with a characteristic timbre (McKusick, 1972).

Infantile Marfan Syndrome

Although infants and young children were among the initial patients in whom MFS was documented, recognition of the disorder in early life is not always an easy matter. On the other hand, certain infants present with a distinct set of clinical manifestations which has led to the suggestion of the existence of a distinct entity, termed infantile or neonatal MFS. A characteristic aged facial appearance due to deep-set eyes, high arched palate, arachnodactyly, pectus deformities, flexion contractures, pes planus, multivalvular involvement, aortic root dilatation and ectopia lentis are among the phenotypic manifestations described (Morse et al., 1990).

INHERITANCE

The autosomal dominant mode of transmission was clearly demonstrated for the first time by Weve in 1931. Since then, scores of pedigrees have been reported which clearly demonstrate autosomal dominant inheritance (Tsipouras et al., 1992). About 80% of individuals documented to have the MFS by reasonably strict criteria inherited the gene from a parent; in the remaining 20% the condition arises as a result of new mutations. Paternal age effect on mutation has been demonstrated in the MFS . According to one study, the average age of fathers of isolated cases was seven years greater than the mean paternal age in the population (Murdoch et al., 1972a). Most infants with the severe infantile form of MFS are isolated cases; for some of these infants the possibility of homozygosity has been raised (Chemke et al., 1984, Schollin et al., 1988).

MOLECULAR BASIS

Brief overview

Over the years several molecules of the extracellular matrix of the connective tissue had been considered as potential candidates in the ætiology of MFS. Among them were the different types of fibrillar collagens, elastin, hyaluronic acid, beta-glucuronidase, elastase and decorin (Tsipouras, 1990). In 1986, Sakai et al., using monoclonal antibodies raised against microfibrils, identified a 350 kDa glycoprotein for which they coined the term fibrillin. Indirect immunofluorescence studies on the skin and cultured dermal fibrobasts of normal individuals and individuals affected with MFS provided the first strong evidence of a structural protein abnormality in Marfan syndrome (Godfrey et al., 1990, Hollister et al., 1990) (Figure 2). In parallel, a consortium of investigators attempted to localize the gene responsible for Marfan syndrome by positional mapping. An exclusion map was constructed that eliminated nearly 75% of the human genome as a likely location of the MFS gene (Blanton et al., 1990). This led to the subsequent assignment of the MFS gene to chromosome 15 (15q15-q23) (Kainulainen et al., 1990; Dietz et al., 1991a; Tsipouras et al., 1991).

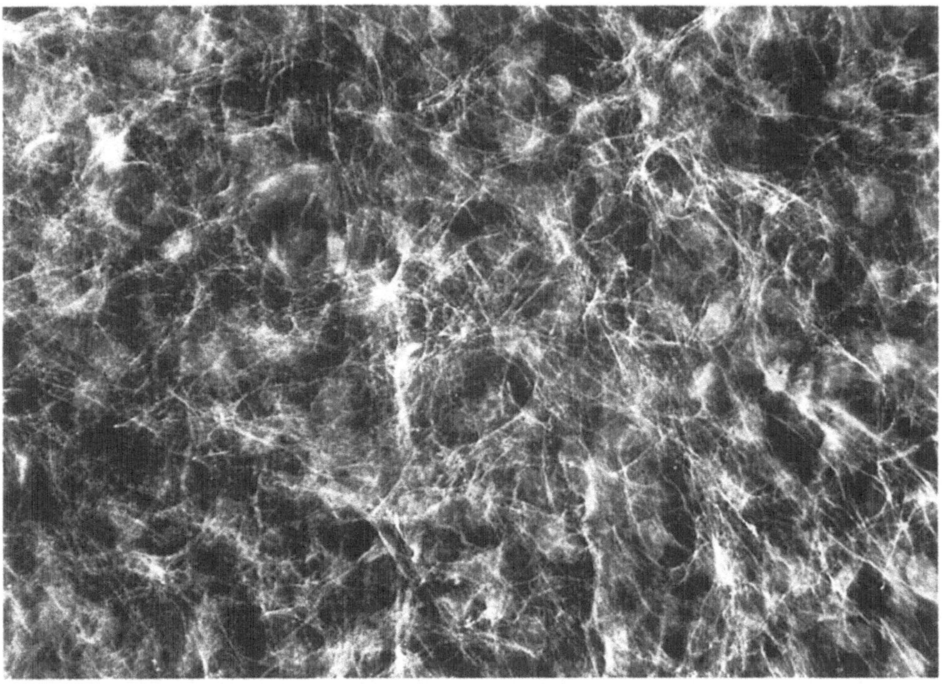

Figure 2 Indirect immunofluorescence of hyperconfluent dermal fibroblast cultures stained with monoclonal antibody to fibrillin. The slides were counter stained with propidium iodide to visualize the cell nuclei. Panel A: dermal fibroblasts from a normal control. Note the normal meshwork of elastin-associated microfibrils. (Courtesy of Dr. Maurice Godfrey and Mr. Jason Cisler, University of Nebraska)

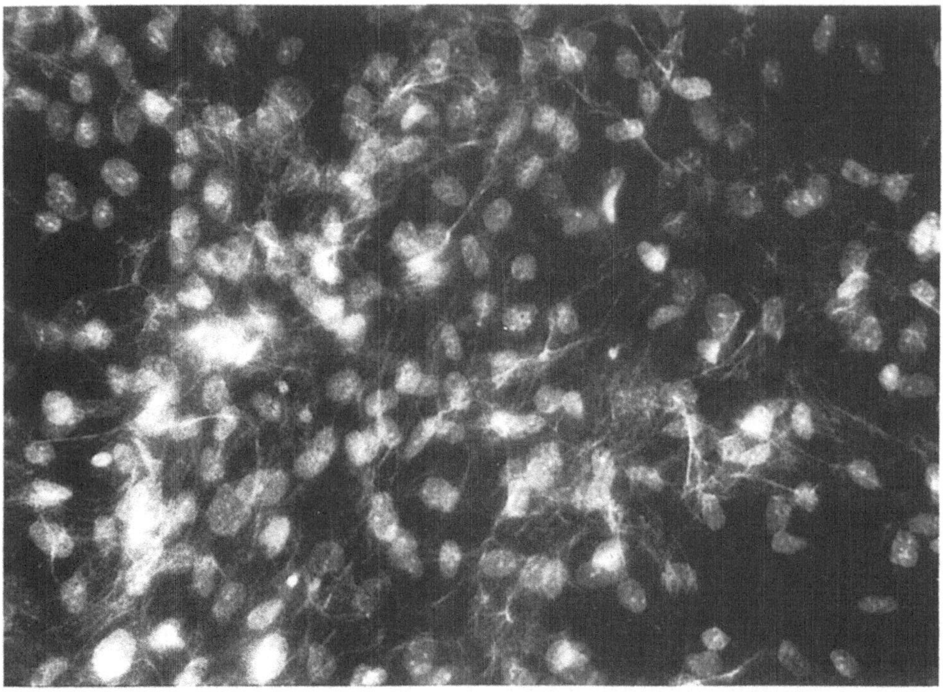

Figure 2 Panel B: dermal fibroblasts from an individual affected with Marfan syndrome. Note the sparsity of the immunostainable microfibrils. (Courtesy of Dr. Maurice Godfrey and Mr. Jason Cisler, University of Nebraska)

In view of the immunohistological evidence implicating fibrillin in MFS, several groups of workers attempted to isolate the gene coding for this protein. The subsequent mapping of a gene for fibrillin to the same chromosome interval as MFS thus designated this gene as a strong candidate in the ætiology of MFS.

Genetic linkage studies in multiplex families using intragenic markers for the fibrillin gene on chromosome 15 (FBN1) clearly showed genetic linkage of MFS to the fibrillin locus (maximum lod score of $\hat{Z}$=25.6 at $\hat{\theta}$=0.00) (Lee et al., 1991, Dietz et al., 1991b; Tsipouras et al., 1992). Genetic analysis strongly indicated the absence of locus heterogeneity in MFS (Tsipouras et al., 1992). However, Boileau et al.(1991), reported a family in which the phenotype segregated discordantly with the FBN1 gene and eight other chromosome 15 markers.

The Fibrillin gene and its product

The term microfibril was originally used to identify morphologically similar extracellular matrix structures displaying a diameter of less than 20nm and lacking the characteristic 67nm banding periodicity of interstitial collagen fibres (Low, 1962). Currently, the microfibrils are divided into two classes according to their average diameter, the larger of the two classes being the 10nm microfibrils also referred to as the elastin-associated microfibrils (Cleary and Gibson, 1983). Immunohistochemical studies have identified microfibrils in the suspensory ligament, pleura, perichondrium, periosteum, meninges, aorta, cartilage, tendon, muscle and many other tissues (Figure 3). The complete macromolecular composition of the microfibrils is as yet unclear, and has been difficult to determine because of the highly insoluble nature of the matrix aggregate. The major protein component appears to be fibrillin. However, four other proteins, with molecular weights of 78kDa, 70kDa, 31kDa and 25kDa, have also been identified (Gibson et al., 1989; Maddox et al., 1989).

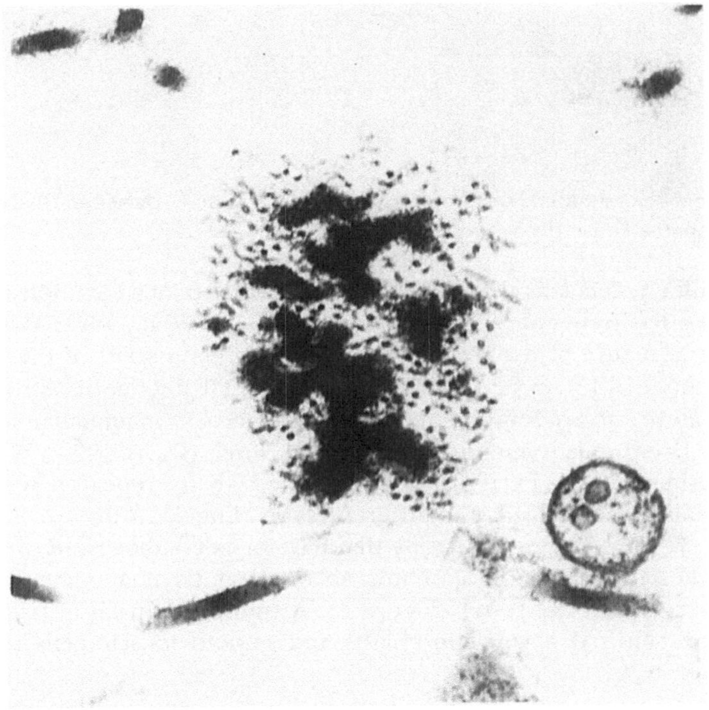

Figure 3 Electron micrographs of microfibrils (magnification 69000x) in developing elastic fibers in bovine ligamentum nuchae at 210 days of gestational age. The elastin deposits among the microfibrils are black due to en bloc staining of the tissue with tannic acid. Panel A: cross-section. (Courtesy of Dr. Elaine C. Davies and Dr. Robert P. Mecham, Washington University School of Medicine)

Fibrillin itself is an acidic glycoprotein with an estimated molecular mass of 350kDa. The amino-acid sequence of the fibrillin molecule has been deduced from the cloned cDNA sequence (Lee et al., 1991; Maslen et al., 1991). The protein has an unusually high cysteine content of 14%, one third of which has the potential to participate in disulphide bonding. In vitro studies suggest that fibrillin is synthesised in a precursor form. Following secretion, the protein is converted to its mature form by proteolysis prior to assembly (Milewicz et al., 1992).

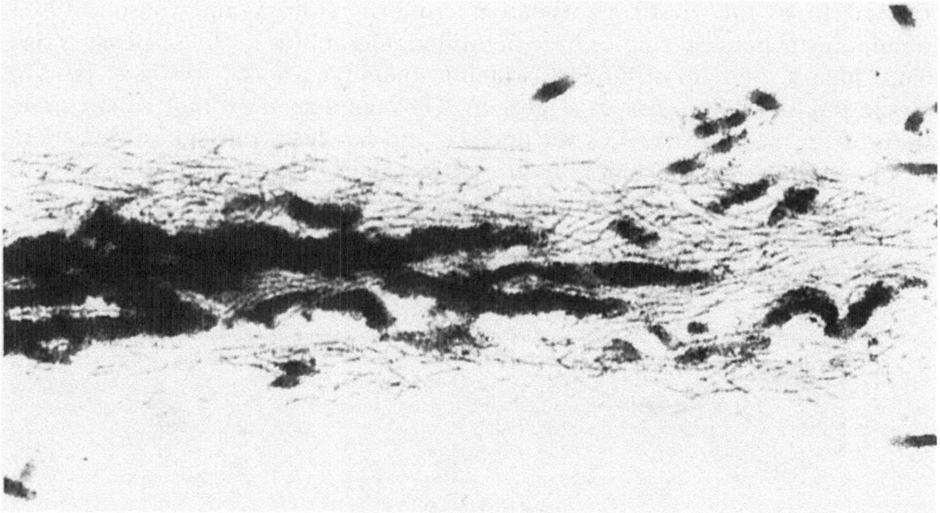

Figure 3 Panel B: longitudinal section. (Courtesy of Dr. Elaine C. Davis and Dr. Robert P. Mecham, Washington University School of Medicine)

The fibrillin molecule is characterised by the presence of a common six cysteine repeat and a less frequent eight cysteine repeat (Lee et al., 1991; Maslen et al., 1991). The structure of the six cysteine repeat is reminiscent of the epidermal growth factor (EGF) peptide (Sakai et al., 1991). EGF-like motifs are distributed widely in nature and are found in molecules as diverse as mammalian LDL receptor and the Drosophila homeotic gene product Notch (Davis, 1990). With respect to matrix proteins, it has been proposed that the EGF-like repeats may serve as localised signals for cell growth and differentiation (Engel, 1989).

From the regions of the FBN1 gene that have been characterised to date, it appears that the genomic organisation mirrors the modular arrangement of the protein product (Lee et al., 1991) (Figure 4). A transcript 10 kb in size has been identified in cultured dermal fibroblasts and smooth muscle cells (Lee et al., 1991).

A surprising outcome of the cloning of the fibrillin gene was that fibrillin is genetically heterogeneous, with evidence for another fibrillin gene on chromosome 5 FBN2. Cross-hybridization of a clone for FBN1 to a genomic library identified clones which mapped to chromosome 5 (5q23-31) (Lee et al., 1991). Comparison of the genomic DNA sequence of regions of FBN2 and FBN1 suggest the two genes have a similar internal structure (Lee et al., 1991). A limited amount of cDNA sequence data has allowed a partial amino-acid sequence to be deduced for the FBN2 gene product (Lee et al., 1991; Maslen et al., 1991).

Preliminary evidence suggests FBN2 may be expressed in cultured dermal fibroblasts (Lee et al., 1991). However, to date no distinct FBN2 gene product has been detected and the function of this protein is unclear. Genetic analysis in families with congenital contractural arachnodactyly, a condition phenotypically similar to MFS (Beals and Hecht, 1971), has established genetic linkage of this condition to

FBN2 (Lee et al., 1991; Tsipouras et al., 1992). It is possible to speculate that the amelioration of the clinical manifestations of congenital contractural arachnodactyly with age may be the consequence of a developmental switch involving the FBN2 and FBN1 genes.

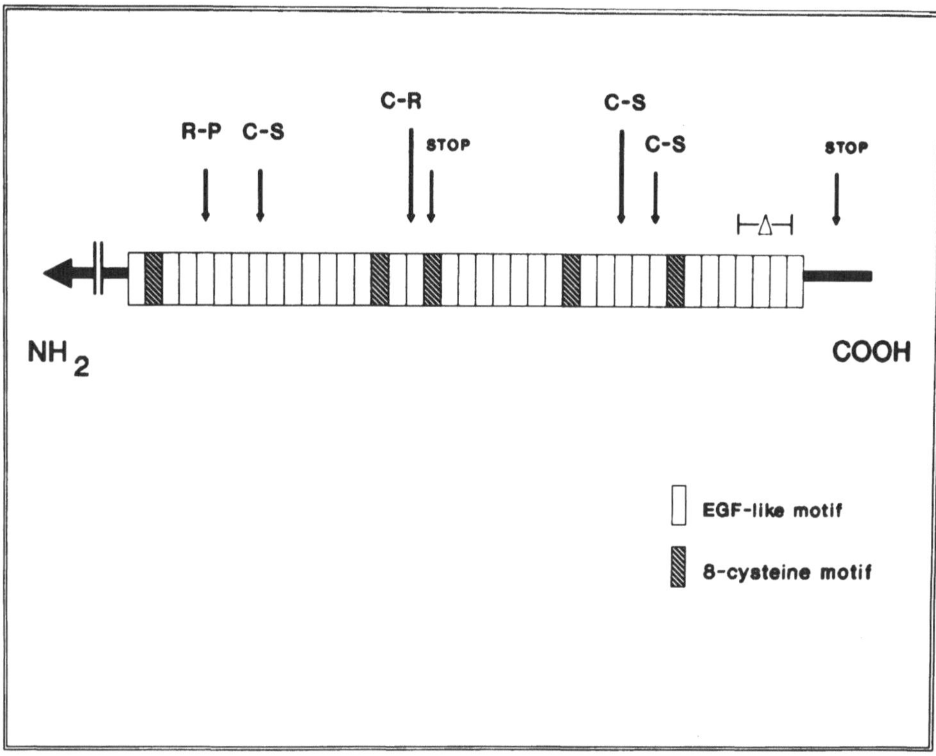

Figure 4 Diagrammatic representation of the Fib15 gene showing the EGF-like repeats and 8-cysteine repeats. The positions of the seven point mutations and two deletions identified to date in MFS patients are indicated by vertical arrows. The one letter amino-acid abbreviation is used to identify amino-acid changes consequent to the point mutations.

Marfan Syndrome and mutations in the Fibrillin gene

Analysis of fibrillin biosynthesis in cultured dermal fibroblasts from individuals with MFS has shown defects in fibrillin synthesis, secretion and incorporation into the extracellular matrix (Milewicz et al., 1992). As has been shown to be the case in other dominantly inherited disorders the defective gene product appears to interfere with the function of the wild-type protein at many different levels.

Further insight into the relationship between the structure and function of the fibrillin molecule is beginning to emerge from the molecular approach, which has facilitated the identification of several distinct mutations in the fibrillin gene. Initial attempts to delineate mutations in the fibrillin gene in individuals affected with MFS revealed no gross deletions or rearrangements, and to date only one genomic deletion has been identified (Kanulainen et al.,1992). The deletion results in the loss of 366 bases of the fibrillin mRNA which consequently produces a truncated protein. This shortened protein appears to be synthesised in normal amounts, secreted efficiently from the cell, and assembled into microfibrils. The microfibrils thus produced are structurally and functionally abnormal.

Seven point mutations (five missense and two nonsense) have so far been identified and characterised (Dietz et al., 1991b, 1992): four of the missense mutations result in cysteine substitutions and one in the substitution of an arginine (Dietz et al., 1992) (Figure 4). As substitution of a cysteine residue would be expected to disrupt local folding and possibly extracellular protein-protein interactions, the consequence of such a mutation can be readily envisaged. Similarly the arginine substitution, which occurs in an EGF-like repeat, might be expected to alter the local secondary structure and thus affect cysteine-mediated bonding. One of the nonsense mutations leads to the production of a fibrillin molecule which lacks the last 116 amino acid residues and is apparently not secreted from the cell (Kainulainen et al., 1992). Immunofluorescence studies of fibroblasts from the individual with this mutation reveals reduced amounts of extracellular fibrillin (Kainulainen et al., 1992). The other nonsense mutation, however, results in a truncated molecule which, although apparently synthesised in reduced amounts, is secreted and does participate in fibrillogenesis. (Kainulainen et al., 1992). Thus even a small amount of structurally abnormal fibrillin is able to elicit a clinical phenotype, thus demonstrating the potential for fibrillin mutations to have a dominant negative effect.

With the exception of the originally reported missense mutation which was found in two unrelated isolated cases of MFS all the mutations reported to date have been unique (Dietz et al., 1991b). In each case, screening of a large number of unrelated individuals affected with MFS for the presence of the mutation has proved fruitless. It is also of interest to note that the mutations identified to date are distributed throughout the entire length of the gene, with no evidence of clustering of mutations in particular regions or corresponding protein domains.

ANIMAL MODEL

An animal model has been provided by a herd of cattle recently reported to be affected by a disorder seemingly analogous to MFS (Besser et al., 1990). The affected animals display long thin limbs, joint laxity, ectopia lentis and aortic dilatation and aneurysm. Autopsy on a calf which died suddenly at age 16 months identified the cause of death to be aortic rupture resulting in hæmopericardium and cardiac tamponade. Electron microscopy revealed that the tunica media of the affected aorta displayed thin and irregular elastic laminae. Immunofluorescent staining of skin biopsy material using fibrillin monoclonal antibodies did not distinguish affected calves from normal. However, it was possible to distinguish affected animals from normal by reduced dermal staining for elastin (Besser et al., 1990).

Although the nature of the genetic defect is as yet unknown, it is possible that this MFS-like condition will provide an excellent model for further study and characterisation of the precise molecular pathology of MFS.

CLINICAL MANAGEMENT

Individuals affected with MFS often require multidisciplinary management by a team of medical specialists. From the clinical point of view, the most serious and potentially fatal complications of the condition are the cardiovascular ones. Death in adults affected with MFS is most commonly due to aortic root disease, in which progressive dilatation results in either aortic regurgitation or aortic dissection and rupture. The median age of death in MFS in the only detailed study was the mid 40s; more than 95% of deaths result from cardiovascular complications (Murdoch et al., 1972b). The cardiological examination schedule depends on the severity of the cardiovascular complications and in individuals with mild features, such as mitral valve prolapse and minimal aortic dilatation, annual examination including an echocardiogram is sufficient. However, if aortic dilatation and regurgitation are significant, more frequent follow-up is necessary. In addition to echocardiography, computed tomography (CT) and magnetic resonance imaging (MRI) have been used for the evaluation of the aorta in MFS. A direct relationship has been observed between aortic root dimension and complications of the MFS. Aortic regurgitation is virtually non-existent in MFS patients whose aortic root dimension is 3.6 cm or less, but occurred uniformly in patients with an aortic root dimension of 6.0 cm or greater, according to one study (Pyeritz et al., 1981a). Patients in whom aortic dilatation extended into the ascending aorta were more likely to have aortic regurgitation and appeared to be at greater short-term risk of progressive dilatation necessitating surgery than patients in whom the aortic dilatation was limited to the sinuses of Valsalva (Roman et al., 1988). Pyeritz et al. (1983) demonstrated that propanolol therapy results in a reduction of the rate of aortic dilatation in affected children and adults. However, the beneficial effects of long term beta-blockade or negative inotropy on the aorta are difficult to ascertain because they have not been subjected to the scrutiny of a controlled drug trial. It is therefore premature to propose treatment of all individuals with the MFS until more information is available.

During the last 10 years, the development of improved surgical techniques for repair of the dilated or dissected aortic root with simultaneous replacement of the aortic valve has been of great importance. An elective repair of an aorta, the root diameter of which is 6 cm or less, is preferable to emergency repair. A modified Bentall procedure with a variety of grafts is used (Bentall and DeBono, 1968). A 10 year survival rate of more than 75% has been reported in one study (Gott et al., 1986). The occurrence of dissection prior to surgery or the presence of aneurysms in multiple aortic segments is associated with a worse postoperative prognosis. Crawford and Coselli (1988) reported a 10 year survival of 56% among the latter group. Replacement of the aortic or mitral valves is a frequent procedure in MFS patients. In children with mitral regurgitation, reconstruction of the mitral valve rather than its replacement is often feasible (Shumway et al., 1988).

Scoliosis is a frequent orthopedic complication in MFS (Pyeritz and McKusick, 1979; Viljoen and Beighton, 1990). Its age of onset varies from early childhood to adolescence. Conservative treatment by bracing or surgical correction with different types of fixation and rodding can be used successfully (Birch and Herring, 1987; Winter, 1990). Pectus deformities are present in more than two-thirds of individuals with MFS. Surgical correction is not warranted for cosmetic reasons only. Other orthopedic problems such as recurrent joint dislocations and subluxations, are managed according to established orthopedic principles.

Regular ophthalmic evaluation is an important aspect of the management of individuals with MFS. Improvement of impaired visual acuity due to myopia is achieved easily with conventional means. Removal of a dislocated lens by surgery should be considered only if satisfactory visual acuity cannot be achieved (Behki et al., 1990; Peyman et al., 1979).

Pregnancy might pose a considerable risk to women affected with MFS (Pyeritz et al., 1981b; Beighton, 1982). In general, women with an aortic root diameter greater than 4 cm (as measured by echocardiography), valvular disease or any evidence of cardiac decompensation are at greater risk, while those with minimal cardiac findings tolerate pregnancy well (Pyeritz, 1981). The continued use of beta-blockers during pregnancy remains controversial.

Although the majority of individuals affected with MFS cope well with their condition, some require psychological counselling as part of their management. In that respect they do not differ from any person affected with a chronic and potentially debilitating condition. The National Marfan Foundation in the United States and similar lay groups in other countries provide valuable psychological and emotional support and disseminate information to their members and other interested individuals.

FUTURE DIRECTIONS

The recent identification of fibrillin as the molecule mutated in MFS heralds a new era in the understanding of this disorder. The possibility now exists to determine fully the molecular pathology of MFS and, in the process, to begin to unravel the complexities of the microfibrillar system. It can be envisaged that the elucidation of the full structure of the FBN1 gene will make it possible to begin to determine the organisation and regulation of this gene. In this regard, the possibility remains that genes flanking the FBN1 gene are also involved in the pathogenesis of MFS. The further identification of fibrillin mutations in patients affected with MFS will allow a genotype-phenotype correlation to be made. In addition to providing clues as to the relationship between structure and function for the various regions of the fibrillin molecule, this may make possible the genotypic diagnosis, both prenatally and postnatally, of MFS. It will also be important to determine the full extent of the fibrillin gene family, the relationship between the various members of this family and the developmental role of the fibrillins.

The identification of fibrillin mutations in individuals with MFS has already begun to provide some clues as to the role played by fibrillin in microfibril assembly and function. The further determination of the molecular composition of microfibrils along with the isolation and characterization of the genes for these proteins and the investigation of the involvement of the various microfibrillar components in the pathology of MFS-like disorders will advance our understanding of the entire microfibrillar array. In the same way, the continued delineation of the precise molecular pathology of MFS-like disorders may lead to improvements in both the diagnosis and treatment of this group of conditions.

It may also be timely to begin to speculate as to the possibility of gene therapy for MFS. Although gene therapy for a dominant condition such as this requires specific repression of the mutant allele rather than simple provision of an absent gene product, the intensity of investigation of the fibrillin gene may provide the necessary background for progress in this area. A recombinant mouse model, constructed by homologous recombination into the FBN1 gene, will definitely become the organism of choice to test gene therapy of MFS.

REFERENCES

Arn, P. H., Scherer, L. R., Haller, J. A., et al. (1989). Outcome of pectus excavatum in patients with Marfan syndrome and in the general population. *J. Pediatr.* **115**:954–958.

Beals, R. K., Hecht, F. (1971). Congenital contractural arachnodactyly. *J. Bone Joint Surg.* **53**–A: 987–993.

Behki, R., Noel, L. P., Clarke, W. N. (1990). Limbal lensectomy in the management of ectopia lentis in children. *Arch.Ophthalmol.* **108**: 809–811.

Beighton, P. (1982). Pregnancy in the Marfan syndrome. *Br. Med. J.* **285**: 464.

Beighton, P., De Paepe, A., Danks, D., Finidori, G., Gedde-Dahl, T., Goodman, R., Hall, J. G., Hollister, D. W., Horton, W., McKusick, V. A., Opitz, J. M., Tsipouras, P. et al. (1988). International nosology of heritable disorders of connective tissue, Berlin, 1986. *Am. J. Med. Genet.* **29**:581–594.

Bentall, H. H., DeBono, A. A. (1968). A technique for complete replacement of the ascending aorta. *Thorax* **23**:338–339.

Besser, T. E., Potter. K. A., Bryan, G. M, Knowlin, G. G. (1990). An animal model of the Marfan syndrome. *Am. J. Med. Genet.* **37**:159–165.

Birch. J. G., Herring, J. A. (1987). Spinal deformity in Marfan syndrome. *J. Pediatr. Orthop.* **7**:546–552.

Blanton, S. H., Sarfarazi, M., Eiberg, H., de Groote, J., Farndon, P. A., Kilpatrick, M. W., Child, A. H., Pope, F. M., Peltonen, L., Francomano, C. A., Boileau, C., Keston, M., Tsipouras, P. (1990). An exclusion map of Marfan syndrome. *J. Med. Genet.* **27**:73–77.

Boileau, C., Alexandre, J. A., Hariti, G., et al. (1991). *Evidence for genetic heterogeneity in Marfan syndrome.* Abstracts of the 11th International Workshop on Human Gene Mapping, London, **18**–22 August 1991: 149.

Chemke, J., Nisani, R., Feigl, A., Garty, R., Cooper, M., Barash, Y., Duskin, D. (1984). Homozygosity for autosomal dominant Marfan syndrome. *J. Med. Genet.* **21**:173–177.

Cheng, T. O., Barlow, J. B. (1989). Mitral leaflet billowing and prolapse: its prevalence around the world. *Angiology* **40**: 7787.

Cleary, E. G., Gibson, M.m A (1983). Elastin-associated microfibrils and microfibrillar proteins. *Int. Rev. Connect. Tissue. Res.* **10**:97–209.

Crawford, E. S. (1983). Marfan's syndrome. Broad spectral surgical treatment of cardiovascular manifestations. *Ann. Surg.* **198**:487–505.

Crawford, E. S., Coselli J. S. (1988). Marfan's syndrome: combined composite valve graft replacement of the aortic root and transaortic mitral valve replacement. *Ann. Thorac. Surg.* **45**:296–302.

Davis, C. G. (1990). The many faces of epidermal growth factor repeats. *New Biologist* **2**:410–419.

Dietz, H. C., Pyeritz, R. E., Hall, B. D., Cadle R. G., Hamosh, A., Schwartz, J., Meyers, D. A., Francomano, C. A. (1991a). The Marfan syndrome locus: confirmation of assignment to chromosome 15 and identification of tightly linked markers at 15ql5-q21.3. *Genomics* **9**:355–361.

Dietz, H. C., Cutting, G. R., Maslen, C. L., Sakai, L. Y., Corson, G. M., Puffenberger, E. G., Hamosh, A., Nanthakumar, E. J., Curristin, S. M., Stetten, G., Meyers, D. A., Francomano, C. A. (1991b). Marfan syndrome caused by a recurrent de novo missense mutation in the fibrillin gene. *Nature* **352**:337–339.

Dietz, H. C., Pyeritz, R. E., Puffenberger, E. K., Kendzior, Jr. R. J., Corson, G. M., Maslen, C. L., Sakai, L. Y., Francomano, C. A., Cutting, G. R. (1992). Title? *J. Clin. Invest.* **89**:1674–1680.

Engel, J. (1989) EGF-like domains in extracellular matrix proteins : localised signals for growth and differentiation. *FEBS letters* **251**:1–7.

Gibson, M. A., Kumaratilake, J. S., Cleary, E. G. H. (1989). The protein components of the 12 nanometer microfibrils of elastic and non-elastic tissues. *J. Biol. Chem.* **264**:4590–4598.

Gibson, M. A., Sandberg, L. B., Grosso, L. E., Cleary, E. G. (1991). Complementary DNA cloning establishes microfibril-associated glycoprotein (MAGP) to be a discrete component of the elastin-associated microfibrils. *J. Biol. Chem.* **266**:7596–7601.

Godfrey, M., Menashe, V., Weleber, R. G., Koler, R. D., Bigley, R. H., Lovrien, E., Zonana, J., Hollister, D. W. (1990). Cosegregation of elastin-associated microfibrillar abnormalities with the Marfan phenotype in families. *Am. J . Hum. Genet.* **46**:652–660.

Gott, V. L., Pyeritz, R. E., Magovern, Jr. G. J., Cameron, D. E., McKusick, V. A. (1986). Surgical treatment of aneurysms of the ascending aorta in the Marfan syndrome. *N. Engl. J. Med.* **314**:1070–1074.

Hollister, D. W., Godfrey, M., Sakai, L. Y, Pyeritz, R. E. (1990). Marfan syndrome: immunohistologic abnormalities of elastin-associated microfibrillar fibre system. *N. Eng. J. Med.* **323**:152–159.

Kainulainen, K., Pulkkinen, L., Savolainen, A., Kaitila, I., Peltonen, L. (1990). Location on chromosome 15 of the gene defect causing Marfan syndrome. *N. Engl. J. Med.* **323**:935–939.

Kainulainen, K., Sakai, L. Y., Child, A., Pope, F. M., Puhakka, L., Ryhanen, L., Palotie, A., Kaitila, I., Peltonen, L. (1992). Two mutations in Marfan syndrome resulting in truncated fibrillin polypeptides. *Proc. Natl. Acad. Sci. USA.* **89**:5917–5921.

Larson, E. W., Edwards, W. D. (1984). Risk factors for aortic dissection: a necropsy study of 161 cases. *Am. J. Cardiol.* **53**:849–855.

Lee, B., Godfrey, M., Vitale, E., Hori, H., Mattei, M. G., Sarfarazi, M., Tsipouras, P., Ramirez, F., Hollister, D. W. (1991). Linkage of Marfan syndrome and a phenotypically related disorder to two different fibrillin genes. *Nature.* **352**:330–334.

Lima, S. D., Lima, J. A. C., Pyeritz, R. E., Weiss, J. L. (1985). Relation of mitral valve prolapse to left ventricular size in Marfan's syndrome. *Am. J. Cardiol.* **55**: 739–743.

Low, F. N (1962). Microfibrils: fine filamentous components of the tissue space. *Anat. Rec.* **142**:131–137 .

Lutman, F. C., Neel, J. V. (1949) . Inheritance of arachnodactyly, ectopia lentis and other congenital anomalies (Marfan's syndrome) in the E. family. *Arch. Ophthal.* **41**: 276–305.

Magid, D., Pyeritz, R. E., Fishman, E. K. (1990). Musculoskeletal manifestations of the Marfan syndrome: radiologic features. *Am. J. Roentgenol.* **155**:99–104.

Marfan, A. B. (1896) . Un cas de deformation congenitale des quatre membres plus prononce aux extremites characterisee par l'allongement des os avec un certain degre d'amincissement. Bull.Mem. *Soc. Med. Hop. Paris* **13**:464–473.

Maslen, C. L., Corson, G. M., Maddox, B. K., Glanville, R. W., Sakai, L. Y. (1991). Partial sequence of a candidate gene for the Marfan syndrome. *Nature* **352**:334–337.

Maumenee, I. H. (1981). The eye in the Marfan syndrome. *Trans. Am. Ophthal. Soc.* 79:684–733.

McKusick, V. A. (1955). The cardiovascular aspects of Marfan's syndrome. *Circulation* **11**:321–342.

McKusick, V. A. (1972). *Heritable disorders of connective tissue.* Fourth edition. Mosby, St. Louis. pp. 61–223.

Mecham, R. P. and Hauser, J. E. (1992). The elastic fibre. In *Cell Biology of the Extracellular Matrix.* Hay, E. D. (ed), 2nd edition. Plenum.

Milewicz, D. M., Pyeritz, R. E., Crawford, E. S., Byers, P. H. (1992) Marfan syndrome: defective synthesis, secretion, and extracellular matrix formation of fibrillin by cultured dermal fibroblasts. J. *Clin. Invest.* **89**:79–86

Morse, R. P., Rockenmacher, S., Pyeritz, R. E., Sanders, S. P., Bieber, F. R., Lin, A., MacLeod, P., Hall, B., Graham, Jr. J. M. (1990). Diagnosis and management of infantile Marfan syndrome. *Pediatrics* **86**:888–895.

Murdoch, J. L., Walker, B. A., McKusick, V. A. (1972a). Parental age effects on the occurrence of new mutations for the Marfan syndrome. *Ann. Hum. Genet.* **35**:331–336.

Murdoch, J. L., Walker, B. A., Halpern, B. I., Kuzma, J. W., McKusick, V. A. (1972b). Life expectancy and causes of death in the Marfan syndrome. *N. Engl. J. Med.* **286**:804–808.

Peyman, G. A., et al. (1979). Management of subluxated and dislocated lenses with the vitreophage. *Br. J. Ophthalmol.* **63**:771–778.

Pyeritz, R. E., McKusick, V. A. (1979). The Marfan syndrome: diagnosis and management. *N. Engl. J. Med.* **300**:772–777.

Pyeritz, R. E., Reider, R., Fortuin, N. J. (1981a). Aortic complications in adult Marfan syndrome are associated with the aortic root diameter. *Clin. Res.* **29**:315A.

Pyeritz, R. E. (1981b). Maternal and fetal complications of pregnancy in the Marfan syndrome. *Am. J. Med.* **71**:784–790.

Pyeritz, R. E. (1983). Propanolol retards aortic root dilatation in the Marfan syndrome. *Circulation* **68** (Suppl.III):365.

Pyeritz, R. E., Fishman, E. K., Bernhardt, B. A., Siegelman, S. S. (1988). Dural ectasia is a common feature of the Marfan syndrome. *Am. J. Hum. Genet.* **43**:726–732.

Roberts, W. C., Honig, H. S. (1982). The spectrum of cardiovascular disease in the Marfan syndrome: A clinicomorphologic study of 18 necropsy patients and comparison to 151 previously reported necropsy patients. *Am. Heart. J.* **104**:115–135.

Roman, M. J., Devereux, R. B., Kramer-Fox, R., Spitzer, M. (1988). Aortic root dilatation in the Marfan syndrome: patterns, familiarity, and short-term clinical courses. *J. Am. Coll. Cardiol.* **11**:74A.

Sakai, L. Y., Keene D. R., Glanville, R. W, Bachinger, H. P. (1991). Purification and partial characterisation of fibrillin, a cysteine rich structural component of conective tissue microfibrils. *J. Biol. Chem.* **266**:14763–14770.

Sakai, L. Y., Keene, D. R., Engvall, E (1986). Fibrillin, a new 350kDa glycoprotein is a component of extracellular microfibrils. *J. Cell. Biol.* **103**:2499–2509.

Schollin, J., Bjarke, B., Gustavson, K. H. (1988). Probable homozygotic form of the Marfan syndrome in a newborn child. *Acta Paediatr. Scand.* **77**:452–456.

Shumway, S. J., Gott, V. L., Reitz, B. A. (1988). A "designer" annuloplasty ring for patients with massive mitral annular dilatation. *Ann. Thorac. Surg.* **46**:695–696.

Tsipouras, P. (1990). Marfan syndrome: light at the end of the tunnel? *Am. J. Hum. Genet.* **46**:643–645.

Tsipouras, P., Sarfarazi, M., Devi, A., Weiffenbach, B., Boxer, M. (1991). Marfan syndrome is closely linked to a marker on chromosome 15ql.5q2.1. *Proc. Natl. Acad. Sci. USA.* **88:**4486–4488.

Tsipouras, P., Del Mastro, R., Sarfarazi, M., Lee, B., Vitale, E., Child, A. H., Godfrey, M., Devereux, R. B., Hewett, D., Steinmann, B., Viljoen, D., Sykes, B. et al. (1992). Genetic linkage of the Marfan syndrome,ectopia lentis, and congenital contractural arachnodactyly to the fibrillin genes on chromosomes 15 and 5. *N. Engl. J. Med.* **326**:905–909.

Viljoen, D ., Beighton, P . (1990) . Marfan syndrome: a diagnostic dilemma. *Clin.Genet.* **37**:417–422.

Weve, H. (1931). Ueber Arachnodaktylie (Dystrophica mesodermalis congenita, typus Marfanis). *Arch. Augenheilk. 104*:1–46.

Winter, R. B. (1990). Thoracic lordoscoliosis in Marfan's syndrome. Report of two patients with surgical correction using rods and sublaminar wires. *Spine* **15**:233–235.

Wood, J. R., Bellamy, D., Child, A. H., Citron, K. M. (1984) . Pulmonary disease in patients with Marfan syndrome. *Thorax* **39**:780–784.

17. LOWE OCULOCEREBRORENAL SYNDROME

ICHIRO OKABE[1] AND ROBERT L. NUSSBAUM[1,2]

[1]Howard Hughes Medical Institute Laboratory; [2]Department of Genetics University of Pennsylvania School of Medicine, 422 Curie Boulevard, Philadelphia, PA 19104, USA

INTRODUCTION

The Lowe oculocerebrorenal syndrome (OCRL), (Mc Kusick #309000), first described in 1952 (Lowe, Terrey, and MacLachan, 1952), is an X-linked disorder characterized by congenital cataract, mental retardation and renal tubular dysfunction. Several hundred cases have been described since the initial report but the exact incidence of this disorder is not known. The primary biochemical defect has been variously ascribed to abnormalities in proline uptake (States, Palmieri, and Segal, 1982) mitochondrial function (Gobernado et al., 1984; Moraes et al., 1991), and proteoglycan synthesis (Yamashina et al., 1983; Yokoi et al., 1981; Fukui et al., 1981). The OCRL locus was first mapped to Xq25-26 by linkage (Silver, Lewis, and Nussbaum, 1987; Wadelius et al., 1989; Reilly et al., 1988; Reilly, Lewis, and Nussbaum, 1990) and cytogenetic methods (Hodgson et al., 1986; Mueller et al., 1991) and then isolated by a positional cloning strategy (Okabe et al., 1992; Attree et al., 1992). Striking homology of the OCRL gene to an enzyme, platelet inositol 1,4,5 tris phosphate-5-phosphatase (Ross et al., 1991), suggests that the primary defect in this disorder may be deranged inositol phosphate metabolism. Current work now focuses on determining the exact function of the product of the gene for OCRL and the pathogenesis of the phenotype. The clinical, laboratory, and genetic features of OCRL have been recently reviewed (Charnas and Nussbaum, 1994).

CLINICAL FINDINGS

Incidence and Inheritance

OCRL is an X-linked disorder that occurs nearly exclusively in males. The disease is rare but appears to be ubiquitous among various nationalities and races (Kornfeld et al., 1975; Ananthakrishna and Biswakumar, 1977; Banerjee, Allen, and McKee, 1982; Yamashina et al., 1983; McSpadden, 1991). Several females have also been reported (Svorc et al., 1967; Sagel, Ores, and Yuceoglu, 1970; Harris et al., 1970; Lythgoe and Ramsey, 1973; Cyvin, Weidemann, and Bathen, 1973; Johnson and Hiles, 1977; Yamashina et al., 1983). In some of these females, however, only a portion of the OCRL phenotype is present while in others, congenital malformations and birth defects not seen in males with OCRL make the diagnosis of OCRL uncertain. One female patient was reported with a deletion of mitochondrial DNA, which raises the possibility that a mitochondrial genocopy for OCRL may exist (Moraes et al., 1991), but her phenotype was that of a progressive multisystem disease only partially overlapping with OCRL. Finally, there are two females reported in the literature in which clear-cut OCRL has arisen *de novo* due to balanced X/autosome translocations involving the Xq25-Xq26.1 region of the X chromosome (Hodgson et al., 1986; Mueller et al., 1991). These patients have an OCRL phenotype indistinguishable from that of hemizygous affected males because the translocations disrupted the OCRL locus on the translocated X chromosome and impeded random X inactivation such that the X chromosome not involved in the translocation is inactivated preferentially.

Ocular Findings

Bilateral congenital cataracts are a major sign of OCRL, present in 100% of affected males. Abnormal lens development begins at 7–9 weeks as a disordered migration of the embryonic lens epithelium (Ginsberg, Bove, and Fogelson, 1981). Lens opacities have been detected clinically *in utero* in some cases as early as 20–24 weeks (Endres et al., 1977; Gaary et al., 1993). Glaucoma develops in about 50% of patients, usually during the first year of life, but may develop as late as the second or third decade (McSpadden, 1991). Corneal keloid formation develops in approximately 25% of the patients, on average at 5 years of age but with a range from 5 months to 18 years (McSpadden, 1991; Cibis et al., 1982). Keloids are often stable and cause no visual difficulty but, if extensive and centrally located, may interfere with vision and can be the major cause of visual disability after age 5 years in patients whose cataracts and glaucoma have been successfully managed. However, some degree of diminished acuity is present in all OCRL patients that is not explainable by cataracts or keloids and may result in strabismus (20%) and nystagmus (69%) (McSpadden, 1991). Overall, about 15% of OCRL patients over 5 years old are blind.

Female carriers can be diagnosed by characteristic lens opacities using the slit-lamp examination (Tongue, 1972; Gardner and Brown, 1976; Cibis et al., 1986; Reilly et al., 1988; Wadelius et al., 1989). In Figure 1A are shown the most common and typical opacities of female carriers. They are punctate, white to gray in color, vary in size from microns to several millimeters, and are distributed in radial or wedge-shaped clusters in the cortex layer but not in the nucleus of the lens (Cibis et al., 1986; Reilly et al., 1988). The distribution of opacities just outside the nucleus suggests that the opacities develop very early in life while the more superficial opacities must arise in adulthood (Cibis, Tripathi and Tripathi, 1990). These opacities are usually greater than 15 in number and may be in the hundreds. A few female carriers show a single subcapsular dense cataract in the posterior pole (Tongue, 1972) (Figure 1B). This form of cataract may be present congenitally and become clinically more apparent with age, thus requiring surgical therapy. Al-

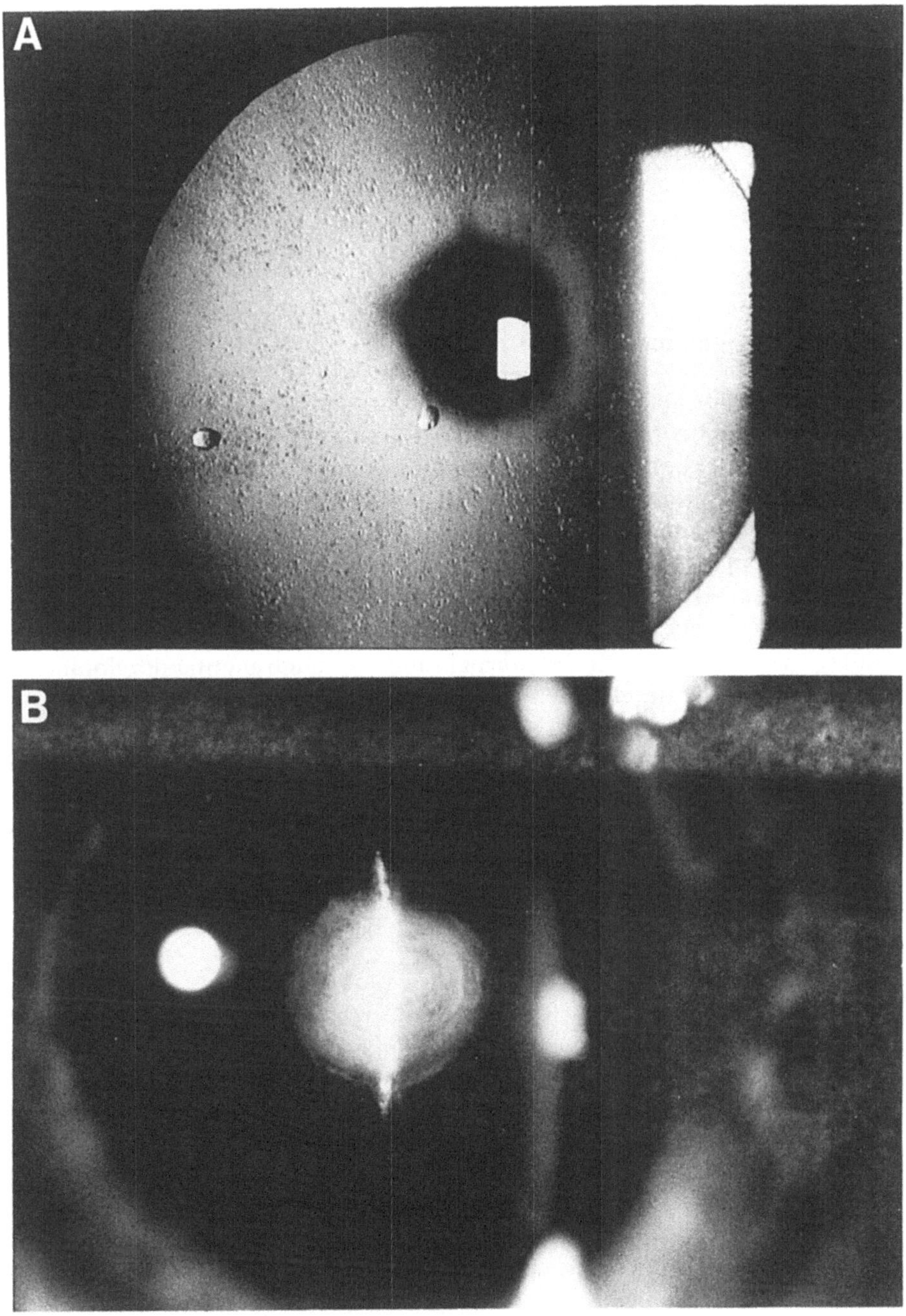

Figure 1 Slit lamp examination of female carriers of Lowe's Syndrome. In panel A is shown the typical pattern of hundreds of micropunctate, grey opacities distributed in a radial distribution in the lens. In panel B is a single dense posterior cataract as seen in some OCRL heterozygotes. (Courtesy of R. A. Lewis, M.D.)

though some controversy exists in the literature concerning the specificity and sensitivity of the lenticular opacities as a method of carrier detection (Gardner and Brown, 1976), reliance on strict diagnostic criteria for both appearance and distribution of the opacities makes carrier detection by slit lamp examination a reliable and sensitive technique (Cibis et al.,1986; Reilly et al., 1988).

Systemic Findings

Growth and development

Average birth weight and length of the patients with OCRL are well within normal range but the rate of growth begins to decline by 3 months of age. By 18 months, weight and height are below normal range, and continue to fall. The growth curve in the patients does not appear to plateau at 16 to 18 years of age as it does in normal men. Growth continues on into early adulthood but mean final weight and height are less than the third percentile of normal values for men (McSpadden, 1991; Charnas et al., 1991).

A wide range of intellectual ability are represented in OCRL patients (McSpadden, 1991; Kenworthy, Park, and Charnas, 1993); 63% have moderate to severe mental retadation, 33% have a mild to borderline form, with a few patients demonstrating intelligence within the normal range. Though mental development is generally delayed, the developmental abnormality is stable, in contrast to the progressive course seen in degenerative disorders.

Some behavior problems are troublesome, especially self-abusive behavior, noncooperative and violent behavior, and eccentric, autistic-like behavior. These behavior problems are both most prevalent and most severe in the 5–13 age group, and tend to decrease in late adolescence and adulthood (McSpadden, 1991; Kenworthy, Park, and Charnas, 1993).

Neuromuscular signs

Hypotonia with decreased muscle mass and decreased deep tendon reflexes are noted in most patients. Hypotonia becomes apparent from a few months of age but is not progressive and tends to improve in association with general growth. The deep tendon reflexes are present at birth but disappear by 1 year of age (Abbassi, Lowe, and Calcagno, 1968). Significant nerve or muscle pathology is absent and suggests that the hypotonia and areflexia are of central nervous system origin (Banerjee, Allen, and McKee, 1982; Charnas et al.,1988).

Seizures occur in 50% of OCRL patients (Charnas, 1989). All three major types of seizure have been reported: febrile, petit mal, and grand mal. The average age of first seizure for these three types of seizure is 2 years, 5 years, 9 years of age respectively. The grand mal type is the most common seizure (McSpadden, 1991). As in the general population, severe, early onset seizures are a poor prognostic sign for intellectual development and seizure control.

Renal dysfunction

Fanconi Syndrome (proximal tubular acidosis, amino aciduria, phosphaturia, and proteinuria) is a principal component of OCRL. The renal tubular dysfunction, in contrast to the cataracts, is not always present at birth and may require a few weeks to months to become apparent. Renal tubular dysfunction is quite variable in its severity or clinical significance, and may not require medical intervention (Charnas et al., 1991). The acidosis is due to bicarbonate wasting (Schoen and Young, 1959; Richards et al., 1965) and leads to the failure to thrive, recurrent infections, and metabolic collapse seen in early case reports when the disease was poorly recognized and untreated. Water resorption is also defective, as reflected in elevated 24 hour volumes and low urine osmolality (Charnas et al., 1991). Aminoaciduria is generalized but variable, ranging from just above the upper limit of normal to 15 times above the upper limit of normal (Payton et al., 1989; Charnas et al., 1991), but the pattern of aminoaciduria is not diagnostic of OCRL (Manz, Bremer, and Brodehl, 1978). Proteinuria is very frequently seen but the amount of urinary losses and age of onset are both highly variable. Proteinuria is usually present in infancy but can first occur later in childhood. When present, protein losses can be substantial (1.38–10.77 gm/m2/d, normal <0.1 g/m2/d). Glycosuria is generally not a feature of the renal tubular dysfunction seen in OCRL. The renal Fanconi syndrome, once established, is usually stable.

Hyperphosphaturia is also frequent but variable, and may lead to osteomalacia, renal rickets, and pathologic fractures if untreated. In approximately half of patients, serum phosphate levels are normal but the fractional excretion of phosphate is elevated. Tubular phosphate wasting often worsens with age (Charnas et al., 1991), and may require phosphate supplementation orally or by nasogastric tube (Redfield et al., 1991). Nephrolithiasis and nephrocalcinosis may occur in OCRL (Schaper and Horstmann, 1963) because of hyperphosphaturia and hypercalciuria, but high urinary volume may be partially protective. Gradual loss of creatinine clearance reflecting progressive renal failure is a feature of OCRL in the second and third decades of life. The average age of onset of renal insufficiency and its severity is, however, not clearly defined.

Skeletal signs

Musculoskeletal complications in OCRL can result either as a primary feature of the disease or as secondary consequences of renal and neuromuscular abnormalities of the illness, i.e., Fanconi Syndrome and hypotonia. Tenosynovitis and arthritis or arthropathy are frequent and appear to be primary complications of OCRL. Joint involvement can present as nontender joint swelling involving both small and large joints, focal nodules on the finger, or bilateral plantar masses. On occasion these masses become painful and require resection (Charnas and Gahl, 1991). It seems likely that these changes are the manifestation of abnormal, excessive growth of fibroblasts from periarticular tissue or tendons. Secondary complications of OCRL include joint hypermobility due to hypotonia or contractures resulting from decreased movement. Rickets is noted in about half of the patients. About half of the patients have had at least one fracture because of diffuse osteopenia (McSpadden, 1991). Osteopenia can be worsened by untreated renal phosphate wasting, but is almost always present, with variable severity despite adequate phosphate replacement, and appears to be a specific manifestation of OCRL. Bone fracture occurs mainly at the arms and legs and increases with age. Scoliosis is also common among the older patients.

Other features

The onset and progression of puberty is generally normal although cryptorchidism is common (Charnas and Gahl, 1991). Peritubular fibrosis and azoospermia presumably reduce fertility (Matin and Sylvester, 1980). OCRL patients lack a strong sexual drive in comparison with other adult males with similar intellectual compromise.

Dental "blue dome" cysts occur during primary tooth eruption and double rows of teeth caused by delayed shedding of primary teeth with simultaneous presence of primary and permanent teeth have been reported (McSpadden, 1991).

Prognosis

Severe renal insufficiency and dehydration or intercurrent infection (Curtin, Joyce, and Ballin, 1967; Matin and Sylvester, 1980; Tripathi, Cibis, and Tripathi, 1980) are the main cause of death in patients with OCRL. Many patients are known to have lived beyond adolescence. The mean age of the 82 patients known to the Lowe's Syndrome Association is 13 years 2 months, with a range of 1 year 7 months to 41 years 3 months (Charnas et al., 1991).

Pathological Findings

Ophthalmic pathology

Lenses are small, flattened, and discoid. There is a diffuse cataract with no differentiation of cortex and nucleus, and pronounced degeneration in the posterior polar region (Curtin, Joyce, and Ballin, 1967; Tripathi, Cibis, and Tripathi, 1980). The anterior subcapsular epithelial hyperplasia and nodular excrescences of the lens capsule are present; however, these are not necessarily specific for OCRL (Zimmerman and Font, 1966; Curtin, Joyce, and Ballin, 1967; Tripathi, Cibis, and Tripathi, 1980; Tripathi, Cibis, and Tripathi, 1986). An anomalous anterior chamber angle with anterior displacement of rudimentary ciliary processes, which could contribute to glaucoma, might be secondary to a microphakia in which zonular fibres drag the ciliary body anteriorly (Zimmerman and Font, 1966; Curtin, Joyce, and Ballin, 1967; Tripathi, Cibis, and Tripathi, 1980; Cibis, Tripathi, and Tripathi, 1990). The small islands or folds of retinal tissue, which might also be dragged anteriorly by the zonules,are present in pars ciliaris retinae (Zimmerman and Font, 1966; Curtin, Joyce, and Ballin, 1967; Garzuly et al., 1973). Otherwise, the retina including optic nerve, sclera, and choroid appears unremarkable in the light microscopic examination (Curtin, Joyce, and Ballin, 1967; Garzuly et al., 1973).

Renal Pathology

Histological examination by light microscopy is usually normal in very young infants but dilatation, atrophy, loss of brush border, and accumulation of proteinaceous material in the tubular lumen appear within the first few months of life (Schoen and Young, 1959; Richards et al., 1965; Witzleben et al., 1968; Van Acker et al., 1967). These tubular abnormalities have been documented in a number of studies, including some in which serial biopsies from the same patients were examined, and affect predominantly the proximal tubules, with some involvement of Henle's loop, distal tubules, and the conducting system occasionally cited as well. Glomeruli are frequently normal in young children, even when proximal tubular lesions are present. After the first few years of life, however, glomerular lesions can be seen. These include thickening of basement membranes, focal fibrosis, and sclerosis. By electron microscopy, swelling of mitochondria in renal tubules was reported in a 22 month old who showed no glomerular abnormalities (Schoen and Young, 1959). When the same child was biopsied 7 years later, distinctive glomerular changes including podocyte fusion and basement membrane thickening were apparent (Witzleben et al., 1968). One report of severe electron microscopic abnormalities in glomeruli and proximal tubules was based on material from a female patient who did not satisfy current clinical diagnostic criteria for OCRL (Sagel, Ores, and Yuceoglu, 1970).

Central nervous system pathology

A variety of pathological findings in the brain have been reported (Richards et al., 1965; Chutorian and Rowland, 1966; Matin and Sylvester, 1980; Banerjee, Allen, and McKee, 1982). The most consistent change in the brain is a loss of or defective formation of myelin. Such changes are not seen in all cases, vary in distribution and extent, and are nonspecific since similar abnormalities have been noted in other metabolic disorders such as maple syrup urine disease, homocystinuria, and phenylketonuria. Multiple tiny cysts without inflammatory changes have also been described in the cerebral white matter but appear to be of no clinical significance (Demmer, Wippold, and Dowton, 1992).

BIOCHEMICAL FINDINGS

Previous studies of OCRL patients and cells have disclosed a number of potential abnormalities. Increased proline uptake in fibroblasts from one patient was reported (States, Palmieri, and Segal, 1982). Since one of the major pathways of proline metabolism is incorporation into collagen, Palmieri et al. (1985) examined collagen formation and found that collagen production was substantially decreased in cultured skin fibroblast from patients. The mechanism and significance of these findings remain unclear.

Metabolic abnormalities of glycosaminoglycans(GAGs) have been reported beginning in 1969 (Suschke and Murken, 1969). Elevated excretion of urinary GAGs was seen in patients with OCRL. However, the results were conflicting, and difficult to reproduce. Akasaki et al. (1978) reported increased low sulfated chondroitin-4-sulfate (C4S) excretion in the urine. Hayashi et al. (1978a;1978b) have shown that low sulfated C4S and heparan sulfate(HS) were elevated in the urine of the patient with elevated excretion of urinary GAGs. This was verified by Kieras et al. (1984) and Wisniewski et al. (1984) except that they found HS excretion was within the normal range. However, Yokoi et al. (1981) reported a decrease in the urinary excretion of HS with no elevation of urinary GAGs in patients.

Given the findings in urine, GAGs synthesis and sulfation have been studied in cultured skin fibroblasts from OCRL patients. On one hand, Donnelly et al. (1984) and Harper et al. (1987) showed that OCRL fibroblasts synthesized normally sulfated GAGs and did not express defects in sulfation of GAGs. Yokoi et al. (1982) reported undersulfation of HS but not CS and DS in fibroblasts of OCRL patients. Yamashina et al. (1983) reported undersulfation of GAGs confined to chondroitin sulfate(CH) and dermatan sulfate(DS) with HS not affected, and this undersulfation of GAGs was attributed to depleted active sulfate (adenosine 3'-phosphate 5'-phosphosulfate; PAPS) pools in OCRL fibroblasts, reportedly resulting from elevated levels of a nucleotide pyrophosphatase (NPPase) activity degrading PAPS (Fukui et al., 1981; Yoshida et al., 1982; Yamashina et al., 1983). They went on to show that the NPPase activities in heterozygotes were intermediate between patients and normal individuals (Yamashina et al., 1983) and the elevated NPPase activities in fibroblast from patients resulted from increased amounts of the enzyme protein (Horie et al., 1988). Elevation of NPPase activities in patients was verified by one other group (Harper et al., 1987). NPPase, however, was eliminated as the primary locus for OCRL by the finding that the NPPase gene maps to 6q22-q23 (Funakoshi et al., 1992). Thus, the occurence and significance of abnormalities of GAGs as primary defects for OCRL still remain obscure.

There are several reports about mitochondrial abnormalities in OCRL (Payton et al., 1989; Ores, 1970; Sagel, Ores, and Yuceoglu, 1970; Gobernado et al., 1984; Moraes et al., 1991). Some of these reports were female patients with phenotypes that overlapped partially with OCRL. All had normal chromosome constitutions and showed morphological mitochondria abnormalities (Sagel, Ores, and Yuceoglu, 1970), or mitochondrial DNA deletion (Moraes et al., 1991). Gobernado et al. (1984) showed a defect in oxidative phosphorylation in mitochondria in a male patient with OCRL. Further studies are needed to determine whether there is a reproducible defect in mitochondrial function in X-linked OCRL but it appears that mitochondrial DNA alteration or deletion may produce partial genocopies of the OCRL phenotype.

MOLECULAR STUDIES

Gene Mapping

Linkage studies of OCRL were first reported in 1982 (Hittner, Carroll, and Prchal, 1982) which ruled out close linkage of OCRL to either the Xg blood group (Xp22.3-pter) or to glucose-6-phosphate dehydrogenase (Xq28). Using restriction fragment length polymorphisms (RFLPs) on eight X chromosome loci, Silver et al. (1987) analyzed four large families with OCRL and identified two loci, DXS10 at Xq26 and DXS42 at Xq24-26, which were tightly linked to OCRL and supported a map assignment to Xq24-q26. Evidence for placing OCRL at Xq24-q26 also came from two female patients with an X;3 translocation breakpoint at Xq25 (Hodgson et al., 1986) and an X;20 translocation breakpoint at Xq26.1 (Mueller et al., 1991). DXS42 and DXS10 have been shown to flank the Xq25 breakpoint by analysis of somatic cell hybrids generated from this t(X;3) patient (Reilly, Lewis, and Nussbaum, 1990). No recombination was found between OCRL and the proximal marker DXS42 in two studies with a combined lod score of 11.8 (Reilly, Lewis, and Nussbaum, 1990; Wadelius, Fagerholm, Pettersson, and Anneren, 1989).

Positional Cloning

Nelson et al. (1991) isolated yeast artificial chromosomes whose human inserts spanned the breakpoint of a female patient with an X;3 translocation. Using these yeast artificial chromosomes as probes, a candidate cDNA, *OCRL-1*, was isolated which is interrupted by both the t(X;3) and t(X;20) breakpoints (Okabe et al., 1992; Attree et al., 1992). Genetic evidence that *OCRL-1* is the gene for OCRL relies on the discovery of mutations in the gene in patients with the disease. The transcript of this gene is absent in both female OCRL patients with X;autosome translocations and is absent or aberrant in size in some unrelated male patients with no detectable rearrangement (Attree et al., 1992). Direct sequencing of reverse transcribed mRNA from male patients with detectable message has revealed point mutations leading to premature termination or missense mutations as well as an exon skipping mutation and a three base deletion of a single amino acid (Leahey, Charnas, and Nussbaum, 1993).

The function of *OCRL-1* is unknown. One clue, however, is that the protein predicted to be encoded by *OCRL-1* has 71% similarity to human platelet inositol polyphosphate-5-phosphatase (Ross, Jefferson, Mitchell, and Majerus, 1991). These data suggest that OCRL may be an inborn error of inositol phosphate metabolism (Irvine, 1992). The role of inositol phosphate is well established as an intracellular messenger for calcium signalling and protein kinase C activation which in turn regulate cellular processes including metabolism, secretion, contraction, neural activity, and cell proliferation (Berridge, 1993). Abnormalities in inositol metabolism have been proposed to have a role in the pathogenesis of cataracts in galactosemia (Cammarata, Tse, and Yorio, 1990), and diabetic complications such as cataracts, peripheral neuropathy, and nephropathy (Greene, Lattimer, and Sima, 1987). The diminution of sulfated proteoglycan concentration is frequently observed in human and animal diabetic states (Epstein, 1989). Thus abnormalities of GAGs in OCRL may be a secondary derangement caused by impaired inositol metabolism. However, it should be stressed that the involvement of inositol phosphate metabolism in the pathogenesis of OCRL remains an unproven hypothesis until the true function of *OCRL-1* is determined and a definitive biochemical defect in inositol metabolism is demonstrated in OCRL.

THERAPY

Current therapies for OCRL are mainly symptomatic treatment, such as extraction of cataracts, management of glaucoma, bicarbonate replacement for renal tubular acidosis, anticonvulsant drugs, orthopedic care for bone fracture, physical rehabilitation, and special education. Phosphate and vitamin D supplementation may also be needed to control rickets. Identification of the gene for OCRL facilitates the understanding of molecular pathogenesis, and will contribute to the development of new strategies for treatment as well as accurate carrier detection and prenatal diagnosis.

REFERENCES

Abbassi, V., Lowe, C. U., and Calcagno, P. L. (1968). Oculo-cerebro-renal syndrome. *American Journal of Diseases of Children*, **115**, 145–168.

Akasaki, M., Fukui, S., Sakano, T., Tanaka, T., Usui, T., and Yamashina, I. (1978). Urinary excretion of a large amount of bound sialic acid and of undersulfated chondroitin sulfate A by patients with the Lowe syndrome. *Clinica Chimica Acta*, **89**, 119–125.

Ananthakrishna, and Biswakumar, I. (1977). Oculo cerebro renal syndrome (a case report). *Indian Pediatrics*, **14(3)**, 223–225.

Attree, O., Olivos, I. M., Okabe, I., Bailey, L. C., Nelson, D. L., Lewis, R. A., McInnes, R. R., and Nussbaum, R. L. (1992). The Lowe oculocerebrorenal syndrome gene encodes a novel protein highly homologous to inositol polyphosphate-5-phosphatase. *Nature*, **358**, 239–242.

Banerjee, A. K., Allen, I. V., and McKee, P. (1982). Oculo-cerebro-renal syndrome: failure to demonstrate specific neuropathological abnormalities in four cases. *Irish Journal of Medical Science*, **15**, 42–45.

Berridge, M. J. (1993). Inositol trisphosphate and calcium signalling. *Nature,* **361**, 315–325.

Cammarata, P.R., Tse, D., and Yorio, T. (1990). Sorbinil prevents the hypergalactosemic-induced reduction in [^{3}H]-myo-inositol uptake and decreased [^{3}H]-myo-inositol incorporation into the phosphoinositide cycle in bovine lens epithelial cells in vitro. *Current Eye Research,* **9 (6)**, 561–568.

Charnas, L. (1989). Seizures in the oculocerebrorenal syndrome of Lowe. *Neurology,* **39** supplement **1**, 362.

Charnas, L., Bernar, J., Pezeshkpour, G. H., Dalakas, M., Harper, G. S., and Gahl, W. A. (1988). MRI findings and peripheral neuropathy in Lowe syndrome. *Neuropediatrics,* **19**, 7–9.

Charnas, L., Bernardini, I., Rader, D., Hoeg, J., and Gahl, W. A. (1991). Clinical and laboratory findings in the oculocerebrorenal syndrome of Lowe, with special reference to growth and renal function. *New England Journal of Medicine,* **324**, 1318–1325.

Charnas, L. R. and Gahl, W. A. (1991). The oculocerebrorenal syndrome of Lowe. *Advances in Pediatrics,* **31**, 75–107.

Charnas, L. R. and Nussbaum, R. L. (1994). The oculocerebrorenal syndrome of Lowe (Lowe syndrome). In *The Metabolic Basis of Inherited Disease,* edited by C.R. Scriver, A. L. Beaudet, W. S. Sly, and D. Valle. New York: McGraw-Hill, in press.

Chutorian, A. and Rowland, L. P. (1966). Lowe's syndrome. *Neurology,* **16**, 115–122.

Cibis, G. W., Tripathi, R. C., and Tripathi, B. J. (1990). Lowe's syndrome - oculocerebrorenal syndrome. In D. H. Gold and T. A. Weingeist (Eds.), *The Eye in Systemic Disease* (pp. 504–506). Philadelphia: J. B. Lippincott.

Cibis, G. W., Tripathi, R. C., Tripathi, B. J., and Harris, D. J. (1982). Corneal keloid in Lowe's syndrome. *Archives of Ophthalmology,* **100**, 1795–1799.

Cibis, G. W., Waeltermann, J. M., Whitcraft, C. T., Tripathi, R. C., and Harris, D. J. (1986). Lenticular opacities in carriers of Lowe's syndrome. *Ophthalmology,* **93**, 1041–1045.

Curtin, V. T., Joyce, E. E., and Ballin, N. (1967). Ocular pathology in the oculo-cerebro-renal syndrome of Lowe. *American Journal of Opthalmology,* **64**, 533–543.

Cyvin, K. B., Weidemann, J., and Bathen, J. (1973). Lowe's Syndrome. *Acta Paediatrica Scandinavica,* **62**, 309–312.

Demmer, L. A., Wippold, F. J. 2d, and Dowton, S. B. (1992). Periventricular white matter cystic lesions in Lowe (oculocerebrorenal) syndrome. *Pediatric Radiology,* **22**, 76–77.

Donnelly, P. V., Reed, P., and DiFerrante, N. (1984). Synthesis and sulfation of glycosaminoglycans in fibroblasts from a patient with Lowe's syndrome. *Connective Tissue Research,* **13**, 89–98.

Endres, W., Schaub, J., Stefani, F. H., Wirtz, A., and Zahn, V. (1977). Cataract in a fetus at risk for Oculocerebrorenal Syndrome. *Klinische Wochenschrift,* **55**, 141–144.

Epstein, S. (1989). Diabetes Mellitus and abnormalities of bone and collagen metabolism. In *Complications of Diabetes Mellitus* (pp. 115–124). Alan R Liss, Inc.

Fukui, S., Yoshida, H., Tanaka, T., Sakano, T., Usui, T., and Yamashina, I. (1981). Glycosaminoglycan synthesis by cultured skin fibroblasts from a patient with Lowe's syndrome. *Journal of Biological Chemistry,* **256(20)**, 10313–10318.

Funakoshi, I., Kato, H., Horie, K., Yano, T., Hori, Y., Kobayashi, H., Inoie, T., Suzuki, H., Fukui, S., Tsukahara, M., Kajii, T., and Yamashina, I. (1992). Molecular cloning of cDNAs for human fibroblast nucleotide pyrophosphatase. *Archives of Biochemistry and Biophysics,* **295**, 180–187.

Gaary, E. A., Rawnsley, E., Marin-Padilla, J. M., Morse, C. L., and Crow, H. C. (1993). *In utero* detection of fetal cataracts. *Journal of Ultrasound Medicine,* **12**, 234–236.

Gardner, R. J. M. and Brown, N. (1976). Lowe's syndrome: identification of carriers by lens examination. *Journal of Medical Genetics,* **13**, 449–454.

Garzuly, F., Jellinger, K., Szabo, L., and Toth, K. (1973). Morbid changes in oculocerebrorenal syndrome. *Neuropediatrie,* **4**, 304–313.

Ginsberg, J., Bove, K. E., and Fogelson, M. H. (1981). Pathological features of the eye in the oculocerebrorenal (Lowe) syndrome. *Journal of Pediatric Ophthalmology and Strabismus,* **18**, 16–24.

Gobernado, J. M., Lousa, M., Gimeno, A., and Gonsalvez, M. (1984). Mitochondrial defects in Lowe's oculocerebrorenal syndrome. *Archives of Neurology*, **41**, 208–209.

Greene, D. A., Lattimer, S. A., and Sima, A. A. F. (1987). Sorbitol, phosphoinositides, and sodium-potassium-ATPase in the pathogenesis of diabetic complications. *New England Journal of Medicine*, **316**, 599–606.

Harper, G. S., Hascall, V. C., Yanagashita, M., and Gahl, W. A. (1987). Proteoglycan Synthesis in Normal and Lowe Syndrome Fibroblasts. *Journal of Biological Chemistry*, **262**, 5637–5643.

Harris, L. S., Gitter, K. A., Galin, M. A., et al., (1970). Oculocerebrorenal syndrome: report of a case in a baby girl. *British Journal of Ophthalmology*, **54**, 78–80.

Hayashi, S., Nagata, T., Kimura, A., et al., (1978a). Urinary excretion of acid glycosaminoglycans and hydroxyproline in a patient with oculocerebrorenal syndrome. *Tohoku Journal of Experimental Medicine*, **126**, 215–223.

Hayashi, S., Nagata, T., Kimura, A., and Tsurumi, K. (1978b). Urinary acid glycosaminoglycans in a patient with oculo-cerebro-renal syndrome. *Tohoku Journal of Experimental Medicine*, **126**, 225–234.

Hittner, H. M., Carroll, A. J., and Prchal, J. T. (1982). Linkage studies in carriers of Lowe oculocerebrorenal syndrome. *American Journal of Human Genetics*, **34**, 966–971.

Hodgson, S. V., Heckmatt, J. Z., Hughes, E., Crolla, J. A., Dubowitz, V., and Bobrow, M. (1986). A balanced de novo X/autosome translocation in a girl with manifestation of Lowe syndrome. *American Journal of Medical Genetics*, **23**, 837–847.

Horie, K., Yano, T., Funakoshi, I., and Yamashina, I. (1988). Elevated nucleotide pyrophosphatase activity in cultured skin fibroblasts from patients with Lowe's syndrome. *Clinica Chimica Acta*, **177**, 41–48.

Irvine, R. (1992). Seeking the Lowe life. *Nature Genetics*, **1**, 315–316.

Johnson, B. L. and Hiles, D. A. (1977). Ocular pathology of Lowe's syndrome in a female infant. *Journal of Pediatric Ophthalmology*, **13**, 204–210.

Kenworthy, L., Park, T., and Charnas, L. R. (1993). Cognitive and behavioral profile of the oculocerebrorenal syndrome of Lowe. *American Journal of Medical Genetics*, **46**, 297–303.

Kieras, F. J., Houck, G. E., French, J. H., et al., (1984). Low sulfated glycosaminoglycans are excreted in patients with Lowe syndrome. *Biochemical Medicine*, **31**, 201–210.

Kornfeld, M., Synder, R.D., MacGee, J., and Appenzeller, O. (1975). The oculo-cerebral-renal syndrome of Lowe: neuromuscular components. *Archives of Neurology*, **32**, 103–107.

Leahey, A. M., Charnas, L. R., and Nussbaum, R. L. (1993). Nonsense mutations in the OCRL-1 gene in patients with the Oculocerebrorenal Syndrome of Lowe. *Human Molecular Genetics*, **4**, 461–463.

Lowe, C. U., Terrey, M., and MacLachan, E. A. (1952). Organic aciduria, decreased renal ammonia production, hydrophthalmos, and mental retardation: a clinical entity. *American Journal of Diseases of Children*, **83**, 164–184.

Lythgoe, C. and Ramsey, M.S. (1973). A possible case of oculocerebrorenal (Lowe) syndrome in a female infant. *Canadian Journal of Ophthalmology*, **8**, 591–593.

Manz, F., Bremer, H. J., and Brodehl, J. (1978). Renal transport of amino acids in children with oculocerebrorenal syndrome. *Helvetica Paediatrica Acta*, **33**, 37–44.

Matin, M. A. and Sylvester, P. E. (1980). Clinicopathologic studies of oculocerebrorenal syndrome of Lowe, Terrey and MacLachlan. *Journal of Mental Deficiency Research*, **24**, 1–8.

McSpadden, K. (1991). *Report of the Lowe's Syndrome Comprehensive Survey*. West Lafayette, IN: Lowe's Syndrome Association.

Moraes, C. T., Zeviani, M., Schon, E. A., Hickman, R. O., Vlcek, B. W., and DiMauro, S. (1991). Mitochondrial DNA deletion in a girl with manifestations of Kearns-Sayre and Lowe syndromes: and example of phenotypic mimicry? *American Journal of Medical Genetics*, **41**, 301–305.

Mueller, O. T., Hartsfield, Jr.,J. K., Gallardo, L. A., Essig, Y.-P., Miller, K. L., Papenhausen, P. R., and Tedesco, T. A. (1991). Lowe oculocerebrorenal syndrome in a female with a balanced X;20 translocation: mapping of the X chromosome breakpoint. *American Journal of Human Genetics*, **49**, 804–810.

Nelson, D. L., Ballabio, A., Victoria, M. F., Pieretti, M., Bies, R. D., Gibbs, R. A., Maley, J. A., Chinault, A. C., Webster, T. D., and Caskey, C. T. (1991). *Alu*-primed polymerase chain reaction for regional assignment of 110 yeast artificial chromosome clones from the human X chromosome: identification of clones associated with a disease locus. *Proceedings of the National Academy of Sciences of the United States of America*, **88**, 6157–6161.

Okabe, I., Attree, O., Bailey, L. C., Nelson, D. L., and Nussbaum, R. L. (1992). Isolation of cDNA sequences around the chromosomal breakpoint in a female with Lowe syndrome by direct screening of cDNA libraries with yeast artificial chromosomes. *Journal of Inherited Metabolic Diseases*, **15**, 526–531

Ores, R. O. (1970). Renal changes in oculocerebrorenal syndrome of Lowe. *Archives of Pathology*, **89**, 221–225.

Palmieri, M. J., O'Hara, J., States, B., and Segal, S. (1985). Decreased procollagen production in cultured fibroblasts from patients with Lowe's syndrome. *Journal of Inherited Metabolic Disease*, **8**, 187–192.

Payton, J. B., Steele, M. W., Wenger, S. L., and Minshew, N. J. (1989). The fragile-X marker and autism in perspective. *Journal of the American Academy of Child and Adolescent Psychiatry*, **28**, 417–421.

Redfield, V. A., Mimouni, F., Strife, F. C., and Tsang, R. C. (1991). Severe rickets in Lowe syndrome: treatment with continuous nasogastric infusion. *Pediatric Nephrology*, **5**, 696-699.

Reilly, D. S., Lewis, R. A., Ledbetter, D. H., and Nussbaum, R. L. (1988). Tightly linked flanking markers for the Lowe oculocerebrorenal syndrome with application to carrier assessment. *American Journal of Human Genetics*, **42**, 748–755.

Reilly, D. S., Lewis, R. A., and Nussbaum, R. L. (1990). Genetic and physical mapping of Xq24–q26 markers flanking the Lowe oculocerebrorenal syndrome. *Genomics*, **8**, 62–70.

Richards, W., Donnell, G. N., Wilson, W. A., Stowens, D., and Perry, T. (1965). The oculocerebrorenal syndrome of Lowe. *American Journal of Diseases of Children*, **109**, 185–203.

Ross, T. S., Jefferson, A. B., Mitchell, C. A., and Majerus, P. W. (1991). Cloning and expression of human 75-kDa inositol polyphosphate-5-phosphatase. *Journal of Biological Chemistry*, **266**, 20283–20289.

Sagel, I., Ores, R.O., and Yuceoglu, A. M. (1970). Renal function and morphology in a girl with oculocerebrorenal syndrome. *Journal of Pediatrics*, **77**, 124–126.

Schaper, G. and Horstmann, W. (1963). Lowe syndrom mit hydrocephalus, nephrocalcinose, und nephrolithiasis. *Monatsschr Kinderheilkd*, **111**, 17–25.

Schoen, E. J. and Young, G. (1959). "Lowe's Syndrome": abnormalities in renal tubular function in combination with other congenital defects. *American Journal of Medicine*, **27**, 781–792.

Silver, D. N., Lewis, R. A., and Nussbaum, R. L. (1987). Mapping the Lowe oculocerebrorenal sydrome to Xq24-q26 by use of restriction fragment length polymorphisms. *Journal of Clinical Investigation*, **79**, 282–285.

States, B., Palmieri, M. J., and Segal, S. (1982). Uptake of proline in cultured cells from patients with Lowe's syndrome. *Biochemical and Biophysical Research Communications*, **109**, 428–433.

Suschke, J. and Murken, J. D. (1969). Uber die Ausscheidung saurer mucopolysaccharide bei einem Kind mit Lowe-syndrom. *Monatsschrift Kinderheilkunde*, **117**, 412–414.

Svorc, J., Masopust, J., Komarkova, A., Macek, M., and Hyanek, J. (1967). Oculocerebrorenal syndrome in a female child. *American Journal of Diseases of Children*, **114**, 186–190.

Tongue, A. C. (1972). Lowe's syndrome: with particular reference to the carrier state. *Transactions of the Pacific Coast Ophthalmological Society*, **53**, 219–227.

Tripathi, R. C., Cibis, G. W., and Tripathi, B. J. (1980). Lowe's syndrome. *Transactions of the Ophthalmological Society UK*, **100**, 132–139.

Tripathi, R. C., Cibis, G. W., and Tripathi, B. J. (1986). Pathogenesis of cataracts in patients with Lowe's syndrome. *Ophthalmology*, **93**, 1046–1051.

Van Acker, K. J., Roels, H., Beelaerts, W., Pasternack, A., and Valcke, R. (1967). The histologic lesions of the kidney in the oculo-cerebro-renal syndrome of Lowe. *Nephron*, **4**, 193–214.

Wadelius, C., Fagerholm, P., Pettersson, U., and Anneren, G. (1989). Lowe oculocerebrorenal syndrome: DNA-based linkage of the gene to Xq4-q26, using tightly linked flanking markers and the correlation to lens examination in carrier diagnosis. *American Journal of Human Genetics*, **44**, 241–247.

Wisniewski, K., Kieras, F. J., French, J. H., Houck, G. E., and Ramos, P. L. (1984). Ultrastructural, neurological, and glycosaminoglycan abnormalities in Lowe's syndrome. *Annals of Neurology*, **16**, 40–49.

Witzleben, C. L., Schoen, E. J., Tu, W. H., and McDonald, L. W. (1968). Progressive morphologic renal changes in the oculo-cerebro-renal syndrome of Lowe. *American Journal of Medicine*, **44**, 319–324.

Yamashina, I., Yoshida, H., Fukui, S., and Funakoshi, I. (1983). Biochemical studies on Lowe's syndrome. *Molecular and Cellular Biochemistry*, **52**, 107–124.

Yokoi, T., Taniguchi, N., and Ikawa, K. (1982). Impaired synthesis of intracellular heparan sulfate in skin fibroblasts of Lowe's syndrome. *Journal of Laboratory Clinical Medicine*, **100(3)**, 461–468.

Yokoi, T., Uozaki, T., Kasei, M., Sato, T., and Taniguchi, N. (1981). Low urinary excretion of heparan sulfate in three patients with Lowe's syndrome. *Clinica Chimica Acta*, **116**, 153–160.

Yoshida, H., Fukui, S., Yamashina, I., Tanaka, T., Sakano, T., Usui, T., Shimotsuji, T., Yabuuchi, H., Owada, M., and Kitagawa, T. (1982). Elevation of nucleotide pyrophosphatase activity in skin fibroblasts from patients with Lowe's syndrome. *Biochemical and Biophysical Research Communications*, **107**, 1144–1150.

Zimmerman, L. E. and Font, R. L. (1966). Congenital malformations of the eye. *Journal of the American Medical Association*, **196**, 96–104.

18. CRYSTALLIN GENES AND CATARACT

MIREILLE CARTIER[1], LAP-CHEE TSUI[1,2], SARAH P. BALL[3] and NICOLETTE H. LUBSEN[4]

[1]*Department of Genetics, Research Institute, The Hospital for Sick Children, 555 University Avenue, Toronto, Ontario M5G 1X8, Canada*
[2]*Department of Molecular and Medical Genetics, University of Toronto, Toronto, Ontario M5S 1A8, Canada*
[3]*Department of Biological Sciences, Washington Singer Laboratories, University of Exeter, Exeter EX4 4QG, United Kingdom,*
[4]*Department of Molecular Biology, University of Nijmegen, Toernooiveld, 6525 ED Nijmegen, Netherlands*

ABSTRACT

Crystallins constitute over 90% of the water-soluble proteins in the ocular lens. They are subdivided into different classes according to their molecular properties. Early biophysical studies indicated that crystallins are important for conferring, as well as maintaining the transparency of the lens. More recent investigations have focused on the isolation and characterization of the genes encoding these proteins; these studies have resulted in the identification of mutations associated with a number of cataracts in mammals, thus strengthening the involvement of the crystallins in lens transparency. Moreover, the phenotype of at least one of these mutants suggests that crystallins are also important for proper lens development.

INTRODUCTION

Crystallins are the major soluble proteins in the eye lenses of vertebrate animals. Most of the crystallins are contained in the fibre cells which occupy the bulk of this multicellular organ (Figure 1; for a review of embryology and morphology of the lens, see Zigman, 1985). The unique spatial arrangement and short range order of these molecules are thought to be important for the maintenance of the remarkable transparency and refractive properties of the lens (Delaye and Tardieu, 1983).

The functional importance of crystallins is also reflected by their evolutionary conservation and their role in disease conditions. Loss of lenticular transparency, commonly known as cataract, can be due to a variety of reasons, such as aging, trauma, or metabolic imbalance, but at least two congenital forms have already been shown to be due to mutations in crystallin genes (see below). Since cataracts

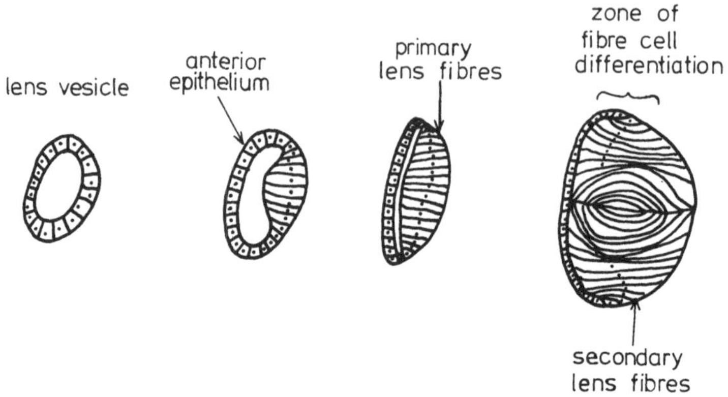

Figure 1 Lenticular development and growth. Development of the lens is initiated from a lens vesicle where the cells from the anterior layer will form the cuboidal epithelial cells and the cells from the posterior region will elongate to form the primary lens fibres. Growth of the lens will continue throughout life, only from the equatorial epithelial cells which will slowly migrate towards the posterior area of the lens and differentiate into fibre cells. These new fibre cells, called the secondary fibre cells, will then push the primary fibre cells to the center of the lens, where they will form the nucleus of the lens.

range from slight, dust-like opacities which do not disturb the vision to complete opacification of the lens, sometimes accompanied by loss of the normal anatomy of the organ, a careful analysis of these inherited cataracts should help in understanding the various roles of crystallins in cataractogenesis.

Different classes of crystallins have been described for a number of vertebrates and, more recently, for some invertebrate species. Most of the initial knowledge about crystallins was derived from the calf, simply because of their accessibility. Three major classes of crystallins, α, β and γ, were originally described. They are distinguished on the basis of their chromatographic properties, with the high molecular weight fraction being the α-crystallins, the intermediate molecular weight fraction, the β-crystallins and the low molecular weight fraction, the γ-crystallins. Although the proteins contained in each of the three fractions are rather heterogeneous, they are antigenically closely related. As we now know, the complexity of the crystallin proteins is not only due to the variety of genes that encode these polypeptides, but is also caused by post-translational modifications of the proteins themselves (for a review of the crystallins, see Harding and Dilley, 1976; Bloemendal, 1981, 1985; Piatigorsky, 1984; Lindley et al., 1985; Wistow and Piatigorsky, 1988; Bloemendal and de Jong, 1991).

While α-, β- and γ-crystallins can be found in almost all vertebrate lenses, unique "taxon-specific" crystallins have been identified for many species (for review, see de Jong et al., 1989; Piatigorsky and Wistow, 1989, 1991). For example, lenses from birds and some reptiles contain a different class of proteins called the δ-crystallins. In humans, there is no evidence for crystallins other than the α-, β- and γ-crystallins.

The various crystallin species are not uniformly distributed within the lens; a specific concentration gradient for each protein is established through temporal and spatial regulation of gene expression as well as post-translational modifications. There is, however, a general increase of protein concentration towards the center (nucleus) of the lens, primarily due to the increasing concentrations of β- and γ-crystallins. As a consequence, the refractive index increases towards the core of the lens, compensating for the changing curvature of the lens (Fernald and Wright, 1983). In addition, since the terminally differentiated fibre cells are never lost from the lens, but just migrate towards the lens center while gradually losing their nucleus and subcellular organelles, older proteins are located at the nucleus of the lens and newly synthesized ones are located in the outer cortex region. The older crystallins in the lens nucleus cannot, therefore, be replaced and must last the lifetime of the organism. Hence, crystallins are very stable proteins, able to resist denaturation by oxidation, UV radiation (sunlight) as well as racemization, all processes that occur with age.

In the following sections, we provide a discussion on each of the three major classes of crystallins found in the human lenses and our current knowledge on the involvement of these proteins in cataractogenesis.

GENERAL PROPERTIES OF THE CRYSTALLINS AND THEIR GENE FAMILIES

α-Crystallins

The α-crystallins are polypeptides of about 20 kDa found as large aggregates with a molecular weight around 800 kDa. The α-crystallins are encoded by two genes, the αA and the αB gene, which, respectively, specify proteins with an acidic (αA2) and basic (αB2) isoelectric point. These two primary gene products share about 60% amino acid identity. Both polypeptide chains can be phosphorylated, giving rise to two other species, αA1 and αB1, respectively (Spector et al., 1985). Further modifications of these proteins may also occur through deamidation (Voorter et al., 1987), N-terminal acetylation (Hoenders and Bloemendal, 1967), O-linked N-acetylglucosamine (Roquemore et al., 1992), C-terminal degradation (Stauffer et al., 1974; van Kleef, Nijzink-Maas and Hoenders, 1974), and racemization (Groenen et al., 1990), leading to a variety of α-crystallin aggregates in the aging lens.

In spite of many years of study, the tertiary structure of α-crystallin remains unknown. On the basis of internal sequence similarities, as well as the exon/intron distribution of the gene, it has been suggested that α-crystallin is a two domain protein (Wistow, 1985). The nature of the α-crystallin aggregates found in the lens is also unclear, with some suggesting that it is an ordered layered oligomer

(Tardieu et al., 1986), while others propose that the protein has a micellar structure (Walsh, Sen and Chakrabarti, 1991; Radlick and Koretz, 1992).

The α-crystallins may also have other functions than that of being an abundant structural protein of the lens. They show sequence homology to the small heat shock proteins (Ingolia and Craig, 1982). In fact, αB-crystallin has been found in other tissues than the lens and its synthesis can be induced by stress (such as heat shock and incubation in Cd^{+2}) in cultured fibroblasts (Bhat and Nagineni, 1989; Dubin, Wawrousek and Piatigorsky, 1989; Klemenz et al., 1991). This protein also accumulates in the brains of humans suffering from Alexander's disease (a degenerative neurological disorder characterized by the presence of Rosenthal fibres or granular inclusions within astrocytes) and in the brains of scrapie-infected hamsters (Duguid, Rohwer and Seed, 1988; Iwaki et al., 1989). More recently, α-crystallins have been found to have chaperone activity (Horwitz, 1992), raising the possibility that they may stabilize other crystallins in the lens.

Genomic organization

As best demonstrated in the hamster, the two α-crystallin genes have the same genomic structure; both have three exons and the introns are located at equivalent positions in the coding sequence, suggesting a close evolutionary relationship between these two genes (Quax-Jeuken et al., 1985b; van den Heuvel et al., 1985). In humans, a partial sequence has been obtained for the αA gene (McDevitt et al., 1986) and a complete sequence for αB (Dubin et al., 1990). Their structure is similar to that of the hamster genes. In addition, the human A-crystallin gene is localized to chromosome 21 region q22.3 (Table 1 and Figure 3; Quax-Jeuken et al., 1985c; Hawkins et al., 1987) and αB-crystallin to chromosome 11, between bands q22.3 and q23.3 (Table 1 and Figure 3; Brakenhoff et al., 1990b).

Expression pattern

During development of the lens, at least in rodents, α-crystallins are the first crystallins to appear. They are present in the lens vesicle before formation of the primary lens fibre cells (McAvoy, 1978; van Leen et al., 1987a, b). The synthesis of α-crystallins commences mainly in the epithelial cell layer and early stages of fibre cell elongation. In the rat, the levels of both αA- and αB-crystallin mRNAs are relatively constant at all ages of growth and development (Aarts, Lubsen and Schoenmakers, 1989; Voorter et al., 1990), suggesting an even distribution of α-crystallins in all cells. Relatively little is known about the developmental regulation of α-crystallin gene expression in humans, except that αA-crystallin mRNA can be detected in the fetal lens (Brakenhoff et al., 1990a).

Table 1 Location of the human crystallin genes

Crystallin gene	*Gene symbol* [a]	*Chromosomal location*	*References*
αA	CRYA1	21q22.3	Quax-Jeuken et al., 1985c Hawkins et al., 1987
αB	CRYA2	11q22.3-23.1	Ngo et al., 1989 Brakenhoff et al., 1990b
βA1/A3	CRYB1[a]	17q11.2-q12	Law et al., 1986 Sparkes et al., 1986
βA2	unassigned	unmapped	
βA4	CRYBA4[a]	22q11.2-13.1	van Rens, Geurts van Kessel and Bloemendal, 1991
βB1	unassigned	probably 22	N.H.L., unpubl.
βB2	CRYB2A[a]	22q11.2-q12	Sparkes et al., 1987 Hogg et al., 1987 Hulsebos et al., 1991
βB3	CRYB3[a]	22q11.2-q12	Hulsebos et al., 1991
γA-F	CRYG1-6	2q33-q35	Tsui et al., 1985 Willard et al., 1985 den Dunnen et al., 1986 Shiloh et al., 1986
γs	CRYG8	3	van Rens et al., 1989

[a] The symbols for the β-crystallins may be changed in the near future to better reflect the name of the protein: βA1/A3 would be CRYBA3/A1, βB2 would be CRYBB2 (and the βB2 pseudogene: CRYBB2P1) and βB3 would be CRYBB3.

β-Crystallins

The β-crystallins appear in the lens as heterogeneous aggregates ranging from dimers (β_L) to octamers (β_H). The calf lens contains seven primary β-crystallin translation products which can be divided into three basic proteins, known as βB1, βB2 and βB3, and four acidic chains, βA1, βA2, βA3 and βA4 (Berbers et al., 1982). The same number of species probably exist in the human lens on the basis of a similar two-dimensional gel separation pattern of embryonic human lens proteins, but an unequivocal assignment cannot be made (W.W. de Jong, personal communication).

Part of the difficulty in the identification of the different β-crystallin species is due to post-translational modifications of these proteins. A portion of the N-terminal arm from both βB1 and βB2 can be removed by proteolytic cleavage (Berbers et al., 1983; McFall-Ngai et al., 1986; Takemoto et al., 1987). The glutamine residues in the N-terminal arms of βA3, βB2 and βB3 are substrates for transamination, in which the N-terminal arm of βB1 (or the C-terminus of αB-crystallin) serves to donate the ε-amino groups from its lysine residues (Lorand, Parameswaran and Velasco, 1991; Groenen, Bloemendal and de Jong, 1992). In some animal species, phosphorylated βB2 can be found (Kleiman et al., 1988). Further, oxidation, cross-linking and deamidation of β-crystallins occur with age, contributing to a complex elution profile consisting of eight (Zigler, Horwitz and Kinoshita, 1980) or more (Thomson and Augusteyn, 1985) components in isoelectric focusing, i.e. gel electrophoresis.

Protein structure

The structure of the calf βB2 dimer has been solved by X-ray crystallography (Bax et al., 1990). The core of each molecule contains four antiparallel β-sheeted (β-barrel) structures known as "Greek key" motifs, organized into two similar domains of two motifs each (Figure 2). The domains are flanked by a N- and a C-terminal arm. The two domains are separated by a connecting peptide in an extended conformation, such that the N-terminal domain of one βB2 chain interacts with the C-terminal domain of the second βB2 chain and vice versa. All known β-crystallin sequences can be modeled in the same tertiary conformation and it is generally assumed that all β-crystallins fold into the same four motif/two domain structure. The most distinct feature for individual β-crystallin molecules is located at their N- and C-terminal arms. In addition to sequence variations, the N-terminal arms differ significantly in length (from about 55 amino acid residues in βB1 to 15 amino acid residues in βA1). The C-terminal arms, varying between 13 and 18 amino acid residues in length, are only found in the basic β-crystallins (for review, see Lubsen, Aarts and Schoenmakers, 1988).

Genomic organization

The complete set of calf β-crystallin sequences has been derived from cDNA clones (Gorin and Horwitz, 1984; Quax-Jeuken et al., 1984; Hogg et al., 1987; van Rens et al., 1991a). βA1 and βA3 are encoded by the same mRNA, using the same reading frame, but with different initiation codons, resulting in a βA1 polypeptide shorter than βA3 by 17 amino acid residues (Quax-Jeuken et al., 1984). This double initiation of the βA1/A3 mRNA has been preserved in evolution and is found not only in calf, but also in chicken and in man (Hogg et al., 1986; Peterson and Piatigorsky, 1986). Further, on the basis of cDNA hybridization patterns,

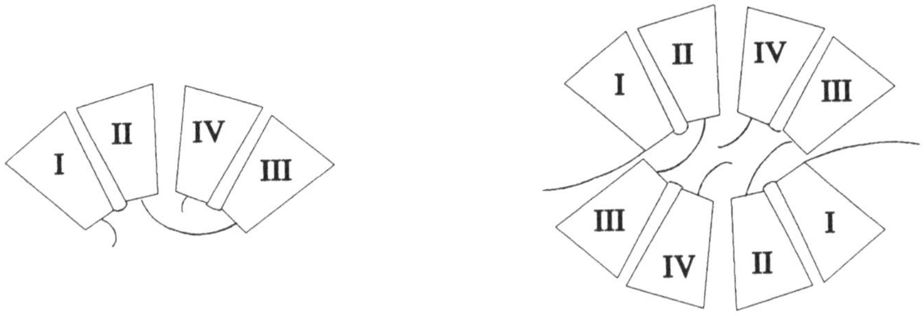

Figure 2 (A) **Figure 2** (B)

Figure 2 Interactions between the different domains of γ- and β-crystallins. (A) depicts the intramolecular interactions of domains II and IV occurring in a γ-crystallin molecule, while (B) illustrates the same domains, from two β-crystallin molecules, interacting in an intermolecular fashion. These representations are free interpretations of the data presented by Blundell et al., 1981 and Bax et al., 1990.

orthologous members of the calf β-crystallin gene family probably exist in human (van Rens et al., 1991b).

The structure of the β-crystallin genes has been derived from representative members in human, rat, and mouse (Inana et al., 1983; den Dunnen et al., 1985b; Hogg et al., 1986). Although the paralogous β-crystallin genes no longer share sufficient sequence identity to recognize each other in hybridization experiments, the genes are clearly evolutionarily related. A typical β-crystallin gene contains six exons. The first exon of the βB1 gene is non-coding, whereas the first exon of the βA3/A1 gene contains the start codon for the βA3 protein. The remainder of the gene structure is remarkably similar among all β-crystallin genes, with the second exon coding for the N-terminal arm and each of the 4 following exons (exons 3 to 6) encoding a Greek key motif. In addition, for the basic chains, exon 6 also encodes the extension of the C-terminal arm.

Because of the similarity in gene structure, it is generally believed that all β-crystallin genes were derived from a single ancestor sequence. Duplication of this ancestral gene and subsequent divergence would have led to the proto-acidic and proto-basic β-crystallin genes which would then have been duplicated further to form the 6-member gene family seen today. In man, all β-crystallin genes except the βA3/A1 and possibly the βA2 are located on chromosome 22 (Table 1 and Figure 3; Law et al., 1986; Sparkes et al., 1986; Hogg et al., 1987; Hulsebos et al., 1991). Three of the genes, βB2-1, βB2-2 and βB3, have been mapped within a distance of 120 kb (Bijlsma et al., 1991), but the βB2-2 sequence probably corresponds to a pseudogene (Brakenhoff et al., 1992). It is of interest to note that only a single copy of the βB2 gene can be found in the rat, although that gene is also closely linked to βB3 (Aarts et al., 1987). Moreover, chromosome duplication has

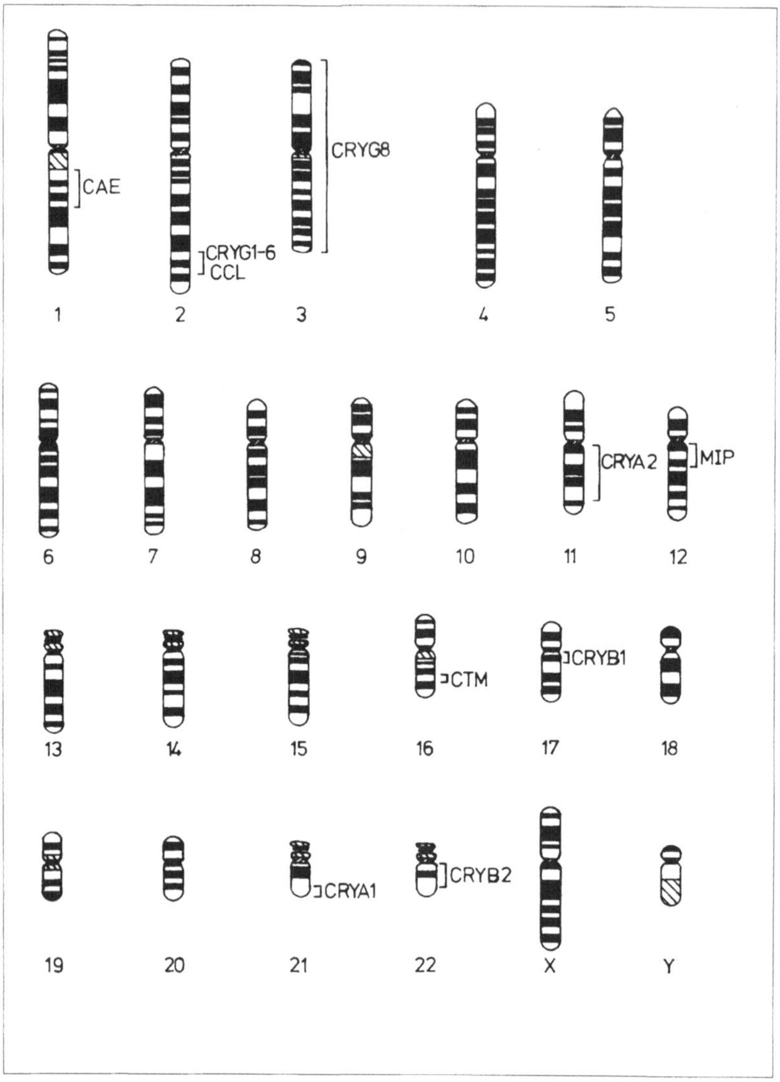

Figure 3 Ideogram of the human chromosomes showing the Giemsa banding patterns. The locations of the major genes expressed in the lens and loci responsible for dominantly inherited congenital cataracts are indicated. Standard gene nomenclature according to Human Gene Mapping 11 (McAlpine et al., 1991) is used. The gene symbols used are as listed in Tables 1 and 2, except for CRYB2 which, here, represents the whole cluster of β-crystallin genes mapped to chromosome 22: βA2, βB2, βB3 and βB4 and MIP, which is the symbol for the major intrinsic protein of the lens fibre membrane.

been proposed as a mechanism for β-crystallin gene duplication; the close proximity of the βA1/A3 gene to the von Recklinghausen neurofibromatosis locus (NF1) on chromosome 17 (Fain et al., 1989) and the co-existence of the βB2 and βB3 genes and neurofibromatosis 2 (NF2) on chromosome 22 (Hulsebos et al., 1991) are thought to be more than coincidental.

Developmental regulation of gene expression

The developmental regulation of the β-crystallin genes has been well documented for the chicken (Ostrer, Beebe and Piatigorsky, 1981; Hejtmancik et al., 1985) and rat (Aarts, Lubsen and Schoenmakers, 1989). The latter study shows that the βB1 and βB3 genes are both active during early lens formation and that their respective transcripts reach their maximum levels around birth. In contrast, the βB2 mRNA only accumulates after birth but remains prominent at least up to one year of age. The βA3/A1 gene shows an intermediate pattern of expression, as its transcript is most abundant between one and three months after birth. Relatively little is known about the developmental profile for the human β-crystallin genes. The pattern may be similar to that of the rat, however, because human βB3 is also expressed early and βB2 is present in later stages (Brakenhoff et al., 1992). Based on the mRNA profile, it may be assumed that there is a differential gradient of β-crystallin species across the lens. For example, the nucleus is expected to contain relatively high levels of βB1 and βB3 whereas the cortex would include primarily βB2. Consistent with this assumption and the fact that the formation of β_H requires the presence of βB1, there is a clear shift of β_H in the rat lens nucleus to β_L in the cortex (Siezen, Anello and Thomson, 1986).

γ-Crystallins

The γ-crystallins are found in the low molecular weight fraction of the soluble proteins from the lens. The 6–7 monomeric proteins of about 20 kDa included in this family can be divided into two branches. The older branch comprises a single protein called γs, which is present in all vertebrate species examined thus far. This protein was once thought to be a monomeric β-crystallin (thus previously named βs), mainly because its N-terminal residue is acetylated, as is the case for the β-crystallins, but not for the other γ-crystallins. Cloning and sequencing studies, however, revealed that the coding sequence and exon/intron organization of the γs gene are more related to those of the γ genes than those of the β genes (Quax-Jeuken et al., 1985a; van Rens et al., 1989). The other branch of the γ-crystallin gene family contains up to six closely related proteins (whose genes are clustered together), which are known as γA to γF. In contrast to the α–, β- and γs-crystallins, these γA to γF-crystallins are not present in all vertebrate species; while these low molecular weight proteins can be found at varying levels in fish, frogs and mammals, they are absent in birds and some reptiles.

Protein structure

The γ-crystallins also contain four highly symmetrical Greek key motifs. The overall structure is similar to that of the β-crystallin, as might be expected from their homology at the primary sequence level. In fact, the three-dimensional structure of the bovine γ-crystallin was the first crystallin structure solved by X-ray crystallography (Blundell et al., 1981). The major difference between the β- and γ-crystallin structures is that the β-crystallin domain interactions are intermolecular, while in the γ-crystallins the domain interactions are intramolecular (Figure 2). As a consequence, γ-crystallins are monomers, while the β-crystallins dimerize. A further difference between β- and γ-crystallins is the lack of N- and C-terminal arms in γ-crystallins. The high internal symmetry and compact shape of the γ-crystallins, together with the placement of surface ion pairs, contribute to the stability of these molecules, which can withstand dehydrated environments. The γ-crystallins are, therefore, better suited for the less hydrated, hard lenses of rodents than the softer, more hydrated lenses of humans and birds.

Genomic organization

The γ-crystallin genes are highly conserved in structure and sequence; each contains 3 exons, with the first exon encoding 3 amino acid residues and the two remaining exons each specifying a domain of two Greek key motifs (Moormann et al., 1983; Lok et al., 1984). At least six γ-crystallin genes have been described for each mammalian species examined. In rat and mouse, five genes (γA, γB, γC, γD and γE) are tandemly linked within a 50 kb DNA segment, while the sixth gene (γF) is located at least 25 kb upstream (Moormann et al., 1985; Quinlan et al., 1987; den Dunnen et al., 1989). The various gene members in this family share between 75 and 90% sequence identity within the coding region, resulting in easy detection of all genes by cross-hybridization.

The human genome also contains six highly related genes (γA-γF), clustered on the long arm of chromosome 2, region q33-q35 (Table 1 and Figure 3; den Dunnen et al. 1985a; Meakin, Breitman and Tsui, 1985). The less related γs gene is localized on chromosome 3 (Table 1 and Figure 3). The relative positions of the γB, γC, γD and γE genes have been established on a 40 kb DNA segment but the exact locations for γA and γF in the gene cluster have yet to be determined. It is of interest to note that the human γE and γF genes both contain an identical in-frame stop codon at the same position within the second exon (Meakin, Breitman and Tsui, 1985). In addition, the γF gene lacks an upstream TATA box which is thought to be important for transcription; the gene also has numerous other mutations downstream from the premature stop codon. Although these two genes are considered as pseudogenes, a low level of γE transcript has been detected (see below). A segment of the γF gene is also found to be duplicated within the gene cluster; this sequence (γG) is non-functional (Brakenhoff et al., 1990a).

Evolution of the γ-crystallin genes

The γs gene is present in virtually all vertebrate species examined. It is probably a direct descendant of a gene recruited by the lens in early vertebrate evolution (Chang and Chang, 1987; van Rens, de Jong and Bloemendal, 1991). The other γ-crystallin genes are present in mammals, fishes and frogs, but absent in birds and higher reptiles. Sequence analysis shows, however, that the mammalian γ-crystallin genes are more related to each other than to any of the genes in fishes and frogs, suggesting that the γ-crystallin gene family evolved separately in the different evolutionary lineages. The γA-γF genes are thus specific to mammals.

As the burst of gene duplications that produced the mammalian γA-γF-crystallin genes occurred recently, these genes are still closely related and clustered together. From sequence comparisons, it has further been speculated that the six γ–crystallin genes (γA-γF) were already present in the common ancestor of man and rodents, but that, in the primate lineage, further reshuffling of this gene family has taken place. Hence, the human γF gene could be a recent duplicate of the γE gene rather than being the homolog (ortholog) of the rodent γF gene, which itself became partially deleted to form γG in the human genome (Brakenhoff et al., 1990a). It is possible that gene conversion also plays a role in the evolution of these γ-crystallin sequences (den Dunnen, Lubsen and Schoenmakers, 1986; Aarts et al., 1988).

Loss of functional gene sequences as well as decreased levels of expression (see below) have occurred after the divergence of primates from rodents. These modifications may be related to the fact that during hominoid evolution a softer lens with accommodative capacity has apparently been favoured, and a soft lens is probably incompatible with a high level of γ-crystallins (the same argument can be put forward to explain the lack of γ-crystallins in the soft and flat bird lenses). Loss of expression of γ-crystallins would, thus, have been an acceptable mutation in the primate lineage.

Based on the structures of β- and γ-crystallins (both have four similar motifs organized into two domains) it may be argued that these two gene families are evolutionarily related. Close examination of the gene structure and DNA sequences indicates that the two families are probably related only through a common ancestral sequence encoding a single "Greek key" motif. It has been proposed that separation of the β- and γ-crystallin families occurred after the formation of a domain structure of the protein and before the formation of the two-domain structure found today (for review, see Lubsen, Aarts and Schoenmakers, 1988). In this assumption, the two motifs in the ancestral domain sequence were separated by an intron which was retained in the β-crystallin genes, but lost before duplication of the domain to form the γ-crystallin gene. It is also possible, however, that the β- and γ-crystallins actually evolved independently from the single ancestral motif sequence.

Regulation of gene expression

All γ-crystallins are specific to the lens; they are located primarily in the nucleus region, presumably to provide the dehydrated environment required. In rodents, the precise spatial and temporal regulation of individual γ-crystallin genes has been well documented and, as expected, most of the genes are expressed at early stages of development (Murer-Orlando et al., 1987; van Leen et al., 1987ab; Goring, Breitman and Tsui, 1992). Transcription of the γA to γF genes is restricted to the terminally differentiated fibre cells and can be readily detected in day 12 embryonic lenses when the primary fibre cells begin to form. The steady state level of mRNA for most γ-crystallin mRNAs increases gradually and reaches a peak at around day 10 after birth. In addition, the γD, γE and γF are relatively more abundant than the other species, whereas the γB gene remains active long after the others have been shut down. A similar expression pattern is found in the calf, where γB is the prominent cortical γ-crystallin (Slingsby and Croft, 1973; Slingsby and Miller, 1983).

There are three to four times less γ-crystallins in the human lens than in the rodent lens (Siezen et al., 1988; Thomson and Augusteyn, 1985). In addition, although the human γA, γB, γC and γD genes are transcriptionally active, only the latter two yield abundant mRNA and protein (Russell et al., 1987; Siezen et al., 1987; Brakenhoff et al., 1990a). The γC and γD genes are relatively active in the fetal lenses but are differentially shut-off after birth. At 22 months, only the γD mRNA can be detected and it remains present till at least 10 years of age (Brakenhoff et al., 1990a). The γE gene is transcribed at a low level; nevertheless, a small amount of its predicted (truncated) protein product has been detected in the human lens (N.H.L., unpubl.). The γF gene lacks a TATA box and is probably silent (Meakin, Breitman and Tsui, 1985).

The exceptional position of the γs gene in the γ-crystallin gene family is also reflected by its expression pattern: in the rat lens, the γs gene is active only after birth (Aarts, Lubsen and Schoenmakers, 1989). However, although the developmental expression of γs has not been followed in man, it is known that the transcript of this gene is present in a 9 week old human fetal lens (N.H.L., unpubl.).

CATARACTS

Since lens opacity can occur for many reasons, it is impossible to define a common pathway of lens disorganization leading to the various forms of cataracts. Nevertheless, numerous studies have described the morphological differences between the normal and cataractous lens (see review by Berman, 1991). The general features of a cataractous lens, aside from the opacity, may include the presence of vesicles, lens fibre swelling, gaps between the fibre cells which may be filled with liquid or debris, and loss of membrane integrity.

At the biochemical level, cataractous lenses generally display high molecular weight proteinaceous aggregates as well as some proteolysis products, unbalanced concentrations of ions such as calcium, sodium and potassium and loss of glutathione and soluble crystallins. The high molecular weight protein aggregates are thought to be caused by the unfolding of the various crystallins, leading to the exposition of hydrophobic groups which may then interact together.

Some of the high molecular weight protein aggregates present in the human cataractous lens have been isolated and fractionated, to reveal peptides of molecular weights around 43 kDa, 20 kDa and 10 kDa (Spector et al., 1979; Garner, Garner and Spector, 1979). While the 43 kDa protein seems to be an extrinsic membrane protein, the 20 kDa is thought to be a crystallin and the heterogeneous 10 kDa fraction is composed of γ-crystallin (as determined by immunological reaction) (Takemoto, Straatsma and Horwitz, 1989). It has been postulated that such 10 kDa fragments may occur from cleavage at the connecting peptide between the two domains of the protein. Indeed, this 10 kDa fragment, which also accumulates with age, has been identified as the C-terminal domain of γD (Srivastava, Srivastava and Silney, 1992). Furthermore, in a study of 13 cataractous lenses, 10 were found to contain γ-crystallin proteins in close association with the membrane (all ten cataracts presenting nuclear sclerosis), while 4 displayed β-crystallins associated with the membrane (two of which had no nuclear involvement) and none showed α-crystallin association with the membrane (Kodama and Takemoto, 1988). From this study, it may be concluded that γ-crystallins can play a role in the cataractogenic process, especially in cases of nuclear cataracts. Such a role may be attributed to their high concentration of thiol groups on the surface of the molecule, which may be oxidized, leading to a destabilization of the protein with concomitant unfolding, and/or participate in intermolecular disulfide bonds with the membrane.

All of these changes, which have been studied mainly in senile cataracts, point to sites for potential mutations in inherited cataracts, such as the crystallin genes and the membrane proteins including the pumps and channels responsible for the osmotic balance. Furthermore, other genes known to play a role in lens development may be implicated in hereditary congenital cataracts.

Human Inherited Cataracts

Inherited cataract is an important cause of visual handicap in childhood accounting for 20% of blindness in children under 15 years, in the United Kingdom. The majority of these conditions are genetically determined. Although some of them are associated with other ocular abnormalities, or as part of a systemic syndrome, most occur as isolated abnormality. There are many different phenotypic forms of inherited cataracts and morphological variation may be observed even within families, reflecting the heterogeneity of the underlying genetic defects and the complexity of lens development (Harding and Dilley, 1976). Autosomal dominant transmission is the most common mode of inheritance, but autosomal recessive and X-linked forms are also known to occur.

Morphologies

Inherited congenital cataracts frequently exhibit a morphology described as pulverulent, ie. an accumulation of dustlike spots. In pulverulent cataracts, the opacities occur in the intermediate layer of the mature lens, in the secondary fibre cells. The inner nuclear region may be clear in zonular pulverulent cataracts or opaque in nuclear pulverulent cataracts, while the outer cortex of the lens is generally clear. As a result, many names appear in the literature, including zonular, lamellar or perinuclear cataract, central or zonular pulverulent cataract, diffuse or total nuclear cataract. In addition, various modes of inheritance have been attributed to these cataracts, suggesting that they may result from different genetic defects.

Other common forms of inherited congenital cataracts may have an anterior polar location in the lens. The opacity is often dense and conical in shape and located either overlying the anterior epithelial layer of an otherwise normal lens or within the epithelial cell layer of the cortex and may be associated with nuclear opacities. Posterior polar locations are rare.

Homogeneously opaque lenses, or total diffuse congenital cataracts, usually represent the end point in the progressive process of opacification, when it is too late to determine the pattern of cataract morphology in the immature state. These cataracts, therefore, probably include many different genetic forms. In animals, it has been shown that the size of the lens during embryonic development has a significant effect in controlling the overall size of the eye (Zwaan, 1979). Since many types of congenital cataract result in retarded growth of the lens, microphthalmia (reduced size of the eye) is inevitable.

Genetic factors

Lens opacification may result directly from aberrant functions expressed in the lens itself, such as alteration of structural proteins (e.g. crystallins) or proteins that serve to protect the lens from damage and to preserve clarity of the lens matrix (e.g. glutathione S-transferase). Alternatively the defect may lie in a metabolic pathway resulting in accumulation and deposition of insoluble material in the lens (e.g. galactokinase activity).

In general, it is anticipated that a gene defect causing a disruption of the structural proteins would have a dominant phenotype whereas an enzymatic defect might be recessive. Therefore cataracts caused by defects in the crystallins and major intrinsic proteins of the lens fibre cells may be predicted to exhibit dominant patterns of inheritance, since the presence of a mutant product is likely to result in loss of lens clarity. On the other hand, proteins involved in protecting the lens from damage are likely to remain effective even at reduced levels of activity, at least during early life, and so congenital cataracts resulting from mutations in these genes will show recessive modes of inheritance. It is possible, however, that in some patients cataracts acquired in later life may be due to a partial deficiency in activity of these protective proteins (semi-dominant).

Linkage studies

The genetic linkage approach may be used to identify the basic genetic defect underlying the various inherited lens opacifications in humans. So far, all of the gene mapping studies on inherited congenital cataract loci have considered only those exhibiting a dominant mode of inheritance. The genetic loci known to cause dominantly inherited congenital cataracts are listed in Table 2 and localized on Figure 3. The possible genetic heterogeneity and the small family size render the mapping approach much less attractive in the case of the recessive forms of cataracts.

Table 2 Positive linkage data for dominantly inherited cataracts

Description of cataract	*Locus symbol*	*Linkage*	*Lod score*	*References*
Zonular, pulverulent (Coppock)	CAE	FY (1q21-q25)	+3.78 at θ = 0.0	Renwick and Lawler, 1963
Zonular, pulverulent (Coppock)	(CAE)	1QH (1q12) FY (1q21-q25)	+2.70 at θ = 0.0 +0.05 at θ = 0.2	Conneally et al., 1978
Posterior, polar	CCDN	HP (16q22.1)	+2.11 at θ = 0.1	Richards et al., 1984
Zonular, central stellate (Marner's)	CTM	HP (16q22.1)	+8.31 at θ = 0.05	Eiberg et al., 1988
Embryonal, nuclear, pulverulent	CCL	CRYG (2q33-35)	+7.58 at θ = 0.0	Lubsen et al., 1987

In 1963, Renwick and Lawler investigated a six generation pedigree, originally described by Nettleship (1909), segregating for a congenital, zonular, pulverulent cataract, otherwise known as a Coppock cataract. Strong evidence was obtained to show close linkage between the cataract locus (CAE) and the Duffy blood group locus (FY) on the proximal long arm of chromosome 1 (lq21-q25) (Figure 3), with a maximum lod score of 3.78 at a recombination fraction of zero (male and female data combined).

Conneally et al. (1978) investigated seven families with dominant congenital bilateral cataracts for linkage with 19 marker loci. While the data were equivocal, one of the families (Family 1) with a zonular pulverulent cataract, similar to that studied by Renwick and Lawler (1963), provided a suggestive lod score (2.70 at a recombination fraction of zero) for linkage to the heterochromatic region on the proximal long arm of chromosome 1 (lQH) (Figure 3). The tests for linkage between the cataract and FY were inconclusive in the same family, however, with maximum lod score of 0.05 at a recombination fraction of 0.2. Nevertheless, it remains possible that this cataract is caused by a mutation allelic with the CAE locus. The lack of evidence for linkage with the chromosome 1 markers in the other families probably reflects genetic heterogeneity in dominantly inherited congenital cataracts.

Strong evidence for linkage was demonstrated between the locus responsible for Marner's cataract (CTM), a zonular, central stellate cataract and the haptoglobin locus (HP) on chromosome 16 (region q22.1, Figure 3) (Eiberg et al., 1988). The maximum lod score obtained was 8.31 at a recombination fraction of 0.05. This same family, originally described by Marner (1949), had already been tested for linkage with other markers, resulting in a suggestion of linkage to the P blood group locus (Mohr, 1954), but the data was questioned by Renwick and Lawler (1963) due to possible mistyping of the P blood group. A linkage has also been suggested between a posterior polar cataract locus (CCDN) and HP, with maximum lod score of 2.1 at a recombination fraction of 0.1 (Richards et al., 1984). It is possible that these two HP-linked cataracts are, although phenotypically distinct, caused by allelic mutations. The basic defects responsible for these cataracts have yet to be defined.

Lubsen et al. (1987) reported a tight linkage between the γ-crystallin gene cluster on chromosome 2 (q33-q35) (Figure 3) and the disease locus in a family with a cataract morphologically resembling the Coppock cataract studied by Renwick and Lawler (1963), except that this cataract was confined to the embryonic nucleus of the lens. The maximum lod score was 7.58 at a recombination fraction of zero, suggesting that this Coppock-like cataract (CCL) could be caused by a defect in the γ-crystallin locus. Preliminary data show that the cataract might be related to overexpression of the truncated γE polypeptide in the CCL patients (N.H.L., unpubl.).

Chromosomal abnormalities associated with congenital cataracts

There are at least two disease syndromes where cataracts appear frequently, potentially as a result of abnormal crystallin gene expression. First, about 60% of patients with Down's syndrome, which is due to trisomy 21, suffer from cataracts. It has been postulated that lens opacity in Down's syndrome may be due to the over-expression of the αA-crystallin gene located on that chromosome. Second, neurofibromatosis type II (NF 2) has been associated with posterior capsular cataracts (Kaiser–Kupfer et al., 1989). Since the NF2 locus is close to the cluster of β-crystallin genes on chromosome 22 and since neurofibromatosis occurs when one copy of this specific chromosomal region is lost, it has been suggested that the appearance of these cataracts may be caused by a deletion of one or more β-crystallin gene(s).

Other gene mapping studies

While positive linkage data have been documented in some family studies, negative findings and inconclusive results have been obtained in many other studies.

Hammerstein and Scholz (1974) found no evidence for linkage between a hereditary central cataract and 11 markers, including FY in a six generation family.

No linkage was detected between the locus responsible for an autosomal dominant nuclear cataract and 21 informative marker loci in a large family studied by Huntzinger, Weitkamp and Roca (1977).

In a study on a single large Italian family with dominantly inherited pulverulent cataracts, there was no evidence for linkage of the locus responsible for the cataracts to FY (Stabile et al., 1983).

Bateman et al. (1986) investigated a large family with dominantly inherited congenital cataracts exhibiting considerable variation between affected members of the family, but with opacification of the embryonal nucleus as a common trait to all the cataracts available for examination. Linkage was excluded conclusively with 11 genetic markers including FY and HP. A further 15 polymorphic loci yielded no evidence for linkage.

Beaumont et al. (1989) and Birdwood et al. (1992) investigated a large English family with autosomal dominant lamellar cataracts for linkage with chromosome lq21-q23 markers and with the γ-crystallin gene cluster. There was no evidence for linkage to either of these regions.

These studies emphasize the high degree of genetic heterogeneity underlying dominantly inherited cataracts. Future studies on the mapping of loci responsible for inherited cataracts should investigate the segregation of alleles of the genes identified as potential candidates for defects in lens clarity, the crystallins and major intrinsic proteins of the lens.

ANIMAL MODELS

Since lens opacity is a phenotype that is readily recognizable, it is perhaps not surprising that there are many animal models for inherited cataracts. In addition to the various cataractous mice (Nakano, Deer, CAT-Fraser, Emory), examples also exist in the rats, guinea pigs, and dogs (reviewed in Gelatt and Das, 1984; Zigler, 1990). Research with animal models offers numerous advantages including the ease with which large amounts of lens material can be obtained, the ability to carry out extensive linkage studies and the possibility to test the role of different mutations in the cataractogenic process through the production of transgenic animals.

We will limit the discussion to three animal models: the Philly mouse, the Elo mouse and the 13/N guinea pig, mainly because they are the ones for which the corresponding mutations have been defined at the molecular level. In all three cases, the mutation is found to be located in a crystallin gene.

The Philly Mouse

The Philly mouse suffers from a dominantly inherited, progressive cataract which appears approximately 15 to 30 days after birth, depending on whether the animal is homozygous or heterozygous for the mutation (Kador et al., 1980; Carper et al., 1982). The cataract first appears as a faint anterior subcapsular opacity around the lens sutures which, within 1 month, develops into a dense anterior subcapsular and nuclear cataract. In the mutant lens, the first cytologically detectable aberration is the appearance of dense bodies in the anterior portion of the newly elongating fibre cells at one week after birth (Uga, Kador and Kuwabara, 1980). This change is followed by the failure of the fibre cells to elongate as well as a general swelling of the lens cells. This swelling initially led to the conclusion that the Philly cataract was due to a defect in osmotic regulation.

Mutation in the βB2-crystallin

A marked decrease in the levels of soluble β- and γ-crystallin proteins was noted in the mutant lens whereas the levels of α-crystallins were apparently normal (Piatigorsky, Kador and Kinoshita, 1980). In addition to this general decrease of soluble crystallin levels, often observed in cataractous lenses, a specific β-crystallin related antigen appeared to be completely lost (Zigler, Carper and Kinoshita, 1981). Subsequent *in vitro* RNA translation studies showed that a 27 kDa β-crystallin protein, now known as βB2, was not produced from homozygous Philly lens RNA (Carper et al., 1982). Sequence comparison of the βB2 cDNA isolated from both normal and mutant lenses revealed that the cDNA found in the Philly lens carried a 12 nucleotide deletion at the 3' end of its coding region (Chambers and Russell, 1991). This deletion would result in a mutant βB2 protein lacking 4 amino acids at the end of the fourth Greek key motif. Consistent with the DNA finding, a protein slightly smaller than βB2 was in fact detected in the Philly lens with a polyclonal antibody directed against βB2, but not with a monoclonal antibody raised against the C-terminal portion of the normal protein (Nakamura et al., 1988). This result suggested that the mutant Philly phenotype may be caused by the presence of an altered βB2 polypeptide.

Altered Properties of the mutant βB2-crystallin

In a normal mouse, the amount of βB2 polypeptide is almost undetectable at 1–5 days after birth but increases markedly thereafter (Carper et al., 1982). This pattern of expression is consistent with the immunocytochemical localization of the βB2-crystallin to the newly elongating fibre cells in the neonates (Carper, Smith-Gill and Kinoshita, 1986). Biochemically, the mutant βB2-crystallin differs from its wild-type counterpart in two ways. First, whereas the wild type βB2 protein is very heat-stable, the mutant protein is heat-labile (Nakamura et al., 1988). Second, the mutant protein is found as high molecular weight aggregates but the wild-type protein is found as an octamer, in the β_H fraction (Russell and Chambers, 1990). Since failure of the fibre cells in the cortical layers to elongate appears to be an early sign of the defect in the Philly lens, βB2-crystallin may be important for normal secondary fibre cell elongation.

The Philly lens also differs from a normal lens in its phase-separation temperature (T_c), or the temperature at which the lens cytoplasm moves from a transparent state (single phase) to a condition of opacity (dual phases). For every lens, this T_c varies with age, since the composition of the lenticular cytoplasm is continually changing. The T_c of the homozygous Philly lens is approximately 17°C higher than that of the wild type lens at day 1 after birth (Clark and Carper, 1987). Thereafter, the T_c of the Philly lens varies in a fashion similar to that of the wild-type lens until around day 27, when the T_c of the Philly mouse suddenly soars, resulting in the appearance of a cataract. Meanwhile, the T_c of the wild-type lens continues to decline. It seems probable that the increase in T_c results from the lack of heat stability and the abnormal aggregation pattern of the mutant βB2 protein, but how this alteration interferes with fibre cell elongation remains to be elucidated.

The Elo Mouse

The Elo (Eye lens obsolescence) mouse carries a dominant and fully penetrant mutation that perturbs embryonic lens development and results in microphthalmia at birth (Oda, Watanabe and Kondo, 1980). Abnormal development of the mutant lens can be detected at day 12 *in utero:* while the posterior cells elongate and fill the lumen in a normal lens, the central posterior cells fail to elongate completely and leave a gap in the lumen of the Elo lens (Oda et al., 1980). In addition, these mutant fibre cells lack the eosinophilic granules normally present and contain numerous lysosomal bodies. These malformed fibre cells subsequently necrose, collapse, and detach from the lens capsule, resulting in the formation of a lens one half to one fifth of the normal size at birth. In this mutant lens, the anterior epithelium, as well as the short equatorial fibre cells, appear normal, suggesting that the defect is limited to the elongating fibre cells.

Cell autonomous defect

When eye rudiments were taken from the Elo mouse and cultured *in vitro,* the lens fibre cells failed to elongate properly, suggesting that the defect was not caused by an extraocular component (Watanabe et al., 1980). In addition, chimeric animals derived from mixing BALB/c and C3H-*Elo/+* embryos were found to have lenses with a spectrum of phenotypes, ranging between wild type, normal-sized but opaque and Elo-like. The severity of the lens defect was proportional to the relative contribution of lens fibre cells from the mutant strain (Yoshiki et al., 1991). In fact, immunohistological analysis showed that the necrotic fibre cells originated from the C3H strain. These studies thus argued strongly that the Elo phenotype was expressed in a cell autonomous fashion.

Mutation in the γ E-crystallin gene

The Elo mutation is located on chromosome 1, close to the -crystallin gene cluster (γA to γF) (Masaki and Watanabe, 1989; Quinlan et al., 1987). The specific involvement of the γ-crystallin locus is further suggested by the marked reduction of the general γ-crystallin mRNA and protein levels when compared to those of β- and γ-crystallins (Masaki and Watanabe, 1988; Quinlan et al., 1987). More recently, a single base pair deletion has been found in the γE-crystallin gene of the Elo mouse (Cartier, Breitman and Tsui, 1992). The deletion, which is located in the middle of the last exon of the gene, would result in a truncated polypeptide missing the last Greek key motif of the normal protein (replaced by 11 amino acids specified by a different reading frame). The causal relationship between this deletion and the Elo phenotype is further supported by the genetic linkage data showing no recombination between the two in 274 meioses. Since mRNA from the mutant allele is readily detectable in the Elo lens, it seems probable that the failure of the lens fibre cells to elongate is caused by the presence of a putative truncated polypeptide. The current hypothesis, therefore, about the Elo phenotype is that its small lens would be due to the presence of a truncated γE polypeptide. Although the molecular mechanism by which this mutant γ-crystallin polypeptide causes a small lens is unknown, this result suggests a role for γ-crystallins in fibre cell differentiation.

13/N Guinea Pig

The 13/N strain of guinea pigs carries a bilateral, congenital nuclear cataract mutation which is a dominant and fully penetrant phenotype (Amsbaugh and Stone, 1984; Stone and Amsbaugh, 1984). The defect seems to be associated with a taxon-specific lens protein, called the ζ-crystallin which has, so far, only been found in some hystrichomorph rodents and camels (Huang et al., 1987; Garland et al., 1991). This 38 kDa protein is present at high levels (up to 10% of the soluble proteins) in normal guinea pig lens and, in somewhat lower amounts, in the liver and kidney. In the homozygous mutant guinea pig the protein is absent from all three organs, while it is present at half the normal levels in the liver and kidney of the heterozygous animal (Huang et al., 1990; Zigler and Huang, 1988). Thus, an association between the inherited cataract condition and the absence of ζ-crystallin has been postulated.

Mutation in the ζ-crystallin

Molecular analysis shows that the ζ-crystallin gene in the cataractous guinea pig harbors a deletion at the splice-acceptor site of exon 7 (Borrás et al., 1990; Rodriguez et al., 1991). As a result, the major mRNA transcript from this gene lacks the entire exon 7 sequence and, since this exon contains 102 base pairs (bp), an internally deleted polypeptide is predicted. In addition, however, the mutation has apparently activated a cryptic splice-acceptor site, giving rise to a minor mRNA species with a frameshift caused by the insertion of a 20 bp intron sequence in place of the missing exon. The presence of these putative mutant polypeptides may explain the fully penetrant and dominant phenotype of the cataract mutation. Neither of the mutant forms are predicted to form the tetrameric structure postulated for the normal protein (Rao and Zigler, 1992).

Enzymatic activity of ζ-crystallin

The presence of ζ-crystallin in liver and kidney implies an enzymatic function for this protein. A distant relationship has in fact been noted between ζ-crystallin and alcohol dehydrogenase (Rodokanaki, Holmes and Borrás, 1989). The ζ-crystallin is characterized, however, by the absence of the oxidation-sensitive thiol groups as well as the lack of superficial segments typical of alcohol dehydrogenase. The absence of these sequences has apparently resulted in the loss of an alcohol dehydrogenase-specific enzymatic activity, while leading to an increased stability for the lens protein (Borrás, Persson and Jörnvall, 1989). Recently, it has been shown that ζ-crystallin is active as another enzyme, an NADPH:quinone oxidoreductase, which may serve to regulate NADPH oxidation or to detoxify quinones in the lens (Rao, Krishna and Zigler, 1992; Rao and Zigler, 1991). It is of interest to note that the truncated polypeptide found in the cataractous lens does not bind NADPH and has no enzymatic activity (Rao and Zigler, 1992), leading to dramatically reduced levels of NADPH levels in the mutant lens (Rao and Zigler, 1990). Therefore, while these data strongly imply a role for the mutant crystallin in the cataractous phenotype, it is not known which defective property (or properties) of this polypeptide is in fact responsible for the diseased state.

Mutagen-Induced Cataract Mice

Studies on the genetic causes of cataract have also been performed with mouse mutants obtained after parental mutagenic treatment (Kratochvilova, 1981; Graw et al., 1986; Kratochvilova and Favor, 1988). Fifteen of the mutants obtained have been tested for allelism and found to segregate into seven different groups (Kratochvilova and Favor, 1992). The largest group comprises five alleles and, based on mapping results from two of the strains, the common mutation locus appears to be located on chromosome 1, close to the γ-crystallin cluster (J. Favor, M.F. Lyon and J. Löster, personal communication). These dominant cataracts may, therefore, be caused by mutations in the γ-crystallin genes.

CONCLUSION

Crystallins are the main soluble proteins of the eye lens. As such, they play a major role in the two specific properties of this organ: accomodation and transparency. It is, therefore, thought that their recruitment by the lens was due to their appropriate biophysical characteristics. This view is further strengthened by the fact that different animal species, with varying lenticular requirements, have recruited numerous proteins as crystallins to fit their particular needs. For example, rodents, which are nocturnal animals, have evolved a hard, dehydrated and spherical lens with high levels of γ-crystallins. Meanwhile, birds, requiring high accommodative power, need a more flexible and hydrated lens, obtained by replacing the γ-crystallins with the δ-crystallins, which allow more hydration.

The study of inherited cataracts has further demonstrated the importance of the crystallins for lenticular function, since four mutations in crystallin genes have been associated with lens disorders so far: the human γ-crystallin cluster associated with the Coppock-like cataract, the murine βB2 and γE implicated in the mutant phenotypes of the Philly and Elo mice and the ζ-crystallin mutated in the cataractous N/13 strain of guinea pigs. Furthermore, two of these crystallin mutations, the ones present in the murine βB2 and γE genes, affect the fibre cell elongation process and may, therefore, implicate these crystallin proteins in the development of the lens. Such results are part of the growing data imparting a function other than structural to the different crystallins. For example, most taxon-specific crystallins were derived from enzymes and still possess an enzymatic activity which may be necessary for proper lens function. Also, as discussed above, the α-crystallins have recently been found to possess some chaperone activity. Therefore, although the three-dimensional structure and biophysical properties of the crystallins are probably the main reasons for the recruitment of these proteins by the lens, their enzymatic activity may have also played an important role in this process and may still be required for the proper function of the organ.

Another point which should also be stressed is the advantage of the genetic approach to investigating the role of crystallins in cataractogenesis. Indeed, although biochemical studies had implicated these molecules in the actual lenticular opacity, it was never possible to demonstrate whether the crystallin involvement was a primary or secondary event. In the case of inherited cataracts, however, if the mutation responsible for the disease lies in a crystallin gene, it is proof that defective crystallins can be a primary cause of cataract. The determination of which mutations cause lenticular opacities will be helpful in gaining a better understanding of the interactions of the different crystallins, both in healthy and diseased states. To that end, it will be important to screen the various crystallin genes, in the different inherited lenticular diseases, for potential mutations.

Although there remain unanswered questions about the crystallins and some of their roles in the development and maintenance of lenticular architecture and function, there is no doubt that these molecules are vital for the proper functioning of this organ.

REFERENCES

Aarts, H. J. M., den Dunnen, J. T., Lubsen, N. H. and Schoenmakers, J. G. G. (1987) Linkage between the βB2 and βB3 crystallin genes in man and rat: a remnant of an ancient β- crystallin gene cluster. *Gene* **59**, 127–135.

Aarts, H. J. M., den Dunnen, J. T., Leunissen, J., Lubsen, N. H. and Schoenmakers, J. G. G. (1988) The γ-crystallin gene families: sequence and evolutionary patterns. *Journal of Molecular Evolution* **27**, 163–172.

Aarts, H. J. M., Lubsen, N. H. and Schoenmakers, J. G. G. (1989) Crystallin gene expression during rat lens development. *European Journal of Biochemistry* **183**, 31–36.

Amsbaugh, D. F. and Stone, S. H. (1984) Autosomal dominant congenital nuclear cataracts in strain 13/N guinea pigs. *Journal of Heredity* **75**, 55–58.

Bateman, J. B., Spence, M. A., Marazita, M. L. and Sparkes, R.S. (1986) Genetic linkage analysis of autosomal dominant congenital cataracts. *American Journal of Ophthalmology* **101**, 218–225.

Bax, B., Lapatto, R., Nalini, V., Driessen, H., Lindley, P. F., Mahadevan, D. et al. (1990) X-ray analysis of βB2-crystallin and evolution of oligomeric lens proteins. *Nature* **347**, 776–780.

Beaumont, C., Garrett, C., Garner, G., Griffiths, B. and Ball, S. P. (1989) Linkage data on dominantly inherited congenital cataracts in families from the South West of England. *Cytogenetics and Cell Genetics* **51**, 959.

Berbers, G. A. M., Boerrnan, O. C., Bloemendal, H. and de Jong, W. W. (1982) Primary gene products of bovine β-crystallin and reassociation behavior of its aggregates. *European Journal of Biochemistry* **128**, 495–502.

Berbers, G. A. M., Hoekman, W. A., Bloemendal, H., de Jong, W. W., Kleinschmidt, T. and Braunitzer, G. (1983) Proline- and alanine-rich N-terminal extension of the basic bovine β-crystallin Bl chain. *FEBS Letters* **161**, 225–229.

Berman, E. R. (1991) *Biochemistry of the eye*, pp. 201–290. New York: Plenum Press.

Bhat, S. P. and Nagineni, C. N. (1989) αB subunit of lens-specific protein α-crystallin is present in other ocular and non-ocular tissues. *Biochemical Biophysical Research Communications* **158**, 319–325.

Bijlsma, E. K., Geurts van Kessel, A. H. M., Westerveld, A. and Hulsebos, T. J. M. (1991) Physical mapping of the βB crystallin genes on chromosome 22. *Cytogenetics and Cell Genetics* **58**, 2045.

Birdwood, S. C., Garrett, C., Garner, G. and Ball, S. P. (1992) The locus causing a form of dominantly inherited congenital lamellar cataracts is not linked to the γ-crystallin gene cluster on chromosome 2q33–q35. *Cytogenetics and Cell Genetics* **58**, 1968.

Bloemendal, H. (1981) The lens proteins. In *Molecular and cellular biology of the eye lens* edited by H. Bloemendal, pp.l–47. New York: John Wiley and Sons.

Bloemendal, H. (1985) Lens research: from protein to gene. *Experimental Eye Research* **41**, 429–448.

Bloemendal, H. and de Jong, W. W. (1991) Lens proteins and their genes. *Progress in Nucleic Acid Research and Molecular Biology* **41**, 259–281.

Blundell, T. L., Lindley, P. F., Miller, L., Moss, D. S., Slingsby, C., Tickle, I. J., et al. (1981) The molecular structure and stability of the eye lens: X-ray analysis of γ-crystallin II. *Nature* **289**, 771–777.

Borrás, T., Persson, B. and Jörnvall, H. (1989) Eye lens ζ-crystallin relationships to the family of "long-chain" alcohol/polyol dehydrogenases. Protein trimming and conservation of stable parts. *Biochemistry* **28**, 6133–6139.

Borrás, T., Jörnvall, H., Rodokanaki, A., Gonzalez, P., Rodriguez, I. and Hernandez Calzadilla, C. (1990) The transcripts of ζ-crystallin, a lens protein related to the alcohol dehydrogenase family, are altered in a guinea-pig hereditary cataract. *Experimental Eye Research* **50**, 729–735.

Brakenhoff, R. H., Aarts, H. J. M., Reek, F. H., Lubsen, N. H. and Schoenmakers, J. G. G. (1990a) The human γ-crystallin genes: a gene family on its way to extinction. *Journal of Molecular Biology* **216**, 519–532.

Brakenhoff, R. H., Geurts van Kessel, A. H. M., Oldenburg, M., Wijnen, J. T., Bloemendal, H., Meera Khan, P. et al. (1990b) Human α-crystallin (CRYA2) gene mapped to chromosome 11q12-q23. *Human Genetics* **85**, 237–240.

Brakenhoff, R. H., Aarts, H. J. M., Schuren, F., Lubsen, N. H. and Schoenmakers, J. G. G. (1992) The second human αB2-crystallin gene is a pseudogene. *Experimental Eye Research* **54**, 803–806.

Carper, D., Shinohara, T., Piatigorsky, J. and Kinoshita, J. H. (1982) Deficiency of functional messenger RNA for a developmentally regulated β-crystallin polypeptide in a hereditary cataract. *Science* **217**, 463–464.

Carper, D., Smith-Gill, S. J. and Kinoshita, J. H. (1986) Immunocytochemical localization of the 27K β-crystallin polypeptide in the mouse lens during development using a specific monoclonal antibody: implications for cataract formation in the Philly mouse. *Developmental Biology* **113**, 104–109.

Cartier, M., Breitman, M. L. and Tsui, L.-C. (1992) A frameshift mutation in the γE-crystallin gene of the Elo mouse. *Nature Genetics* **2**, 42–45.

Chambers, C. and Russell, P. (1991) Deletion mutation in an eye lens β-crystallin. *Journal of Biological Chemistry* **266**, 6742–6746.

Chang, T. and Chang, W.-C. (1987) Cloning and sequencing of a carp βs crystallin cDNA. *Biochimica Biophysica Acta* **910**, 89–92.

Clark, J. I. and Carper, D. (1987) Phase separation in lens cytoplasm is genetically linked to cataract formation in the Philly mouse. *Proceedings of the National Academy of Sciences USA* **84**, 122–125.

Conneally, P. M., Wilson, A. F., Merritt, A. D., Helveston, E. M., Palmer, C. G. and Wang, L. Y. (1978) Confirmation of genetic heterogeneity in autosomal dominant forms of congenital cataracts from linkage studies. *Cytogenetics and Cell Genetics* **22**, 295–297.

de Jong, W. W., Hendriks, W., Mulders, J. W. M. and Bloemendal, H. (1989) Evolution of lens crystallins, the stress connection. *Trends in Biochemical Sciences* **14**, 365–368.

Delaye, M. and Tardieu, A. (1983) Short-range order of crystallin proteins accounts for eye lens transparency. *Nature* **302**, 415–417.

den Dunnen, J. T., Moormann, R. J. M., Cremers, F. P. M. and Schoenmakers, J. G. G. (1985a) Two human γ-crystallin genes are linked and riddled with Alu-repeats. *Gene* **38**, 197–204.

den Dunnen, J. T., Moormann, R. J. M., Lubsen, N. H. and Schoenmakers, J. G. G. (1985b) Intron insertions and deletions in the β/γ-crystallin gene family: the rat βB1-gene. *Proceedings of the National Academy of Sciences USA* **83**, 2855–2859.

den Dunnen, J. T., Jongbloed, R. J. E., Geurts van Kessel, A. H. M. and Schoenmakers, J. G. G. (1986) Human lens γ-crystallin genes are located in the p12-qter region of chromosome 2. *Human Genetics* **70**, 217–221.

den Dunnen, J. T., Lubsen, N. H. and Schoenmakers, J. G. G. (1986) Concerted and divergent evolution within the rat γ-crystallin gene family. *Journal of Molecular Biology* **189**, 37–46.

den Dunnen, J. T., van Neck, J. W., Cremers, F. P. M., Lubsen, N. H. and Schoenmakers, J. G. G. (1989) Nucleotide sequence of the rat γ-crystallin gene region and comparison with an orthologous human region. *Gene* **78**, 201–213.

Dubin, R. A., Wawrousek, E. F. and Piatigorsky, J. (1989) Expression of murine αB-crystallin gene is not restricted to the lens. *Molecular and Cellular Biology* **9**, 1083–1091.

Dubin, R. A., Ally, A. H., Chung, S. and Piatigorsky, J. (1990) Human αB-crystallin gene and preferential promoter function in lens. *Genomics* **7**, 594–601.

Duguid, J. R., Rohwer, R. G. and Seed, B. (1988) Isolation of cDNAs of scrapie-modulated RNAs by subtractive hybridization of a cDNA library. *Proceedings of the National Academy of Sciences USA* **85**, 5738–5742.

Eiberg, H., Marner, E., Rosenberg, T. and Mohr, J. (1988) Marner's cataract (CAM) assigned to chromosome 16: Linkage to haptoglobin. *Clinical Genetics* **34**, 272–275.

Fain, P. R., Goldgar, D. E., Wright, E.C., Andersen, W. L., Willard, H. F., van Tuinen, *P.* et al. (1989) Markers tightly linked to the NF1 locus on chromosome 17. *Cytogenetics and Cell Genetics* **51**, 996.

Fernald, R. D. and Wright, S. E. (1983) Maintenance of optical quality during crystalline lens growth. *Nature* **301**, 618–620.

Garland. D. Rao, V. P., Del Corso, A., Mura, U. and Zigler, S. J., Jr. (1991) ζ-crystallin is a major protein in the lens of *Camelus dromedarius. Archives of Biochemistry and Biophysics* **285**, 134–136.

Garner, W. H., Garner, M. H. and Spector, A. (1979) Comparison of the 10,000 and 43,000 dalton polypeptide populations isolated from the water soluble and insoluble fractions of human cataractous lenses. *Experimental Eye Research* **29**, 257–276.

Gelatt, K. N. and Das, N. D. (1984) Animal models for inherited cataracts: a review. *Current Eye Research* **3**, 765–778.

Gorin, M. B and Horwitz, J. (1984) Cloning and characterization of a cow β-crystallin cDNA. *Current Eye Research* **3**, 939–948.

Goring, D. R., Breitman, M. L. and Tsui, L.-C. (1992) Temporal regulation of six crystallin transcripts during mouse lens development. *Experimental Eye Research* **54**, 785–795.

Graw, J., Favor, J., Neuhauser-Klaus, A. and Ehling, U. H. (1986) Dominant cataract and recessive specific locus mutations in offspring of X-irradiated male mice. *Mutation Research* **159**, 47–54.

Groenen, P. J. T. A., van den Ijssel, P. R. L. A., Voorter, C. E. M., Bloemendal, H. and de Jong, W. W. (1990) Site-specific racemization in aging αA-crystallin. *FEBS Letters* **269**, 109–112.

Groenen, P. J. T. A., Bloemendal, H. and de Jong, W. W. (1992) The carboxy terminal lysine of αB-crystallin is an amine-donor substrate for tissue transglutaminase. *European Journal of Biochemistry* **205**, 671–674.

Hammerstein, W. and Scholz, W. (1974) Familiare form einer"cataracta centralis": klinisch-genetische studie mit koppelungsdaten. *Archiv fur Klinische und Experimentelle Ophthalmologie* **189**, 9–19.

Harding, J. J. and Dilley, K. J. (1976) Structural proteins of the mammalian lens: A review with emphasis on changes in development, aging and cataract. *Experimental Eye Research* **22**, 1–74.

Hawkins, J. W., van Keuren, M. L., Piatigorsky, J., Law, M. L., Patterson, D. and Kao, F.-T. (1987) Confirmation of assignment of the human αA-crystallin gene (CRYAl) to chromosome 21 with regional localization to q22.3. *Human Genetics* **76**, 375–380.

Hejtmancik, J. F., Beebe, D. C., Ostrer, H. and Piatigorsky, J. (1985) α- and β-crystallin mRNA levels in the embryonic and posthatched chicken lens: temporal and spatial changes during development. *Developmental Biology* **109**, 72–81.

Hoenders, H. J. and Bloemendal, H. (1967) The N-terminus of the lens protein α-crystallin. *Biochimica and Biophysica Acta* **147**, 183–185.

Hogg, D., Tsui, L.-C., Gorin, M. and Breitman, M. L. (1986) Characterization of the human β-crystallin gene HuβA3/Al reveals ancestral relationships among the γ-crystallin superfamily. *Journal of Biological Chemistry* **261**, 12420–12427.

Hogg, D., Gorin, M. B., Heinzmann, C., Zollman, S., Mohandas, T., Klisak, I. et al. (1987) Nucleotide sequence for the cDNA of bovine βB2 crystallin and assignment of the orthologous human locus to chromosome 22. *Current Eye Research* **6**, 1335–1342.

Horwitz, J. (1992) Proctor lecture, ARVO, Sarasota.

Huang, Q.-L., Russell, P., Stone, S.H. and Zigler, J. S., Jr. (1987) ζ-crystallin, a novel lens protein from the guinea pig. *Current Eye Research* **6**, 725–732.

Huang, Q.-L., Du, X.Y., Stone, S.H., Amsbaugh, D. F., Datiles, M., Hu, T. S. et al. (1990) Association of hereditary cataracts in strain 13/N guinea-pigs with mutation of the gene for ζ-crystallin. *Experimental Eye Research* **50**, 317–325.

Hulsebos, T. J. M., Bijlsma, E. K., Geurts van Kessel, A. H. M., Brakenhoff, R. H. and Westerveld, A. (1991) Direct assignment of the human βB2 and βB3 crystallin genes to 22q11.2-q12: markers for neurofibromatosis 2. *Cytogenetics and Cell Genetics* **56**, 171–175.

Huntzinger, R. S., Weitkamp, L. R. and Roca, P.D. (1977) Linkage relationships of a locus for congenital total nuclear cataract. *Journal of Medical Genetics* **15**, 113–115.

Inana, G., Piatigorsky, J., Norman, B., Slingsby, C. and Blundell, T. (1983) Gene and protein structure of a β-crystallin polypeptide in murine lens: relationship of exons and structural motifs. *Nature* **302**, 310–315.

Ingolia, T. D. and Craig, E. A. (1982) Four small Drosophila heat shock proteins are related to each other and to mammalian α-crystallin. *Proceedings of the National Academy of Sciences* USA **79**, 2360–2364.

Iwaki, T., Kume-Iwaki, A., Liem, R. K. H. and Goldman, J. E. (1989) αB-crystallin is expressed in non–lenticular tissues and accumulates in Alexander's disease brain. *Cell* **57**, 71–78.

Kador, P. F., Fukui, H. N., Fukushi, S., Jernigan, H. M., Jr. and Kinoshita, J. H. (1980) Philly mouse: a new model of hereditary cataract. *Experimental Eye Research* **30**, 59–68.

Kaiser-Kupfer, M. I., Freidlin, V., Datiles, M. B., Edwards, P. A., Sherman, J. L., Parry, *D.* et al. (1989) The association of posterior capsular lens opacities with bilateral acoustic neuromas in patients with neurofibromatosis type 2. *Archives of Ophthalmology* **107**, 541–544.

Kleiman, N. J., Chiesa, R., Gawinowicz-Kolks, M. A. and Spector, A. (1988) Phosphorylation of β-crystallin B2 (βBp) in the bovine lens. *Journal of Biological Chemistry* **263**, 14978–14983.

Klemenz, R., Frohli, E., Steiger, R. H., Schafer, R. and Aoyama, A. (1991) αB-crystallin is a small heat shock protein. *Proceedings of the National Academy of Sciences USA* **88**, 3652–3656.

Kodama, T. and Takemoto, L. (1988) Characterization of disulfide-linked crystallins associated with human cataractous lens membranes. *Investigative Ophthalmology and Visual Science* **29**, 145–149.

Kratochvilova, J. (1981) Dominant cataract mutations detected in offspring of gamma-irradiated male mice. *Journal of Heredity* **72**, 302–307.

Kratochvilova, J. and Favor, J. (1988) Phenotypic characterization and genetic analysis of twenty dominant cataract mutations detected in offspring of irradiated male mice. *Genetical Research* **52**, 125–134.

Kratochvilova, J. and Favor, J. (1992) Allelism tests of 15 dominant cataract mutations in mice. *Genetical Research* **59**, 199–203.

Law, M. L., Cai, G.-Y., Hartz, J., Kao, F.-T., Hogg, D., Breitman, M. L. et al. (1986) Localization of a β-crystallin gene, the huβA3/Al (gene symbol, CRYBl) to the long arm of human chromosome 17. *Cytogenetics and Cell Genetics* **42**, 202–207.

Lindley, P. F., Narebor, M.E., Summers, L. J. and Wistow, G. J. (1985) The structure of lens proteins. In *The ocular lens* edited by H. Maisel, pp. 123–167. New York: Marcel Dekker.

Lok, S., Tsui, L.-C., Shinohara, T., Piatigorsky, J., Gold, R. and Breitman, M. L. (1984) Analysis of the mouse γ-crystallin gene family: assignment of multiple cDNAs to discrete genomic sequences and characterization of a representative gene. *Nucleic Acids Research* **12**, 4517–4529.

Lorand, L., Parameswaran, K. N. and Velasco, P. T. (1991) Sorting-out of acceptor-donor relationships in the transglutaminase-catalyzed cross-linking of crystallins by the enzyme-directed labeling of potential sites. *Proceedings of the National Academy of Sciences USA* **88**, 82–83.

Lubsen, N. H., Renwick, J. H., Tsui, L.-C., Breitman, M. L. and Schoenmakers, J. G. G. (1987) A locus for a human hereditary cataract is closely linked to the γ-crystallin gene family. *Proceedings of the National Academy of Sciences USA* **84**, 489–492.

Lubsen, N. H., Aarts, H. J. M. and Schoenmakers, J. G. G. (1988) The evolution of lenticular proteins: The β- and γ-crystallin super gene family. *Progress in Biophysics and Molecular Biology* **51**, 47–76.

Marner, E. (1949) A family with eight generations of hereditary cataract. *Acta Ophthalmologica* **27**, 537–551.

Masaki, S. and Watanabe, T. (1988) Suppression of lens γ-crystallin encoded by *len-1* in the mutant mouse with Eye lens obsolescence *(Elo)*. *Experimental Eye Research* **47**, 727–735.

Masaki, S. and Watanabe, T. (1989) Linkage analysis of the mutation locus in the Eye lens obsolescence *(Elo)* mouse. *Genomics* **5**, 259–263.

McAlpine, P. J., Shows, T. B., Boucheix, C., Huebner, M. and Anderson, W. A. (1991) The 1991 catalog of mapped genes and report of the nomenclature committee, Human Gene Mapping 11. *Cytogenetics and Cell Genetics* **58**, 5–103.

McAvoy, J. W. (1978) Cell division, cell elongation and co-ordination of crystallin gene expression during lens morphogenesis in the rat. *Journal of Embryology and Experimental Morphology* **45**, 271–281.

McDevitt, D. S., Hawkins, J. W., Jaworski, C. J. and Piatigorsky, J. (1986) Isolation and partial characterization of the human αA-crystallin gene. *Experimental Eye Research* **43**, 285–291.

McFall–Ngai, M., Horwitz, J., Ding, L.-L. and Lacey, L. (1986) Age-dependent changes in the heat–stable crystallin, βBp, of the human lens. *Current Eye Research* **5**, 387–394.

Meakin, S. O., Breitman, M. L. and Tsui, L.-C. (1985) Structural and evolutionary relationships among five members of the human γ-crystallin gene family. *Molecular and Cellular Biology* **5**, 1408–1414.

Mohr, J. (1954) *A Study of Linkage in Man*. Copenhagen: Munksgaard

Moormann, R. J. M., den Dunnen, J. T., Mulleners, L., Andreoli, P.M., Bloemendal, H. and Schoenmakers, J. G. G. (1983) Strict colinearity of genetic and protein folding domains in an intragenically duplicated rat lens γ-crystallin gene. *Journal of Molecular Biology* **171**, 353–368.

Moormann, R. J. M., den Dunnen, J. T., Heuyerjans, J., Jongbloed, R. J. E., van Leen, R. W., Lubsen, N.H. et al. (1985) Characterization of the rat γ-crystallin gene family and its expression in the eye lens. *Journal of Molecular Biology* **182**, 419–430.

Murer-Orlando, M., Paterson, R. C., Lok, S., Tsui, L.-C. and Breitman, M. L. (1987) Differential regulation of γ-crystallin genes during mouse development. *Developmental Biology* **119**, 260–267.

Nakamura, M., Russell, P., Carper, D. A., Inana, G. and Kinoshita, J. H. (1988) Alteration of a developmentally regulated, heat-stable polypeptide in the lens of the Philly mouse. *Journal of Biological Chemistry* **263**, 19218–19221.

Nettleship, E. (1909) Seven new pedigrees of hereditary cataract. *Transactions of the Ophthalmological Society UK* **29**, 188–210.

Ngo, J. T., Klisak, I., Dubin, R. A., Piatigorsky, J., Mohandas, T., Sparkes, R.S. et al. (1989) Assignment of the αB-crystallin gene to human chromosome 11. *Genomics* **5**, 665–669.

Oda, S.-I., Watanabe, K., Fujisawa, H. and Kameyama, Y. (1980) Impaired development of lens fibres in genetic microphthalmia, Eye lens obsolescence, *Elo*, of the mouse. *Experimental Eye Research* **31**, 673–681.

Oda, S.-I., Watanabe, T. and Kondo, K. (1980) A new mutation, Eye lens obsolescence, *Elo* on chromosome 1 in the mouse. *Japanese Journal of Genetics* **55**, 71–75.

Ostrer, H., Beebe, D. C., and Piatigorsky, J. (1981) β-crystallin mRNAs: differential distribution in the developing chicken lens. *Developmental Biology* **86**, 403–408.

Peterson, C. A. and Piatigorsky, J. (1986) Preferential conservation of the globular domains of the βA3/Al-crystallin polypeptide of the chicken eye lens. *Gene* **45**, 139–147.

Piatigorsky, J., Kador, P. F. and Kinoshita, J. H. (1980) Differential synthesis and degradation of protein in the hereditary Philly mouse cataract. *Experimental Eye Research* **30**, 69–78.

Piatigorsky, J. (1984) Lens crystallins and their gene families. *Cell* **38**, 620–621.

Piatigorsky, J. and Wistow, G. J. (1989) Enzyme/crystallins: gene sharing as an evolutionary strategy. *Cell* **57**, 197–199.

Piatigorsky, J. and Wistow, G. (1991) The recruitment of crystallins: new functions precede gene duplication. *Science* **252**, 1078–1079.

Quax-Jeuken, Y., Janssen, C., Quax, W., van den Heuvel, R. and Bloemendal, H. (1984) Bovine β-crystallin complementary DNA clones. Alternating proline/alanine sequence of βB1 subunit originates from a repetitive DNA sequence. *Journal of Molecular Biology* **180**, 457–472.

Quax-Jeuken, Y., Driessen, H., Leunissen, J., Quax, W., de Jong, W. and Bloemendal, H. (1985a) βs-crystallin: structure and evolution of a distinct member of the γ-superfamily. *EMBO Journal* **4**, 2597–2602.

Quax-Jeuken, Y., Quax, W., van Rens, G., Meera Khan, P. and Bloemendal, H. (1985b) Complete structure of the αB-crystallin gene: Conservation of the exon/intron distribution in the two nonlinked α-crystallin genes. *Proceedings of the National Academy of Sciences USA* **82**, 5819–5823.

Quax-Jeuken, Y., Quax, W., van Rens, G. L. N., Meera Khan, P. and Bloemendal, H. (1985c) Assignment of the human αA-crystallin gene (CRYAl) to chromosome 21. *Cytogenetics and Cell Genetics* **40**, 727–728.

Quinlan, P., Oda, S., Breitman, M. L. and Tsui, L.-C. (1987) The mouse eye lens obsolescence (Elo) mutant: studies on crystallin gene expression and linkage analysis between the mutant locus and the γ-crystallin genes. *Genes and Development* **1**, 637–644.

Radlick, L. W. and Koretz, J. F. (1992) Biophysical characterization of crystallin aggregates: validation of the micelle hypothesis. *Biochimica Biophysica Acta* **1130**, 267–276.

Rao, P. V. and Zigler, J. S., Jr. (1990) Extremely high levels of NADPH in guinea pig lens: correlation with ζ-crystallin concentration. *Biochemical and Biophysical Research Communications* **167**, 1221–1228.

Rao, P. V. and Zigler, J. S., Jr. (1991) α-crystallin from guinea pig lens is capable of functioning catalytically as an oxidoreductase. *Archives of Biochemistry and Biophysics* **284**, 181–185.

Rao, P. V., Krishna, C. M. and Zigler, J. S., Jr. (1992) Identification and characterization of the enzymatic activity of ζ-crystallin from guinea pig lens. *Journal of Biological Chemistry* **267**, 96–102.

Rao, P. V. and Zigler, J. S., Jr. (1992) Mutant ζ-crystallin from guinea-pig hereditary cataracts has altered structural and enzymatic properties. *Experimental Eye Research* **54**, 627–630.

Renwick, J. H. and Lawler, S. D. (1963) Probable linkage between a congenital cataract locus and the Duffy blood group locus. *Annals of Human Genetics* **27**, 67–84.

Richards, J., Maumenee, I. H., Rowe, S. and Lovrien, E. W. (1984) Congenital cataract possibly linked to haptoglobin. *Cytogenetics and Cell Genetics* **37**, 570.

Rodokanaki, A., Holmes, R. H. and Borrás, T. (1989) ζ-crystallin, a novel protein from the guinea pig lens is related to alcohol dehydrogenases. *Gene* **78**, 215–224.

Rodriguez, I. R., Gonzalez, P., Zigler, J.S. and Borrás, T. (1991) A splice-site deletion that causes a defective ζ-crystallin is found in a congenital guinea pig cataract. *Investigative Ophthalmology and Visual Science* **32** (suppl.), 782.

Roquemore, E. P., Dell, A., Morris, H. R., Panico, M., Reason, A. J., Savoy, L.-A. et al. (1992) Vertebrate lens α-crystallins are modified by O-linked N-acetylglucosamine. *Journal of Biological Chemistry* **267**, 555–563.

Russell, P., Meakin, S. O., Hohman, T. C., Tsui, L.-C. and Breitman, M. L. (1987) Relationship between proteins encoded by three human γ-crystallin genes and distinct polypeptides in the eye lens. *Molecular and Cellular Biology* **7**, 3320–3323.

Russell, P. and Chambers, C. (1990) Interaction of an altered β-crystallin with other proteins in the Philly mouse lens. *Experimental Eye Research* **50**, 683–687.

Shiloh, Y., Donlon, T., Bruns, G., Breitman, M. L. and Tsui, L.-C. (1986) Assignment of the human γ-crystallin gene cluster (CRYG) to the long arm of chromosome 2, region q33-q36. *Human Genetics* **73**, 17–19.

Siezen, R. J., Anello, R. D. and Thomson, J. A. (1986) Interactions of lens proteins. Concentration dependence of β-crystallin aggregation. *Experimental Eye Research* **43**, 293–303.

Siezen, R. J., Thomson, J. A., Kaplan, E. D. and Benedek. G. B. (1987) Human lens γ-crystallins: isolation, identification and characterization of the expressed gene products. *Proceedings of the National Academy of Sciences USA* **84**, 6088–6092.

Siezen, R. J., Wu, E., Kaplan, E. D., Thomson, J. A. and Benedek, G. B. (1988) Rat lens γ-crystallins. Characterization of the six gene products and their spatial and temporal distribution resulting from differential synthesis. *Journal of Molecular Biology* **199**, 475–490.

Slingsby, C. and Croft, L. R. (1973) Developmental changes in the low molecular weight proteins of the bovine lens. *Experimental Eye Research* **17**, 369–376.

Slingsby, C. and Miller, L. R. (1983) Purification and crystallization of mammalian lens γ-crystallins. *Experimental Eye Research* **37**, 517–530.

Sparkes, R. S., Mohandas, T., Heinzmann, C., Gorin, M. B., Zollman, S. and Horwitz, J. (1986) Assignment of a human β-crystallin gene to 17 cen-q23. *Human Genetics* **74**, 133–136.

Sparkes, R. S, Hogg, D., Gorin, M. B., Heinzmann, C., Zollman, S., Mohandas, T. et al. (1987) Assignment of a second human β-crystallin gene (CRYB2) to 22q11.2–q12.2. *Cytogenetics and Cell Genetics* **46**, 696.

Spector, A., Garner, M. H., Garner, W. H., Roy, D., Farnsworth, P. and Shyne, S. (1979) An extrinsic membrane polypeptide associated with high-molecular-weight protein aggregates in human cataract. *Science* **204**,1323–1326.

Spector, A., Chiesa, R., Sredy, J. and Garner, W. (1985) cAMP-dependent phosphorylation of bovine lens α-crystallin. *Proceedings of the National Academy of Sciences USA* **82**, 4712–4716.

Srivastava, K., Srivastava, O. P. and Silney, C. (1992) Apparent modification of a 9 kDa polypeptide during aging in human lenses. *Investigative Ophthalmology and Visual Science* **33**, 864.

Stabile, M., Amoriello, A., Capobianco, S., Cavaliere, M. L., Conte, N., De Rosa, C. et al. (1983) Study of a form of pulverulent cataract in a large kindred. *Journal of Medical Genetics* **20**, 419–421.

Stauffer, J. Rothschild, C., Wandel, T. and Spector, A. (1974) Transformation of α-crystallin polypeptide chains with aging. *Investigative Ophthalmology* **13**, 135–146.

Stone, S. H. and Amsbaugh, D. F. (1984) Congenital cataracts in strain 13 guinea pigs: an autosomal dominant human model. *Investigative Ophthalmology and Visual Science* **25**, 606–607.

Takemoto, L., Takemoto, D., Brown, G., Takehana, M., Smith, J. and Horwitz, J. (1987) Cleavage from the N-terminal region of βBp crystallin during aging of the human lens. *Experimental Eye Research* **45**, 385–392.

Takemoto, L., Straatsma, B. and Horwitz, J. (1989) Immunochemical characterization of the major low molecular weight polypeptide (10K) from human cataractous lenses. *Experimental Eye Research* **48**, 261–270.

Tardieu, A., Laporte, D., Licinio, P., Krop, B. and Delaye, M. (1986) Calf lens α-crystallin quaternary structure: a three layer tetrahedral model. *Journal of Molecular Biology* **192**, 711–724.

Thomson, J. A. and Augusteyn, R.C. (1985) Ontogeny of human lens crystallins. *Experimental Eye Research* **40**, 393–410.

Tsui, L.-C., Breitman, M. L., Meakin, S. O., Willard, H. F., Shiloh, Y., Donlon, T. et al. (1985) Localization of the human γ-crystallin gene cluster (CRYG) to the long arm of chromosome 2, region q33-q35. *Cytogenetics and Cell Genetics* **40**, 763–764.

Uga, S., Kador, P. F. and Kuwabara, T. (1980) Cytological study of Philly mouse cataract. *Experimental Eye Research* **30**, 79–92.

van den Heuvel, R., Hendriks, W., Quax, W. and Bloemendal, H. (1985) Complete structure of the αA-crystallin gene. Reflection of an evolutionary history by means of exon shuffling. *Journal of Molecular Biology* **185**, 273–284.

van Kleef, F. S. M., Nijzink-Maas, M. J. C. and Hoenders, H. J. (1974) Intracellular degradation of α-crystallin. *European Journal of Biochemistry* **48,** 563–570.

van Leen, R. W., Breuer, M. L., Lubsen, N. H. and Schoenmakers, J. G. G. (1987a) Developmental expression of crystallin genes: in situ hybridization reveals a differential localization of specific mRNAs. *Developmental Biology* **123**, 338–345.

van Leen, R. W., van Roozendaal, K. E. P., Lubsen, N. H. and Schoenmakers, J. G. G. (1987b) Differential expression of crystallin genes during development of the rat eye lens. *Developmental Biology* **120**, 457–464.

van Rens, G. L. M., Raats, J. M. H., Driessen, H.P.C., Oldenburg, M., Wijnen, J. T., Meera Khan, P. et al. (1989) Structure of the bovine eye lens γs-crystallin gene (formerly βs). *Gene* **78**, 225-233.

van Rens, G. L. M., de Jong, W. W. and Bloemendal, H. (1991) One member of the γ-crystallin gene family, γs, is expressed in birds. *Experimental Eye Research* **53**, 135–138.

van Rens, G. L. M., Geurts van Kessel, A. H. M. and Bloemendal, H. (1991) Localization of the βA4-crystallin (CRYBA4) gene on human chromosome 22 in the region ql 1 .2-q13.1. *Cytogenetics and Cell Genetics* **58**, 2052.

van Rens, G. L. M., Driessen, H. P. C., Nalini, V., Slingsby, C., de Jong, W. W. and Bloemendal, H. (199la) Isolation and characterization of cDNAs encoding βA2- and βA4-crystallins: heterologous interactions in the predicted βA4-βB2 heterodimer. *Gene* **102**, 179–188.

van Rens, G. L. M., Hol, F. A., de Jong, W. W. and Bloemendal, H. (1991b) Presence of hybridizing DNA sequences homologous to bovine acidic and basic β-crystallins in all classes of vertebrates. *Journal of Molecular Evolution* **33**, 457–463.

Voorter, C. E. M., Roersma, E. S., Bloemendal, H. and de Jong, W. W. (1987) Age-dependent deamidation of chicken αA-crystallin. *FEBS letters* **221**, 249–252.

Voorter, C. E. M., de Haard-Hoekman, W. A., Hermans, M. M. P., Bloemendal, H. and de Jong, W. W. (1990) Differential synthesis of crystallins in the developing rat eye lens. *Experimental Eye Research* **50**, 429–437.

Walsh, M. T., Sen, A.C. and Chakrabarti, B. (1991) Micellar subunit assembly in a three-layer model of oligomeric α-crystallin. *Journal of Biological Chemistry* **266**, 20079–20084.

Watanabe, K., Fujisawa, H., Oda, S.-I. and Kameyame, Y. (1980) Organ culture and immunohistochemistry of the genetically malformed lens, in Eye lens obsolescence, *Elo,* of the mouse. *Experimental Eye Research* **31**, 683–689.

Willard, H. F., Meakin, S. O., Tsui, L.-C. and Breitman, M. L. (1985) Assignment of the human γ-crystallin multigene family to chromosome 2. *Somatic Cell and Molecular Genetics* **11**, 511–516.

Wistow, G. (1985) Domain structure and evolution in α-crystallins and small heat-shock proteins. *FEBS Letters* **181**, 1–6.

Wistow, G. J. and Piatigorsky, J. (1988) Lens crystallins: the evolution and expression of proteins for a highly specialized tissue. *Annual Review of Biochemistry* **57**, 479–504.

Yoshiki, A., Hanazono, M., Oda, S.-I., Wakasugi, N., Sakakura, T. and Kusakabe, M. (1991) Developmental analysis of the eye lens obsolescence *(Elo)* gene in the mouse: cell proliferation and *Elo* gene expression in the aggregation chimera. *Development* **113**, 1293–1304.

Zigler, J. S. Jr., Horwitz, J. and Kinoshita, J. H. (1980) Human β-crystallin: I. Comparative studies on the β1, β2 and β3-crystallins. *Experimental Eye Research* **31**, 41–55.

Zigler, J. S., Jr., Carper, D. A. and Kinoshita, J. H. (1981) Changes in lens crystallins during cataract development in the Philly mouse. *Ophthalmic Research* **13**, 237–251.

Zigler, J. S., Jr. and Huang, Q.-L. (1988) ζ-crystallin: its role in formation of congenital nuclear cataracts in strain 13/N guinea pigs. *Investigative Ophthalmology and Visual Science* **29** (Suppl), 317.

Zigler, J. S., Jr. (1990) Animal models for the study of maturity-onset and hereditary cataract. *Experimental Eye Research* **50**, 651–657.

Zigman, S. (1985) Selected aspects of lens differentiation. *Biological Bulletin* **168**, 189–213.

Zwaan, J. (1979) Role of the lens in growth of the eye. *Ophthalmic Research* **12**, 145–146.

19. ANIRIDIA

ISABEL HANSON, TIM JORDAN AND VERONICA VAN HEYNINGEN

MRC Human Genetics Unit, Western General Hospital, Crewe Road, Edinburgh EH4 2XU, UK

INTRODUCTION

Phenotype

Aniridia is a congenital inherited disorder of the eye which derives its name from the noticeable iris hypoplasia seen in most cases. The name is something of a misnomer since there is always some iris tissue present although this may be so minimal as to be invisible to the naked eye. The name also draws attention to the most dramatic feature of the phenotype from the observer's point of view, when in reality the effects on the rest of the eye are far more visually disabling for the patient. The associated eye and visual defects include some or all of the following: corneal opacification, cataract, lens dislocation, ciliary body hypoplasia, foveal dysplasia, optic nerve hypoplasia, strabismus and nystagmus (Hittner, 1989). To this long list of what appear to be mainly primary disorders resulting from maldevelopment of the eye and from altered control of eye movement by the central nervous system, we must add the complication of glaucoma. Very rarely this is actually present at birth, indicating that it may be caused by a primary dysgenesis of the aqueous outflow system but more commonly it has its onset in pre-adolescence or later, when it appears to be due to obstruction of the drainage angle by adhesions between the lens and the trabecular meshwork (Margo, 1983). The wide range of tissues affected has justifiably led to aniridia being described as a panocular disorder (Nelson et al., 1984).

One striking feature of aniridia is the variable expressivity. The phenotype can vary dramatically between individuals even within the same family. For example, the iris hypoplasia ranges from a readily visible almost complete absence of the iris, through enlargement and irregularity of the pupil mimicking a coloboma, to small slit-like defects in the anterior layer seen only on transillumination with a slit-lamp (Figure 1; Hittner et al., 1980). The mildest phenotype is detectable as subtle abnormalities of the iris and macula which can be seen only by fluorescein angiography (Hittner et al., 1992). The effect of aniridia on visual function is also very variable. Some families have been reported in which most affected members only had iris hypoplasia and almost normal visual acuity whereas in other cases the presence of various combinations of defects elsewhere in the eye has caused a severe reduction in acuity (Grove, Shaw and Bourque, 1961; Elsas et al., 1977; Hittner et al., 1980). In addition, all aniridia patients are at risk of developing glaucoma. The incidence of this complication has been variously reported as between 6% and 75% which may reflect differences in the selection of patients stud-

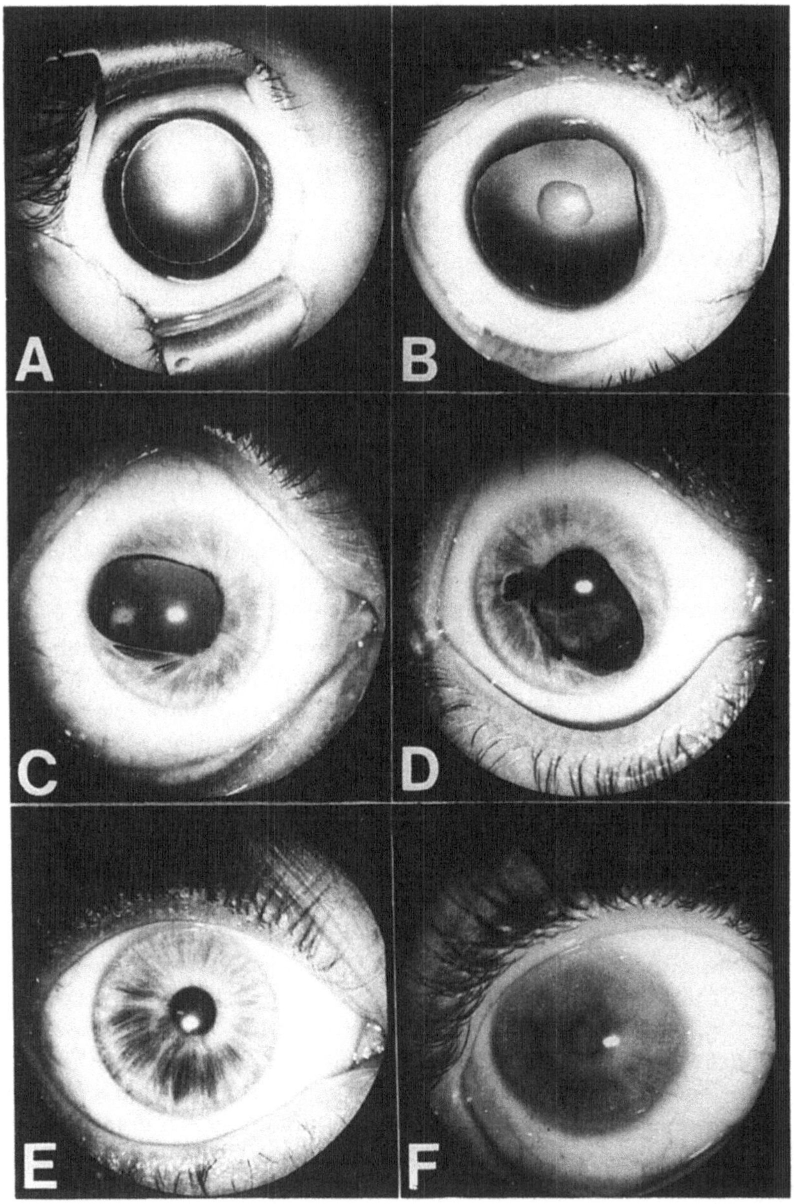

Figure 1 Variable expressivity of aniridia in different members of an affected pedigree. A: virtually complete aniridia. B: small amount of iris visible. C: atypical iris coloboma. D: typical iris coloboma. E: iris thinning with round pupil. F: corneal pannus with iris thinning and round pupil. (*American Journal of Ophthalmology*, 1980, **89**, 531–539). This figure was a generous gift from Dr. H. M. Hittner and is published with permission from the *American Journal of Ophthalmology*. Copyright by The Ophthalmic Publishing Company.

ied and their ages at the time of assessment (Nelson et al., 1984). The glaucoma of aniridia is notoriously difficult to manage and if not treated successfully will destroy any residual vision and can cause intractable ocular pain necessitating removal of the eye.

Genetics

Aniridia is a relatively rare disease, with a population frequency of 1:60,000 to 1:100,000 (Shaw, Falls and Neel, 1960; Shannon et al., 1982). About two thirds of cases are familial, with highly-penetrant autosomal dominant inheritance. The remaining third are sporadic cases with no previous family history. Some of these are new cases of "isolated" aniridia, with no associated anomalies, which will show the usual dominant pattern of inheritance in subsequent generations (Shaw, Falls and Neels, 1960). However, a considerable proportion of these new mutations are associated with cytogenetically visible deletions involving band p13 on the short arm of chromosome 11 (Riccardi et al., 1978; Francke et al., 1979). Most such deletion cases have a constellation of associated anomalies referred to as the contiguous gene syndrome WAGR: a predisposition to Wilms' tumour (a childhood kidney tumour), aniridia, genitourinary abnormalities (particularly cryptorchidism and hypospadias in boys) and often mental retardation. Occasionally other associated anomalies are found with aniridia; these rare syndromes merit special attention to discover the heritability and other allied risks (Nelson et al., 1984).

Aniridia accompanied by cerebellar ataxia and mental retardation is known as Gillespie syndrome (Gillespie, 1965; Nevin and Lim, 1990). This very rare syndrome has been reported in sibs with two unaffected parents, leading to the suggestion of an autosomal recessive mode of inheritance (Gillespie, 1965; Francois, Lentini and de Rouck, 1983; Wittig et al., 1988). However, consanguinity has never been demonstrated and a single parent to child transmission has also been recorded (Crawfurd et al., 1979). It is therefore quite likely that we are in fact dealing with new dominant mutations, and vertical transmission is rarely seen because severely affected patients tend not to reproduce.

MAPPING OF THE ANIRIDIA LOCUS

The Aniridia Locus Maps to 11p13

The iris hypoplasia and associated ocular anomalies seen in aniridia provide no simple immediate biological information about the underlying functional defect. However, the existence of a mutation can be used to identify the gene responsible for these abnormalities. Determination of the genomic position of the mutant gene can lead us by "positional cloning" (Collins, 1992) to the aniridia gene. Genomic position can be deduced either from linkage studies in familial cases or from the analysis of cytogenetic (karyotypic) abnormalities associated with the phenotype. Together, these approaches have provided strong evidence that the aniridia locus maps to chromosome 11, band p13.

Cytogenetics

A key initial observation in mapping the aniridia gene came from cytogenetic analysis of patients with the WAGR syndrome. The visible constitutional interstitial deletions of the short arm of chromosome 11 in these patients had a common region of overlap in band p13 which suggested that 11p13 contained a gene or, more likely, a number of genes which, when hemizygous, caused the WAGR phenotype (Riccardi et al., 1978; Francke et al., 1979).

The 11p13 assignment was strongly supported by evidence from three pedigrees in which aniridia segregated with a chromosomal translocation. Simola et al. (1983) described a family in which three affected members from three generations carried a balanced translocation t(4;11)(q22;p13). Moore et al. (1986) reported a family in which four affected members spanning three generations carried t(11;22)(p13;q12.2). Pettenati, Weaver and Burton (1989) described an affected father and daughter with t(5;11)(q13.1;p13). Wilms' tumour was absent from all three pedigrees, suggesting that Wilms' tumour and aniridia are caused by separate loci within 11p13. The chromosome 11 aniridia locus was designated AN2.

Linkage

A second approach taken to map the aniridia gene was that of linkage, in which affected pedigrees were tested for the segregation of the disease with specific alleles of multiple polymorphic markers of known map position. In one large family all affected individuals carried the same allele of the *ACP1* gene and statistical analysis of the pattern of segregation provided evidence for linkage between aniridia and *ACP1* which maps to 2p25 (Ferrell et al., 1980). The chromosome 2 aniridia locus was designated AN1. In an independent study another large kindred was tested for segregation of aniridia with markers from 2p25 and 11p13. Linkage with 2p25 was excluded in this pedigree while linkage with 11p13 was confirmed (Mannens et al., 1989). It is quite possible for similar phenotypes to be caused by mutations at different loci. However a recent study of the chromosome 2-linked family, including re-evaluation of diagnoses and use of additional markers from 2p and 11p, excluded linkage of aniridia in this family with chromosome 2 and confirmed linkage with 11p13 (Lyons et al., 1992). Since linkage and cytogenetic studies converged on the same small genomic region, this has been the site of intensive effort to identify the aniridia gene(s).

Characterization of the 11p13 WAGR Region

The 11p deletions in different WAGR patients are highly variable in extent, but each must include the genes responsible for the phenotype. Therefore, by determining at the molecular level the proximal (centromeric) and distal (telomeric) limits of the deletions in many patients, it was possible to define the smallest region of overlap (SRO) of the deletions in which the critical genes must lie (Figure 2).

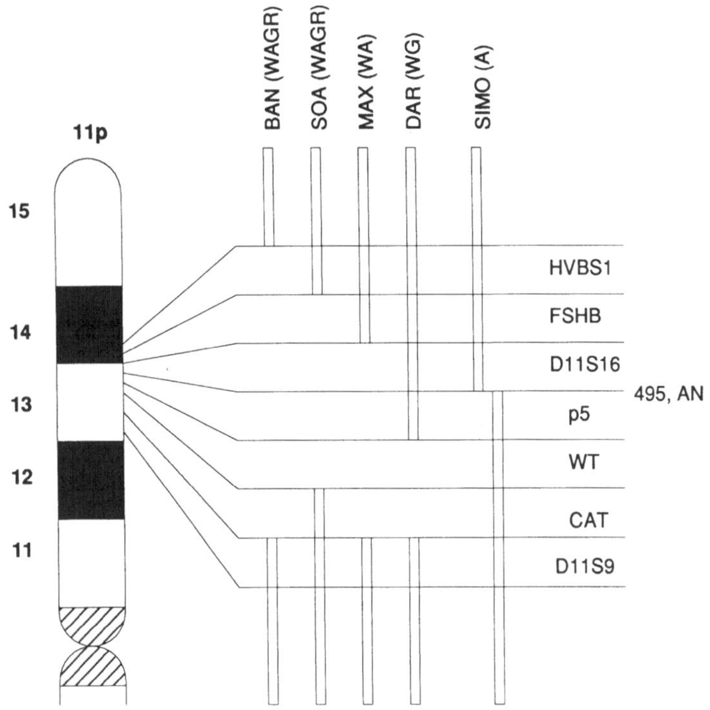

Figure 2 Simplified deletion map of the WAGR region on the short arm of chromosme 11. Vertical bars represent portions of chromosome 11 present in each patient, as defined by the presence or absence of the probes listed to the right. WT, Wilms' tumour locus; AN, aniridia locus. The phenotype of each patient is shown in brackets: W, Wilms' tumour; A, aniridia; G, genitourinary abnormalities; R, mental retardation. BAN and SOA are described in Couillin et al. (1989); DAR in Davis et al. (1988) and Couillin et al. (1989); MAX in Seawright et al. (1989); and SIMO in Junien and McBride (1989).

At first the proximal and distal breakpoints of the deletions were positioned using the relatively low resolution technique of chromosome banding (Riccardi et al., 1978; Francke et al., 1979). Higher resolution was achieved by identifying markers which mapped within the WAGR deletions. This molecular deletion analysis was accomplished either by isolating the deleted chromosome 11 in a somatic cell hybrid or by dosage studies in WAGR cell lines (Seawright et al., 1988; Porteous et al., 1987; Gessler et al., 1989). Initially the markers used were gene products such as enzymes and cell surface markers; later however cloned genes and batteries of random DNA probes isolated from genomic sources enriched in chromosome 11 were predominantly used.

The genes for catalase (*CAT*) and the β-subunit of follicle stimulating hormone (*FSHB*), both of which map to 11p13, were found to be deleted in many but not all WAGR patients, placing them close to, but not within, the SRO (Figure 2; Glaser et al., 1986; Porteous et al., 1987; Seawright et al., 1988). Cytogenetic comparison of the extent of deletions in patients lacking one marker but not the other revealed that *FSHB* was distal to *CAT* with the SRO between the two. The *CAT-FSHB* interval was subdivided into regions which must contain the Wilms' tumour and aniridia genes by determining the positions of the breakpoints in key patients with incomplete WAGR phenotypes. Since aniridia shows complete penetrance, the characterization of the deletion in a patient with reduced catalase activity, Wilms' tumour and genitourinary abnormalities, but not aniridia, placed the Wilms' tumour gene proximal to the aniridia gene (Patient DAR, Figure 2; Davis et al., 1988). The precise position of these breakpoints was defined using random DNA markers. This approach produced a detailed consensus breakpoint map (Junien and McBride, 1989).

Once a high density of markers became available, the construction of a long range restriction map by pulsed-field gel electrophoresis (PFGE) was possible. The high resolution physical map of the 11p13 region produced in this way showed for the first time the distances, in kilobase pairs, between markers, and the positions of the key disease-associated chromosomal rearrangements (Figure 3; Compton et al., 1988; Gessler and Bruns, 1989; Rose et al., 1990). The sites for rare cutting restriction enzymes used for PFGE mapping also mark the position of regions which are relatively rich in unmethylated CpG dinucleotides. Such 'CpG island' sites are often associated with the 5' ends of genes, and are in fact found upstream of both the Wilms' tumour gene (Call et al., 1990; Gessler et al., 1990) and the aniridia gene (Ton et al., 1991).

Cloning of candidate aniridia genes

In theory, disease-associated chromosomal translocations are particularly valuable as tools for positional cloning because the translocation breakpoint often occurs within the gene responsible for the disease phenotype (Gessler, Simola and Bruns, 1989). For this reason, an obvious step towards the identification of the aniridia gene was the characterization of the aniridia-associated translocations. Upon more detailed molecular analysis, the t(11;22) translocation was shown to be associated with a deletion of between 650kb and 850kb of DNA from the 1.4Mb *Not*I fragment in 11p13 (Patient DG-85, Figure 3; Davis et al., 1988; Compton et al., 1988; Rose et al., 1990). Since a stretch of DNA of this size could theoretically encode several genes, the t(11;22) translocation was of limited use in identifying the precise location of the aniridia gene. However, two markers deleted in the t(11;22) translocation were shown to flank the t(4;11) breakpoint and used to position it 100–200kb from the proximal end of the 1.4Mb *Not*I fragment (Patient SIMO, Figure 3; Davis et al., 1989). Another marker mapping just below the breakpoint was used to isolate overlapping DNA fragments from

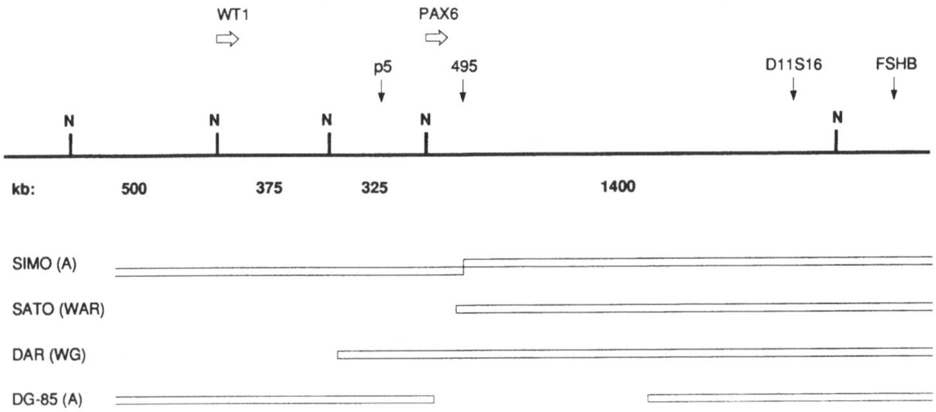

Figure 3 Physical map of the minimal WAGR region in band 11p13 constructed by pulsed field gel electrophoresis (Compton et al., 1988; Gessler and Bruns, 1989; Rose et al., 1990). Sites for the infrequently cutting restriction enzyme *NotI* are shown (N). Positions of some of the probes used to construct the map are represented by vertical arrows. The breakpoints in key patients which helped subdivide the map into regions containing the genes for Wilms' tumour and aniridia are shown below the map. The phenotypes of these patients are designated as for Figure 2. DG-85 is described in Davis et al. (1988), Compton et al. (1988), and Rose et al. (1990); DAR in Call et al. (1990), and van Heyningen et al. (1991); SATO in Ton et al. (1991); and SIMO in Gessler, Simola and Bruns (1989), Davis et al. (1989), and Ton et al. (1991). The Wilms' tumour (*WT1*) and aniridia (*PAX6*) genes are represented by horizontal arrows (Call et al., 1990; Gessler et al., 1990; Ton et al., 1991).

genomic libraries and thus to clone the junction fragment (Probe 495, Figure 3; Gessler, Simola and Bruns, 1989). No sizeable deletion appears to have occurred as both der(4) and der(11) are detected by the 350bp junction fragment. This fragment when hybridised to genomic DNA from different vertebrate species detected conserved fragments in other mammals and chicken. Because expressed sequences are more highly conserved between species, this "zoo blot" approach has proved successful in identifying genes in genomic DNA. When sequenced, the breakpoint region was found to contain an open reading frame flanked by consensus splice sites, features which are characteristic of a coding exon of a gene. Although the breakpoint locus is a strong aniridia candidate on the basis of position, no cognate cDNA clones or sequence data have yet been reported. The role of this proposed gene in aniridia remains uncertain but it is now known that the junction fragment is not deleted in a WAGR patient with classical aniridia whose distal breakpoint is the most centromeric yet known (Patient SATO, Figure 3; Ton et al., 1991).

Meanwhile a second candidate aniridia gene was isolated by Ton et al. (1991) who were approaching the region containing the translocation breakpoint using the techniques of chromosome jumping and walking. Their starting point was a probe (p5, Figure 3) in the 325kb *NotI* fragment immediately proximal to the 1.4Mb *NotI* fragment known to contain the breakpoint. While moving towards the breakpoint, they cloned the junction of the two *Not*I fragments and since this potentially marked a CpG island, it was decided to search for coding sequences in this region. One genomic fragment, conserved across species, was successfully used to isolate cDNA clones from an 8.5d embryonic mouse library and a human adult retinal library. The mouse cDNA was used in turn to isolate a cDNA from a human foetal eye library (Ton et al., 1991).

Sequencing of the cDNA clones revealed that the candidate gene was a member of the *Pax* gene family. *Pax* genes are characterised by the presence of a motif known as a paired box which encodes a 120-amino acid DNA-binding domain. The paired box was first described in the *Drosophila* segmentation gene *paired* (Frigerio et al., 1986; Bopp et al., 1986) and has since been found in other developmentally important *Drosophila* genes (Baumgartner et al., 1987; Bopp et al., 1989). Subsequently, a family of paired-box containing genes have been described in mouse, man, zebrafish and chicken (Deutsch and Gruss, 1991). The expression patterns of the *Pax* genes, as identified by RNA *in situ* hybridisation, and the involvement of these genes in developmental mutations in *Drosophila*, mouse and man has argued for a key role for them in pattern formation (Deutsch and Gruss, 1991; Gruss and Walther, 1992). The vertebrate *Pax* gene family has been best characterised in mouse and so far contains eight members (Walther et al., 1991; Gruss and Walther, 1992). On the basis of predicted amino acid sequence, the candidate aniridia gene is the human homologue of the mouse *Pax-6* gene; indeed the predicted protein products of the mouse (*Pax-6*) and human (*PAX6*) genes differ by only one amino acid. As well as the paired motif, the *PAX6* gene also encodes a paired-type homeobox, another phylogenetically conserved motif found in developmentally important genes and also known to encode a DNA binding activity (Walther and Gruss, 1991; Deutsch and Gruss, 1991).

The candidacy of *PAX6* as the aniridia gene was strengthened by RNA *in situ* hybridisation studies designed to determine the pattern of expression of the *PAX6* gene. This approach revealed that *PAX6* was strongly expressed in the developing eye, at a stage preceeding iris formation, in both human and mouse embryos (Ton et al., 1991; Walther and Gruss, 1991). Interestingly, the domain of expression in the eye was not confined to the rim of the optic cup which would ultimately give rise to the neuroectodermal component of the iris, but also encompassed the neural retina, the lens and the overlying surface ectoderm which contributes towards the cornea. All these structures can be affected in aniridia, and the *in situ* results were therefore consistent with *PAX6* being the aniridia gene. However, to show that a candidate gene isolated by positional cloning is really responsible for the disease it is necessary to demonstrate intragenic mutations in affected individuals. The first genetic evidence that *PAX6* was involved in eye development came from analysis of a mouse mutation called *Small eye.*

Small eye : a mouse model for aniridia

Small eye is a semi-dominant mutation affecting eye development in the mouse. Three alleles have been described, two spontaneously occuring (*Sey* and Sey^{Dey})

and one radiation induced (*Sey*^{H}; Green, 1989). The most obvious phenotypic feature of heterozygous *Small eye* mice is a reduction in eye size. A more detailed examination of the eyes of *Sey*/+ heterozygotes may also reveal cataracts, small vacuolated lenses, adhesions between the lens and cornea, and an abnormally folded neuroretina (Hogan et al., 1986). The eyes of *Sey*^{Dey}/+ and *Sey*^{H}/+ heterozygotes often have colobomata and are generally smaller that those of *Sey*/+ mice but otherwise display a similar range of ocular anomalies (Hogan et al., 1986; Theiler, Varnum and Stevens, 1978).

Mice homozygous for the *Sey* allele have no eyes and imperforate hypoplastic snouts and die shortly after birth because they cannot breathe while feeding (Figure 4; Hogan et al., 1986; 1988). This implicates the product of the *Small eye* gene not only in eye development but in the morphogenesis of the frontonasal region (Hogan et al, 1986; 1988). *Sey*^{Dey} and *Sey*^{H} homozygotes have not been observed and are presumed to die early in gestation. The pleiotropic effects of the *Sey*^{Dey} and *Sey*^{H} mutations suggested that these were deletions which affected a number of genes required for normal growth and viability in addition to eye development. In contrast, the relatively mild phenotype associated with the *Sey* allele led to the proposal that this was most likely to be a point mutation (Hogan, Hetherington and Lyon, 1987).

No comparison was initially made between *Small eye* and aniridia because they appeared distinct at the gross phenotypic level. However, genetic mapping of the *Small eye* mutation revealed that it fell into a linkage group on mouse chromosome 2 which is syntenic with band 11p13 of the human genome (Siracusa et al., 1990; van der Meer-de Jong et al., 1990). It was then shown that markers which are commonly deleted in human WAGR patients are also deleted from mice carrying the *Sey*^{Dey} allele (Glaser, Lane and Housman, 1990). This confirmed that *Sey*^{Dey} was a deletion mutation and, more importantly, suggested that *Small eye* and aniridia were genetically equivalent. Further evidence to support this hypothesis included the similarity between the phenotype of *Sey/Sey* homozygotes and that of a human foetus, presumed homozygous, born without eyes or nose as the result of an aniridia/aniridia mating (Hodgson and Saunders, 1980). In addition, the autosomal semi-dominant mode of inheritance is common to *Small eye* and aniridia, as is the very high penetrance but variable expressivity of the phenotype. Most recently, a careful re-evaluation of the eye phenotype in *Sey*/+ heterozygotes has revealed for the first time iris hypoplasia with a striking resemblance to that seen in aniridia (Jordan et al., 1992).

MUTATION ANALYSIS OF THE *PAX6* GENE

Mouse mutations

Because the *Small eye* mouse was such a plausible model for aniridia and biological material is more readily available from laboratory animals than patients, mutation analysis of the *Pax-6* gene was first carried out in the mouse.

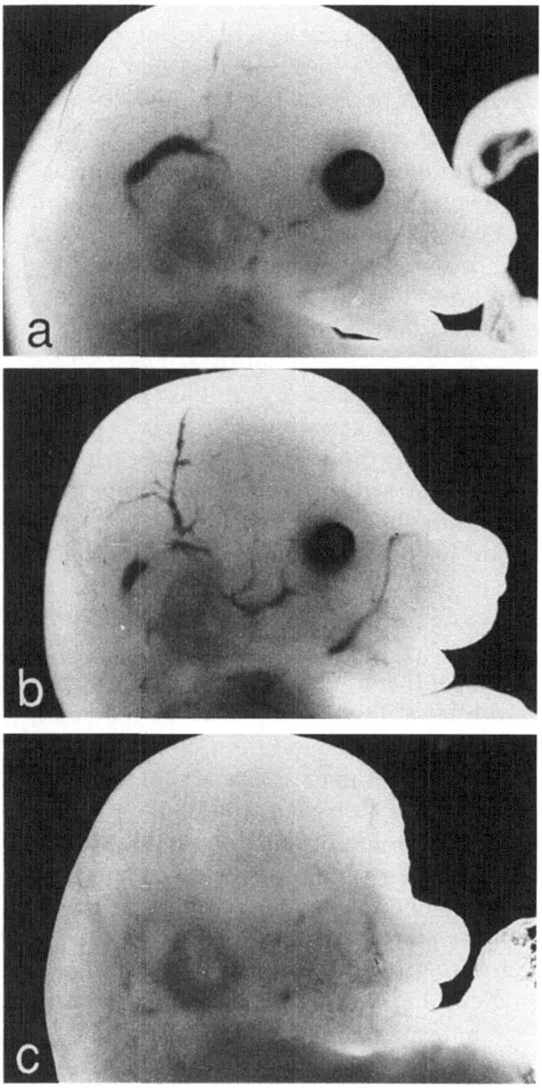

Figure 4 The *Small eye* mouse: photographs of the 15.5 day post coitum mouse embryo. (a) wild type embryo, +/+. (b) heterozygote embryo, *Sey*/+. The eye is reduced in size partly as a result of the smaller lens, and the outline of the optic cup rim is irregular, suggesting delay in the closure of the embryonic fissure. (c) homozygote embryo, *Sey/Sey*. No visible eye is formed and the absence of a nasal cavity alters the shape of the snout.
Reprinted with permission from *Nature* (Hill et al., 1991, **354**, 522–525). Copyright 1991 Macmillan Magazines Limited.

Dosage analysis and interspecific backcrosses demonstrated that *Pax-6* was deleted in the *Sey*^{H} mouse (Hill et al., 1991). A more important goal however was to demonstrate intragenic mutations of *Pax-6* in *Small eye* alleles which were predicted to be point mutations on the basis of phenotype. At this stage little was known about the genomic organisation of *Pax-6* so the most feasible approach to mutation analysis was through mRNA. From the expression pattern it was clear that *Pax-6* was most highly expressed in neural tissue in embryos between 8 and 18 days gestation. Therefore RNA was extracted from the heads of 15d embryos, reverse transcribed, PCR-amplified and analysed for the presence of point mutations by chemical cleavage of mismatch (Hill et al., 1991). This RNA-PCR approach revealed a single base change in the *Sey Pax-6* gene which created a termination codon in the open reading frame. The predicted effect of this would be to truncate the protein product before the homeodomain and thus to impair the function of the Pax-6 protein (Figure 5b).

Additional evidence that *Pax-6* is mutated in *Small eye* came from analysis of the ENU-induced *Sey*^{Neu} mutation (Hill et al., 1991). Although allelism of this mutation with the *Small eye* locus had not been formally demonstrated, genetic mapping suggested that it mapped to the appropriate region of chromosome 2, and the phenotype was virtually indistinguishable from that of *Sey*. RNA-PCR analysis of *Pax-6* in *Sey*^{Neu} revealed an unusually large *Pax-6* transcript which was shown upon sequencing to include an intron. Sequencing of the genomic DNA of *Sey*^{Neu} showed that the invariant G residue of the splice donor site was mutated, thus preventing intron excision from the mRNA. The predicted effect of this mutation would be to terminate translation at a stop codon within the intron and thus to truncate the protein before the serine-threonine rich C-terminal domain (Figure 5b). Although the function of this domain is not known, it is similar in amino acid composition to the trans-activation domain of the transcriptional regulators Oct-1 and Oct-2 (Tanaka and Herr, 1990). Analysis of *Small eye* alleles therefore provided strong evidence for the involvement of *Pax-6* in eye development.

Human mutations

The analysis of human aniridia mutations lagged behind that of mouse mutations because the only material available was lymphoblastoid cell lines which do not express *PAX6* at high levels. Initially the human *PAX6* gene was analysed at the genomic level on an exon-by-exon basis which first necessitated the elucidation of the intron-exon structure of the gene. The genomic organisation of *PAX6* was found to be complex with at least 10 exons spanning about 25kb of the genome (Figure 5a; Jordan et al., 1992). The mutation analysis strategy involved PCR-amplifying a given exon from the genomic DNA of a panel of patients and then resolving the PCR product on a Hydrolink polyacrylamide gel, where heteroduplexes caused by annealing of wild-type products with mutant products have an altered mobility relative to homoduplexes. This method can be used because each patient is expected to have a normal as well as a mutated copy of *PAX6* so

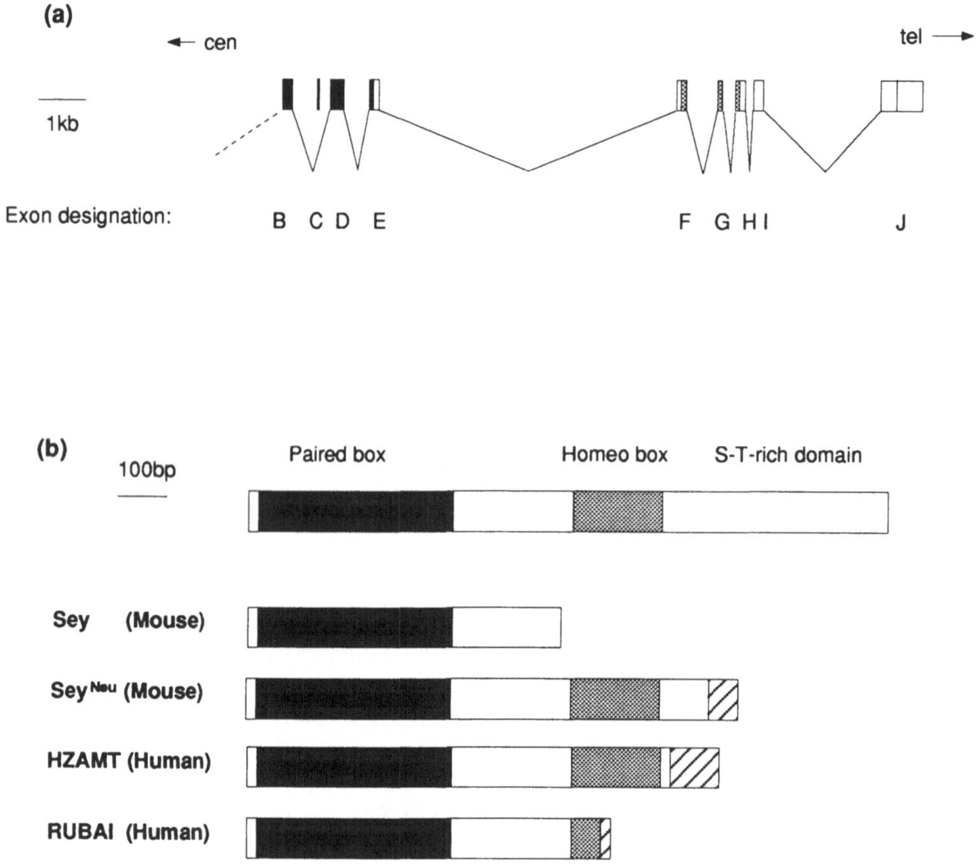

Figure 5 (a) Preliminary genomic organization of the human *PAX6* gene. Exons are represented by rectangles, with regions encoding the paired domain shown in black and regions encoding the homeodomain shown stippled. Since the number of exons encoding the 5' untranslated region is uncertain, as is the position of the translation initiation codon, known exons are arbitrarily designated by letters. Exon C encodes a 42bp sequence which is alternatively spliced into the paired box. Exon J encodes the termination codon (vertical line) and the 3' untranslated region (Jordan et al., 1992).
(b) *PAX6/Pax-6* mutations in *Small eye* alleles and human aniridia patients. At the top is shown the structure of wild-type *PAX6/Pax-6* cDNA with the paired box (black), the paired-type homeo box (stippled) and the region encoding the serine-threonine (S-T) rich domain (Ton et al., 1991; Walther and Gruss, 1991). Below are illustrated the predicted effects on the cDNA structure of the known *Pax-6* (*Small eye*) and *PAX6* (aniridia) mutations (Hill et al., 1991; Jordan et al., 1992). Hatched boxes represent altered open reading frames beyond the mutation site.

that mismatched heteroduplexes are formed when the region containing the mutation is amplified. This approach was used successfully to detect heteroduplexes of anomalous mobility in the PCR product derived from exon H of one aniridia patient. Nucleotide sequencing of the PCR product revealed a two nucleotide insertion which changed the reading frame and would be predicted to cause chain termination just distal to the homeodomain (Patient HZAMT, Figure 5b; Jordan et al., 1992).

A second strategy which has been used to find *PAX6* mutations is that of nested RNA-PCR, which is carried out essentially as described for the *Small eye* analysis but with extra rounds of PCR amplification to compensate for the very low "breakthrough" levels of *PAX6* RNA production in the lymphoblastoid cell lines which are the only available material from the patients. Reverse transcription is followed by two rounds of PCR amplification using two *PAX6*-specific primer pairs, one nested inside the other (Roberts et al., 1991). When this method was applied to the homeobox region of the gene in a panel of patients, an RNA-PCR product of reduced size was detected in addition to the normal product in one patient. Sequencing revealed that the short transcript did not contain exon G, but no deletion was detectable at the genomic level. Therefore the mutation in this patient is most likely a point mutation or sub-visible deletion at the DNA level, probably at or near to a splice site, which causes exon G to be spliced out of the mature *PAX6* mRNA product. The predicted protein product of this mutant allele would terminate after the first third of the homeodomain (Patient RUBAI, Figure 5b; Jordan et al., 1992). These analyses strongly suggest that *PAX6* mutations cause aniridia in at least some patients.

CLINICAL SIGNIFICANCE

Immediate Clinical Applications

These early results of *PAX6* mutation studies open up the possibility of genetic counselling and prenatal diagnosis, both of which are often requested by aniridia patients. The frequency and spectrum of *PAX6* mutations can now be assessed in a large cohort of patients. The existence of well-documented deletion aniridia patients whose phenotypes are clinically indistinguishable from familial cases, who would be predicted to have point mutations, suggests that aniridia results from *PAX6* haplo-insufficiency i.e. loss of one active copy of the *PAX6* gene. Since many different mutations could lead to loss of gene function, we expect to see a wide spectrum of mutations both within the coding region and in control regions outside the transcribed sequence. This implies that each sporadic patient or family will have to be assessed *de novo* before genetic counselling can be given. The way is already open for mutation analysis at the RNA level which will reveal point mutations, small insertions or deletions, or aberrant splicing, all of which can lead to deleterious changes in amino acid sequence (Figure 5b; Hill et al., 1991; Jordan et al., 1992).

In patients where RNA-PCR analysis does not reveal a mutation, it is possible that one copy of the *PAX6* gene is not being expressed. This could result from a deletion of all or part of the *PAX6* gene or a mutation in a control region. Nothing is known at present about the elements which control *PAX6* expression but these will have to be defined, most probably in the mouse, before mutation analysis is possible. Loss of expression of one copy of *PAX6* could be verified by searching in the genomic DNA of patients for heterozygosity at a transcribed site and then showing at the RNA level that only one allele is expressed. A further possibility in patients where no *PAX6* mutation is detectable is that they have mutations in a second aniridia gene. The existence of a second gene is suggested by the t(4;11) breakpoint, which is associated with aniridia but which does not physically disrupt *PAX6* (Gessler, Simola and Bruns, 1989; Ton et al., 1991; Jordan et al., 1992). However the aniridia in this family could ultimately be explained by a "position effect" where the translocated portion of chromosome 4 silences *PAX6* at a distance, perhaps by altering chromatin structure. Again, expressed heterozygosity would be valuable in determining whether the copy of *PAX6* on the translocation chromosome is transcribed.

Another clinically significant aspect of this work is the assessment of the risk of Wilms' tumour in newborns with sporadic aniridia. In contrast to aniridia, both Wilms' tumour and the associated genitourinary abnormalities show reduced penetrance so that even sporadic cases of isolated aniridia may carry a deletion, putting them at risk of developing nephroblastoma. During the mapping of the WAGR region, cosmid clones were isolated for *WT1*, *PAX6* and two intervening markers (Fantes et al., 1992). These can be used in fluorescent *in situ* hybridisation (FISH) analysis of mitotic chromosomes to determine whether or not a deletion is present. If a deletion is indicated, assessment of incipient Wilms' tumour by renal ultrasound should be carried out at regular intervals. We have validated this technique retrospectively in several submicroscopic WAGR deletions. In one unexpected and presumably very rare case a mother with sporadic isolated aniridia gave birth to a son with aniridia who later died as a result of Wilms' tumour which arose in a horseshoe kidney. FISH analysis showed that both were deleted for *WT1* and *PAX6* (Fantes et al., 1992).

A role for *PAX6* mutations in other eye disorders?

A great deal of phenotypic variability has been observed among aniridia patients, even within families who are presumably segregating for the same mutation (Hittner et al., 1980). This variability is particularly striking in the *Small eye* mouse where there are marked differences in eye phenotype even within the same inbred strain (Roberts, 1967). The phenotypic heterogeneity is of clinical importance in assessing familial patterns of inheritance and the associated genetic risks because some individuals are so mildly affected that they may be misdiagnosed (Hittner et al., 1980; Lyons et al., 1992). In view of the heterogeneity amongst aniridia patients it is possible that *PAX6* mutations are also responsible for other disorders which are usually considered to be clinically distinct but may be part of a wide phenotypic spectrum that encompasses aniridia. These include "atypical coloboma" (Figure 1), early onset cataracts and a range of anterior segment defects such as Peters' anomaly, all of which have been observed in families segregating for aniridia and which are consistent with the pattern of *PAX6* expression in the eye (Hittner et al., 1980; Beauchamp, 1978). It will be interesting to see whether *PAX6* is mutated in these disorders and, if so, how the spectrum of mutations compares to that defined for "classical" aniridia.

In the majority of aniridia patients no neurological or brain abnormalities have been noted. It was therefore somewhat surprising to find that *PAX6* is expressed not only in the developing eye but also in the brain and neural tube (Ton et al., 1991; Walther and Gruss, 1991). However in cases of the rare Gillespie syndrome, aniridia is associated with cerebellar ataxia. Might this phenotype also be accounted for by mutation in *PAX6*? It may be significant in this respect that in addition to the eye the cerebellum is the major expressing tissue in adult mice (Ton, Miwa and Saunders, 1992).

THE ROLE OF PAX6 IN EYE DEVELOPMENT

Identifying the affected tissue

The first attempts to understand the molecular pathology of aniridia began by trying to identify which tissues are primarily affected. To decide this, it was necessary to apply what was known about the embryology of the eye. By taking the iris hypoplasia as the typical feature of the disorder and identifying the tissues involved in its embryogenesis, two models were derived. The two tissues believed to contribute to the formation of the iris were the neuroectoderm and the mesoderm, and the two models proposed were the neuroectodermal and the mesodermal, the implication being a primary defect in one or other (Mann, 1957). The posterior layer of the iris arises from the anterior rim of the bilayered optic cup which develops from neuroectoderm, while the anterior layer is formed from what was then thought to be mesoderm but is now known to be mainly neural crest-derived mesectoderm or ectomesenchyme (Johnston et al., 1979; Beauchamp and Knepper, 1984). Looking at the affected ocular tissues throughout the eye it becomes clear that they all receive an ectodermal contribution. The corneal epithelium and lens develop from surface ectoderm while part of the iris and ciliary body, most of the optic nerve and all of the macula develop from neuroectoderm. Although the surface ectoderm and the neuroectoderm can be distinguished as two separate tissues, in the early stages of development (mid to late gastrula through to neural tube stage) they lie in very close proximity at a time when the optic sulcus is first appearing in the neuroectoderm and when the adjacent surface ectoderm is first acquiring a lens forming bias (Couly and Le

Dourain, 1988; Servetnick and Grainger, 1991). The optic sulcus is the first morphological indication of the emergence of the primordial eye from the optic placode on the surface of the developing neural plate. It extends by invagination to form the optic stalk, the end of which then swells to develop the optic vesicle and finally invaginates to produce the optic cup. The posterior layer of the iris, the epithelial layers of the ciliary body and the whole of the neuroretina all develop from the optic cup. The surface ectoderm proceeds through acquisition of a lens forming bias to specification of the lens placode. On contact with the optic vesicle this placode invaginates to form a lens vesicle which eventually separates from the surface ectoderm and moves into the optic cup which is forming simultaneously. The corneal epithelium will form from the remains of the lens placode (O'Rahilly, 1966; Kaufman, 1979). It is therefore quite conceivable that the primary defect in aniridia lies entirely within ectoderm and this idea is supported by the discovery that *PAX6* is expressed in the rim of the optic cup, the developing neuroretina, the lens and the corneal epithelium. It remains to be determined just what *PAX6* is doing in this respect.

PAX6 affects formation of the lens

It is tempting to speculate that *PAX6* is somehow involved in the induction of the lens. The absence of any functional *Pax-6* transcripts i.e. in *Sey/Sey* mice, results in the absence of lens formation in the presence of a correctly positioned optic vesicle which fails to form an optic cup and undergoes involution (Hogan et al., 1986). Existing data show that lens vesicles can form in the absence of the optic vesicle and that the optic vesicle is not capable of producing lenses from non-specified surface ectoderm (Muthukkaruppan, 1965; Grainger, Henry and Henderson, 1988). These findings contradict the results of earlier induction experiments. The most recent experiments in this field point to the existence of a crucial interaction between the neural plate and the adjacent surface ectoderm in order to achieve lens placode specification (Henry and Grainger, 1990). It is interesting to note that many of the so-called free lenses seen in earlier tissue manipulation experiments i.e. lenses occurring in the absence of an optic vesicle, were found in association with the developing nose, since we now know that both the olfactory bulb and the nasal placode express *Pax-6* during development (Saha, Span and Grainger, 1989; Ton et al., 1991; Walther and Gruss, 1991). These findings are evidence of a critical role for *Pax-6* in the formation of the lens, perhaps by the early specification of the lens placode. To understand this and other roles of *Pax-6* will require additional experiments using both traditional tissue manipulation methods and modern molecular techniques. By assembling conjugates of tissues from different stages of development and with different degrees of *Pax-6* expression, it may be possible to define developmental windows and tissue interactions. Somatic cell chimeras in which mixed populations of *Pax-6* expressing and non-expressing cells are juxtaposed in the developing embryo will tell us whether or not the gene acts cell autonomously or if its function can be rescued by the presence of adjacent normal cells (Hatta et al., 1991). The "free lens" evidence argues for the latter. It is to be hoped that such experiments will also provide valuable insights into the molecular events which lie behind the normal development of the eye.

What controls *PAX6* and what are its targets?

Although aniridia can have serious consequences for affected individuals in terms of visual disability, from a developmental point of view the phenotype is surprisingly subtle and localized given that the domain of expression of the *PAX6* gene includes the developing eye, nose, forebrain and neural tube (Walther and Gruss, 1991). The phenotypes of aniridia patients and *Sey/+* mice imply that the eye is especially sensitive to a reduction in *PAX6* levels and that there is a threshold of protein activity which must be exceeded for normal development to proceed (Hill and van Heyningen, 1992). In *Sey/Sey* mice, which completely lack wild type *Pax-6* activity, the structures derived from the optic and nasal placodes fail to mature, but the brain and spinal column appear to be normal (Hill et al., 1991). It may be that neurological defects are present but have been overlooked because they are so much less obvious than the abnormal facial phenotype. Alternatively there may be functional redundancy of Pax gene expression such that absence of *Pax-6* is compensated for by other members of the Pax family which are expressed in overlapping domains during neural tube development (Deutsch and Gruss, 1991).

If, as suggested above, Pax-6 protein activity must exceed certain threshold levels to initiate developmental pathways, the expression of the *Pax-6* gene must be tightly regulated. Pax-6 contains two conserved DNA binding domains and is most likely a transcriptional regulator which will itself control the expression of downstream target genes. At present, nothing is known about these interacting genes and gene products, but one approach to identifying them is to study the expression of other developmental regulators known to be expressed in overlapping temporal and spatial domains. Among these are *Hox-7* and *Hox-8* (*msh*-like homeobox genes believed to define domains within the developing eye, Monaghan et al., 1991); *Emx-1* and *Emx-2* (homeobox genes related to the *Drosophila empty spiracles* gene, Simeone et al., 1992); *XANF-1* (a *Xenopus* homeobox gene expressed in the anterior neural folds, Zaraisky et al., 1992); and the *motch* gene (encoding a transmembrane protein believed to be involved in cell-cell signalling, Del Amo et al., 1992). In addition, it will be very interesting to look for alterations in *Pax-6* expression in mice with other eye mutations such as aphakia (*ak*), coloboma (*Cm*), microphthalmia defective iris (mi^{di}), eye lens obsolescence (*Elo*), dysgenic lens (*dl*), and cataract and small eye (*Cts*) (Green, 1989). By looking at the expression of these genes in the *Small eye* mouse and the expression of *Pax-6* in these mutants it may be possible to detect a hierarchy of interactions. Another more direct method to identify downstream targets involves incubating the Pax-6 protein with linkered genomic DNA fragments and PCR-amplifying the bound fragments to create a library of Pax-6 recognition sequences (Kinzler and Vogelstein, 1989).

Alternative splice forms of transcriptional regulators may be important in the control of target genes. For example, alternative splice forms affecting the structure of the zinc finger DNA binding region of *WT1*, the Wilms' tumour predisposition gene, can alter its binding specificity, thus raising the possibility of interaction with more than one physiological downstream target (Bickmore et al., 1992). Alternative splicing in the paired box of *PAX6*, which results in the insertion of an additional fourteen amino acids into this DNA binding domain, has been demonstrated in mouse, zebrafish and man, but its functional significance remains to be demonstrated (Walther and Gruss, 1991; Jordan et al., 1992; Puschel et al., 1992). RNA *in situ* hybridisation using probes specific for the alternate splice forms may reveal spatial or temporal heterogeneity in their expression.

CONCLUSION

The isolation of the *PAX6* gene and the subsequent discovery of mutations in *PAX6* in aniridia patients illustrates the power of positional cloning techniques to identify the gene underlying a human disease. Although the prospects of developing a treatment for aniridia lie in the distant future there are many immediate benefits from the identification of the aniridia gene. Genetic counselling of aniridia patients will now include the possibility of pre-natal diagnosis, while in atypical cases genetic counselling may be available for the first time. In addition, the elucidation of the role of *PAX6* in aniridia and *Small eye*, the mouse model, will yield fascinating insights into the normal development of the eye.

NOTE :

The complete genomic organisation of the human PAX6 gene has now been deduced (Glaser, Walton and Maas, 1992). Intragenic PAX6 mutations have been characterised in eight additional aniridia patients (Glaser, Walton and Maas, 1992; Hanson et al., 1993)

REFERENCES

Baumgartner, S., Bopp, D., Burri, M. and Noll, M. (1987). Structure of two genes at the *gooseberry* locus related to the *paired* gene and their spatial expression during Drosophila embryogenesis. *Genes and Development* **1**, 1257–1267.

Beauchamp, G. R. (1978). Anterior segment dysgenesis keratolenticular adhesion and aniridia. *Journal of Paediatric Ophthalmology and Strabismus* **17**, 55–58.

Beauchamp, G. R., and Knepper, P. A. (1984). Role of the neural crest in anterior segment development and disease. *Journal of Pediatric Ophthalmology and Strabismus* **21**, 209–214.

Bickmore, W. A., Oghene, K., Little, M. H., Seawright, A., van Heyningen, V. and Hastie, N. D. (1992). Modulation of DNA binding specificity by alternative splicing of the Wilms tumor *wt1* gene transcript. *Science* **257**, 233–237.

Bopp, D., Burri, M., Baumgartner, S., Frigerio, G. and Noll, M. (1986). Conservation of a large protein domain in the segmentation gene *paired* and in functionally related genes of Drosophila. *Cell* **47**, 1033–1040.

Bopp, D., Jamet, E., Baumgartner, S., Burri, M. and Noll, M. (1989). Isolation of two tissue-specific *Drosophila* paired box genes, *Pox meso* and *Pox neuro*. *The EMBO Journal* **8**, 3447–3457.

Call, K. M., Glaser, T., Ito, C. Y., Buckler, A. J., Pelletier, J., Haber, D. A. et al., (1990). Isolation and characterization of a zinc finger polypeptide gene at the human chromosome 11 Wilms' tumor locus. *Cell* **60**, 509–520.

Collins, F. S. (1992). Positional cloning: Let's not call it reverse anymore. *Nature Genetics* **1**, 3–6.

Compton, D. A., Weil, M. M., Jones, C., Riccardi, V. M., Strong, L. C. and Saunders, G. F. (1988). Long range physical map of the Wilms' tumor-aniridia region on human chromosome 11. *Cell* **55**, 827–836.

Couillin, P., Azoulay, M., Henry, I., Ravise, N., Grisard, M. C., Jeanpierre, C. et al. (1989). Characterization of a panel of somatic cell hybrids for subregional mapping along 11p and within band 11p13. *Human Genetics* **82**, 171–178.

Couly, G., and Le Dourain, N. (1988). The fate map of the cephalic neural primordium at the presomitic to the 3-somite stage in the avian embryo. *Development* ***103*** (Supplement), 101–113.

Crawfurd, M. d'A., Harcourt, R.B. and Shaw, P. A. (1979). Non-progressive cerebellar ataxia, aplasia of pupillary zone of iris, and mental subnormality (Gillespie's syndrome) affecting 3 members of a non-consanguineous family in 2 generations. *Journal of Medical Genetics* **16**, 373–378.

Davis, L. M., Stallard, R., Thomas, G. H., Couillin, P., Junien, C., Nowak, N. J. et al. (1988). Two anonymous DNA segments distinguish the Wilms' tumor and aniridia loci. *Science* **241**, 840–842.

Davis, L. M., Everest, A. M., Simola, K. O. J. and Shows, T. (1989). Long-range restriction map around 11p13 aniridia locus. *Somatic Cell and Molecular Genetics* **15**, 605–615.

Del Amo, F. F., Smith, D. E., Swiatek, P. J., Gendron-Maguire, M., Greenspan, R. J., McMahon, A. P., and Gridley, T. (1992). Expression pattern of Motch, a mouse homolog of Drosophila Notch, suggests an important role in early postimplantation mouse development. *Development* **115**, 737–744.

Deutsch, U. and Gruss, P. (1991). Murine paired domain proteins as regulatory factors of embryonic development. *Seminars in Developmental Biology* **2**, 413–424.

Elsas, F. J., Maumenee, I. H., Kenyon, K. R. and Yoder, F. (1977). Familial aniridia with preserved ocular function. *American Journal of Ophthalmology* **83**, 718–724.

Fantes J. A., Bickmore, W. A., Fletcher, J. M., Ballesta, F., Hanson, I. M. and van Heyningen, V. (1992). Non-radioactive *in situ* hybridization for the rapid analysis of submicroscopic deletions at the WAGR locus. *American Journal of Human Genetics* in press.

Ferrell, R. E., Chakarvarti, A., Hittner, H. M. and Riccardi, V. (1980). Autosomal dominant aniridia: probable linkage to acid phosphatase-1 locus on chromosome 2. *Proceedings of the National Academy of Sciences USA* **77**, 1580–1583.

Francke, U., Holmes, L. B., Atkins, L. and Riccardi, V. M. (1979). Aniridia-Wilms' tumor association: evidence for specific deletions of 11p13. *Cytogenetics and Cell Genetics* **24**, 185–192.

Francois, J., Lentini, F. and de Rouck, F. (1983). Gillespie's syndrome. *Ophthalmic Paediatrics and Genetics* **4**, 29–32

Frigerio, G., Burri, M., Bopp, D., Baumgartner, S. and Noll, M. (1986). Structure of the segmentation gene *paired* and the *Drosophila* PRD gene set as part of a gene network. *Cell* **47**, 735–746.

Gessler, M. and Bruns, G. A. P. (1989). A physical map around the WAGR complex on the short arm of chromosome 11. *Genomics* **5**, 43–55.

Gessler, M., Thomas, G. H., Couillin, P., Junien, C., McGillivray, B. C., Hayden, M. et al. (1989). A deletion map of the WAGR region on chromosome 11. *American Journal of Human Genetics* **44**, 486–495.

Gessler, M., Simola, K. O. J. and Bruns, G. A. P. (1989). Cloning of breakpoints of a chromosome translocation identifies the AN2 locus. *Science* **244**, 1575–1578.

Gessler, M., Poustka, A., Cavenee, W., Neve, R. L., Orkin, S.H. and Bruns, G. A. P. (1990). Homozygous deletions in Wilms tumours of a zinc-finger gene identified by chromosome jumping. *Nature* **343**, 774–778.

Gillespie, F. D. (1965). Aniridia, cerebellar ataxia and oligophrenia in siblings. *Archives of Ophthalmology* **73**, 338–341.

Glaser, T., Lewis, W. H., Bruns, G., Watkins, P. C., Rogler, C. E., Shows, T. B. et al. (1986). The β-subunit of follicle stimulating hormone is deleted in patients with aniridia and Wilms' tumor, allowing a further definition of the WAGR locus. *Nature* **321**, 882–887.

Glaser, T., Lane, J. and Housman, D. (1990). A mouse model of the aniridia-Wilms tumor deletion syndrome. *Science* **250**, 823–827.

Glaser, T., Walton, D. S., and Maas, R. L. (1992) Genomic structure, evolutionary conservation and aniridia mutations in the human *PAX6* gene. *Nature Genetics* **2**, 232–239.

Grainger, R. M., Henry, J. J., and Henderson, R. A. (1988). Reinvestigation of the role of the optic vesicle in embryonic lens induction. *Development* **102**, 517–526.

Green, M. C. (1989). Catalog of mutant genes and polymorphic loci. In *Genetic variants and strains of the laboratory mouse*, edited by M.F. Lyon and A.G. Searle, pp 12-403. Second edition. Oxford University Press.

Grove, J.H., Shaw, M.W. and Bourque, G. (1961). A family study of aniridia. *Archives of Ophthalmology* **4,** 105–118.

Gruss, P. and Walther, C. (1992). Pax in development. *Cell* **69**, 719–722.

Hanson, I. M., Seawright, A., Hardman, K., Hodgson, S., Zaletayev, D., Fekete, G., and van Heyningen, V. (1993) PAX6 mutations in aniridia. *Human Molecular Genetics* **2**, 915–920.

Hatta, K., Kimmel, C. B., Ho, R. K. and Walker, C. (1991). The cyclops mutation blocks specification of the floor plate of the zebrafish central nervous system. *Nature* **350**, 339–341.

Henry, J. J. and Grainger R. M. (1990). Early tissue interactions leading to embryonic lens formation in Xenopus laevis. *Developmental Biology* **141**, 149–163.

Hill, R. E., Favor, J., Hogan, B. L. M., Ton, C. C. T., Saunders, G. F., Hanson, I. M. et al. (1991). Mouse Small eye results from mutations in a paired-like homeobox-containing gene. *Nature* **354**, 522–525.

Hill, R. E. and van Heyningen, V. (1992). Mouse mutations and human disorders are paired. *Trends in Genetics* **8**, 119–120.

Hittner, H. M., Riccardi, V. M., Ferrell, R. E., Borda, R. R. and Justice Jr., J. (1980). Variable expressivity in autosomal dominant aniridia by clinical, electrophysiologic and angiographic criteria. *American Journal of Ophthalmology.* **89**, 531–539.

Hittner, H. M. (1989). Aniridia. in "The Glaucomas" edited by R. Ritch, M. B. Shields and T. Krupin, Vol. 2, pp 869-884. Mosby, St. Louis.

Hittner, H. M., Ferrell, R. E., Lyons, L. A. and Kretzer, F. L. (1992). Criteria to detect minimal expressivity within autosomal dominant aniridia families. *American Journal of Ophthalmology* **15**, 700–707.

Hodgson, S. V. and Saunders, K. E. (1980). A probable case of the homozygous condition of the aniridia gene. *Journal of Medical Genetics* **6**, 478–480.

Hogan, B., Hetherington, C. and Lyon, M. (1987). Allelism of Small eyes (Sey) with Dickie's small eye (Dey) on chr2. *Mouse News Letter* **77**, 135–138.

Hogan, B. L. M., Horsburgh, G., Cohen, J., Hetherington, C. M., Fisher, G. and Lyon, M. F. (1986). Small eyes: a homozygous lethal mutation on chromosome 2 which affects the differentiation of both lens and nasal placodes in the mouse. *Journal of Embryology and Experimental Morphology* **97**, 95–110.

Hogan, B. L. M., Hirst, E. M. A., Horsburgh, G. and Hetherington, C. M. (1988). *Small eye (Sey)*: a mouse model for the genetic analysis of craniofacial abnormalities. *Development* (Supplement) **103**, 115–119.

Johnston, M. C., Noden, D. M., Hazleton, R. D., Coulombre, J. L., and Coulombre, A. J. (1979). Origins of avian ocular and periocular tissues. *Experimental Eye Research* **29**, 27–43.

Jordan, T., Hanson, I. M., Zaletayev, D., Hodgson, S., Prosser, J., Seawright, A. et al. (1992). The human PAX6 gene is mutated in two patients with aniridia. *Nature Genetics* **1**, 328–332.

Junien, C. and McBride, O. W. (1989). Report of the committee on the genetic constitution of chromosome 11. *Cytogenetics and Cell Genetics* **51**, 226–258.

Kaufman, M. (1979). Cephalic neuralation and optic vesicle formation in the early mouse embryo. *American Journal of Anatomy* **155**, 425–443.

Kinzler, K. W. and Vogelstein, B. (1989). Whole genome PCR: application to the identification of sequences bound by gene regulatory proteins. *Nucleic Acids Research* **17**, 3645–3653.

Lyons, L. A., Martha, A., Mintz-Hittner, H. A., Saunders, G. F. and Ferrell, R. E. (1992). Resolution of the two loci for autosomal dominant aniridia, AN1 and AN2, to a single locus on chromosome 11p13. *Genomics* **13**, 925–930.

Mann, I. (1957). *Developmental abnormalities of the eye*, Second edition, J. B.Lippincott Philadelphia.

Mannens, M., Bleeker-Wagemakers, M., Blick, J., Hoovers, J., Mandjes, S., van Tol, S. et al. (1989). Autosomal dominant aniridia linked to the chromosome 11p13 markers catalase and D11S151 in a large Dutch family. *Cytogenetics and Cell Genetics* **52**, 32–36.

Margo, C. E. (1983). Congenital aniridia: a histopathologic study of the anterior segment in children. *Journal of Pediatric Ophthalmology and Strabismus* **20**, 192–198.

Monaghan, A. P., Davidson, D. R., Sime, C., Graham, E., Baldock, R., Bhattacharya, S. S. and Hill, R. E. (1991). The *Msh*-like homeobox genes define domains in the developing vertebrate eye. *Development* **112**, 1053–1061.

Moore, J. W., Hyman, S., Antonarakis, S. E., Mules, E. H. and Thomas, G. H. (1986). Familial isolated aniridia associated with a translocation involving chromosomes 11 and 22 [t(11;22)(p13;q12.2)]. *Human Genetics* **72**, 297–302.

Muthukkaruppan, V. (1965). Inductive tissue interaction in the development of the mouse lens in vitro. *Journal of Experimental Zoology* **159**, 269–288.

Nelson, L. B., Spaeth, G. L., Nowinski, T. S., Margo, C. E. and Jackson, L. (1984). Aniridia: a review. *Surveys of Ophthalmology* **28**, 621–642.

Nevin, N. C. and Lim, J. H. K. (1990). Syndrome of partial aniridia, cerebellar ataxia, and mental retardation - Gillespie Syndrome. *American Journal of Medical Genetics* **35**, 468–469.

O'Rahilly, R. (1966). The early development of the eye in staged human embryos, pp 1–42, in "Contributions to embryology" Vol. XXXVIII Carnegie Institution of Washington, Washington, DC.

Pettenati., M. J., Weaver, R. G. and Burton, B. K. (1989). Translocation t(5;11)(q13.1;p13) associated with familial aniridia. *American Journal of Medical Genetics* **34**, 230–232.

Porteous, D. J., Bickmore, W., Christie, S., Boyd, P. A., Cranston, G., Fletcher, J. M. et al. (1987). *HRAS1*-selected chromosome transfer generates markers that colocalize aniridia- and genitourinary dysplasia-associated translocation breakpoints and the Wilms tumor gene within band 11p13. *Proceedings of the National Academy of Sciences USA* **84**, 5355–5359.

Puschel, A. W., Gruss, P. and Westerfield, M. (1992). Sequence and expression pattern of *pax-6* are highly conserved between zebrafish and mice. *Development* 1992, **114**, 643–651.

Riccardi, V. M., Sujansky, E., Smith, A. C. and Francke, U. (1978). Chromosomal imbalance in the aniridia-Wilms' tumor association: 11p interstitial deletion. *Pediatrics* **61**, 604–610.

Roberts, R. C. (1967). Small eyes – a new dominant eye mutant in the mouse. *Genetical Research, Cambridge* **9**, 121–122.

Roberts, R. G., Barby, T. F. M., Manners, E., Bobrow, M. and Bentley, D. R. (1991). Direct detection of dystrophin gene rearrangements by analysis of dystrophin mRNA in peripheral blood lymphocytes. *American Journal of Human Genetics* **49**, 298–310.

Rose, E. A., Glaser, T., Jones, C., Smith, C. L., Lewis, W. H., Call, K. M. et al. (1990). Complete physical map of the WAGR region of 11p13 localizes a candidate Wilms' tumor gene. *Cell* **60**, 495–508.

Saha, M. S., Spann, C. L., and Grainger, R. M. (1989). Embryonic lens induction: more than meets the optic vesicle. *Cell Differentiation and Development.* **28**, 53–172.

Seawright, A., Fletcher, J. M., Fantes, J. A., Morrison, H., Porteous, D. J., Li., S. S.-L. et al. (1988). Analysis of WAGR deletions and related translocations with gene-specific probes using FACS-selected cell hybrids. *Somatic Cell and Molecular Genetics* **14**, 21–30.

Servetnick, M. and Grainger, R. M. (1991). Changes in neural and lens competence in *Xenopus* ectoderm evidence for an autonomous developmental timer. *Development* **112**, 177–188.

Shannon, R. S., Mann, J. R., Harper, E., Harnden, D. G., Morten, J. E. N. and Herbert, A. (1982). Wilms' tumour and aniridia: clinical and cytogenetic features. *Archives of Disease in Childhood* **57**, 685–690.

Shaw, M. W., Falls, H. F. and Neel, J. V. (1960). Congenital aniridia. *American Journal of Human Genetics* **12**, 389–415.

Simeone, A., Gulisano, M., Acampora, D., Stornaiuolo, A., Rambaldi, M. and Boncinelli, E. (1992). Two vertebrate homeobox genes related to the *Drosophila empty spiracles* gene are expressed in the embryonic cerebral cortex. *The EMBO Journal* **11**, 2541–2550.

Simola, K. O. J., Knuutila, S., Kaitila, I., Pirkola, A. and Pohja, P. (1983). Familial aniridia and translocation t(4;11)(q22;p13) without Wilms' tumor. *Human Genetics* **63**, 158–161.

Siracusa, L. D., Silan, C. M., Justice, M. J., Mercer, J. A., Bauskin, A. R., Ben-Neriah, Y. et al. (1990). A molecular genetic linkage map of mouse chromosome 2. *Genomics* **6**, 491–504.

Tanaka, M. and Herr, W. (1990). Differential transcriptional activation by *Oct-1* and *Oct-2*: interdependent activation domains induce *Oct-2* phosphorylation. *Cell* **60**, 375–386.

Theiler, K., Varnum, D. S. and Stevens, L. C. (1978). Development of Dickie's Small Eye, a mutation in the house mouse. *Anatomy and. Embryology.* **155**, 81–86.

Ton, C. C. T., Hirvonen, H., Miwa, H., Weil, M. W., Monaghan, P., Jordan, T., et al. (1991). Positional cloning and characterization of a paired box- and homeobox- containing gene from the aniridia region. *Cell* **67**, 1059–1074.

Ton, C. C. T., Miwa, H. and Saunders, G. F. (1992). *Small eye (Sey)*: cloning and characterization of the murine homolog of the human aniridia gene. *Genomics* **13**, 251–256.

van der Meer-de Jong, R., Dickinson, M. E., Woychik, R. P., Stubbs, L., Hetherington, C. and Hogan, B. L. M. (1990). Location of the gene involving the Small eye mutation on mouse chromosome 2 suggests homology with human aniridia. *Genomics* **7**, 270–275.

van Heyningen, V., Bickmore, W. A., Seawright, A., Fletcher, J. M., Maule, J., Fekete, G. et al. (1990). Role for the Wilms' tumor gene in genital development? *Proceedings of the National Academy of Sciences USA* **87**, 5383–5386.

Walther, C. and Gruss, P. (1991). Pax-6, a murine paired box gene, is expressed in the developing CNS. *Development* **113**, 1435–1449.

Walther, C., Guenet, J. L., Simon, D., Deutsch, U., Jostes, B., Goulding, M. D. et al. (1991). Pax: a murine multigene family of paired box-containing genes. *Genomics* **11**, 424–434.

Wittig, E. O., Moreira, C. A., Freire-Maia, N. and Vianna-Morgante, A. M. (1988). Partial aniridia, cerebellar ataxia and mental deficiency (Gillespie syndrome) in two brothers. *American Journal of Medical Genetics* **30**, 703–708.

Zaraisky, A.G., Lukyanov, S. A., Vasiliev, O. L., Smirnov, Y. V., Belyavsky, A. V., and Kazanskya, O. V. (1992). A novel homeobox gene expressed in the anterior neural plate of the Xenopus embryo. *Developmental Biology* **152**, 373–382.

20. MITOCHONDRIAL DNA MUTATIONS AND THE EYE

MICHAEL D. BROWN, MARIE T. LOTT AND DOUGLAS C. WALLACE*

Department of Genetics and Molecular Medicine, Emory University School of Medicine, 403 Dental Building, 1462 Clifton Road, Atlanta, GA 30322, USA

Ophthalmic dysfunction is one of the most common clinical observations associated with mitochondrial DNA (mtDNA)-based human disease. Ocular pathology in these diseases includes optic atrophy, pigmentary retinopathy, retrochiasmal visual loss, and myopathy involving the extraocular muscles and the eyelids. The wide spectrum of symptoms involving the eye demonstrates the dependence of the ophthalmic system on mitochondrial energy production.

OXIDATIVE PHOSPHORYLATION (OXPHOS)

All tissues depend on chemical energy in the form of adenosine triphosphate (ATP) for cell function and maintenance. The vast majority of cellular ATP is produced by the mitochondrial oxidative phosphorylation (OXPHOS) system which is comprised of five multi-subunit enzyme complexes (Complex I–V) within the mitochondrion inner membrane. Four of these enzyme complexes, NADH:ubiquinone oxidoreductase (Complex I), succinate:ubiquinone oxidoreductase (Complex II), ubiquinol:cytochrome c oxidoreductase (Complex III), and cytochrome c oxidoreductase (Complex IV), are key components of the electron transport chain which transfers electrons from NADH + H^+ and $FADH_2$ to oxygen and concurrently translocates protons from the mitochondrial matrix to the inner membrane space at Complexes I, III and IV. Thus, the process of electron transfer generates an electrochemical gradient across the mitochondrial inner membrane which is utilized by the fifth enzyme complex, ATP synthase (Complex V), to synthesize ATP in the matrix by condensation of ADP and inorganic phosphate. In the rate limiting step, matrix ATP is exchanged for cytosolic ADP by the nuclear-encoded adenine nucleotide translocator.

MITOCHONDRIAL DNA AND OXPHOS

Mitochondrial DNA is essential for aerobic metabolism as it codes for 37 OXPHOS genes including 13 respiratory chain polypeptides, two rRNAs and 22 tRNAs (Figure 1). Functional respiratory chains are assembled from both mitochondrial and nuclear gene products: Complex I contains seven (ND1, 2, 3, 4, 4L,

*Corresponding author

5 and 6) mtDNA-encoded subunits out of approximately 39 total polypeptides; Complex III contains one (cytochrome b) mtDNA-encoded subunit out of approximately 10 total polypeptides; Complex IV contains three (COI, II and III) mtDNA-encoded subunits out of 13 total polypeptides; and Complex V (ATP synthase) contains two (ATPase 6 and 8) mtDNA-encoded subunits out of 12 total polypeptides (Wallace, 1992; Shoffner and Wallace, 1990). Thus, ATP production is coordinately controlled by two physically distinct genetic systems.

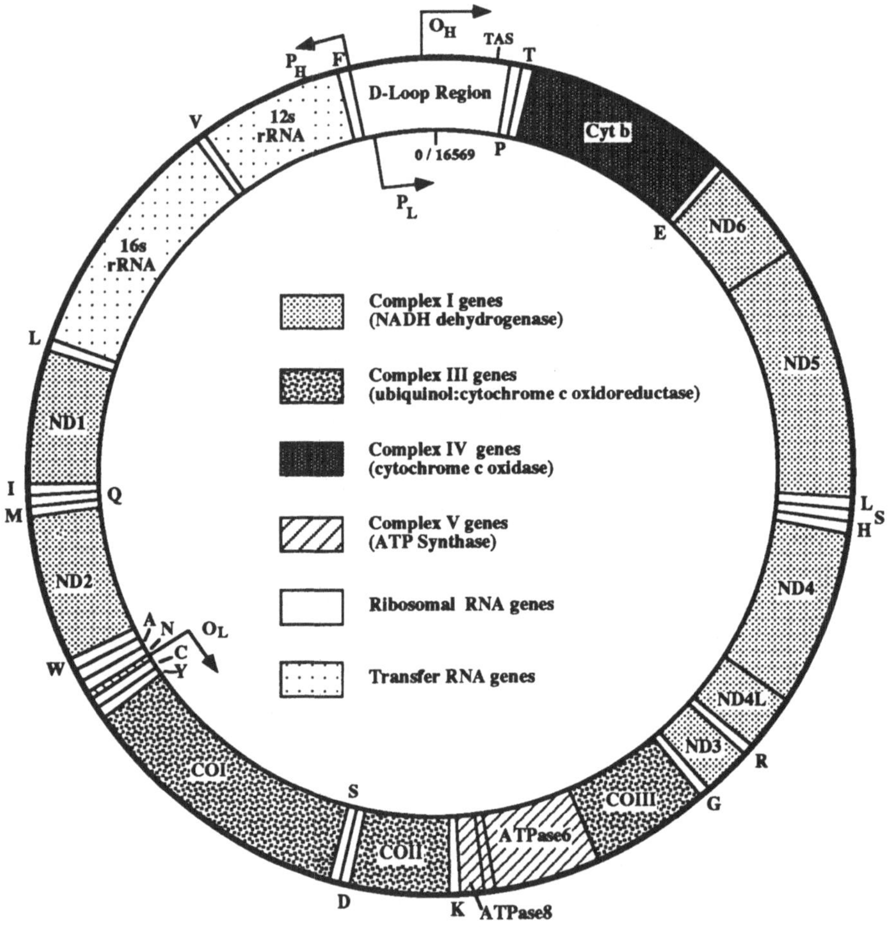

Figure 1 The human mitochondrial DNA. The H (heavy)-strand is shown as thicker outer circle. Both origins of DNA replication (O_H and O_L) and both major promoters (P_H and P_L) are indicated. tRNAs depicted inside the circular genome are transcribed from the L (light)-strand, as is the ND6 gene.

Structurally, the human mtDNA is a closed, circular molecule of 16569 nucleotide pairs (np) located within the mitochondrial matrix (Anderson et al., 1981). It is normally present in thousands of copies per cell and comprises roughly 1% of total cellular DNA. The two strands of the mtDNA show a non-random nucleotide distribution as the heavy strand (H-strand) contains a higher proportion of guanine nucleotides compared to the complementary light strand (L-strand) (Figure 1). The H-strand contains the majority of the coding information as 12 protein-coding, 14 tRNA, and both rRNA genes are found on this strand of the mitochondrial chromosome. The human mtDNA is organized as a highly compact genome as it contains no introns or repetitive DNA spacer sequence elements characteristic of nuclear DNA (nDNA) (Attardi, 1985). A single major non-coding region, the displacement loop (D-loop), is present in the mitochondrial genome and contains many of the sequence elements essential for mtDNA replication and transcription.

MtDNA replication is a bidirectional, asynchronous process involving two origins separated by two-thirds of the genome (Clayton, 1991). H-strand synthesis begins replication and is initiated in the D-loop by an RNA primer produced by the L-strand promoter. H-strand replication proceeds clockwise around the circular molecule, displacing the original H-strand, until the L-strand origin is exposed. The L-strand origin consists of a conserved stem-loop structure within a tRNA-rich region of the mtDNA (Wong and Clayton, 1985). L-strand replication then proceeds counter-clockwise, using the displaced original H-strand as a template. MtDNA replication occurs independently of nuclear DNA replication.

MtDNA transcription occurs bidirectionally from two major promoters which initiate the synthesis of large, polycistronic RNAs from both strands of the mtDNA (Clayton, 1991). These promoters, P_L (L-strand promoter) and P_H (H-strand promoter), are separated by approximately 150 nucleotide pairs in the D-loop region and are independent transcriptional units. Mature mRNAs and rRNAs (12S and 16S) are produced by RNAse-P-like processing of tRNAs, which punctuate the protein-coding and rRNA genes (Figure 1). Mitochondrial mRNAs are distinct from nuclear-derived mRNAs in that they are not capped at the 5' end nor do they contain 5' untranslated sequences (Attardi, 1985; Montoya et al., 1981). Further, most mitochondrial mRNAs lack 3' untranslated sequences except for polyadenylation, which frequently completes translational termination codons (UU<u>A</u>, UG<u>A</u>, U<u>AA</u>) through the addition of 3'-terminal adenine nucleotides (Ojala et al., 1981). Exceptions are the COI, COII, and the ND5 mRNAs which encode complete termination codons and have 3' untranslated sequence prior to polyadenylation (Attardi, 1985; Ojala et al., 1981).

Mitochondrial mRNAs are translated within the matrix using mtDNA-encoded ribosomal RNAs (12S and 16S) and tRNAs. The universal genetic code has been slightly modified in mitochondria, preventing both efficient cytosolic translation of mitochondrial mRNAs and gene transfer between the organelle and the nucleus (Anderson et al., 1981; Wallace, 1982). For example, in human mitochondria, the UGA codon is used for tryptophan (instead of a termination codon), the AGA and AGG codons are termination codons (instead of arginine codons), and the AUA codon is used for methionine (instead of isoleucine).

THE UNIQUE GENETICS OF MTDNA

Five features of the human mtDNA distinguish mtDNA genetics from nuclear genetics (Wallace 1992a,b). First, mitochondria are maternally inherited (Giles et al. 1980; Case and Wallace 1981). At fertilization, the egg contains greater than 10^5 mitochondria while the sperm contribution is less than 0.1% of zygote mtDNA (Gyllensten et al., 1991). Thus, mtDNAs are inherited mother-to-daughter, meaning that pedigrees of mtDNA diseases with multiple affected family members will show exclusive maternal transmission of the phenotype. Second, mtDNAs randomly segregate to daughter cells upon mitosis or meiosis (replicative segregation) (Wallace, 1986). The high copy number of mtDNA genomes allows for the presence of both mutant and wild-type mtDNAs (heteroplasmy) within an organelle or cell. At cell division, the proportion of mutant and wild-type mtDNAs stochastically segregate to daughter cells and, over time, the population of mtDNAs within a cell can drift towards either pure (homoplasmy) mutant or wild-type genomes. Since the magnitude of an OXPHOS deficiency is frequently correlated with the fraction of mutant mtDNAs within a tissue, heteroplasmy due to replicative segregation is an important feature of mtDNA-based disease. Third, phenotypic expression of an mtDNA defect is dependent on a tissue-specific OXPHOS threshold. Each tissue in the body depends on OXPHOS for energy production to a different extent. For example, central nervous system tissue and skeletal muscle are among the most highly oxidative tissues in the body and therefore are highly reliant on OXPHOS. Disease results when the ATP-generating capacity of an organ drops below a minimum OXPHOS threshold necessary for tissue maintenance. Thus, clinical manifestation of an OXPHOS defect is a product of the nature of the mtDNA mutation, the fraction of mutant mtDNA genomes within a tissue, and the OXPHOS requirements of the affected tissue. Fourth, ATP production within a tissue naturally declines with age (Trounce et al., 1989; Yen et al., 1989). This process parallels an increase in somatic mtDNA damage in post-mitotic tissue with age (Linnane et al., 1989; Cortopassi and Arnheim, 1992; Corral-Debrinski et al., 1992). Oxygen radicals are produced at relatively high concentrations at the mitochondrial inner membrane as a result of respiratory chain redox activity, causing the mtDNA to accumulate oxidative damage at a rate 16 times higher than nuclear DNA (Richter et al., 1988). Thus, the age-related decrease of OXPHOS capacity may be correlated with the age-related increase in mtDNA damage. This phenomenon would help explain why some mtDNA diseases are expressed in mid-to-late life and progress with age. Fifth, new mtDNA mutations are fixed at a rate 10–20 times higher than are mutations occurring in nuclear-encoded OXPHOS genes (Neckelmann et al., 1987; Wallace et al., 1987). Because mtDNA OXPHOS genes are essential for aerobic metabolism and subject to equivalent selective pressure as nuclear-encoded OXPHOS genes, deleterious mtDNA mutations are expected to arise with significant frequency.

MtDNA MUTATIONS AND OPHTHALMIC DISEASE

Because the retina, optic nerve and extraocular muscles controlling eye movement are among the most ATP-dependent tissues in the body, it is not surprising that those mtDNA point mutations and deletions/duplications which impair OXPHOS capacity often result in ocular dysfunction. The most common ophthalmic manifestations of mtDNA diseases include optic neuropathy, pigmentary retinopathy, and ophthalmoplegia with ptosis (Table 1).

Table 1 MtDNA mutations and eye disease

MtDNA Mutation	*Phenotype*	*Ocular involvement*
I. Point Mutations		
A. Missense Mutations		
1. Electron transport chain (Complexes I-IV)	LHON	Optic neuropathy
2. ATP synthase (Complex V)	NARP/Leigh's Disease	Pigmentary retinopathy
B. tRNA Mutations		
1. Lysine	MERRF	Occasional optic atrophy, pigmentary retinopathy
2. Leucine	MELAS	Occasional optic atrophy, pigmentary retinopathy
3. Leucine	MMC	None
4. Leucine	MM	None
5. Isoleucine	FICM	None
II. Rearrangements		
A. Deletions		
1. Spontaneous	CPEO	Ophthalmoplegia/Ptosis, occasional pigmentary retinopathy
	KSS	Ophthalmoplegia/Ptosis, pigmtary retinopathy
	Pearson's	Can progress to KSS
2. Heritable KSS[a] ophthalmoplegia/ptosis, bilateral		cataract formation
	Diabetes plus[b]	None
B. Duplications	Variable	Ophthalmoplegia/Ptosis, pigmentary retinopathy

[a] Autosomal dominant transmission
[b] Maternally inherited
LHON, Leber's hereditary optic neuropathy; NARP, neurogenic muscle weakness, ataxia and retinitis pigmentosa; MERRF, myoclonic epilepsy and ragged red fiber disease; MELAS, mitochondrial encephalopathy, lactic acidosis and stroke-like symptoms; MMC, mitochondrial myopathy and cardiomyopathy; MM, mitochondrial myopathy; FICM, fatal infantile cardiomyopathy; CPEO, chronic progressive external ophthalmoplegia; KSS, Kearns-Sayre syndrome; Pearson's, Pearson's marrow/pancreas syndrome; Diabetes plus, diabetes mellitus and hearing loss.

MtDNA Point Mutations and Ophthalmic Disease

Pathogenic mtDNA point mutations have been found in both tRNA genes and protein-coding genes (Wallace, 1992a,b). Transfer RNA mutations are associated with severe mitochondrial myopathies such as Myoclonic Epilepsy and Ragged-Red Fiber disease (MERRF, an A to G transition at np 8344 and a T to C transition at np 8356 in $tRNA^{Lys}$) (Shoffner et al., 1990; Zeviani et al., 1992); Mitochondrial Encephalomyopathy, Lactic Acidosis and Stroke-like symptoms, (MELAS, an A to G transition at nucleotide pair 3243 in $tRNA^{Leu}$) (Goto et al., 1990a); Mitochondrial Myopathy and Cardiomyopathy (MMC, an A to G transition at nucleotide pair 3260 in $tRNA^{Leu}$) (Zeviani et al., 1991); and Fatal Infantile Cardiomyopathy (FICM, an A to G transition at nucleotide pair 4317 in $tRNA^{Ile}$) (Tanaka et al., 1990). Transfer RNA mutations usually alter nucleotides conserved between various species and result in a protein synthesis defect. Despite the heterogeneity of neuro-muscular symptoms associated with these diseases, ophthalmic involvement is relatively uncommon and generally limited to occasional optic atrophy or pigmentary retinopathy (MERRF and MELAS) or retrochiasmal visual loss (MELAS).

In contrast, diseases attributed to mtDNA missense mutations feature ophthalmic involvement (Table 1). Both Leber's Hereditary Optic Neuropathy (LHON) and Neurogenic Muscle Weakness, Ataxia and Retinitis Pigmentosa (NARP), are caused by mtDNA point mutations in OXPHOS polypeptides.

Leber's Hereditary Optic Neuropathy (LHON)

The 1988 discovery of an mtDNA mutation associated with a large number of LHON pedigrees demonstrated that an mtDNA point mutation could cause human disease (Wallace et al., 1988). LHON is a disorder of young adults that results in bilateral loss of central vision due to optic nerve atrophy (Newman, 1991). The average age of onset in LHON patients is 27 years of age and disease progression is rapid, as blindness occurs in one eye and is usually followed by loss of vision in the other within three months time (Newman, Lott, and Wallace, 1991). Loss of vision is usually the only clinical manifestation of LHON, although cardiac abnormalities (preexcitation syndromes, prolonged QT internal) or other neurological symptoms do appear in occasional LHON pedigrees (Newman and Wallace, 1990; Newman, 1991; Ortiz et al., 1992).

Three important features of LHON expression and inheritance are evident from pedigree analysis. First, pedigrees with multiple affected individuals show strict maternal inheritance (Figure 2), an inheritance pattern consistent with the cytoplasmic transmission of the mtDNA. Second, disease expression is variable as not all individuals harboring a homoplasmic LHON mutation manifest clinical symptoms and singleton cases are not uncommon. Third, males are much more frequently affected than females as up to 80–90% of Caucasian LHON patients are

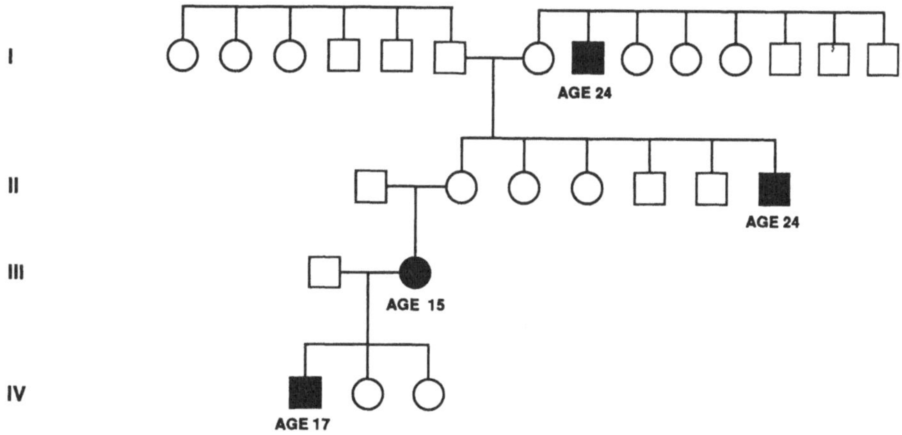

Figure 2 Maternally-inherited LHON pedigree. Pedigree shows incomplete penetrance and male predilection often senn in LHON families. Age of onset is indicated. Individuals depicted have the 14484 and 15257 LHON mutations and are homoplasmic.

male. Thus, although mtDNA mutations have a primary etiological role in the disease, other factors must influence LHON expression. Such factors likely include additional genetic determinants, sex-limiting physiological determinants, and unspecified environmental determinants.

The genetics of LHON have proven to be complex and heterogeneous as twelve mtDNA point mutations have now been associated with LHON (Brown et al., 1992a; Johns, Neufeld and Park, 1992) (Table 2). Eleven of these are missense mutations in genes contributing to respiratory Complexes I and III. The twelfth mutation alters the translational termination codon of the gene encoding cytochrome c oxidase subunit I (Complex IV). These mutations appear to impair respiratory chain function to different degrees, a variation reflected in the risk each mutation represents in causing LHON. Risk can be estimated on the basis of four criteria: (1) nature of the amino acid change, (2) whether or not a mutation can cause blindness by itself, (3) the expressivity of the disease in LHON pedigrees, and (4) the penetrance of the mutation in the general population. Amino acid substitutions are either conservative or radical replacements and, if functionally significant, usually alter evolutionarily conserved amino acids. Expressivity is determined by the proportion of blind individuals in a homoplasmic pedigree and by the proportion of affected males within a pedigree. Since males are more commonly affected than females, more severe LHON mutations might be expected to affect an increased proportion of females. The penetrance of a mutation in the population involves three parameters, heteroplasmy (indicative of a new mutation), the occurrence of a mutation on a single versus multiple mtDNA haplotypes, and the frequency with which a mutation is found in unaffected controls. Based on these criteria, mutations associated with LHON can be classified as high, moderate, or low-risk LHON mutations.

Table 2 mtDNA mutations associated with LHON[a]

			Amino Acid		Expressivity			Penetrance			References
Mutation	% of Patients	Can Cause Blindness By Itself	Change	Conser-vation	% of Relatives Affected	% of Males	Other Clinical Manifestations	Hetero-plasmy	Background Haplotypes	% of Control	
ND1 4160	Rare	Yes	L-P	H	76	54	+	-	1	0	Howell *et al* 1991a
ND4 11778	50	Yes	R-H	H	33-60	~80	+/-	+/-	>1	0	Wallace *et al* 1988
ND1 3460	15	Yes	A-T	M	14-40	33-67	+/-	+/-	NA	0	Huoponen *et al* 1991, Howell *et al* 1991b
Cytb 15257	9	Yes/No	D-N	H	27-80	75-100	-	-	1	0.3	Brown *et al* 1922b, Johns and Neufeld 1991
ND1 3394	Rare	No	Y-H	H	37	100	-	-	1	0.9	Brown *et al* 1992a
CO1 7444	5	Yes/No	Term-K	NA	23-43	~100	-	-	1	1	Brown *et al* 1992c
ND6 14484	~10	?	M-V	M	NA	NA	-	+/-	NA	0	Johns, Neufeld, and Park 1992
ND2 5244	Rare	No	G-S	H	NA	NA	-	+	1	0	Brown *et al* 1992b
Cytb 15812	~5	No	V-M	M	NA	NA	-	-	1	0.1	Brown *et al* 1992b, Johns and Neufeld 1991
ND2 4917	3	No	D-N	H	NA	NA	-	-	NA	3	Johns and Neufeld 1991
ND5 13708	~30	No	A-T	M	NA	NA	-	-	>1	6	Brown *et al* 1992b, Johns and Berman 1991
ND1 4216	~40	No	Y-H	L	NA	NA	-	-	NA	13	Brown *et al* 1992b, Johns and Berman 1991

[a] Mutations are listed in order of estimated severity. H, high; M, medium; L, low; NA, not available; Term, termination codon.

Table 2 mtDNA mutations associated with LHON. Mutations are listed in order of estimated severity. H, high; M, medium; L, low; NA, not available; Term, termination codon.

High-risk LHON mutations

Individuals harboring mtDNA mutations at nps 3460, 4160 and 11778 are at the greatest risk for expressing LHON. All of these mutations cause non-conservative amino acid substitutions and alter amino acids which are conserved among many species. These mutations do not occur together in LHON pedigrees and have not been detected in unaffected controls, indicating that they are sufficiently deleterious by themselves to cause blindness. Further, pedigrees harboring these mutations have a relatively high proportion of affected individuals.

The 4160 (NDl gene, Complex I) mutation appears to be the most severe LHON mutation as roughly 80% of the members of the single 4160-positive LHON pedigree are blind and nearly one-half of the affected individuals are female (Howell et al., l991a). Atypical of most LHON pedigrees, additional neurological symptoms such as dysarthria, ataxia, and severe infantile encephalopathy are prominent in this family. Further, the analysis of OXPHOS enzyme activity from platelet mitochondria revealed a marked reduction in Complex I specific activity in all family members assayed (Parker, Oley and Parks, 1989).

The 3460 (NDl gene, Complex I) and the 11778 (ND4 gene, Complex I) mutations appear to be the next most severe LHON nucleotide substitutions (Wallace et al. 1988; Houponen et al. 1991; Howell et al. l991b). Complete mtDNA sequence analysis of probands carrying either the 3460 mutation or the 11778 mutation has revealed no other mtDNA mutations which appear to play a prominent pathogenetic role in disease expression (Wallace et al., 1988; Brown et al., 1992a). Also, OXPHOS biochemical deficiencies have been detected in individuals harboring these mutations. Majander and colleagues (1991) reported an 80% reduction in the rotenone-sensitive electron transfer activity of Complex I in 3460-positive patients (Majander et al., 1991). A comparable reduction in Complex I enzyme activity has not been observed in 11778-positive patients, however reduced oxidation of Complex I-linked substrates has been detected with polarography (Larsson et al., 1991; Majander et al., 1991). Finally, the 11778 mutation is found on strikingly different mtDNA genotypes, having been detected in African, Asian and Caucasian LHON pedigrees. Thus, this mutation has occurred multiple different times and each occurrence associated with LHON, providing convincing evidence that the 11778 mutation is a primary causal LHON mutation (Singh et al., 1989).

Moderate-risk LHON mutations

LHON mutations at nps 3394, 7444, and 15257 appear to impart an intermediate risk for the expression of LHON. All three mutations have some characteristics of the high-risk LHON mutations such as the radical substitution of highly conserved amino acids and a relatively high fraction of pedigree members who develop optic nerve atrophy. However, each mutation has been detected at a low frequency in the unaffected control population and these LHON mutations are frequently found linked to other LHON mutations. Thus, the np 3394, 7444 and 15257 mutations may represent a relatively high risk when associated with other mildly pathogenic mtDNA mutations and a relatively low risk when not associated with the appropriate mtDNA genotype.

The homoplasmic np 3394 (NDl gene, Complex I) mutation changes a highly conserved tyrosine to a histidine in the N-terminal end of the NDl polypeptide (Brown et al., 1992a; Johns, Neufeld and Park, 1992). This mutation has only been found in association with low-risk LHON mutations. For example, when proband mtDNA from a large 3394-positive pedigree was completely sequenced, only the low-risk mutations at np 4216, 13708 and 14484 were detected (Brown et al., 1992a). No high-risk LHON mutation was found in the genome. However, all affected members of this pedigree were male and the mutation has been detected in nearly 1% of unaffected controls, indicating that the 3394 mutation may be less severe that the 3460, 4160 and 11778 mutations. The predilection of affected males in this pedigree suggests the involvement of an X-linked nuclear allele which influences LHON expression. Such a "two-hit" model involving a mtDNA mutation(s) and an X-chromosomal mutation(s) has been proposed to partially account for the variability of disease expression observed among LHON pedigrees (Bu and Rotter, 1991), although linkage data supporting an X-linked susceptibility locus has been equivocal (Vilkki et al., 1991; Sweeney et al., 1992). Thus, the np 3394 mutation may be a risk-factor for LHON in association with other mtDNA and/or nuclear mutations.

The homoplasmic 7444 (COI gene, Complex IV) mutation is unique among LHON mutations for three reasons: (1) it alters the translation termination codon of an mtDNA gene, (2) it results in the lengthening of a mtDNA-encoded polypeptide, and (3) the mutation impairs Complex IV activity (Brown et al., 1992c). This nucleotide substitution changes the AGA stop codon of the cytochrome oxidase subunit I (COI) gene to an AAA lysine codon, causing a three amino acid extension of the protein. This structural alteration is easily detected by SDS-PAGE of ^{35}S-Met labelled mitochondrial proteins and results in a 40% reduction in Complex IV specific activity. Like the np 3394 mutations, the np 7444 mutation has been found with other LHON mutations and in 1% of the unaffected population. Also, disease expression is sporadic in np 7444-positive pedigrees with males most commonly affected. Thus, the np 7444 mutation may constitute an LHON risk factor when in the appropriate genetic environment.

Perhaps the best example of a moderate-risk LHON mutation whose pathogenicity is influenced by other mtDNA mutations is the homoplasmic np 15257 (cytochrome b gene, Complex III) mutation. This mutation alters a highly conserved aspartate residue in cytochrome b which is located near an invariant heme-binding histidine residue (Johns and Berman, 1991; Brown et al., 1992b). The np 15257-mutation has been found in at least 16 LHON families and is found at a very low frequency (0.3%) in the control population. Pedigrees harboring this mutation show that the disease is moderately penetrant and most of the affected individuals are male. Hence, the np 15257 mutation has many features characteristic of a high-to-moderate risk LHON mutation, yet, in some cases, this mutation alone may be insufficient to cause LHON.

The np 15257 mutation is almost invariably found with a specific subset of lower-risk mtDNA LHON mutations. MtDNA sequence and restriction endonuclease digestion analysis has indicated that patients harboring the np 15257 mutation have very similar mtDNA haplotypes which include the frequently-observed low-risk LHON mutation at np 13708 (ND5, gene, Complex I), and the lower risk LHON mutations at nps 15812 (cytochrome b gene, Complex III) and 5244 (ND2 gene, Complex I) (Brown et al., 1992b). Such similar mtDNA genotypes cluster upon phylogenetic analysis, demonstrating a LHON mtDNA lineage which is defined by the linked np 13708 + 15257 mutations. The np 15812 mutation is found in roughly one-half of the individuals with the 13708 + 15257 genotype and the 5244 mutation has been found in a single individual with the 13708 + 15257 + 15812 genotype. As this lineage sequentially accumulated these four mutations, the number of unaffected individuals decreased and the proportion of blind individuals increased. Blindness occurred in the lineage only after the occurrence of the 15257 mutation. Thus, the np 15257 mutation is the most pathogenic mtDNA mutation in this lineage, but in some cases the risk for developing LHON is apparently increased with the presence of additional mtDNA mutations. The additional mutations interact synergistically with the 15257 mutation to decrease OXPHOS efficacy below an energetic threshold necessary for optic nerve function (Johns and Berman, 1991; Brown et al., 1992a,c).

Low-risk LHON mutations

Six mtDNA mutations appear to impart a low-risk for LHON expression: nps 4216 (ND1), 4917 (ND2), 5244 (ND2), 13708 (ND5) and 14484 (ND6) in Complex I genes and np 15812 in cytochrome b of Complex III (Johns and Berman, 1991; Brown et al., 1992a,b; Johns, Neufeld and Park, 1992). These mutations can alter poorly or highly conserved amino acids and can be found in statistically significant frequencies in LHON families. However, they can also be found in relatively high frequencies in unaffected controls, the exceptions being the heteroplasmic 5244 mutation which has several high-risk LHON mutation characteristics (Table 2) but has only been found in one 15257-positive LHON patient and no controls, and the 14484 mutation which also has not been found in controls. Both of these mutations may therefore represent a more significant risk for developing LHON, but more data is necessary to draw this conclusion. In general, however, the low-risk LHON mutations probably cannot cause LHON as solitary etiological agents, but instead are either non-pathogenic mtDNA polymorphisms linked to a LHON-causing mtDNA genotype or contribute to disease expression as a mutation of low pathogenicity which interacts synergistically with more severe mtDNA mutations. Some of these mutations can be found associated with various other more severe LHON mutations and some are specifically associated with a second LHON mutation, such as the np 15812 mutation which has only been found linked to the more severe np 15257 mutation in LHON patients.

The genetic heterogeneity observed in LHON suggests that optic nerve death results from a generalized electron transport impairment as all LHON mutations affect mtDNA-encoded subunits of the mitochondrial electron transport chain. Because the LHON mutations are of varying severity, each represents a certain risk for disease expression. Thus, LHON expression can be thought of in terms of probability: the more severe the mutation, the greater the risk for blindness and maternal transmission of the phenotype. Conversely, less severe mutations impart a lower risk for blindness and appear more frequently as sporadic LHON cases. Further, risk appears to be influenced by additional genetic and nongenetic factors which either increase or decrease the probability of LHON expression.

Neurogenic Muscle Weakness, Ataxia, and Retinitis Pigmentosa (NARP)

Pigmentary retinopathy is a prominent clinical feature of NARP, a second phenotype due to a mtDNA missense mutation (Table 1). NARP results from a heteroplasmic T to G transversion at np 8993 of the ATPase 6 (Complex V) gene (Holt et al., 1990). This mutation changes a highly conserved leucine to an arginine in the proton channel of ATP synthase, thus inhibiting ATP synthesis (Holt et al., 1990; Tatuch et al., 1992). Like LHON, NARP is maternally transmitted and penetrance is incomplete in an affected family. Unlike LHON, however, disease symptoms often vary markedly among affected family members.

Rapid intrafamilial segregation of heteroplasmic mtDNAs accounts for the variation in penetrance and expressivity seen in NARP pedigrees (Shoffner et al., 1992; Tatuch et al., 1992) (Figure 3). Individuals harboring between 80-90% mutant mtDNAs in their blood are at high risk for developing NARP while family members with less than 80% mutant mtDNAs are at a lower risk for NARP and may remain asymptomatic. NARP symptoms can range from retinitis pigmentosa plus mild neurological problems in mildly affected patients to true retinitis pigmentosa, muscle weakness, ataxia and generalized seizures in severely affected patients. Individuals harboring greater than 90% mutant mtDNAs are at risk for developing the most severe manifestation of this genetic defect: Leigh's disease. Leigh's disease is a fatal infantile disorder characterized by psychomotor retardation, brainstem abnormalities and elevated blood lactate levels. Using both a restriction endonuclease digestion assay and Southern blot analysis, Tatuch et al. (1992) were able to show in one Leigh's patient, that the brain, kidney, liver and cultured fibroblasts all contained greater than 95% abnormal mtDNAs (Tatuch et al., 1992). Thus, the NARP phenotype represents a continuum of symptoms, the severity of which correlates with the proportion of mutant genomes in patient tissue.

The genetics of LHON and NARP illustrate an important feature of mtDNA-based diseases: the variability in genotype/phenotype relationships. LHON is the singular clinical manifestation of multiple mtDNA mutations, acting in various combinations and affecting three different OXPHOS enzyme complexes, while a

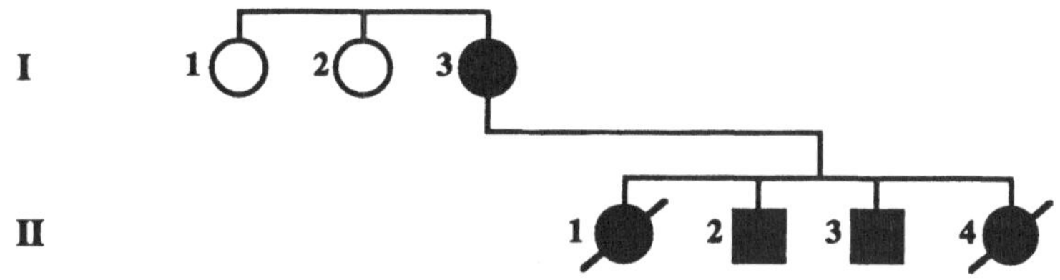

% Mutant mtDNA	**0**	**0**	**86**	**>95**	**>95**	**>95**	**>95**
Age (years)				**2**	**14**	**11**	**1**
RP	-	-	+	**nd**	+	+	+
Leigh's	-	-	-	+	-	-	+
Optic atrophy	-	-	-	-	-	-	+
Migraine	-	-	+	-	-	-	-
M R	-	-	-	+	+	+	+
Ataxia	-	-	-	**nd**	+	+	**nd**
ADD	-	-	-	**nd**	+	+	**nd**

Figure 3 MtDNA genotype relationship to clinical phenotype in a pedigree harboring the ATPase 6 (NARP) mutation at nucleotide 8993. Note variable expressivity resulting from the rapid segregation of the 8993 mutation within the family. RP, retinitis pigmentosa; MR, psychomotor retardation; ADD, attention deficit disorder; nd, not determined.

single, heteroplasmic ATPase 6 point mutation can result in a markedly different phenotype, even within the same pedigree.

MtDNA Rearrangements and the Eye

Large-scale mtDNA deletions and partial duplications are also associated with human disease involving the eye (Table 1) (Wallace, 1992a). Pathogenic deletions are far more common than are duplications and account for a significant fraction of mitochondrial myopathies, a heterogeneous group of diseases hallmarked by the accumulation of structurally and biochemically abnormal mitochondria in muscle. Both types of rearrangements are causally associated with ocular myopathy, observed clinically as Chronic Progressive External Ophthalmoplegia (CPEO) or the more severe disorder, Kearns-Sayre Syndrome (KSS). CPEO is characterized by ophthalmoplegia (paresis of extraocular muscles limiting eye movement), ptosis (droopy eye lids) and mitochondrial myopathy. In addition to CPEO, KSS patients may manifest pigmentary retinopathy, elevated levels of blood lactate and/or CSF protein, ataxia, cardiac conduction abnormalities, dementia, and hearing loss. Other less common ophthalmologic manifestations include corneal opacities, cataracts, and optic atrophy. Pearson Marrow/Pancreas syndrome has also been associated with mtDNA rearrangements. Pancytopenia and pancreatic dysfunction are the hallmarks of this often fatal

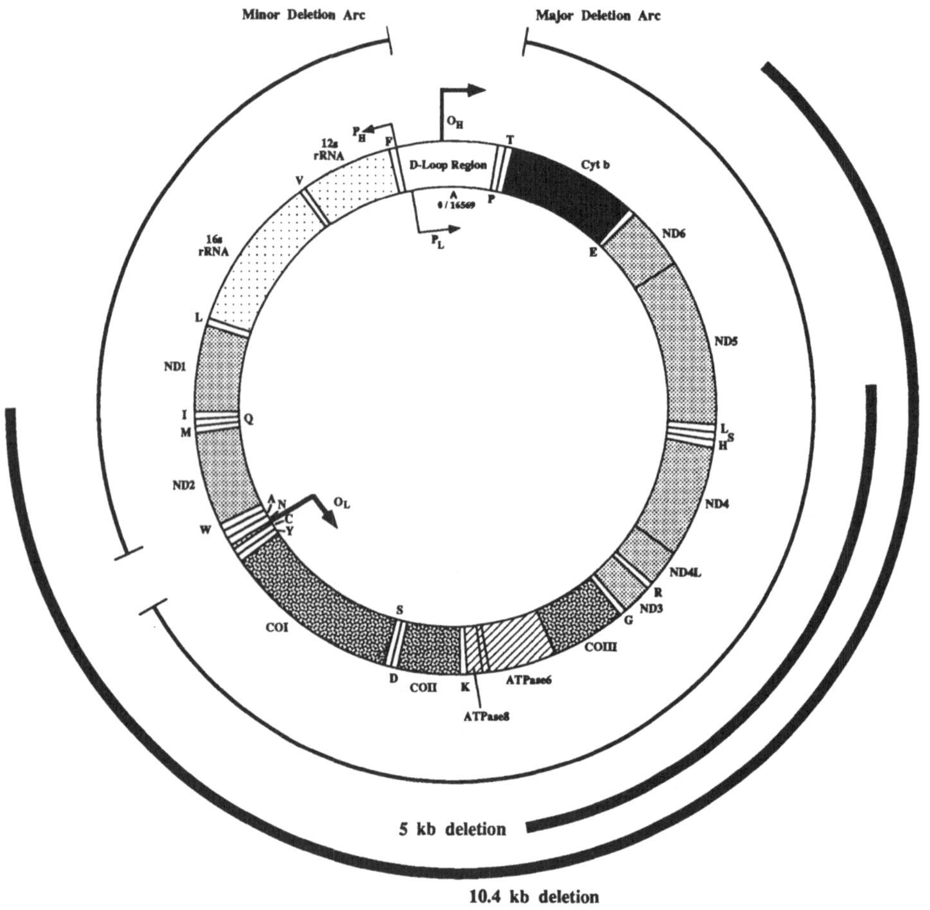

Figure 4 Pathogenic human mtDNA deletions. Deletions typically occur within either the major or minor arcs as defined by the origins of DNA replication (O_H and O_L). Deletions which eliminate O_L, such as the 10.4 kb deletion, are rare. A 5 kb deletion associated with CEOP and KSS is the most common pathogenic mtDNA deletion.

childhood disease. Those children who survive the early stages of Pearson Marrow/Pancreas syndrome often progress to KSS.

MtDNA Deletions

Pathogenic mtDNA deletions can be subdivided based on inheritance patterns. Pedigrees exhibiting familial transmission of deletions are rare and such deletions usually have novel genetic features associated with them which permit inheritance. Spontaneous deletions are comparatively much more common and, as a group, share certain general genetic characteristics.

Spontaneously-occurring deletions account for the majority of CPEO, KSS and Pearson Marrow/Pancreas Syndrome cases seen clinically. Patients harboring spontaneous deletions are observed sporadically as such deletions are somatic mutational events and are therefore not transmitted to successive generations.

Over 120 different spontaneous, pathogenic mtDNA deletions have been mapped and some common genetic features are evident (Wallace et al., 1991). First, although such deletions vary widely in position and size, ranging from roughly 2 kb to 9 kb, the two origins of DNA replication are usually maintained, thus confining spontaneous deletions to the two major arcs of the mtDNA defined by the position of the two origins (Moreas et al., 1989; Goto et al., 1990b; Wallace et al., 1991) (Figure 4). Second, both tRNA and protein-coding genes are commonly deleted (Wallace et al., 1991). Third, sequence analysis of the breakpoints of over 50 spontaneous deletions revealed that these deletions most often occur within directly repeated mtDNA sequences (Shoffner et al., 1989; Schon et al., 1989; Mita et al., 1990; Wallace et al., 1991). Direct repeats are common in mtDNA, in part because of the disproportionate number of H-strand guanine residues. Some direct repeats, which range from three to 13 base pairs in length, are hotspots for deletion formation (Wallace et al., 1991). The best example of such a hotspot is a 13 base pair perfect repeat (5'- ACCTCCCTCACCA 3'), found in the ATPase 8 gene (from np 8470 to np 8482) and the ND5 gene (from np 13447 to np 13459) (Wallace et al, 1991; Wallace, 1992a). This repeat is found at the breakpoint of the most commonly-occurring pathogenic deletion of 4977 base pairs, which has occurred independently over 100 times (Wallace et al., 1991) (Figure 4).

The association of spontaneous deletions with direct repeats suggests a common mutagenic mechanism. Homologous recombination has not been proven to exist in mitochondria and therefore probably does not account for somatic deletion events. However, a "slipped replication" model as proposed by Shoffner et al. (1989) does not rely on strand cross-over, yet is dependent on direct repeats for deletion formation. Under the "slipped replication" model, the upstream, displaced, H-strand direct repeat base-pairs with the complementary downstream L-strand repeat exposed by the DNA replication fork. A strand breakage event just downstream from this pairing generates a 3'-OH for subsequent H-strand synthesis and results in the degradation of the single-stranded H-strand downstream of the breakage event back to double-strand DNA. Thus, slipped replication resolves one wild-type and one deleted mtDNA molecule. Since a minority of deletion events occur in the absence of directly repeated sequences, slipped replication cannot account for all large spontaneous deletions. Other mechanisms, such as illegitimate recombination and topoisomerase II-mediated deletion formation, have been proposed to account for those deletions not associated with direct repeats (Mita et al., 1990).

Patients usually harbor a single deletion which is found in varying heteroplasmic ratios in different tissues of the body. Typically, a high proportion of deleted molecules are present in the highly-oxidative, post-mitotic tissues which are the most affected tissues in these patients. Thus, it is not uncommon to detect 80% deleted mtDNAs in the skeletal muscle of ocular myopathy patients. Lower amounts of deletion are detected in the rapidly-dividing tissues such as the blood and fibroblasts, presumably due to selection against functionally compromised mitochondria. The widespread tissue distribution and lack of genetic transmission indicate that the deletion likely occurred in the ovum or at an early stage of embryonic development. The timing of the deletion event during development and subsequent replicative segregation may account for patient-to-patient differences in deletion tissue distribution and heteroplasmic ratios. This, in turn, explains the variability in clinical expression seen in patients harboring mtDNA deletions as KSS patients are likely to have a higher proportion of deleted mtDNAs in more tissues than CPEO patients.

Disease states resulting from mtDNA deletions progress with age. This is likely due to an increase in the deleted mtDNA to normal mtDNA ratio over time in patient tissues. In a longitudinal study involving KSS patients, Larsson et al. (1990) demonstrated that the proportion of deleted molecules in skeletal muscle increased with age and that this increase in the percentage of deleted molecules was associated with the progression of clinical symptoms. The enrichment of mutant mtDNAs with time can be explained if the smaller deleted molecules had a replicative advantage over larger undeleted molecules, a likely phenomenon given the rate-determining step in mtDNA replication is probably the size of the template molecule (Shoffner et al., 1989).

Determination of the precise pathogenic mechanism of such deletions has been impaired by the lack of correlation between the size and position of the various deletions and the observed OXPHOS defects. In general, patients harbor high levels of deleted mtDNA and low levels of normal mtDNA in affected tissue. Northern blot analysis of mtRNA isolated from cybrid cells containing deleted mtDNAs, *in situ* hybridization studies and *in vitro* transcription data indicate that both wild-type and deleted mtDNA molecules are transcribed (Mita et al., 1989; Nakase et al., 1990; Hayashi et al., 1991; Moreas et al., 1992). However, immunochemical studies and experiments labeling mitochondrial translation products with ^{35}S-Met show that translation is severely impaired in mitochondria containing a high proportion of deleted mtDNAs (Hayashi et al., 1991; Ballinger et al, 1992; Moreas et al., 1992). Since almost all deletions eliminate at least one of the mitochondrial-encoded tRNAs, it is probable that overall mitochondrial translation would be affected by the loss of tRNAs. If so, intraorganellar tRNA complementation might be expected to occur in those mitochondria that harbor a sufficient number of wild-type molecules, as tRNAs transcribed from the normal mtDNAs compensate for those lost in the deleted mtDNAs. This appears to be the case in cybrid cells, as Hayashi et al. (1991) demonstrated that in cybrids containing below 60% deleted mtDNAs, translation occurs in proportion to mRNA levels and that a fusion protein (mRNA coding region spans the deletion breakpoint) can be detected upon SDS-PAGE of ^{35}S-Met labeled mitochondrial proteins. However, in cybrids containing greater than 60% deleted mtDNA, overall mitochondrial translation decreased dramatically, as did synthesis of the fusion protein. Thus, a generalized translation defect due to tRNA deficit is likely in those cells harboring mitochondria with high proportions of deleted mtDNA. Such a pathogenic mechanism would explain why ocular myopathy is a common phenotypic result among patients harboring markedly different deletions.

Pathogenic mtDNA deletions can also be inherited. These deletions are less common than spontaneous deletions and can be further subdivided on the basis of maternal versus Mendelian inheritance. One well-characterized maternally inherited deletion of 10.4 kb was detected in a three generation pedigree exhibiting adult-onset diabetes mellitus and hearing loss (Ballinger et al., 1992). All maternally-related family members studied harbored the heteroplasmic deletion in multiple tissues. Unlike pathology associated with spontaneously-occurring deletions, members of this pedigree did not present with mild or severe forms of ocular myopathy. Also, unlike spontaneously-occurring deletions, this large deletion eliminates the light-strand origin of DNA replication (Figure 4). It is possible that the absence of the light-strand origin decreases DNA replication efficacy and thereby does not allow the smaller deleted molecule a replicative advantage. Thus, progressive enrichment may not occur or be greatly reduced in patient tissues, accounting for the unique pathology and inheritance associated with this deletion.

Several pedigrees exhibiting the autosomal dominantly transmitted propensity for harboring mtDNA deletions have been identified (Zeviani et al., 1989; Zeviani et al., 1990; Cormier et al., 1991). KSS and bilateral cataract formation are common clinical manifestations in many of these pedigrees. In contrast to both spontaneous and maternally-inherited deletions, affected individuals harbor multiple mtDNA deletions. Deletion breakpoints can vary among family members, strongly suggesting that a predisposition to *de novo* deletion formation is inherited, not the deletions themselves. These deletions always spared both origins of DNA replication and occurred within direct repeats. Zeviani et al. (1989, 1990) have identified a mutational hotspot for deletion formation near the $tRNA^{Pro}$/D-loop junction (nps 16070-16080). The non-coding D-loop region contains important sequence elements (including P_L, P_H, and O_H,) that interact in *trans* with nuclear gene products to allow mtDNA replication. Thus, it is quite possible that a mutation in a nuclear-encoded, *trans*-acting factor involved in mtDNA replication promotes deletion formation, perhaps by facilitating slipped replication or intramolecular recombination. Such deletions have been found in both leukocyte and skeletal muscle mtDNA and the number of deletions can increase with time in an individual (Cormier et al., 1991; Servidei et al., 1991). This would explain the progressive clinical course seen in affected family members and is consistent with the constitutive presence of a nuclear-encoded mutant gene product.

MtDNA Duplications

MtDNA duplications have also been associated with progressive ocular myopathy and Pearson Marrow/Pancreas Syndrome (Poulton et al., 1989a,b; Rotig et al., 1990). All patients harbor a single, partially-duplicated mtDNA molecule present in heteroplasmic ratios in multiple tissues. Rotig et al. (1991) have described the maternal transmission of a 26 kb mtDNA molecule from a mother to two daughters. In this pedigree, the mother had ocular myopathy and extremely low levels of rearranged mtDNA while the daughters had mitochondrial myopathy, diabetes mellitus, proximal tubulopathy and ataxia and harbored higher amounts of the partial duplication. As is the case with mtDNA deletions, disease courses are progressive in these patients, possibly due to two sets of DNA origins found on these rearranged molecules. Both the genetic mechanism producing the partial duplications and the pathogenic mechanism producing the phenotype are unknown.

REFERENCES

Anderson, S., Bankier, A.., Barrell, B. G., deBruijin, M. H. L., Coulson, A. R., Drouin, J., Eperon, I. C., Nierlich, D. P., Rose, B. A., Sanger, F., Schreier, P. H., Smith, A. J. H, Staden, R., and Young, I. G. (1981). Sequence and organization of the human mitochondrial genome. *Nature* **290**, 457-465.

Attardi, G. (1985). Animal mitochondrial DNA: an extreme example of genetic economy. *Int. Rev. Cytolog.* **93**, 93–123.

Ballinger, S. W., Shoffner, J. M., Hedaya, E. V., Trounce, I., Polak, M. A., Koontz, D. A., et al. (1992). Maternally transmitted diabetes and deafness associated with a 10.4 kb mitochondrial DNA deletion. *Nat. Genet.* **1**, 11–15.

Brown, M. D., Voljavec, A. S., Lott, M. T., MacDonald, I., and Wallace, D. C. (1992a) Leber's Hereditary Optic Neuropathy: a model for mitochondrial neurodegenerative diseases. *Faseb J.* **6**, 2791–2799.

Brown, M. D., Voljavec, A. S., Lott, M. T., Torroni, A., Yang, C., and Wallace, D.C. (1992b). Mitochondrial DNA Complex I and III mutations associated with Leber's Hereditary Optic Neuropathy. *Genetics* **130**, 163–173.

Brown, M. D., Yang, C.-C., Trounce, I., Torroni, A., Lott, M.T., and Wallace, D. C. (1992c). A mitochondrial DNA variant, identified in Leber's Hereditary Optic Neuropathy patients, which extends the amino acid sequence of cytochrome c oxidase subunit I. *Am. J. Hum. Genet.* **51**, 378–385.

Bu, X. and Rotter, J. I. (1991). X-chromosome linked and mitochondrial gene control of Leber's Hereditary Optic Neuropathy: evidence from segregation analysis for dependence on X chromosome inactivation. *Proc. Natl. Acad. Sci. U.S.A.* **88**, 8198–8202.

Case, J. T. and Wallace, D. C. (1981). Maternal inheritance of mitochondrial DNA polymorphisms in cultured human fibroblasts. Som. *Cell Genet.* **7**, 103–108.

Clayton, D. (1991). Replication and transcription of vertebrate mitochondrial DNA. *Annu. Rev. Cell Biol.* **7**, 453–478.

Cormier, V., Rotig, A., Tardieu, M., Colonna, M., Saudubray, J. M., and Munnich, A. (1991). Autosomal dominant deletions of the mitochondrial genome in a case of progressive encephalomyopathy. *Am. J. Hum. Genet.* **48**, 643–648.

Corral-Debrinski, M., Stepien, G., Shoffner, J. M., Lott, M. T., Kanter, K., and Wallace, D. C. (1991). Hypoxia is associated with mitochondrial DNA damage and gene induction. Implications for cardiac disease. *JAMA* **266**, 1812–1816.

Cortopassi, G. A. and Arnheim, N. (1990). Detection of a specific mitochondrial DNA deletion in tissues of older humans. *Nucl. Acids. Res.* **18**, 6927–6933.

Giles, R. E., Blanc, H., Cann, H. M., and Wallace, D. C. (1980). Maternal inheritance of human mitochondrial DNA. *Proc. Natl. Acad. Sci. U. S. A.* **77**, 6715–6719.

Goto, Y., Koga, Y., Horai, S., and Nonaka, I. (1990a). Chronic progressive external ophthalmoplegia: a correlative study of mitochondrial DNA deletions and their phenotypic expression in muscle biopsies. *J. Neurol.* **100**, 63–69.

Goto, Y., Nonaka, I., and Horai, S. (1990b). A mutation in the tRNA(Leu)(UUR) gene associated with the MELAS subgroup of mitochondrial encephalomyopathies. *Nature* **348**, 651–653.

Gyllensten, U., Wharton, D., Josefsson, A., and Wilson, A. C., (1991). Paternal inheritance of mitochondrial DNA in mice. *Nature* **352**, 255–257.

Hayashi, J.-I., Ohta, S., Kikuchi, A., Takemitsu, M., Goto, Y. I., and Nonaka, I. (1991). Introduction of disease-related mitochondrial DNA deletions into HeLa cells lacking mitochondrial DNA results in mitochondrial dysfunction. *Proc. Natl. Acad. Sci. U.S.A.* **88**, 10614–10618.

Holt, I. J., Harding, A. E., Petty, R. K., and Morgan-Hughes, J. A. (1990). A new mitochondrial disease associated with mitochondrial DNA heteroplasmy. *Am. J. Hum. Genet.* **46**, 428–433.

Howell, N., Kubacka, I., Xu, M., and McCullough, D. A. (199la). Leber's Hereditary Optic Neuropathy: involvement of the mitochondrial NDl gene and evidence for an intragenic suppressor mutation. *Am. J. Hum. Genet.* **48**, 935–942.

Howell, N., Bindoff, L. A., McCullough, D. A. Kubacka, I., Poulton, J., Mackey, D., et al. (199lb). Leber's Hereditary Optic Neuropathy: identification of the same mitochondrial NDl mutation in six pedigrees. *Am. J. Hum. Genet.* **49**, 939–950

Huoponen, K., Vilkki, J., Aula, P., Nikoskelainen, E.K., and Savontaus, M. L. (1991). A new mutation associated with Leber's Hereditary Optic Neuroretinopathy. *Am. J. Hum. Genet.* **48**, 1147–1153.

Johns, D. R. and Berman, J. (1991). Alternative, simultaneous complex mitochondrial DNA mutations in Leber's Hereditary Optic Neuropathy. *Biochem. Biophys. Res. Commun.* **174**, 1324–1330.

Johns, D. R. and Neufeld, M. J. (1992). Cytochrome b mutations in Leber Hereditary Optic Neuropathy. *Biochem. Biophys. Res. Commun.* **181**, 1358 – 1364 .

Johns, D. R., Neufeld, M. J., and Park, R. D. (1992). An ND6 mitochondrial DNA mutation associated with Leber's Hereditary Optic Neuropathy. *Biochem. Biophys. Res. Commun.* **187**. 1551–1557.

Larsson, N.-G., Andersen, O., Holme, E., Oldfors, A., and Wahlstrom, J. (1991). Leber's Hereditary Optic Neuropathy and complex I activity in muscle. *Ann. Neurol.* **30**, 701–708.

Larsson, N. G., Holme, E., Kristiansson, B., Oldfors, A., and Tulinius, M. (1990). Progressive increase of the mutated mitochondrial DNA fraction in Kearns-Sayre syndrome. *Pediatr. Res.* **28**, 131–136.

Linnane, A. W., Marzuki, S., Ozawa, T., and Tanaka, M. (1989). Mitochondrial DNA mutations as an important contributor to ageing and degenerative diseases. *Lancet* **1**, 642–645.

Majander, A., Huoponen, K., Savontaus, M.-L., Nikoskeleinen, E., and Wikstrom, M. (1991). Electron transfer properties of NADH:ubiquinone reductase in the NDl/3460 and the ND4/11778 mutations of Leber's Hereditary Optic Neuroretinopathy. *FEBS Lett.* **292**, 289 – 292.

Mita, S., Schmidt, B., Schon, E. A., DiMauro, S., and Bonilla, E. (1989). Detection of "deleted" mitochondrial genomes in cytochrome-c oxidase-deficient muscle fibers of a patient with Kearns-Sayre syndrome. *Proc. Natl. Acad. Sci. U. S. A.* **86**, 9509–9513.

Mita, S., Rizzuto, R., Moraes, C. T., Shanske, S., Arnaudo, E., Fabrizi, G. M., Koga, Y., DiMauro, S., and Schon, E. A. (1990). Recombination via flanking direct repeats is a major cause of large-scale deletions of human mitochondrial DNA. *Nucl. Acids Res.* **18**, 561–567.

Montoya, J., Ojala, D., and Attardi, G. (1981). Distinctive features of the 5'-terminal sequences of the human mitochondrial mRNAs. *Nature* **290**, 465–470.

Moraes, C. T., DiMauro, S., Zeviani, M., Lombes, A., Shanske, S., Miranda, A. F., Nakase, H., Bonilla, E., Werneck, L. C., Servidei, S., Nonaka, I., Koga, Y., Spiro, A. J., Brownell, K. W., Schmidt, B., Schotland, D. L., Zupance, M., DeVivo, D. C., Schon, E. A., and Rowland, L. P. (1989). Mitochondrial DNA deletions in progressive external ophthalmoplegia and Kearns-Sayre syndrome. *N. Engl. J. Med.* **320**, 1293–1299.

Moraes, C. T., Ricci, E., Petruzzella, V., Shanske, S., DiMauro, S., Shon, E.A., and Bonilla, E. (1992). Molecular analysis of he muscle pathology associated with mitochondrial DNA deletions. *Nat. Genet.* **1**, 359–367.

Nakase, H., Moraes, C. T., Rizzuto, R., Lombes, A., DiMauro, S., and Schon, E.A. (1990). Transcription and translation of deleted mitochondrial genomes in Kearns-Sayre syndrome: implications for pathogenesis. *Am. J. Hum. Genet.* **46**, 418–427.

Neckelmann, N., Li, K., Wade, R. P., Shuster, R., and Wallace, D. C. (1987). cDNA sequence of a human skeletal muscle ADP/ATP translocator: lack of a leader peptide, divergence from a fibroblast translocator cDNA, and coevolution with mitochondrial DNA genes. *Proc. Natl. Acad. Sci. U. S. A.* **84**, 7580–7584.

Newman, N. J. (1991). Leber's Hereditary Optic Neuropathy. *Ophthalmic. Clin. N. Am.* 4, 431–447.

Newman, N. J., Lott, M. T., and Wallace, D. C. (1991). The clinical characteristics of pedigrees of Leber's Hereditary Optic Neuropathy with the 11778 mutation. *Am. J. Ophthalmol.* **111**, 750–762.

Newman, N. J. and Wallace, D. C. (1990). Mitochondria and Leber's Hereditary Optic Neuropathy. *Am. J. Ophthalmol.* **109**, 726–730.

Ojala, D., Montoya, J., and Attardi, G. (1981). tRNA punctuation model of RNA processing in human mitochondria. *Nature* **290**, 470–474.

Ortiz, R. G., Newman, N. J., Manoukian, S. U., Diesenhouse, M. C., Lott, M. T., and Wallace, D. C. (1992) Optic disk cupping and electrocardiographic abnormalities in an American pedigree with Leber's Hereditary Optic Neuropathy. *Am. J. Opthalmol.* **113**, 561–566.

Parker, W. D., Oley, C. A., and Parks, J. K. (1989). A defect in mitochondrial electron transport activity (NADH-coenzyme Q oxidoreductase) in Leber's Hereditary Optic Neuropathy. *N. Engl. J. Med.* **320**, 1331–1333.

Poulton, J., Deadman, M. E., and Gardiner, R. M. (1989). Tandem direct duplications of mitochondrial DNA in mitochondrial myopathy: analysis of nucleotide sequence and tissue distribution. *Nucl. Acids Res.* **17**, 10223–10229.

Poulton, J., Deadman, M. E., Ramacharan, S., and Gardiner, R. M. (1991). Germ-line deletions of mtDNA in mitochondrial myopathy. *Am. J. Hum. Genet.* **48**, 649–658.

Richter, C., Park, J.-W., Ames, B. N. (1988). Normal oxidative damage to mitochondrial and nuclear DNA is extensive. *Proc. Natl. Acad. Sci. U.S.A.* **85**, 6465–6467.

Rotig, A., Cormier, V., Blanche, S., Bonnefont, J. P., LeDeist, F., Romero, N., Schmitz, J., Rustin, P., Fischer, A., and Saudubray, J. M. (1990). Pearson's marrow-pancreas syndrome. A multisystem mitochondrial disorder in infancy. *J. Clin. Invest.* **86**, 1601–1608.

Schon, E.A., Rizzuto, R., Moraes, C. T., Nakase, H., Zeviani, M., and DiMauro, S. (1989). A direct repeat is a hotspot for large-scale deletion of human mitochondrial DNA. *Science* **244**, 346–349.

Servidei, S., Zeviani, M., Manfredi, G., Ricci, E., Silvestri, G., Bertini, E., Gellera, C., Di Mauro, S., Di Donato, S., and Tonali, P. (1991). Dominantly inherited mitochondrial myopathy with multiple deletions of mitochondrial DNA: clinical, morphologic, and biochemical studies. *Neurol.* **41**, 1053–1059.

Shoffner, J. M., Fernhoff, M. D., Krawiecki, N. S., Kaplan, D., Holt, P., Koontz, D. et al. (1992). Subacute necrotizing encephalopathy: oxidative phosphorylation defects and the ATPase 6 point mutation. *Neurol.* (in press).

Shoffner, J. M., Lott, M. T., Voljavec, A.S., Soueidan, S. A., Costigan, D. A., and Wallace, D.C. (1989). Spontaneous Kearns-Sayre/chronic external ophthalmoplegia plus syndrome associated with a mitochondrial DNA deletion: a slip-replication model and metabolic therapy. *Proc. Natl. Acad. Sci. U. S. A.* **86**, 7952–7956.

Shoffner, J. M., Lott, M. T., Lezza, A. M., Seibel, P., Ballinger, S. W., and Wallace, D. C. (1990). Myoclonic epilepsy and ragged-red fiber disease (MERRF) is associated with a mitochondrial DNA tRNA(Lys) mutation. *Cell* **61**, 931–937.

Shoffner, J. M. and Wallace, D. C. (1990). Oxidative phosphorylation diseases. Disorders of two genomes. *Adv. Hum. Genet.* **19**, 267–330.

Singh, G., Lott, M. T., and Wallace, D. C. (1989). A mitochondrial DNA mutation as cause of Leber's Hereditary Optic Neuropathy. *N. Engl. J. Med.* **320,** 1300–1305.

Sweeney, M. G., Davis, M. B., Lashwood, A., Brockington, M., Toscano, A., and Harding, A. E. (1992). Evidence against an X-linked locus close to DXS7 determining visual loss susceptibility in British and Italian families with Leber's Hereditary Optic Neuropathy. *Am. J. Hum. Genet.* **51**, 741–748.

Tanaka, M., Ino, H., Ohno, K., Hattori, K., Sato, W., Ozawa, T., Tanaka, T., and Itoyama, S. (1990). Mitochondrial mutation in fatal infantile cardiomyopathy. *Lancet* **336**, 1452.

Tatuch, Y., Christodoulou, J., Feigenbaum, A., Clarke, T. R., Wherret, J., Smith, C., et al. (1992). Heteroplasmic mtDNA mutation (T->G) at 8993 can cause Leigh disease when the percentage of abnormal mtDNA is high. *Am. J. Hum. Genet.* **50**, 852–858.

Trounce, I., Byrne, E., and Marzuki, S. (1989). Decline in skeletal muscle mitochondrial respiratory chain function: possible factor in ageing. *Lancet* **1**, 637–639.

Vilkki, J., Ott, J., Savontaus, M. L., Aula, P. and Nikoskelainen, E. K. (1991). Optic atrophy in Leber Hereditary Optic Neuroretinopathy is probably determined by an X-chromosomal gene closely linked to DXS7. *Am. J. Hum. Genet.* **48**, 486–491.

Wallace, D. C. (1982). Structure and evolution of organelle genomes. *Microbiol. Rev.* **46**, 208–240.

Wallace, D. C. (1986). Mitotic segregation of mitochondrial DNAs in human cell hybrids and the expression of chloramphenicol resistance. *Som. Cell Molec. Genet.* **12**, 41–49.

Wallace, D. C. (1992a). Diseases of the mitochondrial DNA. *Annu. Rev. Biochem.* **61**, 1175–1212.

Wallace, D. C. (1992b). Mitochondrial genetics: A paradigm for aging and degenerative diseases. *Science* **256**, 628–632.

Wallace, D. C., Bunn, C. L., and Eisenstadt, J. M. (1975). Cytoplasmic transfer of chloramphenicol resistance in human tissue culture cells. *J. Cell Biol.* **67**, 174–188.

Wallace, D. C., Lott, M., Torroni, T. and Shoffner, J. (1991). Report of the committee on human mitochondrial DNA. *Cyto. Cell. Genet.* **58**, 1103–23.

Wallace, D. C., Singh, G., Lott, M. T., Hodge, J. A., Schurr, T. G., Lezza, A. M. S., et al. (1988). Mitochondrial DNA mutation associated with Leber's Hereditary Optic Neuropathy. *Science* **242**, 1427–1430.

Wallace, D. C., Ye, J. H., Neckelmann, S. N., Singh, G., Webster, K. A., and Greenberg, B. D. (1987). Sequence analysis of cDNAs for the human and bovine ATP synthase beta subunit: mitochondrial DNA genes sustain seventeen times more mutations. *Curr. Genet.* **12**, 81–90.

Wong, T. W. and Clayton, D. A. (1986). DNA primase of human mitochondria is associated with structural RNA that is essential for enzymatic activity. *Cell* **45**, 817–825

Yen, T. C., Chen, Y. S., King, K. L., Yeh, S. H., and Wei, Y. H. (1989). Liver mitochondrial respiratory functions decline with age. Biochem. *Biophys. Res. Commun.* **165**, 944–1003.

Zeviani, M., Bresolin, N., Gellera, C., Bordoni, A., Pannacci, M., Amati, P., Moggio, M., Servidei, S., Scarlato, G., and DiDonato, S. (1990). Nucleus-driven multiple large-scale deletions of the human mitochondrial genome: a new autosomal dominant disease. *Am. J. Hum. Genet.* **47**, 904–914.

Zeviani, M., Amati, P., Bresolin, N., Antozzi, C., Piccolo, G., Toscano, A., and DiDonato, S. (1991a). Rapid detection of the A-G (8344) mutation of mtDNA in Italian families with myoclonus epilepsy and Ragged-Red Fibers (MERRF). *Am. J. Hum. Genet.* **48**, 203–211.

Zeviani, M., Gellera, C., Antozzi, C., Rimoldi, M., Morandi, L., Villani, F., Tiranti, V., and DiDonato, S. (1991b). Maternally inherited myopathy and cardiomyopathy: association with mutation in mitochondrial DNA tRNALeu(UUR). *Lancet* **338**, 143–147.

Zeviani, M., Muntoni, F., Savarese, N., Serra, G., Tiranti, V., Carrara, F., et al. (1992). A MERRF/MELAS overlap syndrome associated with a new point mutation in the mitochondrial DNA $tRNA^{Lys}$ gene. *Eur. J. Hum. Genet.* **1**, 80–87.

GLOSSARY

*methods ^ terms # abbreviations

^3' end: The end of a nucleic acid strand that contains a free 3' hydroxyl group on the sugar portion of the nucleotide. DNA synthesis proceeds from this end of the strand when there is a DNA template (DNA polymerase), an RNA template (reverse transcription), or nonspecific addition (e.g. terminal deoxynucleotide transferase)

^5'end: The end of a nucleic acid strand that contains a free 5' hydroxyl group on the sugar portion of the nucleotide. In DNA cleaved with enzymes, this hydroxyl group is usually phosphorylated but can be removed and replaced by ^{32}P using a combination of phosphatase and kinase. Like DNA replication, both transcription and translation of nucleic acid molecules proceed in a 5' to 3' direction along the DNA or RNA strand.

^acceptor splice site: The junction between the 3' end of an intron and the 5' end of an adjacent exon.

***agarose gel electrophoresis:** The separation of charged molecules (e.g. negatively charged nucleic acids) using an electric field across an agarose matrix (*see* acrylamide gel electrophoresis). Agarose is a polysaccharide prepared from seaweed and can be used for the separation of small as well as very large nucleic acids. Because the gel is formed by hydrogen-bonding of the sugar chains, agarose gels are easily disrupted by heat and/or denaturing agents.

^allele: An allele is an alternative form of a single gene or other locus. A single allele is inherited separately from each parent. In situations where a genetic marker locus is defined by a nucleic acid sequence or probe, an allele may be defined as a single nucleotide difference that causes a change in a restriction endonuclease site (*see* RFLP), a conformation alteration (*see* SSCP, DGGE), a mismatch between the sequences from the two chromosomes (*see* heteroduplex analysis), a deletion or insertion of DNA, or variations in the length of a microsatellite repeat (*see* CA repeats).

^allele-specific oligonucleotide: An appropriately selected oligonucleotide can be labelled and used for hybridisations with sufficient stringency (specificity) that single nucleotide differences as well as small rearrangements can be detected in genomic DNA. When used in conjunction with PCR, allele-specific oligonucleotides can be attached to a membrane and used to type a large set of alleles rapidly for a given DNA sequence, such as HLA antigen genes.

***Alu PCR:** The use of sequences from Alu repeats (*see* below) as primers for PCR (*see* polymerase chain reaction). Generally the sequences are selected from Alu segments that have evolved after speciation so that human sequences can be detected independently of DNA from other organisms. Because there are so many Alu repeats within the genome, short DNA sequences between adjacent Alu repeats can be amplified from many regions of the genome. The technique has been used to identify and characterize radiation hybrids, somatic cell hybrids and YACs containing human genomic DNA.

^Alu repeats: There are an estimated 3-5 x 10^5 repetitive elements in mammalian DNA that are part of a family referred to as Alu repeats. These repetitive elements are 300 bp in length and were first recognized when genomic DNA was cut with the restriction endonuclease, AluI. They are scattered throughout the genome and because subfamilies of the Alu repeats have arisen since mammalian speciation, specific Alu primers can be used to selectively amplify human genomic DNA from a mixture of yeast or rodent DNA (*see* Alu PCR).

^antisense strand: *see* sense strand

#ASO: *see* allele-specific oligonucleotide

***autoradiography:** The detection of radioactively labeled material using X-ray film or a comparable photographic emulsion. Can be used for in situ hybridisation of chromosomes, cells, and tissue slices as well as for detection of radioactively labeled bands or spots from dot blots, Northerns, and Southerns.

^avidin: A tetrameric protein derived from egg whites that tightly binds biotin (*see* streptavidin/biotin).

#b: *see* base

^bacteriophage: A virus that specifically infects bacteria. The most commonly used bacteriophages in molecular biology are derived from lambda phage. Typically, these virus-derived vectors can hold 5–20 kb of inserted, foreign DNA. Although less ideal for manufacturing large amounts of an inserted DNA than plasmids or cosmids, bacteriophage are an important host for cloning because the DNA can be packaged *in vitro* into phage particles that enter bacteria with very high efficiency. The genetic information from the bacteriophage also cause infected cells to lyse, allowing for simple and sensitive screening of large number of plaques (*see* plaque).

^base: A nitrogenous base used in nucleic acids consisting either of a purine (adenine, guanine) or pyrimidine (thymine, cytosine (DNA) or uracil (RNA)) residue. A nucleotide base is one of the monomeric units within nucleic acid and consists of a phosphoric acid group attached to a deoxyribose (DNA) or ribose (RNA) sugar and a purine or pyrimidine base. A 500 base RNA refers to a nucleic acid strand comprised of 500 nucleotides.

^**base pair:** A pair of hydrogen-bonded nucleotide bases (i.e a purine and pyrimidine each attached to a ribose and phosphate groups). The sizes of double-stranded nucleotide chains are often indicated by the # of base pairs that are shared. A 500 base pair (bp) DNA fragment consists of two strands of DNA, each containing 500 nucleotides and a total molecular weight of approximately 330 kilodaltons.

^**beta-galactosidase:** A bacterial enzyme that cleaves-galactose. The activity of this enzyme can be monitored by using indicator substrate molecules such as X-gal or Blue-gal and the appearance of a blue color. The expression of beta-galactosidase introduced into bacteria as part of a recombinant plasmid is used to distinguishing non-recombinant and recombinant plasmids or cosmids. The enzyme is often coupled to different promoters and used as an indicator gene in transfections and transgenic animals.

^**biotin:** *see* Streptavidin/biotin

#**bp:** *see* base pair

^**breakpoint:** The location at which there is a break in the continuity of a chromosome allowing two previously noncontiguous pieces to become joined (*see* crossing over, recombination, translocation).

***bubble PCR:** Amplification of specific fragments of DNA when only one primer of known sequence can be constructed. The genomic DNA is cleaved into pieces (either specifically with restriction endonucleases or nonspecifically) and then special DNA adaptors are ligated onto the exposed ends. The special adaptors, known as bubble adaptors, have mismatching internal sequences but matching ends that allow for stable ligation. The bubble primer is made to be identical (not complementary) to one of the strands of DNA within the bubble. When PCR is performed, the sequence-specific PCR primer allows the Taq polymerase to synthesize a complementary strand of DNA. The polymerase synthesizes the complement to one of the strands within the bubble. In the next round of primer annealing and amplification, the bubble primer is able to anneal to this newly synthesized DNA and allow a new complementary strand to be made. Without the first synthesis step directed by the sequence-specific primer, the bubble primer is unable to anneal and prime DNA synthesis. Using this technique, nonspecific amplification of DNA fragments that contain identical ligated adaptor molecules is avoided.

^CA repeats: There are hundreds of thousands of microsatellite repeats within the mammalian genome. These dinucleotide, trinucleotide and tetranucleotide repeats include (CA)n, (CAA)n, (CAAT)n, (AT)n, and others. The most commonly used repeat for PCR-based genotyping and linkage mapping is the CA repeat. Thousands of CA-repeat based markers have been isolated and mapped to the human genome. For many of these repeats, there is considerable variations (alleles) among individuals which allows them to be used for linkage studies.

#cDNA: *see* complementary DNA

***cDNA synthesis and cloning:** cDNA synthesis begins with either total or polyA+RNA and an oligonucleotide primer that will anneal to the 3' polyA tract of the RNA (oligodT) or a set of random primers that will anneal/hybridise along the length of the RNA. Reverse transcriptase is used to synthesize a complementary DNA (*see* above). By a variety of methods, the RNA of the RNA/DNA hybrid is replaced by DNA so that a double-stranded DNA contains the original genetic information that was encoded in the RNA. There are numerous variations in the method, including the use of primers that are already attached to a vector, using PCR to amplify the double-stranded DNA, different techniques to prime the second strand synthesis, and different strategies for inserting this synthesized DNA into a vector. Once the process of building and incorporating the cDNA into the vector is completed, then the recombinant material is introduced into E. coli either by transfection (for bacteriophage vectors), transformation (introducing unpackaged DNA into cells) or electroporation (creating transient holes in the cellular membrane with an electric shock so that the DNA can enter the cells). Cells are spread and grown so that individual colonies can be established. Antibiotic selection is used to ensure that only cells that have taken up the vector will grow. The individual colonies can then be screened by a variety of methods to identify those that contain a specific cDNA of interest.

^cell cycle: The sequence of stages that occur between successive cell divisions. The period of mitosis is referred to as the M phase, followed by a growth period (G_1), a DNA synthesis period (S) and a second growth period, (G_2) before entering the M phase again. Cells that have stopped dividing and have entered a stationary phase are said to be in G_0.

^centiMorgan: A unit of measure of recombination frequency between two markers such that one centimorgan equals a 1% chance that a recombination event would occur in a single generation. While there is a rough correlation of physical distances (Mb) and genetic distances (cM), with 1 Mb equivalent to about 1cM in man, there are regions of reduced or increased recombination that limit direct comparisons.

^centiray: A unit of measure similar to centimorgan but the centiray is based on the probability or frequency of DNA breaks induced by radiation. This measure of genetic distance between markers is used in radiation hybrid mapping and more closely correlates with physical distance than recombination frequencies. In general, one centiray equals approximately 50 kb of DNA.

^chemiluminescence: The production of light by a chemical reaction. An enzyme (e.g. horseradish peroxidase) that can catalyze a chemiluminescent reaction is covalently linked to a molecule (avidin or an antibody) that also binds a DNA "probe" (such as a specific DNA labelled with biotin or digoxygenin). Specific detection of DNA fragments complementary to the probe can be achieved after hybridisation by the addition of appropriate substrate molecules and the detection of light on photographic film. Although it is a more complicated procedure than autoradiography, the method has the advantage of creating stable reagents, high sensitivity, rapid exposure times, and avoidance of radioactivity.

^chromosome: A self-replicating genetic element within the nucleus of a eukaryotic cell that is comprised of a single double-stranded molecule of DNA and a large number of proteins. Each chromosome contains a centromere where spindle fibers attach during cell division and telomeres at the ends of each chromosome that are required for replication. All of the genetic information in the nucleus of human cells is contained within a set of 23 pairs of chromosomes which range in size from 50 to 250 Mb (average 150 Mb), comprising a total of three billion base pairs of haploid DNA. The size of the human mitochondrial chromosome is 16,569 base pairs.

***chromosome banding:** When cells in metaphase are lysed and spread on glass slides, the chromosomes are easily visible by light microscopy. Because the structure and composition of the chromosomes are not homogeneous, different dyes (such as Giemsa and quinacrine) can be used that will create a standardised banding pattern along the length of each chromosome. These patterns can be used to identify specific chromosomes, establish a physical "ruler" along the chromosome by which genes can be mapped, and aid in the detection of chromosomal rearrangements such as deletions and translocations.

***chromosome painting:** Alu-PCR (*see* Alu-PCR) can be used to amplify human-specific genomic DNA from flow-sorted chromosomes or somatic cell hybrids (*see* flow cytometry and somatic cell hybrids), so that a hybridisation reagent can be produced that will specifically label all or part of specific chromosomes. FISH analysis can be carried out (*see* fluorescence *in situ* hybridisation) by using fluorescence detection of the labeled PCR material that will "paint" all or part of a chromosome. These reagents are now commercially available and are widely used to detect subtle chromosomal rearrangements that were previously undetectable by standard cytogenetic methods including chromosome banding.

#cM: *see* centiMorgan

^coding region: The coding region is that portion of the mRNA transcript (the processed RNA synthesized from the gene) that contains the genetic information for the amino acid sequence of the protein product. *see* exon, intron, splicing, messenger RNA, open reading frame.

^codon: The nucleotide triplet in mRNA that encodes for 1) the beginning of peptide synthesis (initiation codon), 2) a specific amino acid to be added to the peptide chain or 3) the termination of peptide synthesis (termination codon).

^colony: A population of cells that have arisen from a single progenitor cell. In plasmid, cosmid or yeast libraries, individual colonies can be identified by spreading the library on agar plates. Libraries generated in bacteriophage vectors can also be spread on plates to identify homogeneous, clonal populations of viruses which cause discrete spots of bacterial lysis, known as plaques.

^complementary DNA (cDNA): The strand of DNA that is synthesised using a specific or non-specific primer to an RNA template (RNA-dependent DNA polymerase such as reverse transcriptase) or to a DNA template (DNA-dependent DNA polymerase). The sequence of this synthesized DNA is determined by the rules of base pairing. The term, cDNA, is frequently used to specifically identify DNA that has been made to be complementary to messenger RNA or polyA+RNA.

^contigs: A group of clones that provide overlapping coverage of one or more regions of a genome.

***contig mapping:** The process of ordering the clones from a library to establish a genomic map based upon the relative order and overlap of these smaller genetic elements.

^cosmids: A type of cloning vector that is derived from plasmids but contains the COS site of lambda bacteriophage so that the recombinant vector can be packaged and transfected into E. coli. Cosmids are generally used for cloning genomic DNA in the size range of 25-40 kb because of their greater efficiency at entering bacteria by transfection compared to the transformation of bacteria with plasmids containing large inserts.

^Cot-1 DNA: Much of the DNA within the mammalian genome is repetitive (*see* repetitive DNA) and this DNA can be paritally purified by denaturing total genomic DNA and allowing it to reanneal to itself. Because repetitive sequences are present in many copies, the majority of the DNA that initially reanneals will contain these repetitive elements. DNA that has reannealed under certain defined conditions is called Cot-1 DNA and contains the most highly repetitive DNA within the genome. This material can be used in hybridisations to compete out repetitive DNAs on target molecules, allowing unique sequences that are represented in the hybridisation probes to be detected.

^CpG island: The distribution of nucleotides is not random throughout the genome and regions that are rich in C followed by G dinucleotides (CpG) are relatively rare. Studies have shown that these CpG islands are often near the promoter regions of genes. The identification of CpG islands is made with rare cutting restriction endonucleases that recognize these rare sequences. The presence of CpG islands is one of the indicators of potential genes in a large segment of genomic DNA.

#cR: *see* centiRay

^crossing over: Crossing over occurs when two paired chromosomes exchange matching regions of their DNA. *see* recombination

^dATP, γ-^{32}P or α-^{32}P labelled: Deoxyadenosine triphosphate with the γ- position phosphate containing ^{32}P is used for 5'end-labeling of DNA or oligonucleotides using T4 polynucleotide kinase. α -^{32}P-dATP is used for labelling DNA molecules by incorporation of the nucleotide into the polynucleotide chain using a DNA polymerase.

***denaturing gradient gel electrophoresis (DGGE):** This type of polyacrylamide gel electrophoresis method distinguishes DNA fragments (100–500 bp) of similar sizes but of different nucleotide sequences and is used in mutation detection. The denaturation or melting properties of different regions of a double-stranded DNA molecule are very sensitive to the nucleotide composition and sequence. The partial and/or complete denaturation of the DNA strands dramatically alters the mobility of the molecules in a polyacrylamide gel. A denaturing gradient is established by a gradient of a denaturant such as formamide within the gel (DGGE) or less commonly by a thermal gradient (TGGE). PCR-generated fragments are the source of DNA for DGGE. Additional sequences (GC clamps) are often introduced on to the ends of the fragment of interest in order to improve the sensitivity and effectiveness of the detection method.

^deoxyribonucleic acid: A covalent polymer of deoxynucleotide subunits (*see* nucleotide).

^digoxygenin: A steroid derived from the digitalis plant that can be covalently linked to DNA by incorporation of derivatized nucleotides. The labelled DNA can then be detected by a variety of methods (chemiluminescence, enzymatic stain, fluorescence) using a specific antibody conjugated to an appropriate indicator molecule or enzyme.

#DGGE: *see* denaturing gradient gel electrophoresis

#DNA: *see* deoxyribonucleic acid

^donor splice site: The junction between the 3' end of an exon and the 5' end of an adjacent intron.

***dot blotting:** A method of detecting and/or quantitating the presence of a specific nucleic acid fragment by hybridising a labeled probe to DNA or RNA that has been spotted and fixed to a membrane surface.

#dsDNA: double-stranded DNA

***electroporation:** A method for permeabilizing cell membranes to allow for the passage of large molecules such as DNA into the cell. The cells are suspended in a special cuvette and exposed to a brief electrical pulse that causes transient holes in the membrane. The electrical conditions vary considerably among different cell types. The technique can be used for eukaryotic cells or bacteria.

#EST: *see* expressed sequence tag

^exon: A region of a gene that is transcribed and is present in the final messenger RNA (mRNA) transcript. The exons contain the 5' and 3' untranslated portions of the mRNA as well as the sequences that code for the amino acid sequence of the protein product (*see* intron, splicing, coding region, messenger RNA).

^expressed sequence tag: Equivalent to the sequence tagged site (STS) except that the unique sequence that is used for construction of the primers is within an exon of a gene.

^expression vector: A special vector (*see* Vector) that allows for the expression of a foreign gene in an appropriate host. Expression vectors are used when a gene can only be recognized by detection of its protein product (immunologically or functionally), or as a means of manufacturing the protein for biochemical and cell biology studies.

^expressivity: The range of altered features (phenotypes) associated with a specific genetic constitution of an individual (genotype).

***field inversion gel electrophoresis:** A special type of agarose gel electrophoresis that uses an asymmetric alternating electric field to move the DNA through the gel matrix. The intermittent reversal of the electric field allows for the separation of much larger DNA fragments then can normally be achieved by standard agarose gel electrophoresis. The method is effective up to about 1 Mb of DNA. To separate larger DNA fragments, pulsed-field gel electrophoresis method is more often used. *see* agarose gel electrophoresis, pulsed-field gel electrophoresis.

#FIGE: *see* field inversion gel electrophoresis

#FISH: *see* fluorescence in situ hybridisation

***flow cytometry:** Cells or subcellular fractions such as chromosomes can be analysed and sorted by dispersing individual elements into minute, electrostatically-charged water droplets, determining the optical or fluorescent properties of the element within the droplet, and then sorting the droplet using a controlled magnetic field. By this approach, individual chromosomes can be isolated to 80–90% purity and then used for the construction of chromosome-specific genomic libraries.

^fluorescence: The detection of specific molecules based upon their ability to emit light at a higher wavelength than the light absorbed by the molecule. Because of the separation between the excitation and emission wavelengths, one can use selective filters to detect molecules with high sensitivity and low background. Examples of fluorescent molecules used in molecular studies include ethidium bromide, and derivatives of fluorescein, phthalocyanines, and rhodamine.

***fluorescence *in situ* hybridisation:** Commonly referred to as FISH, fluorescence *in situ* hybridisation is the localization of one or more markers to specific chromosomes using hybridisation of a probe to metaphase or interphase chromosomes arrayed on a glass slide and detection with one or more fluorescent dyes. In addition to providing physical mapping data, FISH can also be used to detect microdeletions in selected regions as well as aneuploidy and chromosomal rearrangements.

#G_0: The stationary growth phase of the cell cycle (*see* cell cycle).

#G_1: The growth phase after mitosis during the cell cycle (*see* cell cycle).

***gel electrophoresis:** *see* agarose gel electrophoresis, polyacrylamide gel electrophoresis

***gel retardation assays:** A method to identify proteins that bind to specific nucleotide sequences within the promoter regions of genes. Detection of specific binding is achieved by demonstrating that the mobility of a labeled DNA fragment in a gel is specifically decreased by the binding of one or more proteins to an internal sequence. Unlabelled oligonucleotides that contain the target sequ-ence can be used to competitively block this binding and allow the labeled DNA to migrate as if the binding protein was not present.

^gene: A fundamental unit of genetic structure residing in a single location within the genome that encodes for information that is capable of being expressed by one or more cells during the lifetime of the organism.

^genetic map: The linear ordering and spacing of markers and genes along the chromosomes based upon the frequency of recombination between the linked elements.

^genome: The entire set of genetic information contained within a single (haploid) set of chromosomes.

^heterogeneous nuclear RNA: RNA that is found within the nucleus and consists of unprocessed and partially processed transcripts, as well as small nuclear RNAs. Heterogeneous nuclear RNA is rapidly processed by polyadenylation and splicing out or removal of the introns so that the final transcript, the messenger RNA, only contains the exon regions of the gene and can be exported to the cytoplasm.

^heterozygosity: 1) The condition of having dissimilar alleles at a given locus. 2) A measure of the number and distribution of alleles in the population associated with a marker *see* polymorphism information content

#hnRNA: *see* heterogeneous nuclear RNA

^horseradish peroxidase: An enzyme that, when combined with the appropriate substrate, creates a chemiluminescent or insoluble product that can be visually detected. By coupling the enzyme with an appropriate molecule, e.g. an antibody or avidin, it can be used for the detection of specific DNA fragments or proteins (*see* chemiluminescence, in situ hybridisation).

#HRP: *see* horseradish peroxidase

***in situ* hybridisation:** In situ hybridisation (ISH) is a method for directly visualizing the cellular or subcellular location of one or more DNA or RNA sequences using hybridisation with a labeled probe and a suitable detection method. *In situ* hybridisation is used for the localisation of one or more markers to specific chromosomes arrayed on a glass slide (*see* Fluorescence in situ hybridisation). ISH is also used to identify the location of cellular or viral transcripts or viral DNA on histologic preparations of cells and tissue and to evaluate the developmental and/or tissue-specific expression of specific genes.

^**intron:** That portion of a gene that in transcribed and present in the initial transcript (hnRNA) but is then removed by splicing to produce the final transcript (mRNA). The sequences within the intron are generally not used for the coding of the protein, but some genes have alternative splicing in which some introns are removed only part of the time, thus creating more complex final transcripts and heterogeneous protein products (*see* exon, splicing, coding region).

***jumping libraries:** A genomic library that has been constructed by cutting genomic DNA with a rare-cutting restriction endonuclease. The large fragments (100–200 kb) are allowed to circularize and ligated. The ligated circles are then cut with a second enzyme and the fragments are cloned. Clones are selected for the presence of the first, rare-cutting restriction endonuclease site. These clones contain small fragments that were originally separated by the length of the original fragment. They provide the means for "jumping" across a distance that is greater than can be normally contained within a single cosmid or phage clone.

^**karyotype:** The display of a complete set of chromosomes from a given individual or cell (*see* Banding).

#**kb:** A thousand bases or base pairs

***library:** A collection of recombinant molecules or clones (foreign DNA inserted within a vector and propagated in cells) that represents a heterogeneous source of DNA or RNA. Libraries can contain the population of transcripts (mRNA) within a cell or tissue (*see* cDNA cloning), the genomic DNA from an entire organism, DNA from a specific chromosome (i.e. chromosomes sorted by flow cytometry or somatic cell hybrids), or DNA from a limited region of a chromosome (e.g. a microdissected portion of a chromosome or from a YAC containing a large genomic fragment). Subtracted libraries are those in which the population of clones has been constrained by the elimination of those clones that are in common between two populations of DNA or RNA. Libraries are said to be normalised when steps have been taken to minimise the difference in copy numbers of unique elements. For example, cDNA libraries may contain clones representing transcripts that have less than 0.001% abundance within a cell and other transcripts that constitute more than 10% of the mRNA population.

***library screening:** The identification of specific clones (either genomic or cDNA) from a library (*see* above) using either a hybridisation probe (such as a previously cloned DNA, an oligonucleotide or a PCR product) or a detection method for a specific gene product (such as an antibody to a protein product, a specific enzymatic or biological assay).

^**linkage:** The greater than chance co-inheritance of two or more markers/ genes. When two genes are linked, they are sufficiently close to each other that the probability of being inherited together is greater than 50% (the probability of independent assortment).

***linkage mapping:** *see* genetic mapping

^**locus:** The position that a gene or marker occupies in a chromosome.

^**lod score:** *see* discussion of Linkage in Chapter 1(II)

#LOH: *see* loss of heterozygosity

^**loss of heterozygosity:** Heterozygosity refers to the state of having pairs of dissimilar alleles. Loss of heterozygosity can occur as a result of structural deletions, mitotic recombination or nondisjunction. In some instances the cell is left with a single copy of the gene and in other circumstances the cell contains two copies of the same parental chromosome, leading either to a hemizygous (1 copy) or a homozygous state (2 copies). An example of loss of heterozygosity is when a retinal cell containing a mutation in the Rb (retinoblastoma) locus loses its normal Rb allele by a nondisjunction event. The retinal cell becomes hemizygous for the Rb mutation and is vulnerable to clonal expansion and the development of a malignant state.

^**lymphoblastoid cells:** *see* transformation

#M: The mitosis phase of the cell cycle (*see* cell cycle).

^**marker:** A position on a chromosome which can be identified and monitored by the inheritance of a trait or gene product or by a unique attribute of the DNA sequence.

#Mb: A million nucleotides or nucleotide base pairs.

***microdissection of chromosomes:** A method for isolating limited regions of a specific chromosome by cutting the chromosome with a micropipette under direct vision with a light microscope. The selected DNA region is either cloned directly or amplified as small fragments using PCR and cloned to create a region-specific genomic library.

^**microsatellite repeats:** *see* CA repeats

^**messenger RNA:** The RNA in the cell that encodes the processed transcript from a gene and is the template for protein synthesis. Most, but not all, of the messenger RNA in eukaryotic cells is polyadenylated (*see* polyA+RNA).

#mRNA: *see* messenger RNA

***nick translation:** A method of labeling specific, pure fragments of double-stranded DNA by introducing nicks into the molecules with DNaseI followed by template-specific synthesis of DNA incorporating radioactive or derivatized nucleotides. The resulting mixture of small fragments represent the entire sequence of the original template and can then be used as a hybridisation probe (*see* random primer labeling).

^nondisjunction: The failure of two homologous chromosomes to separate properly during cell division and miosis.

***Northern blot:** A method for the analysis of RNA. An RNA mixture is fractionated by size using denaturing agarose gel electrophoresis. Denaturing the RNA is necessary to eliminate secondary structure and to allow the RNA molecules to migrate in accordance with their size. These gels generally employ formaldehyde or glyoxal, and rarely, methyl mercuric hydroxide, as the denaturants. The RNA is then transferred to a filter or support so as to preserve the size fractionation achieved by gel electrophoresis. Specific RNA species can then be detected and assessed for size and amount using a hybridisation probe.

^nucleoside: A nitrogenous base (purine or pyrimidine) attached to a ribose or deoxyribose sugar.

^nucleotide: The basic subunit of DNA (a deoxynucleotide) or RNA (a ribonucleotide). Each nucleotide is comprised of a nitrogenous base (adenine, guanine, or cytosine in DNA and RNA; thymine in DNA and uracil in RNA), a sugar (deoxyribose for DNA and ribose for RNA), and a phosphoric acid molecule that serves as the backbone linkage between nucleotides.

***nucleotide sequencing:** There are a number of methods for establishing the nucleotide sequence of DNA. The most commonly used techniques employ enzymatic synthesis of DNA using a specific primer, a DNA polymerase, and a mixture of deoxynucleotides (dATP, dCTP, dGTP, dTTP) that contain a small amount of a modified deoxynucleotide (dideoxynucleotide) that can be incorporated into the growing DNA strand but then blocks further elongation of the chain. Four sets of reactions are normally carried out for each template (the DNA to be sequenced) and primer, with each reaction containing one of the four possible dideoxynucleotides (ddATP, ddCTP, ddGTP, ddTTP). Each reaction mix contains a collection of DNA strands that were all synthesized from a common starting point but are terminated at varying points along the length of the template DNA wherever a specific dideoxynucleotide has been incorporated. The mixtures are then separated on adjacent lanes of a high resolution, denaturing polyacrylamide gel that can distinguish DNA molecules differing by a single base in length. The pattern on the gel can be detected by autoradiography or by fluorescence detection if the primers are suitably labelled. The combined pattern of bands provides a direct correspondence to the nucleotide sequence of the template DNA.

^**open reading frame:** A region of genomic DNA that potentially encodes for an amino acid sequence of a protein . The region is characterized by a stretch of nucleotide sequence that begins with an initiation codon, followed by a series of triplets that would encode for different amino acids, and ends with a termination codon that would end peptide synthesis.

#**ORF:** *see* open reading frame

#**PCR:** *see* polymerase chain reaction

^**pedigree:** A description or diagram of the ancestral history of a family. Females are depicted as circles, males as squares, individuals whose sex is unknown are shown as diamonds. Solid or blackened objects indicate individuals expressing the trait or disease. A diagonal line through the object indicates that the person is deceased. Heterozygous female carriers of X-linked conditions are denoted by a dot in the center of the circle. Consanguinous matings are denoted by a horizontal double line connecting the two individuals. The birth order (oldest to youngest) is generally presented from left to right beneath the parents and twins are shown by a common branch point. An arrow is used to indicate the proband (or propositus) of the family.

^**penetrance:** The proportion of individuals sharing a common genotype who show a particular phenotype. If all gene carriers display at least some evidence of the phenotype, the penetrance is said to be complete. A genotype may show complete penetrance but highly variable expressivity such that it is difficult to recognize individuals with subtle manifestations of the genotype.

#**PFGE:** *see* pulsed-field gel electrophoresis

^**phage:** *see* bacteriophage

^**physical map:** A linear array indicating the order and physical distance (i.e. base pairs) between two or more markers or genes on a chromosome. Distances can be established by a variety of methods including nucleotide sequence, contig mapping, overlapping hybrids, and measurements from *in situ* hybridisation.

#**PIC:** *see* polymorphism information content

^**plaque:** A plaque is an individual patch of lysed bacteria on a culture plate that is the result of clonal expansion of a single bacteriophage particle (*see* bacteriophage).

^**plasmids:** An extrachromosomal genetic element found in bacteria that is capable of self-replication. Naturally occurring plasmids were first identified by their ability to confer antibiotic resistance to bacteria. Plasmids are double-stranded circular DNA's that range in size from 1 to 200 kb. Most plasmids that are used as cloning vectors range in size from 2.5 to 7 kb and contain a number of unique restriction sites for insertion of foreign DNA, drug-selection gene(s), and other DNA sequences that allow for replication in the bacterial host, gene expression or *in vitro* transcription.

***polyacrylamide gel electrophoresis:** The separation of charged molecules (e.g. negatively charged nucleic acids) using an electric field across a polyacrylamide matrix or gel. For molecules of a given charge density (the same number of charges per unit size), the separation is based upon size and/or conformation of the molecules. Polyacrylamide is a covalently crosslinked polymer that is resistant to most denaturing agents such as heat, urea, formamide. Polyacrylamide gels which contain the denaturing agent, urea, are typically used for high resolution analysis of single-stranded DNA fragments such as for nucleotide sequencing.

^**polyA$^+$ RNA:** RNA that contains a 3' tail of adenine nucleotides (polyA). The enzymatic addition of adenine nucleotides occurs shortly after transcripation as part of the processing of the primary RNA transcript in the cell nucleus. This RNA represents the majority of the messenger RNA within a cell. Poly A$^+$ RNA is purified from total cellular RNA by selection with oligo dT (an oligonucleotide consisting of 15–30 thymidine residues) that can be bound to cellulose, membranes, or beads.

***polymerase chain reaction:** A method (patented by Perkin Elmer Cetus Corporation) for amplifying a defined segment of double-stranded DNA by using two DNA primers, one matching a unique sequence at the 5' end of the sense strand and the other corresponding to the sequence of the 5' end of the antisense strand (Figure 1). When the DNA template is denatured by heat and allowed to cool, the primers can specifically anneal to the complementary DNA strands and serve as starting points for DNA synthesis by a thermostabile DNA polymerase (e.g. Taq polymerase). As a result, two new strands of DNA are synthesised that are complementary to the original DNA template and two double-stranded molecules are generated where previously there was only one. The process of denaturation, primer annealing and DNA synthesis is then repeated and both DNAs can serve as templates. With each successive round or cycle, there is a doubling of the synthesised material. Within a small number of cycles (30–35) the initial template can be amplified 10^5–10^6 fold. As the process continues, a nearly homogeneous population of DNA fragments of a defined length are synthesised (Figure 1). These fragments can then be cloned or directly analysed for mutations or by nucleotide sequencing. For Alu-PCR, the primers are derived from human-specific regions of the highly repetitive Alu sequences. For reverse transcription-PCR or RT/PCR, the initial step consists of synthesising a complementary DNA (cDNA) to an RNA template using reverse transcriptase and then using the cDNA as the template for a standard polymerase chain amplification.

^**polymorphism:** The occurrence of two or more alternative genotypes (or alleles) within a population such that the frequency of the least common allele is at least 1% in the population (i.e. at least 2% of the population are heterozygotes).

^**polymorphism information content:** A measure of the informativeness of a marker based upon the relative frequencies of the different alleles in the population and the proportion of genotype combinations (matings) that are informative (i.e. the parental alleles can be distinguished). A marker that detects only two alleles, one of which is present in over 90% of the population, would have a low PIC value and would not be very useful for linkage mapping. A marker that detects multiple alleles, none of which is present in more than 50% of the population, would have a high PIC value and would be more informative in a linkage analysis.

^**primer:** An oligonucleotide or pre-existing polynucleotide chain that can serve as the starting point for the addition of new deoxyribonucleotides by DNA polymerase or reverse transcriptase.

^**proband:** The individual in a family who is the first to be identified or to seek the advice of a genetic counsellor and after whom other family members are ascertained. Also referred to as the propositus (*female*, proposita) (*see* pedigree).

^**probe:** Single-stranded DNA or RNA molecules representing a specific nucleotide sequence that are used to detect complementary sequences in a complex nucleic acid mixture. The probe molecules are modified so that they can be detected either by autoradiography, fluorescence or by secondary binding of detector molecules (e.g. conjugates of antibodies or avidin with enzymatic indicators) (*see* markers).

^**promoter:** That portion of the gene that is involved in the control of gene expression, including tissue and developmental specificity and regulation. The promoter includes the site at which the RNA polymerase binds and initiates the transcription of RNA from the gene.

***pulsed-field gel electrophoresis:** This method of agarose gel electrophoresis separates large DNA molecules ranging in size from 100 kb to 10 Mb. The technique is used for physical mapping of genomic DNA using long-range restriction endonucleases and for the analysis of YACs and yeast chromosomes. The cells to be analysed are embedded in agarose plugs, lysed and de-proteinised with proteolytic enzymes because very large DNA molecules are highly vulnerable to shearing in aqueous solutions. The extracted DNA molecules are then separated in an agarose gel using alternating directions of electrical current that are at varying angles to the final direction of DNA migration. This alternating application of electric fields causes the DNA molecules to shift their orientation and conformation so that they can gradually pass through the gel matrix and be separated on the basis of size and shape (*see* FIGE).

***radiation hybrid mapping:** This method of physical mapping of markers relies on the breaking up of chromosomal DNA within cells by application of X-irradiation and then rescue of chromosomal fragments within a somatic cell hybrid. A radiation hybrid usually contains multiple fragments from different chromosomes. Statistical methods similar to those used for genetic mapping (linkage analysis) are used to establish whether two or more markers are physically linked within a retained fragment. Because the size of the retained fragments can be regulated by the radiation dose, markers as close as 50 kb apart can be separated and mapped. The map unit used for radiation hybrid mapping is the centiRay, cR.

***random primer labeling:** There are several methods of generating labeled DNA that can be used as a probe for hybridisation. Random primer labeling involves the use of small oligonucleotides (e.g. hexamers) that can anneal to the template DNA and serve as the starting points for DNA synthesis using a DNA polymerase and nucleotides that are radioactively labelled (^{35}S, ^{32}P,^{33}P). The probe can then be detected using autoradiography. If the nucleotides are modified to contain biotin, then a detection method based upon the binding of biotin and an avidin-conjugate can be used (horseradish peroxidase stain or chemiluminescence) Other DNA labeling methods include nick translation, incorporation of ^{32}P on to the 5' end of the probe or primer using T4 polynucleotide kinase, or the addition of radioactive or derivatized nucleotides on to the 3' end of a DNA probe using terminal deoxynucleotide transferase.

^recombination: The generation of new combinations of genes or markers in the progeny that differ from those in the parents as a result either of independent assortment of chromosomes or of crossing-over between homologous chromatids in meiosis. Crossing-over occurs when two paired chromosomes (chromatids) exchange matching regions of their DNA.

^repetitive DNA: DNA sequences that are present in multiple copies in the genome. The most well-known repetitive elements in the mammalian genome are Alu elements which are approximately 300 base pairs long and the LINE-1 elements (long interspersed nuclear elements, also known as L1 repeats) that range in size from 1 to 6 kb. There are also hundreds of thousands of short repetitive elements known as microsatellite repeats (*see* microsatellite repeats, Cot-1 DNA).

^restriction endonuclease: A large number of enzymes have been isolated from bacteria and other prokaryotes that have the ability to recognize specific sequences of DNA and direct the double-stranded cleavage of the DNA in a reproducible fashion. The length and sequence of the recognized DNA plays an important role in determining how frequently the restriction endonuclease cuts the genome. Enzymes can cut the DNA so as to create blunt-ended molecules, double-stranded DNAs with 5' or 3' single stranded overhangs, or even cut the DNA at a location several bases away from the recognition site. Restriction endonuclease digestions are frequently used to map DNA, identify restriction fragment length polymorphisms, characterise cloned segments of DNA, and to facilitate the joining or ligation of DNA fragments for the purpose of cloning.

^restriction fragment length polymorphism: Variation in the length of a specific DNA restriction fragment that is detectable in different individuals in a population. The variation in DNA fragments are due to the gain, loss or displacement of specific restriction endonuclease sites. DNAs from different individuals are cut with a restriction endonuclease, fractionated by agarose gel electrophoresis and analyzed by a Southern blot. Polymorphisms are detected by the presence of different sized bands that hybridise to a unique DNA probe (Figure 1).

***reverse transcription:** The synthesis of a complementary DNA to an RNA template using a primer (such as oligo dT) and an RNA-dependent DNA polymerase such as reverse transcriptase.

#RFLP: *see* restriction fragment length polymorphism

^ribonucleic acid: A covalent polymer of ribonucleotides (*see* nucleotide).

#RNA: *see* ribonucleic acid

***RNase protection:** A method to detect and quantitate small amounts of a single-stranded RNA transcript using a labelled, homogeneous RNA probe that is complementary (antisense) to the target RNA transcript. The target and probe are hybridised together in solution and the unhybridised probe is degraded with ribonuclease. The probe that hybridises to the target creates a double-stranded RNA that is resistant to certain ribonucleases. The protected, labeled RNA can then be measured directly or purified by polyacrylamide gel electrophoresis. It is a more sensitive method for RNA measurements than a Northern blot because the unhybridised probe is removed enzymatically (*see* Northern blot).

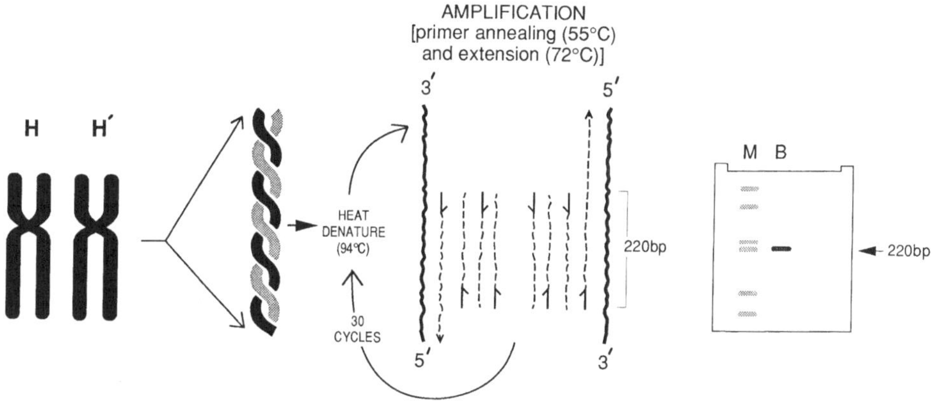

Figure 1 Amplification of specific target sequences using the polymerase chain reaction. A complex DNA mixture such as total genomic DNA is denatured by heat (94°C) and annealed to specific single stranded primers (ca. 20 bases long) at a stringent temperature (55°C) that reduces non-specific annealing. The two primers are spaced 0.05–5 kb apart and are complementary to the sense and antisense strands respectively that flank the region to be amplified. A heat-stable polymerase (e.g. Taq polymerase) is added which replicates the DNA in a 5' to 3' direction, commencing at each primer on opposite strands. The mixture is denatured and the cycle repeated up to 40 times. The resultant amplification is $(1+X)^n$ where X is the amplification efficiency of each cycle (about 0.85) and n is the number of cycles, although the reaction usually reaches a plateau after 20–30 cycles (about 10^6-fold amplification). B, amplified product; M, size markers.

#RT: *see* reverse transcription

***S1 nuclease mapping:** A method to identify the end of an RNA transcript by hybridising a labeled DNA fragment to a complementary RNA. The single-stranded ends of the molecules that do not match are digested with S1 nuclease . If the DNA contains more than the complement to the entire RNA strand, then the DNA fragment will be shortened by the enzyme, leaving only those regions protected by the RNA. The change in size can be determined on a sequencing (denaturing) polyacrylamide gel and the length of the original RNA molecule can be calculated.

#S phase: The DNA synthesis phase of the cell cycle (*see* cell cycle).

^sense strand: Also known as the coding strand, the sense strand is the strand of duplex DNA whose sequence corresponds to the nucleotide sequence of the mRNA transcribed from that gene. The complementary strand to the sense strand is often referred to as the antisense strand of DNA. The antisense strand of genomic DNA is the template for RNA transcription.

^sequence tagged site: A specific location in the genome that is defined and identified by means of flanking PCR primer sequences that specifically amplify that locus. The site can therefore be identified using PCR within genomic clones, somatic or radiation cell hybrids. The PCR products represent the sequence tagged sites, STSs, and are used for the construction of genetic and physical maps. The presence of a given STS in two or more genomic clones is an indication that the clones overlap and form part of a contig (*see* PCR, physical maps, contig mapping).

^shuttle vectors: Genetic elements that allow for the propagation of foreign DNA in more than one cell type. For example, a plasmid that allows for propagation of a foreign gene in bacteria and can also be stably transferred to a yeast host or to a eukaryotic cell for gene expression (*see* vector).

***single strand conformation polymorphism:** Also called single strand conformation analysis, this method detects single base differences between two segments of DNA (usually under 400 bp in length) by observing the subtle differences in mobility of the DNA when the single-stranded molecules are allowed to form intra-strand base pairing as they migrate through a non-denaturing polyacrylamide gel. The differences in mobility are due to the slightly different folding patterns shown by partially base-paired single-stranded DNA molecules as a result of nucleotide differences.

***somatic cell hybrid mapping:** Somatic cell hybrids are cell lines that contain the majority of their DNA from one species and a smaller amount of DNA (a few chromosomes or chromosomal fragments) from another species. Panels of these hybrid cells, containing a broad distribution of human chromosomes or chromosomal fragments, can be readily used to localise new markers or genes to specific human chromosomes or chromosomal regions.

***Southern blot:** The detection of specific DNA fragments after immobilisation on a membrane by hybridisation with a labelled probe. The technique was named after its developer, E.M. Southern. The mixture of DNA fragments are first separated by agarose gel electrophoresis and then transferred or blotted to a membrane so as not to disturb the physical array of fragments. The filter is then hybridised to the probe and washed so that only those probe molecules that have hybridised to precisely complementary DNA fragments are retained. Autoradiography or other detection methods are then used to visualise the retained probe DNA (*see* agarose gel electrophoresis, probe, restriction fragment length polymorphism).

^splicing, intron/exon: The process by which the initial transcript of the gene (called heterogeneous nuclear RNA or hnRNA) is edited to remove the intron regions and appropriately connect the exons that make up the final mRNA transcript (*see* intron, exon).

#SSCP: *see* single strand conformation polymorphism

#ssDNA: single-stranded DNA

^streptavidin/biotin: Streptavidin is a commercially available protein derived from *streptomyces avidini* that contains four binding sites for biotin molecules. It behaves similarly to avidin, a slightly larger protein derived from egg whites. Streptavidin or avidin are readily conjugated to other molecules that allow for the immobilisation or detection of biotinylated molecules. DNA or RNA can be biotinylated by photochemical linkage or by the incorporation of biotinylated nucleotides by enzymes such as DNA or RNA polymerases.

#STS: *see* Sequence Tagged Site

^TATA Box: A nucleotide sequence, e.g. TATAAAA, within the promoter region, usually located 19–27 bp from the transcription start site of eukaryotic genes.

^transcription: The synthesis of RNA that is complementary to the antisense strand of a DNA template by a DNA-dependent RNA polymerase.

***transfection:** 1) The process of introducing foreign DNA into cells by packaging the DNA into particles that use viral mechanisms of infecting cells, but are not able to establish a continuing infection (i.e. propagation of infectious particles). 2) The introduction of DNA into cultured eukaryotic cells either to assess transient gene expression or to obtain stable integration.

***transformation:** 1) The introduction of foreign DNA into cells by transiently altering the permeability of the cell membrane to allow the molecules to enter. Transformation can be achieved by chemicals in the medium, attached to the DNA (e.g. calcium phosphate, lipofectin), or by an electric shock (*see* electroporation).
2) Transformation is also used to describe a change in the behavior of a cell that allows the cell to proliferate and escape from growth control. An example of this transformation is observed when the retinoblastoma gene is inactivated and the cells begin to proliferate and form a malignancy. Another example of cellular transformation is the transformation of blood lymphocytes by Epstein-Barr virus to create lymphoblastoid lines that can be continually grown in culture as a replenishable source of genomic DNA.

***transgenic animal:** A transgenic animal is one that expresses foreign genetic material in addition to its own. It is a means of studying the expression of a synthetic or foreign gene in an intact animal. Foreign genetic material may be introduced by fusing mouse embryonic stem (ES) cells containing a modified gene with mouse blastocysts and implantation into a foster female which results in the producion of chimaeric offspring. Alternatively, a foreign gene is introduced into an animal (e.g. mouse, rat, pig, cow) by injecting the DNA directly into the pronucleus of a fertilized egg. The injected egg is implanted into a pseudo-pregnant female to allow for normal development. Single or multiple copies of the foreign gene (transgene) integrate in a random fashion within the host genome. These transgenes can be expressed with appropriate developmental- and tissue-specificity depending on 1) the promoter elements that are included with the coding region and 2) the effects of the flanking sequences around the insertion site. Transgenic animals are used for studies of gene expression and function and for the creation of animal models of genetic disease. Because the transgenes are integrated into the animal's genome, the genetic material can be stably propagated through the animal's progeny. (*see* Chapter 1(II) and Figure 5 of that Chapter)

^**translation:** The cellular process by which an mRNA sequence directs the ordered assembly of amino acids into a peptide chain.

^**translocation:** A chromosomal aberration in which regions of one or more chromosomes are displaced to another chromosomal segment. The point at which the two previously noncontiguous DNA segments are joined is referred to as the translocation breakpoint.

^**variable number tandem repeats:** DNA segments that contain variable numbers of short tandemly repeated segments with a particular sequence. Genomic DNA is cut with a restriction enzyme outside the repeated sequence, the specific fragment is size fractionated using gel electrophoresis and then detected by hybridisation. Because of the variable number of repeats, the fragments demonstrate different lengths (polymorphisms) in different individuals which can be used as alleles for that locus.

^**vector:** A genetic element that is used to carry a foreign piece of genetic information into a cell and will support the propagation of that foreign DNA within the cell. Vectors can refer to plasmids, cosmids and phage which support foreign DNA in bacteria, as well as YACs in yeast and viral elements that introduce DNA into mammalian cells.

#VNTR: *see* variable number tandem repeats

***Western blot:** A method that is analogous to Southern and Northern blotting, in which proteins are fractionated by size using denaturing polyacrylamide gel electrophoresis and then transferred to a membrane or filter. The detection of a specific protein is achieved using an antibody that has been raised against all or part of that protein. Detection of the bound antibody can be achieved by radioactive and nonradioactive methods.

^**yeast artificial chromosome:** An artificial chromosome that contains large fragments of foreign DNA (0.1-1.5 Mb) within a linear vector molecule that can stably propagate itself and segregate within the yeast *Saccharomyces cerevisiae.* A YAC contig (*see* contig) can define a large region of the genome with a limited number of clones. The use of these vectors is hampered by occasional instability of the inserted DNA, cloning artifacts such as rearrangements and joining of unrelated smaller pieces of DNA (chimaeric clones), and difficulties in obtaining large amounts of the purified YAC free of background yeast chromosomes. YACs provide a source of DNA for the identification of genes, Alu-PCR, fluorescence *in situ* hybridisation and chromosome painting, limited genomic libraries, and contig mapping of large genomes (*see* Alu-PCR, chromosome painting, cloning, contig mapping, fluorescence *in situ* hybridisation, and library).

#YAC: ***see*** **yeast artificial chromosome**

zinc finger proteins: Proteins that contain one or more segments that are capable of binding Zn to two closely-linked cysteine residues followed by two histidine residues. The binding of Zn creates loops or fingers that are capable of binding DNA and potentially regulating transcription.

Colour Plate I (A): show a normal optic disc and retinal vessels. See page 155.

Colour Plate I (B): shows the optic disc pallor and attenuation of retinal vessels in the fundus of an RP patient. Typical intra-retinal bone spicule pigmentary deposits are visible around the edges of the photograph. See page 156.

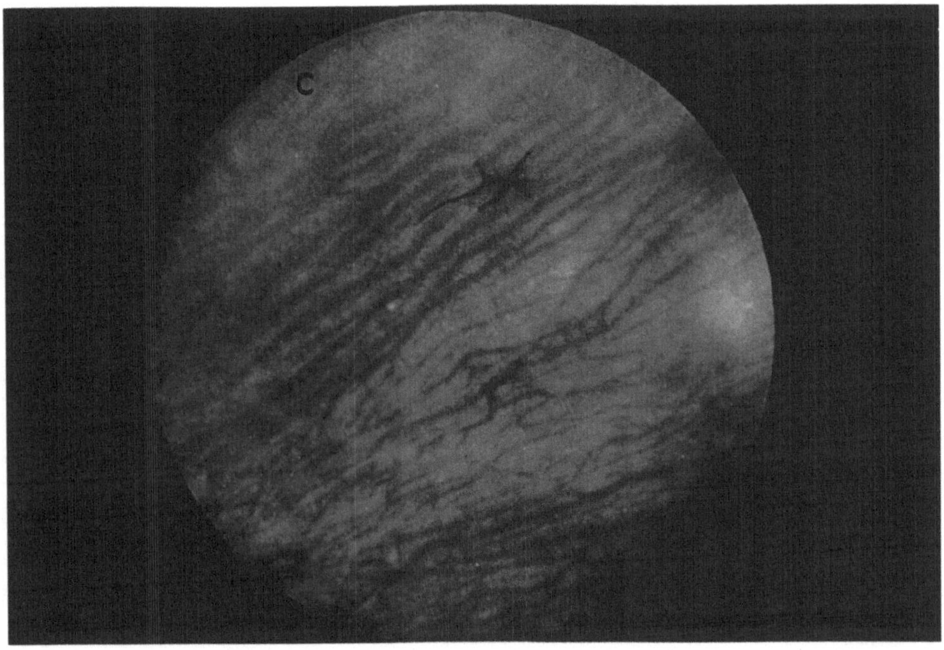

Colour Plate I (C): shows large bone spicule pigment deposit in the mid-peripheral retina of an RP patient. See page 156.

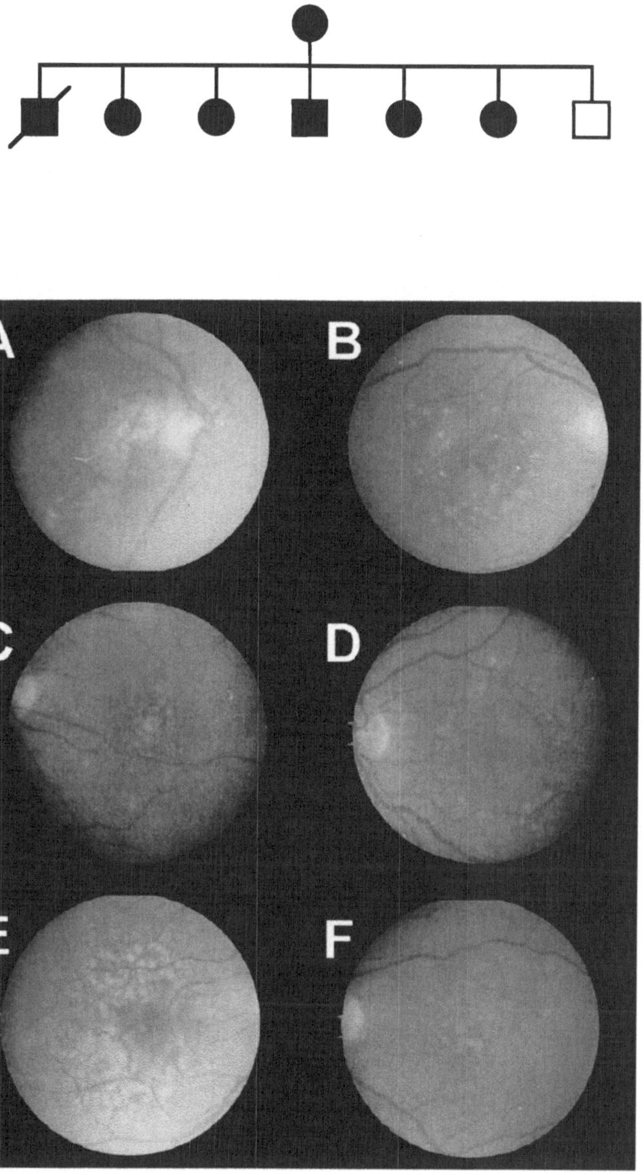

Colour Plate II Ophthalmoscopic similarity of macular disease in relatives affected with familial macular degeneration. The pedigree is shown at the top. Closed symbols indicate individuals with macular degeneration. Fundus photographs of the mother (A) and five affected children (B-F) are shown in decreasing order of age. The drusen are soft, confluent and limited to the central macula. The peripheral fundus is normal. Three patients (A, B and D) have some geographic atrophy in the center of the affected areas. The view of the fundus in patient A is limited by cataract. See page 177.

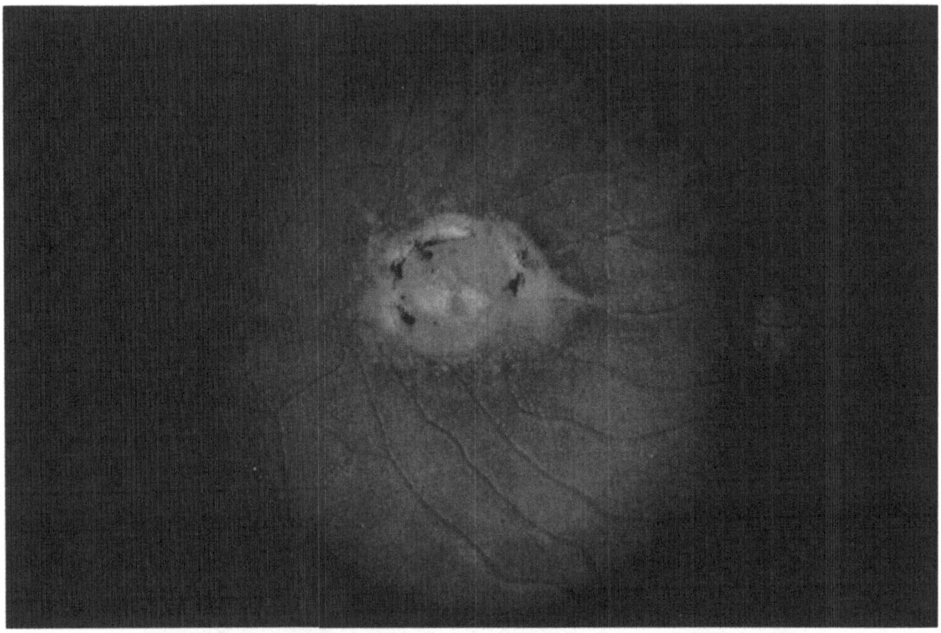

Colour Plate III North Carolina macular dystrophy. Typical features of the disease are the central crater with a shiny base surrounded by a gliotic rim. See page 183.

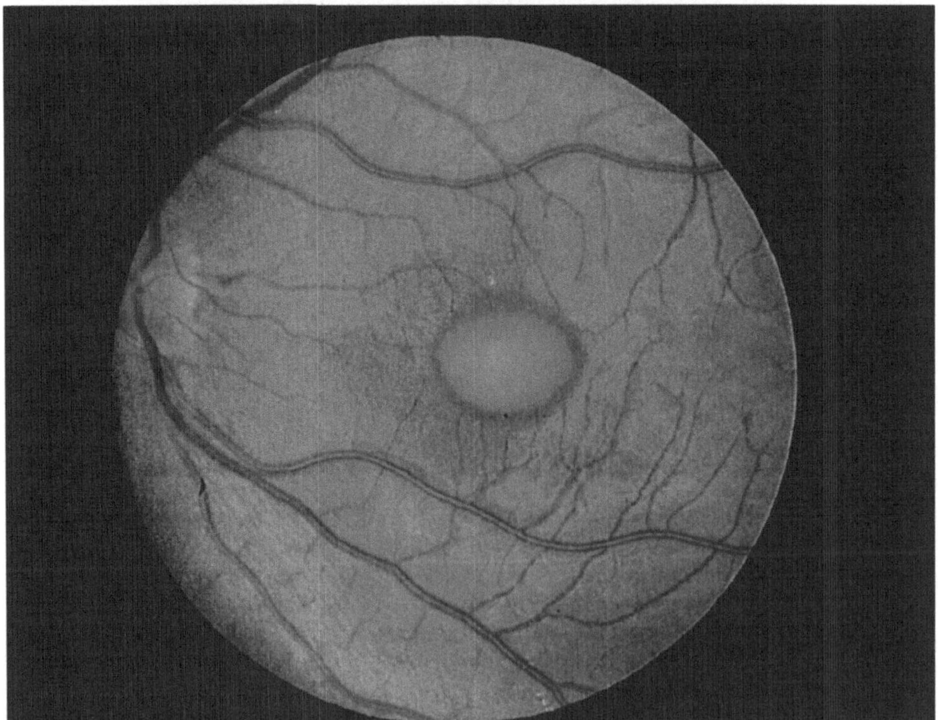

Colour Plate IV Best's vitelliform dystrophy. A classic "egg yolk" deposit of lipofuscin-like material is present beneath the retinal pigment epithelium of the central macula. See page 184.

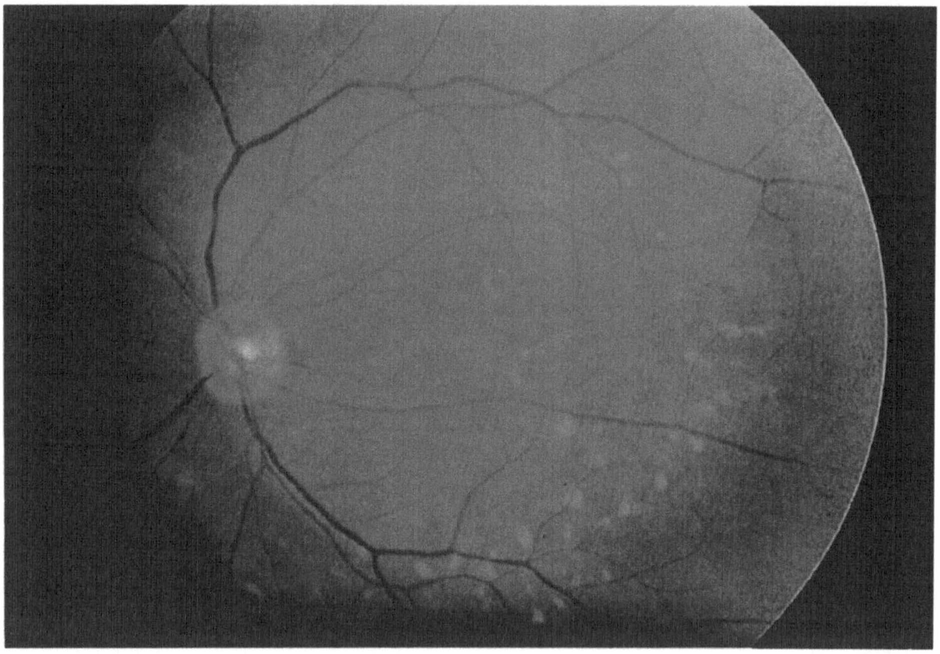

Colour Plate V Butterfly dystrophy. There is a yellowish subfoveal lesion with five radiating arms. In addition, round pigment deposits that are larger than typical drusen are present just posterior to the major vascular arcades. From Nichols, et al., *Nature Genetics* (1993); used with permission. See page 185.

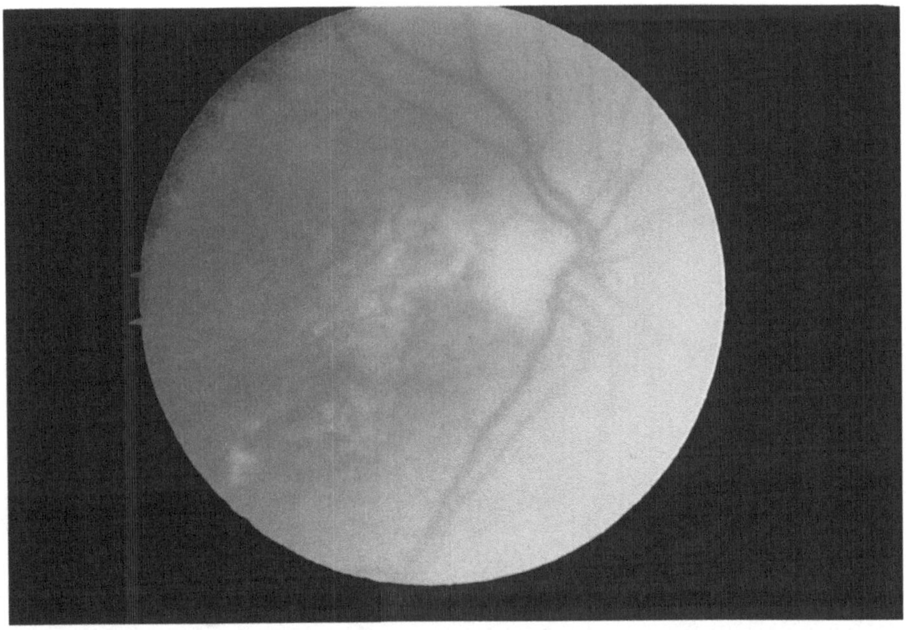

Figure 2 A

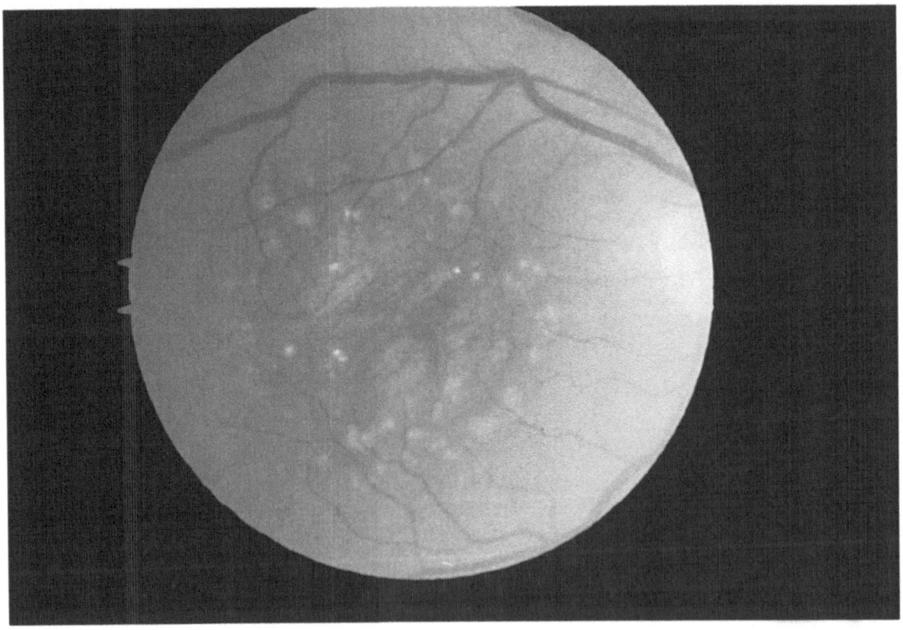

Figure 2B

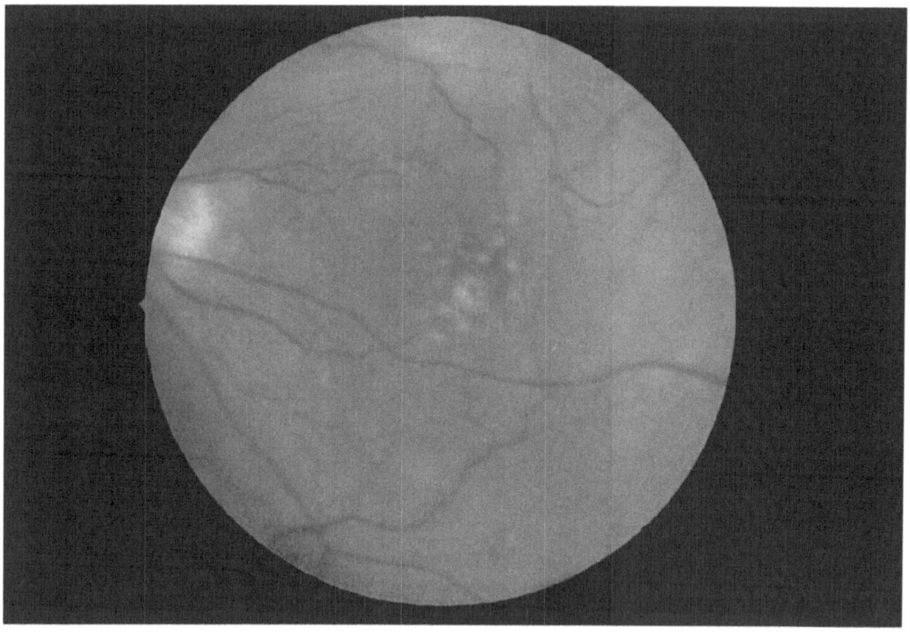

Figure 2C

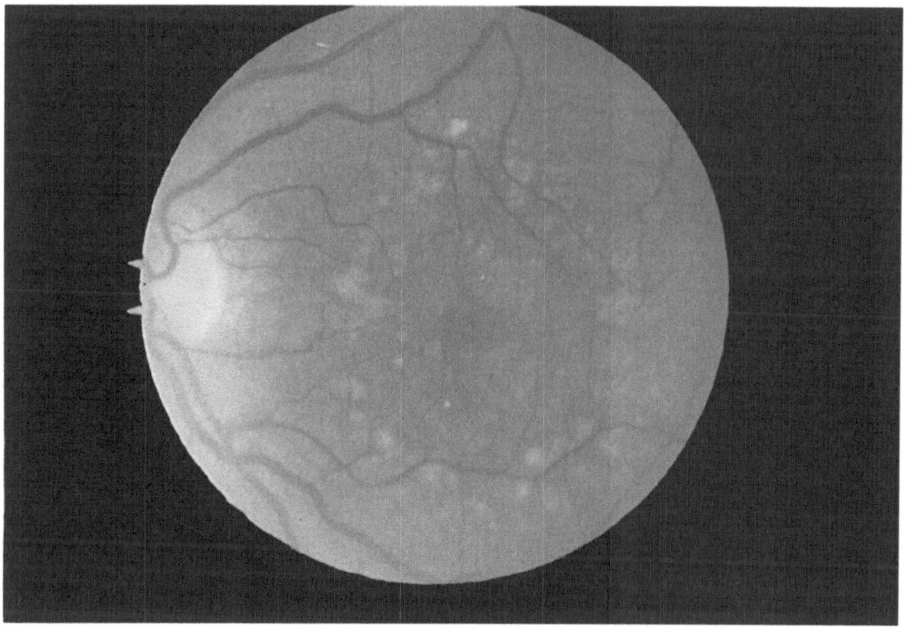

Figure 2D

Colour Plate VI Photographs of the fundi of 4 independent patients with Usher syndrome showing the range of involvement. The first two panels are from type I patients and the other two are from type II patients. See pages 364 and 365.

INDEX

3' end, 7, 491
5' end, 7, 491
11-*cis* retinal, 161, 219
Abetalipoproteinaemia, 63–64
Absorption spectrum — colour pigments, 219, 222, 244
Acceptor splice site, 491
Achromatopsia, 220
Agarose gel electrophoresis, 491
Age-related macular degeneration (AMD), 173–191
  definition of AMD, 173–174
  heritability of AMD, 179–181
  heterogeneity of AMD, 174–179
  therapy, 190
Ageing, 179, 361, 413, 425, 472, 484
Åland eye disease, 288–290
Albinism, 89–113
  classification, 96
  definition of human albinism, 93–95
  diagnosis, 112–113
  ocular albinism, 89, 110–112
    autosomal recessive ocular albinism (OA2), 111–112
    X-linked recessive OA (OA1), 110–111
  oculocutaneous albinism (OCA), 89, 95–108
    brown OCA, 107
    minimal pigment OCA, 97–98
    mutations in human TYR gene, 98–102
    platinum OCA, 98
    rufous OCA, 107–108
    temperature sensitive OCA, 98
    tyrosinase negative OCA, 97
    tyrosinase positive OCA (OCA2), 103–108
    tyrosinase related OCA (OCA1), 96–103
    yellow OCA, 97–98
  treatment, 113
Allele, 491
Allele-specific oligonucleotide (ASO), 158–160, 491
Alu PCR, 492
Alu repeats, 5, 492
Aminoaciduria, 403, 405
Amino acid sequence, 309, 314, 389
Amniocentesis, 211
Angelman syndrome, 105–106
Animal models — albinism, 89–113, 104
  aniridia, 452–455
  cataract, 429–434
  choroideremia, 316
  CSNB, 293
  platelet storage disease, 109–110
  Marfan syndrome, 392
  retinal degeneration, 123–139
  Usher syndrome type II, 377
Aniridia, 445–462
  AN1, AN2, 448
  clinical features, 445–447
  cloning of candidate aniridia genes, 450–452
  cytogenetic abnormalities, 448–449
  frequency, 447
  genetics, 447
  homozygote (AN2), 453
  linkage studies, 447–448
Anophthalmia, 453–454
Anterior segment defects, 321, 459
Antibodies, 204, 289, 334, 387–388, 392
Antisense strand, 492
Appaloosa horses, 293
Arrestin, 128, 130–131, 163
ASO, 158–160, 491–492
Ataxia, 359, 363, 447, 480–481, 486
ATP, 469
ATP synthase (Complex V), 469–470, 480
Atrophy, of choroid and RPE, 304
Audiograms, 362
Auditory evoked responses, 361–362
Autism, 323
Autoradiography, 492
Avidin, 492
Axonemes, 371

Bacteriophage, 492
Bardet-Biedl syndome, 368
Base, 1, 492
Base pair, 493
Behaviour disorders, 402

Best's disease — *see* Macular dystrophy, vitelliform
Beta-galactosidase, 493
Bietti's crystalline dystrophy, 66–67
Biotin, 493
Bipolar cells, 74
Blindness, 259, 321, 367, 400, 477, 479–480
Blood-retinal barrier, 280
Blue-cone monochromacy, 71, 220, 242
bp, 493
Breakpoint, 493
Bruch's membrane, 175–176, 178
Bubble PCR, 493
Bull's eye dystrophies, 68–69
Butterfly dystrophy of the fovea, 66, 185–187

CA repeats, 6, 494
Calcium channel, 40, 42
Calcium signalling, 40, 408
Candidate gene, 182, 184, 189, 266, 315, 329, 372, 374–376, 407
Cap site, 7–8
Carriers — in choroideremia, 62, 304
  in Norrie disease, 324
  in OCRL syndrome, 399–401, 408
  in X-linked congenital stationary night blindness, 287–288
  in X-linked juvenile retinoschisis, 347–348, 350–351
  in X-linked retinitis pigmentosa, 57, 260–261, 269
Catalase genes, 449–450
Cataract, 260, 321, 367, 399, 401, 404, 408, 413–435, 445, 453, 459, 485
  aging, 413, 425
  anterior polar cataract, 426
  congenital, 399
  Coppock cataract, 427–428
  dominant inherited cataract — inherited forms in man, 420, 425
  embryonal cataract, 427
  hereditary central cataract, 429
  lamellar cataract, 425–426, 429
  linkage studies in human inherited cataracts, 427–429
  Marner's cataract, 427–428
  morphology, 426
  mutagen-induced, 433
  posterior capsular cataract, 428
  posterior polar cataract, 426–428
  pulvurulent cataract, 426, 429
  total cataract, 426
  zonular cataract, 426–427
Cattle model, Marfan syndrome, 392
cDNA — *see* Complementary DNA
Cell cycle, 206, 494
CentiMorgan, 11, 494
Centiray, 495
Centromere, 2
CEPH families, 14, 349, 375
Chaperone activity, 416
Chediak-Higashi syndrome, 109
Chemical cleavage of mismatch, 189
Chemiluminescence, 495
Chimaeric mice, 16–17, 432, 460, 511–512
Choriocapillaris, 62, 67, 69–70, 175, 304
Chorionic villus sampling, 208
Choroid, 62, 66–67, 70, 303–304, 316
Choroideremia, 61–63
Choroideremia (CHM), 61–63, 303–316, 341
  chromosomal abnormalities with choroideremia, 305–307
  clinical aspects, 303–304
  cloning CHM gene, 307–311
  contiguous gene deletion syndromes, 304–305
  critical region, 305
  fundus changes in heterozygous females, 62
  genomic structure CHM gene, 308–309
  geranylgeranyl transferase, 315–316
  murine homologue of CHM, 308, 314–315
CHML gene, 308, 312, 314, 316, 375
Chromosomal jumping, 306
Chromosomal localisation
  dominant inherited cataract genes in man, 420
  hCHML gene, 315
  major human genes expressed in lens, 420
  NDP gene, 327
  phototransduction genes, 131
  RS gene, 350
  RP2, RP3 genes, 263

Usher syndrome genes, 360, 373
Chromosomal walking, 306
Chromosome 1, human, 375–378
Chromosome 2, human, 417
Chromosome 3, human, 157–165, 169
Chromosome 4, human, 135
Chromosome 6, human, 138–139, 165–169, 181, 184–188
Chromosome 7, human, 168
Chromosome 8, human, 168–169
Chromosome 11, human, 98–102, 183, 372–374, 417, 447–452
Chromosome 13, human, 197–211
Chromosome 14, human, 370–372
Chromosome 15, human, 106–107, 387–392
Chromosome 17, human, 417
Chromosome 21, human, 417
Chromosome 22, human, 417, 429
Chromosome X, human, 217–254, 259–272, 277–295, 303–316, 321–334, 339–353, 399–408
Chromosome banding, 495
Chromosome painting, 495
Chromosomes, 2, 202, 495
Chronic granulomatous disease, 264
Chronic progressive external ophthalmoplegia (CPEO), 481–484
Cilia, 371
Ciliary body hypoplasia, 445
Ciliary processes, 322
Ciliary zonules, 386
cM — *see* centiMorgan
Coats' disease, 367
Cochlea, 361–362
Coding region, 496
Codon, 7, 496
Coloboma, 445–446, 459
Colony, 496
Colour matching, 218, 222, 224, 246–248
Colour pigments — long-wavelength, 219
  medium wavelength, 219
  normal variation, 223
  short-wavelength, 219
Colour pigment genes — 225, 228, 236
  defects, 241
  fusion genes, 226–229, 233
  stoichiometry, 234–236, 239–240
  structure, 235–236
Colour sensation, 219
Colour vision, 217–254, 281, 288
  variations in normal colour vision, 223–225
  defects, 70–71, 217–254
Complementary DNA (cDNA), 22, 136, 494, 496
Complementation, genetic, 484
Complex I-V, respiratory chain, 469
Compound eye, *Drosophila*, 30
Compound heterozygotes, 98
Cone defects — congenital, 70–71
Cone dystrophies, 68–69, 181
Cone photoreceptors, 154, 217, 243
Cone pigments — *see* colour pigments
Cone-rod dystrophies, 68–69
Congenital achromatopsia, 220
Congenital adrenal hypoplasia, 288
Congenital cataract, *see under* cataract
Congenital contractural arachnodactyly, 390–391
Congenital photoreceptor defects, 70–75
Congenital retinal folds, 347
Congenital rod monochromacy (achromatopsia), 220
Congenital stationary night blindness (CSNB), 71–72, 271–272, 277–295
  autosomal dominant, 278–279, 294
  carriers, X-linked, 287–288
  classification, 278–279
  clinical manifestations, 279–281, 291–292
  complete type, 283, 285
  CSNB1, 289, 294
  CSNB2, 289, 294
  dark adaptation, 291
  electrophysiological evaluation, 281–288
  histopathology, 292
  incomplete type, 284–286, 289, 294
  inheritance, 279
  mapping genes for CSNB, 293–294
  Nougaret variety, 279
  pathogenesis, 292–293
  photopic responses, 286
  X-linked, 271–272, 279–280
Connective tissue, 383, 387
Contigs, 378, 496
Contig mapping, 496
Contiguous gene deletion syndrome,

304, 368–369, 377, 447
Copper-binding regions, 99
Cornea, 66, 321, 399, 445, 481
Cosmids, 329, 352, 496
Cot-1 DNA, 496
CpG island, 9, 264–265, 450, 452, 497
cR — *see* centiRay
Crossing over — 10, 497
critical, 375
unequal, 220, 227–228
Crystallins, 413–435
α-crystallins, 415–417
αA-crystallin gene in Down's syndrome, 428
chaperone activity, 416
chromosomal localisation, 416–417
developmental expression, 416
gene organisation (αA and βB), 415–417
heat shock protein homology, 416
β-crystallins, 417–421
evolution, 419
gene organisation, 418
interactions, 419
post-translational modification, 418
protein structure, 418
βB2-crystallin, 430–431
altered properties in Philly mouse, 431
mutations in Philly mouse, 430
γ-crystallins, 421–424, 428, 432, 434
evolution, 423
gene organisation, 422
in cataract, 425
linkage studies in man, 428
protein structure, 422
γE-crystallin gene — mutation in Elo mouse, 432
δ-crystallin, 415, 434
ζ-crystallin, 432–433
enzymatic activity, 433
mutation in 13/N guinea pig, 432–433
aggregation in cataract, 425
enzymatic activity, 433–434
post-translational modification, 414, 418
taxon-specific, 414, 432, 434
Cybrids, 484
Cyclic GMP, 124, 129, 163
Cyclic GMP-phosphodiesterase, 38, 123–136, 162–163, 170
structure, 128
abnormalities in rd mouse, 128–130
gene structure, β-subunit, 134
Cyclic GMP-gated ion channel, 162–163
Cysteine repeats, 389–390
Cytochrome c oxidase (Complex IV), 470, 475, 478
Cytogenetic analysis, 201, 326, 447–448
Cytoskeletal proteins, 43

Dark adaptation, 60, 71, 73–75, 167, 259, 288, 290–291, 346
dATP, γ-^{32}P or α-^{32}P labelled, 497
Deafness, 304, 323–324, 359–361, 481, 485
Deletions — 264–265, 305–306, 328–329, 324, 449, 458
chromosomal, 200–201, 288, 304, 447
in mt DNA, 481–485
smallest region of overlap (SRO), 449–450
submicroscopic, 203, 264–265, 268, 307, 309–310, 329
Dementia, 323–324, 481
Denaturing gradient gel electrophoresis (DGGE), 186, 189, 497
Deoxyribonucleic acid (DNA) — 2, 497
binding domains, 452
fractionation, 20
methods of modifying, 20–21
repair genes, 376–377
repetitive, 5–6
sequence, 158–159, 190, 203, 210, 332, 329, 407, 457
Detached retina, 321–322
Deuteranomaly, *see* Trichromacy
Deuteranopia, *see* Dichromacy
Development — of the eye, 452, 455, 459, 461
Developmental switch, 391
Diabetes mellitus — 485–486
cataracts in, 408
Diagnosis — 112, 189, 324, 332, 334, 341, 350, 458
in inherited retinal disorders, 53–76
differential, 325, 347
Dichromacy, 218, 220–221
deuteranopia, 220–221, 225
molecular genetics, 225–226
protanopia, 220–221, 226

molecular genetics, 226
tritanopia, 220
Differentiation — cellular, 205–206, 334
Digoxygenin, 497
DGGE — *see* Denaturing gradient gel electrophoresis
D-loop — mt DNA, 471
DNA — *see* Deoxyribonucleic acid
Dog, Irish setter, 135
Dominant drusen, 179, 183
Dominant negative effect, 392
Donor splice site, 498
Dopachrome tautomerase, 91, 93
Dot blotting, 498
Down's syndrome — cataract, 428
dsDNA, 498
Down's syndrome, 428
cataract in, 428
αA-crystallin gene, 428
Drosophila melanogaster, 29–46
allele-dependent photoreceptor degeneration, 45
compound eye structure, 30–32
mutants — *ninaE*, 45
ninaC, 34, 43, 46
norpA, 34, 36, 41
paired, 452
rdgA, 41, 45
rdgB, 34, 41
rdgC, 34, 44
retinal degeneration, 29–46
trp, 34, 39–40, 46
Drusen, 174, 176, 179, 183, 344
Duplications — mt DNA, 481, 485–486
Dystrophin, 289

E1A protein, 205
E7 protein, 205
Ectopia lentis, 385
Elastin, 393
Electro-nystagmography, 363
Electro-oculogram (EOG), 65, 68–69, 345
Electron transport impairment, 480
Electrophoresis, 18, 20, 499
Electroretinogram (ERG), 30–31, 42, 45, 56, 59, 65–66, 68, 71–75, 124, 167–168, 201, 259, 281–285, 287, 289–290, 346–347, 366
negative ERG, 283, 287, 289
Schubert-Bornschein, 283–284
Electroporation, 498
Elo mouse — *see* mouse mutants
Embryonic stem (ES) cells, 15–17, 511
Emx-1, Emx-2 — *see* homeobox genes
Enhancer sequences, 10
Epidermal growth factor motif, 390, 392
EST, *see* Expressed sequence tag
Esterase D, 200
Ethical issues, 211
Evolution, of crystallins, 419, 423
Evolutionary conservation, 308, 451–452
Exons, 7, 498
Expressed sequence tag (EST), 498
Expression vector, 498
Expressivity, 347, 445–446, 458, 475, 480–481, 498
Extracellular matrix molecules, 387
Eye development, 452, 455, 459, 461

Familial exudative vitreo-retinopathy (FEVR) — *see* X-linked FEVR
Fanconi syndrome, 403
Fatal infantile cardiomyopathy (FICM), 474
Fibre cell elongation, 431, 434
Fibrillin genes, 387–392
Marfan syndrome and mutations in fibrillin gene, 391–392
Fib5 gene, 390–391
Fib15 gene, 388
Fib17 gene, 390
Fibroblast growth factor 3 (FGF3), 372
Field inversion gel electrophoresis, 307, 499
FIGE — *see* Field inversion gel electrophoresis
FISH — *see* Fluorescence *in situ* hybridisation
Fleck retina of Kandori, 75, 278–279
Flow cytometry, 19, 499
Fluorescein angiography, 56, 62, 66, 68, 74, 280, 346, 445
Fluorescence, 499
Fluorescence *in situ* hybridisation, 452, 458, 499
Founder effect, 102, 339, 309
Fovea, 93–94, 186, 288, 342, 366–367, 445
French Acadian population, 372
Fundus albipunctatus, 74, 278, 281
Fundus flavimaculatus, 67–68

Fundus reflectometry, 292
Fundus, retinal, 278, 280, 304, 364–366

G_0, 499
G_1, 499
Galactokinase, 426
Galactosemia, 408
Ganglion cells, 72
Gel electrophoresis, 499
Gel retardation assays, 500
Gene, 5–9, 500
Gene — anatomy, 6–10
Gene conversion, 423
Gene expression - 424, 310, 333, 459
color pigment genes, 242–243, 248–249, 253
γ-crystallins, 424
Pax-6, 452, 459–460
Gene structure — colour opsins, 236
CHM, 307–309
crystallins, 415–418, 422
cyclic GMP-PDE β-subunit, 134
PAX6, 455–456
peripherin/RDS, 166
mouse rds, 184–185
NDP gene, 329–332
RB1, 203–204
TRP1, 92
tyrosinase, 92
Fib15, 391
Gene therapy, 39, 43–44, 395
Genetic code, 1
Genetic counselling, 351, 457, 462
Genetic distance, 5
Genetic map, 12, 351, 500
Genetic mapping, 10–17
genetic map lengths, 12
genetic markers, 14
Genetic silencing, 203
Genitourinary anomalies, 324, 447
Genome, 500
Genome — anatomy, 2–6
Genotype-phenotype correlations, 269, 333, 480
Gillespie syndrome, 447, 459
Glaucoma, 321, 399, 404, 445
Glutathione-S-transferase, 426
Glycerol kinase deficiency, 288
Glycosaminoglycans (GAGs), 406, 408
Goldmann-Favre disease, 347
Guanylate cyclase, 162–163
Guinea pig, 13/N, 432–433
Gyrate atrophy, 310

H-strand, mt DNA, 471
Haemorrhages, in retinoschisis, 346
Hallgren syndrome, 368
Haplo-insufficiency, 137, 457
Haplotype analysis, 267–268
Haplotype association, 351
Heat shock proteins, 416
Hermansky-Pudlak syndrome, 108–109
Heteroduplex analysis, 158, 189, 455
Heterogeneity — clinical, 359–379
genetic, 14, 169, 262, 269, 350, 359–379, 388, 390,
Heterogeneous nuclear (hn) RNA, 7, 500
Heteroplasmy, 472, 480, 483, 485–486
Heterozygosity, 500
Histopathology — Norrie disease, 323
CSNB, 292
retinoblastoma, 197
juvenile retinoschisis, 341
hnRNA — *see* Heterogeneous nuclear RNA
Homeobox genes — 452, 455, 461
Emx-1, 461
Emx-2, 461
Hox-7, 461
Hox-8, 461
XANF-1, 461
Homology — human-mouse, 144
Homoplasmy, 472, 474, 478
Horseradish peroxidase, 500
HRP — *see* Horseradish peroxidase
Human crystallin genes — chromosomal localisation, 417
Human mitochondrial (mt) DNA, 470
Hybridisation — *in situ*, 500
subtractive, 130, 136
differential, 130, 136
Hybrids — somatic cell, 137, 139, 200, 266, 314, 327, 449
Hyperphosphaturia, 403
Hypopigmentation, 89–113, 288

Imaging, CT and MRI, 393
Imprinting, 3, 106, 200
Inborn error of metabolism, 408
Inner nuclear layer, 282
Inner plexiform layer, 282
Inositol phosphate metabolism, 36–37, 40, 407–408

Inositol polyphosphate-5-phosphatase, 407
Interference, 11
Inter-specific crosses, 15, 132, 144, 455
Introns, 7, 501
Iris hypoplasia, 445, 453
Iris pigment, 63, 94, 288
Isoprenylation site, 135

Junk DNA, 6
Jumping libraries, 501
Juvenile retinoschisis, *see* X-linked juvenile retinoschisis

Karyotype, 501
Kearns-Sayre syndrome (KSS), 481–484

Lactate, 474, 481
Large T antigen, 205
Laurence-Moon syndrome, 368
Leber's hereditary optic neuropathy (LHON), 474–480
  clinical features, 474
  high-risk mutations, 476–477
  moderate-risk mutations, 476–479
  low-risk mutations, 476, 479–480
Leigh's disease, 480
Lens — abnormalities in aniridia, 453
  development, 399, 406, 413–414, 421
  developmental regulation, 421
  dislocation, 445
  induction, 460
  major human genes expressed in, 420
  structure, 415
Library, 501
Library screening, 501
Light — effects on retinal degeneration, 38, 40–41, 43–45, 176, 190
Linkage analysis, 10–14, 106, 139, 145, 157, 165, 181–182, 189, 200, 209–210, 262, 266, 289, 293–294, 326–327, 348–349, 352, 372, 374, 387–388, 391, 407, 427–429, 447–448, 502
Lipofuscin, 65, 67, 183–184
Locus, 502
Locus-control region (LCR), 220, 243
Lod score, 12–13, 502
LOH — *see* Loss of heterozygosity
Loss of heterozygosity, 200, 206, 502
Lowe oculocerebrorenal syndrome (OCRL), 399–408
  biochemical findings, 406–407
  clinical features, 399–405
    fertility, 404
    growth, 402
    infections, 404
    ocular features, 400–402
    systemic features, 402–404
  linkage studies, 407
  pathological features, 404–405
  positional cloning, 407–408
  therapy, 408
L-strand, mt DNA, 471
Lymphoblastoid cells — *see* transformation

M phase, cell cycle, 502
Macromelanosomes, 109, 111, 288
Macula, 64–65, 259, 341, 445
Macular degeneration, 167, 242, 342, 344, 347
  see Age-related macular degeneration
Macular dystrophy, 59, 138, 169,
  adult vitelliform, 65, 168, 187
  North Carolina, 145, 170, 179, 181
  pseudo-inflammatory (Sorsby's), 69–70, 179, 183
  vitelliform, 64–66, 170, 179, 181, 183
MAO — *see* monoamine oxidase
Mapping functions, 11
Marfan syndrome (MFS), 383–395
  animal model, 392–393
  clinical features, 383–386
    cardiovascular features, 385, 393
    musculoskeletal features, 383–384, 394, 404
    ocular features, 385–386
  infantile, 386
  inheritance, 386
  linkage studies, 387–388
  management, 393–394
  mortality, 393
  mutations in fibrillin gene, 391–392
Marker, 502
Maternal inheritance, 472, 475
Maximum likelihood, 13
Mb, 502
McLeod syndrome, 264
Melanin synthesis, 90–91

Melanoma, 202
Mental retardation, 109, 264, 304, 324, 368, 402, 447
Mesoderm, 459
Messenger RNA, 502
Metastasis, 199
Microcephaly, 324
Microdeletions — *see* Deletions
Microdissection of chromosomes, 502
Microfibrils, 387–389
Microinjection, 17, 164
Microphthalmia, 321, 426, 431
Microsatellite — markers, 6–14, 165, 189, 351
  repeats — *see* CA repeats
Mitochondrial abnormalities, 176, 399, 406, 482–486
Mitochondrial DNA, 3, 469–486
  deletions, 482–485
  duplications, 485–486
  mutations and ophthalmic disease, 473–486
  Oxphos, 469–472
  polymorphisms, 479
  replication, 471, 483–485
  transcription, 471
Mitochondrial encephalopathy (MELAS), 474
Mitochondrial genome, 3–5, 470–471
Mitochondrial myopathy, 474, 480–486
Mizuo phenomenon, 278
Monoamine oxidase (MAOA, MAOB), 328
Monochromacy, 71, 218–220
  blue cone monochromacy, 220
  congenital rod monochromacy, 71, 220
Mosaicism, 199
Mouse — genetic map, 14–17
Mouse models — *see* Animal models
Mouse mutants — albino (*c*), 101–103
  aphakia (*ak*), 461
  beige (*bg*), 109
  bouncy (*bc*), 139
  brown (*b*), 92
  cataract and small eye (*Cts*), 461
  coloboma (*Cm*), 461
  cribriform degeneration (*cri*), 123
  dysgenic lens (*dl*), 461
  extreme dilution (c^e), 112
  Eye lens obsolescence (*Elo*), 431–432, 461
  Himalyan (c^h), 101
  microopthalmia (*mi*), 139
  microphthalmia defective iris (mi^{di}), 461
  motch, 461
  muscular dystrophy (*mdx*), 17
  nervous (*nr*), 123
  pearl (*pe*), 293
  Philly (*Phil*), 430–431
  pink-eyed dilution (*p*), 104, 106
  purkinje cell degeneration (*pcd*), 123
  retinal degeneration (*rd*), 15, 39, 123–135
  retinal degeneration-3 (*rd-3*), 123, 377
  Retinal degeneration-4 (*Rd-4*), 123
  retinal degeneration slow (*rds*), 15, 58, 123, 136–139, 160, 166–167
  shaker-1 (*sh-1*), 374
  situs inversus viscerum (*iv*), 371
  slaty (*slt*), 91, 93
  Small eye (*Sey*), 15, 452–455
  vitiligo (mi^{vit}), 123
Mouse retinal degenerations, 123–151
mRNA — *see* Messenger RNA
Mucin, 333
Muller cells, 73–74, 341
Mus spretus, 15, 132, 144, 455
Muscular dystrophy — Duchenne type, 264, 288, 290
Mutagen-induced cataract, 433
Mutagenesis, 39, 43–44
Mutations — 3, 35, 102, 203, 220, 235, 312, 325, 340, 351, 432, 453, 455–456, 462
  analysis, 332
  clustering, 99–100, 313
  criteria for pathogenicity, 188
  dominant versus recessive, 36
  deletion, 9, 138, 166–167, 186–187, 204, 312, 392, 407, 473
  duplication, 473
  exon skipping, 407
  expansion, 9
  frameshift, 9, 98, 186, 312, 457
  hot-spot, 485
  identification, 210
  insertion, 137, 204, 312
  missense, 7–9, 98, 138, 168, 185, 187, 204, 210, 333, 392, 407, 473–475, 480
  in mt DNA, 469–486

new, 197, 200, 386, 447
nonsense, 7, 98, 134–135, 168, 187, 204, 312, 392, 407, 455
pathways, 229–230
null, 137, 167–168
spectrum and phenotypic effects, 187
splice site, 9, 204, 312, 333, 455, 457
in stop codon, 478
temperature sensitive, 98, 101
tRNA, 473–474
Myelin, 405
Myoclonic Epilepsy and Ragged-Red Fiber Disease (MERRF), 474
Myopia, 72, 261, 269, 271–272, 280, 283, 288, 293, 304, 386
Myopic degeneration, 260

NADH dehydrogenase (Complex I), 470, 475, 477–479
NDP gene, *see* Norrie disease
Neovascularization, 174
Neuroectoderm, 459–460
Neurofibromatosis type II (NF2), 428–429
cataract in, 428
chromosome 22, 429
Neurogenic muscle weakness, ataxia and RP (NARP), 474, 480–481
Neurological disorder — in Norrie disease, 323–324
in NARP, 480–481
in LHON, KSS, 474
Neurotransmission — in CSNB, 293
Nick translation, 503
Night blindness, 154, 259–260, 271, 277–295, 304, 366
Nondisjunction, 503
Norrie disease (NDP), 294, 321–334, 347
clinical features, 321–324
cytogenetics, 326–327
differential diagnosis, 324–325
gene isolation, 329
histology, 323–324
linkage studies, 326
mutation analysis, 332–333
pericentric inversion, 327
physical fine mapping of NDP gene, 328–329
deletion studies, 328–329
long range restriction mapping, 329
structure of NDP gene, 329–332
North Carolina macular dystrophy — *see* under Macular dystrophy
Northern blot, 311–312, 329, 333, 503
Nucleic acids — detection, 25–26
hybridisation, 26–27
manipulation, 19
production of pure, 19, 22–23
separation, 18–19
Nucleoside, 503
Nucleotide, 503
Nucleotide pyrophosphatase, 406
Nucleotide sequencing, 503
Nutritional deficiency, 176
Nystagmus, 93, 220, 260, 280, 287–288, 399, 445

Ocular albinism (OA), 89, 110–112
autosomal recessive ocular albinism (OA2), 111–112, 288
X-linked recessive OA (OA1), 109–111
Ocular myopathy, 485–486
OCRL gene, *see* Lowe oculocerebrorenal syndrome
Oct-1, Oct-2, 455
Oculocutaneous albinism (OCA), 89, 95–108
brown OCA, 107
minimal pigment OCA, 97–98
mutations in human TYR gene, 98–102
platinum OCA, 98
rufous OCA, 107–108
skin cancer, 93
temperature sensitive OCA, 98
tyrosinase negative OCA, 97
tyrosinase positive OCA (OCA2), 103–108
tyrosinase related OCA (OCA1), 96–103
yellow OCA, 97, 100
Oculocerebrorenal syndrome, *see* Lowe oculocerebrorenal syndrome
Oguchi's disease, 73–74, 278, 280
Olfactory marker protein (OMP1), 374
Ommatidium, 32
Oncogenes, 197
Open reading frame, 329, 332, 504
Ophthalmoplegia, 473, 481
Opsin, 34, 219, 222

Optic neuropathy, 473–481
Optic nerve fibre misrouting, 95, 105, 110, 288
Optic nerve hypoplasia, 445
Oregon eye disease, 289
ORF — *see* Open reading frame
Ornithine aminotransferase (OAT), 310
Oscillatory potentials, 282–284, 287
Osteosarcoma, 206
Outer plexiform layer, 289
Oxidative damage to DNA, 472
Oxidative phosphorylation (OXPHOS), 469–472
Oxygen radicals, 472

P locus, 106
P-element, 39
Paired box, 452
Paralogous regions, 374
Paternal age effects, 386
Pattern dystrophy, 66, 168, 179, 183, 185, 189
Pattern formation, 452
Pax gene family, 452
PAX6 gene in aniridia, 452–462
Pax-6 gene in mouse, 452–455
PCR — *see* Polymerase chain reaction
Pearson marrow/pancreas syndrome, 481, 483, 485–486, 473
Pedigree, 504
Penetrance, 197, 199, 203–204, 325, 347, 369, 458, 475, 480, 504
Peripherin/rds gene (mouse), 136–138
Peripherin/RDS gene (human), 58–61, 66, 137–139, 162, 165–168, 183, 187
  mutation in adult vitelliform macular dystrophy, 169, 187
  mutations in autosomal dominant macular dystrophy, 169, 187
  mutations in autosomal dominant RP, 165–168, 187
  mutations in human, 58–61
  mutation in retinitis punctata albescens, 169, 187
Peter's anomaly, 459
PFGE — *see* pulsed-field gel electrophoresis
Phage, 504
Phagosomes, 185
Phenocopies, 68
Philly mouse — *see* mouse mutants
Phosducin, 130
Phosphatase, 44, 129
Phospholipase C gene — mutants in Drosophila (*norpA*), 36–39
Phosphatidylinositol transfer protein, 42
Phosphorylation, 42, 44, 205
Photophobia, 94, 220
Photoreceptor — 30–32, 153, 282, 304, 371
  degeneration in Drosophila, 29–46
  in normal mouse, 138
  dystrophies — central, 64–70
    peripheral, 54–64
  defects, congenital, 70–75
  inner segment, 161
  outer segment, 58, 60, 72, 123–124, 161, 185, 316
  R1-6, 31–32, 45
  R7/8, 31–32, 45
  vertebrate, 126
Phototransduction - invertebrates, 36–37, 42, 44
  vertebrates, 128, 155, 160
Physical map, 352, 451, 504
Phytanic acid, 63
PIC — *see* Polymorphism information content
Pigment — yellow p. in macula/lens, 222
Pigment dystrophy, 168
Pigmentation of retina, 89–113, 155, 260, 304, 244, 366
Plaque, 505
Plasmids, 505
Platelet storage pool deficiencies — *see* Animal models
Pleiotropy, 368, 453
Polyacrylamide gel electrophoresis, 505
PolyA$^+$ RNA, 505
Polymerase chain reaction (PCR), 21, 159, 165, 186, 203, 210–211, 246, 455, 505
Polymorphism, 101, 166, 188, 242, 479, 506
Polymorphism information content, 506
Position effect, 265, 458
Positional cloning, 157, 329, 447
Prader-Willi syndrome, 105–107
Preimplantation screening, 211

Prenatal diagnosis, 269, 350–351, 395, 408, 457, 462
Prenatal screening, 207–208, 211
Presumed ocular histoplasmosis syndrome (POHS), 174, 189
Primary colour, 218
Primer, 506
Proband, 506
Probe, 506
Proline uptake, 406
Promoter, 9, 92, 204, 313, 332, 506
Protanomaly, *see* Trichromacy
Protanopia, *see* Dichromacy
Protein kinase C, 408
Protein kinase II, calcium/calmodulin-dependent (Camk2a), 139
Protein phosphatase 2A, 129
Protein prenyltransferase, 315
Proteolytic cleavage, 389
Provirus, 134
Pseudogenes, 422
Pseudoinflammatory macular dystrophy — *see* under Macular dystrophy
Psychiatric disorder, 323, 368
Psychological counselling, 394
Ptosis, 473, 481
Pulsed-field gel electrophoresis (PFGE), 19, 328, 450–451, 506

Rab 3A, 315–316
Rab 3A GDP dissociation inhibitor (GDI), 315
Rab geranylgeranyl transferase, 315
Radiation hybrid mapping, 507
Random primer labeling, 507
Rayleigh colour matching, 221
Receptor potential — Drosophila, 36, 39, 41
Recombinant inbred (RI) strains, 15
Recombination, 10–12, 328, 507
  recombination fraction, 11–12
Recoverin, 162–163
Red/green colour blindness, *see* dichromacy/trichromacy
Refsum disease, 63
Regulatory sequences, 9, 92, 134, 137, 164, 204, 313, 332
Repetitive DNA, 5, 137, 165, 507
Replicative segregation, 472
Replication slippage, 313, 483
Rescue of mutant phenotype, 134, 137
Respiratory chain, 469
Restriction endonuclease, 508
Restriction fragment length polymorphisms (RFLP), 3–4, 101, 140, 157, 326, 207–208, 348, 407, 508
Retina — light micrograph of normal mouse, 125
  light micrograph of *rd/rd* mouse, 127
Retinal degeneration (*rd*) gene, 123–136
  characterisation of *rd* gene, 131–132
  isolation of *rd* gene, 130–136
  site of *rd* mutation, 134–135
Retinal degeneration — in blue-cone monochromacy, 220
  and cone pigment genes, 242
Retinal degeneration mutants — Drosophila, 29–51
  mutants of *ninaC* gene, 43–44
  mutants of *ninaE* gene, 33–36
  mutants of *norpA* gene, 36–39
  mutants of *rdgB* gene, 41–42
  mutants of *rdgC* gene, 44
  mutants of *trp* gene, 39–40
Retinal degenerations in mouse, 123–145
  cyclic GMP levels, 123–127
  rd mouse, 124–139
  rds mouse, 136–139
Retinal degeneration slow (*rds*) gene, 123
  human counterpart of mouse *rds* gene, 138–139
  identification of *rds* gene, 136–138
Retinal detachment, 339, 346, 386
Retinal development, 341
Retinal disorders — human inherited, 53–88
Retinal fundus — Best's vitelliform dystrophy, 184
  in retinitis pigmentosa, 156
  macular degeneration, 177
  normal, 155
Retinal pigment epithelium, 62, 65–69, 74, 168, 174–175, 183, 185, 259–260, 282, 316
Retinitis pigmentosa (RP) human, 35, 54–61, 138, 153–170, 184, 259–272, 292, 341, 359, 363, 480
  autosomal dominant RP (ADRP),

54–56, 58, 138, 145, 153–170
mutation of gene on chromosome 7, 168
mutation of gene on chromosome 8, 168
mutations of peripherin/rds gene, 165–168
mutations of rhodopsin gene, 158–165
autosomal recessive RP, 54–55, 57, 169–170, 371
expressivity, 56
functional studies, 55–58
genetic studies, 54–55
prevalence, 262
sector RP, 55–56, 59–60
sine pigmento, 366, 377
sporadic, 54–55, 260
tapetal reflex, 57, 261
type I (diffuse) RP, 55, 59, 165
type II (regional) RP, 55, 60, 165
unilateral RP, 56
X-linked RP, 54–55, 57–58, 145, 169–170, 259–272, 371
clinical features, 260–261
genetic mapping, 262–268
RP2 gene, 262, 266–268, 294
RP3 gene, 262, 264–265
genotype-phenotype correlations, 269
prevalence, 262
RP2 gene, 58, 266–268
RP3 gene, 58, 264–265
Retinitis punctata albescens, 74, 139, 166, 168, 187
Retinoblastoma (Rb), 197–211, 323
bilateral, 198, 209
cytogenetic analysis of Rb tumours, 201–204
family history, 197, 209
genetics of retinoblastoma, 197–199
prenatal diagnosis, 207–208
risk assessment, 208–210
trilateral, 207
unilateral, 198, 209, 211
Retinoblastoma (RB1) gene, 199–208
functions of RB1 gene, 205–206
identification of causative mutations, 210–211
isolation of RB1 gene, 202–203
pRB protein, 205–206
RB1 mutations in other tumours, 206–207
risk assessment, 208–209
Retinoma, 198, 203
Retinoschisis — autosomal dominant, 347
autosomal recessive, 347
Retinoschisis, *see* X-linked juvenile retinoschisis (RS)
Reverse transcriptase/polymerase chain reaction (RT/PCR), 312, 333, 407, 455, 457–458
Reverse transcription, 508
RFLP — *see* Restriction fragment length polymorphism
Rhabdomere, 32, 40–41, 44
Rhodopsin, 33-37, 41, 44, 58–60, 72, 74, 128–129, 138–139, 153, 157–165, 220, 293–294
in mouse, 139
mutants in Drosophila (*ninaE*), 33–36
mutations in autosomal dominant RP, 58–60, 158-165
transgenic mice with ADRP mutation, 164–165
Rhodopsin kinase, 128–129, 163
Ribonucleic acid, 508
alternative splicing, 462
heterogeneous nuclear, 7
protection, 508
splicing, 7
Risk — in age-related macular degen eration, 190
in Leber's hereditary optic neuropathy, 475
in retinoblastoma, 208–209
RNA — *see* Ribonucleic acid
RNA protection, 508
Rod defects — congenital, 71–75
Rod monochromatism, *see* monochromacy
Rod outer segment, 58, 60, 72, 123–124, 161, 185, 316
ROM-1 gene, 61, 183, 185, 372
RT — *see* Reverse transcription

S1 nuclease mapping, 509
S phase, 509
Sense strand, 509
Sequence tagged site, 510
Shuttle vectors, 510
Signal sequence, 333
Single strand conformation polymorphism (SSCP), 18, 138, 158–159, 166,187, 189, 203, 246, 312–313, 334, 510

Small eye mouse (Sey), 452–455
Somatic cell hybrid mapping, 510
Sorsby's fundus dystrophy, 69–70
Southern blot, 510
Spectral sensitivity of opsins, 244–245
Splicing, 7
Splicing, intron/exon, 510
SSCP — *see* Single strand conformation polymorphism
ss DNA, 511
Stargardt's disease, 67, 70, 183
Stickler disease, 347
Streptavidin/biotin, 511
STS — *see* Sequence tagged site
Synaptic vesicle transport, 316
Synteny, 453

TATA box, 511
Telomere, 3
Trans-activation domain, 455
Transcription, 511
  breakthrough, 457
Transcription factors, 204, 206, 455, 461
Transducin, 128, 130, 139–140, 142, 161
Transfection, 511
Transformation, 511
Transgenic animal, 15–17, 103, 134, 137, 164, 188, 190, 395, 511
Translation, 512
Translocation, 512
  balanced, 202, 448, 450, 458
  breakpoint, 450–451
  in cone dystrophy, 181
  X/autosome, 304, 306–307, 309, 324, 326, 399, 407
Trichromacy, 218, 221–223
  anomalous, 221, 249–252
  deuteranomaly, 221, 226, 241,
    molecular genetics, 226
  protanomaly, 221, 226
    molecular genetics, 226
  tritanomaly, 221
Tumour — analysis, 211
  multifocal, 202
  non-ocular, 206–207, 211
  progression, 207
  suppressor, 197
  virus, 205
Tyrosine transport protein, 106
Tyrosinase, 90–93, 112
  mouse tyrosinase (c-locus) mutations, 102–103
  tyrosinase gene (TYR), 92
    mutations in human TYR gene, 98–102
    tyrosinase related protein-1 gene (TRP1), 92–93
    tyrosinase related protein-2 gene (TRP2), 93

Ubiquinol:cytochrome c oxidoreductase (Compex III), 470, 475, 478–479
Usher syndrome (US), 359–379
  brain abnormalities, 369
  classification, 359–360
  clinical features, 359–369
  hearing loss, 360–362
  inheritance, 369–370
  Usher type I, 170, 359–375
    linkage to chromosome 11, 372–375
    linkage to chromosome 14q, 370–372
    in Poitu-Charentes region, 370
  Usher type II, 145, 170, 315, 359–370, 375–378
    linkage to chromosome 1q, 375–377
  Usher type III, 360–361
  Usher type IV, 368
  speech, 360–361
  sperm, 371
  vestibular function, 363

Variable number tandem repeats, 512
Vector, 22–23, 512
Vestibular hair cells, 371
Vestibular responses, 360, 363
Visual evoked responses, 366
Visual fields, 154, 259–260, 281, 366,
Visual pigment — *see* colour pigments, opsin, rhodopsin
Vitamin A, 41, 64
Vitamin E, 64
Vitelliform macular dystrophy (Best's discase) — *see under* Macular dystrophy
  adult vitelliform macular dystrophy, 65–66, 168–169, 187
  atypical macular dystrophy, 181
Von Willebrand factor, 333
VNTR — *see* Variable number tandem repeats

Wagner disease, 347
WAGR syndrome, 447–448, 451
  deletions leading to mapping of aniridia locus, 448–450
Western blot, 512
Wilms' tumour, 447–448, 450, 458

X-inactivation, 3, 203, 260, 287–288, 304, 324, 326, 399
X-linked familial exudative vitreoretinopathy, 325
X-linked juvenile retinoschisis (RS), 339–353
  clinical features, 341–348
  epidemiology & prevalence, 339–341, 351
  linkage studies, 348–352
  prenatal diagnosis, 350–351
  therapy, 346

Yeast artificial chromosome, 266, 329, 352, 407, 513
YAC — *see* Yeast artificial chromosome
Young-Helmholtz theory, 227–228

Zinc finger proteins, 513